Bridge and Highway Structure Rehabilitation and Repair

About the Author

Dr. Mohiuddin A. Khan is a consulting engineer. He helped to develop the *NJ State Bridge Design Manual,* and is coauthor of the FHWA-approved *Handbook of Scour Countermeasures.* Dr. Khan is registered as a licensed professional engineer in five states and is a chartered engineer from England. As a bridge engineer, he is familiar with both U.S. and European practices. He has recently designed curved girder bridges and VMS/CMS sign structures. Dr. Khan is Chairman of Structures Group for ASCE, Philadelphia Section, served on ASCE Methods of Analysis Committee for six years, and is a member of the ASCE National Seismic Effects Committee. He teaches bridge inspection and design to graduate classes and practicing engineers at Drexel, Widener, and Temple Universities. Dr. Khan is conducting research in deck vibrations, integral abutments, and seismic design of bridges for the New Jersey Departments of Transportation and at Temple University in Philadelphia.

Bridge and Highway Structure Rehabilitation and Repair

Mohiuddin A. Khan

McGraw Hill

New York Chicago San Francisco Lisbon London Madrid
Mexico City Milan New Delhi San Juan Seoul
Singapore Sydney Toronto

The McGraw·Hill Companies

Cataloging-in-Publication Data is on file with the Library of Congress.

McGraw-Hill books are available at special quantity discounts to use as premiums and sales promotions, or for use in corporate training programs. To contact a representative please e-mail us at bulksales@mcgraw-hill.com.

Bridge and Highway Structure Rehabilitation and Repair

1 2 3 4 5 6 7 8 9 0 WDQ/WDQ 1 9 8 7 6 5 4 3 2 1 0

ISBN 978-0-07-154591-4
MHID 0-07-154591-3

This book is printed on acid-free paper.

Sponsoring Editor
Larry S. Hager

Acquisitions Coordinator
Alexis Richard

Editorial Supervisor
David E. Fogarty

Project Manager
Virginia Howe, Lone Wolf Enterprises, Ltd.

Copy Editor
Jacquie Wallace

Proofreaders
Mary Riker and Afton Woodson

Indexer
Leona Woodson

Production Supervisor
Richard C. Ruzycka

Composition
Lone Wolf Enterprises, Ltd.

Art Director, Cover
Jeff Weeks

Contents

PREFACE xiii

ACKNOWLEDGMENTS xvii

SYMBOLS, NOTATIONS, AND TERMINOLOGY xix

SECTION 1 ADMINISTRATIVE ISSUES 1

Chapter 1 Overview of Bridge Engineering 1

1.1 REHABILITATION AS A MODERN INDUSTRY - 1

1.2 A FRESH APPROACH TO MAINTENANCE - 4

1.3 THE NEED TO KEEP BRIDGES FUNCTIONAL - 7

1.4 MAINTENANCE OF A SUPERSTRUCTURE - 12

1.5 MAINTENANCE/REPLACEMENT OF THE SUBSTRUCTURE - 13

1.6 OVERVIEW OF MAINTENANCE PROCEDURES - 16

1.7 A STRATEGIC PLAN TOWARD REPAIR AND REHABILITATION - 21

1.8 A LOOK AHEAD FOR PROGRESS IN MAINTENANCE - 23

1.9 THE MAINTENANCE AND PROTECTION OF TRAFFIC DURING CONSTRUCTION - 28

1.10 SUCCESS OF WELL-MAINTAINED STRUCTURES - 31

1.11 DEFICIENT STRUCTURES - 33

1.12 PRESERVING AESTHETICS - 34

1.13 MAINTAINING THE ENVIRONMENT - 35

BIBLIOGRAPHY - 37

Chapter 2 Diagnostic Design and Selective Reconstruction **39**

2.1 MAINTENANCE ENGINEERING - 39

2.2 THE REHABILITATION PROCESS - 44

2.3 PROGRESSIVE DESIGN PHASES FOR CONTINUITY - 52

2.4 THE ROLE OF REDUNDANCY AND FRACTURE CRITICAL MEMBERS - 60

2.5 THE ROLE OF GOVERNMENT AGENCIES IN MAINTAINING INFRASTRUCTURE - 61

2.6 COMBINING OLD AND NEW TECHNOLOGIES FOR REHABILITATION - 66

BIBLIOGRAPHY - 73

Chapter 3 Bridge Failure Studies and Safety Engineering **75**

3.1 HISTORY OF DISASTERS AND SAFETY MANAGEMENT - 75

3.2 THE ROLE OF FORENSIC ENGINEERING - 78

3.3 MANY ASPECTS OF FAILURES - 81

3.4 A DIAGNOSTIC APPROACH - 81

3.5 A HISTORICAL PERSPECTIVE OF RECENT FAILURES - 85

3.6 DESIGN DEFICIENCY AND PREVENTIVE ACTIONS - 92

3.7 FATIGUE FAILURES AND SUGGESTED PREVENTIVE ACTIONS - 94

3.8 CONSTRUCTION DEFICIENCY AND SUGGESTED PREVENTIVE ACTIONS - 95

3.9 VESSEL COLLISION OR FLOATING ICE AND SUGGESTED PREVENTIVE ACTIONS - 99

3.10 TRAIN ACCIDENTS CAUSING BRIDGE DAMAGE AND PREVENTIVE ACTION - 101

3.11 VEHICLE IMPACT AND PREVENTIVE ACTION - 102

3.12 BLAST LOAD AND PREVENTIVE ACTION - 103

3.13 FIRE DAMAGE TO SUPERSTRUCTURES AND PREVENTIVE ACTION - 104

3.14 SUBSTRUCTURE DAMAGE DUE TO EARTHQUAKE AND PREVENTIVE ACTIONS - 105

3.15 WIND AND HURRICANE ENGINEERING - 109

3.16 LACK OF MAINTENANCE AND NEGLECT - 110

3.17 UNFORESEEN CAUSES LEADING TO FAILURES - 112

3.18 A POSTMORTEM OF FAILURES - 114

3.19 THE STUDY OF MODES OF FAILURE - 116

3.20 STEPS TO AVOID FAILURES - 117

3.21 REITERATING NEEDED PREVENTIVE MEASURES - 124

BIBLIOGRAPHY - 128

SECTION 2 STRENGTHENING AND REPAIR WORK **133**

Chapter 4 An Analytical Approach to Fracture and Failure **133**

4.1 THEORETICAL CONCEPTS USED IN DEVELOPING COMPUTER SOFTWARE - 133

4.2 STRESS ANALYSIS - 135

4.3 REVIEW OF ELASTIC ANALYSIS - 137

4.4 ANALYSIS OF SLAB BEAM BRIDGES - 138

4.5 METHODS OF ANALYSIS OF THE SUPERSTRUCTURE - 140

4.6 EFFECT OF BOUNDARY CONDITIONS ON BRIDGE BEHAVIOR - 143

4.7 FULL COMPOSITE (ARCHING AND DOME) ACTION IN SLAB AND BEAMS - 146

4.8 NUMERICAL AND COMPUTATIONAL MODELS - 151

4.9 ANALYSIS OF APPROACH SLAB RESTING ON GRADE - 152

4.10 NONLINEAR ANALYSIS IN STEEL AND CONCRETE - 159

4.11 SINGLE SPAN LIVE LOAD ANALYSIS - 161

4.12 SELECTION OF STEEL GIRDERS - 162

4.13 REVIEW OF COMMON FAILURE THEORIES - 169

4.14 PLASTIC BEHAVIOR OF STEEL SECTIONS - 172

4.15 PLASTIC BEHAVIOR OF STEEL NON-COMPOSITE SECTION - 173

4.16 PLASTIC BEHAVIOR OF COMPOSITE SECTIONS - 174

4.17 SHEAR DESIGN FOR REINFORCED CONCRETE AND PRESTRESSED BEAMS - 180

BIBLIOGRAPHY - 182

Chapter 5 Load and Resistance Factor Rating and Redesign **185**

5.1 RATING AND REDESIGN METHODS - 185

5.2 UTILIZING ULTIMATE LOAD BEHAVIOR OF MATERIALS - 190

5.3 LRFD SERVICE LOAD REQUIREMENTS - 194

5.4 FATIGUE AND FRACTURE - 204

5.5 SELECTION OF TRUCK LIVE LOADS - 209

5.6 DETAILED AASHTO LOAD COMBINATIONS FOR RATING AND DESIGN - 216

5.7 CONSTRUCTION LOADS AND LOAD COMBINATIONS - 223

5.8 LRFD LOAD COMBINATIONS FOR STRENGTH, SERVICEABILITY, AND EXTREME CONDITIONS - 226

5.9 LRFD T-BEAM BRIDGE DESIGN - 228

5.10 SOFTWARE FOR SUPERSTRUCTURE - 232

5.11 SOFTWARE FOR SUBSTRUCTURE - 233
BIBLIOGRAPHY - 234

Chapter 6 Applications of Bridge Design and Rating Methods **237**

6.1 INTRODUCTION - 237
6.2 LIMIT STATES DESIGN EQUATION - 237
6.3 LEGAL LOADS - 240
6.4 SIMPLIFIED FORMULA - 246
6.5 RATING PROCEDURES FOR CONCRETE AND STEEL BRIDGES - 255
6.6 RATING OF SECONDARY STRUCTURAL MEMBERS - 260
6.7 EXAMPLE OF LOAD RESISTANCE FACTOR RATING (LRFR) - 261
6.8 DESIGN OF A DECK SLAB - 266
6.9 RATING PROCEDURE FOR REINFORCED CONCRETE T-BEAM BRIDGE - 274
6.10 RATING OF PRESTRESSED CONCRETE GIRDER - 276
BIBLIOGRAPHY - 279

Chapter 7 Bridge Widening and Deck Replacement Strategy **281**

7.1 SELECTING BRIDGE WIDENING OR REPLACEMENT - 281
7.2 SUPERSTRUCTURE WIDENING - 284
7.3 REHABILITATION/RETROFIT OR REPLACEMENT DECISIONS - 285
7.4 REPLACEMENT PLANNING OR A NEW BRIDGE AT A DIFFERENT LOCATION - 295
7.5 CONSTRUCTABILITY ISSUES - 303
7.6 SUPERSTRUCTURE REPLACEMENT PROCESS - 308
7.7 CASE STUDIES OF ALTERNATES FOR REPLACEMENT OR NEW BRIDGES - 317
7.8 PRACTICAL CASE STUDIES - 318
7.9 SELECTED CASE STUDIES - 328
BIBLIOGRAPHY - 331

SECTION 3 REPAIR AND RETROFIT METHODS **133**

Chapter 8 Inspection, Rating and Health Monitoring Techniques **335**

8.1 BRIDGE MANAGEMENT SYSTEM (BMS) - 335
8.2 BRIDGE INSPECTION LEADING TO DIAGNOSTIC DESIGN - 344
8.3 TYPES OF INSPECTIONS - 349

8.4 BRIDGE COMPONENTS FOR INSPECTION AND STRUCTURAL EVALUATION - 351

8.5 SUBSTRUCTURE COMPONENTS - 357

8.6 FHWA CONDITION AND SUFFICIENCY RATINGS - 359

8.7 DUTIES, QUALIFICATIONS, AND TRAINING OF INSPECTORS - 364

8.8 OVERLOAD PREVENTION - 366

8.9 SUPERSTRUCTURE LOAD CARRYING CAPACITY - 367

8.10 CONDITION RATING FOR EXTREME EVENTS - 368

8.11 STRUCTURAL HEALTH MONITORING (SHM) AS A RATING AND DIAGNOSTIC TOOL - 372

8.12 MONITORING USING NONDESTRUCTIVE EVALUATION (NDE) AND MODERN NONDESTRUCTIVE TESTING (NDT) METHODS - 380

8.13 ADAPTIVE MANAGEMENT AND MAINTENANCE FOR BRIDGES OVER WATERWAYS - 383

8.14 CONCEPT STUDY REPORT, PLANS, AND RECOMMENDATIONS - 387

8.15 VALUATION OF UNKNOWN FOUNDATIONS - 390

8.16 DISASTER MANAGEMENT - 390

8.17 TRAINING AND REGISTRATION NEEDS - 391

BIBLIOGRAPHY - 391

Chapter 9 Conventional Repair Methods **397**

9.1 SCOPE OF CONCRETE AND STEEL REPAIRS AND RELATED WORK - 397

9.2 ASSOCIATED TASKS AND TEAMWORK - 406

9.3 SHORT-TERM REPAIRS IN LIEU OF REPLACEMENT - 412

9.4 PRIORITIZATION OF BRIDGE DECKS - 414

9.5 SCOPE OF REPAIRS - 417

9.6 ALTERNATIVE METHODOLOGY AND TECHNIQUES OF REPAIRS - 419

9.7 REHABILITATION OF STEEL AND PREVENTION OF CORROSION - 426

9.8 GALVANIC TECHNOLOGY - 431

9.9 CORROSION IN THE POST-TENSIONING REGIONS OF BEAMS - 433

9.10 DEVELOPING NEW CONSTRUCTION PRODUCTS - 433

9.11 DRAINAGE - 434

BIBLIOGRAPHY - 434

Chapter 10 Concrete Repair Methods **439**

10.1 INTRODUCTION TO COMMON REPAIR PROCEDURES - 439

10.2 DETAILS OF REPAIRS - 440

10.3 SUPERSTRUCTURE REPAIRS - 452

10.4 DECK SLAB REPAIRS - 453

10.5 PAY ITEMS AND QUANTITIES FOR DECK REPLACEMENT PROJECTS - 462

10.6 ACI RECOMMENDED METHODS OF REPAIR - 463

10.7 PRESTRESSED CONCRETE BRIDGES - 466

10.8 BRIDGES LOCATED ON RIVERS - 467

10.9 CONCRETE REPAIR/RESTORATION OTHER THAN FOR DECKS - 468

BIBLIOGRAPHY - 471

Chapter 11 Advanced Repair Methods **477**

11.1 INNOVATIONS MADE IN RECENT YEARS - 477

11.2 DEVELOPMENT IN DESIGN CODES AND MARKET READY TECHNOLOGY - 483

11.3 USE OF MODERN CONSTRUCTION MATERIALS - 485

11.4 USE OF RECYCLABLE MATERIALS - 486

11.5 USE OF FIBER REINFORCED POLYMER (FRP) - 488

11.6 ADVANCEMENTS IN CONCRETE TECHNOLOGY - 494

11.7 ADVANCEMENTS IN STEEL MATERIALS TECHNOLOGY - 500

11.8 ADVANCEMENTS IN CONSTRUCTION TECHNOLOGY - 502

11.9 ACCELERATED BRIDGE CONSTRUCTION TECHNOLOGY - 505

11.10 PREFABRICATION TECHNOLOGY - 509

11.11 REHABILITATION OF STEEL BRIDGES - 513

11.12 RESEARCH IN NEW TECHNIQUES FOR MONITORING, NEW MATERIALS, AND VIRTUAL DESIGN - 517

BIBLIOGRAPHY - 521

Chapter 12 Protection of Bridges against Extreme Events - 529

12.1 EXTREME EVENT DAMAGE FROM FLOOD SCOUR - 529

12.3 CONDITION OF EXISTING BRIDGES ON RIVERS - 532

12.4 PREVENTIVE ACTION AGAINST BANKS AND FOUNDATION SCOUR - 535

12.5 HEC-18 COUNTERMEASURES MATRIX (CM) - 549

12.6 DESIGN GUIDELINES - 554

12.7 CONSTRUCTABILITY ISSUES - 557

12.8 SCOUR OF PILES, PILE GROUPS, AND CAISSONS - 560

12.9 SCOUR AT WINGWALLS - 561

12.10 SCOUR AT CULVERTS - 562

12.11 SCOUR MEASURING EQUIPMENT - 562

12.12 DEBRIS ACCUMULATION - 562

12.13 EXTREME EVENT OF EARTHQUAKE DAMAGE - 563

12.14 SEISMIC ASSESSMENT - 569

12.15 THE ROLE OF SEISMIC DESIGN CODES - 570

12.16 SEISMIC DESIGN - 576

12.17 COMMON RETROFIT CONCEPTS AND CODE APPLICATIONS - 578

12.18 STEEL AND CONCRETE BRIDGE DETAILING REQUIREMENTS - 582

12.19 RETROFIT AND STRENGTHENING - 586

12.20 DEVELOPMENTS IN PASSIVE DAMPING SYSTEMS - 595

12.21 FOUNDATION RETROFIT - 596

12.22 DISASTER MANAGEMENT - 599

BIBLIOGRAPHY - 601

APPENDICES

APPENDIX 1 - SCOUR COUNTERMEASURES - 611

APPENDIX 2 - ESTIMATE OF UNIT CONCRETE REPAIR COSTS (2009) - 615

APPENDIX 3 - QUICK REFERENCE TO AASHTO LRFD 2008 SPECIFICATIONS - 617

APPENDIX 4 - QUICK REFERENCE TO AASHTO LRFR MANUAL - 621

INDEX 623

Preface

INTRODUCTION

During the author's career as a consulting engineer and teacher of structural engineering subjects, it was observed that few textbooks are available on bridge design or inspection and rating. Bridge engineering background information needs to be presented beyond AASHTO LRFD specifications and the manual for condition evaluation and LRFR for highway bridges. What is especially needed is coverage of recent state-of-the-art developments.

A bridge engineer has to face many aspects—structural and otherwise. This book is about problem solving and providing an insight into important issues. As a bridge engineer and teacher of modern bridge engineering, I find that there is a need for a book to address new technology on planning, detailed design, and rehabilitation aspects. Topics such as seismic retrofits and scour countermeasures need to be summed up in the form of a book for the benefits of students and practicing engineers.

Although AASHTO LRFD and state codes have also been updated to include some of the changes, they take time to catch up with innovations and incorporate new procedures. They serve as guidelines and do not aim at educating the reader with fundamentals. Some state codes are more enterprising in developing the technology compared to federal codes. The subject matter covered by these voluminous specifications is extensive, diverse, in-depth, mathematical, and at times not so easy to understand by bridge engineers. This book will serve basically as a companion reference manual to the codes and specifications, with emphasis on new topics, and it will focus on both traditional and nontraditional design problems.

Practicing engineers continuously find the need for a book which can simplify the presentation and make it more palatable for office use. A book should address the state of art of bridge engineering, highlight major issues, offer necessary explanations, provide sample solved examples for the day-to-day design issues, and use case studies of practical problems.

If the developments in the subject are reported by this book and understood by teachers and students at the university level, future practicing engineers will have a jump-start and the purpose of the book will be well served.

OBJECTIVES

The issues are basically presenting an overall perspective of the multi-billion dollar transportation industry, the resulting advantages of which to the society appear to surpass those offered by other disciplines, except perhaps the medical profession. Indeed, transportation is a complement even to the medical profession, by facilitating an ambulance's trip to the hospital in the shortest time, without being stuck in a traffic jam or delayed by a roadway closed for repairs.

The book's theme and scope are about engineering the rehabilitation and repairs of existing highways bridges in an efficient and cost-effective manner. Its special features are about facing "ground realities" or "river realities" when "fixing the bridge." No more mathematics should be required than is actually needed, since mathematical technique is a means to an end and is not an end in itself.

Innovative ideas for structural planning and precast connection details developed by design and sometimes construction teams over the years will be presented, making this book unique.

NEED TO INCORPORATE GROWING TECHNOLOGIES

There is a need to update the growing volume of knowledge of this important and practical subject. Existing books seldom include rehabilitation aspects exposed to the dynamic forces at work and may not fully appreciate the need for teamwork or sensitivity to client's and the user's needs. More important is speaking the engineer's language and taking a series of small steps rather than making one big leap. Also, recommending one or more outdated books to the students does not seem to solve the problem. There are relatively few books on similar topics and on developments in modern technology on design and construction disciplines. However, the available books seem to contain information from a different era, information which is already obsolete and has becomes less relevant than the conditions for which it was originally written.

Maintenance of highways include the much neglected highway structures, such as cantilever and overhead sign structures, special loads resulting from variable message signs, noise walls, precast modular retaining walls, and providing relief bridges and scour countermeasures at bridges and embankments.

For bridges located on waterways, use of deeper foundations and adequate bridge openings are being recommended. An earlier handbook on protection against flood scour (a major cause of bridge failures in the United States), which I coauthored for NJDOT jointly with City University of New York, has now been approved by FHWA for use by consultants. I had also developed Sections 45 and 46 of *NJDOT Bridge Design Manual* pertaining to seismic design and bridge scour.

I find that it is the right time to rewrite the contents of the subject and address the latest changes in the code requirements (such as important changes in the format of SI&A Datasheet and that the method of rating needs to be based on the new LRFR Method). The latter method has generated development of new LRFD software, such as LEAP, SAM, Merlin-Dash, PENNDOT and other LRFD based software for seismic and scour analysis. Also, a book should fully present the language of ongoing design, construction culture, and design development phases.

TRACK RECORD OF SUCCESSFUL BRIDGE DESIGN AND CONSTRUCTION

The bridge engineer needs to be groomed to a technical decision making position, starting from a junior engineer to a senior engineer and to a team leader responsibilities. In the design environment, there is a need to develop an engineering sense to address the issues and resolve them in an engineering sense in the limited time available.

The audience for whom the book is intended includes practicing structural and bridge engineers in United States and abroad, government agency engineers, planners, highway and traffic engineers, geotechnical engineers, hydraulic engineers, transportation specialists, contractors, product vendors, senior engineering students, and university professors.

The book is also intended for those with the following job titles: engineering managers, project managers, design engineers, construction supervisors, instructors, and final year students.

In addition, the book is intended for those who might be affiliated with the following professional associations and organizations: American Society of Civil Engineers, American Society of Highway Engineers, American Concrete Institute, AASHTO, FHWA, TRB, American Institute of Steel Construction, and state board professional engineering licensing agencies.

KEY BENEFITS

Experience in the design and construction supervision of bridges located in the United States, Europe, and Asia can be utilized in developing a book that can be recommended simultaneously to students and bridge engineers alike. Some of the prestigious bridge projects, where I have successfully participated as a designer include curved steel bridges at Washington National Airport, continuous span curved bridges currently under design at Washington Dulles Airport, the elevated Monfayette Expressway Project in Western Pennsylvania with high piers, prestressed concrete box bridges for Jeddah-Mecca Expressway, numerous rehabilitations of highway transit bridges in New York City, Boston, and Philadelphia and the recently completed integral abutment bridge on Route 46 over Peckman River in New Jersey.

In addition, the book will address

1. Recent advancements in analytical and design techniques, such as the application of load and resistance factor design (LRFD) methods and the need and availability of specialized software.
2. Changes in AASHTO design codes for bridges and highways and in the design manuals and guidelines of each state.
3. Revisions to design criteria based on recognition of earthquake vulnerability and bridge foundation scour from floods.
4. Increase in the volume of vehicular traffic on highways results from enhancement in automobile industry production and with their marketing motto "one car for each family". This has resulted in an overload of existing bridges and highways, increased wear and tear and in the number of accidents. The highway network should be able to accommodate increases in ADT and ADTT. Frequent traffic counts may be necessary for adjusting direction of flow in the network.
5. Modern materials technology, the developments in new types of concrete, steel and other construction and repair materials.
6. Developments in construction methodology, the use of long span cranes, hauling of long span girders and improvements in erection on sites.
7. Changes in the architecture of bridges, need for planning of wider highway lanes for use by wider and heavier trucks and the enhancement of roadside facilities for long distance travel by cars, and distribution of goods in every nook and corner of the country.
8. Updates on important developments in bridge and highway maintenance and use of management techniques of bridges and highways based on science and technology.

Professional engineers will learn to create a more efficient design process and construction managers will learn how to save time and money with a better bridge management system. The book incorporates the essence of latest developments by making use of the experience and expertise acquired in implementing major bridge projects.

Recent innovations in telecommunications, automobile, aircraft, and telephone industries seem to have influenced the overall format of engineering disciplines. It is important for any book to generate sufficient interest in the subject matter being utilized by university students. A realistic approach is likely to help the students so that they are mentally prepared for the design tasks ahead, rather than being focused only on the design formulae.

A fresh approach to the topics is reflected in the contents. The book will be used by the student, the teacher, and the engineer as guidelines, resulting in time saved and the dangers of misconceptions avoided. Both academic and professional classes would find the contents as appropriate and relevant.

The book may be used both as a text book or a desk reference.

There is a wealth of information available in research publications, in the proceedings of international conferences, technical journals, and periodicals, where there are papers by learned authors and speakers from all over the world. Notable among the conferences are SEI Proceedings, New York City Bridge Conference, Pittsburgh Bridge Conference, and specialty conferences and seminars. Journals, which provide latest articles on the related issues including vendor products from the world of construction, are *Civil Engineering, New Civil Engineer, Roads & Bridges,*

Structural Engineer, Structure, and ASCE journals *Bridge Engineering, Journal of Structures Division,* and *Modern Steel Construction.* My book contains much needed information on innovations presented in all the technical resources.

Many chapters and the solved examples will contain references to the applicable codes and specifications. This is also a legal requirement—the engineer needs to be well versed with codes.

Acknowledgments

It gives me a great pleasure to express my gratitude to the following individuals who took keen interest in the development of the book and devoted their valuable time to review the chapters:

Nazhat Aboobaker, Ph. D., P.E., Adjunct Professor at Widener University, Philadelphia for encouragement and suggestions.

Chris Menna, P. E., Head of Bridge Design Department, City of Philadelphia, for patiently going over the contents of the book and for his recommendations.

Others who helped in the progress of publishing and printing include Larry Hager, Roger Woodson, Leona Woodson, and of course Virginia Howe for her dedication during the production phase.

Without them the book could hardly be possible.

The references from the many AASHTO, FHWA, NCHRP, ASCE-SEI publications, conference proceedings and from authors on the subject (listed in the bibliography) are also acknowledged.

Symbols, Notations, and Terminology

Greek Alphabet

A	α	Alpha	N	ν	Nu
B	β	Beta	Ξ	ξ	Xi
Γ	γ	Gamma	O	o	Omicron
Δ	δ	Delta	Π	π	Pi
E	ϵ	Epsilon	P	ρ	Rho
Z	ζ	Zeta	Σ	σ	Sigma
H	η	Eta	T	τ	Tau
T	τ	Theta	Υ	υ	Upsilon
I	ι	Iota	Φ	ϕ	Phi
K	κ	Kappa	X	χ	Chi
Λ	λ	Lambda	Ψ	ψ	Psi
M	μ	Mu	Ω	ω	Omega

Greek Symbols

θ	angle, angle of twist, co-ordinate
δ	deflection, displacement
Δ	elongation or deflection; also change in magnitude
ρ	ratio of nonprestressed reinforcement in a section
γ	shear strain
$\gamma_{xy}, \gamma_{xz}, \gamma_{y}$	shearing strain components in rectangular coordinates
$\gamma_{r\theta}$	shearing strain in polar coordinates
τ	shearing stress
$\tau_{xy}, \tau_{xz}, \tau_{yz}$	shearing stress components in rectangular coordinates
P	single load
$\sigma_x, \sigma_y, \sigma_z$	tangential stress in polar coordinates
$\epsilon_t, \epsilon_\theta$	tangential unit elongation in polar coordinates
θ_t	unit angle of twist
ϵ	unit elongation
$\epsilon_x, \epsilon_y, \epsilon_z$	unit elongations in x, y, and z directions

Nomenclature

C_m	a factor relating an equivalent uniform moment diagram to the actual diagram
A	area
A_f	area of compression flange
A_w	area of web
f_a	axial stress; also, allowable axial fatigue stress
d_b	bar diameter
d	beam depth; also, distance between center of beam flanges
M_n, M_{nt}	bending and twisting moments per unit length of a section of a plate perpendicular to n direction
M	bending moment
M_x, M_y	bending moments per unit length of sections of a plate or beam perpendicular to x and y axes, respectively
$M_{y/d}$	bending moment that causes first yielding
f_b	bending stress
f_c	buckling stress for columns; also, compressive stress in plates or tubes; also, bending stress in concrete; also, theroetical elastic lateral bucking stress
u, v, w	components of displacements
M_{cr}	cracking moment of concrete
M_d	dead load moment
D	diameter
d	depth, diameter
e	eccentricity
k	effective length factor for a compression member
b_e	effective plate width
P_e	Euler load of column
FS	factor of safety for static stress
b_f	flange width
D	flexural rigidity of a plate or shell
q	intensity of a continuously distributed load
L	length
M_l	live load moment
n	modular ratio (ratio of modulus of elasticity of steel to that of concrete)
E_c	modulus of elasticity of concrete
E_s	modulus of elasticity of steel
G	modulus of elasticity in shear
E	modulus of elasticity in tension ad compression
M_x	moment caused by loads perpendicular to x axis
N_x, N_y	normal forces per unit length of sections of a plate perpendicular to x and y directions, respectively
M_p	plastic moment
Z	plastic section modulus
a	plate length or stiffener spacing; also depth of compressive region in concrete slab of composite beam; also, torsional flange bending constant

b^{l}	plate width between stiffeners
b	plate width; also, effective width of concrete slab
r, θ	polar coordinates
J	polar second moment of area
υ	Poisson's ratio
p	pressure
r_x, r_y	radii of curvature of the middle surface of a plate in xz and yz planes, respectively
r	radius of gyration
x, y, z	rectangular coordinates
I	second moment of area
S	section modulus; also, stress in cyclic loading
Q	shear force
V	shear force
Q_x, Q_y	shearing forces parallel to axis per unit length of sections of a plate perpendicular to x and y axes, respectively
Q_n	shearing force parallel to z axis per unit length of section of a plate perpendicular to n direction
f_v	shear stress
f'_c	specified compression strength of concrete
V	strain energy
f	stress
T	temperature, torque
R_m	a term used in required percentage of steel expression for flexural members
t_c	thickness of concrete slab
t_f	thickness of flange
t_w	thickness of web
h	thickness of a plate or a shell
J	torsional constant
T	torsional moment; also, tensile force in composite beam; also, proof load of bolts; also membrane torsion
I_{cr}	transformed moment of inertia of cracked concrete section
M_{xy}	twisting moment per unit length of section of a plate perpendicular to x axis
M_u	ultimate moment
w_c	unit weight of concrete
l_u	unsupported length of a compression member
V	volume
W	weight, load
γ	weight per unit volume
b_b	width of bottom flange
f_y	yield stress (yield point or yield strength, whichever is applicable)

Other notations as introduced in text or used in references listed in Bibliography.

Bridge Terminology

Abutment—Abutment has dual purpose, earth retaining and supporting all or part of the bridge. The two abutments define the beginning and end of bridge and serve as anchors.

Approach—Approach or approach slab interfaces with the abutment to provide continuity to the roadway. The roadway is capped by a slab resting on unyielding soil to prevent differential settlement after a heavy downpour.

Backwall—Backwall or the stem separates the approach slab from the bridge. They interface with wingwalls.

Barrier—Barrier is a railing or reinforced concrete parapet which serves as a protective dwarf wall placed over the deck to guide vehicles and prevent collisions.

Beam—Beam or girder is a very important primary member which supports the deck. It spans over the river width or the interchange. It is supported by pier and abutments and is held in position by bearings resting over substructure. It is made of timber, aluminum, steel, reinforced or prestressed concrete.

Bearing—Bearing is a mechanical device which permits expansion, contraction and rotation of beams. They also transfer heavy reactions over the superstructure over a wider area.

Bracing—Bracing is a secondary member placed in between beams to resist wind load in horizontal direction.

Deck—Deck is the most important part of superstructure. It supports the moving loads and can be built in reinforced concrete, timber or open grid steel. It is usually made composite with the supporting beams.

Deck joint—A joint in transverse or longitudinal direction helps to release thermal stress in the deck slab. They are placed before the abutment backwall. Deck Joint consists of steel angles with vertical leg anchored into the deck concrete to shield the corners of concrete edges from any damage resulting from the impact of wheels.

Diaphragm—Diaphragm is a secondary member which interconnects the beam in transverse direction and enables lateral distribution of vertical and horizontal loads. They are usually spaced at twenty feet or less.

Embankment—Embankment is body of earth which transitions from deck elevation to the lower ground and is given a gentle slope using fill material for that purpose.

Footing—Footing serves as the fixed feet of bridge. It is the most important part of the bridge family. Shallow footings transfer the loads from the bridge over the wider area of compacted sub-soil. Deep footings are bearing or frictional piles or caissons driven to the required depth deep into soils.

Haunch—Haunch is a concrete transition member placed over the flange of beam and under the deck slab. Like a pedestal, its height is adjusted to achieve the required deck elevation. Haunches over 4 inch depth require reinforcement.

Integral abutment—Abutment wall resting on piles in which top stub is made integral with the deck slab and the approach slab. Bearings or deck joints are not required. Special boundaries of beam serve as a longitudinal frame. Structural performance of bridge under seismic conditions is also improved.

Maintenance—Routine or regular activities, which are intended to preserve and maintain a structure's original serviceability and functionality.

Parapet—Parapet is a barrier placed at the edges of deck slab. It also supports lighting poles or sign panels.

Pedestal—Pedestal supports the bearings as very short concrete columns. They maintain the required bridge seat elevations and their adjustable heights in concrete are varied to suit.

Pier—Pier is a wide column which supports the intermediate parts of the longer span bridge and shares the loads with abutments. They may be placed in the middle of a river.

Pile—Pile serves as a stiff nail driven into a hard soil media. Acting in a group, it provides stability to the bridge.

Rehabilitation—Comprehensive repair of a bridge structure's most deteriorated elements that are intended to restore and significantly extend its original serviceability and functionality.

Repair—Activities, usually isolated to a portion of one element of a structure, that are necessary to restore serviceability or functionality due to distress from things such as vehicle impact damage, observed scour or severe localized deterioration.

Sidewalk—Sidewalk is an elevated deck for use by the pedestrian and is placed at the side of a bridge deck. It is usually protected by railing from the traffic.

Scupper—Scupper is used for rapid deck drainage. It is a vertical pipe located at the deck inlet at the face of parapet and connected under the bridge substructure into a manhole of drainage system.

Sheeting—Sheeting is timber or steel planks driven in the soil to allow construction of footing. They serve as temporary retaining walls and are usually left in place.

Sign structure—Sign structure supports the sign panels which provide directions to the drivers about the exact location. Overhead, cantilever, butterfly types are used. It is usually placed well before the exit ramp or intersection.

Underdrain—Underdrain is a water drainage pipe usually located behind abutment. It prevents water table from rising to the deck elevation.

Wearing surface—Wearing surface protects the deck from abrasion and cracking from friction and impact of the fast moving trucks and vehicles.

Wingwall—Wingwall is located on the sides of the bridge width. The two splayed or return walls provide stability to abutment, by confining the earth behind the abutment and preventing it from moving.

1

Overview of Bridge Engineering

Section 1

Administrative Issues

Section Overview

- Practice and procedures for diagnostic design, life cycle costs, deficiencies, and studies of bridge failures.
- Developments in theory and code methods such as LRFR and LRFD.
- Improving aesthetics and ensuring security.
- Need for preserving historical bridges and safety use in engineering.

1.1 REHABILITATION AS A MODERN INDUSTRY

1.1.1 Overview of Bridge Engineering

Maintenance of transportation facilities is a major industry. Transportation is an important requirement in people's daily lives. Both urban and rural areas are linked by highways and bridges. Global civilization is now dependent upon the use of automobiles and the public transport system. This giant industry maintains infrastructure for safe travel on roads, highways, and interstate highways. Highway maintenance costs hundreds of billions of dollars annually.

Bridge engineering is the marriage of many diverse disciplines including architecture. This variety of disciplines requires some knowledge of the chemistry of natural and man-made materials, composites, metallurgy, structural mechanics, statics, dynamics, statistics, probability theory, hydraulics, and soil science, among other topics. Modern design seems to be reflected in the electronic CAD drawings required in the field for construction and repairs.

Christina Rosetti offered the following poem describing the kind of colorful rainbow bridges she would like to see:

There are bridges on the rivers,
As pretty as you please.
But the road that bridges earth to heaven,
Is prettier far than these.

All constructive ideas seem to benefit mankind!

Engineers have constructed nearly one million highway and railway bridges in the U.S. alone. These include pedestrian, equestrian, and small garden bridges or the hidden culverts. The process has evolved over hundreds of years and is reflected in a wide variety of geometry, structural forms, shapes, and materials.

1.1.2 Benefits of Rehabilitation

1. Improved structural performance based on:
   - Redundancy—Examples are the use of continuous spans, provision of alternate load paths, and reduction of fatigue in fracture critical members.
   - Fatigue performance—Examples are the use of HPS with increased ductility and energy absorption, use of prestressing to reduce tension stress range, and elimination of details more critical than AISC category C.
   - Resistant design against extreme events—such as for earthquakes and foundation scour.
2. Serviceability: Examples are provisions for deck replacement and widening, durability, maintainability, and inspectability.
3. Economy: Examples are the reduction of life cycle costs, and use of efficient design and construction methods, structural details, HPC, prestressed or steel grid decks, weathering steel, and HPS.
4. Constructability: Examples are ease of fabrication, weldability, transportation lengths to eliminate field splices, and ease of erection with minimum disruption to traffic.
5. Diagnostic design: It is project specific and site specific. It differs from new bridge design in a number of ways by requiring:
   - Structure condition evaluation and load rating
   - Alternative analysis and computer applications
   - Use of new repair materials and state of the art rehabilitation techniques
   - Staged construction
   - Modern construction techniques
   - Decision making models such as decision matrix, life cycle costs, and risk analysis.

Table 1.1 shows a list of topics for bridge and highway rehabilitation and related issues covered in this book. In this chapter, an overview of the factors leading to successful maintenance is presented, and a brief history of some of the old and best-kept bridges is provided.

1.1.3 Rehabilitation Is a Means to an End and Not an End in Itself

The bridge is only a small part of the highway. Rehabilitation activities comprise many associated civil engineering disciplines and tasks depending upon the complexity of the project, namely:

- Constraint management including scope, schedule, and budget
- Community outreach and public involvement
- Acquiring the right of way
- Surveying
- Utilities relocation
- Maintenance and protection of traffic or stream flow
- Temporary signals, signing and striping
- Drainage
- Approaches and sidewalks
- Geotechnical engineering
- Inspections and monitoring
- Environmental concerns and acquiring permits.

Hence bridge engineer needs to work as part of a team and not independently.

Table 1.1 List of topics addressed in each chapter of this book.

S. No.	Some of the Topics Addressed	Chapter Number
1	Overview of Bridge Engineering	1
2	Diagnostic Design and Selective Reconstruction	2
3	Bridge Failure Studies and Safety Engineering	3
4	An Analytical Approach to Fracture and Failure	4
5	Load and Resistance Factor Rating and Redesign	5
6	Applications of LRFD and LRFR Methods	6
7	Bridge Widening and Deck Replacement Strategy	7
8	Inspection, Rating, and Health Monitoring Techniques	8
9	Conventional Repair Methods	9
10	Concrete Repair Methods	10
11	Advanced Repair Methods	11
12	Bridge Protection of Bridges Against Flood Scour and Earthquake	12

Rehabilitation and repair topics are constantly changing and represent an ongoing process. A list of publications is placed at the end of each chapter for reference purposes.

The influence of physical location is of paramount importance:

1. Overland bridges may be described as those:

- Located over intersections, with alignment perpendicular to lower level roadway
- Located over intersections, with alignment skew to lower level roadway: In the U.S., the majority of bridges have skew geometry in plan even though urban planning of roads is east-west and north-south.
- Elevated bridges in valleys and gorges, over 30 feet in height
- Located over railroads perpendicular to multiple railway tracks
- Located over railroads skew to multiple railway tracks
- Ramps connecting lower level roadways to higher level roadways
- Flyovers, cloverleaf, and spaghetti intersections
- Third level overpass over an existing bridge.

2. Bridges over waterways include the following:

- Bridges over small rivers
- Bridges over wide rivers
- Elevated bridges over non-tidal rivers
- Elevated bridges over tidal rivers
- Coastal bridges connecting adjacent islands
- Floating bridges over wide rivers.

They have the following issues:

- Underwater construction for footings of abutments and piers—Waterway bridges have deep foundations. They are difficult and more expensive to construct and maintain than land bridges. Neither underwater inspections nor over-water inspections are easy compared to land bridges.
- Flood frequency, water levels, overtopping of water
- Foundation erosion
- Aggradation and degradation.

3. Construction over water would also require:
   - Deep foundation excavation
   - Cofferdams
   - Barges
   - Special supports.

1.1.4 Environmental Issues

1. Thermal loads: They should be applied as daily recurrent cycles for long-term behavior of structures. The method of construction should account for hot weather concreting and cold climate concreting procedures. For each case, deck expansion joints and expansion bearings are provided to release thermal stresses.
2. Wind loads, tornados, and hurricanes: Structures need to be modeled correctly, as and where applicable.
3. Progressive collapse: Seismic zones have large variations. Dynamic analysis methods shall use the classification of seismic zones and return periods. In the case of a collision, earthquake, or flood erosion, the total bridge should not be a write-off and most of the bridge should be salvaged.

   The old practice was to use two rows of bearings on the pier making each span simply supported or partially continuous over the piers. There is an in-built advantage in this approach so that progressive collapse does not carry over to the adjacent span.

1.1.5 Mathematical and Computing Issues

Improved mathematical methods: Mathematical methods for every type of bridge are presently limited. They need to be systematically organized to simulate and compute short-term construction conditions and long-term aspects of repeated loads. Mathematical methods shall adjust for different geometry, substructure type, construction method, construction loads, and all special design issues.

1.1.6 Constructability Issues

One of the practical difficulties in rehabilitation design is to incorporate the vast variation in construction sites, materials of construction, geometry of the bridge, loads, access, topography and location. Construction methods and construction-related analysis and design are linked to bridging requirements. No two sites are alike. The topography may be flat or hilly. Connecting ends of bridges may have very different elevations. The erection of precast bridges over busy highways is a modern day challenge.

1.2 A FRESH APPROACH TO MAINTENANCE

1.2.1 Changing Technology

In order to maintain old and historic bridges, it is important to understand the transition of structural development from old to new technology. For example, use of arch structures made of stone, masonry, or timber was popular in the olden days. Covered timber truss bridges had fewer maintenance problems.

Development of metallurgical processes produced wrought iron and cast iron in larger capacities during the mid-1800s. It soon led to the use of metal trusses. Except in the areas of plentiful and large timber trees, timber trusses quickly lost their popularity in the late 1800s. The earlier use of trusses for small or medium spans has been overtaken with the current use of composite slab-beam structures.

The success of metal bridges has continued, from wrought iron, cast iron, and mild steel to high performance steel. Even in reinforced and prestressed concrete bridges, higher grade steel rods and high tensile steel strands are being used.

Rehabilitation requirements for both older and historic categories demand that safety, durability, and economy be selected as the main criteria with optimization of materials in superstructures, substructures, and foundations serving as secondary criteria.

Figure 1.1 shows the relationship between the percentage decrease in the useful life of a bridge versus time.

It appears that in the last two decades, considerable advancement in an increased array of construction materials, details, components, structures, and foundations has taken place. Hence, it is an opportune time to inculcate thought processes, develop structural solutions, and introduce modern technology such as remote health monitoring.

The dotted line shows continued use of a bridge, but with higher maintenance costs. Deferred/no maintenance, low maintenance, and regular maintenance alternatives need to be considered.

1.2.2 Strategic Topics and Their Advancement

1. Rehabilitation technology: It is more diverse than the original design effort. It involves inspection, interpretation of data, selection of repair and rehabilitation methods, in addition to analysis, computer aided design, and application of AASHTO and state codes of practice.
2. AASHTO's objectives: A plan was developed by the AASHTO Highway Subcommittee on Bridges and Structures in June 2005 to help achieve "safe, cost-effective, low-maintenance, long-life structures." Some of their recommendations, such as accelerated bridge construction, serve as strategic topics and require further attention. Other topics are added here as deemed necessary by practitioners.
3. Technology developments at the state level: It is important to specialize in developing structural solutions for specific issues that may relate to local traffic, topography, geography, environmental loads, and extreme events. Currently, their diversity restricts them from being adequately addressed in the national level design specifications or in the maintenance manuals. In resolving local specific issues in a timely manner, existing bridge design manuals in each of the 50 states serve as necessary supplements to The American Association of State Highway and Transportation Officials (AASHTO) Load and Resistance Factor (LRFR) and Load and Resistance Design (LRFD) design and construction specifications.

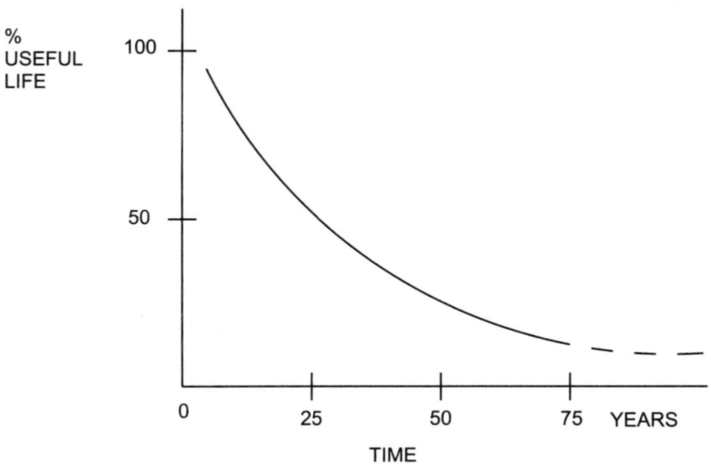

Figure 1.1 The relationship between the percentage decrease in the useful life of a bridge versus time. The dotted line shows continued use of a bridge, but with higher maintenance costs. Deferred/no maintenance, low maintenance, and regular maintenance cases are considered.

Developing effective methods of repair and rehabilitation should result in minimum cost of maintenance for the innumerable situations. This would lead to selection of the most promising and emerging preservation methods to ensure continuity. One of the goals is to focus on safety, serviceability, and economy.

4. New methods of prefab design: Development of LRFR and LRFD guidelines may apply to prefabricated elements, drainage guidelines, and methods for scour protection, seismic design, software applications, and check lists in QA/QC documents.
5. Innovative applications: New materials include FRP composites, SIKA CarboDur and SIKA Wrap or equivalent, accelerators, air entraining admixtures, water reducers, super plasticizers, pozzolans, emulsions, anti-washout admixture for underwater concrete, use of HPC and HPS, scour countermeasures, seismic retrofit, etc.
6. New aids in planning: The use of GIS and imaging technology, stay-in-place formwork, precast deck panels, and orthotropic decks is becoming popular.

1.2.3 New Traffic Loads

In Europe and many other countries, bridges built prior to the 18th Century still exist and continue to serve traffic needs. Older bridges were originally designed to carry much lighter carriages, carts, and wagons, but now heavy trucks, automobiles, and military vehicles are using them. The aim of the new generation of bridges is to provide for heavier truck loads. Such old bridges also need to be rated and posted for maximum live load.

A fresh approach to the evaluation of maintenance issues will help in mitigating the inevitable deterioration of bridges, thus leading to minimal discomfort. An insight into the needed approach is highlighted in this book. Such an approach is likely to reduce costs and construction durations. It may also result in increased standardization, introduce uniformity of construction and lead to longer-lasting, low-maintenance structures.

1.2.4 Explaining the Proposed Rehabilitation Process

Current practice: Inspection reports serve as the eyes and ears of the design engineer. Recommendations coming from the field assessment are evaluated by rating analysis programs. FHWA rating criteria is generally used. Alternatives based on repair methods and cost consideration are studied before preparing rehabilitation designs and drawings. However, newer monitoring methods have emerged using remote sensors and robotics. Examples of other innovations are HPC and HPS materials, isolation bearings and scour countermeasures, LRFR methods, etc.

The following questions need to be re-addressed in light of changing technology:

Why repair and rehabilitate?

Safety, continuity of use, and failure prevention are the primary reasons. The following considerations are of paramount importance:

1. Improving traffic conditions, geometry, sight distance, and clearances.
2. Addressing environmental concerns.
3. Increasing load carrying capacity.
4. Correcting deficiencies.
5. Providing for possible future widening.
6. Minimizing costs to be incurred.

When to repair and rehabilitate?

Repairs should directly follow the recommendations presented in inspection reports. Emergency repairs are generally required immediately after an emergency or after extreme events, such as vessel collision, flood scour, or earthquake. Chapter 3 analyzes various types of failures, their causes, and methods of preventing failures.

What components of bridge to repair and rehabilitate?

Deck, deck joints, or bearings are subjected to deicing salts and constant wear and tear and need the greatest attention. Usually the substructure is to some extent over designed and is less likely to need repairs, except for repairs resulting from erosion or earthquakes.

Who is eligible to repair and rehabilitate?

An experienced or licensed professional engineer is qualified to oversee the delicate tasks.

Where to repair and rehabilitate on a priority basis?

Localized component repairs are needed. Selective criteria can be applied based on location and on a priority basis. Bearings retrofit is one example.

Which bridges to repair and rehabilitate first?

Those which require emergency work have a higher priority. Condition evaluation of each item is based on the list provided by the recording and coding guide. Interstate bridges, those on military routes, or those serving schools, fire stations, and hospitals should have top priority. Evaluation of defects should be confirmed by in-depth inspection or nondestructive testing.

How to initiate repair and rehabilitation?

1. The first step is field inspection and structural health monitoring.
2. The second step is preparing an inspection report.
3. The third step is computing the condition rating and sufficiency rating, for funding approval.
4. The fourth step is analysis and load rating (both inventory and operating ratings). Additional ratings for extreme loads such as scour and seismic vulnerability may be required.
5. The fifth step is preparing a rehabilitation report.
6. The sixth step is implementing diagnostic design procedures.
7. The seventh step is selecting methods of repair, retrofit, rehabilitation, or replacement.
8. The eighth step is preparing contract documents and selective reconstruction.

Is there a choice between rehabilitation and replacement?

With thousands of bridges to be fixed, economics, inconvenience to the public during reconstruction, or sentimental/historical reasons can discourage replacement. Replacement is expensive and causes interruption in service during the construction period. Environmental concerns and permit requirements will be greater for new bridges, especially those with four or more lanes. A sufficiency rating, diagnosis of deficiencies, and cost benefit analysis need to be performed to determine the course of action.

Planning and design procedures for replacement are covered by LRFD Design Specifications and will only be briefly discussed in this book.

1.3 THE NEED TO KEEP BRIDGES FUNCTIONAL

1.3.1 Maintenance Engineering

A bridge needs to be reconstructed after each complete maintenance cycle (Figure 1.2). General considerations include a variety of interdisciplinary approaches, including:

1. Extensive planning considerations, such as
    - Funding and cost
    - Functional requirements
    - Right of way
    - Maintenance and protection of traffic and staged construction
    - Relocation of utilities

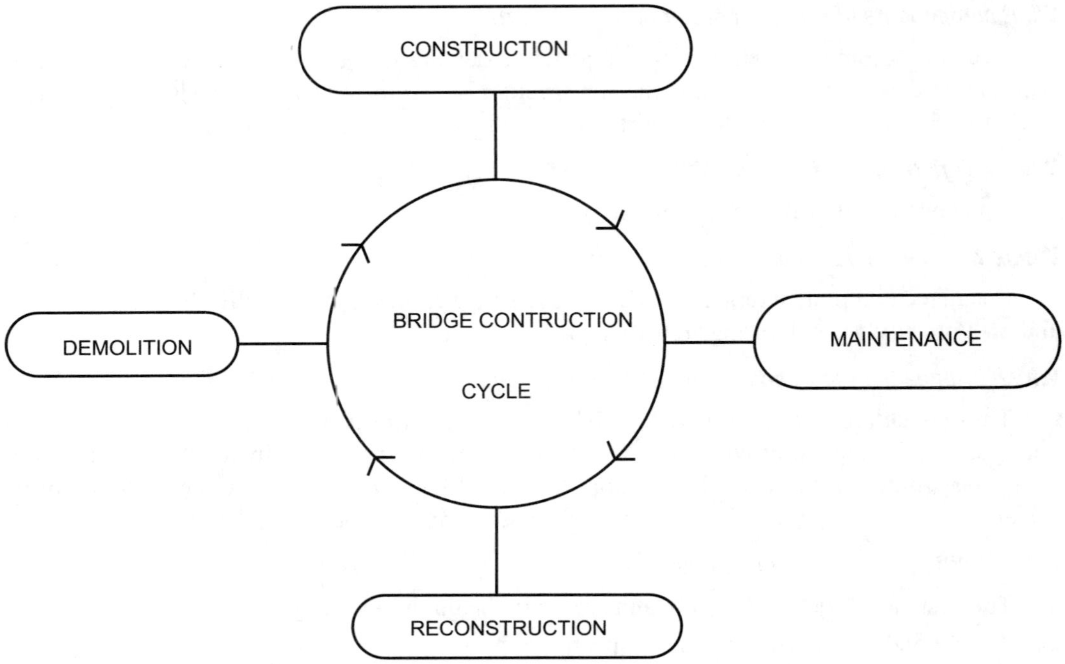

Figure 1.2 The inevitable cycle in the life of a bridge.

- Improving soil conditions
- Use of standard geometry, alignment, and profile
- Vertical and horizontal clearances
- Constructability
- Use of alternates analysis
- Aesthetic requirements
- Environmental permits
- Public involvement
- New technology and innovative methods
- Future maintenance and inspection access
- Development of rehabilitation and replacement schemes
- Performing value engineering.

1.3.2 Types of Materials for Bridge Girders

1. The following types of materials are used for bridge girders:

Type 1–Timber: Solid sawn, glulam.

Type 2–Concrete: Made of cast in place or precast construction:

- Slab bridge, T-beam, reinforced and highly redundant prestressed concrete AASHTO I-girders and box girders, and segmental beams
- Plain concrete or reinforced concrete
- Prestressed concrete, post-tensioned or pre-tensioned.

Type 3–Steel: Rolled steel and fabricated plate girders, multi-girder and through bridges, box girders, trusses, arches, and rigid frames.

- Old bridges were built with mild steel used at the time of construction. Commonly used grades of steel were A7, A373, A36, A440, A441, A572, and A709. Currently grades 50, 50W, 70W (and 100W) are being used.

- Weathering steel: Using ASTM A242, A588 or A709 "W" grade steel does not require painting, except at the ends.
- Special steel: Using special chemical composition steel or with special heat treatment, such as quenched and tempered plate (usually A514 or A517 grade).
- Hybrid steel section: This is used to describe a section, which is composed of more than one type of steel (e.g., web is composed of grade 50W steel and the flanges are composed of 70W).

Type 4–Aluminum, wrought iron, or cast iron. Wrought iron and cast iron were commonly used before 1900.

Type 5–Stone or brick masonry.

2. Selection of modern materials: A choice can be made for arriving at optimum solution for the use of available construction materials.

Steel versus concrete:

Structural steel has the advantage of being a permanent material with repetitive uses extending over decades. When recovered from the demolition of a disused bridge, it is salvaged and sent to steel mills and foundries for re-rolling. With the expansion in transportation systems, there are supply shortages, resulting in higher price as the demand for steel increases.

Recent trends are for greater use of precast prestressed concrete girders in small and medium span ranges, although steel girders are applicable to any span length. HPS 70W steel welded girders are increasingly being used for long spans. Both steel and reinforced concrete are essential materials for bridge construction as indicated by the use of steel in the form of steel reinforcing bars, prestressing strands, rolled steel joists, or welded plate girders.

- Epoxy coated reinforcing steel bars—Substructure rebars are generally plain uncoated bars. Galvanized steel rebars are used in marine environments. Both epoxy coated and galvanized steel bars are more expensive than plain bars. While local bond properties between concrete and uncoated bars is known, tests need to be performed to determine local bond of different types of chemical used in coatings.
- Modern concrete materials—Composites, fiber reinforced polymer concrete, and exodermic decks are being successfully used.
- Recycling of construction materials—Good quality aggregate has the potential to be cleaned and reused for all types of concrete construction. Recycled concrete aggregate from debris can be washed and reused. So far the practice has been to use the debris in landfills and to some extent in pavement construction.
- Normal weight concrete
- Lightweight concrete
- High performance concrete
- High performance steel
- Aluminum
- Glue laminated timber: Glulam hardwood bridges are economical for small spans.

The following precautions are also applicable to timber bridges:

- Timber friction piles can be damaged while driving in firm ground.
- Timber pile piers in rivers should not be subjected to debris accumulation.
- Timber bridges are posted for low speed.

1.3.3 Span Classification

In practice, most bridges are a single span with two or three lanes over narrow rivers. Piers are needed for continuous spans over highways, valleys, or wide rivers. The minimum single span for a bridge is 20 feet, below which a culvert is normally used. Pedestrian bridges with lighter live loads can be smaller in length than 20 feet.

Table 1.2 Shows the overlap of bridge types for a variety of span lengths.

S. No.	Approximate Range of Span Length (feet)	Bridge Type Selection
CONCRETE		
1	20–50	RC/PS slab
2	50–140	PS box
3	30–150	PS I/bulb girder
4	140–200	PT box girder
5	130–750	PT segmental box
6	6–36	Precast concrete culvert
STEEL		
7	40–110	Composite beams
8	60–300	Steel plate girder
9	100–600	Steel box girder
10	500–1700	Steel arches
11	500–1800	Steel truss
FLEXIBLE CABLE		
12	500–1500	Cable stayed box
13	600–2000	Steel cable stayed
14	1000–6500	Steel suspension

For practical considerations, the selection of bridge types may be classified broadly as:

1. Small span (20 to 40 feet): Economical examples are reinforced concrete slab and T-beam bridges, precast prestressed cellular deck bridges, timber bridges, prestressed concrete adjacent and spread box beam bridges. Bridges with steel stringers have relatively higher life cycle costs for small spans when compared to timber or modern precast concrete bridges.
2. Medium span (over 40 to 120 feet): Economical examples are prestressed concrete adjacent and spread box beams and steel girder bridges.
3. Long span (over 120 to 240 feet): Economical examples are steel girder bridges (50W or hybrid 70W and 50W grades), steel deck and through trusses, and prestressed concrete arches.
4. Very long span (over 240 feet): Economical examples are steel arches, prestressed concrete segmental boxes, cable stayed, and suspension cable bridges.

1.3.4 Deck Geometry, Components, and Materials

1. Geometric shape of deck: Common deck shapes are rectangular, skew, or curved. There are separate AASHTO bridge design specifications for curved girder bridges. Super elevation and sight distance requirements would reduce accidents.
2. Deck joints, deck grooves, longitudinal and cross slopes, drainage inlets.
3. Deck material: Timber, concrete, and steel have been largely used for decks. Concrete is widely used for all types of traffic. FRP (fiber reinforced polymer) decks are now being used. Exodermic (grillage) and orthotropic decks are also being used.
4. Prefabricated decks and composite steel-concrete I-beams are popular for small and medium spans. Examples of proprietary products are CONSPAN and former INVERSET techniques.
5. Deck width is based on traffic volume and number of lanes. It should match with approach roadway width and acceleration and deceleration lane widths.

1.3.5 Girder Shapes

The most commonly used girder shapes are (Table 1.2):

- I-shaped girders
- Adjacent or spread box beams
- Composite beams
- Through girders.

1.3.6 Typical Tasks Associated with Rehabilitation (see Section 1.7)

1. Site surveying and in-depth inspection.
2. Geotechnical investigation.
3. Redesign criteria/application of LRFD method
   - Live load analysis
   - Hydraulic and scour analysis
   - Seismic analysis
   - Structural calculations for sizing and connections.

1.3.7 Expected Life of a Bridge

Some bridges in the U.S. have survived for over 150 years with proper maintenance. New design techniques assume a shorter life of a bridge or just the superstructure of 75 to 80 years. With changes in demographics and urban congestion, intersections may need to be re-planned or widened.

Planning considerations and design philosophy about performance of construction materials and joints, etc. have changed since the time when older bridges were designed.

Some bridges may become functionally obsolete sooner and extreme events, accidents, or fatigue may increase the life cycle costs and cause them to be replaced even before the stipulated 75 years.

1.3.8 Published References and Resource Applications

An approximate allocation of design time needed is as follows:

1.	AASHTO Code	25 percent
2.	State design codes (text)	15 percent
3.	State and in-house standard details	15 percent
4.	Textbooks, handbooks, and HEC-18, HEC-23, etc.	10 percent
5.	Computer software/math CAD and Excel spread sheets	20 percent
6.	Technical specifications	10 percent
7.	Construction specifications	5 percent
		100 percent

AASHTO Codes consist of important subjects covering bridge design specifications, sign support structures, and design specifications for curved girder bridges.

1.3.9 Contract and Bid Documents

Comprehensive bid documents for plans, specifications, and estimates (PSE) are required:

1. Construction plans.
2. Specifications.
3. Estimates.
4. Construction schedule.

1.3.10 Post-design Tasks

1. Responses to contractor's queries prior to selection.
2. Bid approval.
3. Construction coordination, field meetings.
4. Answering request for information (RFI).
5. Design change notices (DCN).
6. As-built drawings preparation.

1.4 MAINTENANCE OF A SUPERSTRUCTURE

1.4.1 Maintenance versus Replacement

1. The superstructure is subjected to greater wear and tear from traffic while the substructure is less affected. As a result, substructures are not replaced as often as superstructures. Hence, the life cycle costs and rehabilitation efforts are greater for the superstructure components.
2. Replacement of superstructure components is more frequent than for the substructure. This is discussed in detail in a later chapter. Replacement can apply to a component or to the entire superstructure:
   - Parapet replacement
   - Deck overlay replacement
   - Deck joint replacement
   - Deck and parapet replacement
   - Deck, parapet, and girder replacement
   - Deck drainage replacement
   - Bearing replacement
   - Entire superstructure replacement.
3. Maintenance applies to:
   - Existing concrete deck repairs
   - Deck protective systems
   - Deck drainage
   - Bearings retrofit
   - Concrete structure repairs
   - Steel girders rehabilitation.

1.4.2 Parapets and Railing Designs for Crashworthiness

A wide variety of railing styles can be seen on bridges throughout the U.S. Railing not meeting current standards requires upgrading, and upgrading of non-crash tested railing is necessary. Parapet repair or replacement using New Jersey Barrier (Figure 1.3) is also required. The following details are required:

1. Details showing removal of existing concrete.
2. Dimensions for placement of new concrete.
3. Treatment of the parapet at expansion joints.
4. Parapet transition details.
5. Typical sections with reinforcing steel.
6. Joint spacing.
7. Limits for purpose of measurement and payment.
8. Pay item of the work to be included with each item.

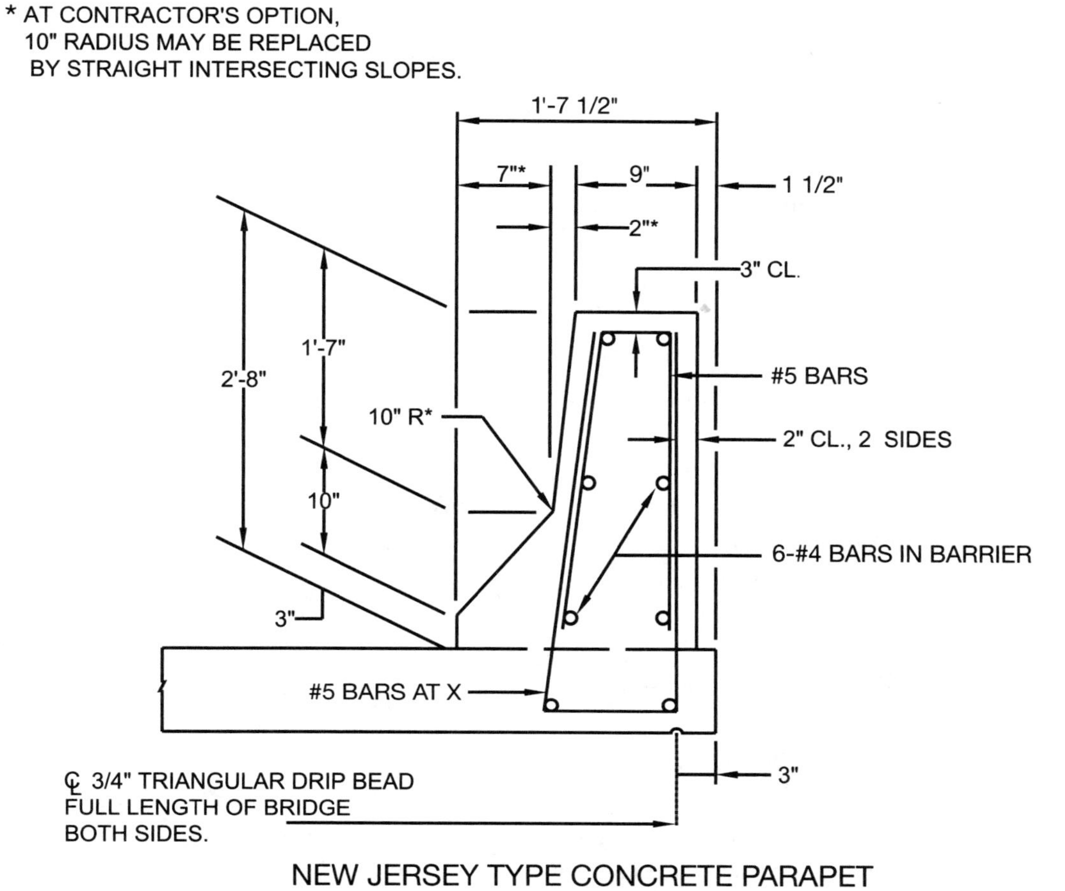

Figure 1.3 Crash tested barriers require regular maintenance.

1.5 MAINTENANCE/REPLACEMENT OF THE SUBSTRUCTURE

1.5.1 Common Bearing Types

1. Rocker and roller bearings
2. Elastomeric pad
3. Multi-rotational
4. Isolation bearing
   - New bearings are preferred over trying to repair existing bearings.
   - Elastomeric pad bearings involve the least maintenance.
   - If one bearing is defective, replacing the entire row may be considered due to uniform performance.
   - Use of a single row of bearings for continuity is the modern trend.
   - The provision of vertical curves increases vertical clearance and allows navigation.

The following types of bearings are seen on U.S. bridges:

Type 1

- Steel roller
- Steel rocker.

Type 2

- Steel sliding on phosphor bronze
- Steel sliding on steel
- Steel sliding on lubrite.

Type 3

- Pot bearing with P.T.F.E. (Ex.Teflon).

Type 4

- Multi-rotational (pot bearing) guided
- Multi-rotational (pot bearing) unguided
- Multi-rotational (disc bearing) guided
- Multi-rotational (disc bearing) unguided.

Type 5

- Elastomeric with P.T.F.E. (Ex. Teflon)
- Elastomeric, fabric type with P.T.F.E. (Ex. Teflon)
- Elastomeric, steel laminated
- Elastomeric, fabric laminated
- Elastomeric, steel laminated w/Ext. load plate
- Elastomeric, steel laminated w/lead core
- Elastomeric, laminated with P.T.F.E. (Ex. Teflon).

1.5.2 Substructure Replacement

In the past, there has often been an over design by using gravity and massive wall type abutments, piers, and their foundations. This had an in-built advantage in that when it came to replacement, only the superstructure was replaced.

It is an unusual situation if the substructure needs to be replaced while the superstructure is in satisfactory condition. In some cases, superstructures can be lifted off the bearings and reused. In most cases, the entire bridge is replaced, except when using the roll in-roll out technique which can preserve the superstructure.

1.5.3 Abutment Types

1. Cantilever wall:
   - Full height abutment
   - Mid-height abutment
   - Stub and semi-stub abutments (shown in Figure 1.4).
2. Spill-through abutment.
3. Modern types are:
   - Integral abutments
   - Semi-integral abutments
   - MSE (mechanically stabilized earth) wall abutments.

1.5.4 Pier Types

Multiple bents and flared caps are aesthetically pleasing. Common shapes are shown in Figure 1.5 and include:

1. Solid wall
2. Hammerhead

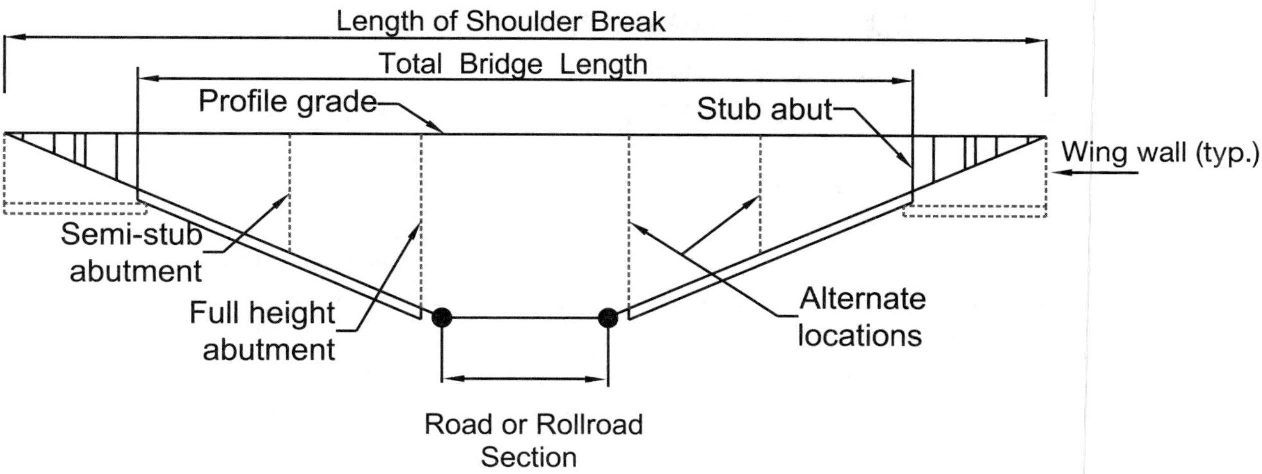

Figure 1.4 Alternates with full height, semi-stub, stub, semi-integral, and integral abutments.

3. Multiple column bent
4. Modern types are:
   - Multiple pile bent
   - Integral pier.

1.5.5 Foundation Types

1. Spread footing.
2. Drilled shaft or caisson wall.
3. Pile foundation (end bearing or friction piles)
   - Steel H pile or W sections
   - Steel pipe pile
   - Concrete pile or steel encased
   - Prestressed concrete pipe
   - Steel sheet piles.

For the selection of the foundation, the expertise of a geotechnical engineer will be utilized.

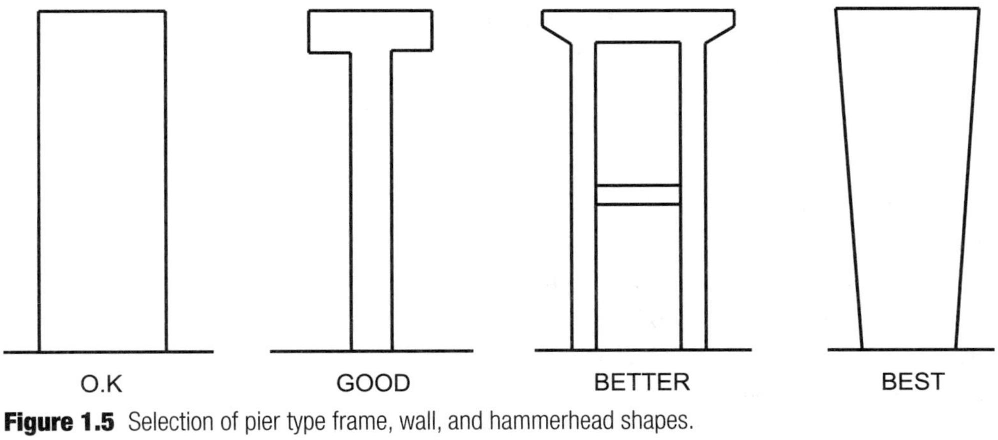

Figure 1.5 Selection of pier type frame, wall, and hammerhead shapes.

1.6 OVERVIEW OF MAINTENANCE PROCEDURES

1.6.1 Classifying the Type of Bridge

1. Importance: A bridge is a necessary but small part of a highway system. Its importance is linked to:
   - The network of roads it serves
   - The volume of traffic it carries
   - Its relative importance to meet funding criteria and funding priority.
2. Inspection classification: For inspection purposes, many highway agencies have the following classifications:
   - Inventory: A bridge included in an inventory file when it carries moving loads.
   - Collapsed: A bridge that once satisfied the inventory bridge definition, but is now closed due to collapse.
   - Closed: A closed bridge that once satisfied the inventory bridge definition, but is now temporarily closed for any reason except collapse. Secondary uses such as pedestrian traffic may be allowed.
   - Abandoned: A bridge that once satisfied the inventory bridge definition, but is now permanently closed.
   - Deleted: A bridge which has been deleted from the inventory.
   - Temporary: A bridge that is used to maintain traffic during a modification or replacement.
3. Functions: In the U.S., bridges are located on one of the following networks and are classified as such:

   Type 1–Interstate

   Type 2–Arterial

   Type 3–Collector

   Type 4–Local

   Forty percent of all bridges serve local roads. Thirty-three percent serve interstate or arterial highways. Interstate bridges allow higher speeds. Interstates have express lanes but are thoroughfares with limited access and exits. Interstate and arterial bridges carry almost 90 percent of average daily traffic (ADT) for rural and urban areas. Twenty-seven percent of bridges serve collectors. Collectors collect and distribute traffic between arterials and local roads.

   They are typically two lane roads and provide for shorter trips at lower speeds.
4. Type of traffic: Each type has special requirements for varying live load impacts. Deck surfacing is made of timber, concrete, or steel deck:

   Type 1–Highway bridges carrying vehicular traffic

   Type 2–Transit and railroad bridges carrying train traffic

   Type 3–Pedestrian bridges

   Type 4–Equestrian bridges

   Type 5–Airport bridges carrying aircraft
5. Ownership

   Ownership governs design criteria and procedures for maintenance or reconstruction:

   Local government owned—51 percent

   State government owned—48 percent

   Federal government owned—1 percent

Figure 1.6 A scenic countryside concrete bridge with a waterfall.

6. Construction materials

Old bridges contain older types of materials while new bridges contain more modern materials. Science and technology offer us the opportunity to develop high performance materials that make bridges more durable and easy to maintain. Providing a longer lifespan can justify a greater initial investment.

- New timber materials: Laminated or composite timber, bamboo sheets
- New concrete materials: Normal weight, lightweight, high performance concrete with corrosion inhibitor overlays
- Metal: High performance weathering steel, aluminum
- Composites and fiber reinforced polymers
- Recycled materials.

The girders commonly used for small or medium spans are made of timber, steel I beam, reinforced concrete, prestressed concrete I shapes, or box sections. Arches have been constructed in masonry, timber, steel, and concrete. Steel cables and trusses are used for long spans. Aluminum and composites used for pedestrian bridges result in lightweight decks.

7. Geometry

Structural analysis is based on bridge geometry:

Type 1–Normal right angle plan

Type 2–Skew plan

Type 3–Horizontally curved plan

Type 4–Bridge on curved vertical alignment

8. Span length

Construction issues are based primarily on span lengths:

Type 1–Short span < 50 feet

Type 2–Medium span 50 to 500 feet

Type 3–Long span > 500 feet

9. Structural system

The design of a bridge is related to the structural system. Beam, truss, and arch configurations may be used for medium span lengths:

Type 1–Slab

Type 2–Through

Type 3–Slab-beam

Type 4–Truss

Type 5–Arch

Type 6–Cable stayed

Type 7–Segmental

Type 8–Suspension

1.6.2 Geometric Requirements for Surface Elements

1. The components of surface elements are lane, shoulder, sidewalk, approaches, and ramps.

2. A typical travel lane is 12 feet wide. For staged construction, it can be less than 12 feet (but not less than 10 feet). Minimum width of a vehicle is 4 feet between wheel centers and generally 6 feet overall. A vehicle with wide load is required to display warning sign "WIDE LOAD."

3. The minimum width of a shoulder is 3 feet between the edge of the travel lane and the concrete barrier and less than 3 feet between the edge of the temporary lane and the concrete barrier during staged construction. Small shoulder widths serve as buffer zone to avoid accidents. Standard shoulder width is 10 feet with a minimum 6 feet width for emergency.

4. For safety reasons, a sidewalk is generally provided on both the sides of the roadway. Even during staged construction a provision for temporary pedestrian bridge and utility support is usually required (Figure 1.7). Sidewalks are elevated by 8 inches from the outer edge of the shoulder or the outer edge of the lane. For heavy traffic, a safety fence is required. The typical width of a sidewalk is 5 feet.

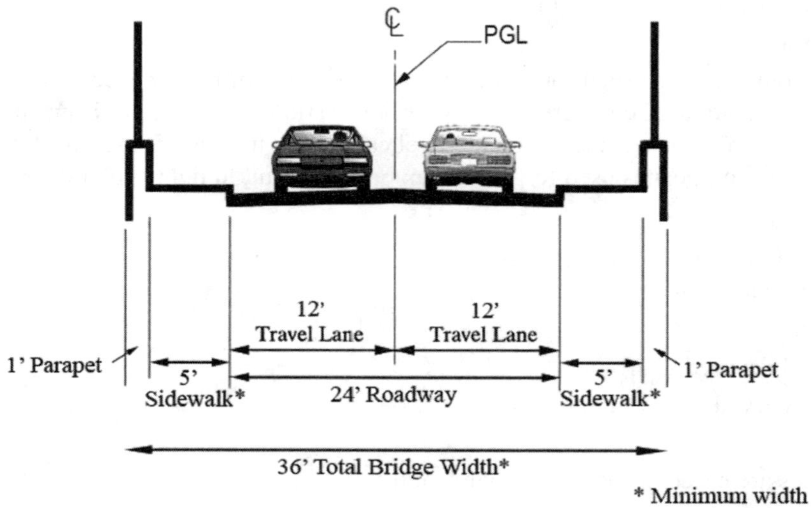

Figure 1.7 Proposed bridge typical section.

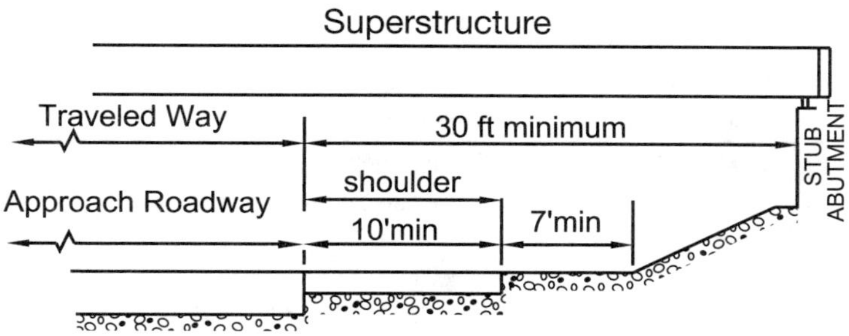

Figure 1.8 Minimum horizontal clearances to abutment walls measured from the edge of lane.

5. Entry or exit ramps connect two levels of traffic moving approximately at right angles. For safety reasons both entry and exit ramps are located adjacent to the right lane which carries slower traffic. A ramp has traffic moving in a single curved direction while a bridge has traffic moving in both directions. An acceleration lane is for transition from a slow speed entry ramp merging into fast moving traffic. Similarly, a decelerating lane serves as a transition between a fast lane and slow speed exit ramp.

1.6.3 Horizontal and Vertical Under Clearances

1. If existing horizontal or vertical clearances are not adequate, the existing bridge needs to be replaced with a new bridge that has higher clearances. As an alternate, posting for vertical over clearance or under clearance is required in keeping with agency requirements. AASHTO specifications have defined minimum horizontal and vertical clearances to bridge substructure and superstructure. These may be modified by state and local codes.

2. The minimum horizontal clearance between the edge of the lane and the concrete face of the abutment or pier is applicable. A commonly used minimum horizontal clearance is 30 feet to abutment face from the edge of the travel lane (Figure 1.8 and Table 1.3) and 16 feet 6 inches for minimum vertical clearance (Table 1.4) from the top of the road surface and minimum 23 feet from the top of the rail (Figure 1.9). Older bridges were designed for lower

Table 1.3 Minimum horizontal clearances.

Serial Number	Bridge Type	Located Over	Minimum Horizontal Clearance	Remarks
1	Highway	Intersection	30′-0″	AASHTO code governs
2	-do-	Waterway	N/A	Due to soil erosion, horizontal clearance is difficult to maintain
3	-do-	Railroad		-do-
4	Railroad	Intersection	14′–0″	AREMA code governs
5	-do-	Waterway	Varies for each river. Determined by Coast Guard	Due to soil erosion, horizontal clearance is difficult to maintain
6	-do-	Railroad	14′–0″	-do-
7	Pedestrian	Intersection	14′–0″	Less maintenance due to low live load fatigue
8	-do-	Waterway	N/A	-do-
9	-do-	Railroad	14′–0″	-do-

Table 1.4 Minimum vertical clearances.

Serial Number	Located over	Minimum Vertical Clearance	Remarks
1	Interstate highway	16′–6″ (5.03 m)	Freight trucks with unusual height to utilize selected routes
2	State highway	15′–6″ (4.72 m)	Height caters for most of trucks
3	Local street	14′–6″ (4.42 m)	Minimum truck traffic required on local streets
4	Pedestrian	17′–6″	Comfort of walking with the least noise from traffic above
5	Waterway	Usually determined by Coast Guard	Applicable only for navigable rivers
6	Over railroad	23′–0″	Trains with special freight height

height trucks than those on highways today. Any deviation needs to be posted as warning signs to prevent accidents.

3. Vertical clearance requirements: Minimum requirements are based on the importance of the highway. It would be uneconomical to design all bridges to a single horizontal or vertical clearance requirement rather than based on their importance and frequency of use. Some bridges may have additional levels for carrying traffic, e.g., George Washington Bridge, New Jersey.

In such cases there would be top level, bridge deck level, and lowest level such as highway, railroad, or navigable river in a direction at a right angle.

4. Each highway agency can modify clearances to a certain acceptable extent from those laid down by AASHTO. These variations need to be approved by AASHTO before the state

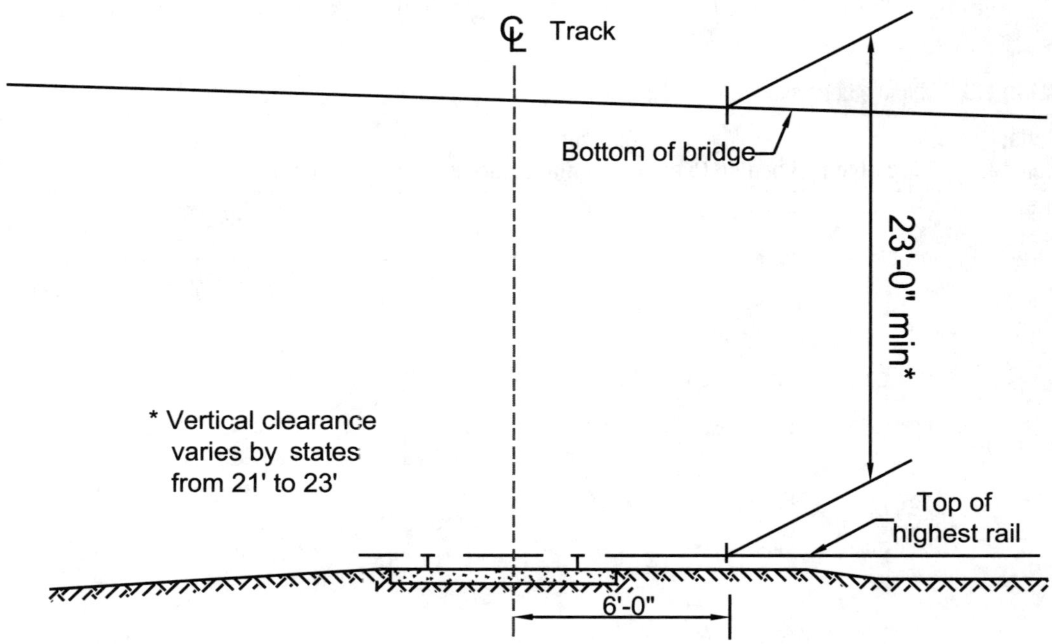

Figure 1.9 Vertical under clearance requirement for a railroad.

highway code is implemented. The changes can be either less or more based on experience, judgment, and special conditions present in that state. For projects funded by federal programs, AASHTO specifications must be followed. The highway agency can approve design modifications when planning a new bridge based on prevailing clearances of bridges located on that highway.

1.6.4 Physical Parameters for Reconstruction

Figure 1.10 shows that in addition to material, span length, geometry, and vertical under clearance, average daily traffic (ADT) volume will have an influence on selecting the type of bridge for reconstruction.

1.7 A STRATEGIC PLAN TOWARD REPAIR AND REHABILITATION

1.7.1 Adequate Funding Availability

Adding new highways and structures requires considerable funding, both for short-term construction and for long-term maintenance. Unfortunately, it is difficult to guarantee availability of sufficient maintenance funds, say 20 years after construction. As more bridges are added to the current pool, the greater the increase in future maintenance costs.

Highway agencies need to ration their funding between new bridge construction and maintaining existing ones, with the latter taking priority. One way to control costs is to improvise, and another is to reduce maintenance costs per bridge with help from modern engineering and technical know-how. Typical funding sources are addressed in Chapter 2.

1.7.2 Normal Maintenance Procedures

In the U.S., bridge inspectors and structural engineers are responsible for performing inspections and preparing reports for rehabilitation (see heading number 1.2.4). The following steps are followed:

1. An inspection report is prepared describing the condition of the bridge, its past history, diagnosis of defects, methods of repair or retrofit, and cost estimates.
2. Structural inventory sheets and field data are obtained from two yearly inspections by inspection teams using the PONTIS system. Remote sensors are also used.

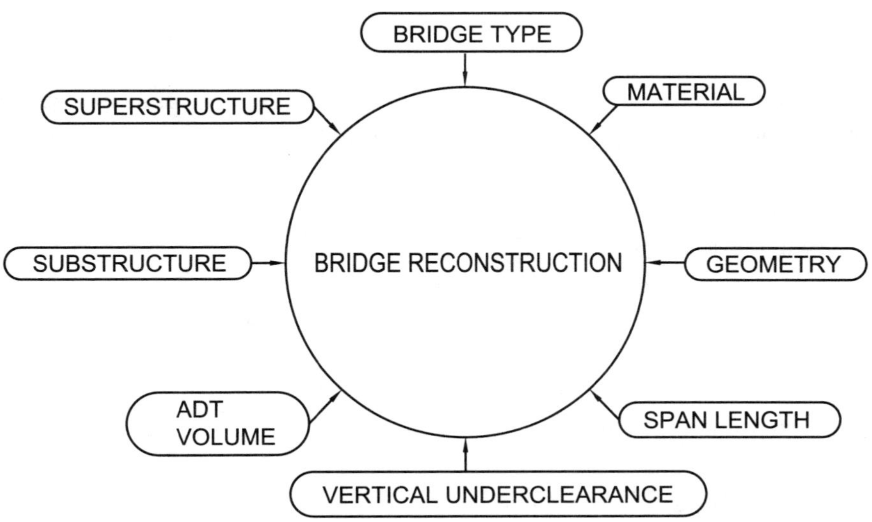

Figure 1.10 Physical parameters to be considered in reconstruction.

3. Diagnosis of defects and classification of deficiencies are based on procedures provided in the AASHTO Recording and Coding Guide. The sufficiency rating method is used.
4. Rating analysis is carried out for acceptable live load trucks. Inventory ratings and operating ratings are computed based on formulae for load capacity.
5. Condition rating and sufficiency rating are evaluated.
6. In addition to live load ratings, ratings for extreme events such as scour sufficiency rating and seismic rating are computed using special methods.
7. A structural team prepares diagnostic design and construction documents based on recommendations from the inspection report.
8. A rehabilitation report is prepared for record purposes and includes a feasibility analysis and approximate costs.

1.7.3 Developing a Maintenance Manual for the Owner

1. Since many times the highway agency is overseen by a non-technical person, the importance of the engineering aspects of maintenance may not be fully appreciated. If a prescribed manual is made available in simple terms highlighting the importance of rehabilitation, methodology, and life cycle costs, it can make decision making for funding easier and more timely.
2. For minimizing future life cycle costs, a maintenance manual (akin to that of a maintenance manual provided with a car at the time of purchase) should be developed and provided for future use by maintenance engineers.
3. Most highway agencies appoint consulting teams for inspection and rehabilitation of thousands of bridges each year. In such situations, a maintenance manual is required for uniformity of methods for all the bridges under the control of the agency.
4. Methods for repairing an old bridge are different from the newer types due to the materials used, span length, lane width, etc. A maintenance manual should address the uniqueness of each material. It will be unique for each type of bridge since the technology used for timber, steel, or concrete bridges would be different.
5. Sections 2 and 3 of this book address the theoretical and practical concepts and procedures for the rehabilitation, repairs, and computer aided design, which form the basis of a maintenance manual. Guidelines given in the manual will be different for each bridge and will depend upon the type of bridge, its function, and performance.

1.7.4 Maintenance Manual Contents

1. A maintenance manual may be included in the scope of work of the structural engineer or the owner's representative. A maintenance checklist needs to be provided which includes:
   - Inspection methods
   - Structural health monitoring
   - Rating and structural evaluation
   - Diagnostic design, repair, or retrofit methods including repainting.
2. Contract documents: The following documents shall be made available with the proposed maintenance manual for future use and reference:
   - As-built drawings
   - Ready access to construction history and any special features to be monitored
   - Technical specifications
   - Actual strengths of concrete, steel, and other materials from laboratory tests
   - Details of vendor supplied products such as bearings and connections.

Experience has shown that many of the construction documents may be lost over time. Databases of scanned copies should be maintained as electronic files.

3. Checklist

   A checklist of bridge inventory items for structural evaluation should include:

   - Deck repairs or replacement
   - Deck joint repairs or replacement
   - Parapet or railing repairs or replacement
   - Repainting of girders.

4. Records of emergency repair methods, including accidents, fire, extreme events, etc.
5. Maintenance schedule.

1.8 A LOOK AHEAD FOR PROGRESS IN MAINTENANCE

1.8.1 Upgrading Knowledge Database/Theoretical Approach

1. Progress in the structural mechanics and applied mathematics areas has not been significant in the past 50 years. Development of closed form solutions is desirable to:
   - Identify and maintain consistent reliability indices within LRFD for all bridge and highway structure elements, including calibration to reflect local materials and practices.
   - Identify and calibrate the service limit states.
   - Integrate information from maintenance and operations into code development and vice versa.
   - Identify load distribution for foundation elements.
2. For simple spans, design formulae for moving live loads are developed by the author in Chapter 4. Similarly, rating formulae are provided in Chapter 6. Additional concepts are provided through design techniques covered in Section 2.
3. Load and resistance factor rating (LRFR) methodology has been adopted by different states. There is always a need to improve the durability of transportation structures within the context of the LRFD specifications.

1.8.2 Accessing Databases of Bridge Failures

Applying failure theories and introducing preventive methods: The life of structures can be increased by studying the reasons for failures and developing a methodology to minimize them. Details are provided in Chapter 3 of this book.

1.8.3 Improved Seismic Resistant Design and Erosion Science

Methods to monitor foundations against extreme events to protect and/or strengthen foundations against scour, earthquake, and impact damage are required.

The objectives are to strengthen the soil, liquefaction mitigation, and to protect foundations against saltwater and corrosion. Retrofit techniques for improved foundation performance under extreme events need to be developed.

Details for scour and seismic retrofits are provided in Chapter 12 of this book.

1.8.4 Optimizing Structural Systems by Use of Enhanced Concrete and New Materials

1. Deck overlays, high-performance concrete, high-performance and corrosion-resistant steels, and fiber reinforced polymer composites have demonstrated significant initial and long-term cost savings, and more efficient construction, resulting in less traffic disruption. Geo-materials, geo-synthetic products, and ground improvement techniques can be utilized. Some of the newer materials like self consolidating concrete and ultra high performance concrete for superstructures can increase efficiency.

2. Construction materials may be classified as traditional, new, or emerging. The goal is to fully employ these, and even newer optimized materials, by exploiting the potential and properties of materials, including life-cycle performance. Their advantages and limitations need to be understood. The objectives are:
   - Assessment of real and perceived barriers to deployment of various elements of optimized structural systems.
   - Development of appropriate state criteria for the use of these materials, details, components, and structures for adoption into the LRFD specifications.
   - Quantification of the impact of increased traffic volume and loads, nondestructive tests, methods for protection, preventing salt ion intrusion, and new materials and techniques for deck construction and repairs.

1.8.5 Use of Construction Systems Such as Accelerating Bridge Construction (ABC) and Design-Build Construction (DBC)

Many states have legislation that mandates minimizing traffic disruption during construction. Innovative construction methods, materials, and systems are needed for reducing on-site construction time. Accelerated bridge construction results in projects being completed more quickly and therefore the impact to users may be lessened.

Projects can be completed while maintaining traffic capacity including little impact to peak traffic. Selection of the most effective techniques, the most promising emerging techniques, and benefit/cost parameters is required to indicate when accelerated construction is appropriate.

Accelerated bridge design and construction research will advance technology by developing improved prefabricated structural systems using enhanced details, materials, and foundation systems, using modern technology such as high capacity cranes and tractor trailers.

The organization of a combined contractor-consultant team has reduced some of the friction and lack of coordination between contractors and consultants. Traditional systems have been refined by contractors playing a greater role as team members, which has resulted in more realistic construction related design, better use of contractor's resources, and a faster turnout in construction.

1.8.6 Extending Service Life

1. Using efficient repair methods and improved retrofits.

   It is important to understand the processes that decrease the serviceability of existing structures and to determine the optimum time to apply the preservation methods. There is a need to develop approaches to preserve, maintain, or rehabilitate the existing system by managing these processes. An outcome would be to develop strategies that would extend the service life of the existing inventory of bridges and highway structures.
2. Main load carrying members are the core of any bridge. They require :
   - Repair and strengthening methods, improved retrofit and replacement of bearings. The majority of retrofits are in the form of upgrading the bearings, scour countermeasures, and use of crash tested parapets. Details are provided in Section 2 of this book.
   - Methods to eliminate expansion joints and bearings
   - Corrosion mitigation techniques including coatings.
3. Substructures would require methods for strengthening piers and abutments.
4. Deployment of the most promising emerging preservation methods would be required.

   These concepts are provided in design techniques covered in Section 2 of this book.
5. Modernizing structural details such as using jointless decks.

   Integral and semi-integral abutment bridges are being used extensively to eliminate the problematic deck joint. Jointless bridge systems result in more durable bridges since combined longitudinal frame behavior is offered by moment connections between the ends of beams and pinned connections with the structural approach slab resting on grade.

1.8.7 Improved Bridge Management

1. Monitoring by using NDT, SHM and remote sensors: The potential exists for the development of early problem detection and warning systems and the use of NDT facilities. Enhanced nondestructive evaluation (NDE) and visual techniques can result in increased structural reliability. Monitoring systems assist in more efficient management of existing bridges and highway structures. Cost savings can be achieved through efficiently managing existing bridges by implementing bridge monitoring systems. The present biennial bridge inspection interval could be transitioned to longer periods through the use of enhanced monitoring.
2. The management objectives are:
   - To understand what information should be collected from prototype structural components and to characterize the condition, or health, of both the superstructure and the substructure. Through efficient asset management, the service life of bridges and highway structures can be extended.
   - Promising cost-effective technologies and enhanced monitoring strategies need to be identified.
   - Deployment of multiple integrated health assessment systems is needed.
3. Resources for bridge inspection are becoming scarcer as inspection budgets are strained by our aging bridge inventory. These circumstances require more intensive inspection, but at a lower cost. With the recent increase in available monitoring and computing technology, it is feasible to develop and deploy intelligent bridge monitoring systems.
4. Implementation of effective monitoring systems results in reduction of man-hours and development of optimum inspection and repair schedules. Effective monitoring systems can:
   - Assess long-term performance and increase system reliability
   - Improve the credibility of inspections and subsequent ratings through less subjective data
   - Improve uniformity of data, enabling the development of better decision-making tools
   - Improve and augment visual assessment, and provide early detection and warning.

   Details are provided in Chapter 6 of this book.

1.8.8 Enhancing Knowledge Base

1. Applying knowledge and use of FEM by software development, CADD, and graphics:

   The quality of a technical workforce must be maintained to meet the rapid development of new technology and the preservation of existing technology when required. Improvement in relationships between the various bridge industry sectors, i.e., owners, consultants, industry, and academe, can assist in maintaining a quality workforce.
2. The current workforce needs to be trained in planning and leadership development.
3. Long term economic benefits can be achieved by disseminating bridge engineering knowledge and developing new and more effective approaches consistent with the evolving bridge-engineering community and emerging technology.
4. Research ideas can be found through journals, seminars, and conferences.
5. Managing pools of knowledge presented in textbooks, journals, conference proceedings, and seminars: For the highways agency, research is a key source of new ideas, knowledge, tools, and technologies. We need to deliver innovative solutions for our customers. It also enables us to set technical standards and provide expert advice so that our bridges are safe and reliable.

1.8.9 Continuing Education and Leadership Training of Engineers

1. The survival of bridge engineering as a flourishing profession requires:
   - Continuous education consistent with industry needs
   - Increased dissemination of technical information across networks
   - Integration of existing and new information.
2. The training objectives are:
   - Identification of good professional and undergraduate training strategies
   - Identification of strategies for the establishment of a bridge-engineering knowledge database
   - Assessment of the current level of collaboration between academe and the rest of the community,
   - Identification of successful mentoring and succession planning strategies.

1.8.10 Preservation Techniques for Historical Bridges

Bridges that are landmarks serve as tourist attractions. Their pictures appear on postcards that are purchased and mailed to other countries by visitors. Their maintenance is special and has been possible through advancements in remote sensors, robots, coatings technology, and high strength steel and concretes.

Rehabilitation requires a specialized approach known as a "preservation design."

1.8.11 Improving the Security of Bridges

1. Recent natural disasters and the increasing threat of terrorism highlight the need for effective monitoring and for rapid recovery of the use of our bridges and highway structures.

   Once again modern technology is providing a sophisticated approach to security such as the use of wide band Internet networks for all security systems, digital CCTV surveillance systems, access control systems, and biometric devices.
2. To avoid high cost investments in installing security systems, it is important to analyze the risk assessment. For important bridges carrying high ADT the design criteria needs to consider structural response to applicable blast loads similar to subjecting the bridge to a high magnitude (safe shutdown) earthquake. For continuous bridges sudden and progressive collapse can be avoided by retrofits such that the resulting damage of a bomb blast can be isolated.
3. Deployment of the performance standards for security design of major bridges and development of a performance-based specification and accompanying design manual is required.

1.8.12 Use of Aesthetics in Planning

Bridges serve as national landmarks. *A thing of beauty is a joy forever.* Details are provided in heading number 1.12 under "Preserving Aesthetics."

1.8.13 Contributing To FHWA National Policy

1. Project decisions affecting technical, cultural, and cost issues are being made without receiving adequate input from bridge engineers. There is a need to expand the role of the bridge engineer in transportation development and in social and policy development.
2. Enhanced contributions of bridge engineers to transportation policy decisions can result in:
   - More practical input to context-sensitive design approaches
   - Enhanced utilization of transportation systems through nationwide uniformity in size and weight restrictions
   - A balanced view on environmental project requirements.

3. National policy objectives as defined by FHWA are:

- To understand the functioning and decision-making consequences affecting transportation systems
- To develop strategies in which bridge engineers more effectively contribute to transportation-policy decisions.
- To develop recommendations to AASHTO on oversize/overweight vehicles and the long-term impact of construction on the environment
- To develop strategies to enhance public involvement of bridge engineers, including outreach to all stakeholders.

1.8.14 Implementing Diagnostic Design Procedures

1. Preliminary diagnostic design

The steps are:

- Preparing plans indicating type, size, and location (TS & L)
- Preparing plans indicating geometry, alignment, profile, span length, and beam spacing
- Preparing a report on accessibility to the site, estimates of cost, and material availability.

2. Final Diagnostic Design

The steps are:

- Preparing plans, specifications, and estimates (PS & E)
- Adjusting maximum truck load: After rehabilitation, the number of fatigue cycles and fatigue stress level are expected to be lower since the remaining useful life of the substructure is reduced to less than the original estimated life of 75 years. A fatigue evaluation study needs to be carried out for the remaining fatigue life for the level of live loads (Table 1.5).
- Posting the bridge for a lower live load.

1.8.15 Compliance with Codes and Standards

1. When a bridge is being considered for rehabilitation, it should be reviewed for compliance with current standards. Existing vertical clearance, horizontal clearance, load capacity, free board, seismic capacity, lane width, and shoulder width should be compared to current standards.

2. Hydraulic and seismic history should also be reviewed. If the existing features are non-standard, consideration should be given to improving them under rehabilitation or by replacing the bridge. If improvements cannot be made or the improvements that can be made will not be up to current standards, a non-standard feature justification will be required. This should be taken into account when making the rehabilitation versus replacement decision.

Table 1.5 Truck weight for fatigue evaluation.

S. No.	Deck Function	Deck Type	Truck Type	Remarks
1A	Pedestrian use	Timber	H5	
1B	-do-	Lightweight	H5	
1C	-do-	Normal concrete	H5	
2A	Passenger cars only	Lightweight	H10	
2B	-do-	Normal concrete	H10	
3A	Full service	Lightweight	HS20	Place load restrictions
3B	-do-	Normal concrete	HS20	-do-

3. Current AASHTO manual for condition evaluation of bridges
4. Current AASHTO policy on design standards: interstate system
5. Current state geometric design policy for bridges

1.9 THE MAINTENANCE AND PROTECTION OF TRAFFIC DURING CONSTRUCTION

1.9.1 Projected Traffic Count

1. Building bridges is all about solving traffic issues.

   Maintaining and protecting traffic is the primary planning issue. Table 1.6 shows an increase in maintenance with increased traffic volume. When an existing bridge is to be replaced or is to undergo major rehabilitation, the decision whether to maintain traffic in the proximity of the existing bridge or to detour traffic must be made. This decision is based upon consideration of many conditions, engineering feasibility, cost effectiveness, ADT/truck traffic, and the impact on the local economy, emergency services, the environmental impact, and obtaining right-of-ways. Adequate public coordination must be performed in order to minimize adverse impacts.
2. Construction stages planning

   Traffic counts are required for average daily traffic (ADT) volume. A traffic count needs to be performed to assess the impact on traffic flow during construction. Warning signs must be placed weeks in advance so that the users may select an alternate route to avoid congestion. Local authorities should be contacted to determine if they have any restrictions regarding lane closures.

   Prior to developing staging plans, the agency's traffic operations department will provide the maximum allowable lane closure hours in each direction and the maximum number of lanes that can be closed at one time. An eight to 10 hour night window is required for the contractor to properly complete his work. Extra hours are permitted for weekend work.
3. Traffic count is needed for
   - Planning of the number of lanes
   - Fatigue analysis of girders
   - Detour purposes
   - Posting any weight or speed restrictions
   - Widening may be required due to high ADT, congestion and traffic jams.
4. Using the full detour (bridge shut down) option during reconstruction

   Full detour (shown in Figure 1.11) or partial detour plans on adjacent local roads or other traffic control schemes, if applicable, must be approved by local authorities.

Table 1.6 Classification of daily traffic.

S. No.	Traffic Volume	ADT	Location	Remarks
1	Very low	—	Rural area bridges serving low density population	Low fatigue and minimum maintenance
2	Low	< 4000	—	—
3	Medium	> 4000 < 25000	Bridges serving medium density population	Average fatigue and maintenance
4	High	> 25000	—	—
5	Very high	—	Interstate bridges during peak traffic	High fatigue and maximum maintenance*

* May require increased inspection frequency

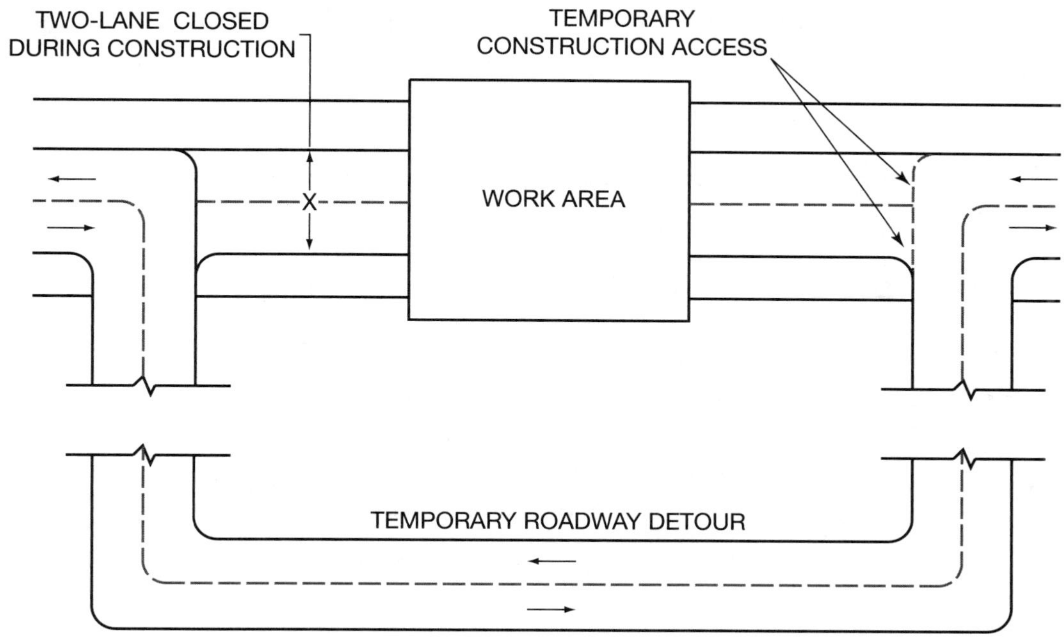

Figure 1.11 Diverting traffic on local roads.

Structural drawings showing construction in each stage should conform to traffic control plans.

A set of applicable standard traffic control plans is to be used as a basis for developing the final traffic control plans. These plans shall be customized to reflect site conditions and the ability of the shoulder to withstand traffic.

Plans must comply with MUTCD (Manual on Uniform Traffic Control Devices), Work Zone Traffic Control Guidelines and AASHTO LRFD specifications. All nonstandard signs shall be sized according to MUTCD with letter heights and alphabet size given for each line.

If shutting down will result in unacceptable delays from detours and/or if there is too much local opposition to shutting down, construction needs to be carried out in more than one stage.

1.9.2 Planning for the Maintenance and Protection of Traffic (MPT)

MPT will be based on a hierarchy of options:

1. Half-width staged construction (shown in Figure 1.12)
2. Schedule nighttime work hours to expedite construction
3. Construct new bridge adjoining existing bridge (use existing bridge for traffic maintenance)
4. Prescribe temporary stream crossing and approaches (e.g., stream crossing made of multiple pipes/fill material)
5. Prescribe temporary bridge and approaches for bridges carrying high traffic volume
6. Provide bridge on new alignment
7. Temporary measures to keep bridge open:
    - Structure is in advance stage of deterioration; partial lane closure may be adopted by posting lower load limits.
    - Continue using on temporary basis if there is high traffic volume and/or there is no money immediately available for full replacement.

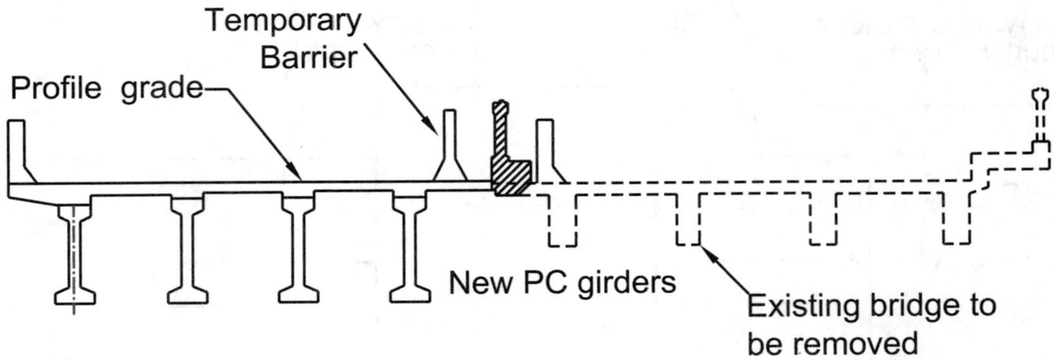

Stage 1: Lane closure, demolition, new construction, signing and striping.

Stage 2: Divert trafic from existing lanes to new construction.
Demolish and reconstruct remaining bridge. Signing and striping.
Remove temporary barriers and divert traffic.

Figure 1.12 Half-width two-staged contruction which is the preferred alternative to a detour.

1.9.3 Construction Staging Plans

The plans shall include cross-sections of the bridge for each stage of construction.

1. Fewer stages will provide less time for completion. Time required for the completion of concrete work for each stage is nearly that of a bridge. Two main stages are preferred over three or four, although there may be sub-stages.
2. An off-peak traffic hours construction schedule and/or incentive/disincentive clauses may be prescribed for bridges carrying extremely high traffic volumes.
3. During installation of countermeasures, small cranes or pile driving equipment may be parked on a lane or shoulders. A lane closure would then be required. Coordination with the state traffic operations and local transportation officials will be necessary.
4. Reducing the period of construction through accelerated bridge construction (ABC).

1.9.4 Involvement of Stakeholders and Public Outreach

Rehabilitation issues require the blessings of all stakeholders. Their cooperation is required in order to minimize adverse impacts on community. The following aspects may be considered:

1. Define problem and assess community needs.
2. Concept development: Ranges of possible "solutions."
3. Understanding the community: Public relations outreach and communication through town hall meetings.
4. Develop general solutions: Consensus on vision and process.
5. Develop traffic scheme: Application of MUTC and Work Zone Traffic Control Guidelines is required.
6. Temporary measures to keep bridge open:
   - Structure is in advance stage of deterioration; partial lane closure may be adopted by posting lower load limits.
   - Continue using on a temporary basis if there is high traffic volume and/or no money is immediately available for full replacement.

- Shutting down will result in unacceptable delay and detour.
- There is too much local opposition to shutting down.

These alternatives must be carefully considered as to their practicality, overall cost, delay of traffic, and impact to the surrounding community. In some cases, the type of bridge work will be driven by the fact that there is only one practical solution to managing the traffic.

7. Use of a temporary bridge during construction: Examples are the portable Bailey bridge as successfully used by military engineers and the Mabey panel bridge.

1.10 SUCCESS OF WELL-MAINTAINED STRUCTURES

1.10.1 An Historical Perspective

1. Civil engineering structures are indispensable to a civilized society. So great is the importance of structures that in recognition, modern marvels are displayed in pictures and on postcards. They include the Pyramids, the Roman Coliseum, The Taj Mahal, the Great Wall of China, and many other landmarks such as the Empire State Building, Leaning Tower of Pisa, and Sydney Opera House.
2. Romans were the first great bridge builders. Their stone arch bridges are still in use today. Other building materials used were timber, masonry, cast iron, wrought iron, and steel. London's Tower Bridge, Venice's Rialto Bridge, and Prague's Charles Bridge are well known examples. It appears that the use of weaker materials had restricted their use in the past. Historic bridges however show ingenuity and the future of the industry is bright. Much stronger building materials such as composites and HPS 100W are on the horizon.
3. Structural engineering serves the community needs and the bridge is an important part of structural engineering infrastructure. It displays and demonstrates the wisdom of sages. Every structure, renowned or not, seems to provide immense benefit to society.
4. Bridge engineering facilitates transportation and is an industry in itself. It is rated high as a scientific profession in terms of benefits and level of complexity. It is made up of many trades, including land development, excavation, fill, drainage, pile driving, sub-surface and overhead utility pipes, masonry, cast-in-place and precast concrete components, steel reinforcing bars, carpentry, metal work, welding, deck overlay, traffic signs, lighting, macadam surface, and railroad construction.
5. More recent examples in the Americas of structures using similar construction technology as that used for bridges include the Hoover, Boulder, and Grand Coulee dams, sky scrapers, nuclear power plants, and the Panama and Eerie canals.

1.10.2 The Golden Age of Bridge Building Between the First and Second World Wars

1. Well known examples are:
   - The George Washington, 1931
   - The longest steel arch Bayonne, 1931
   - Sydney Harbor, 1932
   - Tri-borough, 1936
   - Golden Gate, 1937
   - Blue Water Bridge linking Michigan with Ontario, 1938
   - Tacoma Narrows Bridge (which collapsed in November 1940): Details of failure under wind are given in Chapter 3.
2. Suspension bridges fell out of favor in Europe in the 19th Century because they were vulnerable to wind. American bridge builder John Roebling studied the failure of these bridges and determined that adding more weight, guides, trusses, and stays to suspension bridges

would make it less likely for the wind to move them. Roebling went on to build the Niagara Gorge suspension bridge in 1854 and the famous Brooklyn Bridge in New York.

Except for the ill-fated Tacoma Narrows Bridge that had design flaws, the above long life bridges have been successfully maintained on a regular basis.

3. Similarly, the French constructed the tallest bridge, Millau Viaduct, to connect Paris with the Mediterranean. Akashi-Kaikyo in Japan emerged as the longest span bridge. The proposed intercontinental Bering-Strait Bridge would link North America, Asia, and Europe.

1.10.3 The Challenges of Maintaining Bridges on the Mighty Mississippi and Missouri Rivers

1. The bridges of the Mississippi provide the people and industries of the region a vital link for interstate travel and commerce from Minnesota to Louisiana. The river itself provides a border between states. For the agricultural bounty of the midwest it serves as a nautical highway to the sea. A variety of bridge construction can be seen in cable stayed bridges and pneumatic caissons.

2. Constructing the well-known structures has been a challenge for engineers, however, maintaining such huge bridges has been an even greater challenge.

Notable examples are:

- Gateway Arch Bridge in St. Louis
- Super bridge: Clark Bridge in Alton, Illinois
- New Greenville
- The Bridge at St. Louis (due to Captain James Buchanan Eads)
- Minneapolis Stone Arch (due to James J. Hill)
- The ill-fated Route I35W bridges.

Similarly, it has not been easy to harness or maintain the Missouri River for crossing purposes.

1.10.4 Arterial Highway Systems of the Northeast United States

1. The New York and New Jersey region transportation system is one of the largest arterial systems in the world and includes navigable rivers in addition to bridges and tunnels. It serves many modes of transportation and users, such as:

- Pedestrians
- Marathon runners
- Bicycles
- Automobiles
- Taxis
- HOVs
- Buses
- Rapid transit
- Ferry users.

2. It is hard to imagine the chaos the Big Apple traffic would face and the resulting adverse effects on the economy of the northeast region without regularly maintained long span bridges, such as:

- Brooklyn
- Harlem River
- Williamsburg

- Queensboro
- Manhattan
- Outer Bridge
- Goethals
- Bayonne
- Verrazano Narrows
- Bronx/Whitestone and Throgs Neck
- Hell Gate Railroad Bridges.

1.10.5 Other Major U.S. Bridges

Some of the major bridges in the U.S. which continue to survive environmental effects through regular maintenance are:

1. Chesapeake Bay Mackinaw Straits, which separates the upper and lower peninsulas of Michigan
2. Al Zampa Memorial over Carquinez Strait, which separates San Pablo and Suisun Bay in the San Francisco Bay area
3. Bridge over Lake Pontchartrain in Southern Louisiana, one of the longest bridges ever built
4. Steel City Pittsburgh, Pennsylvania known as the City of Bridges, with a wide variety of bridges
5. Long span steel bridges in Louisville, Kentucky.

1.11 DEFICIENT STRUCTURES

1.11.1 Foundation Failures

The following causes of failures need to be rectified:

1. For widened structures, differential settlements of new and old components in widening shall be considered. It is possible that the existing foundation has settled. New footings may need to be placed on piling or drilled shafts in an attempt to prevent differential settlement. The widened section should be designed so that superstructure deflection for the new and old deck is identical.
2. Scour analysis: Foundations need to be investigated for scour. The investigation consists of determining on what the substructures are founded and the foundation depth, and deciding whether potential scour will endanger the substructure's integrity. Local scour and stream meander need to be considered.

1.11.2 Seismic Resistance

If an existing bridge does not meet current AASHTO or state design specifications, seismic retrofit needs to be considered.

1.11.3 Hydraulic Inadequacy

Environmental or Coast Guard concerns may push the rehabilitation versus replacement decision in the direction of rehabilitation while hydraulic inadequacies and poor stream alignment may push the decision toward replacement.

1.11.4 Soil Conditions

Any signs of foundation settlement may push the decision toward requiring the replacement of structure.

Chapter 3 addresses the issues contributing to the failure of bridges, in particular those which were least maintained or neglected, and the impact failures have on the future approach to maintenance of existing structures.

1.12 PRESERVING AESTHETICS

1.12.1 Planning to Ensure Safety and Operation

1. One of the most significant design factors contributing to the aesthetic quality of the structure is unity, consistency, or continuity. These qualities will give the structure an appearance of a design process that was carefully thought out. Sound planning also leads to safety and effective operation at intersections.

2. Aesthetics is required for rehabilitation: A bridge should have a pleasant appearance. As the old saying goes, "A thing of beauty is a joy forever." It should have a visual relationship with the surrounding area and also have visual impact. The famous structural engineer Hardy Cross laid down the criteria of a beautiful bridge: "The first requirement of a beautiful bridge is that it must stand up long enough for us to look at it."

New bridge facades should preferably blend with the appearance of existing bridges in the vicinity.

3. Bridges should have an open appearance and avoid abrupt changes in elevation or curvature. Abrupt changes in beam depth should be avoided when possible. Whenever sudden changes in the depth of the beams in adjacent spans are required, care should be taken in the development of details at the pier.

- Avoid mixing structural elements, for example, concrete slab and steel beam superstructures or cap and column piers with wall type piers.
- In general, continuous superstructures shall be provided for multiple span bridges. Where construction joints cannot be avoided, the depth of spans adjacent to the joints preferably should be the same. The use of very slender superstructures over massive piers needs to be avoided.
- Lighting can make a big difference in the aesthetics of a bridge (Figure 1.13).
- Use of precast mechanically stabilized earthwork (MSE) walls.

For abutment, wing wall, and retaining walls, MSE walls are gaining popularity due to their elegant styles, low cost, and quick construction.

4. Normally it is not practical to provide cost premium aesthetic treatments without a specific demand; however, careful attention to the details of structure lines and forms will generally result in a pleasing structure appearance.

Figure 1.13 Lighting and its reflection in the water is important to the aesthetics of a bridge.

- No patchwork in concrete or dissimilar steel painting.
- Use of innovative ideas and new technology: Lightweight and weather-resistant transparent noise varrier sheets incorporate polyamide filaments that hold broken sheet in place in the event of impact by a car or truck.

5. Cost considerations related to aesthetics

The aesthetics of the structure can generally be accomplished within the guidelines of design, requiring only minor project cost increases. If form liners are being considered, the depth of the projections should be as deep as possible in order to have the desired visual effect. Using shallow depths provides very little visual effect or relief when viewed from a distance. The depth of the form liner shall not be included in the measurement of the concrete clear cover.

1.12.2 Pedestrian and Equestrian Traffic

In practice, aesthetic requirements are more important in pedestrian and foot bridges located in parks and highways than for other bridge types. Aesthetic planning of rural area bridges is different than that for urban areas. Each structure should be evaluated for aesthetics.

1.12.3 Long Spans

The architecture of long span bridges, such as cable stayed or suspension cable bridges, has always attracted attention. They are pleasant to look at and also serve as cultural icons. Like movable bridges they are a subject in themselves requiring specialized analysis and design. Modern structural engineering and construction technology have contributed to ever increasing longer spans (world's longest span Akashi-Kaikyo Bridge) in addition to addressing safety, security, and maintenance issues.

Existing long spans using steel and concrete have used the following types of structural systems:

1. Suspension cables with towers, their foundations, anchorages, and approach spans.
2. Cable stayed bridges.
3. Truss bridges.
4. Arch bridges.
5. Segmental bridges.

Figure 1.16 shows aesthetic planning of curved approaches to a cable stayed bridge.

1.12.4 Rural Settings

Planning of rural area bridges for aesthetics will be different from urban area bridges. While girder types are more common in urban settings, arches and deep trusses are easier to blend in rural areas.

1.13 MAINTAINING THE ENVIRONMENT

1.13.1 Environmental Concerns

1. The purpose of reconstruction is to benefit transport facility users. The process of short term construction and long term impacts should not be adverse to road users or local residents. A large number of environmental concerns must be addressed related to water, including:
   - Maintaining water quality
   - Providing fish passage
   - Avoiding wetlands contamination
   - Avoiding stream encroachment

- Improving drainage
- Construction impact on floodplain
- Soil erosion and sediment transport: Minimizing the erosion of native substrate due to sediment transport after the installation.

2. Issues related to ecology:

- Preservation of vegetative species: Ecology (flora and fauna), minimizing impacts to natural vegetation by controlling construction access points
- Re-vegetation of disturbed areas with species may be required
- Landscape preservation
- Preservation of endangered species.

3. Maintaining air and water quality:

- Maintaining air quality and avoiding air contamination, pollution
- Noise mitigation from construction vehicles, pile driving, concreting, excavation, welding, etc.
- Relocation hazards of underground and bridge supported utilities
- Reactions with acid producing soils.

4. Related Issues:

- Historical and cultural aspects, aesthetics, "landmark"
- Temporary works and scaffolding
- Traffic disruption
- Socio-economic aspects, relocation
- Right-of-way issues
- Permitting considerations and implementing EPA/DEP procedures.

1.13.2 Action Required by Environmental Engineer

1. To minimize adverse environmental impacts, address these environmental concerns:

- Develop a baseline survey to define current environmental issues
- Develop an assessment of the impact of proposed repairs on air pollution or on the water environment
- Develop considerations or measures to avoid or mitigate adverse impacts
- Eliminate impacts on wetlands by using top to bottom construction and temporary bridge
- Develop alternatives to minimize impacts.

2. Environmental related items:

- Requirements for paint removal and containment and disposal of contaminants shall be incorporated as per the current State Department's policy.
- The need for bridge mounted sound walls shall be determined in conjunction with the need for sound walls on the adjoining roadway.

1.13.3 Permit Requirements

1. The proposed construction should neither damage an existing wetland nor adversely affect the historical significance of the bridge itself or its surroundings, except as permitted through the environmental evaluation process. For bridges located on streams, a *flood hazard area general permit* is required.

2. Engineering data and documentation are required for permit approval. As per regulations, the following reports/proforma need to be submitted:

- EA (environmental assessment): An EA is required when the significance of the environmental impact is not clearly established.
- EIS (environmental impact statement): EIS documents need to be prepared when a replacement or new bridge (usually with four or more lanes) has significant impact on natural, ecological, or cultural resources including on endangered species, wetlands, flood plains, groundwater, fauna, and flora. An EIS is required when there are impacts on properties protected by the DOT Act or the Historic Preservation Act. Significant impacts on noise and air quality need to be avoided.
- CE (categorical exclusions): An action that does not have a significant effect on the environment falls under CE. Examples provided by FHWA are reconstruction or modification of two-lane bridges, adding pedestrian or bicycle lanes, widening for shoulders, installation of signs, etc.

BIBLIOGRAPHY

Accelerated Bridge Construction

Bashir, Tariq M. Nysdot Lozano and Genaro Lozano, *Roll-In Roll-Out Reduces Construction Impact to Local Traffic*, HNTB Corporation.

High Performance Steel Bridge Concepts, American Iron and Steel Institute, J. Muller International, November 1996.

Khan, Mohiuddin Ali and Richard W. Dunne, "Recent Developments in ABC Technology," *Paper Presented to FHWA Conference on Accelerated Bridge Construction*, Baltimore, MD, 2008.

Lushko, Brian, "Accelerated Bridge Construction," *Structure Magazine*, August 2007.

Precast Slab System Workshop Summary Report, Minnesota Department of Transportation and Federal Highway Administration, September 2005.

Prefabricated Bridges, American Association of State Highway and Transportation Officials Technology Implementation Group, May 2002.

Ralls, Mary Lou, Ronald Medlock and Sharon Slagle, "Prefabricated Bridge National Implementation Initiative," *NCBC Concrete Bridge Conference*, Nashville, TN, October 2002.

Aesthetic Bridges

Aesthetic Bridges User's Guide, Maryland State Highway Administration, January 2005.

Barratt, Stephen D., "New York's First Transparent Noise Barrier," *Bridges Magazine*, July/August, 2006.

Long Span Structures

Khan, Mohiuddin Ali, "An Experimental Study of Flat Cable Nets," *First Kerensky International Conference on Tension Structures*, Institute of Structural Engineers, American Society of Civil Engineers, American Society of Civil Engineers and International Association of Bridge and Structural Engineering, London, 1988.

Khan, Mohiuddin Ali, "Factors Affecting Structural Behavior of Suspension Cables Nets," *International Conference Proceedings*, American Society of Civil Engineers/Structural Engineering Institute, New York, NY, 2005.

Khan, Mohiuddin Ali, "Structural Applications of Suspension Cables and Cable Nets," *International Conference Proceedings*, American Society of Civil Engineers/Structural Engineering Institute, Seattle, WA, 2003.

Khan, Mohiuddin Ali and H. Tottenham, "Elastic Non-linear Analysis of Prestressed Cable Networks Using Equivalent Continuum Theory," *IASS Pacific Symposium Proceedings*, Tokyo, October 1993.

Khan, Mohiuddin Ali and H. Tottenham, "Static Behavior of Prestressed Cable Cooling Towers," *Second International Conference on Civil Engineering Proceedings*, Pahlavi University, Shiraz, Iran, 1976.

Textbooks and Handbooks on Bridge Engineering

Barker and Puckett, *Bridge Engineering*, Second Edition, Wiley & Sons, New York, NY, 2006.

Helsel, Bill, National Society for the Preservation of Covered Bridges World Guide Steering Committee, *World Guide to Covered Bridges*, National Society for the Preservation of Covered Bridges, 1989.

Park, Sung H., *Bridge Inspection and Structural Analysis*, Park, Trenton, NJ, 1980.

Park, Sung H., *Bridge Superstructure Design and Rehabilitation*, Park, Trenton, NJ, 2005.

Tonias and Zhao, *Bridge Engineering*, McGraw-Hill, New York, NY, 2006.

2

Diagnostic Design and Selective Reconstruction

Section 1
Administrative Issues

2.1 MAINTENANCE ENGINEERING

2.1.1 Maintenance Principles

1. Maintenance engineering is both an art and a science. Availing of the state of art and modern techniques in repair and rehabilitation will be required. It is an essential part of bridge engineering which comprises of a host of other tasks and disciplines, including:

- Administration and related tasks such as bridge management, marketing, public relations, ethics, budgetary control, accounting, contract law, land acquisition, and right-of-way jurisdiction
- Technical tasks such as analysis, diagnostic design, development of procedures, training, mentoring, continuing education, construction coordination, and construction supervision.

2. Among the many facets of engineering, maintenance requires a more specialized approach than original design. Maintenance policy principles are basically governed by design codes and guidelines. The following guidelines need to be followed:

- In the case of replacement, the approach should be the three B's—to *"Build Back Better."*
- Only identified deficiencies of the superstructure need to be fixed.
- Selective reconstruction depends upon performance of the diagnostic design.
- The most common maintenance issues concern deck repairs.
- Maintenance is to a large extent based on the average daily traffic volume and also on location in relation to hospitals, schools, or military

routes. Hence, all important and weight restricted bridges need increased inspections. Maintaining a healthy interstate and local transportation system requires diagnosis based on regular inspections by a team of qualified inspectors and repairing the bridge to an acceptable standard.

2.1.2 An Engineering Approach to Rehabilitation

1. Many activities accompany rehabilitation. In addition to recommending optimal rehabilitation procedures and developing solutions to the long term needs of the crossings and the manner in which to address them, public perception is important. What may be most important to the public are traffic delays caused by disruptions for recurring repairs.

   When the public is involved early in a project, their suggestions can make planning easier. If they fully understand the reasons for proposed delays due to staged construction or detours, they are more likely to accept delays when they occur. Often times, the public helps to shape and form the way work is performed. A thorough public outreach program is therefore paramount to a successful project.

2. A bridge widening feasibility study to address future improvements may also be carried out. Approval of reports by the owner is followed by preparing preliminary and final design plans and providing post design services for the initial near-term repairs.

   A team effort is required to prepare a concept study for the near-term repairs in a report documented with plans, including a condition assessment of the bridge and approach roadways.

3. The objectives are to restore serviceability and original functionality following distress from severe localized deterioration, flood or vehicle impact damage, wind, earthquake, and observed scour. Rehabilitation/repair tools such as preventive maintenance actions, life cycle costs evaluation, and bridge management systems will be applied for near-term and long-term tasks. Practical considerations may be summarized as follows:
   - For increasing the width of the bridge in the long term, the deck slab in a through girder bridge is made wider and the supporting floor beams longer.
   - For bridges located on waterways, underwater inspections are needed to evaluate the current conditions of foundations based on which existing abutments and piers will be restored.
   - Environmental impact should be minimized, and river pollution should be prevented during construction.
   - The historic appearance and shape of members should be maintained.
   - Traffic detours should be kept to a minimum.
   - Most near-term repairs should be salvaged for reuse in the long term.
   - Safety of the bridge through increased redundancy and carrying out structural analysis for the new system with cables for live loads and speed should be ensured.
   - Public needs regarding bridge closures and construction schedules should be addressed through outreach.
   - Near-term repairs include those that can be accomplished with minimal impact to traffic
   - Coordination with utilities and roadway agencies to identify acceptable reconstruction solutions is required.

2.1.3 Engineering Maintenance

1. Engineering maintenance is another name for need-based mitigation and reconstruction (Figure 2.1). It is a combination of both art and science for protecting, strengthening, upgrading

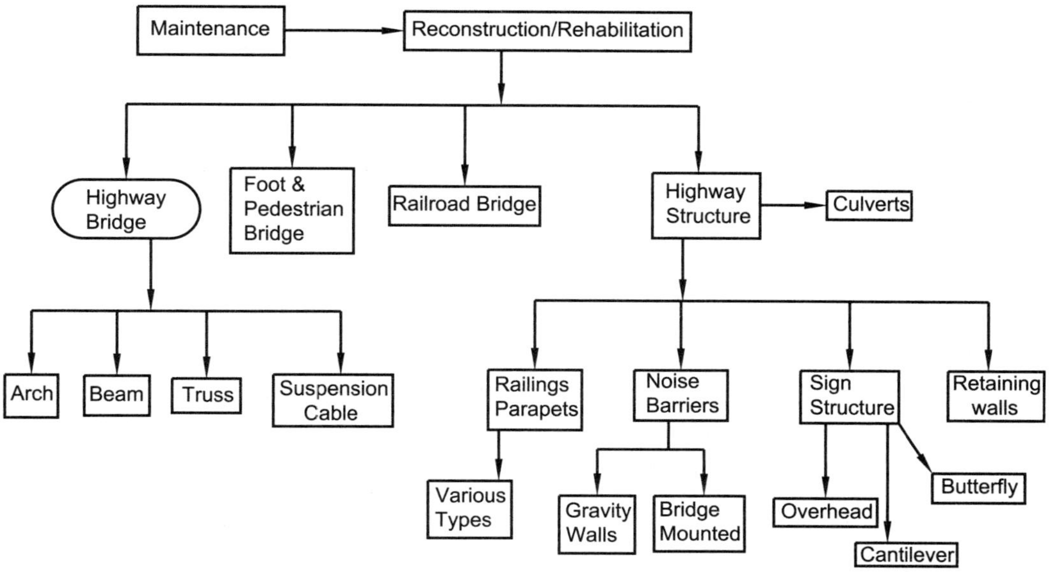

Figure 2.1 A bridge structural maintenance flowchart.

and improving the performance of bridges through repairs and retrofits. Timely maintenance helps to correct things that may have gone wrong during the design or construction phase.

2. FHWA defines maintenance as *"Routine and/or regular activities, including preventative, maintenance actions, which are intended to preserve and maintain a structure's original serviceability and functionality."*

A well known definition of reconstruction or rehabilitation is *"Comprehensive or major repairs of a structure's most deteriorated elements to restore and significantly extend its original serviceability and functionality."*

3. Rehabilitation work covers many areas (Figure 2.1) and can be described as both major and minor repairs. Minor rehabilitations address non-structural repair or improvement of certain bridge elements. Examples are concrete surface repair, deck overlays, joint and bearing restoration, minor repair to primary steel members, secondary member steel repair, and restoration of steel members by adding cover plates and high strength bolts.

2.1.4 Defining the Objectives of Bridge Rehabilitation

1. With the aging of the national highway infrastructure, state and local governments are spending a chunk of their budgets on bridge rehabilitation. Bridge rehabilitation is project specific since no two bridges are alike and all are located in different traffic conditions. Rehabilitation design is diagnostic and the diversity and complexity of the issues make it different from conventional new bridge design.

2. Objectives of rehabilitation: The objectives of rehabilitation are round-the- clock access for road users, rideability, inspectability, condition evaluation, and maintainability. It requires restoring structural members which are deficient. Different engineering solutions such as repairs, retrofit, and refurbishing may be used.

Basic objectives are to ensure safety by correcting deficiencies, providing comfort to users, maintaining the environment, and serviceability (summarized in Figure 2.2).

It also means routine or incidental work necessary to maintain function of the bridge deck with improved traffic conditions, increased load capacity, and low cost. Various structural solutions are discussed in subsequent chapters.

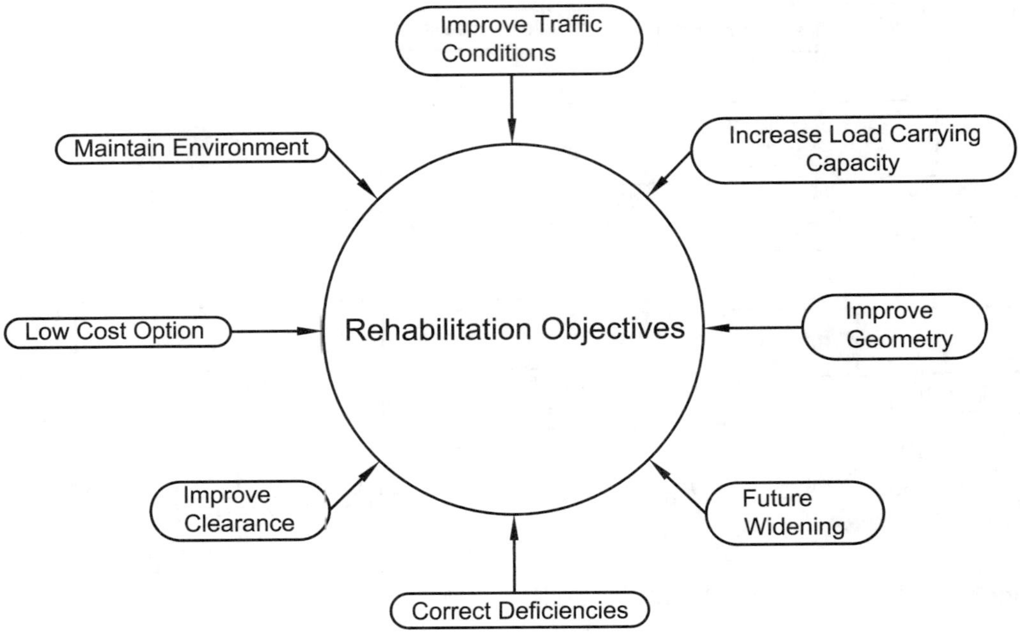

Figure 2.2 Bridge rehabilitation objectives.

3. Bridges need to be maintained regularly for safety and security reasons (Figure 2.3).

There are legal requirements for conforming to and complying with the standard procedures.

The advantage of following standard procedures, in addition to safety and security, is uniformity of construction for the numerous bridges rebuilt each year.

These procedures or specifications are prescribed by federal agencies such as AASHTO and FHWA. They cover analysis methods, applied loads, and construction materials such as steel, concrete, timber, and aluminum. In addition, design methods for substructures and foundations are addressed.

Figure 2.3 Fascia girders need to be repainted to prevent corrosion.

4. It may require reconstruction of part of the bridge other than the deck or even replacing the entire bridge. Initially, theoretical analysis, application of design codes, and preparation of construction drawings in the design office are required, followed by field work based on technical specifications and special provisions.
5. As discussed in Chapter 1, major rehabilitations involve major structural repairs or the replacement of primary bridge elements. Examples include:
    - Pier and pier cap replacement
    - Deck replacement
    - Superstructure replacement
    - Bridge widening
    - Strengthening and retrofit.

2.1.5 Mitigation Solutions

1. Another way of looking at maintenance is defining it as "mitigation."

    Mitigation technology is an alternate to rehabilitation, providing alternate solutions through one or more of the following approaches (Figure 2.4):
    - Minimizing the impact by limiting the degree or magnitude of the action and its implementation
    - Rectifying the impact/deterioration by restoring the impacted environment
    - Eliminating the impact/deterioration over time by preservation and maintenance operations during the life of the action
    - Compensating for the impact/deterioration by replacing or providing substitute resources, alternate solutions, and environmental controls.
2. The following items need to be considered:
    - Improve traffic conditions
    - Improve safety

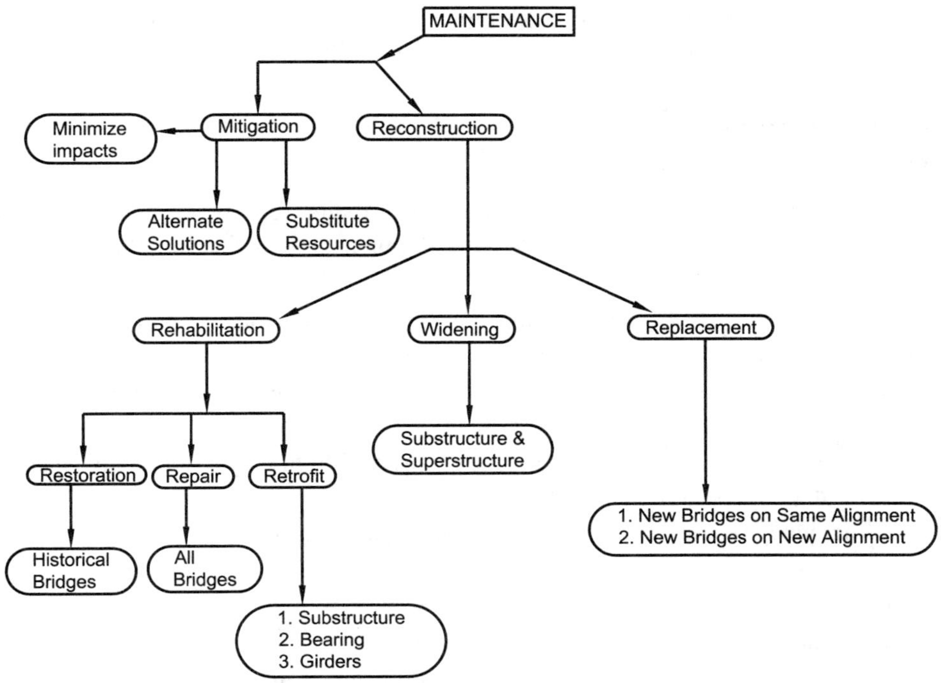

Figure 2.4 Pillars of engineering maintenance.

- Address environmental concerns
- Increase capacity
- Improve geometry—sight distance
- Correct deficiencies
- Future widening
- Improve horizontal and vertical under clearances
- Correct joints and bearing problems
- Improve deck and drainage.

2.1.6 Structural Solutions

1. In general, most recurring maintenance problems would require unique structural solutions for restoration and strengthening. Although rehabilitation is usually associated with older bridges, it may be required for newer bridges when planning, design, or construction mistakes are made. The capacity of existing bridges built for a lighter live load are fully tested when heavier vehicles are permitted.

Fewer maintenance problems are likely to occur when live loads are small (such as with pedestrian or cyclist loads) compared to those by repeated heavier trucks or permit loads.

2. Rehabilitation of bridges is a far more diverse and challenging subject than a new design based merely on code compliance. For maintenance of an existing bridge, there are fewer alternatives available to the designer than when designing a new bridge.

Some common rehabilitation examples include:

- Replacing a collision damaged fascia girder
- Replacing deck joints and bearings by jacking the superstructure
- Repairing cracks in a deck slab or in an earthquake damaged concrete pier
- Strengthening unknown foundations by underpinning with mini piles to prevent soil erosion
- Bridge performance can be upgraded by seismic retrofit, scour countermeasures, and by widening to provide additional lanes, shoulders, or sidewalks.

2.2 THE REHABILITATION PROCESS

2.2.1 Inspection Reports and Planning Issues

Recommendations are provided in inspection and rehabilitation reports for emergency repairs, which should be implemented as soon as possible. Options include the following courses of action:

1. No reconstruct option is to dismiss the problems for some good reasons. The scope of work, type of design, construction effort, and cost are evaluated before a decision is made.

2. In extreme cases, the bridge may be shut down indefinitely. A temporary detour is followed.

3. In other cases, some lanes may be closed to reduce the risk of failure by reducing live loads. A partial detour in one direction only may be used.

4. The client's emergency repair funds make immediate repairs possible.

2.2.2 Quality Planning

1. Rehabilitation generally involves hundreds of man hours of diagnostic design and several years of work before the bridge is fully restored. Over the years, an engineering system has evolved which requires creativity, innovation, ingenuity, constructability, cost effectiveness, and public involvement. Design procedures are usually completed in phases, using intuition, thumb rules, case studies, and practical experience (Figure 2.5).

2. As stated earlier, maintenance involves inspection, interpretation of data, selection of repair methods, analysis, computer aided design, and application of AASHTO and state codes of practice. Only a brief description of various aspects is given here.
3. Plan for geometry, minimum skew or curvature, adequate sight distance, sufficient horizontal or vertical clearances, and adequate opening over waterways. For important bridges repainting is required on a regular basis (Figure 2.6).
4. Use modern high strength and corrosion resistant materials.
5. Include structural design aspects such as minimum deflection and vibration of girders, use of jointless decks, keeping deck surfacing un-cracked, unrestricted bearing movements, and ductility of joints.
6. Meet the functional requirements such as providing an adequate number of lanes to prevent overload, and posting of warning signs and directions located ahead of the bridge.
7. Provide facilities for ease of maintenance such as a provision for inspection chambers, structural health monitoring by remote sensors, and nondestructive testing.
8. The construction industry has also benefited from the use of new machinery, cranes, and tractor trailer vehicles for freight. Precast concrete technology and pre-assembled replacement bridges offer quick and reliable solutions by minimizing delays and reducing construction time.

   Experience has shown that if any of these are lacking, indirect costs in terms of structural damage, accidents, or delays (which were not provided for in the original budget) will accrue.
9. The technology of maintenance varies according to the uniqueness of defects and quality control. Quality assurance and quality control are supported by modern technology as seen in the quality of maintenance chart shown in Figure 2.7.

   Planning deficiencies should be avoided since they lead to a functionally obsolete condition. The goal should be:

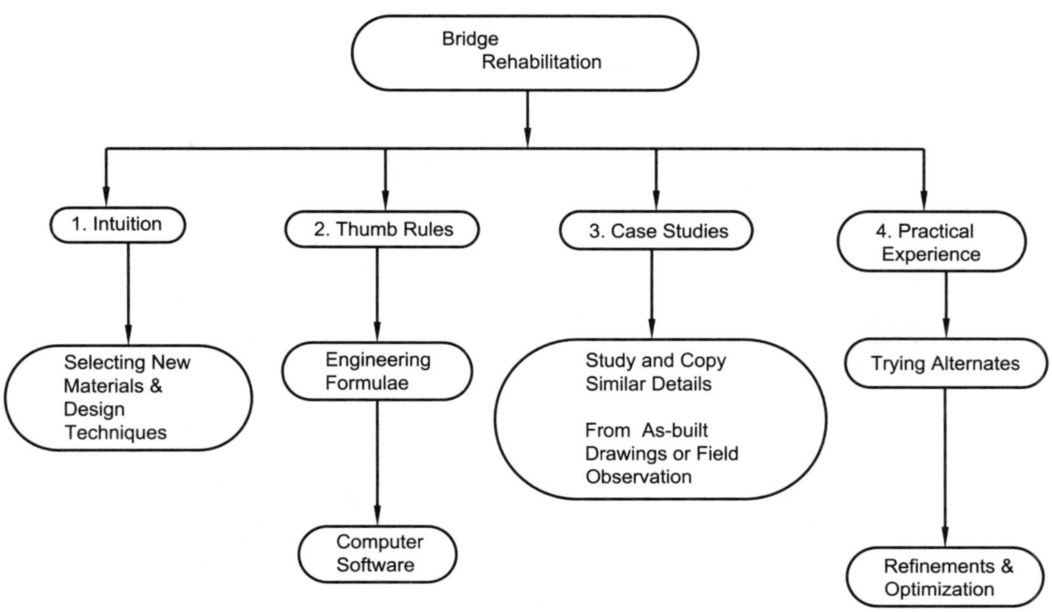

Figure 2.5 Ingredients of successful rehabilitation and design.

Figure 2.6 The Delaware River's Calhoun Street truss bridge connects Pennsylvania and New Jersey.

- To develop low cost solutions
- To improve traffic conditions
- To increase capacity
- To improve geometry and sight distance
- To improve horizontal and vertical under clearances
- To provide for future widening.

10. Pre-reconstruction planning and design of public projects should consider increasingly diverse needs and opinions of users within our society. New skills to communicate and incorporate feedback in planning are required so that the project best meets the recurring needs of society.

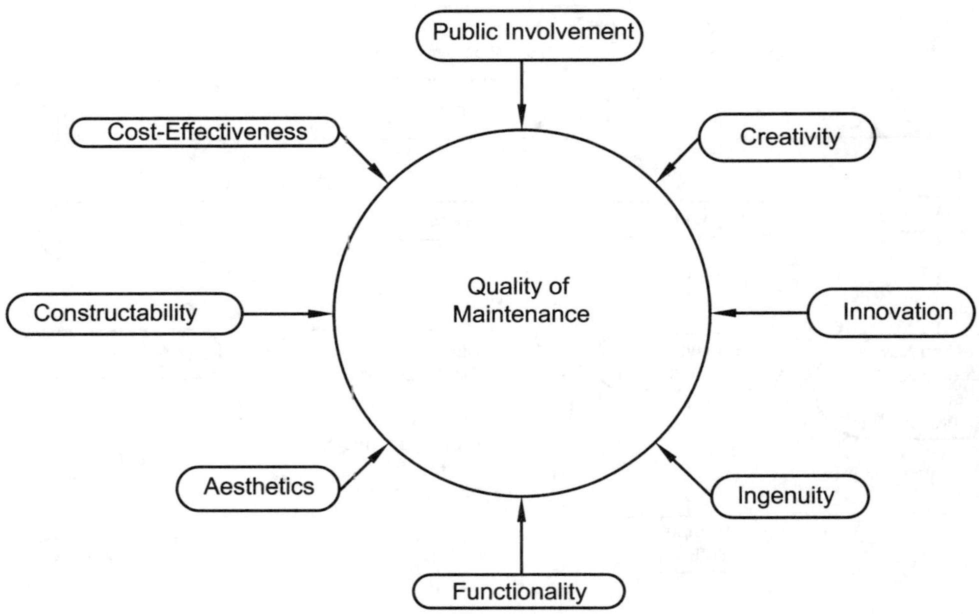

Figure 2.7 Factors contributing to quality maintenance.

2.2.3 Engineers to Supervise Bridge Repairs and Retrofits

1. Background: The subject of bridge repair and reconstruction has developed as a craft. It still maintains some of the ad hoc planning approach. The early engineering reconstruction practice was based on apprenticeship. It was performed mainly in the field under an experienced artisan or "guru." Builders did not require a formal education. A concept or hidden theory was built into the construction process and was applied indirectly, without expressing it in a mathematical format.

   Practical experience requirements for engineers are still applicable. A professional engineer's license cannot be granted before many years of supervised training.

2. Experience: The success of any craft requires successful field practice verification and time tests. Practical experience gained in the field has become one of the major strengths of bridge design. It provides guidelines for a healthy continuity of design and generates structural refinements.

   The word "engineer" was coined a long time ago. It is used both as a noun and sometimes as a verb, "to engineer." In the U.S., an engineer is generally referred to as a consulting engineer. A resident engineer usually represents the owner on the construction site. Due to the unpredictability of deficiencies, a maintenance engineer is required to be a licensed professional engineer.

   These days the intuitive practice and in-built logic of age-old craftsmanship is being interpreted in the light of familiar scientific principles. The formation of the Institution of Civil Engineers in many countries has set the education standards for practicing engineers. The training and continuing education process is making steady progress to keep up with technology advancements.

3. A common approach: Due to the high risk factor for failure, a similar level of repair is needed for bridges as is normally required for regular maintenance of a dam, nuclear power plant, or even an aircraft structure. Research may be required to select the best repair material or method. Repairs, whether long-term or short-term, need to be carried out under the guidance of a professional or licensed engineer following initial diagnostic design.

2.2.4 Efficient Planning and Regular Inspection Lead to Minimum Maintenance

Maintenance engineering: It includes planning, diagnostic design, and selective reconstruction.

1. Efficient planning: If sufficient thought has gone into the planning of a new bridge there will be fewer problems down the road and maintenance will be minimized. Proper investment at the construction stage will minimize subsequent repair and rehabilitation cost.

$$\text{Total cost} = \text{Initial cost} + \text{Life cycle cost}$$

$$\begin{aligned}\text{Life cycle cost} = {} & \Sigma\,(\text{Cost of routine inspections} + \text{Maintenance and retrofits}) + \\ & \Sigma\,(\text{Repairs from extreme events} + \text{Cost of demolition})\end{aligned}$$

   The total cost is computed over the life of the bridge.

   Extreme events may or may not apply within its life. There can be unforeseen events including accidents resulting from vehicle and vessel collision, floods and scour, earthquake damage, fire and bomb blasts, etc.

2. Life cycle costs are linked to concrete spalls, beam corrosion, and other deficiencies (Figure 2.8). If the initial cost does not cover all structural requirements, the life cycle cost for repair and rehabilitation will be much higher.

   The most sensitive elements of a large project are life cycle costs and value engineering (Figure 2.8).

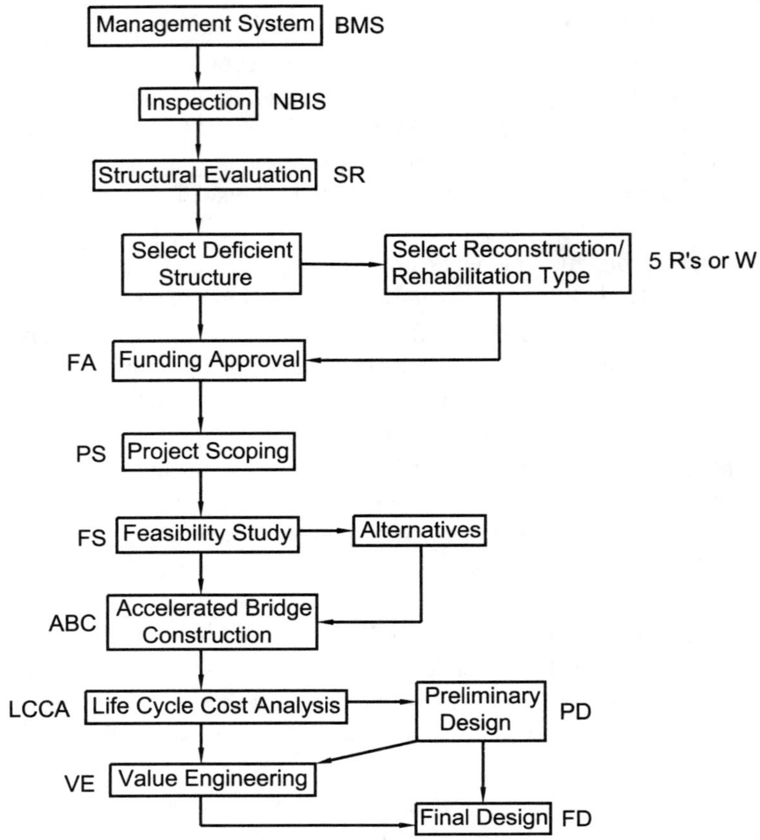

Figure 2.8 The pre-reconstruction planning and design process.

2.2.5 Simplified Structural Options (5 Rs, Widening, or Demolish)

1. A well-maintained bridge is the goal of the engineer. To a certain extent, choice for the type of maintenance is between the five Rs, namely:

Repair, Retrofit, Restore, Rehabilitate, or Replace, with further option to widen (Figure 2.9) or demolish.

Unlike the design tasks, the scope of maintenance work is varied.

2. The rehabilitation of highway structures other than bridges is also considered equally important. These include bridge approaches, sign structures, railings, parapets, fences, noise barriers, and culverts, and each is a subject in itself.

2.2.6 Basic Maintenance Activities

1. Keeping bridges in perfect condition involves both diagnostic design and selective reconstruction. Design related activities are based on:

- NBIS inspections
- Interpretation of data
- Selection of repair and rehabilitation methods
- Analysis
- Computer aided design
- Application of AASHTO and state codes of practice.

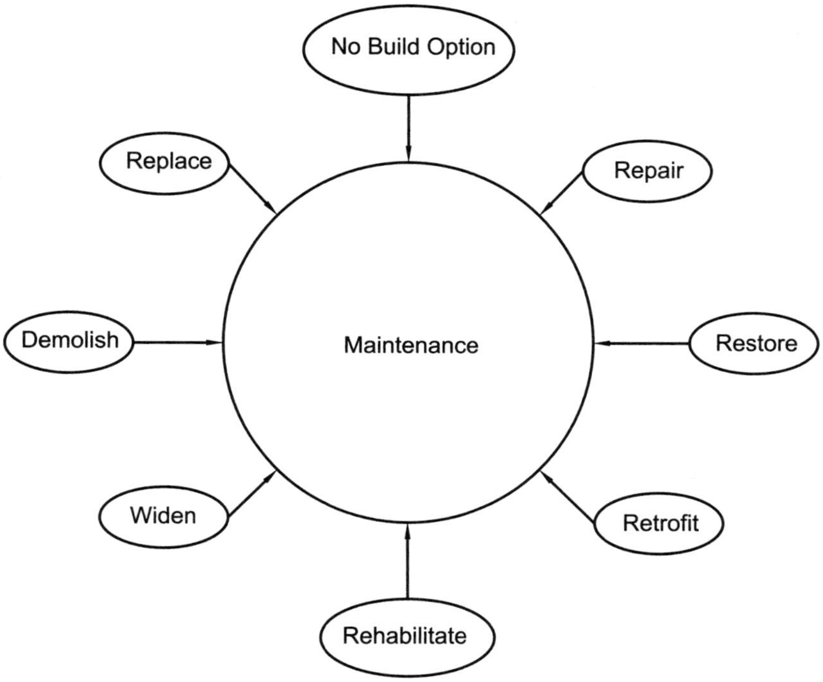

Figure 2.9 Types of structural options.

2.2.7 Design Aspects to Minimize Maintenance

The following issues are common to many states:

1. Elimination of deck joints by using integral and semi-integral abutments.
2. Fiber wrapping columns to increase seismic resistance.
3. Wider shoulders versus number of deck drains.
4. Use of culvert versus small span bridge.
5. Use of low water crossing in place of a bridge.
6. Transverse post-tensioning of deck slab.
7. Reduced deck thickness due to arching action.
8. Elimination deck vibration at longitudinal joints due to traffic.
9. Use of grade beams under approach slab.
10. Design of bridges with sharp skew angles.
11. Design of scour countermeasures.
12. Settlement of approach slabs.

2.2.8 Selection of Concrete Repair Materials

1. Latex modified and micro-silica modified concrete overlays.
2. Asphalt overlays.
3. Silica fume overlays.
4. Use of precast concrete approach slabs.
5. Improved concrete mix design for extremely hot weather.
6. Use of cast-in-place versus precast concrete.
7. Use of HPC.

8. Use of precast concrete railings in place of steel railings.
9. Use of chloride for cold weather causing deck cracking.
10. Improved deck curing techniques.
11. Use of concrete inhibitor aggregates.
12. Use of granite aggregates versus limestone aggregates.
13. Use of Class C fly ash.
14. Shotcrete and gunnite repairs are ineffective.

2.2.9 Selection of Steel

1. Use of stainless steel.
2. Use of weathering steel with painted beam ends.
3. Galvanized joint hardware.
4. Use of epoxy coated rebars and thicker epoxy coats.

2.2.10 Selection of Composite and Elastomeric Materials

1. Elastomeric bearing pads.
2. Composite reinforcing bars.
3. Expansion joint material for strip seal.
4. Epoxy injection.
5. Use of silicone joint with polymer nosing.
6. FRP sheet reinforcement.

2.2.11 Useful Life of an Overlay

Estimated duration of concrete decks, use in years:

1. Overlay thickness 1.25 to 2.5 inches and corresponding scarification depth of 0.25 to 1 inch.
2. LMC, 19 years (maximum life).
3. Low slump dense concrete, 18 years.
4. Asphalt with membrane, 17 years.
5. Fly ash concrete, 17 years.
6. Silica fume concrete, 16 years.
7. Standard concrete mix, 14 years.
8. Asphalt without membrane, 12 years.
9. Plasticized dense concrete, 11 years.
10. Thin bonded epoxy, 10 years (maximum life).

2.2.12 Effective Deck Drainage

1. Use of deck waterproofing membranes.
2. Extension of deck drains below bottom of girder flanges.
3. Prevention of clogging of deck drains.

2.2.13 Construction-Related Aspects for Rehabilitation

1. Method used for joints.
2. Use of staged construction versus detours for reconstruction.
3. Night time deck and parapet pours.

4. Slip forming parapets.
5. Engineered fills for backfills behind abutments.
6. Repair spalling concrete.
7. Hydro-removal of concrete for patching.

2.2.14 Corrosion Protection Strategies for Increased Life

1. Epoxy rebar, top mat only.
2. Epoxy rebar, top and bottom mats.
3. Low permeability concrete.
4. Corrosion inhibitor.
5. Surface sealer.

2.2.15 Deck Cracking and Efflorescence

1. High negative moment over piers.
2. Deficient rebars detailing.
3. Shrinkage cracks due to high water/cement ratio during curing.
4. Excess cement paste in concrete.
5. Excess number of shear connectors.
6. Small aggregate sizes.
7. Inadequate bar cover.

2.2.16 Cost Effective Preventive Strategies

1. Address problems related to deferred maintenance.
2. Improved procedures for over 100 maintenance issues.
3. Avoid field welding for fracture critical tension members.
4. Resolve MPT issues during reconstruction.
5. Ensure environmental protection during maintenance.
6. Implement effective management techniques such as planning, scheduling, monitoring, and reporting.
7. Train bridge engineers and technicians through bridge management training courses.

2.2.17 Potential for New Applications

Further information needs to be developed for the following:

1. Deck crack sealing with high molecular weight methacrylate.
2. Deck crack sealing with silane/siloxanes.
3. Precast prestressed concrete beam repair.
4. Special coatings on prestressed concrete beam end, e.g., paint or sealer.
5. Substructure concrete sealers.
6. Composite wrapping of substructure caps.
7. Repairing undermined footings.
8. Debris removal after culvert demolition.
9. Culvert sediment removal.
10. Approach slab patching.
11. Deck surface sealing with boiled linseed oil.

2.3 PROGRESSIVE DESIGN PHASES FOR CONTINUITY

2.3.1 Routine, Diagnostic, and Preservation Design

1. Original or routine design (shown in Figure 2.10): New bridge design is based on the latest codes of practice from AASHTO and the state DOT. The latest technology is utilized. There are many options available for superstructure planning and design. Structural solutions and the type of foundation to be used are broad based.

Major funding is required. However, life cycle costs can be controlled through judicious planning and design. In original design, no load posting is required.

AASHTO LRFD specifications deal primarily with routine design of new members and not with the redesign of deficient or "rogue" members. At present there are no separate specifications for diagnostic design. Although there are many types of defects, the practical issues related to all types of defects need to be addressed on their own merits, perhaps on a case-by-case basis. A comprehensive maintenance code would therefore be desirable.

2. Diagnostic design: Older bridges were designed to different criteria, using different materials and for different loads, a long time ago. Fatigue and environmental constraints such as corrosion have given rise to structural deficiencies. To identify the deficiencies and the degree of damage, a diagnosis based on inspection or remote health monitoring is carried out on a regular basis. Some major and some minor repairs may be required, based on the diagnosis.

Smart solutions need to be based on diagnostic type design. Diagnostic design steps required for a typical pre-reconstruction planning process are shown in Figure 2.10.

An example requiring diagnostic design is when a lower strength of concrete exists. This can be known by taking core samples from abutment or wingwall and testing the samples in a laboratory. Older bridges may have used lower strengths (below 3000 psi) while new bridges the require Class A concrete (4000 psi or higher) in some states. Similarly, current requirements of 60,000 psi for yield of reinforcing bars may not be met, since in olden days yield strength was not even 40,000 psi. Such bridges may be strengthened if they are not beyond repair.

Another example requiring diagnostic design is at zones of high fatigue in a steel beam subject to reversal stress from moving loads. This would require flange plates to be added. If new plates are not added, allowable stresses in bending and shear may be lowered in keeping with the fatigue deficiency. Load posting to a lower live load may be required.

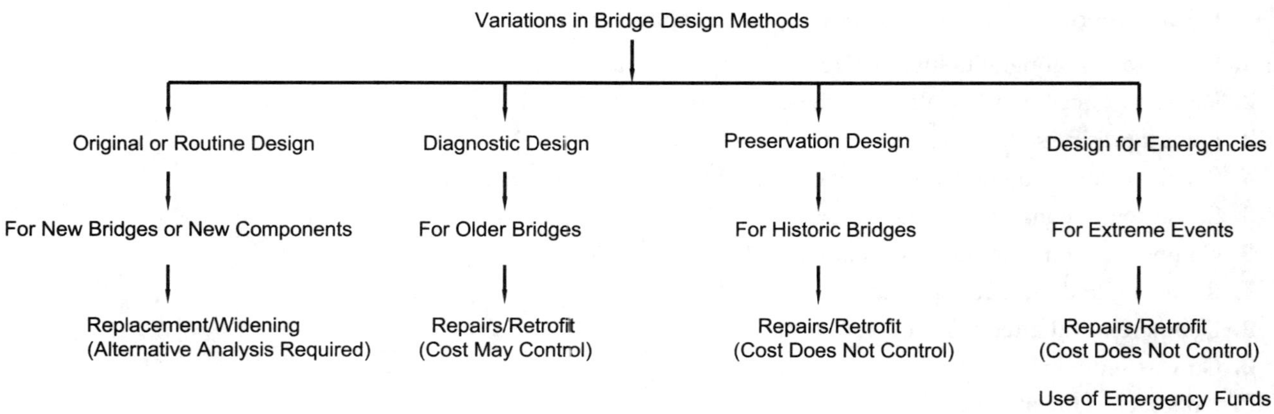

Figure 2.10 Progressive design phases that may occur in the life of a bridge.

3. Selective type of reconstruction: Selective reconstruction is tailored to suit existing field conditions. Based on an alternatives study, value engineering is usually carried out. Since structural solutions and type of foundation to be used are site specific and bridge oriented, unique applicable solutions need to be considered.

Funding for the diagnostic design and repair would be less than the cost of replacement. However, life cycle costs cannot be controlled due to limitations in the original design as any deficiency is likely to be recurring.

Repair, retrofit, and rehabilitation requirements would require deficient members to be repaired or replaced. Selected primary ailing members need to be checked for new conditions and load combinations. If the old member is retained, selective reconstruction or a "fix" is required. Load posting may be required during diagnostic design.

Consider physical constraints such as matching new and old components: A newly constructed width in an existing bridge may have a different response to shrinkage and creep strain of concrete.

4. Preservation design: Certain important bridges which are on the Historic Register have special requirements for repairs and retrofit. It is a special case of diagnostic design and many features are common. While funding based priorities are generally used in diagnostic design and the rehabilitation process is different, funding is less of an issue for the preservation of historically significant bridges.

In diagnostic design, in place of an expensive solution it is preferred to replace the bridge. However, in a preservation design an expensive solution is not a deterrent, and restoration is carried out for aesthetic, historical, and sentimental reasons. Hence, the methods of structural solutions and the type of repair technology in each case tends to be different. More sophisticated methods may be adopted for the repairs of a historic bridge since members or their appearance need to be preserved.

Similarly, inspection frequency would be higher for preservation design than for diagnostic design. In preservation design, load posting may be frequently required.

According to AASHTO LRFD specifications the minimum life of a bridge is assumed to be 75 years, which incidentally coincides with the life span of the generation of engineers who may be regarded as its worldly creators. It appears that historic bridges and most of the older bridges have survived longer than originally estimated as a result of regular inspections.

5. Innovative repair, strengthening, and retrofit techniques listed in Chapters 7 through 12 cover both diagnostic and preservation approaches to designing for maintenance.

6. Steps in a reconstruction process: Steps required for reconstruction include the 5 R's and/or widening (W) as shown in Figure 2.9.

- Bridge Management System (BMS) is used for National Bridge Inspection System (NBIS) based on which sufficiency rating (SR) is computed.
- Bridge or highway structure is evaluated for deficiency or is declared as functionally obsolete.
- The type of maintenance based on the 5 R's or widening is selected.
- Funding approval from FHWA or highway funding program is obtained.
- Project scoping is developed based on which feasibility study for preferred and other alternates is carried out. The FHWA method for accelerated bridge construction and life cycle cost analysis is applied. Bridge security and environmental issues are considered.

For large or complex projects, Value Engineering is performed. If necessary, the preliminary design is modified. Final design and contract documents are prepared for award of the reconstruction contract. Constructability and MPT are considered.

2.3.2 Survey of Structural Deficiencies and the Need for Rehabilitation

In this chapter, the need for diagnostic and preservation design and maintenance principles is discussed. The role of the federal government and states for overseeing and funding is addressed. While original construction is based on routine design, selective reconstruction is based on diagnostic and preservation design. Design issues for the historic and covered bridges are highlighted.

1. Bridge is not structurally deficient: Only minor repairs may be required. It appears that a percentage over 5 percent deficient is a cause of concern in terms of monitoring and rehabilitation, or even replacement.
2. Bridge is structurally deficient: Structurally deficient bridges (SDB) may be defined as those with deteriorated conditions and which may be subjected to load restrictions for safety considerations. Either its design or existing condition has impacted its ability to adequately carry its intended traffic loads. It has a span > 20 ft and has not had major reconstruction in the past 10 years. Figure 2.11 shows the percentages of bridges in the U.S. which are structurally deficient.

   SDB does not necessarily mean an unsafe bridge, but if not repaired, in time such bridges are likely to become unsafe. A recent survey conducted in 2008 (Table 2.1) points out the following alarming percentages (in excess of 5 percent) of SDBs in some of the 50 states:

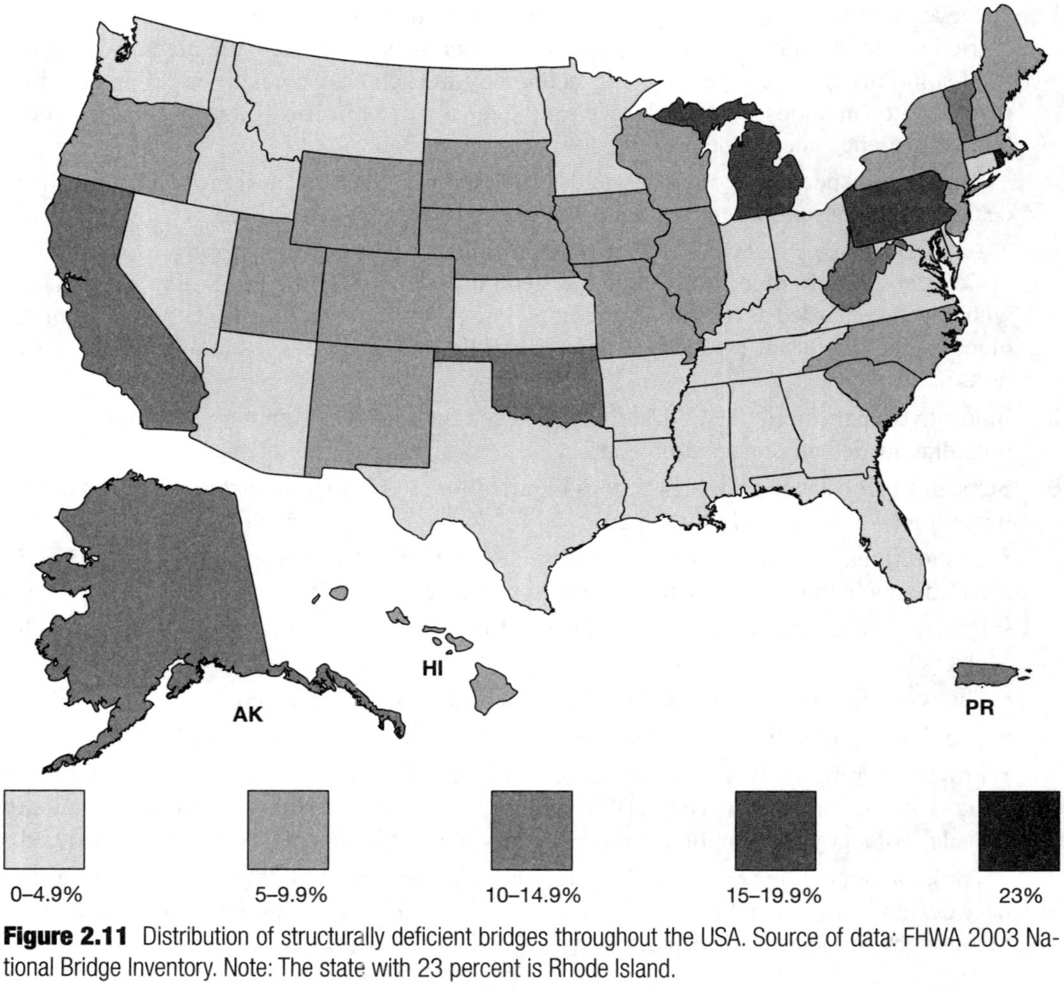

Figure 2.11 Distribution of structurally deficient bridges throughout the USA. Source of data: FHWA 2003 National Bridge Inventory. Note: The state with 23 percent is Rhode Island.

Table 2.1 The increasing number of SDB's in the U.S.

State	Estimate of Total Number of Bridges	SDB	Approximate Percent Deficient (Rounded %)
Pennsylvania	31,704	8140	26%
Oklahoma	22,723	5435	24%
Iowa	24,797	4763	19%
Missouri	24,140	4332	18%
California	23,971	3517	15%
Ohio	27,998	2862	10%
Mississippi	16,575	2830	17%
Kansas	25,500	2707	11%
Illinois	26,710	2615	10%
Nebraska	15,000	2294	15%
North Carolina	17,783	2272	13%
New York	17,361	2128	12%
Indiana	18,494	2030	11%
Texas	50,474	1871	4%
Alabama	15,827	1769	11%
Virginia	20,842	1755	8%

An estimate of total repair costs made by ASCE in 2005 shows it would cost over $10 billion for repairs of all SDB's. The number of SDB's increases with time and continued usage.

The condition of bridges and the number of SDB's is reported on a two-year inspection cycle. Note that Pennsylvania has over 8,000 SDB's; however, Texas, with the largest number of bridges on its inventory (over 50,000), has fewer SDB's (Table 2.1).

2.3.3 Identify Deficiencies

1. In evaluating rehabilitation, every component, as well as structural capacity, deck geometry, scour, seismic adequacy, and current deficiencies, needs to be assessed. Projects should correct bridge deficiencies that contribute to accident clusters and cause a functionally obsolete bridge.

Ignoring the impact/deterioration altogether and not taking any action for a long time will lead to functionally obsolete bridges (Figure 2.12).

2. For structural solutions, a complete rehabilitation for removing all deficiencies, or justifying their retention, is necessary. It includes the work required to restore the structural integrity of portions of the original bridge deck, as well as the installation of a deck protective system.

3. Functionally obsolete bridges: A functionally obsolete bridge (FOB) has a reduced ability to adequately meet traffic needs and is below the accepted design standards. The following factors contribute to the increase in FOB's:

- Structures in the advanced stage of deterioration
- Low traffic volume and/or no money available for repairs
- Too much local opposition to change
- Safety issues
- Unacceptable delays and detours.

The two obvious solutions to these issues are closing an FOB or replacing it.

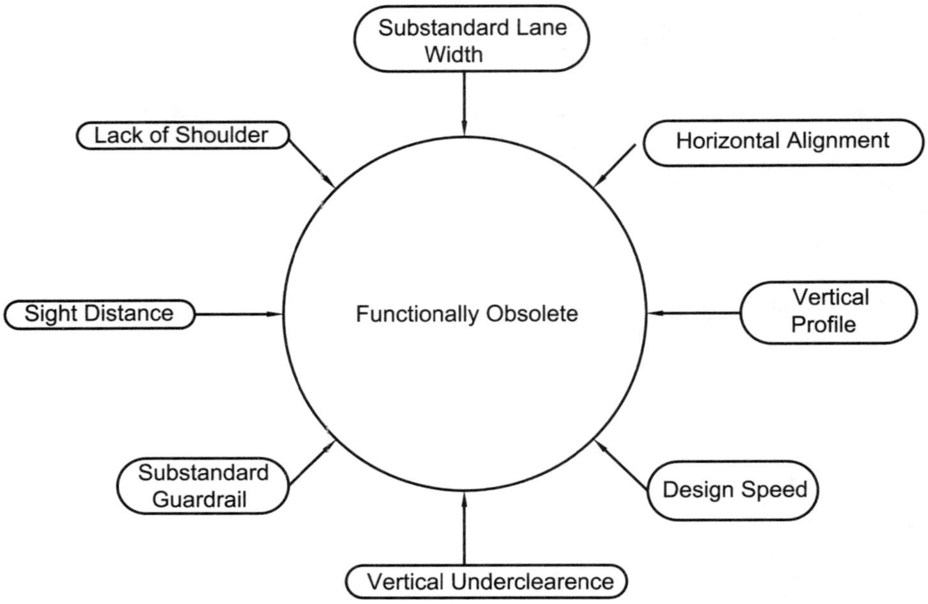

Figure 2.12 Deficiencies causing functionally obsolete bridges.

2.3.4 Survey of Bridge Maintenance Repair and Rehabilitation (BMRR) Practices

A comprehensive survey carried out by FHWA in the year 2000 brought to light a host of issues. The most common rehabilitation activities are related to:

1. Deck slab, bearings, and deck joints.
2. Cleaning, washing, and flushing; replacement of deck joints.
3. Deck patching and overlays.
4. Deck crack sealers and surface sealers.
5. Painting steel to prevent corrosion (Figure 2.13).

Figure 2.13 A weather resistant paint system protects steel in a corrosive environment on a bus route.

6. Bearing cleaning.
7. Bearing lubrication.
8. Deck drain/scupper cleaning.

Most overlays of asphalt and concrete are expected to provide nearly 20 years of satisfactory service. Deck slabs are expected to provide 25 years of repair-free service using corrosion protection strategies.

2.3.5 Progressive Design Phases

1. Routine design methods are basically applicable to new bridges, replacement bridges, and for widening of existing bridges. For existing bridges with deficiencies, diagnostic design methods are required. For historic and older bridges preservation design is needed. The important phases in the life of a bridge are transformation from a new to an older bridge and then perhaps to a historic bridge for the lucky ones. Others get unceremoniously demolished. Only those structures which are well maintained and possess optimum proportion and reserve of strength have withstood the test of time. For successful maintenance, any reconstruction effort must be based on diagnostic design and planning, made possible through the skills of trained personnel.

   As bridges get older, deficiencies are likely to emerge sometimes under heavier loads than expected. Old bridges still need to perform as well as any new bridge. This requires constant maintenance by correcting any deficiencies, performing repairs, and replacing any deficient members. Once again there are legal requirements for following standard procedures for maintenance as laid down by the highway agency, state, and federal agencies.
2. The structural health of every bridge may not be identical. Performance of each bridge depends upon its age, traffic volume, geometry, span length, load intensity, type of material, etc. Deficiencies may also arise from substandard design and lack of quality control in construction. "Diagnosis" will be determined by inspection and structural health monitoring. Hence, the type of diagnostic design will be different in each case and will aim to remove the diagnosed deficiency. A repair or replacement strategy will be implemented depending upon the condition of the bridge at any given time.

   Methods of repairs, retrofits, and strengthening are discussed in Chapters 7 through 12.
3. Engineering tasks for rehabilitation start with "diagnostic" design. It is a specialized but restricted application of "routine design." Guidelines for diagnostic design are:
   - It should meet FHWA criteria of highway for life.
   - The bridge needs to be rated for live load used in the original design if known but checked for new loads. It should be posted for a live load based on new design criteria. The subject encompasses both field and office procedures. The AASHTO Manual of Condition Evaluation of Bridges is applicable.
   - Remaining useful life needs to be evaluated for economical justification of continued use.
   - Accuracy of design will be based on the accuracy of the inspection report. It requires inspection to identify all the deficiencies and perform diagnosis and structural evaluations. Correct diagnosis is the basis of the rehabilitation process. It is made possible by inspection, monitoring, coding guide, load rating, analysis, and the applications of modern practice for rehabilitation and repair. Types of reconstruction depend upon evaluation of physical conditions and inspection-based diagnosis.
   - Each difficulty needs to be resolved. For example, future widening may not be easy for through type girder bridges. Also, where the fascia girder is weaker than the interior girder, replacement or strengthening of the fascia is required.

2.3.6 A Survey of Common Structural Deficiencies for Diagnostic Design

1. Condition evaluation: Documents required for preliminary assessment of the structure's condition include:
- In-depth inspection report
- FHWA Recording and Coding Guide for the structure inventory. The old SI&A sheet codes for condition evaluation or bridge closure have now been replaced by a PONTIS data sheet.
- Rehabilitation report
- Sufficiency rating
- Underwater inspection report
- Scour rating
- Seismic rating.

2.3.7 Investigate Structural Deficiencies

1. It is important to investigate internal structural deficiencies in the structure. Structural deficiencies include one or more of the following:
- Lack of redundancy in the structural system: Figure 2.14 shows a high degree of redundancy in a modern continuous bridge.
- Poor condition of superstructure and deck
- Poor condition of substructure
- Low structural capacities
- Bearings malfunction
- Corrosion
- Fatigue and fracture of beam connections.

2. Substandard geometry: Factors to consider in improving substandard geometry are:
- Lane and shoulder width
- Maximum profile grade
- Minimum horizontal radius
- Super-elevation rate and cross slope
- Stopping sight distance and K value
- Clearance—vehicular/navigational
- Level of service and design speed
- Substandard deck geometry

Figure 2.14 The author's design of a two-span precast highly redundant bridge. (U.S. Route 50 in southern N.J.)

3. Superstructure rehabilitation issues (as discussed in Chapter 1):
- Deck reconstruction and design
- Deck protective system
- Steel superstructure rehabilitation
- Deck joints: Joint types, compression seals, strip seals, modular and expansion joints, joint replacement, deck joint rehab, joint design
- Deck drainage: Scuppers, inlets
- Bridge railings: Bridge railing rehab
- Barrier design: Median barrier and parapets
- Bearings retrofit and design: Evaluation and strengthening of rocker and roller types, elastomeric pads, sliding, multi-rotational, dampers, and seismic isolation bearings
- Girder retrofit: Composite sections, shear connectors, web stiffeners, cross frame design, design of splice
- Bearings retrofit and replacement: Restrainers, seat width improvement
- Special procedures for through bridges
- Utilities relocation: Design of hangers and pipe supports.

4. Substructure rehabilitation issues:
- Geotechnical issues and foundation design: Foundation types, underpinning and rehabilitation, mini piles, pin piles, pile groups, caissons
- MSE walls, modular walls, restoring abutments and piers, underpinning methods
- Abutment and wingwall repairs
- Pier jacketing
- Column strengthening
- Seismic retrofit
- Scour countermeasures retrofit.

2.3.8 Geometric Issues

Both AASHTO Standard 16th Edition and AASHTO LRFD specifications address the following standard issues for bridges to remain functional:

1. Design speed.
2. Lane and shoulder width.
3. Maximum profile grade.
4. Minimum horizontal radius.
5. Super-elevation rate and cross slope.
6. Stopping sight distance and K value.
7. Clearances—vehicular/navigational.
8. Level of service.

Design exceptions as required need to be approved by the state.

2.3.9 Parameters Affecting Functional Obsoleteness

The following parameters causing functional obsoleteness need attention:

1. The deck geometry.
2. Load carrying capacity.
3. Clearances.
4. Approach roadway alignment.

2.4 THE ROLE OF REDUNDANCY AND FRACTURE CRITICAL MEMBERS

2.4.1 Definitions

Redundancy is a desirable structural quality to have in any structure. It reduces the risk of failure and increases safety. It means that in an assembly of members prior to the collapse of an overstressed member, the load carried by that member will be redistributed to adjacent members or elements, which do have the capacity to temporarily carry additional load. This redistribution of peak stress to members with lower stress would prevent the collapse of the structure and is referred to as redundancy. This is a kind of bonus offered by the configuration of members acting together as an assembly.

There are three types of redundancies which may be described as:

1. Structural redundancy: Structural redundancy is defined as that redundancy which exists as a result of the continuity within the load path. Any statically indeterminate structure may be said to be redundant. For example, a single span is statically determinate and cannot distribute load or stress to another span. It is therefore non-redundant. A continuous two-span bridge has structural redundancy.

   However, AASHTO conservatively classifies exterior spans as non-redundant where the development of a fracture would cause two hinges which might be unstable.
2. Load path redundancy: Load path redundancy refers to the number of supporting elements, usually parallel, such as girders or trusses. For a structure to be non-redundant, it must have two or less load paths (i.e., load carrying members), like the ones which only have two beams or girders. The failure of one girder will usually result in the collapse of the span, hence these girders are considered to be non-redundant and fracture critical.
3. Internal redundancy: With internal redundancy, the failure of one element will not result in the failure of the other elements of the member. The key difference between members which have internal redundancy and those which do not is the potential for movement between the elements. Plate girders, which are fabricated by riveting or bolting, have internal redundancy because the plates and shapes are independent elements. Cracks which develop in one element do not spread to other elements.

   Conversely, plate girders fabricated by rolling or welding are not internally redundant, and once a crack starts to propagate, it may pass from piece to piece with no distinction unless steel has sufficient toughness to arrest the crack. Internal redundancy is not ordinarily considered in determining whether a member is fracture critical but as affecting the degree of criticality.

2.4.2 Fracture Critical Members (FCMs) Linked to Redundancy

Inspection and maintenance of FCMs are important in avoiding a collapse. Some load carrying bridge members are more critical to the overall safety of the bridge and, thus, are more important from a maintenance standpoint. Although their inspection is more critical than other members, the actual inspection procedures for FCMs are no different.

The AASHTO manual "Inspection of Fracture Critical Bridge Members" states that "Members or member components (FCM's) are tension members or tension components of members whose failure would be expected to result in collapse." To qualify as an FCM, the member or components of the member must be in tension and there must not be any other member or system of members which will serve the functions of the member in question should it fail. The alternate systems or members represent redundancy. Once an FCM is identified in a given structure, the information should become a part of the permanent record file on that structure. Its condition should be noted and documented on every subsequent inspection. The criticality of the FCM should also be determined to fully understand the degree of inspection required for the member and should be based upon the following criteria:

1. Degree of redundancy.
2. Live load member stress: The range of live load stress in fracture critical members influences the formation of cracks. Fatigue is more likely when the live load stress range is a large portion of the total stress on the member.
3. Fracture toughness: The fracture toughness is a measure of the material's resistance to crack extension and can be defined as the ability to carry load and to absorb energy in the presence of a crack.
   - FCM's designed since 1978 by AASHTO standards are made of steel meeting minimum toughness requirements.
   - On older bridges, coupon tests may be used to provide this information.
   - If testing is not feasible, the age of the structure can be used to estimate the steel type which will indicate a general level of steel toughness.

   Welding, overheating, overstress, or member distortion resulting from collision may adversely affect the toughness of the steel. FCM's that are known or suspected to have been damaged should receive a high priority during the inspection, and more sophisticated testing may be warranted. A bridge that receives proper maintenance normally requires less time to inspect. Those with FCM's in poor condition should be inspected at more frequent intervals than those in good condition.
4. Fatigue prone design details: Certain design details have been more susceptible to fatigue cracking. The thoroughness of a fracture critical member inspection should be in the order of their susceptibility to fatigue crack propagation.

2.4.3 Method of Analysis and Mathematical Approach

The method of analysis is based on whether the bridge is redundant (statically indeterminate) or non-redundant (statically determinate). The latter can be analyzed by simple laws of equilibrium and boundary conditions, while the former also requires stress-strain, strain-displacement relationship, and compatibility equations. It appears that the difference in the theoretical approach governs structural behavior and hence inspection and maintenance requirements.

2.5 THE ROLE OF GOVERNMENT AGENCIES IN MAINTAINING INFRASTRUCTURE

2.5.1 Administration of Infrastructure

1. Administrative responsibilities: Effective administration would ensure effective maintenance and eventually prevent failures. However, bridge safety is directly linked to the basic funding required for maintenance purposes. It is also the engineer's bread and butter to design and maintain bridges. Failures, if correctly diagnosed, lead to improvements in design and maintenance procedures. Besides design codes regularly being updated, it is also important to understand the role of overseeing and to ensure that adequate funding is provided by the federal and state agencies.
2. The federal/state role: Since potential failures are a hazard to public safety, they fall under the jurisdiction of federal government agencies such as:
   - The Federal Highway Administration (FHWA)
   - The National Transportation Safety Board (NTSB)
   - Federal Emergency Management Agency (FEMA).

   The Federal Highway Administration (FHWA) is the main agency to oversee highway bridges for:
   - Maintenance
   - Safety
   - Reestablishing mobility
   - Reconstructing bridges after a catastrophic failure.

The National Transportation Safety Board (NTSB) is the entity that usually investigates the causes of bridge failures. It has the general authority under 49 U.S.C. §1131 to investigate selected highway accidents in cooperation with state authorities.

The Coast Guard (CG) and the Army Corps of Engineers (COE) have the responsibility of clearing and reopening the waterways after floods or a vessel collision. The CG is the authority that will declare the river safe for navigation once river debris has been removed. The COE is the agency responsible for clearing federal navigation channels and assisting in the removal of river debris with a barge-based crane operation.

3. Federal Highway Funding Programs are the main source for funding of bridge repairs.

FHWA's Emergency Relief Program (ER)

The ER program is also administered through the state DOT's. The ER program provides funding for bridges damaged in natural disasters or that were subject to catastrophic failures. The program provides funds for emergency repairs immediately after the failure to restore essential traffic, as well as for long-term permanent repairs. The ER program is considered to have a good track record in getting traffic alternatives such as detours, transit, or ferryboat service in place. It also uses innovative contracting to accelerate the rebuilding of any damaged federal-aid highway facilities.

Rehabilitation is warranted by inclusion of an existing bridge in an approved funding program.

Interstate preventive maintenance (IPM) projects cater to:

- Accidents caused by deficiencies
- Corrosion prevention by painting
- Sealing of cracks
- Deck joint repairs
- Highway capacity improvement
- Safety improvement or other structural improvement programs.

4. Examples of special situations for funding are:

- The need for providing scour countermeasures
- Seismic retrofit of bearings and connections
- Condition of the bridge: According to Bridge Management System Coding Manual, Publication 100A. For example, in Pennsylvania a condition rating of 6 or less would require the need for rehabilitation.
- If a bridge is structurally deficient or functionally obsolete, with a sufficiency rating of 50, it may receive funds through Federal Critical Bridge (FCB) Funds for either replacement or rehabilitation.
- If a bridge is structurally deficient or functionally obsolete, with a sufficiency rating between 50 and 80, it may receive funds through Federal Critical Bridge (FCB) Funds for rehabilitation only.
- All the deficiencies and problems listed in inspection reports must be addressed and resolved.

5. The Federal-Aid Highway Program is funded by the Highway Account of the Highway Trust Fund (HTF). These are several large "core" formula-driven programs through which highway funds are apportioned to the state DOT's, namely:

- Interstate Maintenance Program (IMP)
- National Highway System (NHS)
- Surface Transportation Program (STP)

- Congestion Mitigation and Air Quality Improvement Program (CMAQ)
- Highway Bridge Replacement and Rehabilitation Program (HBRR)
- Discretionary programs under the control of FHWA or earmarked directly by Congress, such as the Safe, Accountable, Flexible, and Efficient Transportation Equity Act—A Legacy for Users (SAFETEA-LU).
- States can "flex" funds from other federal-aid highway programs to increase spending on bridges.
- In addition, there is nothing to prevent a state from spending its own funds on bridge projects beyond the minimum local matching share.

6. The Highway Bridge Program (HBP) is the primary federal program to fund the replacement or rehabilitation of structurally deficient or functionally obsolete bridges. HBP is also referred to as the Highway Bridge Replacement and Rehabilitation Program (HBRR). HBRR is the primary source of federal funds for replacement, reconstruction, and capital maintenance. HBRR funds are apportioned to the states by a formula based on each state's relative share of the total cost to repair or replace deficient highway bridges. Plans for the spending of these funds are under the control of the state DOT's.

Each state is guaranteed use of ¼ to 10 percent of total program allocations. The federal share under HBP is 80 percent, except that for Interstate bridges, where the federal share rises to 90 percent.

These funds are usually not spent on new bridges, but are available for:

- Systematic preventative maintenance
- Rehabilitation to restore structural integrity or to correct major safety defects
- Replacement of low-water crossings, and bridges made obsolete by certain COE projects and not rebuilt with COE funds
- Painting, seismic retrofitting, anti-scour measures, and deicing applications
- Total replacement of a structurally deficient or functionally obsolete highway bridge with a new facility constructed in the same general traffic corridor.

7. A funding application report will address the following issues:

- Geometry, number of lanes, horizontal and vertical underclearance
- Deck condition: Concrete strength, cracking, corrosion detection by half-cell method, delaminations, spalls, salt content above and below reinforcement layers, and air content
- Deck drainage, substructure drainage, and drainage disposal
- Safety railings.

2.5.2 The Role of State DOTs

Although the Federal-Aid Highway Program provides federal money to highways and bridges, the money itself is normally under the control of the states. The state departments of transportation (DOTs) have to comply with detailed federal planning guidelines on where and on what the money will be spent. Their functions are:

- To let the contracts
- To oversee the project development and construction process
- To provide for the inspection of bridges.

The options available to each state are:

- Increase funding to perform immediate repairs
- Implement weight restrictions

- Install high-tech sensors and train additional inspectors
- Close down some lanes, if feasible
- Close down the bridge adversely affecting travel, trade, and commerce.

Funding can be increased by:

- Increases in highway trust funds
- Issuing bonds and taking out debts
- Raising gasoline taxes
- Hiking tolls on roads and bridges
- Shifting funds from other non-transportation allocations.

2.5.3 Project Scoping Process

Scoping is the first major stage of the project where most important decisions are made. The end products of this stage are:

1. Project objectives.
2. Design criteria.
3. Feasible alternatives.
4. A reasonable cost estimate.
5. Historic preservation.
6. To identify key environmental issues, e.g., wetlands, endangered species, protected streams, contaminated soil, asbestos, lead based paint, noise, etc.

The information needs to be assembled and analyzed in this stage. It must be of sufficient detail to demonstrate that the project is defined by these "scoping products." Analysis should show that the project scope is appropriate. Only then should it be progressed to the next stage of project production.

As described in Chapter 1, the scope of work applies to the following aspects:

1. Life cycle cost evaluation.
2. Maintenance and protection of traffic.
3. Hydraulic and scour studies.
4. Seismic retrofits.
5. Environmental considerations and acquiring permits.
6. Performing value engineering for optimum project cost.

2.5.4 Project Funding

The decision to continue maintaining a bridge or to demolish and replace it is based on safety considerations and the cost required to overcome deficiencies. Sources of funding for most public works are as follows:

1. FHWA provides 80 to 100 percent funding on selected bridges.
2. States will provide the remaining percentage of funds.
3. Local governments would not generally provide for new bridges, but for rehabilitation.
4. Private funding may sometimes be available but is unspecified.

Results from investigations given below are to be included in scoping document.

2.5.5 Scoping for a Rehabilitation Project

1. Addressing specific deficiencies: The scoping document may also serve as a design approval document.

- Obtaining and examining bridge inventory, load rating data, and the latest inspection report considering the overall condition of the bridge and the specific condition of the major structural elements
- The year constructed and design loading provides clues to the potential serviceability of a rehabilitated structure
- Identifying geometry, materials used, and details that may limit potential alternatives
- Obtaining and examining record plans, structure width, type of construction, materials used, and fabrication methods employed.

2. Verifying documented Information includes the following steps:

- Verifying data to assure that the information in the bridge inventory and inspection system and on the record plans is accurate
- Visiting the project site. This is not meant to be an in-depth bridge inspection, rather, a verification visit to assist in a feasibility assessment.

3. Evaluating the hydraulic adequacy of the structure, if applicable, includes the following steps:

- Identifying susceptibility to flooding, scour, and damage from floating ice and debris
- Performing a hydraulic assessment
- Performing some preliminary engineering activities prior to the closure of scoping activities
- The technical activities for this phase are focused on the feasibility, including a list of reasonable alternates and their cost estimates.
- General considerations that help define the feasibility of each alternate are required.

4. Determining reasonable costs and a schedule for the most feasible alternate includes the following steps:

- Providing project specific programming information
- Comparing general requirements of work to other projects of similar size and type and estimating a reasonable cost for work
- Preparing an approximate schedule.

5. Summarizing recommendations of scoping activities includes the following steps:

- The information gathered and the conclusions reached through these activities should be presented in the project's scoping document.
- Any unfeasible alternates should be eliminated.

6. Projects requiring painting: Repainting of steel bridges is required at regular intervals of 15 to 20 years, especially for structures located over rivers. To remove and dispose of existing lead-based paints, repainting costs have escalated in recent years and in some cases have become nearly comparable to the cost of a new bridge. The steps are:

- The cost effectiveness of fresh paint, which may only last 20 years, needs to be investigated. When combined with rehabilitation costs, the total expenditure may justify its replacement, especially for functional improvements such as providing additional lanes, sidewalks, or increased live load capacity.
- Any paint containing lead or toxic materials must be indicated, either on the plans or in a special provision, to advise the contractor for taking necessary precautions.
- If lead or toxic material content in the paint system is not known, samples need to be sent for analysis. Information on the prospective bidder's test results will be included in bidding documents, whether or not lead is present in the paint.

2.6 COMBINING OLD AND NEW TECHNOLOGIES FOR REHABILITATION

2.6.1 Preservation Design for Historic and Older Bridges

1. A specialized approach to design is required for effective maintenance of historic bridges and may include a diagnosis of deficiencies. It is based on the findings of a field inspection and structural health monitoring. Selective reconstruction in the form of repair, retrofit (structural strengthening) is the result.

2. Preservation design is a special type of diagnostic design: Preservation design is required for bridges that are listed on the National Historic Register. They have a high importance factor since their preservation is a sentimental consideration, depicting past history or culture, and cost is not usually a major factor. Their importance emanates from their ability to serve as a living museum.

3. "Preservation" design generally applies to historic bridges. It maintains existing shapes and sizes, and optimization is not a consideration. To maintain the historical integrity of each bridge, help is needed from sophisticated technology and special design methods, in addition to those deployed for diagnostic design.

To develop retrofitting or reinforcing systems, the process involves advanced numerical modeling and simulation of the loading regime. The same analysis-based computer programs will compute both rating and redesign separately.

4. Federal law protects historic bridges. Special attention is required for their rehabilitation or improvement. The director of the Division of Historical Resources (the Department of State) serves as the State Historic Preservation Officer (SHPO). The SHPO and state DOT are responsible for determining what effect any structural changes will have on a historic bridge.

5. According to the National Park Service, the structures and places that are part of its National Historic Landmarks Program "possess exceptional value or quality in illustrating or interpreting the heritage of the United States."

Historical bridges increasingly make significant impacts as community landmarks. They can be as simple as freeway overpasses decorated with American flags. They are symbols of local attractions as distinctive as the pedestrian bridges or as famous as the Golden Gate Bridge. In a small way, they promote tourism and are depicted on postcards that are mailed the world over by tourists.

6. Guidelines for historic bridge maintenance and rehabilitation based on the Secretary of the Interior's Standards for Rehabilitation may be quoted as follows:

- "The original character-defining qualities or elements of a bridge, its site, and its environment should be respected. The removal, concealment, or alteration of any historic material or distinctive engineering or architectural feature should be avoided.
- Distinctive engineering and stylistic features, finishes, and construction techniques or examples of craftsmanship that characterize an historic property shall be preserved.
- Deteriorated structural members and architectural features shall be retained and repaired, rather than replaced. Where the severity of deterioration requires replacement of a distinctive element, the new element should match the old in design, texture, and other visual qualities and where possible, materials. Replacement of missing features shall be substantiated by documentary, physical, or pictorial evidence."

7. A historic bridge serves as a "landmark" and is subjected to obtaining the following permits as applicable, which is a most time-consuming process:

- Construction impact
- Floodplain impact
- Wetlands impact

- Documentation
- Move to other location

2.6.2 LRFD Code Compliance

1. Inspection reports form the basis of interpretating of field deficiencies, structural evaluation, rating, selection of rehabilitation method, analysis and computer aided design, and application of AASHTO and state codes of practice. Generally, only a few components of a bridge need replacement, while many others are retained. Sometimes this may result in a mismatch due to differences in old and new materials.
2. Inspection of 50- to 100-year-old structures has shown that they were designed using old AASHO code, which was in effect at that time. Both design criteria and construction techniques have changed since those days. We are dealing with many thousands of older bridges. Design practices in the olden days were based on intuition, thumb rules, case studies, and limited practical experience. They were later followed by hand calculations using simple formulae. Modern day design requires code compliance and proficiency in computer analysis and software use. Stiffness matrix and finite elements methods are currently used for analysis and ultimate load methods for design.
3. A mixed technology scenario is sometimes necessary for widening and rehabilitation. Table 2.2 summarizes the variations in old and new materials and methods. Notable differences compiled by the author show applicable new live loads and new design criteria.
4. At the time of their design, there were no criteria for heavy permit loads and live loads were much lighter. Also, there was no ultimate load design, fatigue resistant details, or security considerations. Not all long span bridges for example would meet new security criteria. There are practical difficulties in upgrading and bringing up to date an old bridge to conform to newer technology.
5. Multi-girder construction has taken over non-redundant through girder systems. Working stress design has been replaced by load resistance factor design. Use of materials having steel yield strength of 30,000 psi is to be replaced by 50,000 or 70,000 psi of HPS 70W grade. Similarly, concrete crushing strength has increased from 2500 psi to 5000 psi of HPC.
6. The history of failures has shown that major failures are due to floods, earthquakes, or fatigue. It is important to check the safety for both scour and seismic criteria given in AASHTO LRFD Code 2007 and recommend adequate retrofits. An important part of seismic response is the substructure stiffness and foundation capacity. Also, the effect of scour on foundation capacity may reduce seismic response since existing pier footings, which have become exposed, may cause settlement during a seismic event.

   Hence, old bridges require added attention and special procedures need to be developed for their rehabilitation. Repairing old structures to meet new AASHTO criteria requires planning, use of modern materials, and innovative techniques.

2.6.3 Restoration Methods for Historic Bridges

1. Bridges which are listed in federal and state historic registers have slightly different criteria than routine type repairs. All new modifications need to conform to the original features and to shapes and sizes of members and connections, which are now likely to be extinct. Such repairs and retrofits tend to be more expensive.
2. Each historic bridge needs to be treated for its uniqueness on an individual basis. Repairs of suspension cable bridges, truss bridges, and arch bridges present special problems. For example, the deck stiffening of the Golden Gate Bridge is an interesting case in point.
3. Relocation factors:
   - Due to deficiencies in original site selection, foundation settlement may have taken place. It may be desirable to move the bridge to a new location. In such cases, new foundations

Table 2.2 Notable differences in old and new materials and methods.

	Old/Existing Construction	Replacement	Widening	Rehabilitation
Inspection and structural evaluation	N/A for original construction	Def ciencies must be established	Not a deficient bridge	Modern inspection methods with ultra high speed 3D laser techniques
Materials	Traditional materials used	Modern materials used	Partly traditional and partly modern	Special materials used
Construction techniques	Traditional construction used	Formwork, pile bent, precast components	-do-	Modern techniques used
Geometry and vertical underclearance	Substandard	16 feet 6 inches standard cross slope and grade	Substandard	Limited changes
Structural system	Non-redundant through girders	Redundant system	Widening not possible	Non-redundant through girders
Live loads	Live load to be checked by rating. bridge may be posted.	HL-93	Live load to be checked by rating.	Live load to be checked by rating.
Seismic design	N/A	Seismic zone and multiple spans to be considered in design	N/A (same as existing)	Seismic retrofits may be required
Scour analysis/ countermeasures design	River characteristics have changed	Based on hydraulic studies	Improves narrow openings and reduces scour	River maintenance is required for bridges over scour critical rivers
Foundation scour	Riprap	Gabions, grout bags, sheet piles	Gabions, grout bags	Gabions, grout bags
Bridge security	N/A	Full security	Limited security	Limited security
Environmental issues	Monitoring agencies such as EPA were not required	Stream encroachment and other construction permits required	Stream encroachment and other construction permits required	Some air, water and noise control requirements need to be complied with

need to be constructed, and the entire superstructure needs to be transported as one piece requiring a single operation.

- The superstructure may be dismantled piece by piece and reassembled at another location. One example is that of the historic London Bridge, which was relocated to the U.S. at a considerable expense. In other situations when huge reservoirs are created to prevent inundation, selected bridges downstream may be removed to an alternate site.
- Application of high capacity cranes, dollies, and heavy duty trucks needs to be investigated for feasibility of moving and transporting the superstructure to a new location.

2.6.4 Structural Data for Preservation Design

1. The majority of historic bridges are over 100 years old. They are special bridges (such as cable stayed with towers) or made of wrought iron, cast iron, and mild steel, with through Pratt, Pony, or Warren truss superstructures and stone masonry arches. Wrought iron structures represent an important era in the development of structures in the U.S. Figure 2.15 shows a historic bridge.

2. Decks are generally made of timber or corrugated metal. Test results of the mechanical properties of wrought iron show that average yield strength was 24.5 ksi but minimum was as low as 18 ksi; while average ultimate strength was less than 40 ksi and minimum was 26 ksi. Over the years there have been average reductions in the cross sectional area of 25 percent.

3. For historic bridges, preservation is the preferred option. Structural procedures include the following tasks:
   - Preparing records of inventory
   - Preparing CADD drawings and detailed documentation of members and connections
   - Research into alternate materials and refurbishment methods
   - Structural analysis and load capacity rating
   - Replacing old deck and railing with new
   - Replacing deteriorated components with new components
   - Conversion into pedestrian bridges if necessary, while preserving the bridge
   - Foundation strengthening or design and construction of new foundations
   - Installing sensors for SHM.

2.6.5 Restoration of Historic Footbridges

1. Footbridges promise increased challenges to designers by incorporating more elegant and modern materials, thoughtful designs, and complex engineering. Even railway bridges can be included in this trend of illustrating and interpreting local heritage. Footbridge design has become an aesthetically pleasing way for neighborhoods and communities to create a desired landmark.

   To strengthen masonry arches, a successfully used *reinforcing anchor system process* involves:
   - Maintaining the historical integrity of the bridge
   - Simulation of the loading regime to specify a retrofitted reinforcing system
   - Numerical modeling
   - Designing the number of bars to be used and their location
   - Dry or low volume wet diamond drilling techniques
   - At selected locations, where a structure is in need of reinforcement, a steel bar enclosed in mesh fabric sleeve is inserted
   - A non-polymer grout is injected into the sleeve under low pressure.

2. Preserving historic railings that may not meet current standards on bridges that are listed or are eligible to be listed in the National Register of Historic Places presents a special challenge in maintaining the appearance of the bridge. Original railing on a historic bridge is not likely to meet:
   - Current standards for combination traffic and pedestrian railings, e.g., a minimum height of 42 inches and the minimum 6-inch limit for openings in the railing
   - Current crash test requirements
   - Current standards for railing height and for combination traffic and pedestrian railings.

3. Options for upgrading the railing include the following:
   - Replace the existing railing with an approved, acceptable railing of similar appearance.
   - Place approved traffic railing inboard of existing railing, leaving the existing railing in place to act as dummy railing.
   - Where an existing railing is especially decorative, remove the current railing and mimic into a new acceptable railing.

- Design a special railing matching the appearance of the existing railing with higher strength material to meet the crash test requirements.

It may not be necessary to crash test the new railing if the geometry and calculated strength equals or exceeds a crash tested traffic railing.

4. Historic Shelby Street Bridge in Nashville, Tennessee has been retained as a footbridge. One hundred year old Shelby Street Bridge originally was built in 1909. Problems with concrete used in its construction led to repairs in the 1920s and again in 1960. In 1998, the bridge was declared unfit for traffic and was slated for demolition. However, because of its historic nature, the bridge was not torn down, but was converted into a pedestrian bridge linking entertainment venues and the coliseum on either side of the Cumberland River.

The bridge was repainted in 2003 with an inorganic, zinc-rich primer, a cycloaliphatic amine cured epoxy middle coat, and a fluorourethane topcoat. The bridge often is lit with lights of various colors, often reflecting the season. A gray topcoat was selected because it shows the color of the lights better. At 3150 feet, it is one of the longest pedestrian bridges in the world.

2.6.6 Restoring Historic Masonry Arch Bridges

1. A two lane precast concrete arch bridge in Lancaster County, PA replaced a masonry bridge built in 1917, but retained the original structure's character.

A safety inspection of the original 1917 bridge revealed deterioration and structural deficiencies too extensive for repair. Design was sensitive to surrounding rural landscape. The county was able to quickly secure the Pennsylvania Historic Museum Commission's (PHMC) approval for the bridge replacement. The new concrete bridge mimics the features of the original historic structure and matches the stone-masonry-like façade of a nearby farmhouse.

2. Restoration of the Wisconsin Avenue Bridge in Washington, D.C.

The new arch bridge needed to be simulated. Based on the analysis of a three dimensional simulation model, reinforcement configurations to resist the stresses were finalized. It was successfully restored by using a proprietary system.

The method allowed 2.5-inch diameter holes to be drilled in the arch. Stainless steel 1-inch diameter anchor bars were embedded into a polyester sock material. Cementitious grout was pumped into the sock causing it to inflate and encase the steel in 6000 psi compression grout.

3. The historic 19th Century bridge located in Valley Forge National Park in Pennsylvania is a wrought-iron through truss built in 1886 and was rehabilitated using advanced composite materials. It spans 19 meters and carries one lane of vehicular traffic. Based on the deterioration of stringers that supported a timber deck, a rehabilitation project was initiated. The solution chosen was to remove the stringers and in their place install a glass fiber-reinforced polymer (GFRP) slab. The weight of the slab was roughly half the weight of the existing stringers, thereby reducing the dead load effects on the remaining structure. A wood wearing surface was installed on top of the GFRP slab to restore the deck to its initial condition. In 1998, a diagnostic load test was performed on the bridge using a pre-weighed truck. Strains and deflections of the GFRP slab and truss elements were recorded for two different truck weights and for various load passes.

4. Evaluation of historic bridges constructed in the first half of the 20th Century: Construction Technology Laboratories, Skokie, IL used advanced NDT methods to provide vital structural and material information necessary for successful rehabilitation. Laboratory testing addressed such topics as the extent of deterioration and chloride concentration in the concrete, as well as identification of aggregates and cement constituents, so repair concrete could be developed to match existing concrete. A combination of visual inspection, nondestructive methods, and coring followed by laboratory testing is used in all cases. NDT minimizes invasive investigation and disruption to bridge operations.

2.6.7 Some Well-Known Historic Bridges

1. The ancient Chinese bridge "Rainbow Truss" was built of timber logs and bamboo strings in a fashion of interlocking arches and cross beams, it is unique in both design and construction. The configuration of the arches causes it to induce much less shear force and bending moment than a beam of the same span. Its construction with small diameter round or rectangular timber logs is an advantage due to large quantities of small diameter logs readily available. Various timber bridges following the model of Rainbow Truss were later built with rectangular members as seen today in North America.
2. The Dragon Bridge of Li Chun (also known as An Ji Bridge) in Ancient China, an open spandrel, segmental stone arch bridge, is probably the oldest surviving bridge dating back to seventh century A.D. It was the first major segmental bridge when constructed and was later refurbished.
3. The John A. Roebling Bridge constructed in 1867 was the first suspension bridge to span the Ohio River near Kentucky. The main span is 1100 feet long. Based on load tests for four truck types and three bus types and compared to element level analysis, the posted weight limits on the bridge were 17 tons for two axle trucks and 22 tons for three or more axles.
4. A large number of extant lenticular (fish belly or parabolic) iron truss bridges were built in New England by the Berlin Bridge Company prior to 1900. There are still 43 pony truss, 27 through truss, and three Warren truss bridges made of cast iron which have survived.
5. The Bear Tavern Road Bridge in Mercer County, New Jersey, located on a quiet street near the author's office, includes two king through trusses built in 1882. The steel shape is a pratt truss. The floor beams have been replaced, and the bridge is functioning without a reduction in the original live load.

Figure 2.15 An example of a historic pedestrian bridge in Hunterdon County, N.J.

6. Market Street Bridge in Chattanooga, Tennessee is the second longest double leaf bascule span in the U.S. The steel truss main span is 300 feet long. There are a total of 16 spans, including three arch spans. It is a landmark located in downtown Chattanooga and has undergone major repairs, reopening in 2007.
7. The Calhoun Street Bridge on the Delaware River was placed on the National Historic Register in 1975. The two-lane bridge with a sidewalk is only 20 feet wide. It has an open steel grid deck and seven simply supported spans of pratt truss. Damaged truss members will be repaired by straightening. It is load posted and does not allow truck traffic.

Table 2.3 Locations and construction dates of covered bridges by state.

State	Number	Ranking	Earliest	Latest
Alabama	14		c1850	1934
California	13		1862	1984
Connecticut	5		1841	1976
Delaware	2		c1870	1870
Georgia	17		c1840	1975
Illinois	9		1854	1987
Indiana	93	4th	1838	1922
Iowa	12		1869	1969
Kentucky	13		c1835	c1925
Maine	12		1857	1990
Maryland	6		c1850	1880
Massachusetts	17		1832	1985
Michigan	6		1832	1980
Minnesota	1		1869	
Mississippi	1		1966	
Missouri	5		1868	1980
New Hampshire	57	5th	1827	1987
New Jersey	1		1866	
New York	29	7th	1823	1982
North Carolina	5		c1860	c1910
Ohio	143	2nd	1829	1986
Oregon	51	6th	1914	1989
Pennsylvania	227	1st	180?	1988
South Carolina	1		1909	
Tennessee	6		1876	1977
Vermont	100	3rd	c1820	1982
Virginia	9		c1800	1920
Washington	6		1905	1976
West Virginia	17		1852	1911
Wisconsin	2		1876	1962
30 States	880			

2.6.8 Revival of Covered Bridges Is Possible

Covered bridges are structural icons representative of the New England countyside. A covered bridge is a timber, steel, or reinforced structure supporting a deck surface that carries loads over an obstruction (e.g., a river). Its structural components are protected from the elements by various coverings: walls, roofs, and decks.

Covered bridges have been built in many different situations and in widely varied settings. There used to be many covered railroad bridges, but only a few still remain in the U.S. According to the World Guide, there are only eight North American railroad covered bridges still standing.

Timber structures were covered simply to help protect the timber from the ravages associated with periodic wetting. Timber truss structures without coverings would often fail after 10 to 20 years of service. Coverings quickly proved their worth by greatly extending the life of the structure, so much so that the use of timber structures without coverings was only accepted for a brief period of time. The author has composed a short poem on the need for maintenance:

A bridge is not so inanimate,
It owes its fitness to man;
It obeys all the acts of God,
Will sway much like a fan;
And with diligent looking after,
Will serve you all it can.

As reported in literature, Table 2.3 from World Guide gives details of historic covered bridges built in timber or masonry with as many as 880 surviving in 30 states.

Apart from a sense of aesthetics, the advantages are protection from deicing salts of the deck slab and deck joints and reduced thermal effects.

BIBLIOGRAPHY

Analysis and Design

Design Manual—Part 4, Structures, Pennsylvania Department of Transportation, September 2007.

International Handbook of Earthquake Engineering: Codes, Programs and Examples (IHEE), Edited by Mario Paz, Kluwer Academic Publishers, Dordrecht, 1995.

Khan, Mohiuddin Ali and M. ElGaaly, *Proceedings of Two Day Seminar*, Drexel University, Philadelphia, PA, 2002.

Khan, Mohiuddin Ali and H. Nassef, *Proceedings of Two Day Seminar*, American Society of Civil Engineers, North Jersey Section, Trenton, NJ, 2004.

LRFD Bridge Design Specifications, American Association of State Highway and Transportation Officials, Washington, DC, 1998.

PCI Bridge Design Manual, Chapter 8.12 Strut & Tie Modeling of Disturbed Regions, Precast/Prestressed Concrete Institute, Chicago, IL.

Rehm, Kelley C., "The Future of Bridge Design, States Implement Load and Resistance Factor Design Specifications to Meet and October Deadline," *ASCE Journal of Bridge Engineering*, American Society of Civil Engineers, January/February 2007.

Structural Standards Manual, Maryland State Highway Administration, Baltimore, MD, 2006.

Repair and Rehabilitation

Bane, R., "Strenthening Masonry Arch Bridges," *ASHE Scanner*, Summer 2007.

Bowman, Mark D., "Brittle Fracture of the Blue River Bridge," *Structures 2004—Building On The Past: Securing The Future,* ASCE Structural Engineering Institute, Nashville, TN, May 2004.

Davis, Allen G., Carlton A. Olsen and Kevin A. Michols, "Evaluation of Historic Reinforced Concrete Bridges," *Structures 2001—A Structural Engineering Odyssey,* ASCE Structural Engineering Institute, Washington, DC, May 2001.

Dexter, R., K. Johns and R. Di Bartolo, "Smart Sign Support," *Civil Engineering*, Volume 17, pp. 62–65, September 2000.

Dunker, K. and L. Dajin, "Foundations for Integral Abutments," *ASCE Practical Periodical on Structural Design an Construction*, Volume 12, Issue 1, pp. 22–30, February 2007.

Historic Bridges Evaluation, Preservation and Management, edited by Hojjat Adeli, CRC Press, April 2008.

Life Cycle Cost Analysis Primer, Office of Asset Management, Federal Highway Administration, Washington, DC, August 2002.

Miller, Ann B., et. al., "Final Report: A Management Plan for Historic Bridges in Virginia," Virginia Transportation Research Council, Charlotteville, VA, January 2001.

Movable Bridges, Operating Complex Systems, New Jersey Department of Transportation Structural Design Bureau, Trenton, NJ.

Perrow, Charles, *The Next Catastrophe: Reducing Our Vulnerabilities to Natural, Industrial, and Terrorist Disasters*, Princeton University Press, Princeton, NJ, March 2007.

Priority Repair Procedure, New Jersey Department of Transportation Bureau of Structural Evaluation, Trenton, NJ, January 2008.

Rong, Alex, "Structural Analysis of an Ancient Arch Bridge Using Modern Technology," *Structures 2004—Building On The Past: Securing The Future,* ASCE Structural Engineering Institute, Nashville, TN, May 2004.

Shenton, III, Harry W., Michael J. Chajes, William W. Finch, Jr., Scott Hemphill and Rick Craig, "Performance on a Historic 19th Century Wrought-Iron Through-Truss Bridge Rehabilitated Using Advanced Composites," *Advanced Technology in Structural Engineering Proceedings of Structures Congress 2000*, American Society of Civil Engineers, Philadelphia, PA, May 2000.

Steel Bridge Design Handbook, Chapter 23, Corrosion, National Steel Bridge Alliance.

"Survey Results of Bridge Maintenance Practices in the Midwest," Federal Highway Administration, Washington, DC, May 2000

Umphrey, J., Derek Beck, G. Ed Ramey and Mary L. Hughes, "Rapid Replacement of Four GDOT Bridge Decks," *Practical Periodical on Structural Design and Construction*, Volume 12, No.1, pp. 48–58, February 2007.

Zhou, Y. E., "Fatigue Problems in Steel Bridge Structures," *Structures 2004—Building On The Past: Securing The Future*, ASCE Structural Engineering Institute, Nashville, TN, May 2004.

3

Bridge Failure Studies and Safety Engineering

3.1 HISTORY OF DISASTERS AND SAFETY MANAGEMENT

3.1.1 Unexpected Material Deterioration and Failure

Engineering is usually about avoiding failures and investigating why failures occur and ways to fix the problem. There is a need to understand the conditions giving rise to past failures and ways to avoid such failures so that loss of life can be minimized. Historical events and selected case studies demonstrate the causes of each type of failure. Future design codes can make use of the deficiencies identified in order to develop guidelines for safe practice. If failures are interpreted correctly, a great deal of information for correct analysis, anticipated behavior, detailed design, and construction can be obtained to help formulate accurate design guidelines.

Failures occur in different forms in a material. Physical forms of failure can be seen as infinitely large deformation and metallurgical disintegration of elements. It can be localized cracking without collapse or discontinuity or total separation in a component.

At failure, critical sections for plastic hinges are located at the midspan of beams or under the concentrated load where deflection or bending moment is highest. It can also be at a support where shear force, reaction, or negative bending moment is the highest.

Failures are encountered on construction sites and are not just confined to the collapse of structures. Deaths and injuries to construction workers by far exceed the number of fatalities of the bridge users in failure events. Structural design methods related to construction loads and equipment need to be refined.

Physical causes are varied such as erosion, reversal of stress, impact, vibrations, wind, and extreme events. Usually, it is a combination of dead load stress combined with one or more external transient forces resulting in a compound critical stress. If dead load stress is already high and approaching the elastic limit of members, any applied force or stress will exceed the allowable limit and lead to failure. Scour evaluation reflects the scour sufficiency rating or coding. Scour rating and evaluation needs to be based on a more refined analysis.

3.1.2 Aftermath of a Bridge Failure

1. Shutdown of approaches to traffic.
2. Emergency relief work by police such as calling hospital ambulances and helicopters.
3. Provision of a detour or alternate route.
4. Emergency repairs and retrofits, if applicable.
5. Forensic engineering to resolve litigation issues.
6. Reconstruction.
7. Deficient bridges need to be replaced by efficient bridges designed by using the latest criteria. Issues include:
    - Solving design challenges
    - Availing of benefits provided by new materials
    - Deploying new techniques such as extending the span length by beam splicing and post-tensioning
    - Rapid reconstruction.

3.1.3 Studying the Reasons for Failure

1. Past failure studies have shown that failures occur due to a variety of reasons. The primary causes of failure and the numerous secondary causes contributing to failure need to be investigated. Primary effects may not all be dangerous by themselves, but when combined with secondary effects, their cumulative action can trigger a collapse.

    Lessons need to be learned from each failure. It appears that much water has flowed *over the bridge* since the disaster at Schoharie Creek in New York State. If such failures can be prevented or even minimized, the engineer has done his duty for the community. Failure of a bridge due to flood is shown in Figure 3.1.
2. There is an abundance of information available about failures that are attributed to design limitations and construction quality, in addition to those caused by extreme events.
3. Eliminating the root causes leading to failure: One of the duties of any engineer is to minimize the possibility of failure. Engineering maintenance is not just patching concrete. It means preventing failure by managing, disciplining, and applying structural mechanics to the structural domain of bridge components, made up of single or composite materials.

    It is important to understand both the mechanics and mechanism behind a failure and the applicable theory of yielding and fracture so that structural integrity is maintained and future designs are made safer.

3.1.4 Steps to Failures

No list can be complete since many failures or even near failures in the past have not become public knowledge. A structural review of hundreds of failed bridges provides important data for:

1. Those bridges which collapsed.
2. Those which get demolished due to their poor condition, since allowing them to continue would result in imminent failure.
3. The wide variety of failures can be classified as:
    - Old bridge failures—Such failures are expected due to eventual disintegration of materials.

Figure 3.1 School children visiting the site of a railway bridge failure in India in 2005. The bridge was washed by floods causing the deaths of 114 people.

- Landmark or historic bridges have improved inspection and maintenance and last longer than other old bridges of the same era.
- Failure of new bridges—These are relatively few but occur due to design or construction errors.

4. Design deficiency leading to bridge collapse: In 2007, a newly built prestressed concrete curved girder bridge collapsed in Karachi, Pakistan during rush hour killing over a dozen people and injuring dozens more (Figure 3.2).

Figure 3.2 Immediately after recent collapse of the bridge in Karachi, ambulances assemble to carry the injured to hospitals.

5. Failure during construction: A surprising number of failures have occurred during construction due to a lack of skilled labor, failure of lifting equipment and cranes, fabrication errors, or instability during erection.

3.1.5 Detailed Objectives

Failures are like open books.

There is an old saying that *"failures are the pillars of success."* It is based on deductive reasoning that if you identify, eliminate, or minimize the reasons for failure step by step, you have reduced the risk and probability of failure, thereby leading to eventual success.

Primary objectives of our study are to identify all causes of failure and to understand the technical issues and reasons behind failure.

The knowledge and experience gained from failure analysis can be applied:

1. To save lives and property.
2. To study failure mechanisms, formation of plastic hinges, and modes of failure.
3. To develop analytical methods and efficient design codes; perform post design checks.
4. To study actual load combinations leading to collapse.
5. To study the behavior of materials.
6. To refine ultimate load methods of design including load factors, distribution factors, and resistance factors.
7. To improve construction practices and to implement reconstruction methods correctly.
8. To resolve disputes for insurance claims during forensic engineering.

3.2 THE ROLE OF FORENSIC ENGINEERING

3.2.1 Combining Science, Engineering, and Legal Requirements

1. Forensic engineers explain how and why failures occur. The forensic process gets to the roots of the problem. It gives a clear insight into structural behavior and lack of maintenance, providing an independent account of deficiencies that lead to loss of life and/or damage. Conducting forensic engineering involves:
   - Expert witnessing
   - Technical knowledge
   - Detective skills
   - Legal aspects of the damage caused.
2. The basic procedures of conducting a forensic engineering investigation are applied following a failure. Ultimately, based on the knowledge gained from past experiences, the engineer is better prepared technically to successfully face the challenges that will arise in the future.
3. A forensic engineer knows that if design and reconstruction criteria were correctly implemented a failure could be prevented. As a routine, bridges get demolished from sabotage, fatigue and fracture, old age, inherent defects, or a lack of compliance to normal functions. In addition, they may fail unexpectedly or due to continued neglect.
4. Collapse due to earthquake could be delayed when standard details, such as ductile moment resistance connections are used. The expected life of 75 years or more for modern bridges and their components may not be achieved without regular inspection, structural evaluation, preventive action, and timely rehabilitation.

3.2.2 Documentation for Various Phases of Investigation

1. The structural forensic engineer, who is called in following a collapse, plays a crucial role in determining what the first steps of an investigation should be. He is the most qualified

to recognize perishable evidence and its potential value. The decisions made will directly affect the evidence upon which investigations will depend.

2. For examination by the judicial process, the design engineer will maintain important records for the bridge such as contract documents including feasibility studies, site selection, selection of bridge type, and value engineering.

   In particular documents must refer to:

   - Design criteria used and compliance with AASHTO and state codes
   - Design assumptions, calculations, and computer output—Evidence of client approval for use of computer software
   - Contract drawings
   - Special provisions of technical specifications which are not covered by state standard specifications
   - Approved shop drawings prepared by contractors
   - A record of requests for information (RFI) and design change notices (DCN) during construction
   - As-built drawings.

3.2.3 Methodology for Forensic Analysis

Accidents at construction sites can often result in serious injuries or death from defective or collapsing scaffolds and falls through roofing structures. Other cases include electrocutions, ladder injuries, defective machinery (cranes, hoists, conveyors, tractors, forklifts, etc.), malfunctioning tools, and injuries or death from collapse of walls or floors.

If the failure involves a construction site accident, the investigation will include:

- Workplace safety and liability
- Compliance with OSHA

The first steps after failure: The first steps following a collapse are critical (Figure 3.3). The structural forensic engineer who is called in following a collapse is responsible for documenting

Figure 3.3 Failure of the I-35W bridge in Minneapolis, Minnesota on August 1, 2007.

this vital information. Immediate issues that a structural forensic engineer may be faced with when arriving at a collapse site and in the ensuing days include issues such as:

- Safety
- Preserving perishable evidence
- Reserving samples
- Documentation
- Interviews
- Document gathering
- Preliminary evaluation of causes of failure.

3.2.4 Formal in-Depth Inquiry into the Defects Contributing to Failure

The condition of a failed bridge needs to be thoroughly examined. The following are some of the deficiencies a forensic engineer will typically encounter in his investigations:

1. Inability to define loads accurately, such as magnitude and unpredictable level of stress distribution from settlement.
2. Limited redundancy in the structural system.
3. Inability to fully include plastic behavior or composite action between the concrete deck slab and repeated beams, arching action, creep and shrinkage strain distribution in the deck slab.
4. Lack of information on fracture mechanics in general and lack of understanding of fracture of new materials, in particular.
5. Inelastic behavior of connections and joints, splices, gusset plates, bolts, and welds.
6. Complex behavior as a unified assembly of uncombined multiple components of mixed (old and new) materials and structural systems, resulting from rehabilitation or widening methods.
7. Delamination and reduction in strength of the concrete deck due to deicing salts (as observed from chain drag test).
8. Malfunction and locking of old bearing assemblies due to lack of maintenance, freezing of expansion bearings, large thermal forces causing compression, and local buckling of truss members and flanges.
9. Inability to prevent scour at pile top.
10. Inability to fully incorporate different types of soil interaction at abutments.
11. Lack of drainage behind abutments and pressure build-up behind abutments.
12. Inadequacy of Rankine, Coulomb, or Mononobe-Okabe theories for non-homogenous soil conditions for wing walls and stub abutments resulting in unstable design.

3.2.5 Forensic Engineering as a Diagnostic Tool

An area where forensic science is required is failure from blast loads. Accidental explosions have resulted from numerous sources, including high explosives. These impulsive events call for specialized forensic analysis methodology. The damaging loads exist for only short durations, making conventional static analysis inappropriate for backing out possible root causes. Examining structural load path transfer during a blast is required in order to provide additional support to portions of the structure under attack.

An important part of an explosion forensic analysis is to use damage indicators from the surroundings to determine the strength of the blast wave. The damage indicators are:

1. Deformed structural members.
2. Deflections in metal panels.
3. Debris throw.

The final result from analyzing many indicators leads to an understanding of the manner of explosion and magnitude of the explosion source energy. Correlation methods used are:

1. Semi-empirical damage correlations to single-degree-of-freedom analysis.
2. Semi-empirical damage correlations to dynamic nonlinear finite element analysis.
3. Use of multiple damage indicators to identify the manner of explosion.

3.2.6 Preparing Judicial Reports

An investigation report will clearly identify the reasons behind the failure and will cover any administrative and technical lapses or force majeure.

Knowledge of state and federal laws addressing the rights of victims affected by the disaster will be required. Training in global bridge engineering to include forensic engineering is desirable for designers to appreciate the consequences of failures resulting from their actions or inactions. Continuing education seminars to create interest in objective designing need to be made mandatory.

3.3 MANY ASPECTS OF FAILURES

1. Table 3.1 lists additional causes in light of more recent events. They may be identified on the basis of old and new technology.

   Additional causes of failures may be listed as:

   - Joints and connection failures (I-35W bridge failure in Minnesota)
   - Tornados and hurricanes (Louisiana disaster from Hurricane Katrina)
   - Bomb blast and vandalism (a bridge collapse in Manchester, New Hampshire and the 1992 A406 flyover in England)
   - Ice damage (author's structural solutions as structural engineer for timber fender collapse at the navigable Delaware and Raritan River bridges for the New Jersey Turnpike Authority)
   - Earthquake damage to bridges in Pakistan (author was a member of the U.S. AID Team which compiled a reconstruction report after the 2005 earthquake)
   - Scour collapse of Peckman's River Bridge from Hurricane Floyd (the six-lane collapsed bridge on Route 46 was replaced with an integral abutment bridge, using deep pile foundations and shielded with sheet piling).
2. Failures during construction and due to earthquakes have been much higher than those shown by the earlier studies and need to be taken seriously in designing for construction loads or for seismic events.
3. Design and detailing errors need to be minimized with QA/QC procedures and checking.
4. In the past, less attention has been paid to extreme events and construction conditions. However, AASHTO LRFD Bridge Design Specifications have included both extreme events and construction loads in design. Further research is needed in these relatively new disciplines.

3.4 A DIAGNOSTIC APPROACH

3.4.1 Comparative Study of Failures

The identification and diagnosis of failures is the starting point for meeting rehabilitation objectives and drafting a comprehensive code of practice for design. The author has carried out in-depth studies of such causes and their prevention from many independent sources. Only five major sources are listed here:

1. According to Jean Louis Briaud of Texas A&M, a great number of bridges continue to fail due to flood, collision, and overload. Bridges with narrow waterway openings and erodible soils are most susceptible to bridge collapse.

   Other frequent principal causes are design, detailing, construction, and material defects.

Table 3.1 Influence of technology level on bridge failures.

Bridge Component	Old Technology	New Technology	Remarks
Deck slab	Open steel grid or steel floor beam supported or low strength concrete	HPC, Exodermic and FRPC	Concrete deck is replaced every 15 or 20 years
Overlays for protection	Bitumen or screed for concrete deck	Latex modified concrete, corrosion inhibitor aggregate concrete	Wearing surfaces added (FWS) as required
Girders or trusses	Made of cast iron, wrought iron, or mild steel with low yield strength	Made of Grade 50 steel, HPS 70W and HPS 100W	Hybrid girders being used
Structural system	Use of non-redundant through trusses	Use of redundant multiple girder system	Composite action due to shear connectors
Joints and connections	Riveted connections	High strength bolts and welds	Detailing procedures revised in subsequent codes
Parapet and railing	Non-crash tested	Crash tested	Through girders also used as parapets in old system
Seismic resistance	Rigid connections in substructure	Use of ductile moment resisting frames for piers	Substructure detailing procedures changed
Bearings	Rocker and roller	Elastomeric pads or multi-rotational	New bearings allow thermal changes and seismic movements
Corrosion protection	Lead-based paint	Weathering steel with selected paint system	Paint costs have increased as percentage of total cost
Foundations	Shallow or on timber piles	Deep foundations, steel piles or drilled shaft	In-depth soil information is required
Protection against scour	Use of riprap	Gabions, sheet piles and articulated concrete blocks	Additional cost of countermeasures is incurred
Design Aspects			
Live load	H-15 and H-20	HS-20, HS-25, HL-93, and permit trucks	Majority of old bridges are posted
Strength design	Allowable stress or load factor	Load and resistance factor design (LRFD)	New designs are economical or safe
Load combinations for analysis	Extreme conditions not considered	Collision, seismic analysis, and scour analysis considered	Use of computer software has made possible over a dozen load combinations
Inspection methods	Visual	Visual and SHM	Frequency of inspections is increased
Rating methods	Load factor	Load and resistance factor rating (LRFR)	Scour vulnerability and seismic vulnerability introduced

2. Wardhana and Hadipriono studied over 500 bridge structure failures in the United States. The age of the failed bridges ranged from one year (or during construction) to 157 years. The most frequent causes of bridge failures were floods and collisions.

   Bridge overload and lateral impact forces from trucks, barges/ships, and trains constitute 20 percent of the total bridge failures. In the U.S. alone, over 36,000 bridges are either scour critical or scour susceptible.

3. Campbell R. Middleton of the University of Cambridge Engineering Department has organized comprehensive databases summarizing the failure records according to the chronology of events, country of occurrence, reasons of failure, and bridge type.
4. Bjorn Akesson in his 2008 book "Understanding Bridge Collapses," estimates scour as responsible for about 50 percent of bridge failures. The book presents useful information on structural details and deficiencies of 20 well-known bridges that failed between 1847 (Dee Bridge on River Severn, England) and 2003 (Sgt. Aubrey Cosens VC Memorial Bridge in Canada) through analysis of failures. Other causes of failures listed in his book are lack of material strength and maintenance, accidents, fatigue, brittle fractures, buckling, wind loading, aerodynamic instability, fire, and poor anchorage capacity. While failures of very old bridges built before 1940 appear to be of academic interest due to a different level of technology, (much lower intensity and frequency of truck live loads), the failures in the past 75 years (erstwhile generation of bridges) are of practical interest. A 70-year life span is the current minimum design life of each generation of bridges based on fatigue, as suggested by AASHTO LRFD Specifications. Many of the bridges listed by Akesson failed much earlier.
5. Jana Brenning in the Scientific American Journal was one of the earliest to address the causes of failures in 1993. Since then more information has been compiled. Technological advances in information systems have a great impact on data collection and analysis.

Water, salt, stress, and corrosion can make a bridge structurally deficient. A decrease in the load rating will result in imposing weight limits.

3.4.2 Understanding the Multiple Structural and Fracture Causes

The most common causes of bridge failure include:

1. Overstress of girders from section loss (Figure 3.4), design defects, and deficiencies: See Section 3.6, Design Deficiency and Preventive Actions, for historical events and recommended preventive actions.

Figure 3.4 The deterioration of girders is a potential cause of bridge failures.

Pre-collapse deficiencies can include incorrect assumptions, data errors, incorrect analysis, noncompliance with code guidelines, incorrect connection details, and mistakes in drawings. Other frequent principal causes are design, detailing, and use of substandard materials.

2. Long-term fatigue and fracture (see Section 3.7, Fatigue Failures and Suggested Preventive Actions): Failures can be due to fatigue of steel or concrete girders from repeated reversal of stress. Brittle fracture results in the unplanned loss of service, very costly repairs, concern regarding the future safety of the structure, and potential loss of life.
3. Failures during construction (see Section 3.8, Construction Deficiency and Suggested Preventive Actions): Include constructability issues such as construction inspection, construction supervision, quality control, use of substandard materials, and deficient design of temporary works. Heavy construction loads and construction defects such as poor workmanship, substandard materials, inadequate concrete curing, imperfections in steel, lack of fit, and lack of quality control are other possible causes.
4. Accidental impact from ships and vessels (see Section 3.9, Vessel Collision or Floating Ice and Suggested Preventive Actions).
5. Accidental impact from trains and defects in geometry such as vertical under clearance (see Section 3.10, Train Accidents Causing Bridge Damage and Preventive Action).
6. Accidental impact from vehicles (see Section 3.11, Vehicle Impact and Preventive Action).
7. Failures due to blast loads (see Section 3.12, Blast Load and Preventive Action).
8. Failures due to fire damage (see Section 3.13, Fire Damage to Superstructures and Preventive Action): Fire may result from accidental spraying of gasoline, any stored material under the bridge catching fire, overturned vehicles, lightning, or vandalism.
9. Failures due to earthquakes (see Section 3.14, Substructure Damage Due to Earthquake and Preventive Actions): Includes failure due to earthquake from limited bearing seat width or plastic hinge formation.
10. Failures due to heavy winds, tornados, and hurricanes (see Section 3.15, Wind and Hurricane Engineering).
11. Failures due to lack of inspection (see Section 3.16, Lack of Maintenance and Neglect). Lack of maintenance or inspection leads to:
    - Malfunction of bearings
    - Corrosion of steel: Lack of painting. Failures due to corrosion of steel girders caused by evaporation and condensation of river water.
12. Failures due to unforeseen events in spite of maintenance (see Section 3.17, Unforeseen Causes Leading to Failures).
    - Ice damage of piers and failure of timber fenders (unexpected loads and load combinations: *Force majeure*
    - Poor deck drainage: *Negligence*
    - Soil settlement: *Force majeure*
    - Freezing of bridge surface: *Negligence*
    - Failure due to gussets or connections: Design deficiency.
13. Failure due to experimentation: Some failures may occur due to experimentation with new types of materials or new systems such as undefined and unpredictable material properties in cast in place or precast construction. Although this approach may be unavoidable in certain disciplines such as space exploration, caution is necessary.

Any one of the above factors may contribute to bridge failure or may trigger a collapse, but failures actually occur due to a combination of loads, of which the principal or additional cause can be one of those listed above.

Although load combinations have been defined by AASHTO LRFD Bridge Design Specifications, they do not include some of those listed above.

Table 3.2 Recent international bridge failures, and the nature of each collapse.

Location	Year	Description	Nature of Collapse
Shershah Bridge Karachi, Pakistan (see Figure 3.5)	2007	Ten people died.	Bridge less than two weeks old
Southern China	2007	Many casualties	Bridge over river hit by a ship in fog
Kashiwazaki City, Nigata, Japan	2007	Many casualties	Due to earthquake
Laval, Quebec, Canada	2006	Autoroute 19 overpass collapsed killing five and injuring six.	Not available
India	2005	A rail disaster killed 114 people.	Flood washed a rail bridge away
Southern Spain	2005	A section of a highway bridge collapsed killing six people.	Under construction
Daman, India	2003	Long span suspension bridge	Bridge collapse over a river
Central China	2002	Two bridges killing a combined 19 people	Not available
Lisbon, Portugal	2001	Collapse caused a tour bus to plunge into a river, killing more than 50 people.	Bridge collapse over a river
Seongsu Bridge, Seoul, South Korea	1994	Collapse killed 32 and injured 17.	Not available

3.5 A HISTORICAL PERSPECTIVE OF RECENT FAILURES

3.5.1 Bridges Not Located on Rivers

Bridges have failed the world over and continue to do so. Most failures can be avoided with efficient monitoring and timely maintenance. Some examples of recent bridge failures in the U.S. are given below where full or partial failures have resulted from:

1. Fatigue and fracture (numerous railway and highway bridges).
2. Corrosion and web cracking of steel girders (I-95 curved bridge located near the Philadelphia Airport).
3. Collision damage due to limited vertical under clearance (North Jersey Bridge).
4. Fire and excessive heat (I-95 bridge northeast of Philadelphia).
5. Earthquake movements (bridge failures in California).
6. Excessive wind (Tacoma Narrows Bridge).

Examples of more recent failures outside the U.S.: A brief description of casualties is provided in Table 3.2.

3.5.2 Examples of Foundation Scour

1. Failure due to foundation scour and settlement from soil erosion is a threat to bridge structures.

   The majority of scour-related failures can be avoided by providing modern countermeasures at their foundations to prevent erosion.

2. Scour from Riverine Flow

   Foundation scour is a major cause of bridge failures. Foundation settlement is caused from erosion and weak soil conditions. When a pier tilts (Figure 3.6), there is potential for a bridge to collapse without warning. This is a definite safety issue for travelers.

   Scour likely reduces the capacity of existing foundations due to the removal of scoured material. Section 46 of the NJDOT LRFD Design Manual for Bridges and Structures, developed

Figure 3.5 Sudden collapse of Shershah curved girder bridge in Karachi, Pakistan.

by the author, requires precautionary scour protection at all times based on a scour analysis. Countermeasures are needed based on HEC-23.

Other examples of bridge failure include a bridge in Ellis County, Texas in 2004 and another major failure in New York (Schoharie Creek bridge located on the NY state thruway) in 1987.

3. Scour from tidal flow: Tidal scour analysis using procedures outlined in HEC-18 and HEC-25 is required.

- Tidal scour is more common than the fluvial flood flows of shorter durations. The destructive nature of scour evolves in cumulative smaller steps rather than occurring in a single flood event.

Figure 3.6 Abutment settlement, backwall collapse, and bearing failure.

Table 3.3 Recent bridge failures in the U.S., including the nature of each collapse.

Location	Year	Description	Nature of Collapse
Interstate 35W Minneapolis, MN	2007	Twelve people died.	Bridge collapse over a river
Webbers Falls, OK over the Arkansas River (see Figure 3.10)	2002	A 500-foot section of the I-40 bridge collapsed, killing 14 people.	A barge struck one pier of bridge, causing partial collapse.
Walnut St. Bridge, Harrisburg, PA	1996	Some injuries	Bridge collapse over a river
Tennessee River Bridge, Clifton, TN	1995	Some casualties	Bridge collapse over a river
Schoharie Creek Thruway Bridge, Amsterdam, NY	1987	A total collapse of the bridge killed 10 people	Bridge collapse over a river
Sunshine Skyway Bridge, Tampa, FL	1980	A ship hit the bridge during a storm, 35 people were killed in the collapse.	Bridge collapse over a river

- Scour is induced due to wave action.
- Scour is induced due to tidal currents.
- Effects of the interaction of simultaneous fluvial and tidal currents may be present.
- The effects of littoral drift can increase lateral migration and affect long term erosion. Bridges located on tidal waterways are also subjected to contraction and local scour.

4. The physical factors affecting tidal scour include:
- Peak tidal velocities
- Variations between flood velocity and ebb velocity
- Range of tidal amplitude between neaps and spring tides
- Locations and shapes of scour under different flow conditions
- Cumulative effect of a series of tides.

Bridges on rivers (subjected to floods or accidents) seem to be affected the most. Historic failures due to flood scour in the U.S. and abroad: Table 3.4 lists some of the recorded scour failures in the U.S., Canada, Germany, Britain, Austria, Portugal, India, China, South Korea, and Australia.

3.5.3 Diagnostic Case Studies by Author

Temporary underpinning and replacement design of New Jersy's Route 46 bridge on Peckman's River. Due to Hurricane Floyd in 1997, overtopping of bridge occurred. Much of Route 46 Peckman's River area was fully flooded. A replacement bridge was designed by the author using integral abutments with a single row of piles. Abutment settlement occurred and heavy cracking of approaches took place (Figures 3.6 and 3.7). Temporary pile bents (with piles over 90 feet long) were driven in front of abutments to transfer the load from the abutments. In addition to replacing the damaged bridge, the approach slab had to be reconstructed.

Planning recommendations for Peckman's Bridge:

1. The direction of the abutment skew is now parallel to the meandered direction of river flow to minimize scour.

2. Based on hydraulic analysis, the opening size has been increased to minimize overtopping flood.

3. Use of integral abutments and integral approaches make the bridge more resistant to longitudinal forces.

4. Steel girders have been replaced by prestressed spread box beams to prevent corrosion.

Table 3.4 Historic failures due to flood scour.

U.S. Bridges	Location	Year	Details of Failure
Interstate 29 West Bridge	Sioux City, IA	1962	Natural hazard (flooding scour)
Bridge near Charleston, SC	Cooper River, SC	1965	Natural hazard (flooding scour), pier failure
Interstate 17 Bridge	Black Canyon, AZ	1978	Natural hazard (flooding scour)
Schoharie Creek Bridge	Near Fort Hunter, NY	1987	Flooding and storm led to collapse of two spans after scouring of a pier.
Twin I-5 Bridges (Arroyo Pasajero River)	Coalinga, CA	1995	Scour of bridge foundations
Tennessee River Bridge	Clifton, TN	1995	Scour
Walnut Street Bridge (Susquehanna River)	Harrisburg, PA	1996	Scour and ice damage
Hatchie River Bridge	Near Covington, TN	1999	Scouring and undermining of the foundations
Interstate 20 bridge on Salt Draw River	Near Pecos, TX	2004	Scour from floodwaters after two days of heavy rain
Lee Roy Selmon Expressway	Tampa Bay, FL	2004	Scour hole developed under a concrete pier causing bridge to drop.
Rural Bridge (Beaver Dam Creek)	Near Shelby, NC	2004	Natural hazard (flooding scour) causing bridge washout
Canadian Bridges			
Bridge over a river	British Columbia	1981	Flood scour and tree debris in water destroy bridge.
German Bridges			
Bridges over Weser River	Bremen	1947	Flooding, floating ice, and ships led to the collapse of the bridges.
Bridge over Mosel River	Near Koblenz	1947	Flooding and floating ice led to collapse of the bridge.
Esslingen Bridge	Esslingen	1969	Water entering sheet pile wall caisson and flooding during construction
Bridges in Germany	South and East Germany	2002	Extensive flooding in South and East Germany due to scour
British Bridges			
Drimsallie Bridge	Inverness, Scotland	1973	One span of bridge collapsed due to washout of abutment in flood—foundation scour
Glanrhyd Railway Bridge over River Towy	Near Llandeilo, Wales	1987	Flooding, bridge collapsed as a train drove over it—flood scour.
Multispan masonry arch Ness viaduct	Inverness, Scotland	1989	Heavy floods washed multispan masonry arch bridge away, just after a freight train had passed over it—flood scour
Five-span bridge at Forteviot, (May River)	10 km south of Perth, Scotland	1993	Flooding, erosion of the gravel bed beneath the downstream face of the shallow founded pier, concrete bag scour protection washed away—flood scour.

Table 3.4 Historic failures due to flood scour *(continued)*.

Austrian Bridges	Location	Year	Details
Motorway Bridge	Near Salzburg	1959	Scour due to flooding
Two-span truss bridge over Traun River	Between Linz and Selzthal	1982	Scour led to loss of pier and partial collapse of bridge girder.
Five-span box girder motorway bridge over Inn River	Near Kufstein	1990	Scouring led to settlement and major damage of distorted superstructure.
Bridge in Braz	Braz, Vorarlberg	1995	Express train plunged into ravine after mudslide destroying rail bridge—scour due to floods
Bridges in Austria	Various locations	2002	Flooding in Thurnberg, Engelstein, Salzburg and other cities —scour
Portuguese Bridges			
Bridge over river	Lisbon	2001	Bridge collapse caused a tour bus to plunge into a river
Indian Bridges			
Bridge between Jabalpur and Gondia	Madhya Pradesh	1984	Flooding scour
Bridge (Nalgonda district)	Near Veligonda	2005	Floodwaters from two breached reservoir tanks upstream washed away the embankment leaving only the rail tracks dangling in the air. Train derailed and plunged into a rivulet due to breaches on the track.
Railway bridge	India	2005	Flood washed a rail bridge away—foundation scour.
Long span suspension bridge over river	Daman	2003	Flood velocity
Chinese Bridges			
Two bridges	Central China	2002	Floods
South Korean Bridges			
Bridge over river	Seoul	2004	Foundation scour
Australian Bridges			
River bridge 40 km west of Charters Towers	Queensland	2005	A man was checking floodwaters when the bridge he was standing on collapsed—flood scour.

5. Deep sheet piling sections were used to minimize scour of abutment and wingwalls.
6. Both superstructure and substructure designs are based on the LRFD method. Figure 3.8 shows the completed bridge.
7. Replacement design of the Rancocas Creek Bridge by the author (see Figure 3.9) shows the parapet type used by Burlington County, NJ.

Flooding of Rancocas Creek and occasional overtopping of the bridge showed that the bridge was functionally obsolete. It had become dangerous and its hydraulic opening needed to be increased. Unlike the spread box beams of Peckman's River Bridge, adjacent box beams were used which helped to reduce life cycle costs by avoiding corrosion and repainting.

Figure 3.7 Installing temporary pile bents on Peckman's River Bridge.

3.5.4 Preventive Action against Foundation Scour

Performing scour analysis and use of effective countermeasures, such as deep foundations and river training, may be required in addition to conventional shielding. Design of countermeasures will be based on HEC-23 procedures.

Scour countermeasure options include river training measures, dredging, driving sheet piles around the piers and abutments, and filling the gaps with riprap. Alternate measures such as grout bags, cable-tied blocks, and AJAX type blocks will be considered, based on tidal flow conditions.

Figure 3.8 Overhead sign structure and integral abutment designed by author.

Figure 3.9 Author's design of aesthetic, strong, and easy-to-repair concrete parapet.

Wide cracks at piers need to be pressure grouted with epoxy grout, as recommended in the inspection report. Design of required repairs and retrofits to substructure will consider:

- Modern countermeasures technology requires new construction techniques. The contractor performing such tasks will need properly trained construction crews.
- Providing cofferdams, sheet piling, and bed armoring will require temporary construction work. Alternatives for using quick construction should be considered.

Figure 3.10 Failure of Webbers Falls bridge in Oklahoma in 2002.

- The available flow width may be reduced due to construction of cofferdams.
- Velocities through the remaining opening will increase, thereby increasing scour in the channel and around the structure.
- Any river pollution from construction material needs to be monitored. Regular cleaning of the channel may be required.
- Approvals for stream encroachment permits will be necessary.
- The effect of driving sheeting or bed armoring on existing utility needs evaluation.
- Countermeasures may extend into adjacent property limits. Any construction easement needs to be carefully evaluated and permits obtained.
- Tidal conditions will affect working methods and working hours for construction.
- The health and safety of construction personnel will be a concern due to water depth.

3.6 DESIGN DEFICIENCY AND PREVENTIVE ACTIONS

3.6.1 Incorrect Design Assumptions or Modeling Errors

Table 3.5 shows reasons for failures from assumptions which are usually made in analysis. Failures result when assumptions do not truly represent the behavior of the superstructure in the field. Figure 3.11 shows a newspaper cartoon published in Karachi, Pakistan after the 2007 collapse of a bridge.

3.6.2 General Preventive Measures

1. Provide space for bearing inspection chambers.
2. Apply load and resistance factors based on LRFD methods.
3. Ensure seismic retrofit against minor and recurring earthquakes.
4. Provide scour countermeasures.
5. Effectively monitor through remote sensors.
6. Study failure mechanisms of different types of structural systems.
7. Maintain quality control and personnel safety during construction.
8. Develop and make available codes for rehabilitation of mixed structural systems should be developed and made available.

Figure 3.11 The impact of engineering decision on bridge users: A cartoon shows a family in Karachi offering prayers before crossing a bridge. The public is cautiously aware of bridge safety conditions.

Table 3.5 History of failures due to design deficiency.

U.S. Bridges	Location	Year	Reasons for Failure
Tacoma Narrows suspension bridge	Washington State	1940	Insufficient bending and torsion stiffness, aerodynamic instability
King's River Slough Bridge	Near Fresno, California	1947	Overloading from an agricultural train
Elbow grade bridge, timber truss	Willamette National Forest	1950	Parts of truss underdesigned
Silver bridge, chain suspension bridge	Ohio River	1967	Fatigue
Bridge over Kaslaski River	Illinois	1970	Not anchored against uplift—design error
Syracuse bridge	New York	1982	Torsional buckling due to lacking lateral support
Oakland highway bridge	California	2007	Design deficiency
Canadian Bridges			
Second Narrows Bridge	Vancouver, British Columbia	1958	Web buckling of transverse girder due to design error
The Autoroute 19 Overpass bridge	Laval, Quebec	2006	Design deficiency
German Bridges			
Continuous truss bridge over Leda River	Near Leer	1960	Earth pressure horizontal load not considered —design error
A2 bridge	Near Lichtendorf, Schwerte	1968	Moving supports due to creep, shrinkage, and low temperature-pier head destroyed, settlement of bridge—design failure
Rodach River bridge	Near Redwitz	1973	Bridge collapses under overload from ready-mix concrete mixer—design failure
Vorland Bridge	Hochheim	1973	High temperature resulting in support plates moving – design failure
Zeulenroda Bridge	East Germany	1973	Buckling of steel box section flange plate due to lack of stiffeners
Austrian Bridges			
Reichsbrücke	Vienna	1976	Lack of reinforcement in pier footing
Indian Bridges			
Assam bridge	Assam	1977	Heavy train—overloading
Punjab Province bridge	Punjab	1977	Packed coach on bridge—overloading
Dombivli Railway Station foot overbridge	Dombivli	2004	Faulty design, not enough resistance—failed during construction
South Korean Bridges			
Seongsu Bridge	Seoul	1994	Design deficiency

9. Cracks in substructure due to foundation settlement needs to be prevented by underpinning. (see Figure 3.12).
10. Develop codes for new materials such as FRP decks should be developed. New techniques of repairs as discussed in this book need to be incorporated in the codes.
11. Implement greater vendor and construction engineer participation in revising and developing design codes.

3.7 FATIGUE FAILURES AND SUGGESTED PREVENTIVE ACTIONS

3.7.1 Fatigue of Members and Connections/ Controlling Level of Truck Traffic

1. Load-induced fatigue damage assessment: In 2005 ASCE Proceedings, J. Robert Connor and John W. Fisher of Lehigh University, Bethlehem, Pennsylvania, and William J. Wright of FHWA reported on brittle fractures in steel and preventive maintenance strategies.

   Cumulative fatigue damage of uncracked members and fasteners that are subjected to repeated variations or reversals of load-induced stress needs to be assessed. The lists of detail categories and illustrative examples to consider in a fatigue damage assessment are shown in AASHTO specifications and the AISC Handbook.

   If cracks have already been visually detected, a more complex fracture mechanics approach for load-induced fatigue is required instead of the procedure outlined here.

   Generally, upon visual detection of cracking, the vast proportion (perhaps over 80 percent) of the fatigue life has been exhausted and retrofitting measures should be initiated.

   Infinite fatigue life or remaining finite fatigue life needs to be evaluated.
2. Calculated fatigue stress range: The factored live load stress range, produced by the method given in the LRFD code is considered an approximate method of analysis.

3.7.2 Suggested Preventive Action Against Excessive Fatigue

Adequate man hours need to be spent in analysis and design to conform to all aspects of AASHTO LRFD code. To avoid failures:

1. Durability requirements need to be addressed on a scientific basis.
2. AASHTO code provisions and other relevant codes need to be followed.
3. Fatigue failures are most critical when there is no evidence of fatigue cracking leading up to

Figure 3.12 Foundation settlement cracks at the junction of the abutment and wingwall create a potential danger of collapse.

the fracture. Hence, they occur without warning and the details are essentially non-inspectable. The following preemptive retrofit strategies appear to be highly desirable:

- Avoid use of high carbon brittle steel
- Avoid poorly executed welding leading to high residual stress level
- Avoid bad detailing
- Avoid dynamic loads that cause high strain rates and reversal of stress
- CAD drawings should have sufficient details. Emphasize increasing the strength of joints by adding bolts and the strength of the girder web and flanges by adding plates, etc.
- QA/QC procedures should ensure adequate checking of criteria, method of analysis, design details, and technical specifications.

3.8 CONSTRUCTION DEFICIENCY AND SUGGESTED PREVENTIVE ACTIONS

3.8.1 Lack of Quality Control or Construction Supervision

Failure or early demolition of a bridge needs to be avoided at all costs. The most vulnerable stage is during construction. During construction, critical items are inadvertently overlooked, thereby leading to failure. Critical items include:

1. Inadequate design of formwork or its premature removal.
2. Inadequate bracing.
3. Improper sequence of concrete placement.
4. Improper sequence of erection.
5. Improper placement of reinforcing bars.
6. Incorrect profiles of post-tensioning tendons.
7. Welding deficiencies in steel connections.
8. Incorrect thickness of gusset plates.

Table 3.6 shows reasons for failures during construction or by negligence in the field. It is an area of weakness where expertise in construction techniques is desirable.

Many failures seem to happen during construction. One of the difficulties is that construction practice varies from state to state and from job site to job site. Site organization is based on selection of one general contractor, who in turn selects several sub-contractors who have specialized in a particular trade such as concreting, formwork, steel fabrication, bearings, reinforcing steel, etc. Construction procedures need to be streamlined.

3.8.2 Method of Construction

1. Current specifications do not adequately cover construction related design and temporary loads. Future construction codes should address issues created by the use of the latest technology such as new construction loads. Technical specifications may also be made comprehensive to give minutest details of construction procedure.
2. Accelerated bridge construction: Modern construction technology seems to be pulling the train on design methods. Precast technology is a world apart from traditional wet construction methods. Self-propelled modular transportation (SPMT) has enabled the transportation of long span assembled girders without the need for splices, resulting in increased factory production.

The connection design used for precast construction is different than that used in traditional construction. Joint strength must be be tested in a structures laboratory.

3.8.3 Suggested Preventive Action against Construction Failures

1. Design of temporary works such as scaffolding supports and formwork supporting a deck during concreting needs to be in accordance with AASHTO temporary works design manual

Table 3.6 History of failures during construction (constructability issues).

U.S. Bridges	Location	Year	Construction Deficiencies
Hinton truss bridge	West Virginia	1949	Insufficient design capacity of cantilever arm during construction phase
Sullivan Square Viaduct motorway bridge	Boston, Massachusetts	1952	Instability of scaffolding during construction
Buckman Bridge	Jacksonville, Florida	1970	Voided pier filled with sea water during construction—expansion of pier—partial collapse of bridge
Motorway bridge (Arroyo Seco River)	Near Pasadena, California	1972	Scaffolding collapsed under weight of fresh concrete
Concrete 5-span box girder bridge near	Near Rockford	1979	Large cracks, failure of epoxy-filled joint (not enough hardened to take design shear force)
Multiple span box girder bridge	East Chicago, Indianapolis	1982	Scaffolding collapsed under weight of fresh concrete
Prestressed concrete precast box girder bridge	Saginaw, Michigan	1982	Temporary support elements too weak during construction
Walnut Street Viaduct over I-20	Denver, Colorado	1985	Failure of pier head during construction sent eight bridge girders onto road
El Paso bridge	El Paso, Texas	1987	Inadequate scaffolding during construction
Motorway bridge	Near Seattle, Washington	1988	Girders not tied together by diaphragms, domino effect during construction
Box girder bridge	Los Angeles, California	1989	Collapsed when scaffolding was removed during construction
Baltimore bridge	Baltimore, Maryland	1989	Prestressing not in place, asymmetric loading during construction
Truss bridge	Concord, New Hampshire	1993	Stiffener mounted at wrong place during construction
3-span 3-girder composite bridge	Near Clifton, Tennessee	1995	Executed construction sequence different from the one planned
Marcy bridge (Utica-Rome Expressway project)	Marcy, New York	2002	Global torsional buckling during concreting, bridge not braced properly
Imola Avenue Bridge	Napa, California	2003	3- 100-ton hydraulic jacks to raise falsework failed to support poured-in-place concrete deck slab
Bridge near Pawnee City	Near Pawnee City, Nebraska	2004	Failure of falsework caused bridge collapsed during concrete pouring
I-70 Bridge	Denver, Colorado	2004	Bracings, fastened to bridge with bolts, came loose as girder collapsed—construction failure
Canadian Bridges			
Dawson Creek suspension bridge (Peace River)	British Columbia	1957	Movement of anchorages on footings which were not fixed properly—sub-standard construction
Second Narrows Bridge (Gerber hinge)	Vancouver	1958	Bad construction details detected, but no action taken —construction failure
Arch bridge over Rideau River	Ottawa	1966	Scaffolding collapsed under weight of fresh concrete —construction failure
3-span arch bridge	Elwood	1982	Lateral buckling of scaffolding due to insufficient lateral supports—construction failure
Composite bridge near Sept-Iles	Near Quebec	1984	Failure during construction from faulty calculations —design errors

Table 3.6 History of failures during construction (constructability issues) *(continued).*

German Bridges	Location	Year	Details
Motorway bridge	Near Frankenthal	1940	Failure of lifting equipment during construction
Hindenburg bridge over Rhine River	Cologne	1945	Collapse during refurbishment
Motorway composite bridge	Near Kaiserslautern	1954	Insufficient stiffness of top members about weak axis —construction failure
Nordbrücke bridge over Rhine River	Düsseldorf	1956	Insufficient crane capacity to carry double load—construction failure
Continuous motorway bridge	Near Limburg	1961	Settlement of temporary foundations, load redistribution, scaffolding collapse—construction failure
Heidingsfeld motorway composite bridge	Heidingsfeld	1963	Temporary concrete support plates underdesigned—construction failure
Vorland Rees-Kalkar plate girder bridge	Between Rees and Kalkar	1966	Temporary supports underdesigned—construction failure
Bridge near Wennigsen, Niedersachsen	Near Wennigsen, Niedersachsen	1971	Scaffolding collapsed under weight of fresh concrete —construction failure
Steel box girder bridge over Rhine River	Koblenz	1971	Plate buckling of bottom chord in compression—cantilevered construction failure
Continuous Hangbrücke over Laubachtal	Near Koblenz	1972	Scaffolding collapsed under weight of fresh concrete —construction failure
Steel box girder bridge	Zeulenroda	1973	Plate buckling of bottom chord—cantilevered construction failure
Bridge over Leubas River	Near Kempten	1974	Scaffolding collapses under weight of fresh concrete —construction failure
Brohltal bridge, segmental construction	Brohltal	1974	Incremental launch construction led to concrete crushing when low prestressing cable positions are over support, settlements
Timber truss	Bad Cannstatt	1977	Construction sequence not thought out—construction failure
13-span Rottachtal bridge	Near Oy	1979	Incremental launch, large cracks, inversed position of gliding plate (top/bottom)—construction failure
Bridge near Dedensen	Near Dedensen	1982	Lateral buckling of construction support girder during removing of lateral supports
Simple span, steel truss bridge	Road bridge	1982	Temporary support elements too weak—construction failure
Bridge on DB Lohr-Wertheim railway line	Near Kreuzwertheim	1984	Use of uncertified lifting bars and too weak bolt nuts —construction failure
Composite Czerny Bridge	Heidelberg	1985	Use of wrong bolts—construction failure
New (composite) Grosshesselohe bridge	Munich	1985	Ignorance of load case "displacement of mobile scaffolding"—construction failure

(Table continues on next page)

Table 3.6 History of failures during construction (constructability issues) *(continued)*.

German Bridges	Location	Year	Details
A3 motorway bridge (Main River)	Near Aschaffenburg	1988	Critical load case during incremental launch not included, shear failure during construction
Approach bridge (beam-and-slab)	Cologne-Wahn Airport	1995	Scaffolding collapsed under weight of fresh concrete —construction failure
British Bridges			
Barton Bridge	Lancashire, England	1959	Buckling of temporary props—construction failure
Cleddau Bridge	Milford Haven, Wales	1970	Incremental launch of long span, box girder plate buckling over support—construction failure
Loddon Bridge	Berkshire, England	1972	24 m span collapsed during placing of concrete due to failure of falsework—construction failure
Austrian Bridges			
Fourth Danube Bridge (plate box girder bridge)	Vienna	1969	Plate buckling of bottom chord in compression—construction failure
Soboth prestressed concrete bridge	Soboth	1970	Collapsed during cantilevered construction, prestressing bars badly put in place
Prestressed concrete bridge over Tauern motorway	Gmünd	1975	Concrete resistance not yet achieved, construction not in accordance with design
Rheinbrücke bridge over Rhine River	Near Höchst, Vorarlberg	1982	Scaffolding collapses under weight of fresh concrete —construction failure
Spanish Bridges			
Highway bridge	Southern Spain	2005	Under construction
Indian Bridges			
Bihar district bridge	Bihar	1978	Construction failure
Japanese Bridges			
Prestressed concrete bridge	Avato, Japan	1979	Incremental launch, when cantilevers coming from two sides were to be joined, differences in length appear. Temporary construction to correct it led to collapse of both cantilevers—construction failure
Tokyo West bridge over Tama River	Tokyo West	1984	Scaffolding removal sequence was not well-thought-out —construction failure
Hiroshima bridge	Hiroshima	1991	Stability problem, sliding—construction failure
Australian Bridges			
Westgate Bridge over Yarra River	Melbourne	1970	Plate buckling due to weak splicing of longitudinal stiffeners —construction sequence was not well-thought-out
Loddon River bridge	Near Victoria	1972	Scaffolding collapsed under weight of fresh concrete —construction failure

and OSHA standards. Lack of quality control for materials testing and an unrealisticly quick construction schedule need to be avoided.

2. New foundation construction techniques: Many problems and substandard performance of foundations observed in structures on expansive soils occur from faulty construction practices. The construction equipment and procedures that are used depend on the foundation soil characteristics and soil profiles. Construction techniques that promote a constant moisture regime in the foundation soils should be used during and following construction.
3. Self-consolidating concrete (SCC) for use in drilled shaft applications: When conventional concrete is used in congested drilled shafts, coarse aggregates may bridge between reinforcing bars, which may lead to segregation of the concrete between the inside and outside of the reinforcing cage. SCC is feasible for use in congested drilled shaft applications.

3.9 VESSEL COLLISION OR FLOATING ICE AND SUGGESTED PREVENTIVE ACTIONS

3.9.1 General

Bridge piers located on navigable rivers are likely to be hit in fog or in darkness usually from barges or ocean going ships. Damage to timber fenders which shield the piers may also be caused by floating ice at high velocities. The EOR must assemble the following information:

1. Characteristics of the waterway including:
   - A nautical chart of the waterway
   - Type and geometry of the bridge
   - Preliminary plan and elevation drawings depicting the number, size, and location of the proposed piers, navigation channel, width, depth, and geometry
   - Average current velocity across the waterway.
2. Characteristics of the vessels and traffic including:
   - Ship, tug, and barge sizes (length, width, and height)
   - Number of passages for ships, tugs, and barges per year (prediction for 25 years)
   - Vessel displacements
   - Cargo displacements (deadweight tonnage)
   - Draft (depth below the waterline) of ships, tugs, and barges
   - The overall length and speed of tow.
3. Accident reports.
4. Bridge importance classification.

Table 3.7 shows failure details for a large number of impacts from ships, which is also a cause of concern for the shipping industry.

3.9.2 Design Vessel

The design of all bridges over navigable waters must be checked for possible vessel collision. Conduct a vessel risk analysis to determine the most economical method for protecting the bridge. The number of vessel passages and the vessel sizes are embedded as an integral part of the vessel collision risk analysis software.

1. The Florida DOT's MathCAD software for conducting vessel collision risk analysis may be used. The software computes the risk of collision for several vessel groups with every pier. When calculating the loads and load factors probability, the overall length of each vessel group is used instead of the length overall (LOA) of a single design vessel.
2. Widening of bridge on navigable waterway: Major widening spanning navigable waterways must be designed for vessel collision. Minor widening spanning navigable waterways will be considered on an individual basis for vessel collision design requirements.

Table 3.7 History of failures due to accidents or impact from ships (human error).

U.S. Bridges	Location	Year	Failure Details
Swing bridge	Boston-Charlestown, Massachusetts	1945	Ship impact hit half-open swing bridge
John P. Grace Memorial Bridge	(Cooper River), South Carolina	1946	Ship forced by wind into bridge deck
Bridge near Charleston	(Cooper River), South Carolina	1965	Ship impact, error of ship captain
Chesapeake Bay Bridge	Annapolis, Maryland	1970	Military ship lost control and hit the bridge during stormy weather, five spans collapse, 11 other spans damaged
Sidney-Lanier Bridge	Brunswick, Georgia	1972	Ship impact, misunderstanding between captain and staff
Chesapeake Bay Bridge	Annapolis, Maryland	1972	Ship impact, two spans collapse, five other spans damaged
Lake Pontchartrain bridge	Lake Pont	1974	Ship impact, captain slept
21-span, Pass Manchac Bridge	Louisiana	1976	Ship impact, error of ship captain
Benjamin Harrison Memorial Bridge (James River)	Near Hopewell, Virginia	1977	Ship impact, failure of ship guidance electronics
Bridge over Passiac River	Union Avenue, New Jersey	1977	Ship impact caused two spans collapse
Southern Pacific Railroad Bridge	Berwick Bay, Louisiana	1978	Ship impact caused steel truss of 70 m to falls into water and sink
Sunshine Skyway Bridge	Near St. Petersburg, Florida	1980	Ship impact, not enough care by captain in bad weather
Herbert C. Bonner Bridge (Oregon Inlet)	North Carolina	1990	Ship impact, four piers damaged, five spans collapsed
Truss bridge	Near Mobile, Alabama	1993	Ship impact
Queen Isabella Causeway	South Padre Island, Texas	2001	Four barges and a tugboat struck the bridge
Interstate 40 Bridge over the Arkansas River	Webber Falls, Oklahoma	2002	Ship collides with one of the piers, bridge collapses on 150 m of length
Canadian Bridges			
Fraser River Swing Bridge	New Westminster/ Vancouver	1975	Ship impact caused 120 m spanto collapses—accident
Australian Bridges			
Tasman Bridge over Derwent River	Hobart, Tasmania	1975	Ship impact, inexperienced captain
Recent Chinese Bridges			
Bridge over river	Southern China	2007	Bridge was hit by a ship in fog
Swedish Bridges			
Almo Bridge	Near Gothenburg, Sweden	1980	Ship impact on steel arch due to lack of visibility in bad weather

3. Span length: The length of the main span between centerlines of piers at the navigable channel must be based upon Coast Guard requirements, the vessel collision risk analysis (in conjunction with a least-cost analysis), and aesthetic considerations.
4. Soil conditions: Soil depth that is subject to local and contraction scour must be calculated. The soil model must utilize strength characteristics over this depth and compared to that of redeposited soil.

3.9.3 Suggested Preventive Action against Ship Collision

1. Provide adequate lighting around bridge substructures to prevent accidents during nighttime or fog.
2. Provide and maintain timber fenders. Timber fenders should be provided around piers to resist the impact of ships on bridge substructures.
3. Schedule Coast Guard crews to monitor bridges at nighttime.
4. Improve training of ship captains and crew in navigation under bridges.

3.10 TRAIN ACCIDENTS CAUSING BRIDGE DAMAGE AND PREVENTIVE ACTION

3.10.1 Train Impact

Table 3.8 shows failure details of numerous train impacts or derailments.

Table 3.8 History of bridge failures due to accidents or impact from trains (human error).

U.S. Bridges	Location	Year	Failure Details
Alabama Rail Bridge	Alabama	1979	Train impact
Turkey Creek Bridge	Sharon Springs, Kansas	2002	Train brakes became very hot, setting timber bridge on fire
West Grove Bridge	Silver Lake, Kansas	2004	Bridge collapsed after derailment of 40 of the coal train's 137 cars
German Bridges			
2-span bridge (over railway line)	Near Eschede	1998	Derailed train impact under bridge—accident
Wuppertal Schwebebahn bridge over Wupper River	Wuppertal Schwebebahn	1999	Maintenance personnel left maintenance equipment on tracks; train derails
British Bridges			
St. John's Bridge	London	1957	Derailed train rams steel pier of 350 ton girder bridge, bridge collapses on train
Indian Bridges			
Bridge over Beki River	Northeast of India	1977	Derailed train impact on bridge—accident
Australian Bridges			
Bridge near Granville Station over railway line	Sydney	1977	Derailed train impact under bridge

Table 3.9 Failures due to accidents or impact from vehicles (human error).

U.S. Bridges	Location	Year	Reasons of Failure
Goose River Bridge	Rockport, Maine	1947	Truck impact on truss
Truss bridge over Milwaukee River	Trenton, Wisconsin	1980	Truck impact on main truss
Truss bridge	Shepherdsville, Kentucky	1989	Litter collector was higher than bridge clearance
Historic Tewksbury Township pony truss bridge	Hunterdon County, New Jersey	2001	Truck struck the bridge abutment and caused its collapse
Highway 14 overpass over I-45	60 miles south of Dallas, Texas	2002	Truck slammed into overpass, causing bridge collapse
I-95 bridge	Bridgeport, Connecticut	2004	Car collided with oil tanker, causing fire
German Bridges			
2-span bridge over motorway A2	Near Dortmund	1979	Truck high speed impact on piers —accident
2-span composite bridge over motorway A3	Near Duisburg	1979	Crane on back of lorry frees itself and impacts bridge deck—overload accident
2-span continuous composite bridge over motorway A1	Near Sittensen	1979	Truck looses control, impact on pier —accident
Suspension bridge over Dortmund-Ems Canal	Near Münster	1980	Truck impact on hanger due to ice—design failure
Bridge over Mittlerer Ring	Munich	1981	Self-erected skip on dump truck impact—overload accident
British Bridges			
Bridge over M62 motorway	Near Manchester, England	1975	Impact of crane on road below—accident
Historic wood and metal bridge	Codsall Railway Station, Staffordshire, England	2005	The bridge collapsed after being hit by a maintenance vehicle—accident.

3.10.2 Suggested Preventive Action against Train Collision

1. Slowing down of trains to prevent derailment.
2. Better coordination between train drivers and station masters at bridge approaches.
3. Railway engineers to study causes of failures and incorporate changes in operations criteria.

3.11 VEHICLE IMPACT AND PREVENTIVE ACTION

Table 3.9 shows failure details of numerous truck and vehicle impacts on superstructures or substructures that can be minimized by taking necessary precautions.

3.11.1 Minimum Vertical Under Clearance Is a Bottleneck for Truck Traffic

AASHTO requirements are to use 16 feet 6 inches vertical under clearance. Many old bridges in the U.S. and abroad do not meet this requirement. Newer model trucks at times get stuck under the girders causing serious damage to the bridge and the truck.

3.11.2 Vehicular Accidents/Damage to Superstructures and Substructures

Vehicular accidents can cause damage to parapets and railings. Girders and piers get hit by vehicles due to insufficient horizontal and vertical under clearance. Vehicular accidents may also be caused by poor deck drainage and water accumulation after a heavy downpour. Some causes include:

1. There may not be enough scuppers installed on the deck.
2. The magnitudes of cross slope (< 4 percent) and longitudinal slope (< 2 percent) may not be adequate.
3. The local intensity of rainfall may be underestimated.
4. In winter months, ice layers forming on the concrete deck may create a slippery surface for vehicle wheels.
5. In addition, deicing salts may cause early deterioration and cracking of a deck, making a bumpy ride for motorists.

3.11.3 Suggested Preventive Action against Vehicle Collision

1. Ensuring adequate drainage of decks during storms.
2. Efficient de-icing after snowfall and posting warning signs for bridge deck freezing.
3. Improving traffic signs at approaches.
4. Avoiding centrifugal forces due to a sharp radius, or improving sight distance.
5. Discouraging drunk driving with severe penalties.
6. It is important that vehicles and the bridge be properly maintained.
7. Also, drivers should not be under the influence of drugs or suffering from drowsiness due to lack of sleep.

3.12 BLAST LOAD AND PREVENTIVE ACTION

3.12.1 Introduction

1. Bridges provide vital links for the nation's economy, defense, and quality of life. In addition to natural hazards, the transportation infrastructure in the U.S. is vulnerable to physical, biological, chemical, and radiological attacks.

   To keep the system safe and operational under all circumstances systems and technologies should be developed to prevent, detect, respond to, and remediate all attacks. The design of bridges for blast and impact is important. Multiple hazards affect the stability and service life of structures.
2. University of Missouri-Columbia (UMC) has a unique National Center for Explosion Resistant Design. They are engaged in research projects to establish prototypical functionality and architectural standards for blast-resistant barricade systems through applied research, design, and test efforts.
3. To begin addressing this concern, FHWA has initiated a state pooled grant program entitled "Validation of Numerical Modeling and Analysis of Steel Bridge Towers Subjected to Blast Loading." The research program includes a combination of blast testing against scaled steel tower components and parallel numerical computations using a variety of analytical tools.
4. Inelastic dynamic progressive collapse analysis of truss structures: In a truss, dynamic effects may arise not only due to external dynamic loads, but also due to a sudden reduction in the stiffness and load carrying capacity of a member. Resulting dynamic redistribution of internal forces may lead to progressive collapse of the entire truss structure.

The sudden reduction in the load carrying capacity of a member may occur due to post buckling and subsequent inelastic cyclic member behavior.

If a truss with redundant members is subjected to a slight quasi-static overload such that the failure is initiated by buckling of a member, post buckling and subsequent cycles of inelastic force-deformation behavior of truss members may create adverse member force redistribution that leads to complete collapse of the structure.

3.12.2 Suggested Preventive Action against Failure from Bomb Blast

1. Load combinations in AASHTO code should be clearly defined, and a suitable method of analysis should be adopted.
2. Attention should be paid to multi-hazard design procedures to maintain needed safety levels.
3. Life cycle costs should be included for making appropriate retrofit decisions.
4. Improving security at bridges by surveillance cameras, preventing parking in the vicinity of bridges, installing fire controlling devices near bridge parapets, and more vigilant policing are required.

3.13 FIRE DAMAGE TO SUPERSTRUCTURES AND PREVENTIVE ACTION

3.13.1 Vehicular Accidents

Overturning of trucks due to slippery decks may cause spraying of the deck with gasoline and fire. Also, vehicular accidents can be the cause of fire.

Table 3.10 shows historic details of bridge failures due to fire.

3.13.2 Repairing the Fire Damaged Notre Dame Bridge

On April 12, 2003 a fire ignited underneath an important bridge in the City of Manchester, New Hampshire. Although the fire was intentionally set and caused major interruptions to traffic, telephone lines and some city fire department communications, there is no evidence that this was an act of terrorism. Although there are emergency plans in place and precautions to prevent additional threats, disasters (natural and man-made) are still going to occur.

Table 3.10 Bridge failures due to fire (accidental spill of cil or vandalism).

U.S. Bridges	Location	Year	Details of Failure
Two U-section bridges (Floyd River)	South of Le Mars, Iowa	1941	Fire due to collision of two vehicles
Notre Dame Bridge	Manchester, New Hampshire	2003	Arson
I-95 bridge	Northeast of Philadelphia, Pennsylvania	2004	Fire started due to accidental burning of used tires dumped under the bridge piers
Wooden bridge spanning Rio Hondo flood control channel	Pico Rivera, California	2005	Fire began in combustibles beneath the bridge and spread to wooden infrastructure—Arson
British Bridges			
Tubular Britannia Bridge	Menai Straits, Whales	1970	Timber roof impregnated with tar paints helped fire, steel box girder superstructure deflections of up to 0.75 m
Flyover over A406, Staples Corner	London, England	1992	IRA bomb exploded underneath, causing serious damage to roads and nearby buildings—vandalism

3.13.3 Suggested Preventive Action against Fire Damage

1. Emergency fire extinguishing equipment should be provided near bridges which are located more than one hour away from a fire station.
2. Transverse and longitudinal wind analysis and uplift with the correct intensity need to be considered in the design.
3. Member sizes and joints must be designed for the AASHTO LRFD load combination strength III.
4. Timber bridges should be painted with fire resistant paint and regularly sprayed.

3.14 SUBSTRUCTURE DAMAGE DUE TO EARTHQUAKE AND PREVENTIVE ACTIONS

3.14.1 Progressive Collapse of Piers

1. Bridges are known to move sideways and collapse under their own weight during seismic events. The sway collapse mechanism of bridges depends upon the type of substructure (Figure 3.13).

   During earthquakes which last for more than a few seconds, active and passive resistance, for instance, of integral and semi-integral abutments on piles would not be the same as abutments with expansion bearings. Similarly, the response of piers with pile or frame bents is different than that of say a hammerhead piers. Earthquake damage records show that earthquakes can cause foundation settlement due to liquefaction, damage piers, and cause cantilever mechanisms.
2. Ground deformation: One of the major causes of destruction during an earthquake is the failure of the ground surface. The ground may fail due to fissures, abnormal or unequal settlement, or complete loss of soil shear strength. A loose saturated sand deposit when subjected to vibration (or cyclic loading) tends to compact and decrease in volume. If drainage is unable to occur, the pore water pressure increases. Based on the effective stress principle the shearing strength of saturated sand is given by the well known shearing strength equation:

$$\tau = (\sigma n - u)\tan\varphi$$

During longer shaking of loose saturated sand deposits, increased pore pressure (u) becomes equal to the overburden stresses (σn) and the ground may lose its shearing strength resulting in settlements and tilting of structures. Loss of strength during cyclic loading occurs in clays also, but loss of strength does not occur until after large strains have developed.

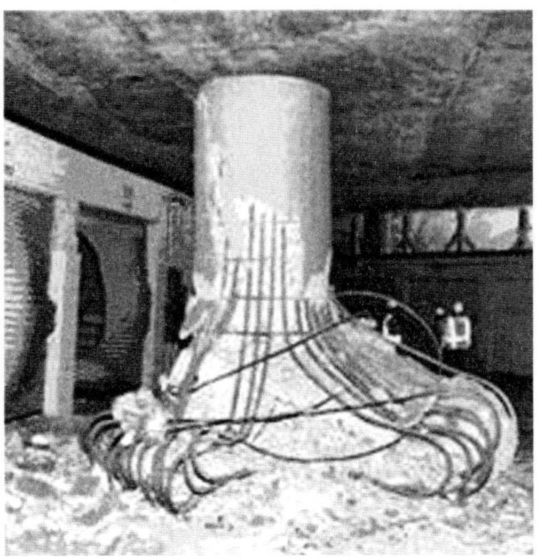

Figure 3.13 Foundation failure as a result of earthquake.

3.14.2 Seismic Ranking and Prioritizing

1. Retrofit priorities are defined in FHWA Seismic Retrofitting Manual for Highway Bridges and will depend upon:
   - Structural vulnerability (Some components are more vulnerable than others, such as girder connections, bearings, seat width, piers, abutments, and soils.)
   - Seismic and geotechnical hazards
   - Importance factor.
2. The rating system:
   - The quantitative part consists of the seismic rating (bridge ranking) and is based on the structural vulnerability or seismic hazard.
   - The qualitative part consists of an overall priority index consisting of importance factor, remaining useful life, non-seismic deficiencies and redundancy.

3.14.3 Seismic Retrofit Goals

The primary goal of seismic retrofitting is to minimize the risk of unacceptable damage during an earthquake. Damage is unacceptable if it results in the collapse of all or parts of the bridge or loss of use of this vital transportation route.

Bearings, sliding plates, and anchor bolt nuts may exhibit moderate to severe rust and material loss.

The old approach for continuous girders was to provide two lines of bearings on each pier. It is important that all bearings allow movement during a seismic event and therefore will require regular maintenance. At the time of original old bridge constructions, there were no seismic criteria in effect. Hence, it is important to analyze the seismic vulnerability of bridge using three-dimensional finite element software, such as SAP 2000 or ADINA in order to minimize the risk of:

- Structural damage
- Loss of life
- Collapse of all or part of bridge
- Loss of use of a vital transportation route (essential route).

AASHTO minimum seat width requirement to be satisfied—formula revised to include skew of support S measured from line normal to span L.

$$N = (200 + 0.0017L + 0.0067H)(1 + 0.000125S^2)$$

3.14.4 Seismic Retrofit Process

1. The following issues need to be addressed:
   - Preliminary screening—inventory
   - Detailed evaluation
   - Computed vulnerability rating
   - Seismic ranking evaluation
   - Design of retrofit measures.
2. Evaluate and upgrade the seismic resistance of existing bridges.
   Examples of seismic retrofits are:
   - FRP wrapping
   - Column strengthening and jacketing
   - Substructure stabilization/repairs/foundation improvement
   - Bearing strengthening/use of restrainers
   - Seat width improvement/bearing seat retrofit
   - Bearing replacement/using elastomeric pads or isolation bearings

- Using dampers
- Strengthening of members and connections.

3.14.5 Seismic Resistant Design for Movable Bridges

Bridges need to be analyzed in both open and closed positions, and for positions in between. AASHTO movable bridge specifications specify that the seismic load used for the open position may be reduced by 50 percent if the bridge is in that position for less than 10 percent of time.

Counterweights on bascule bridges represent large seismic inertias. The effect of the large mass on the girders should be explored. When the span is raised, the span drive braking machinery becomes effective. Seismic acceleration forces on brakes need to be investigated.

Several methods of seismic retrofit are outlined for bearings and expansion joints within the FHWA Retrofit Manual. Primarily, the bearings, joint restrainers, and minimum seat widths for seismic Zone 2 criteria retrofit need to be addressed.

Because of stringent operational and mechanical tolerance requirements, movable bridges need to be evaluated not only for stress, but more importantly for displacements. Efficient functioning of the expansion bearings therefore is important. In bascule bridges, there are heavy counterweights that significantly affect the seismic behavior of the long structure.

3.14.6 Fragility Analysis of R.C. Bridge Pier Considering Soil-Structure Interaction

Seismic fragility methodology for highway bridges: Bridge fragility curves, which express the probability of a bridge reaching a certain damage state for a given ground motion parameter, play an important role in the overall seismic risk assessment of a transportation network.

3.14.7 Case Studies of Seismic Failures

Table 3.11 gives details of seismic failures for numerous bridges located in USA and abroad. Although earthquakes have been known to cause damage for hundreds of years, it is only since the Santa Barbara and Norma Prieta earthquakes in California that interest in design and studies by NCHRP have been initiated.

3.14.8 Some Case Studies of Recent Collapsed Bridges

1. Gujarat India Earthquake: Bhuj suffered major damage. Widespread liquefaction and lateral spreading of soils have been reported in Rann of Kutch (India) and in many parts of the Southeast Sindh. Craters several feet wide developed on and around Badin-Kadhan road.

   Fault rupture results from ground vibration due to the upward transmission of the stress wave from rock to the softer soil layers. These stress waves are body waves that reach the surface at an angle depending upon the distance of the surface point from the epicenter or point on the surface over the origin. These body waves may generate two other surface waves that are confined to elastic-half-space and are known as "Raleigh wave" and "love wave." The seismograph may also record the ground motions of these waves, which are complex in nature.

   Rocks in the region are primarily Jurassic to Cretaceous age sedimentary and volcanic rocks. The earthquakes in India and Pakistan are the result of the compression thrust of the Eurasian Plate with the Indian Plate. The neotectonic geology of Kutch (Malik, et al, 2000) consists of a series of folds and faults with a general WNW/ESE trend.

   The kinetic energy of the waves is dissipated in the earth's crust with distance from the source, and its magnitude is registered at various intensities at the locations through which the body waves pass.

2. Balakot Bridge failure in the Pakistan Earthquake of 2005: Several bridges that were not designed for seismic resistance were severely damaged in the northern region of Pakistan during the 7.6 intensity earthquake. Balakot Bridge suffered the greatest damage and the main highway was shut down.

Table 3.11 History of bridge failures due to earthquake (design issues).

U.S. Bridges	Location	Year	Details of Failure
I-S and Antelope Valley Freeway interchange	Near San Fernando, CA	1971	Natural hazard (Sylmar Earthquake)
Cypress Freeway	Oakland,, California	1989	Natural hazard (Loma Prieta Earthquake)
Section of east span of San Francisco-Oakland Bay Bridge	San Francisco, California,	1989	Natural hazard (Loma Prieta Earthquake)
Motorway bridge	Junction Antelope Valley	1992	One span collapses during earthquake
Interstate 5 Bridge	Los Angeles, California	1994	Earthquake measuring 6.6 on the Richter scale
Japanese Bridges			
Showa Bridge	Showa	1964	Intensity 7.5 Niigata Earthquake, movement of the pier foundations natural hazard (earthquake)
Nishinomiya Bridge	Nishinomiya	1995	Hyogo-Ken Nanbu Earthquake, separation of the two supporting piers caused by the lateral ground displacements
Hanshin elevated expressway (Kobe-Osaka highway)	Hanshi	1995	Hanshin Earthquake (7.2 on Richter Scale), five sections of expressway were tossed aside
Bridge in Kashiwazaki City	Niigata	2007	Due to Niigata-Chuetsu-Oki Earthquake
Pakistani Bridges			
Various bridges	North of Pakistan and Azad Kashmir	2005	Intensity 7.6 Pakistan Earthquake

3.14.9 Suggested Preventive Action against Earthquake Failures

1. Some seismic retrofits include the use of isolation bearings, snubbers, and restrainers at discontinuous beams over piers; seismic detailing; the provision of ductile joints; and the adoption of seismic criteria in design. Member sizes and joints must be designed for the AASHTO LRFD load combination strength VI. Details of seismic retrofits are given in Chapter 10.
2. Earthquake engineering should consider the vulnerability of column failure and incorporate design against progressive collapse. The need for ductile detailing to prevent column damage has been emphasized in NCHRP 12-49. In design, ductility needs to be linked to specific collapse mechanisms and any cantilever action resulting from loss of support. A progressive collapse approach may increase bridge safety.
3. Conclusions for seismic behavior: It has been shown that ground response during earthquake depends upon the soil conditions underlying a site. A gross instability of the soil may develop, resulting in large permanent movements of the ground surface and associated distortion of supported structures.

The intensity of the earthquake movements should be studied in the forms of displacements and accelerations, as useful to the seismic engineering profession. Therefore, it is important to record destructive earthquakes by means of accelerograph. From the recorded accelerations the response spectra for different location areas may be calculated and the results may be applied in the seismic design of foundations and structural engineering.

3.15 WIND AND HURRICANE ENGINEERING

3.15.1 Failures Due to Heavy Winds/Hurricanes

Louisiana State University is developing new courses to create a hurricane engineering minor with plans to broadly disseminate information to engineering faculty and the profession at large. A book is also being developed titled *Hurricane Engineering: Planning, Analysis, Design, Response, and Recovery of Civil Engineering Systems*. Table 3.12 shows a brief list of failures from wind and tornados.

3.15.2 Case Studies of Wind Failures

1. Tay Bridge in Scotland

Historically, the 1 1/12-mile long bridge was used by Queen Victoria for travel to and from Balmoral Castle. The tall truss columns were made of cast iron while the girders were made of wrought iron. The bridge was subjected to sway from wind thereby causing fatigue. On the night of the disaster, high winds caused maximum sway. A train passing over the bridge was subjected to full wind pressure. The lever arm from wind forces acting on the superstructure and the full width of the train increased bending at the base of the column. Increased tension on anchor bolts caused them to fracture and break. Wind loads were not considered in design, and the forces could not be distributed due to a lack of wind bracing. The failed bridge was replaced by the well-known Forth Road Bridge.

2. The Tacoma Narrows Bridge

The original Tacoma Narrows Bridge, at all stages of its short life, was very active in the wind. Its nickname of Galloping Gertie was earned from its vertical motions in even very modest winds. Its collapse on November 7, 1940 has attracted wide attention at the time and ever since.

3.15.3 Suggested Preventive Action Against Failure from Wind and Hurricanes

1. A short history of "Galloping Gertie":

- Twisting motion: The bridge had galloped, but not twisted, prior to November 7, 1940. At maximum twist, one sidewalk was 28 feet higher than the other. A 600-foot section fell into Puget Sound. After the main span fell, the side spans sagged 45 feet.

Table 3.12 History of bridge failures due to wind and tornados.

U.S. Bridges	Location	Year	Details of Failure
2-span truss bridge over Mississippi	Chester, Illinois	1944	Uplifting wind load not considered – design error
Eric Bridge	Cleveland, Ohio	1956	Natural hazard (wind)
Hood Canal Bridge	Washington	1979	Wind and storm
Interstate 10 Bridge	Phoenix, Arizona	1979	Natural hazard (wind and storm)
1900 built Kinzua Viaduct steel bridge	North Central Pennsylvania	2003	Tornado speed of 140 km/h produced complex pattern of high-velocity winds
McCormick County bridge (Little River)	East of Mount Carmel, South Carolina	2004	Debris from Hurricane Jeanne stacked against bridge's support piles
Interstate 10 Bridge	Escambia Bay, Pensascola, Florida	2004	Natural hazard—Hurricane Ivan
German Bridges			
A1 cable-stayed bridge over Nordelbe River	Hamburg	1970	Wind vibrations—design failure

2. Interpretating the Tacoma Narrows Bridge failure: Cable supported bridges are subject to:
   - Wind-induced drag (the static component)
   - Flutter (the instability that occurred at the Tacoma Narrows)
   - Buffeting (where gusts "shake" the bridge).
3. Adequate aerodynamic performance is required with respect to each of these effects:
   - For modest span bridges, drag generally controls the strength required to resist wind.
   - Flutter becomes critical when the wind acting on the structure reaches a critical velocity that triggers a self-excited unstable condition. The task in design is to assure that it has a very low probability of occurrence. This can be achieved by providing a stiff structure and/or an aerodynamically streamlined superstructure shape.
   - The magnitude of buffeting response under higher probability wind conditions must be controlled. It influences fatigue of the bridge materials as well as users' comfort.
4. Addressing these issues in an engineering context requires the use of wind tunnel models. Current practice is converging on the use of such models for the aerodynamic properties of the bridge shape only. The mechanical properties of the bridge, and the final wind evaluation, are performed using computer models that incorporate the wind tunnel results.

3.16 LACK OF MAINTENANCE AND NEGLECT

3.16.1 Lack of Effective Inspection and Rehabilitation Systems

Regular inspection of transportation facilities is critical to public safety.

1. Older bridges were not designed to meet the new design criterion.
2. More and heavier trucks are using our roads everyday, increasing the rates of deterioration of bridges and pavements.
3. Growing traffic on our waterways is increasing the probabilities of a barge/bridge collision that can result in a disaster.
4. Aging highway signage and high mast lighting (luminaries) are becoming structurally unsound, usually due to failing connections, and are dropping to the roadways with large potential for damages and the risk of loss of life.
5. While the federal government did establish minimum guidelines and requirements for bridge inspections, it has not established similar requirements for signs or luminaries. As a result, some states do not inspect signs or luminaries. Similar requirements for bridge pier protection, highway signs, and luminaries should be in place.

Table 3.13 shows a list of failures caused by miscellaneous reasons including oversight or neglect.

Table 3.13 History of bridge failures due to lack of maintenance or neglect (management issues).

U.S. Bridges	Location	Year	Details of Failures
Bridge over King's Slough River	Near Fresno, California	1947	Overloading from agricultural train
3-span bridge in Lafayette Street	St-Paul, Minnesota	1975	Brittle failure of new steel
Point Pleasant Bridge	West Virginia	1967	Fatigue crack in eye bar chain suspension bridge
Fulton Yates Bridge	Near Henderson, Kentucky	1976	Overloading during refurbishment
K&I Railroad Bridge	Louisville, Jefferson County, Kentucky	1979	Vehicle exceeding weight limit
Connecticut Turnpike Bridge (Mianus River)	Near Greenwich	1983	Corrosion of joint hangers (Gerber-joint), constraint stresses due to skew

Table 3.13 History of bridge failures due to lack of maintenance or neglect (management issues) *(continued)*.

U.S. Bridges	Location	Year	Details of Failures
Sergeant Aubrey Cosens VC Memorial Bridge	Latchford, Ontario, Canada	2003	Corrosion of hangar pins supporting floor beams of long span bridge
Shannon Hills Drive bridge	Shannon Hills, Arkansas	2004	Pedestrian bridge collapsed from the weight of crane—overloading
Sappa Creek bridge	Northwest of Norcatur, Kansas	2004	Overloading from heavy grain trucks over the bridge
Laurel Mall pedestrian bridge connecting parking and shopping areas	Laurel, MD	2005	Bridge was attached by metal bolts and brackets which had corroded
Lakeview Drive Bridge, Interstate 70	Washington County, PA	2005	Corrosive road salt draining onto side of a concrete girder overpass
Interstate 35W over Mississippi River	Minneapolis, Minnesota	2007	Lack of redundancy—structurally deficient bridge not maintained.
Canadian Bridges			
Duplessis bridge, plate girder composite bridge, 2-span, (St. Maurice River)	Between Montreal and Quebec	1951	Brittle failure of new steel—sub-standard construction material. Lack of testing of materials—management error
Wood trestle bridge	Near McBride, British Columbia	2003	The bridge collapsed under a CN Rail freight train. Inspectors discovered severe rotting in the wood bridge in 1999—problems with railway's inspection and maintenance program
British Bridges			
Suspension bridge	Near Bristol, England	1978	Unexpectedly heavy truck traffic, hanger failures, construction errors—overloading
Ynys-y-Gwas Bridge	West Glamorgan, Whales	1985	Segmental construction with thin mortar joints, highly permeable mortar at joints allowed moisture, chlorides and oxygen ready access to tendons, corrosion of longitudinal tendons at the segment joints—Deterioration
Austrian Bridges			
Timber foot bridge	Near Zell am See, Pinzgau	1974	Rotten piers not detected during inspection —deterioration
Reichsbrücke over Danube River	Vienna	1976	Water enters pier and freezing-thawing cycles destroy unreinforced pier, shear failure, lacking inspection and maintenance
Timber foot bridge	Vorarlberg	1976	Rotten piers, no inspection—deterioration
Indian Bridges			
Damanganga River bridge	Daman	2003	Irregularity of the administration in repairs, should have been replaced for 15 years
Australian Bridges			
King Street Bridge over Yarra River	Melbourne	1962	Brittle failure of new steel, lack of testing of materials —management error

3.16.2 Preventive Action against Negligence and Lack of Maintenance

1. Increasing inspections and using Structural Health Monitoring (SHM) methods.
2. Performing rating analysis.
3. Adopting structural health monitoring methods.
4. Timely rehabilitation is required.
5. Using weathering steel to reduce maintenance and painting cost.

3.17 UNFORESEEN CAUSES LEADING TO FAILURES

3.17.1 Ice Damage of Piers with No Timber Fender Shielding

1. AASHTO equation for dynamic ice pressure: Timber fenders may be damaged by moving ice sheets or crushed by compressive forces exerted by frozen layers of ice. A 50-year return period may be considered.

$$F = C_n \, p \, t \, w$$

The author studied ice loads for timber fender design for bridges located on Delaware River and Raritan River Bridges for the New Jersey Turnpike Authority.

where F is the ice force in lbs

C_n = Nose inclination coefficient

p = Ice pressure in psi

t = Ice thickness in inches

w = Pier width/projected fender width

AASHTO specifications give a range of values of ice pressure from 100 to 400 psi based on the conditions of ice incidence (provided by the Department of the Army Cold Regions Research and Engineering Laboratory (CRREL). Timber fenders and sheeting were modeled as structural members for analysis and were designed as timber structures.

$C_n = 1.0$ for inclination of nose to vertical = 0 degrees to 15 degrees

$C_n = 0.75$ for inclination of nose to vertical = 15 degrees to 30 degrees

Thickness of ice, t = 10 to 18 inches

w = width of pier

2. Static ice pressure (US Army Corps of Engineers): In 1995, Haynes suggested a minimum thermal expansion force of 5.0 kips/ft.

3.17.2 Malfunction of Bearings/Thermal Stresses

1. For continuous spans a thermal analysis is required. Due to dirt accumulation, expansion bearings may malfunction, generating thermal forces that increase with the length of the span.
2. Piers are usually not designed for bending moments caused by the daily recurring longitudinal thermal force.

3.17.3 Examples of Unexpected Failures and Suggested Preventive Actions

1. Repairing the fire damaged Notre Dame Bridge: Thomas A. French of Manchester, New Hampshire describes that on April 12, 2003 a fire ignited underneath an important bridge in the City of Manchester, New Hampshire. In the year and a half since the fire, the investigation has determined the official cause to be arson.

2. Cracks in the I-95 bridge: A significant crack was discovered on an I-95 bridge over the Brandywine River in Delaware. The steel girder bridge carries six lanes of traffic just north of downtown Wilmington. The crack was located on the fascia girder at midspan of the bridge's main span. The entire bottom flange was found to be fractured, with the crack extending upwards to within 0.3 meters of the upper flange. The circumstances leading up to the crack, the cause of the crack, review of the repair strategy, and the results of load tests performed prior to and during the repair are described in a paper by Michael Chajes, Dennis Mertz, Spencer Quiel, Harry Roecker, and John Milius of the University of Delaware and DMJM Harris.
3. Failure of I-35W Highway bridge over the Mississippi River in Minneapolis, Minnesota: The following is the summary of recommendation of the National Transportation Safety Board dated January 15, 2008: On August 1, 2007 there was a failure in the superstructure of the 1000-foot-long deck truss portion of the 1900-foot-long bridge (Figure 3.14).

   Approximately 110 vehicles were on the portion of the bridge that collapsed, and 17 vehicles fell into the water. As a result of the bridge collapse, 13 people died, and 145 people were injured. The bridge was originally opened to traffic in 1967. The steel deck truss portion of the bridge consisted of two parallel main trusses connected through transverse floor trusses supporting the reinforced concrete deck. It was considered to be fracture-critical because the load paths in the structure were non-redundant, meaning that a failure of any one of a number of structural elements in the bridge would cause a complete collapse of the entire bridge. This type is also referred to as a non-load-path-redundant bridge.
4. The National Transportation Safety Board had a concern regarding certain elements of the bridge (gusset plates), which has prompted issuance of the following safety recommendation:

   The ends of the beams in the main trusses were connected by riveted gusset plates at 112 nodes (joints) along the deck truss portion.

   Additional load: As part of renovations, the average thickness of the concrete deck was increased from 6.5 inches to 8.5 inches, and the center median barrier and outside barrier walls were increased in size. These changes added significantly to the overall weight of the structure.

Figure 3.14 An aerial view of the failure of the I-35W bridge over the Mississippi River in Minneapolis, Minnesota.

Roadway construction was being conducted on the deck truss portion of the bridge when the bridge collapsed. Machinery and paving materials were being parked and stockpiled on the center span.

5. The Safety Board is concerned that, for at least the I-35W bridge:
   - The bridge was designed with gusset plates that were undersized.
   - The design error was not detected when plans were created. Because of this design error, the riveted gusset plates became the weakest member of this fracture-critical bridge. Normally due to low cost, gusset plates are expected to be stronger than the beams they connect.
   - The methods used in calculating load ratings and the inspections conducted through the National Bridge Inspection Standards (NBIS) program are not expected to uncover original mistakes in gusset plate designs or calculations.

6. Because of this accident, the Safety Board cannot dismiss the possibility that other steel truss bridges with non-redundant load paths may have similar undetected design errors. Consequently, before any future major modifications or operational changes are contemplated, owners should ensure that the original design calculations for this type of bridge have been made correctly.

7. The National Transportation Safety Board makes the following recommendation to the Federal Highway Administration: All non-load-path-redundant steel truss bridges within the national bridge inventory require that owners conduct load capacity calculations to verify that the stress levels in all structural elements, including gusset plates, remain within applicable requirements whenever planned modifications or operational changes may significantly increase stresses.

3.18 A POSTMORTEM OF FAILURES

3.18.1 The Need for Applying Safety Engineering

1. Developments in safety engineering require that engineers should be trained in the basic concepts of identifying and controlling hazards. It involves recognition, diagnosis, and implementation of a control selected from one or more options.

2. Murphy's Law seems to apply to bridges as well, especially when the number of bridges runs into the millions. *For some bridges, whatever can possibly go wrong will.*

An engineer's goal is to prevent hazards and prevent the fulfillment of Murphy's Law.

In the planning, design, construction, and maintenance of bridges there are many single and chains of activities that can contribute to a disaster. In the numerous decision making tasks required in planning and design, engineers may inadvertently create hazards at sites, in equipment, in operation, or by neglecting the environment.

Such activities are related to quality of materials, products, equipment, processes and the environment. Any error in calculation will be in the direction of maximum harm. *Where hindsight is missing, the risk of failure is increased.*

3. A hazard may be defined as *a condition or circumstance that can lead to adverse or harmful consequences.* Hazards are seldom created deliberately, but are usually created unintentionally or inadvertently. Sources of hazards are:
   - Human limitations
   - Errors in judgment
   - Poor assumptions
   - Pressure to meet commitments or unrealistic schedules
   - Over ambitious incentives and penalties offered to the contractor to finish the job
   - Poor communication

- Duress and job stress
- Lack of knowledge to perform a certain task efficiently.

Hence, factors that contribute to unforeseen incidents need to be reduced or eliminated through efficient planning and design:

Planning is the process of developing a product such as a bridge, formulating a program of action, or structuring an orderly arrangement of activities for design and construction.

Design is the extension of planning when more detail and specific information is required.

3.18.2 Design Errors

1. Converting units of measure such as feet into inches.
2. Failure to include a correct factor of safety.
3. Use of a static force in place of a dynamic force.
4. Selection of the correct material.
5. Constructability issues resulting from design.
6. An incorrect estimate of the service life of a product.
7. Neglecting environmental effects such as thermal forces, vibration, corrosion, and abrasion.

3.18.3 Production Errors

1. It is not always possible to construct the way it is shown on the drawing.
2. Changing fasteners or connectors because the sizes specified are not available can weaken a structural joint.
3. Local overstress during shipping, erection, and handling.

3.18.4 Maintenance Errors

1. Inadequate maintenance or repair.
2. Poor access for inspection such as for bearings.
3. Lack of funds to carry out timely repairs.

3.18.5 Communication Errors (Figure 3.15)

1. Poor detailing.
2. Lack of construction coordination through responses to RFI (request for information).
3. DCN (design change notice) for resolving constructability issues.
4. Adjustments in dimensions due to site geotechnical conditions.
5. Access issues.
6. Availability of materials, etc. may contribute to failures.

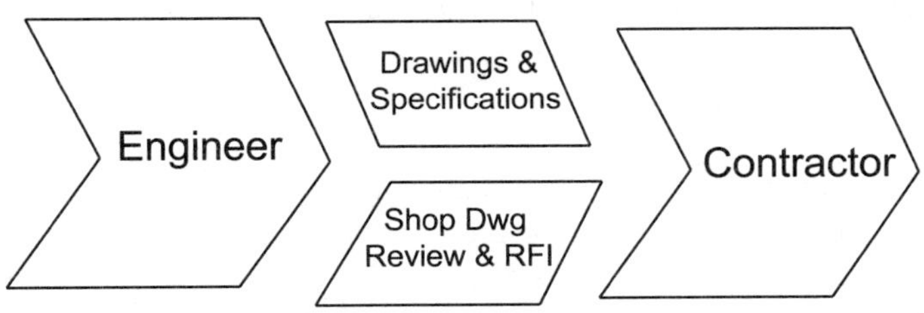

Figure 3.15 The four components of communication.

It is important that effective communication be maintained and weekly meetings be held between the consultant, client, and the contractor to resolve immediately the imminent day-to-day issues. Full-time two-way communication needs to be maintained by full use of the modern media such as fax, e-mail, letter, and cell phone.

3.18.6 Hazard Control Principles

Engineering sense requires that the following intuitive approach be adopted by the management team:

1. Recognize defects and deficiencies through inspection and monitoring.
2. Define preventive action and estimate of costs.
3. Allocate of resources to match the needs.
4. Assign responsibility for preventing action such as redesign and repairs.
5. Adopt quality assurance and quality control procedures.
6. Implement in a timely manner.

3.19 THE STUDY OF MODES OF FAILURE

3.19.1 Mechanics of Failure

In practice, failures occur in different forms in a material. They are likely to be different for steel, concrete, and timber bridges. Physical forms of failure can be seen as infinitely large deformation and metallurgical disintegration of elements. It can be localized cracking without causing collapse or discontinuity and total separation of a bridge component. Common types of failures are:

1. Yielding (metals—crushing, tearing or formation of ductile or brittle plastic hinges).
2. Buckling (metals, web buckling).
3. Crushing (concrete).
4. Fracture and fatigue (metals and concrete—reduced material resistance, local hairline cracks, minor or major cracks in the deck slab, girders or abutments, reversal of stress in welds and connections, vibrations).
5. Rupture (shearing).
6. Large deformations (metals and concrete—impact, sway, violent shaking during seismic events, erosion of soil in floods, settlement due to expansive soils).
7. Stress concentrations (concrete deck slabs with sharp skew).
8. Corrosion (metals and concrete—reduction in material area).

3.19.2 Forms of Failure

For composite beams, plastic hinges form at midspan and for continuous beams at supports. Visible tension yielding occurs in the bottom flange and at the continuous supports at the top flange accompanied by cracking at the surface of the slab.

In prestressed concrete beams, collapse may occur due to principal tensile stress at anchorages or breakage of the corroded strands.

3.19.3 Physical Causes

1. Usually it is a combination of more than one type of force that causes failure. For example, dead load stress must always be combined with one or more external transient forces to apply a compound critical stress. If dead load stress is already high and approaching the elastic limit, any applied force or stress will exceed the allowable limit and lead to failure.

2. As discussed earlier, construction difficulties, wind, hurricane, tornado, flood, support settlement, earthquakes, and tsunamis are some of the major environmental forces responsible for failures. Physical causes are varied, such as vibrations, wind, extreme events, reversal of stress, impact, erosion, and violent shaking during earthquakes.
3. Wind, earthquakes, and floods are acting at random in multi-dimensions. In the mathematical model, they are resolved in three directions at right angles. The vertical component of wind or an earthquake may be smaller than the two horizontal components, but can cause uplift of a bridge deck over the bearings and therefore is important. The magnitude of a seismic vertical component will depend upon the distance from the epicenter of the earthquake. Seismic forces acting at right angles to each other in a given plane can be resolved with maximum stress occurring in the resultant direction.
4. For modifications to existing structures, specifications used for original design need to be checked against the latest LRFD specifications.
5. Studies have revealed the following causes (also addressed in Section 3.4):
   - Construction problems and difficulties seem to be the biggest issue.
   - Lack of timely inspection, maintenance, or neglect: The expected life of 75 years or more for modern bridges and their components may not be achieved without effective inspection, structural evaluation, and timely rehabilitation.
   - Design deficiencies, bridge design code violations, and in-depth analysis.
   - Truss types and non-redundant structural systems are most vulnerable.
   - Use of inferior material in some cases such as cast iron when wrought iron was available. Today, HPS 70W steel plates fabricated with high strength welds are available.
   - Acts of man, such as vandalism.
   - For bridges located on rivers, impact from ships and soil erosion from floods account for most failures.
   - Acts of God such as accidents, fire, explosions, and extreme weather events.
6. Any one of the above factors may contribute to bridge failure or may trigger a collapse. However, failures can only occur due to a combination of loads of which the principal or additional cause can be one of those listed above. The load combinations have been defined by AASHTO LRFD Bridge Design Specifications, 2004. AASHTO load combinations do not include accidents, fires, or vandalism. However, AASHTO rating specifications address details of inspection and the Manual on Rehabilitation addresses repair and maintenance.

3.20 STEPS TO AVOID FAILURES

3.20.1 Signs and Events Leading to Failure

In situations where considerable uncertainty exists in the ability to reliably inspect critical details and in the prediction of accumulated damage, preemptive retrofit strategies appear to be highly desirable.

1. It is too late when collapse and loss of life have already happened (replacement required).
2. Appearance of minor cracks and signs of possible incremental collapse (immediate action required such as bridge temporarily closed for investigation or partial lane closure).
3. Damage and unserviceability (bridge is closed for major repairs or retrofit).
4. Identification of any structural deficiency from inspections (action is required).

Figure 3.16 shows the role of extreme events, overload, or accidents, etc. in evaluating the vulnerability ratings of bridges. The rating shows the extent to which a bridge is vulnerable to failure. The likelihood of unexpected failure needs to be minimized.

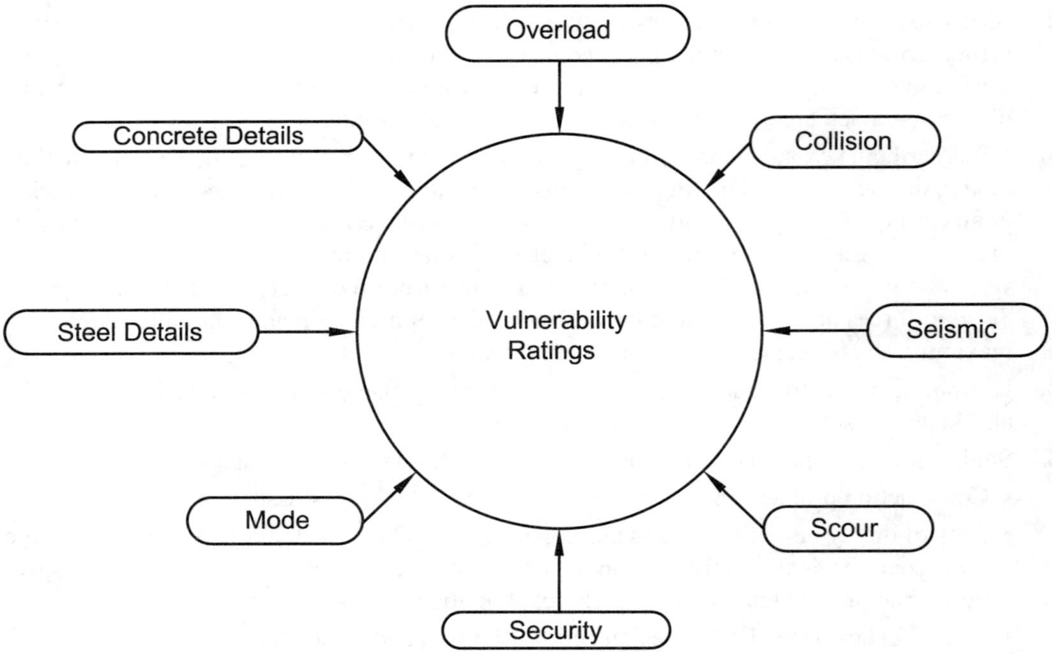

Figure 3.16 Factors affecting vulnerability rating.

Underlying principles for avoiding failures dictate that:

1. There will be regular or routine maintenance

   There should be no problem with pre-approved budgeting and availability of maintenance funds by the owner for emergency or routine maintenance.
2. Engineering planning for original construction shall be sound.
3. There will be no design mistakes:
   - There is a need to critically examine state-of-the-art, theoretical concepts, current design methods, and the quality of construction and maintenance.
   - Design errors can occur due to a lack of expertise on the part of the engineer. Nonperformance of independent checks (for design concepts or structural details shown on drawings) through lack of a quality control program is a recipe for disaster.
   - It should be possible to minimize or prevent failures by developing and implementing checklists for maintenance adequacy and quality design requirements.
   - Bridges are subject to displacement, movement, and sway of the substructure due to floods.
   - If design and reconstruction criteria are correctly implemented, many of the failures can be prevented.
4. Collapse can be reduced by using standard details and, flexible and ductile moment resistance connections.
5. There will be no construction defects: Construction errors generally arise due to a lack of adequate supervision when meeting tight deadlines.
6. Some failures may occur due to experimentation with new types of materials or new systems such as undefined and unpredictable material properties in cast in place or precast construction.

7. Future design code specifications or construction procedures need to consider reasons for failures as guidelines in developing a comprehensive approach. When implemented, such measures are likely to minimize the frequency of failures.
   - Durability requirements need to be addressed on a scientific basis.
   - AASHTO code provisions and other relevant codes need to be followed.
   - The engineer needs to refresh his knowledge of dynamic analysis, hydraulics, and materials science through continuing education. These disciplines would help to combat challenges of an early failure.
8. Fatigue, scour, and seismic analysis should be carried out in detail. Design criteria must include specialized subjects such as fragility analysis and ship collision analysis. All designs should be refined to include fracture mechanics principles for fracture of steel, concrete, and composites. Thermal analysis and bearing design to prevent malfunction are necessary.
9. Although the past studies in general were able to identify immediate causes of failures, their diagnosis was restricted to case studies only. Most of the older bridges which failed did not have as-built drawings to evaluate their strengths in bending and shear.
10. Results of failure studies are not easy to analyze due to a large number of variables contributing to failure. Comparisons made between failure types revealed similar trends of failures occurring during the bridge's service life. Also, lack of maintenance and human-induced external events occurred frequently.
11. Forensic engineering investigations of failures are partly based on technical reasons and to some extent they narrow down the blame either the designer or the contractor. However, there may be reasons other than individual responsibility.
12. Design criteria used did not include scour of foundations or earthquakes. Design live loads were lighter than the modern vehicles, resulting in less fatigue.
13. Mandatory insurance of bridges may induce highway agencies to take a greater interest in research and development of safety procedures (Figures 3.17 to 3.19).

3.20.2 Regular Inspection of Transportation Facilities

Inspection is critical to public safety. A problem of growing damages to bridges identified over time has resulted in increasing the minimum design stress for H-20 loading.

1. Older bridges were not designed to meet the new criterion.
2. More and heavier trucks are using our roads every day, increasing the rates of deterioration of bridges and pavements.

Figure 3.17 A truck falls in a river due to a bridge failure in 2004.

Figure 3.18 Failure of the Oakland Bridge.

Figure 3.19 Quebec Bridge collapse.

3. Growing traffic on our waterways is increasing the probabilities of a barge/bridge collision that can result in a disaster.
4. Aging highway signage and high mast lighting (luminaries) are becoming structurally unsound, usually due to failing connections, and are creating the potential for damage and loss of life.
5. An evaluation of risk and a vulnerability rating based on inspections

3.20.3 Use of Effective Monitoring Methods

1. Underwater inspection or instrumentation as a bridge management tool: Figure 3.20 shows visual monitoring followed by flood watch and follow-up monitoring of scour critical bridges.
2. Use of modern techniques: The use of wireless and remote sensors enables the movements of bridges to be monitored around the clock. This is most desirable in flood situations. Modern sensors, when installed on scour critical bridges, minimize the possibility of sudden collapse and serve as a warning for a bridge to be closed.

   Modern techniques are further discussed in Chapter 9 on scour.
3. Investigating unknown foundations: Figure 3.21 shows the need for geotechnical procedures in evaluating unknown foundations.

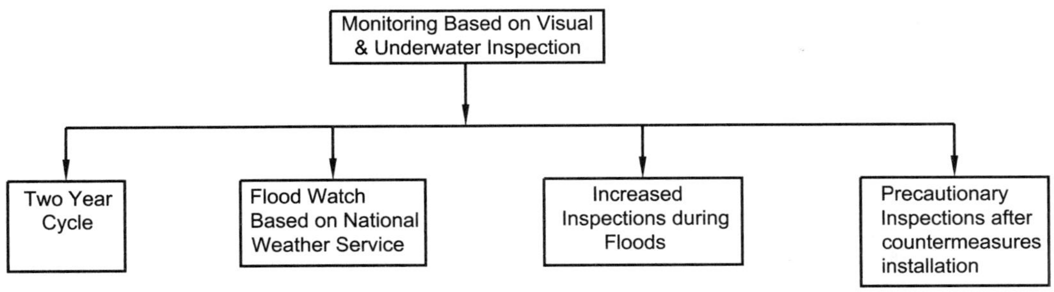

Figure 3.20 Diagram for monitoring scour critical bridges.

4. Scour safety evaluation: The scour sufficiency rating needs to be evaluated. A scour analysis is carried out to evaluate scour depths based on FHWA Publication HEC-18. A scour report consists of a detailed field survey, substructure information, scour analysis results, hydraulics related findings, and countermeasure recommendations. Harnessing the river to reduce flood velocities may also be used.

   Common countermeasures are riprap, gabion baskets, concrete blocks, and sheet piling.

3.20.4 Structural Planning and Solutions

Planning is a major area of concern. A mistake made in planning can seldom be overcome. Bad planning decisions will ultimately lead to bad results. The following planning principles shall be observed:

1. The bridge footprint must be located in a suitable site ensuring stable foundations.
2. Sharp skew angles in the deck slab must be minimized to avoid stress concentration in the acute corners.
3. Selection of a structural system with redundancy and adequate construction materials.
4. The optimum number of spans, continuity over supports, and equal spans need to be considered.
5. Provision for future widening of the highway below must be considered. This may require provision of a median, space for future lanes, an acceleration and deceleration lane, emergency shoulders, a sidewalk, and zebra crossing.
6. If a slab-beam bridge is selected, the number of girders and spacing shall be optimum. Girder spacing shall preferably be in the range of 6 to 12 feet. A higher degree of redundancy needs to be considered. In place of through girders that are non-redundant, a minimum of four girders is required. However, an odd number of girders is preferred for stage construction since the middle girder serves as a spine beam reducing deck deflection.

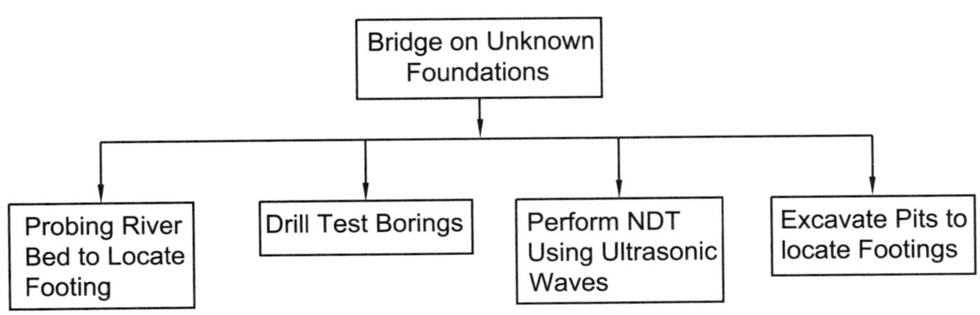

Figure 3.21 Geotechnical procedures for locating unknown foundations.

Both fascia and interior girders need to be designed. If the fascia is lighter than the interior, good practice will be to adopt the heavier interior girder design throughout. This is necessary for future widening, otherwise the fascia girder and part of the deck will need replacement.

7. Longitudinal joints shall be avoided. In place of transverse deck joints, integral abutments may be preferred. Deck joints usually have maintenance issues due to higher impact from vehicles.
8. Walls of the abutment and pier should be placed at 90 degrees to the direction of traffic flow to avoid any skew. Minimum vertical clearance for new bridges shall be 16 feet 6 inches. The horizontal distance to the face of the abutment and pier needs to be a minimum 30 feet from the edge of the travel lane. This distance may be reduced if guard rails are provided in front of the abutment or pier.
9. Geotechnical borehole results must be applied in selecting shallow or deep foundations. Weaker soils require deep foundations, such as piles, drilled piers, or caissons. For firm soils with a bearing capacity of 2 ksf and above, shallow footings should be considered.
10. Traffic count is required to determine the number of lanes, with traffic projections for 20 years ADTT. A provision for deck widening will be included.

3.20.5 Steps to Avoid Superstructure Failures

1. Durability requirements need to be addressed on a scientific basis.
2. The engineer needs to become familiar with the principles of related engineering disciplines, such as dynamic analysis, hydraulics, and materials science.
3. AASHTO LRFD code provisions and other relevant FHWA guidelines to be followed:
   - Strength and serviceability load combinations for constructability that are not covered in AASHTO LRFD code are presented in Chapter 5. Application of these loads will minimize the number of construction accidents.
   - Strength and serviceability load combinations for ship collisions that are not covered in AASHTO LRFD code are presented in Chapter 5.
   - Collapse can be reduced by using standard details and ductile moment resistance connections.
   - The expected life of 75 years or more for modern bridges and their components may not always be achieved without regular inspection, structural evaluation, and timely rehabilitation.
4. Deficient modeling: The following are some of the unknowns a bridge designer will typically encounter:
   - Inability to define loads accurately, such as magnitude and unpredictable level of stress distribution from settlement
   - Limited redundancy in structural system
   - Inability to fully include plastic behavior of composite action between the concrete deck slab and repeated beams, arching action, creep, and shrinkage strain distribution in the deck slab
   - Lack of information on fracture mechanics in general and lack of understanding of fracture of new materials in particular
   - Inelastic behavior of connections and joints, splices, gusset plates, bolts, and welds
   - Complex behavior as a unified assembly of uncombined multiple components of mixed (old and new) materials and structural systems, resulting from rehabilitation or widening methods
   - Delamination and reduction in strength of concrete deck due to deicing salts (as observed from chain drag test)

- Malfunction and locking of old bearing assemblies due to lack of maintenance; freezing of expansion bearings; large thermal forces causing compression and local buckling of truss members and flanges
- Inability to prevent scour at pile top
- Inability to fully incorporate different types of soil interaction at abutments
- Lack of drainage behind abutments and pressure build-up behind abutments
- Inadequacy of Rankine, Coulomb or Mononobe-Okabe theories for non-homogenous soil conditions for wing walls and stub abutments, resulting in unstable foundation design.

5. Providing space for bearing inspection chambers.
6. Codes for rehabilitation of mixed (old and new) structural systems should be developed to enhance life and prevent early collapse.
7. Greater vendor and construction engineer participation in revising and developing design codes for countermeasures is encouraged.
8. Study of the failure mechanism of different types of structural systems.
9. Maintaining quality control and personnel safety during construction.

3.20.6 Steps to Avoid Substructure Failures

1. The latest techniques of repair and rehabilitation of substructures need to be incorporated in codes.
2. Fragility analysis of reinforced concrete pier to include soil-structure interaction: A procedure needs to be developed to evaluate effective soil strains, as a function of dynamic shear forces generated at the planar surface of a river bed due to flowing water. Uncertainties in soil properties can be included in the vulnerability assessment framework.

3.20.7 Steps to Avoid Foundations and Piles Failures

1. Pile design: The ultimate bearing capacity (UBC) of axially loaded piles must be limited to the compressive and/or tensile loads determined for reduced capacity for projected scour.
2. Load redistribution must not be permitted when the axial pile capacity is reached; rather, axial capacity must be limited to the ultimate limit as established by L-pile analysis.
3. Lateral soil-pile response must be determined by concepts utilizing a coefficient of subgrade modulus provided or approved by the geotechnical engineer. Pile group effects must be considered.
4. Substructure stabilization, repairs, and foundation improvement through underwater inspection need to be carried out for all scour critical bridges.
5. Use of a dynamic screening tool for pile bents: An evaluation procedure developed by the Alabama DOT may be employed. It is a screening tool in macro and micro flood charts.
6. Preliminary or general checks such as bridge bent being located in water with scour possible, including:
   - Checking bent piles for possible kick-out or plunging failure
   - Checking bent piles for buckling failure
   - Checking the bent for transverse to bridge centerline pushover failure from combined gravity and flood water loadings
   - Foundation settlement from scour and weak soil conditions: When a pier tilts, there is potential for a bridge to collapse without warning. This definitely is a safety issue for travelers.
7. Install countermeasures for risk reduction on the following lines:
   - River training measures
   - Placing guide banks to move scour away from the abutment foundation

- Installing spurs or bendway weirs at a bend that is migrating toward a bridge abutment. Spurs will redirect the flow away from the abutment.
- Hydraulic countermeasures: Placement of armoring such as riprap around exposed foundation
- Structural countermeasures: Underpinning of footings that were undermined by using grout or grout bags.

3.21 REITERATING NEEDED PREVENTIVE MEASURES

3.21.1 Proposed Solutions

1. Correct diagnosis: The history of failures shows that they are not a rule, but they are not an exception either. They may not be common, but are recurring. Failures may occur at irregular intervals and for different reasons. Like all grim realities, every bridge which is constructed will eventually collapse. Some will take longer than others.

Bridges are subject to displacement, movement, and sway due to the changing positions of live load, impact, varying temperatures, and transient environmental forces such as wind, floods, and earthquakes. It is prudent to take necessary precautions and promptly fix any deficiency. To avoid adverse effects resulting from failures, a diagnostic approach in design is required.

2. Technological advances in information systems have had a great impact on data collection and analysis.

3. There is a need for continuous inspection through structural health monitoring.

All the man-hours for bridge inspections may not be usefully spent if the bridge management system in effect has not prevented failures. It is therefore important to examine the recurring lapses in our inspection methods, structural health monitoring, and advance warning systems.

4. Engineering agreements may be used to perform a wide variety of safety inspection work for bridges, culverts and other structures. Inspection related work may include:

- Routine and in-depth inspections
- While two year inspection cycles keep track of changes in structural conditions, circumstances may change overnight for the worse. Failure of the Minnesota I-35 bridge is a case in point.
- Special inspections (fracture critical members, damage investigations, etc.)
- Underwater inspections
- Bridge capacity analysis and ratings
- Special studies and testing
- Providing space for bearing inspection chambers.

5. There is uncertainty about the fate of thousands of bridges classified as structurally deficient. Using vulnerability assessment methods, some have been assessed as more vulnerable to failure than others. Such bridges may be located in a seismic zone or on a scour critical river where the probability of failure is higher than in non-seismic zones or where foundations are not subjected to erosion.

3.21.2 Design Related Actions

1. Due to in-built safety factors in design, the majority of failures occur not due to a single cause but due to the combination of two or more extreme conditions. For example, thin gusset plates in a steel truss are more vulnerable to failure when corroded than when they are galvanized.

Greater vendor and construction engineer participation in revising and developing design codes needs to be introduced and implemented.

2. Fire resistant design: In 2008, the National Institute of Standards and Technology (NIST) in Gaithersburg, Maryland recommended the use of a load combination with extreme fire conditions in their Report on the (2001) Collapse of World Trade Center Building 7. This collapse was caused primarily due to fires. Fire damage to several bridges has shown:
   - Longer span bridges with thinner web plates are more vulnerable to local buckling and need to be designed to resist fire. Currently, temperature forces are calculated for an increase between 68° and 112° F. Thermal stress is added to flexural stress under a dead load and the cumulative effect may exceed the allowable.
   - The failure mechanism under fires needs to be studied. At high temperatures allowable shear and bending stress decrease. Also, the load path is likely to shift and loads may become eccentric to the center line of web causing additional moments not allowed for in design.
   - High temperatures cause heat corrosion. Hence, corrosion resistance needs to be improved. Another method of increasing fire resistance is to use a non-flammable spray paint.
   - The author developed Section 45 (Scour at Bridges) and Section 46 (Seismic Design and Retrofit) for the NJDOT LRFD Bridge Design Manual. Based on research on New Jersey scour critical bridges, a "Handbook of Scour Countermeasures" was developed by the author jointly with City University of New York and was approved by FHWA. Based on studies of material response to fire, a fire resistant design criteria and design method for steel and prestressed girders needs to be developed.
3. Additional topics which require attention are:
   - Applying a more realistic LRFR method for rating
   - Applying load and resistance factors based on the LRFD method
   - Studying failure mechanisms of different types of structural systems
   - Maintaining quality control and personnel safety during construction
   - Ensuring seismic retrofit against minor and recurring earthquakes
   - Providing scour countermeasures
   - Developing and making available codes for rehabilitation of mixed structural systems
   - Developing codes for new materials such as FRP decks
   - Considering new techniques of repairs similar to those discussed in this book need to be considered for inclusion in the codes.
4. Provide redundancy in design with more than one load path.
5. Risk management.

3.21.3 Construction Related Activities

Preventing accidents during construction: The construction industry has the highest rate of accidental deaths compared to say mining or mountain climbing. OSHA (Occupational Safety and Hazards Administration) maintains records of the recurring events. While original construction is done only once, maintenance is continuous. Hence, construction issues and difficulties are always present. Examples are:

1. Scaffolding failures.
2. Crane failures.
3. Subsidence during deep excavations or land slides.
4. Fire, combustion, and respiratory problems.
5. Underwater diving hazards.
6. Unprotected welding operations.
7. Wind gusts.
8. Explosions.

9. Hazardous and inflammable materials and explosives.
10. Extreme hot and cold weather conditions.
11. Failure of steel ropes, chains, slings, or hoisting gear.
12. Lack of a fall protection net.
13. Electrical shock.
14. Malfunction of conveyors, equipment, and instruments.
15. Slippery surfaces.
16. Lack of helmets and safety gear.
17. Fear of heights and dizziness.
18. High noise level.
19. Poor health, hearing, or eyesight, absent mindedness, fear, and anxiety.
20. Lifting of heavy objects and materials.
21. Exposure to radiation, chemicals, and lead paint.
22. Working in poor lighting or a confined space.
23. Lack of ventilation.
24. Pipe burst.

3.21.4 Priority Issues During Construction

1. Eliminate hazards.
2. Minimize hazard level.
3. Introduce safety devices, fire extinguishers, first aid and access to hospitals.
4. Provide warnings.
5. Provide safety gear and equipment.
6. Provide head, eye, and hearing protection.
7. Provide safety training to workers.
8. Training in lifting methods and recommended weight limit.
9. Implement safety procedures.

Accelerated bridge construction needs to be promoted. This will minimize long duration construction projects. The use of precast decks and frames, transport tractor trailers, and high-capacity cranes is desirable.

The use of lightweight aggregate concrete for reducing dead weight moments and self-consolidating and fiber reinforced polymer concrete are helpful.

Other innovations as discussed in this book need to be investigated, as well.

3.21.5 Improving Knowledge Databases

Progress should be made in the development of applied mechanics and structural mechanics theorems to boundary value problems in obtaining more accurate and closed form solutions.

Application of refined theories such as:

1. Assessment of arching action in deck slabs at supports and boundaries as recommended by AASHTO LRFD Specifications. The author's research results justify the inclusion of membrane forces in deck slab and T-beam action. (The author's investigation is explained in Chapter 4.)
2. Analysis of initial curvature (camber) in girders on dead and live load deflections and stress distribution.
3. Analysis of shear deflection for deck and girder overhangs in piers.

4. Influence lines method redefined in terms of maximum deflection, shear force, and reactions for single span beam.
5. Application of theories of yielding in steel structures such as cantilever sign structures.

There is a need for continuing education in the specialized area of software development. With the large variety of bridge structures and structural systems, there is a need to introduce graduate level courses at universities, as well.

3.21.6 Risk Management

The probability of failure can be linked to the degree of risk. For example, designing a suspension cable, cable stayed or a deep arch bridge located near an airport for direct impact from a plane may increase significantly the bridge's cost. With the probability of such an extreme event occurring being extremely small, any risk reduction by making the bridge stronger is not required. Three types of risk levels are considered:

1. Risk reduction is required.
2. Risk reduction is optional.
3. Risk reduction is not required.

$$\text{Risk} = \text{Frequency} \times \text{Severity of potential loss}$$

Frequency is the probability of occurrence of an event such as once a year or once in 100 years. Severity of loss may be loss of life, injury, or a financial loss.

Loss control is the controlling of conditions which contribute to loss.

Risk management consists of identifying risk, analysis of risk, minimizing or eliminating risk, providing varying levels of resources, and administering the risk management process.

The objectives of risk management are to:

1. Fulfill social responsibility such as public service, public image, and public relations.
2. Maintain stability of benefits such as continued use of a bridge.
3. Ensure continuity of growth in commerce and industry from transportation.

Risk reduction is a function of the relative importance of a bridge. The higher the importance of a bridge, the higher the importance of risk reduction.

The importance factor of a bridge is classified as:

1. Bridge on a military route.
2. Bridge serving a hospital.
3. Bridge located on a school route.
4. Bridge located on a major highway.
5. A long-span or high-cost bridge.

Risk reduction requires increased monitoring, maintenance, and resource allocation for correcting any deficiency. Hence,

$$\text{Risk Reduction} = \text{k} \times \text{Importance Factor}$$

The factor k is based on average daily traffic (ADT). By neglecting the condition of a deficient bridge, the probability of the number of lives lost will increase.

Table 3.14 shows a matrix for the assessment of risk. No risk reduction is required for minor damage or when probability is categorized as improbable.

A paper by Henry Petroski, *To Engineer is Human: The Role of Failure in Successful Design,* American Society of Civil Engineers proceedings, 2003 addresses the role of assumptions and oversight in analyzing failures. To quote a nineteenth century English poet describing the famous Westminster Bridge in London,

Table 3.14 Risk assessment based on probability.

	Probability				
Hazard Severity	**Occurring at Regular Intervals**	**Very Likely**	**Occurring at Irregular Intervals**	**Not Likely**	**Improbable**
Bridge on verge of collapse	Risk reduction required	Risk reduction required	Risk reduction required	Optional risk reduction	No risk reduction
Bridge condition critical	Risk reduction required	Risk reduction required	Optional risk reduction	No risk reduction	No risk reduction
Marginal damage	Optional risk reduction	Optional Risk reduction	No risk	No risk reduction	No risk reduction
Minor damage	No risk reduction	No risk reduction	No risk reduction	No risk reduction	No risk reduction

Earth has not anything to show more fair,
When engineers endeavor to take good care!

With due apologies to the poet, the second line is added here to underline the importance of maintenance.

To summarize, nature does not forgive any error in planning, and a little hindsight would help. Failures are, however, the pillars to success.

BIBLIOGRAPHY

Analysis of Recent Bridge Failures in the United States

Briaud, Jean-Louis, "Failure of Bridges," American Society of Civil Engineers, 2003.

Brauer, Roger L., *Safety and Health for Engineers*, 2nd Edition, John Wiley & Sons, Inc., Hoboken, NJ, 2006.

Petroski, Henry, *To Engineer is Human: The Role of Failure in Successful Design*, Vintage Books, New York, NY, 1992.

Additional References

Abram, J., Nick Nicholson, E. B. Williamson and K. A. Marchand, "Regular Inspection of Transportation Facilities is Critical To Public Safety," *2006 Structures Conference*, American Society of Civil Engineers, St. Louis, MO, May 2006.

Akesson, Björn, *Understanding Bridge Collapses*, Taylor and Francis Group, London, 2008.

Baird, J., R. A. Nolte and D. Selke, "Blast Mitigation for Structures," *2006 Structures Conference*, American Society of Civil Engineers, St. Louis, MO, May 2006.

Biezma, M. V. and F. Schanack, "Collapse of Steel Bridges," *Journal of Constructed Facilities*, Vol. 21, No. 5, pp. 398–405, September/October 2007.

Billah, K. Y. and R. H. Scanlan, "Resonance, Tacoma Narrows Bridge Failure, and Undergraduate Physics Textbooks," *American Journal of Physics*, Vol. 59, No. 2, February 1991.

Bureau of Transportation Statistics, "National Transportation Statistics," *ASCE Proceedings*, Washington, DC, 2005.

Chajes, Michael, Dennis Mertz, Spencer Quiel, Harry Roecker and John Milius, "Steel Girder Fracture on Delaware's I-95 Bridge Over the Brandywine River," *2005 Structures Congress and Forensic Engineering Symposium*, American Society of Civil Engineers, New York, NY, April 2005.

Chopra, A. K. and C. Chintanapakdee, "Accuracy of Response Spectrum Estimates of Structural Response to Near-Field Earthquake Ground Motions: Preliminary Results," *Structural Engineering World Congress*, Elsevier Science Ltd., Oxford, England, 1998.

Conner, J. R., J. W. Fisher and W. J. Wright, "Recent Brittle Fractures in Steel Bridges and Preventative Mitigation Strategies," *2005 Structures Congress and Forensic Engineering Symposium*, American Society of Civil Engineers, New York, NY, April 2005.

Duwadi, S. R. and M. Lwin, *Securing the Highway Infrastructure—An R&D Program*, Federal Highway Administration, McLean, VA.

Duwadi, S. R., *Surveillance and Security Technologies for Bridges and Tunnels*, Federal Highway Administration, McLean, VA.

Ettouney, M. and S. Alampalli, "Blast Hazard Considerations within a Multihazards Environment: An Application to the Theory of Multihazards," *Proceedings of the 2006 Structures Congress*, Brad Cross, John Finke, Editors, May 18–21, 2006, St. Louis, MO.

French, Thomas A. and Hoyle, Tanner & Associates, "Repairing the Fire Damaged Notre Dame Bridge," *ASCE Proceedings 2005*, Manchester, NH.

Government Accountability Office, *Highway Emergency Relief: Reexamination Needed to Address Fiscal Imbalance and Long-Term Sustainability*, GAO-07-245.

Government Accountability Office, *Federal-Aid Highways: Trends, Effect on State Spending, and Options for Future Program Design*, August 2004.

Hart, Gary C., C. G. Ekwueme and Paulos B. Tekie, "Analysis of the Collapse of a Bridge Falsework," *Proceedings of the 2005 Structures Congress and the 2005 Forensic Engineering Symposium*, April 20–24, 2005, New York, NY.

Hodgson, III, Donald, Anton K. Schindler, Dan A. Brown and Mary Stroup-Gardiner, "Self-Consolidating Concrete for Use in Drilled Shaft Applications," *Journal of Materials in Civil Engineering*, Volume 17, Issue 3, pp. 363–369, May/June 2005.

Ibrahim, Sheikh, "Load Rating Evaluation of Gusset Plates in Truss Bridges," *FHWA Bridge Design Guidance No.1*, February 2008.

Iwan, W. D., "Drift Spectrum: Measure of Demand for Earthquake Ground Motions," *Journal of Structural Engineering*, American Society of Civil Engineers, 123(4), 1997, pp. 397–404.

Kanai, K., T. Tanaka and S. Yoshizawa, "Comparative Studies of Earthquake Motions on the Ground and Underground (Multiple Reflection Problem)," *Bulletin of the Earthquake Research Institute*, Vol. 44, March 1966, pp. 609–643.

Kirk, Robert S. and William J. Mallett, *"Highway Bridges: Conditions and the Federal/State Role,"* Transportation Policy Resources, Science, and Industry Division, 2007.

Kirk, R. S., "Repairing and Reconstructing Disaster-Damaged Roads and Bridges: The Role of Federal-Aid Highway Assistance," *CRS Report RS22268.*

Kirk, R. S. and W. J. Mallett, *Highway Bridges: Conditions and the Federal/State Role,* August 10, 2007, Transportation Policy Resources, Science, and Industry Division.

Kohli, V., Jeremy Eden, Fulcrum Corporation, M. L. Levitan and Charles P. Siess, Jr., *New Curriculum for a Planet at Risk,* LSU Hurricane Center, Louisiana State University, 2006.

MacMurdo, J. (1823), Paper Relating to the Earthquake Which Occurred in India in 1819, Trans. Literary Soc. Bombay 3.

Malik J. N., P. S. Sohoni, S. S. Merh and R. V. Karanth, *Proceedings of the Hokudan International Symposium and School on Active Faulting*, K. Okumera, H. Goto, and K. Takada, Editors, 2000.

Malla, R. B. and P. Agarwal,"Inelastic Dynamic Progressive Collapse Analysis of Truss Structures," *Proceedings of the 2006 Structures Congress*, Brad Cross, John Finke, Editors, May 18–21, 2006, St. Louis, MO.

Meymand, P. J. (1998), *"Shaking Table Scale Model Tests of Nonlinear Soil-Pile-Superstructure Interaction in Soft Clay,"* Ph.D. Dissertation, Dept. of Civil & Environmental Engineering, University of California, Berkeley, California.

Middleton, C. R., *Bridge Forum,* University of Cambridge Department of Engineering, 2007, England.

Musick, Nathan, "Trends in Public Spending on Transportation and Water Infrastructure, 1956 to 2004," Congress of the United States Congressional Budget Office, August 2007.

Nielson, B. G., Clemson University and R. DesRoches, *Seismic Fragility Methodology for Highway Bridges,* Georgia Institute of Technology, Atlanta, GA.

Poston, R. D. and J. S. West, "Investigation of the Charlotte Motor Speedway Bridge Collapse," *Proceedings of the 2005 Structures Congress*, April 2005, New York, NY.

Ray, J. C., *Validation of Numerical Modeling and Analysis of Steel Bridge Towers Subjected to Blast Loadings*, U.S. Army Engineer Research and Development Center.

Ratay, R. T., "A Course in Forensic Structural Engineering," *Proceedings of the 2005 Structures Congress and the 2005 Forensic Engineering Symposium*, April 20–24, 2005, New York, NY.

Reese, L. C., W. R. Cox and F. D. Koop (1974), "Analysis of Laterally Loaded Piles in Sand," *6th Offshore Technology Conference*, OTC 2080, Vol. 2, Houston, TX, pp. 473–485.

Seed, H. B. and K. L. Lee, 1966 "Liquefaction of Saturated Sands During Cyclic Loading," *Journal of Bridge Engineering*, American Society of Civil Engineers, Vol. 92, No. SM 6, pp. 105–134.

Seed, H. B., I. H. Idriss and F. W. Kiefer, (1969), "Characteristics of Rock Motion During Earthquakes," *Journal of the Soil Mechanics and Foundations Division*, American Society of Civil Engineers, 9 (SM-5), p. 1204.

Seed, H. B., "Practical Seismic Design of Structures," *Seminar, Structures Group*, Metropolitan Section, American Society of Civil Engineers, 1997, New York, NY.

Shenton III, H. W., Michael J. Chajes, William W. Finch, Cameron P. Chasten and Chien-Ming Chu, "Field Test/Fatigue Investigation of the Summit Bridge," *Proceedings of the 2006 Structures Congress*, Brad Cross, John Finke, Editors, St. Louis, MO.

Smith, T. D. and R. Slyh, "Side Friction Mobilization Rates for Laterally Loaded Piles from the Pressuremeter, The Pressuremeter and Its Marine Applications," *Second International Symposium ASTM STP 950*, J. L. Briaud and J. M. E. Audibert, Editors, American Society for Testing and Materials, 1986.

United States Coast Guard and Army Corps of Engineers Activities, "Metropolis & Beyond," *Proceedings of the 2005 Structures Congress and the 2005 Forensic Engineering Symposium*, April 2005, New York, NY.

U.S. Department of Transportation, Federal Highway Administration and Federal Transit Administration, *2006 Status of the Nation's Highways, Bridges, and Transit: Conditions and Performance*, Chapter 2, Washington, DC, 2007.

U.S. Department of Transportation, Federal Highway Administration and Federal Transit Administration, *2006 Status of the Nation's Highways, Bridges, and Transit: Conditions and Performance*, Exhibit 9-8, Washington, DC, 2007.

U.S. Department of Transportation, Research and Innovative Technology Administration, Bureau of Transportation Statistics, National Transportation Statistics, Tables 1–27, Washington, DC.

Wang, S. T., "Lessons Learned from Civil Engineering Failures," *Proceedings of the 2005 Structures Congress and the 2005 Forensic Engineering Symposium*, April 20–24, 2005, New York, NY.

Wardhana, K. and F. C. Hadipriono, "Analysis of Recent Bridge Failures in the United States," *Journal of Performance of Constructed Facilites*, Volume 17, Issue 3, pp. 144–150, August 2003.

Wesevich, J. W. and D. B. Olson, "Explosion Forensic Analysis," *Proceedings of the 2005 Structures Congress and the 2005 Forensic Engineering Symposium*, April 20–24, 2005, New York, NY.

Williamson, E. B. and K. A. Marchand, "Recommendations for Blast-Resistant Design and Retrofit of Typical Highway Bridges," *Proceedings of 2006 ASCE Structures Congress*, University of Texas.

Zhou, Y. E. and Tom Feroli, "Investigation Analysis of Accident/Postmortem and Lessons Learned of Girder Fractures of I-895 Bridge over US 1 and Patapsco River, MD," *ASCE Proceedings*, 2005.

National Transportation Safety Board Publications

NTSB Report HWY-06-MH-024; *Ceiling Collapse in the Interstate 90 Connector Tunnel*, Boston, MA, July 10, 2006.

NTSB Report HAB-06-01; Highway Accident Brief—*Passenger Vehicle Collision with a Fallen Overhead Girder Eastbound on Interstate 70 at the Colorado State Route 470 Overpass,* Golden, CO, May 15, 2004.

NSTB Report HAR-04-05; Highway-Marine Accident Report—*U.S. Towboat Robert Y. Love Collision with the I-40 Highway Bridge near Webbers Falls, Oklahoma*, May 26, 2002.

NTSB Report HAR-94-03; *U.S. Towboat Chris Collision with the Judge William Seeber Bridge,* New Orleans, LA, May 28, 1993.

NTSB Report HAR-94-02; *Tractor-Semi Trailer Collision with Bridge Columns on Interstate 65,* Evergreen, AL, May 19, 1993.

NTSB Report HAR-90-03; *Collapse of the Harrison Road Bridge Spans,* Miami town, OH, May 26, 1989.

NTSB Report HAR-90-01; *Collapse of the Northbound U.S. Route 51 Bridge Spans Over the Hatchie River Near Covington, Tennessee,* April 1, 1989.

NTSB Report HAR-89-04; *Collapse of the S.R 675 Bridge Spans Over the Pocomoke River Near Pocomoke City, Maryland,* August 17, 1988.

NTSB Report HAR-88-02; *Collapse of New York Thruway (I-90) Bridge Over the Schoharie Creek, Near Amsterdam, New York*, April 5, 1987.

4

An Analytical Approach to Fracture and Failure

Section Overview

- Developments in theory, design, rating, and code methods used in computer software.
- Strength and serviceability methods for structural evaluation and rating, which form the basis for repair and resonstruction.
- AASHTO load combinations for live loads and extreme conditions are extended to construction conditions.

Section 2
Strengthening and Repair Work

4.1 THEORETICAL CONCEPTS USED IN DEVELOPING COMPUTER SOFTWARE

4.1.1 Analysis for Engineering Maintenance

One of the continuous duties of an engineer is maintenance. Maintenance is not just patching concrete. It includes preventing failure by managing, disciplining, and applying structural mechanics to the structural domain of bridge components, made up of single or composite materials.

Chapter 3 covered reasons for failure. It is important to understand both the mechanics/mechanisms behind a failure and the applicable theory of yielding and fracture so that future designs can be made safer. Retrofit of bridge components, widening or replacement usually require use of computer software. Theories of both elasticity and plasticity are required to understand the behavior of a bridge or a given member. In general, at failure nonlinear behavior due to large deformations or due to formation of plastic hinges (nonlinear stress) will take place.

In this chapter, facilitated computer-based analysis based on understanding of the component behavior. In addition the review of fundamental concepts, finer points in analysis such as arching action in deck slabs, theories of yielding of steel (due to Tresca and Von Mises), modified compression field theory (MCFT), and simplified bending and shear formulae under moving loads for single span girders are addressed. It will be noted that the more accurate the analytical methods, besides safety, the greater the economy in design.

Hand calculations are kept to a minimum since in the design office the culture of obtaining numerical solutions is shifted to software usage.

4.1.2 Diagnostic and Preservation Approach to Analysis

Criteria for effective performance can best be satisfied by a diagnostic and preservation approach. All diagnostic or preservation design is based on analysis of the deficient structure and the remodeled or rehabilitated structures.

No reliable design is possible without accurately knowing the deformations, bending moments, shear forces, and foundation reactions under real and projected environmental conditions. Practical considerations will not be overlooked when analyzing or redesigning for rehabilitation. Diagnostic analysis will address a variety of issues, including:

1. Meeting any security needs against bomb blasts for maintaining important bridges.
2. Creating minimum environmental impacts.
3. Meeting constructability requirements.
4. Achieving the expected serviceability.
5. Low initial construction or life cycle costs.
6. Easy inspectability.
7. Preserving aesthetics.
8. Resisting extreme conditions of design for earthquakes and flood scour.

4.1.3 Diagnostic and Preservation Methods

For new or replacement design, the theoretical approach is based on conventional theories laid down in Chapters 3 and 4 of AASHTO LRFD specifications. The application of theoretical tools used for rehabilitation is also similar; for example, the stiffness matrix and finite elements method are still applicable. However, for bridge rehabilitation the unknowns are greater. Hence, the objectives or requirements and the physical procedures in the field are site specific and need to be carried out on a case-by-case basis.

Examples of diagnostic and preservation methods are:

1. Analysis for staged construction: Parapet or girder replacement would require lane closure and partial bridge shut down. Since structural behavior is modified, a new analysis of the superstructure would be required.
2. Jacking the beam ends at abutments and piers: For bearing repairs and replacement, it is customary to apply the load upwards by using hydraulic jacks. Jacks are placed directly under the beams or a jacking beam is installed. Since the direction of load causes tension in the slab, grid beam, analysis is required to limit the bending and shear stresses and to control vertical deflection. Jacking load is applied in successive increments of 1/16 to 1/8 inch.
3. Deck replacement: An analysis of girders is required for concrete pour sequence and for new loads from a deck slab. Both non-composite and composite cases need to be considered.
4. Seismic retrofit: Placing of isolation bearings would change the structural behavior of the members during a seismic event. Seismic analysis is required to design bearings.
5. Providing scour countermeasures: Scour analysis based on HEC-18 and HEC-23 methods would be required.
6. Rehabilitation for movable bridges, curved bridges, segmental and cable-stayed bridges, erection loads, painting loads, etc. present additional challenges. For analysis, the mathematical model should match field conditions rather than be based on unrealistic assumptions.
7. Underpinning of foundations: To prevent the settlement of foundations, a dead load analysis of the bridge substructure and superstructure is required.
8. Post widening behavior: It is important that the remodeled bridge behave as a single structure rather than two or more separate structures.
   - New concrete materials for widening have different shrinkage and creep strains than the old concrete.
   - A new foundation is likely to settle more than the existing foundation.
   - Old fascia girders need to be analyzed as interior girders with increased live loads.
   - Longitudinal joints need to be checked for seismic response in the transverse direction.

4.1.4 Relevant Theoretical Concepts

A bridge engineer needs to be familiar with the mathematical approach developed in the past two centuries and how it should be applied to practical problems. A review of some analytical methods is presented, and some newer topics are addressed. These include topics such as:

1. Nonlinear methods of analysis.
2. Arching action in slabs.
3. Shear deflections.
4. Shear behavior of high strength concrete beams.
5. Application of theories of yielding and fracture.
6. Use of finite element methods.
7. Simplified formulae for moving loads.

4.2 STRESS ANALYSIS

4.2.1 A Review of the Basic Theory of Elasticity

1. Stress management may be defined as measuring the minutest change in length as a ratio of the original length, multiplying strain by modulus of elasticity or rupture of the material and making sure that the resulting stress does not exceed the bond between atomic particles.
2. Elongation gives rise to strain, which results in stress. Excessive elongation or shortening observed at a given cross section is likely to cause separation of the atomic level bond between particles. Failures are a result of stresses and strains exceeding the allowable resistance of materials.

   Strains can be axial strain or bending strain. Axial strain results from a linear change in length while bending strain is due to deflection, rotation, and curvature at the section being considered. Hence, all failures emanate from a deformation, which is a physical change in length or rotation.
3. As stated, a change in length gives rise to a finite strain, which in turn causes a certain type of stress. Hence, axial force gives rise to axial stress, shear force to shear stress, bending moment to bending stress, and torsion to equivalent shear stress.

When bending stress and shear stress act in the same plane (e.g., in a vertical plane or deep beam) they can be combined together as a principal stress acting in the resultant principal plane. In a column subjected to bending moment, axial stress can be added to bending stress since both act in the same direction.

At failure, the critical state of stress can be due to elongation, which is causing maximum principal stress, principal strain, shear stress, and shear strain energy or total strain energy. Deductive reasoning concludes that external work done will have exceeded the internal strain energy.

1. Review the fundamentals of analysis: The objective is to evaluate correct deflection and stress under load and make the design safe enough for the life of the structure.
2. Analyze the superstructure (deck slab, parapets, and beams) and the substructure (bearings, abutment, and piers) separately.
3. Analyze the superstructure and the substructure as a combined structure (in particular, integral abutment bridges or arches).
4. Use reduced allowable strength in design due to fatigue stresses: Specifications review is usually based on AASHTO LRFD and state codes.

4.2.2 External Effects Leading to Member Sizing

1. A review of the fundamentals of analysis is presented here. Knowledge of applied mechanics can be utilized by applying laws of equilibrium, compatibility, strain displacement relations and boundary conditions. Whether it is the finite element method or partial differential

equations approach, it is mathematically convenient to assume Cartesian coordinates in three dimensions.

2. By performing analysis, we translate the physical concept into a mathematical procedure which serves as the basis of sizing the member or finalizing its design. It determines external force effects on the structure, which for equilibrium is resisted by internal forces within the material.
3. Stress resultants on a structural member can be idealized as bending moments acting in two planes at right angles and a torsion in the third plane. Out of the three moments, one is torsion and the other two are bending moments, all acting at right angles to each other and located in planes at right angles.

Similarly, the shear forces acting in two planes at right angles and an axial force in the third plane are complementary to each other. Out of the three forces, one is axial force and two are shear forces, all acting at right angles to each other and located in planes at right angles. This reference system in setting up a mathematical model is widely used in structural mechanics due to its simplicity in locating moments and forces in a structural member.

It may be further noted that axial force can be compressive or tensile. Compressive force may result in global or local buckling when compressive stress is very high. Tensile stress may result in tearing of material due to direct tensile stress.

Magnitude and distribution of forces and moments can be analyzed by the following approaches:

1. Idealizing or selecting a structural behavior—Such as representing a straight line for a beam, a convex curve for an arch, or a concave curve for a cable. The line element, whether straight or curved, in each case can be assumed to act at the centroid of the cross section.
2. Mathematical model—Such as a thin plate as compared to a thick plate.
3. Reference system for measurements—Such as rectangular (cartesian x, y, z coordinates) or curvilinear (r, θ coordinates).
4. Analytical tools—Such as stiffness methods, strain energy method, or differential equations.
5. Numerical model—Such as solving a system of algebraic equations using matrix inversion, iteration, or the step-by-step Gaussian elimination method.
6. Computational model—Such as programming computers with high speed, large storage capacities, and virtual memory, which would enable solving a large number of unknowns resulting from multiple load combinations or post-processing of results.
7. Manual computation methods—Such as checking computed results by empirical methods.

A designer is also an analyst and must have a clear concept about the fundamentals of the subject. Hence, to keep up with technological advancements, the knowledge base for applying the principles and theorems of statics and dynamics to structural behavior needs to be supplemented by continuing education or training courses.

4.2.3 Resistance to Applied Loads

Resistance can be defined as a resisting force or stress. It is the maximum quantifiable value beyond which cracking or failure of material will occur. A bridge can only be insured if it meets the legal requirement of Resistance > Applied load.

The three recognized methods of design are presented here. They evolved to meet society's need for a reliable infrastructure:

1. Elastic or working stress is the oldest method.
2. The load factor design (LFD) method served as an approximate ultimate load method. It has now been replaced by the more refined LRFD method for highway bridges. For railway and transit bridges, the LFD method is still being used.

3. The LRFD method deals with uniform design, with the objective of equal probability of failure everywhere on the bridge. It uses a probability-based reliability theory. The aim is to achieve a probability index β of 3.5 to 4.

   β = (Mean value of resistance − Load)
   and is also known as a measure of standard deviation.

The load resistance factor method has a more advanced ultimate load approach than the LFD method and in design aspects such as using a wide range of load combinations, selection of load factors, and resistance factors. Its use is now mandatory by AASHTO.

4.2.4 Structural Systems

The structural behavior of primary members in a bridge is a function of span length and the selected material. A primary system can be idealized as:

1. Beam action—applicable to small and medium spans of composite slab-beam systems.
2. Truss action—applicable to medium and long spans of through trusses braced transversely by floor beams.
3. Arch action—applicable to medium and long spans of curved compression members.
4. Suspension cables—applicable to very long spans of cable supported bridges.

4.2.5 Substructure Type

Substructure systems fall into the following general categories:

1. Solid wall full-height abutments and piers (long cantilever behavior).
2. Stub or partial-height abutments (short cantilever behavior).
3. Frame type abutments and piers (short column bent frames).
4. Pile bent type piers (long column bent frames).

4.3 REVIEW OF ELASTIC ANALYSIS

4.3.1 Fundamental Equations

Elastic analysis for dead load, live load, and other applicable loads forms the starting point of the factored design method. Using different factors for each type of load, elastic moments, shear forces, and reactions are either exaggerated or modified to simulate ultimate load effects. Analysis in all cases must comply with the laws of equilibrium and stability.

Well-known laws of equilibrium may be stated as:

Sum of vertical forces, $\Sigma V = 0$ (4.1a)

Sum of horizontal forces, $\Sigma H = 0$ (4.1b)

Sum of moments, $\Sigma M = 0$ (4.1c)

Beam bending: The most commonly used beam bending equation for bending stress, resulting from a bending curvature (or deflection) and stiffness (or shape, size, and material property) is well known:

$$f/y = E/R = M/I \tag{4.2}$$

where

$$1/R = (d^2w/dx^2)/(1 + (dw/dx)^2)^{3/2} \tag{4.3a}$$

f = stress, y = distance of fiber from neutral axis, E = modulus of elasticity, R = radius of curvature, M = bending moment, I = moment of inertia.

Since (dw/dx) is small, $(dw/dx)^2$ is negligible

$$I/R = (d^2w/dx^2) \tag{4.3b}$$

as given by Newton's law of curvature

$$\text{Slope} = EI\ dw/dx$$

where w is the vertical deflection due to bending

$$\text{Bending Moment } M = EI\ d^2w/dx^2 \tag{4.4a}$$

$$\text{Shear Force } V = dM/dx = EI\ d^3w/dx^3 \tag{4.4b}$$

$$\text{Load intensity } q = dV/dx = EI\ d^4w/dx^4 \tag{4.4c}$$

Beam torsion: in a similar format to equation 4.2, the general torsion equation is given by:

$$\tau/r = C\ \theta/L = T/J \tag{4.5}$$

where τ = shear stress, r = radius of a point on x-section of beam, C = shear modulus, θ = angle of twist w.r.t. center line of x-sec., L = span, T = torsional moment.

This equation expresses shear stress resulting from a torsional moment to torsional rigidity and twist of sections. Generally, torsion is accompanied by bending moment due to self weight. They simultaneously cause deflection, twist, and strain in a member.

Analysis used as basis of design: For equilibrium, compressive force C = tensile force T.

The distance between the point of application of the two opposite forces (commonly known as lever arm) when applied by compressive or tensile force will be equal to applied moment.

Bending moment → Bending stress < Allowable material bending stress

Shear force → Shear stress < Allowable material shear stress

Reaction → Bearing stress < Allowable material bearing stress

Compressive force → Compressive stress < Allowable material compressive stress

Buckling → Compressive stress > Allowable material compressive stress

Tensile force → Tensile stress < Allowable material tensile stress

Axial force → Axial stress and buckling < Allowable axial and buckling stress.

Torsion → Shear stress < Allowable shear stress.

Allowable stress for a given material is usually obtained from laboratory tests on specimens. Stresses acting in the same plane are combined.

Physical parameters: Analytical results are used for selecting member sizes:

1. Stress criteria based on the magnitude of flexure and shear forces.
2. Deformations based on serviceability and limiting live load deflections.

Every physical parameter should be included in any analysis, namely:

1. Geometry and curvature.
2. Deck width, number of lanes, and overhang.
3. Girder spacing.
4. Materials—reinforced concrete, steel, aluminum, prestressed concrete, timber, or masonry.
5. Plan aspect ratio of deck slab and thickness.
6. Skew angle.

4.4 ANALYSIS OF SLAB BEAM BRIDGES

4.4.1 Analytical Approach to Composite Bridge Decks

1. Drastic changes in distribution coefficients: The two theories based on idealization of composite deck and girder action are:
   - The simplified beam theory, using transverse distribution factor—The bridge deck is idealized as a series of T-beams (line girders composite with deck slab) in the longitudinal

direction. With the use of diaphragms in the transverse direction, a load distribution factor is applied. The longitudinal beam is assumed stiffer than the transverse diaphragm. This approach is commonly used in the U.S. The primary mode of deformation is by bending and torsion.

The simplified beam theory uses AASHTO LRFD design code for two or more design lanes loaded (AASHTO Table 4.6.2.2.2.b-1). A set of tables for distribution coefficients for computing bending moments are given in AASHTO 2007. The distribution formula Eq. (4.6a), takes into account spacing, span, and beam stiffness such as:

$$DF = 0.075 + (S/9.5)^{0.6} (S/L)^{0.2} (Kg/12 L t_s^3)^{0.1} \tag{4.6a}$$

where: S = beam spacing, L = span, t_s = thickness of slab.

In older AASHTO codes, the distribution of loads in the transverse direction was based on a simpler equation for the distribution factor. The model used for original analysis was a typical longitudinal interior girder, which was generally simply supported between abutments or not fully continuous (with two lines of bearings over the piers), namely

$$DF = \text{Girder spacing}/5.5 \tag{4.6b}$$

Spacing generally varied between 6 and 12 ft and the empriicaldistribution factor variation differed widely for each spacing. A similar approach is used for DF of timber bridges (AASHTO Table 4.6.2.2.2.a-1).

- In the grid or grillage theory in which the transverse spanning deck strips (and diaphragms) allow full two-way bending action. Transverse diaphragms are idealized as truss members. The bridge deck is idealized as a grid in plan (line girders in longitudinal directions and line diaphragms in transverse directions). The grid or grillage theory is commonly used in European countries.

2. Comparison between the simplified beam theory and grillage theory.

With a longer span and small spacing, longitudinal girder bending will be controlled. For small spans and wider spacing, transverse distribution of moments and forces will be higher (approaching a grillage type bending). A finite element analysis for the bridge deck using the grillage model will give a more accurate estimate of girder deflection, moment, shear distribution, and diaphragm forces.

- In both the girder and grillage theories, deck slabs are analyzed as a thin plate. However, for lower L/d ratios, such as a 9-inch thick deck slab with a continuous span of 6 feet, for example, it is likely to behave as a thick plate.
- Due to composite action between girders and the deck slab, arching action may result, thereby increasing the compressive forces in the top flange of girders.

3. Compact rolled steel joists (compactness defined by Iyc/Iyy) have proven to perform better transverse stress distribution than the fabricated plate girders. Use of non-compact plate girder sections will only allow partial moment and shear distributions compared to girders with compact sections. Due to the relative thickness of the web and the gentle transition between the rolling of flange and web profile, there are fewer instances of rolled steel joist failures than those of plate girders resulting from local web or flange buckling.

4. Basic assumptions for analysis:

- Diaphragms can be modeled as co-existing beam or truss in the transverse direction. Truss action is by X-frame, and K-frame configurations of the diaphragm need to be considered.
- The concrete deck may be assumed isotropic, i.e., having equal strength in all directions. Certain types of grid decks may be orthotropic. Poisson's ratio for the slab and beam system for one-way action is neglected in the longitudinal beam theory.
- Vertical shear deformation in the deck slab is considered negligible.
- Wheel loads may be approximated as an equivalent patch load acting under the wheel width. A 45 degree distribution in both directions may be assumed, located away from the

wheel load and distributed across the depth of slab.

4.5 METHODS OF ANALYSIS OF THE SUPERSTRUCTURE

4.5.1 Analytical and Computer Modeling Methods

For monolithic slab and beam bridges, AASHTO Section 4.6.1 shows a simplified line model of a superstructure, neglecting modeling in transverse distribution if the span length exceeds 2.5 times the deck width. Primary flexural members may be idealized as:

1. Line girder—Stiffness matrix, moment distribution, strain energy, and area-moment methods.
2. Finite Strip—AASHTO Section 4.6.1 describes limitations of the strip method for skew slabs. It is suitable for slab and beam bridges spanning in direction parallel to traffic. Width requirements for equivalent strips for concrete, steel, and timber decks are specified in AASHTO Table 4.6.2.1.3.1-1.

   Empirical equations for deck bending moments and deflections are given in detail in AASHTO Section 4.6.2.1.8.
3. Grillage.
4. Continuum (finite element modeling).
5. Frames.

An alternative to using a computer program is to develop equations for maximum bending moment, shear forces, and reactions based on which upper bound values can be generated using Excel spreadsheets or Mathcad software. The author has developed such an approach (Sections 4.8 and 4.9).

Table 4.1 shows mathematical modeling, grid idealization of girder and flange, and box beam and flange and strip idealization. When using finite elements, the side aspect ratio of a rectangular finite element is < 5. For beam models the number of nodes shall not be less than five.

The mathematical model of foundations shall represent soil properties, elastic properties, piles, and soil-pile interaction. Since it is difficult to obtain an accurate model, upper bound and lower bound solutions can be considered.

The elastic theory using small deflections: The slab and beam system is subjected to small live load deflections. Hooke's Law is based on small linear deflections when measured relative to the thickness of the member.

The comfort level while traveling on a bridge with heavy trucks is also a consideration in maintaining small deflections and small accelerations. Damping of vibrations will depend upon the type of deck surface.

In the case of a simple beam, bending is accompanied by shear force. Pure shear force usually does not exist by itself and is complementary to bending moment. Also in beam bending, the moment is equivalent to two equal and opposite axial forces, with one acting as compression on one side of a neutral axis and the other acting as tension on the other side.

4.5.2 Analysis of Trusses

Trusses may be analyzed as a special case of plane frame or space frame analysis. Out-of-plane buckling of slender compression members needs to be evaluated. Both steel and concrete trusses are common. Three types of bridge trusses are used:

- Trusses supporting deck slab: Analysis shall consider composite action.
- Through trusses with transverse floor beams spanning across the bottom flanges of trusses and braced at the top. Floor beams are composite with deck slab.

Table 4.1 Types of mathematical models.

Coordinate Dimensions of Mathematical Model	Numerical Model	Type of Structural Theory
1	Stiffness method/flexibility method Slope deflection/moment distribution Strain energy method	Beam theory
2	Grillage analysis Finite difference/Finite elements Harmonic analysis Finite strip analysis Classical plate solutions	Beam, two-dimensional thin plate theory
3	Grillage analysis Finite difference/finite elements Harmonic analysis Finite strip Classical plate solutions	Three-dimensional theory of elasticity for beams, thick and simplified thin plate theory

- Pony trusses acting as through bridges and unbraced at the top. A through truss bridge has a low degree of redundancy and greater risk of collapse.

4.5.3 Analysis of Box Beams

Three-dimensional analysis is required for transverse frame action and for modeling of boundary conditions. Both St. Venant's torsion stress and torsional warping stress need to be considered. The following well-known stiffness matrix method is used by most bridge analysis software. Using matrix notations, $\{F\} = [k]\{\delta\}$.

4.5.4 Three Dimensional/Triaxial Behavior of Flat Plates and Overhead Sign Structures

A slab (or plate) offers the greatest benefits to the user, whether it is a bridge deck supporting moving vehicles, a roof slab offering protection from the elements of nature, or a floor slab in a residential building. Hence, stress distribution in slabs is of paramount importance in structural engineering.

A thin plate element is subjected to beam type bending in two directions and the accompanying twisting moments and shear. A plate may be idealized as a series of beams placed in two directions at right angles which interact with each other according to Poisson's ratio, i.e., a fraction of the load is also shared by beam bending at right angles to primary bending.

S. P. Timoshenko has expressed bending of a plate or slab using partial differential equations as follows:

Equation 4.4a for line element of a beam in the x direction can be generalized for a (plate) surface element by allowing secondary curvature in the y direction as:

$$-M_x = D\,(\partial^2 w/\partial x^2) + \upsilon\, D\,(\partial^2 w/\partial y^2) \tag{4.6a}$$

where flexural rigidity D is similar to EI value of the beam. Moment of inertia of plate includes Poisson's ratio υ, and while beam $I = b\,h^3/12$, Plate $I = h^3/12\,(1 - \upsilon^2)$ for unit width where υ is Poisson's ratio. A plate will be subjected to maximum number of moments and forces and is truly a 3-dimensional problem. As given by Timoshenko, theory of plates and shells Chapter 1, Equation 3.

For unit width, $D = E\,h^3/12\,(1 - \upsilon^2)$

$$-M_y = D\,(\partial^2 w/\partial\, y^2) + \upsilon\, D\,(\partial^2 w/\partial\, x^2) \tag{4.6b}$$

$$-M_{xy} = M_{yx} = (1 - \upsilon)\, D\,(\partial^2 w/\partial x\, \partial y) \tag{4.6c}$$

For load intensity resulting from one-directional beam bending can be generalized for two-directional plate bending as

$$q\,(x, y) = D\,[(\partial^4 w/\partial\, x^4) + 2\,(\partial^4\, w/\partial^2 x\, \partial y^2) + (\partial^4\, w/\partial\, y^4)] \tag{4.7}$$

The above equation is known as Lagrange Biharmonic Equation for plate bending and associated twisting.

4.5.5 Application of Finite Element Method

Modeling of concrete decks and piers is based on finite elements using computer methods. FEM is a powerful computational method. Like the finite differences method, deflections are calculated for elements of deck supported by grid beams or the wall of a pier by an indeterminate analysis for each position of the unit load. While the finite differences method uses line elements in two directions, the finite element method uses surface elements and therefore has a higher degree of accuracy.

Using Energy Principle:

$$\text{Work Done} = (A) \times (C) = (B) \times (D)$$

(B) and (C) are moving unit load quantities at two points X and Y. A and D are the reciprocal deflections at Y and X.

Finite Element Modeling:

$$\text{Stiffness matrix } [k] = \int_{vol} [B]^T [D]\, [B]\, \{\delta\} \tag{4.8}$$

Plate elements with 3, 4, and 8 nodes or beam elements are used. Energy theorems and Castigliano's Principles are applied. The well-known method was originally developed by Clough and Wilson at the University of California Berkeley.

Due to availability of high speed and large storage computers, FEM has been successfully applied to the modeling of bridges.

Recent developments in FEM:

1. Boundary Elements Method (BEM)—Developed by Carlos Brebbia and Hugh Tottenham, it uses the variational principle, which is transformed into a sequence of FEM equations. The number of equations is smaller than for FEM.

   Accurate predictions of stress concentrations near skew boundaries are possible. Presently approximate shear correction factors for skew angles range between 60 and 90 degrees.
2. Probabilistic Finite Element Method (PFEM)—Developed by T. Belytschko at Northwestern University, this method eliminates various spurious modes that are associated with FEM modeling of nodal forces. The Lagrange Variational Principle is used. Probabilistic characteristics of stress-strain, strain-displacement, and random character of applied loads and outer boundary are accounted for, resulting in more accurate solutions.

4.5.6 Substructure Stiffness Methods

Piers are usually made of reinforced concrete. Hammerhead, multi-frame, and pile bent methods are in use. Each type uses the stiffness method of analysis.

Abutments are designed as retaining walls subjected to vertical loads from the bridge and lateral loads from braking forces, active and passive earth pressures, longitudinal and transverse wind forces, and thermal and seismic forces. Settlement due to erosion is a possibility for scour critical bridges.

4.6 EFFECT OF BOUNDARY CONDITIONS ON BRIDGE BEHAVIOR

4.6.1 Longitudinal Beam Modeling

Modeling is not complete without the true boundary conditions of beams for the fixed and expansion bearings on unyielding supports. Usually bridge deck boundary conditions can be identified as simply supported, fixed, free, or continuous. Beams are modeled as simply supported on abutments and as continuous over piers. Deflections, bending moments, and shear forces are computed on that basis.

However, actual load distribution from the slab is not always uniform or rectangular, as is usually assumed. Interaction or composite behavior with the slab alters local and global behavior of the beam depending upon the type of connection.

In analysis, the approach has been to neglect the effect of composite behavior or interaction of slab and supporting beams.

In the majority of slab and beam systems, the magnitude of total compressive stress at the top of the slab near supports is the cumulative effect of arching action and compresssion due to bending. For a single span, formation of plastic hinge is at the midspan of the beam only as a result of excessive tension. At supports, composite action or T-beam action is neglected, and rectangular section is assumed. Plastic hinge also develops at supports of the beam at the ultimate load stage, due to fixity moment, and eccentricity of slab-beam connection has little effect.

Single panel behavior: Through girder bridges with a single lane and most pedestrian bridges fall into this category. The following three types are used in practice:

1. Thick deck slab bridges without beams (or with concealed beams) spaning longitudinally.
2. Others are supported by two upstand or through girders and floor beams span transversely. Both the deck slab and floor beams are constructed monolithic with the deck slab continuous in a longitudinal direction.
3. Composite multi-girder systems with the deck slab made monolithic with supporting beams by using shear connectors (increasingly used).

   Multiple panel behavior: A large majority of bridges are constructed with shear connectors between the beam and slab to ensure a unified behavior. Two-directional bending is possible for small spans with wider girder spacing.

4.6.2 Mathematical Modeling for Composite System

The mathematical modeling problem for composite systems is multifold:

1. Shape effect—shape effect in slabs alters the boundary conditions with beams influencing stress distribution.
2. Skew effect—skew effects at the edges of the slab.

   Skew slabs in bending were studied experimentally by C. P. Seiss at the University of Illinois. Arching action in bridge decks has been the subject of recent study by Csonka and others and is also addressed in the latest LRFD AASHTO code for the design of bridges (Figure 4.1).

   Effects of skew are important at corners since acute angled corners generate local stress concentration. Due to non-symmetry of live load near skewed corners, uplift of simply supported beam ends may occur. These practical considerations need to be taken into account in reinforcement detailing and anchor design at bearings.
3. Bridge decks curved in the plan.

   Methods of analysis of curved bridges such as V-load are described in detail in AASHTO LRFD, Section 4.6.2.2. AASHTO Curved Bridge Code requires a curved deck analysis only if

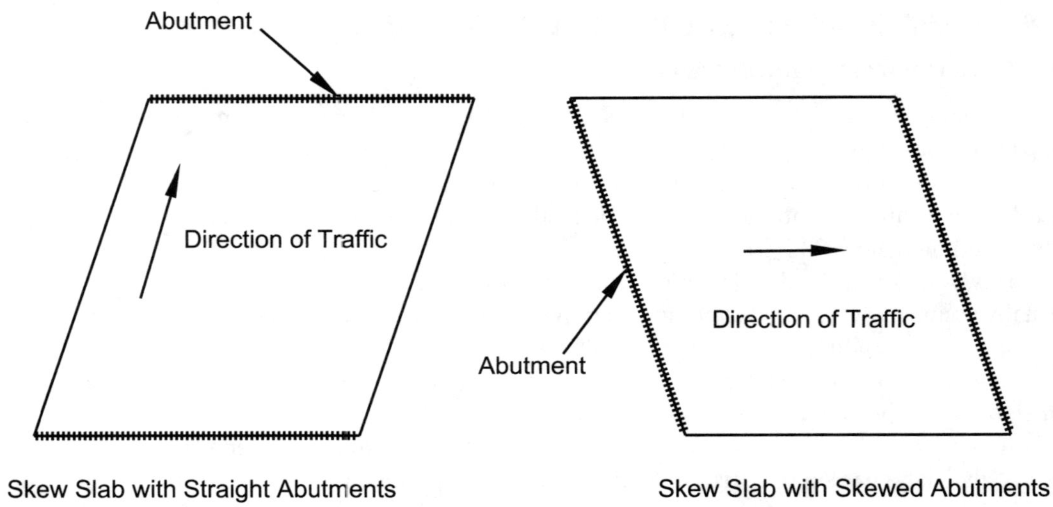

Figure 4.1 Varying boundary conditions at skew slab edges.

radius of curvature has certain sharpness. Plan of curved decks with and without skew edges (Figure 4.2). Software such as SAP 2000, DESCUS, BSDI or STRUDL can be used.

4. Vertically curved decks.

When alignment of a highway is on a vertical curve, the bridge deck geometry follows the curvature. It can be a hogging curvature (an arch) or sagging curvature (an inverted arch). Since the degree of curvature affects sight distance and speed while traveling, super

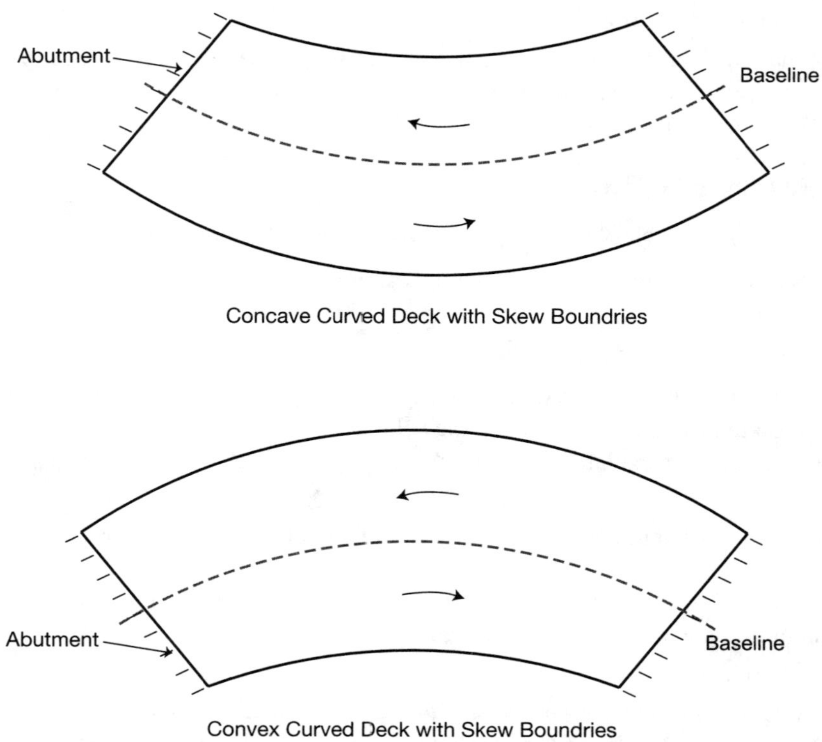

Figure 4.2 Plan of concave and convex decks.

elevation is generally restricted to 8 percent. In analysis, arch geometry effects are generally neglected compared to primary beam action.

4.6.3 Composite Action between the Deck Slab and Beams Using Classical Theory and Numerical Methods

The widely used approaches are:

1. Composite action resulting from deflecting beams—bridge decks are subjected to an additional deflection at the boundaries due to bending of girders.
2. Composite action resulting from geometry—eccentric connection of slab and beam.
3. Composite action resulting from type of connection—shear connectors or encasement (see Figure 4.3).
4. Composite action resulting from composite materials—concrete deck with precast concrete or steel beams:
   - Use of stud shear connectors for steel beams and stirrups for prestressed concrete beams. Section properties are based on combined slab T or Ell between beam sections.
   - Encasing steel beams in concrete without shear connectors. Section properties are based on combined slab and beam sections.
   - Non-composite action for dead loads only but composite action for dead and live loads. Section properties are based on beam section only.

For arching action, slab and beam flange should have the same curvature, and the slab should not lift up independently (see Figure 4.4).

Galerkin series for partial composite action: In 1953 B. G. Galerkin proposed the following solution for slabs supported on elastic beams:

Relative stiffness of beam to slab parameters = γ, δ in x and y direction.

Plate panel size $a \times b$, $\gamma = EI/aD$, $\delta = EI/bD$

$$w = q\,[\gamma\,(16x^4 - 24\,a^2\,x^2 + 5\,a^4) + \delta\,(16y^4 - 24\,b^2\,y^2 + 5\,b^4)]\;1/384D\,(\gamma + \delta) + \Sigma\,A_n \operatorname{Cosh}(n\pi y/a) \operatorname{Cos}(n\pi x/a) + \Sigma\,B_n \operatorname{Cosh}(n\pi x/b) \operatorname{Cos}(n\pi y/b) + \Sigma\,C_n y \operatorname{Sinh}(n\pi y/a) \operatorname{Cos}(n\pi x/a) + \Sigma\,D_n\,x \operatorname{Sinh}(n\pi x/b)\,(\operatorname{Cos} n\pi/y/b) \tag{4.9}$$

where δ, γ, x, and A_n to D_n are constants to be evaluated from boundry conditions and n = 1, 3, 5, …, we satisfy differential equation $q(x, y) = D\,[(\partial^4 w/\partial x^4) + 2\,(\partial^4 w/\partial^2 x\,\partial y^2) + (\partial^4 w/\partial y^4)]$ and conditions of symmetry. Refer to theory of plates and shells by Timoshenko, Chapter 6.

Using x = a/2 and y = b/2, the edge conditions similar to those given below are obtained.

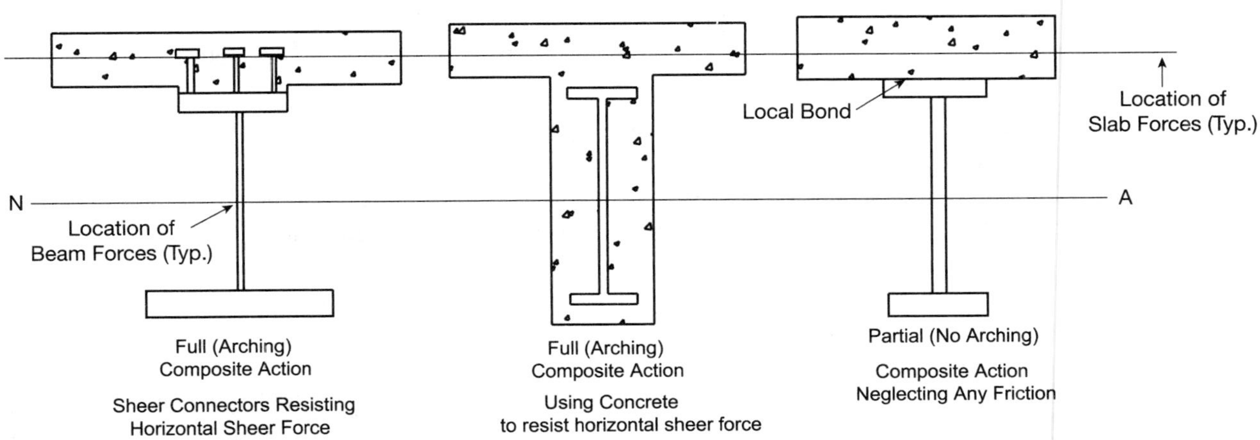

Figure 4.3 Full and partial composite action between slab and beam due to eccentric type of slab-to-beam connections.

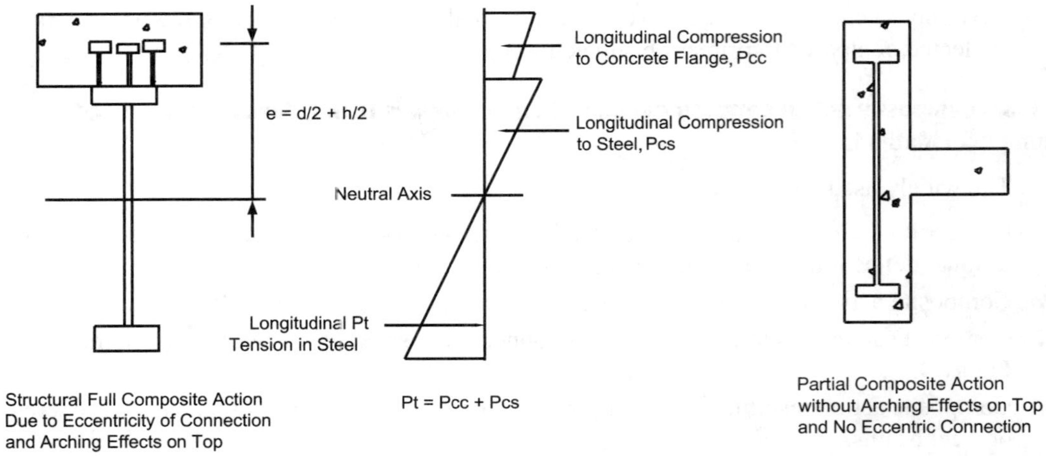

Figure 4.4 Full and partial composite action between slab and beam due to geometry of connection.

$$M_y = 0, [(\partial^2 w / \partial y^2) + \upsilon (\partial^2 w / \partial x^2)]_{y=b/2} = 0$$

$$IV_y = \frac{d^2M}{dx^2}, D [(\partial^3 w / \partial y^3) + (2 - \upsilon) (\partial^3 w / \partial x^2 \partial y)]_{y=b/2} = E_b I_b (\partial^4 w / \partial x^4)_{y=b/2} \quad (4.10)$$

For simply supported conditions,

$$E_b I_b = \infty$$

For bending of a square plate supported by four identical beams,

$$\delta/\gamma = 1, A_n = B_n \text{ and } C_n = D_n$$

The unknown coefficients A_n are eliminated by equating to zero edge moments.

Taking only four terms, n = 1, 3, 5 and 7 in above equation (4.9), four linear equations for C_1, C_3, C_5 and C_7 are obtained.

The results are compared in Table 4.2 with author's solutions.

When EI = 0, we have the case of beamless edges.

It is seen that Poisson's ratio has greater effect at the edges than at the middle of plate.

4.7 FULL COMPOSITE (ARCHING AND DOME) ACTION IN SLAB AND BEAMS

4.7.1 Introduction

In bridge decks, any curvature of slab resulting from geometry or boundary conditions would cause added planar forces to accompany primary bending. Neglecting arching action can give rise to underestimation of compression and cracking in decks, which is one of the reasons for early deterioration of concrete and subsequent replacement of deck slabs. The primary reasons for added membrane action are:

1. Convex vertical curved decks and supporting beams, due to rising curved alignment (usually rise is restricted to 8 percent).
2. Concave vertical curved decks and supporting beams, due to falling curved alignment.
3. T-beam action, when the slab of the composite T-beam located at the top is in full compression and the bottom of beam is in tension. AASHTO specifications assume a minimum flange width of (12 ds + br), 1/3rd of span length or spacing of beams, to act in full compression.
4. Curvature of slab due to deflected curved shape under heavy live loads (relative to deck's original flat surface).
5. Double curvature of slab at corners of slab and beam ends, due to diagonal bending.
6. Differential bending of the beam and slab in different directions (normally at 90 degrees). Both the slab and beam bending in different directions.

Table 4.2 Comparative solutions using the differential equations approach for two-way bending of a square panel (Poisson ratio = 0).

	Relative Slab: Beam Stiffness $\gamma = 0$ (Beamless deck)		Relative Slab: Beam Stiffness $\gamma = \infty$ (Stiff beams/simply supported)	
Method of Solutions	Central Slab Deflection × qL^4/D	Central Slab Moment × qL^2	Central Slab Deflection × qL^4/D	Central Slab Moment × qL^2
Galerkin solution for partial composite boundary	0.0287	—	0.00456	0.03684
S. J. Fuch's* solution for partial composite boundary	0.0282	0.1055	0.004062	0.03941
Khan's solution for partial composite boundary	0.0285	0.1056	0.004048	0.03604
Khan's solution for full composite boundary	0.0285	0.1056	0.004049	0.03606

* Proceedings of ASCE, Vol. 79, Separate No. 199, 1953.

7. In a through two-girder system, the ends of the slab panels composite with the floor beam are connected in the tension zone of the through girder and are subjected to overall tension.

8. Stresses arising from sharp skew edges accompanied by stress concentrations.

4.7.2 Arching Action in Deck Slabs

Arching action in slab and beam systems was studied as early as 1953 by B. G. Galerkin in Moscow and later by Allen and Severn at Bristol University and by Khan and Kemp at the University of London. Recently investigations were conducted by Csagoly, Hewitt, and Klinger. AASHTO LRFD recommends incorporating arching action in the detailed design of deck slabs.

AASHTO C9.7.2.1 states that arching action occurs in deck slabs and an internal compressive dome is created. Near the boundaries, bending stress is accompanied by axial or planar stress. When the magnitude of axial stress exceeds the bending stress in tension, the net stress is overall compression. Dome effect appears to be greater at the corners of the slab panel, where the deck rotates in a diagonal direction. Dome action would occur at the beam supports due to slab bending in the diagonal direction in opposite panels of multiple panels. The deflected shape of a deck slab can be idealized as an arch over supporting beams and usually a reverse arch at midspan. Both dead load and live load deflected shapes are additive and are similar reverse curves. Beam action changes to arch action with the full depth of the slab acting in compression at supports.

At supports such as piers and abutments, additional stiffness is provided to composite sections in transverse directions by diaphragms, even though they are not directly in contact with concrete. It minimizes one-way bending of the deck slab and introduces two-way bending local to stiffened composite section. A transverse strip of slab over supports tends to act as a concealed or secondary beam. To prevent cracking of concrete, additional reinforcing bars in strips in the direction of the diaphragms would reduce any overstress. A similar effect is observed in flat or beamless slabs in building floors, where column strips are designed as concealed beams.

Transverse and longitudinal bending of the slab will give rise to diagonal bending of slab strips between diagonally opposite bearings. Diagonal bending will be smaller if the aspect ratio of the panel (L/a) is large.

Near the corners of panels, the diagonal deflected shape of the slab will be in the form of a dome with predominant local compression and rapid transition to tension at midspan (Figure 4.5). Detailing of reinforcing bars at present does not take into account true stress distribution. This inadequacy in detailing is partly responsible for slab cracks, a nagging maintenance problem in bridges.

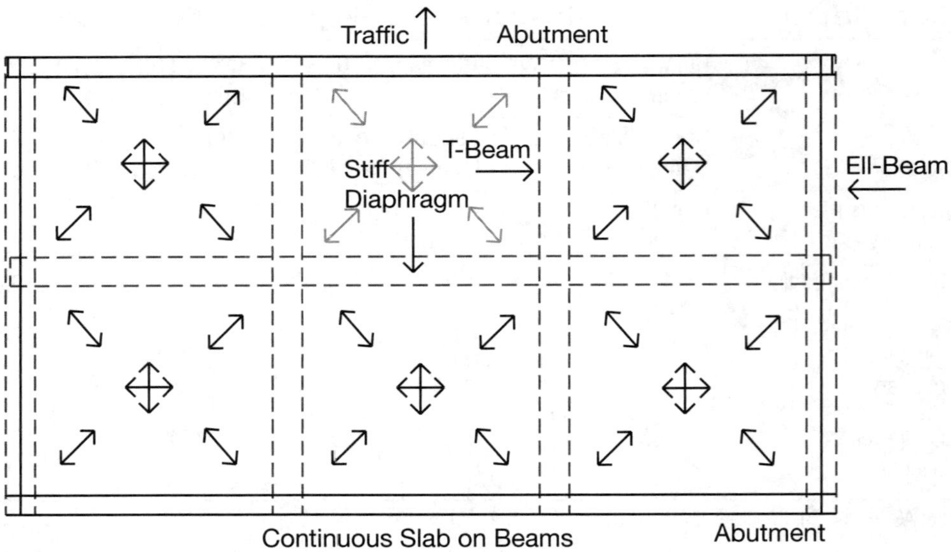

Figure 4.5 Dome or arching actions at boundary of continuous deck slabs (biaxial and diagonal bending).

The magnitude of dome action will depend upon:

1. Actual boundary conditions of the slab caused by beam depth and resulting eccentricity between the slab neutral axis and beam neutral axis.
2. In an internal panel, boundary conditions will approach to fixity depending upon the relative stiffness of slab to beam, with the shear connectors preventing any in-plane displacements.
3. Aspect ratio (span of slab to span of supporting beam at right angles).

Compression results from membrane or planar stress which is currently neglected in deck slab design. Membrane force would result in reduced deflections and would alter stresses in shear connectors which provide composite action. Deck slabs curved in plan and composite with curved beams would experience a greater effect of membrane forces in the slab and axial force in the curved beam.

The biharmonic equation for planar stress needs to be applied simultaneously with the plate bending equation to compute membrane forces.

$$N_x (\partial^2 w/\partial x^2) - 2N_{xy} (\partial^2 w/\partial x \, \partial y) + N_y (\partial^2 w/\partial y^2) = 0 \tag{4.11}$$

By using Airy stress function, the field equation for deck slab analysis becomes:

$$(\partial^4\phi/\partial x^4) + 2(\partial^4\phi/\partial^2 x \, \partial y^2) + (\partial^4\phi/\partial y^4) = 0 \tag{4.12}$$

where $\phi = \psi/D$

The arching action in a slab in a given panel gives rise to pure compression zones near supports. This effect is analogous to beam-column behavior in which compressive stress is accompanied by bending stress. Compressive stress is higher than bending stress at supports, and net stress is compression. It changes to tension below the neutral axis at midspan.

Unlike floors in buildings where two-way slab bending is predominant, bridges have unidirectional bending except when stiff diaphragms are present.

The following are advantages of arching action:

1. Midspan deflection is lower.

2. Continued cost benefits due to savings in reinforcement.
3. Structural slab thickness can be less.
4. Improved detailing procedure will result.

Description of full composite action:

1. The curvature of the slab in transverse and longitudinal directions is governed by relative stiffness at the boundary of the slab and girder. The greater the eccentricity of the slab and beam, the greater the arching or dome effect.

   The load path for gravity dead and live loads is from slab to beam. The slab deflects in the transverse direction, and the beam deflects in the longitudinal direction. The resulting deflected shape of the slab is similar to a shallow arch, with the highest point on the beam centroid and lowest point on the slab centroid.
2. Arching action or dome boundary effects would occur due to full composite action between slab and supporting beam. The phenomenon has been referred to as arching action, membrane action, or shear lag effects in slabs (Figure 4.6). It is due to structural mechanics but is not related to the material composite behavior that may result from the use of dissimilar or composite materials of the slab and beam.

Boundary effects: Due to the deflected shape of an arch, compressive forces additional to primary bending are introduced at slab boundaries as planar forces (Figures 4.7). For a comprehensive analytical approach, the following effects at the boundaries of the slab need to be considered:

1. Dual deflections of the slab and supporting beam at a right angle.
2. Reverse curvature of the slab near the boundaries, with sagging at midspan and hogging at supports.
3. At midspan, the top part of the beam above neutral axis is in a longitudinal compressive zone. Slab longitudinal edge, which is connected to the top of the beam, will be fully immersed in the much deeper compressive zone of the beam. Due to compatibility, the longitudinal force will cause a compressive planar force in the slab.
4. Slab curvature at the support (which is generally neglected) will depend upon the eccentricity of the beam and slab centroids. If centroids of the beam and slab coincide (for example in a through girder type connection) arching action is negligible.
5. In hybrid steel girders, flange steel and web steel strengths would alter the shape of the beam

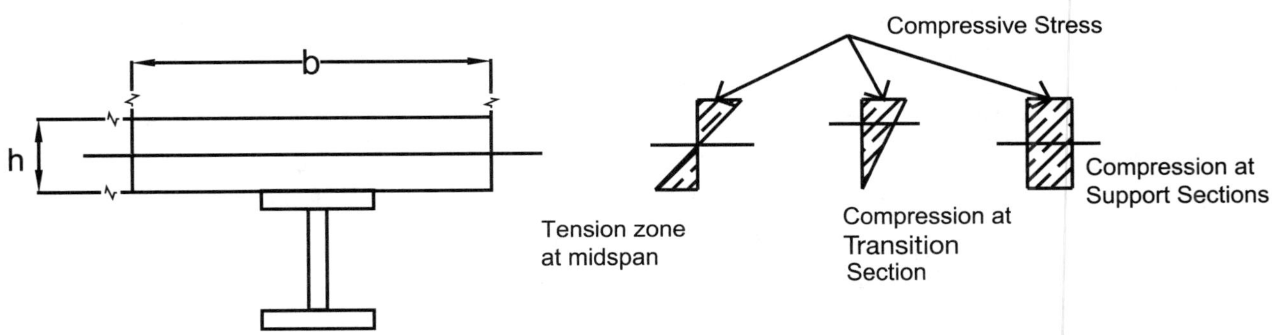

Figure 4.6 Progressive distribution of compressive stress across slab thickness due to arching action.

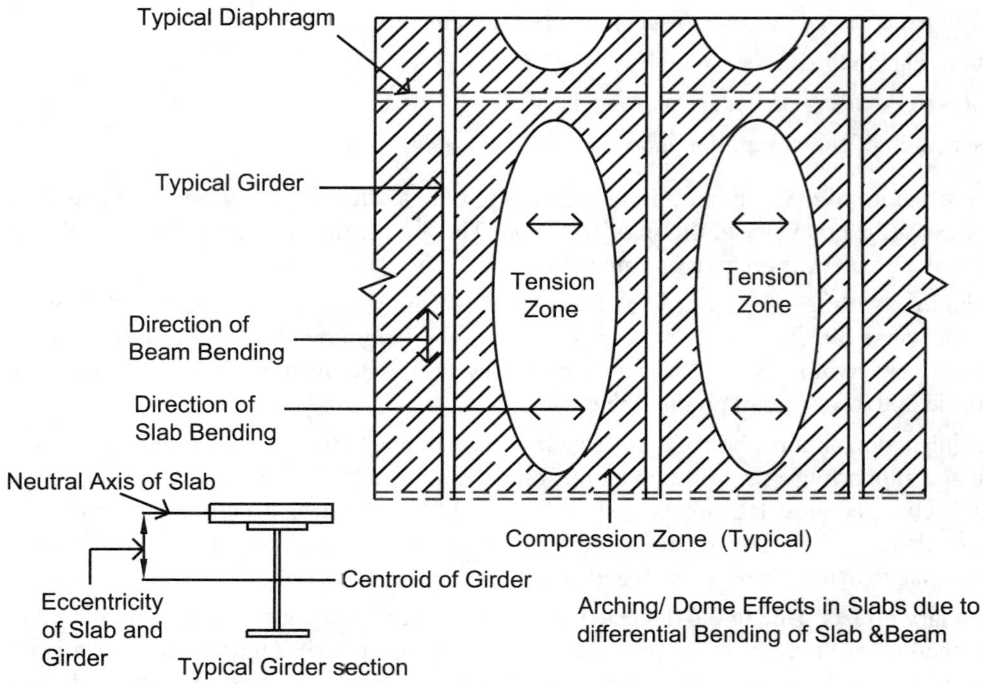

Figure 4.7 Tension and compression zones in continuous slabs.

compression forces at the junction between the flange and web. If the flange is partly immersed in deck slab concrete, the magnitude of compression in the slab will be different.

6. The transfer of planar shear in the slab is accompanied by vertical shear and bending in the slab.

4.7.3 Summary of T-Beam Analysis

Table 4.3 shows a comparative study of full composite action as discussed earlier in Section 4.6.3. Figure 4.8 shows the bending moment and shear force distribution in an element of a deck slab. Simplified formulation of a theoretical model can either be based on the classical plate and shell theory using a differential equations approach or by using a finite element model that would allow for bending and in-plane forces.

Boundary conditions can be developed as equilibrium, compatibility, stress-strain, and strain-

Table 4.3 A comparative study of full composite action in R.C. slabs with rectangular beams.

		Deflection at Midspan of Slab		Bending Moment at Midspan		Bending Moment at Support	
γ Relative Slab: Beam Stiffness	Poisson's Ratio υ	Galerkin Series	Khan	Galerkin Series	Khan	Galerkin Series	Khan
0 (Beamless Deck)	0.25	0.0257	0.0262	0.1109	0.1106	0.1527	0.1514
1 (Elastic Beams)	0.25	0.0117	0.0119	0.0691	0.0684	0.0559	0.0558
4 (Elastic Beams)	0.25	0.0067	0.0068	0.0539	0.0532	0.0117	0.0195
∞ (Stiff Beams Leading to Simply Supported Condition)	0.25	0.0041	0.0041	0.0460	0.0451	0	0
0 (Beamless Deck)	0.30	0.0249	0.0260	0.1090	0.1116	0.1404	0.1502

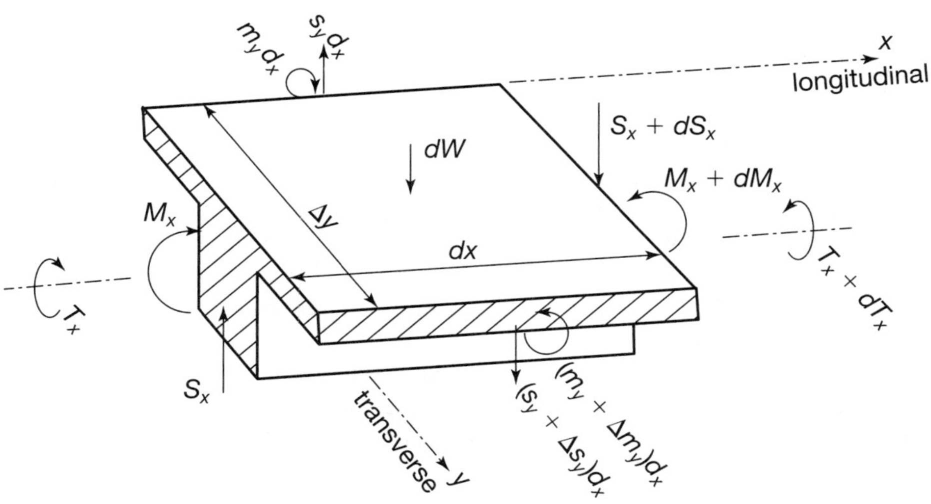

Notations:

dW = applied load acting on the slab-beam element

S_x = shear force, acting on beam

s_x = shear force per unit width, acting on slab

M_x = bending moment, acting on beam

m_y = bending moment per unit width, acting on slab

T_x = torsion on beam, acting about x axis

Figure 4.8 Forces on an element of composite T-beam.

displacement relations. Resulting sets of linear equations can be solved as banded matrices or by Gaussian elimination. Curved and skew beams may contribute to compressive membrane forces, especially in the vicinity of slab boundaries.

By incorporating the true stress distribution in a slab, it may be possible to reduce its thickness and the percentage of reinforcing bars in transverse and longitudinal directions, resulting in overall economy in design. With in-depth parametric study, recommendations on deck design and supporting beams can be made for future design codes.

Thick slab effects: It is assumed that for arching action to develop, a slab is thin compared to its span (typically $L/h < 40$). However, when girder spacing is small, say 5 ft, and the thickness of the slab is 12 inches. Most studies for arching action were for thin building slabs, such as 4 inches thick spanning over 15 feet with L/h ratio = 15/0.33 = 45.

L/h ratio = 5 only. Deep beam effect would occur. Thin plate theory is no longer applicable. For finite element analysis, the use of special elements needs to be considered. Nonlinear stress distribution across thickness will result. The arching effect due to stiff boundaries and the eccentric connection of the slab and beam will be better known.

4.8 NUMERICAL AND COMPUTATIONAL MODELS

4.8.1 Introduction

For slab and beam systems with simple boundary conditions, only limited closed form solutions using partial differential equations have been derived using the classical theory of elasticity. The availability of computer methods such as FEM has shifted the focus away from closed form solutions for some time, especially for complex geometric conditions.

Computer solutions using numerical models are available with some limitations.

4.8.2 Commonly Used Methods

1. Matrix inversion.
2. Matrix partitioning and solving banded matrices.
3. Gaussian elimination method.
4. Gauss-Seidel iteration methods.

The deflection function in response to applied load can be expressed in the form of algebraic polynomial function, such as R. H. Wood's series for slab and beam systems.

Harmonic analysis such as the Fourier series in terms of assumed deflection function with unknown coefficients: Well-known Navier series, Ritz series, Galerkin series, and Allen and Severn series have been used for thin plates subjected to vertical deflection and for stress function representing membrane stress.

The finite difference method derived from a Taylor series is based on a differential equations approach in which slopes (dw/dx) and (d^2w/dx^2) and other derivatives of deflections are replaced by algebraic equations, assuming successive differences in deflections at selected nodes.

1. Deflection function is assumed as y = f(x).

   A change in assumed deflection curve over a small distance h can be expressed as:

$$f(x + h) = f(x) + h\, f'(x) + h^2/2!\, f''(x) + \ldots\ldots\ldots + h^n/n!\, f^n(x) \tag{4.13}$$

   Neglecting higher order terms, the first derivative represents slope as a rate of change of deflection. $f'(x) = [f(x + h) - f(x)] / h$

2. Successive derivation will give expressions for bending curvature, vertical shear, and rate of loading at selected nodes.
3. The resulting system of algebraic equations can then be solved in the form of matrices.

4.9 ANALYSIS OF APPROACH SLAB RESTING ON GRADE

4.9.1 Alternate Approaches Using Differential Equations or Finite Element Methods

Approach slabs are required when ADTT is high (greater than 500 trucks). The new method is to support the end of the approach slab on a groove located behind the backwall so that settlement will not take place. The factors affecting stresses are:

1. Thickness of approach slab—Bridge and highway design manuals for each state have developed thicknesses varying from 12 to 18 inches. However, the thickness of the slab needs to be developed from a finite element analysis.
2. End conditions or method of support of approach slab—15 to 30 feet length in direction of traffic is generally used.

   Three types are possible:

   - Approach slabs on grade
   - Approach slab at bridge end simply supported on a ledge or groove at the top of the backwall and on grade at highway end.
   - Approach slab at integral abutment bridge end cast in place (integral) with concrete of backwall and deck slab. On highway end simply supported by an ell beam resting on grade. Eighteen-inch-thick slab is generally used.

3. Subgrade reaction of the soil-supporting slab—Usually subgrade material is selected by the highway engineer and an estimate of quantities is included in the highway costs. However, for integral abutments, selected fill material is used. The material is compacted in layers.

   The coefficient of subgrade reaction "k" values varies between 200 and 800 lbs/sq. inch/inch.

   When the bridge is located on a stream, the rise and subsequent fall of the water table may cause the subgrade to settle. Small diameter perforated pipes for drainage are commonly placed at a level where buildup of water will not take place.

4. Properties of concrete—Usually the same concrete is used as for deck slabs, with fc′ values between 4.0 and 5.0 ksi. This helps in rapid construction.

5. Thermal expansion and contraction—Longitudinal and transverse joints are provided at regular intervals. The edges of slab joints are armored.

The S. P. Timoshenko and Woinowsky-Krieger Theory of Plates and Shells proposes the following mathematical method for elastic subgrade reaction for soil spring stiffness k:

Lagrange general biharmonic equation for plate bending may be written as:

$$D\,(\partial^4 w/\partial x^4) + 2D\,(\partial^4 w/\partial^2 x\,\partial y^2) + D(\partial^4 w/\partial y^4) = (q - wk) \tag{4.14a}$$

Dividing by D,

$$(\partial^4 w/\partial x^4) + (\partial^4 w/\partial^2 x\,\partial y^2) + (\partial^4 w/\partial y^4) = (q/D - wk/D) \tag{4.14b}$$

where q is intensity of vertical load.

w_0 = Deflection of edges of bottom plate, intensity of foundation reaction = $k\,(w_0 - w)/D$

k = modulus of foundation in lb/sq in/in. k varies between 100 and 800 based on Casgrade's soil classification but needs to be verified by borehole tests.

4.9.2 Closed Form Solutions

The format can be developed in terms of fourth order differential equations. The following expressions for deflections of slab on grade are based on the Navier solution and are of practical interest in application to design of the approach slab. They can be programmed using Excel® spreadsheets:

1. For an approach slab of width a, length b simply supported on all four edges,
Apply the Navier solution to obtain the expression of deflection of plate as:

$$\text{For rigid foundations, } w = \Sigma_{m=1}^{m=\infty}\,\Sigma_{n=1}^{n=\infty}\,A_{m\,n}\,(\sin m\pi x/a)\,(\text{Sin } n\pi y/b)$$

where $\beta = \dfrac{m^2}{a^2} + \dfrac{n^2}{b^2}$

Apply the Navier solution to obtain the expression of distribution of load acting on the plate as:

$$q = \Sigma_1^{m=\infty}\,\Sigma_1^{n=\infty}\,a_{m\,n}\,(\sin m\pi x/a)\,(\text{Sin } n\pi y/b)\ \ldots.. \tag{4.14c}$$

Similarly, for reaction of subgrade:

$$p = kw = \Sigma\,\Sigma\,k\,A_{m\,n}\,(\sin m\pi x/a)\,(\text{Sin } n\pi y/b)$$

$$A_{m\,n} = a_{m\,n}/[\pi^4 D\,(m^2/a^2 + n^2/b^2) + k]$$

For uniformly distributed load of intensity q:

$$w = (16\,q\,/\,\pi^2\Sigma^{m=\infty}{}_{m=1,3,5}\,\Sigma^{n=\infty}{}_{=1}\,(\sin m\pi x/a)\,(\text{Sin } n\pi y/b)/$$

$$mn[\pi^4 D\,(m^2/a^2 + n^2/b^2) + k]\ \ldots. \tag{4.14d}$$

When k = 0, deflection reduces to that given by Navier equation for deflection of a uniformly loaded plate. Refer to the theory of plates and shells by Timoshenko, Chapter 8.

4.9.3 HL-93 Trucks versus Heavier Cooper Railway Train Live Load

Shear forces transmitted by wheels of railway wagons are very high and punching shear failure is possible. In railway tracks, sleepers and ballasts are used to distribute wheel loads at 45 degrees to subgrade.

For permit loads and HS-25 trucks or tandem trucks, slab thickness will be governed by punching shear, and thickness used should be adequate. Reinforcing bars should be designed for bending moments from FEM analysis. Usually bending shear is not a problem for large width slabs.

- Comparison of analysis of approach slab on grade using differential equations and finite element methods
- Exposed pile analysis using L-pile or COMP624P program
- Fatigue analysis for category B
- Sheet piling analysis
- Location of section for maximum BM

In structural analysis applied to the bridge framework of composite slab-beam girders with substructures, many assumptions are made. The mathematical model used is as accurate as the validity of the assumptions. Effects of some assumptions are addressed here.

4.9.4 Shape Effects for Steel or Aluminum Sign Structures

Sign structures with hollow tubes: Selection of closed hollow tubular sections rather than open sections is recommended for resisting torsion. Members' torsional capacity is evaluated and can be used for primary design. Three thickness ranges are considered:

If r_1 = Internal radius, r_2 = External radius,

r = Radius of solid shaft of equivalent area, identical unit weight and maximum stress

$T_{solid} = \tau \pi r^3/2$

$T_{hollow} = (\tau \pi/2)(r_1^4 - r_2^4)/r_1$

Equating maximum stress, $T_{hollow}/T_{solid} = (r_1^4 - r_2^4)/r_1 r^3$ (4.16)

Equating areas, $(\pi/2) r^2 = (\pi/2)(r_1^2 - r_2^2)$

$T_{hollow}/T_{solid} = (r_1^2 + r_2^4)/r_1 r$

$= r_1/r (1 + 1/n^2)$, if $r_1/r_2 = n$ (4.17a)

Also $r^2 = r_1^2 - (r_1/n)^2$

$r_1/r = n/\surd(n^2 - 1)$

$T_{hollow}/T_{solid} = (n^2 + 1)/[n\surd(n^2 - 1)]$; (4.17b)

$n = 1.25, T_{hollow}/T_{solid} = 2.563/(1.25 \times 0.75) = 2.734$

$n = 1.1, T_{hollow}/T_{solid} = 2.21/1.1 \times 0.458 = 4.384$

A hollow beam can carry 4.384 times greater torsion than an equivalent solid beam of the same weight and is ideal for members subjected to torsion.

4.9.5 Initially Curved Shallow Beams with Camber (Linear Strain)

It is common practice to design beams with upwards camber to negate dead load deflection. Camber introduces compressive stress in bottom flange especially at midspan. This simple approach has advantages similar to prestressing to introduce compression. Net deflection in practice will be due to live load only. An initially vertical curved beam is used. However, the effect of camber or initial curvature needs to be considered in analysis as follows:

If the depth of the cross section is small compared to the radius of the curvature, stress distribution is linear (Figure 4.9). This is the case for most of the beams used in practice. However, for deep beams with initial camber stress, distribution will be non-linear (Figure 4.10).

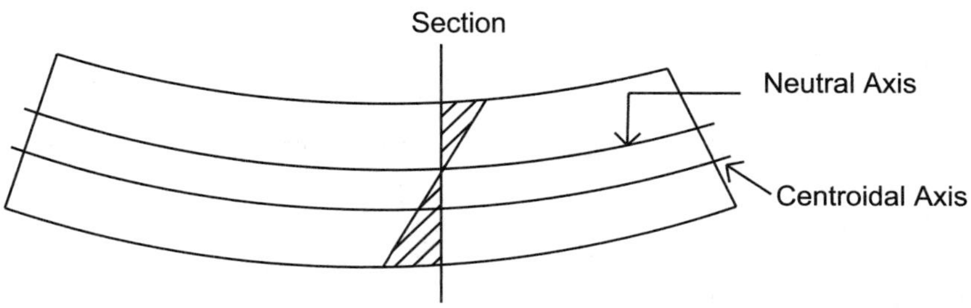

Figure 4.9 Linear stress distribution for shallow beams.

If R1 r1 is initial radius of neutral axis, and R2 is final radius,

$$\text{Net strain 'e' under load is } e = (1/R2 - 1/R1)\, y \qquad (4.18a)$$

Bending equation may be expressed as

$$\sigma/y = E\,(1/R2 - 1/R1) = M/I \qquad (4.18b)$$

A correction factor in strain and stress values needs to be applied.

4.9.6 Deep Beams with Small Initial Curvature (Non-linear Strain)

It is assumed that a neutral axis does not coincide with a geometric centroid. The centroidal axis of a cross section is assumed at distance n from the neutral axis (Figure 4.10).

If R1 is radius of curvature to neutral axis,
R2 is radius of curvature to centroidal axis

$$n = R2 - R1$$
$$R1 = d/\,[\log_e (R2 + d/2)/(R2 - d/2)] \qquad (4.19a)$$

Equation 4.18b may be expressed as $\sigma = \dfrac{M_y}{nA\,(R_1 + y)}$ (4.19b)

$$[\log_e (R2 + d/2)/(R2 - d/2)] = d/R2\,[1 + 1/3\,(d/2R2)^2 + 1/5\,(d/2R2)^4 + \ldots] \qquad (4.19c)$$

A nonlinear equation will lead to non-linear stress distribution across the depth of a deep beam. Currently this non-linear effect is being neglected by AASHTO for deep beams.

4.9.7 Shear Stress Distribution

For a rectangular beam $A = bd$, $I = bd^3/12$

$$A\,\check{y} = b\,(d/2 - y) \cdot (d/2 + y) \cdot ½ = [(d/2)^2 - y^2] \cdot b/2 \qquad (4.19d)$$
$$\tau_{xy} = W\,[(d/2)^2 - y^2]/bd^3 \text{ at a distance y from N.A.}$$

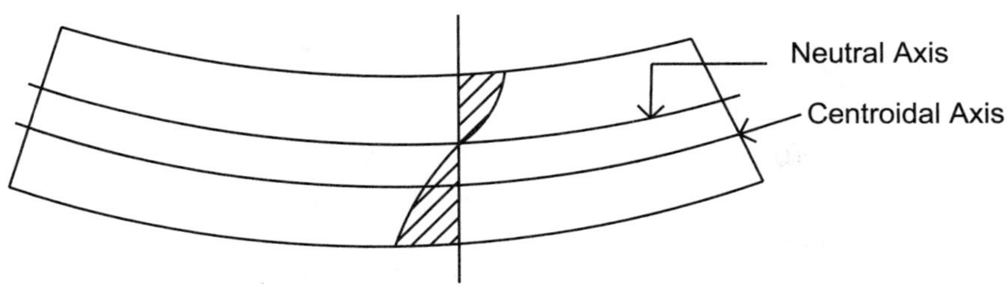

Figure 4.10 Nonlinear stress distribution for deep beams.

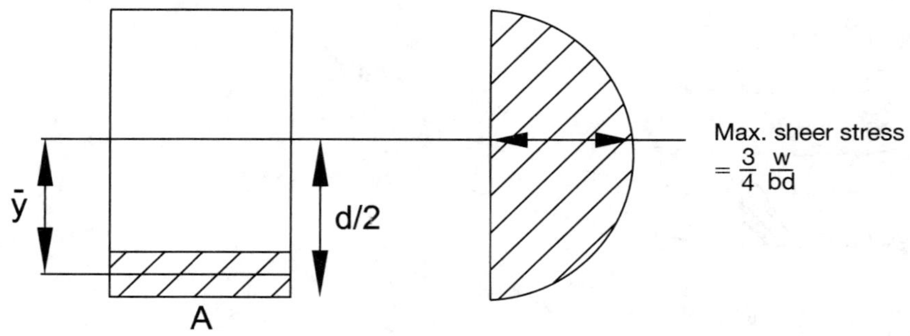

Figure 4.11 Parabolic shear stress distribution across a rectangular beam section.

$$\text{Actual shear stress } \tau_{xy} = F\,A\,\bar{y}/I \cdot b \tag{4.19e}$$
$$\text{Approximate shear stress} = \tau_{xy} = F/A_w$$

It can be shown that:

$$\text{for } y = \frac{d}{2},\ \tau_{xy} = 0$$
$$\text{for } y = 0, (\tau_{xy})_{max}$$
$$= \frac{3}{4}\frac{w}{bd}$$

Total deflection: Consider a 100-foot girder subjected to total distributed load W.

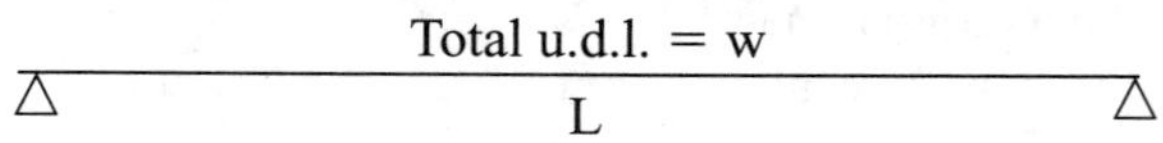

Max. deflection due to bending = $5\,WL^3/384\,EI$
$I = bd^3/12;$

$$\text{Max. deflection due to shear} = \frac{3WL}{20bdG} = \frac{6WL}{20bdE},\ \text{assuming } G = \frac{E}{2}$$

$$\text{Total deflection} = \frac{WL}{2bdE}\left(\frac{5}{16}\frac{L^2}{d^2} + \frac{3}{5}\right) \tag{4.20}$$

4.9.8 Estimating Additional Deflection due to Shear

There is higher shear deflection on cantilever spans than on continuous spans (Figure 4.13).

1. Consider a 5-foot Overhang: W = 1.0 KIP, $I = 12 \times 9^3/12 = 729$ inch units, w = 0. 33 kips/ft

Deflection due to bending = $WL^3/3EI + wL^4/8EI =$
$1.0 \times 5^3 \times 1728/3\ EI + 0.33 \times 25 \times 25 \times 1728/8EI =$
$125 \times 1728/(3 \times 3000 \times 729 + 5/3/8 \times 3000 \times 729)$
$= 0.033 + 0.02$

Total deflection due to shear = $6\ WL/5bd\ G + (3/5)\ wL^2/bdG$

$G = E/2(1 + \gamma)$; $E = 3.0 \times 10^6$ psi; $\gamma = 0.15$; $G = 3000/2.3$ ksi

Deflection = $6 \times 1.0 \times 5 \times 2.3 \times 12/5 \times 12 \times 9 \times 3000 = 0.059''$

$$+\frac{3}{5}\times\frac{0.33\times 5^2\times 2.3}{12\times 9\times 3000}\times 12 = 0.0004'' + 0.0005 = 0.0009''$$
$$= 0.009/0.059 = 15\%$$

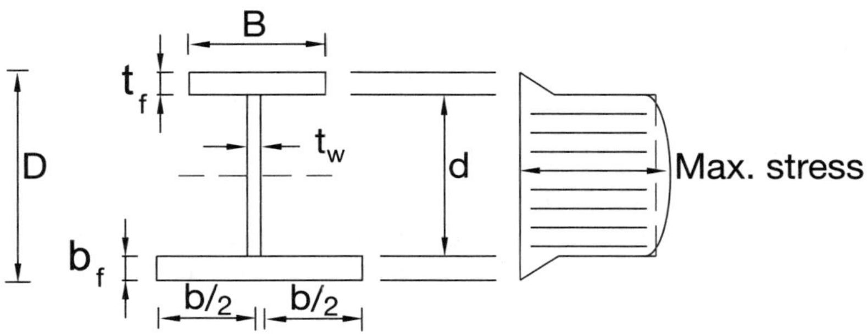

Figure 4.12 Shear stress distribution across an I-beam section.

Percent increase in this case, for shear deflection. Currently shear deflection is being neglected by AASHTO for shallow beams.

2. Shear deflection at the cantilever end of hammerhead piers: Shear deflection of deep beams is much higher than that of overhangs. The greater the depth, the greater the deflection.
3. Predicting vibrations on lighter spans: Light, slender span designs are more popular and easier to build, thanks to advances in material technology and construction techniques. As modern-day designs of pedestrian bridges have become lighter in form, migrating away from boxy, steel truss bridges to arch or cable-stayed spans with thinner decks, vibrations have become more of a concern.

   Typically, structural members are designed to vibrate at a frequency that would not interfere with those caused by a person running or walking across the bridge. Studies have shown that while pedestrians expect a bridge that looks lighter to exhibit more vertical movement, lateral movements tend to be uncomfortable for the user.

   In Figure 4.12 $\tau_{xy} = \frac{V}{8I}\left[d^2 + \frac{B}{t_w}\left(D^2 - d^2\right)\right], I = \frac{BD^3 - bd^3}{2}, \sigma_b = \frac{12M.y}{BD^3 - bd^3}$

4. Engineers need to compute lateral forces from sway: Computational analyses and insight from studies of pedestrian bridges with live loads exist. Member sizes used allow designers to predict the dynamic behavior of a footbridge to a certain extent.

   Damping effects: There are still many factors that make these predictions difficult. Damping depends on a number of parameters such as:

1. Materials used.
2. Complexity of the structure.

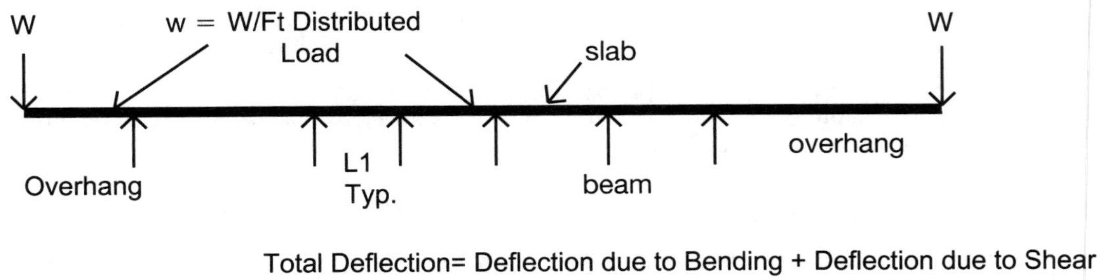

Figure 4.13 Slab and beams load diagram.

3. Type of surfacing.
4. Bearing conditions.
5. Railings.
6. Modeling of random live loads.

There are many possible varying locations and numbers of people on a bridge to be applied and many ways to model them, all of which will offer different results. While one person walking across the bridge will cause minimal vibration, several individuals jumping on the deck can create higher oscillations. A vibration analysis will be needed, especially for longer spans. Use of tested computer software is required.

4.9.9 Definitions Based on Structural Behavior

Various physical parameters which represent theoretical concepts govern analytical results. There should be no ambiguity about their definitions given below.

1. Redundant and non-redundant sections: As defined by the theory of indeterminate structures, use of a two-girder system or through trusses will cause the least redundancy or least indeterminacy. The disadvantage is that the probability of failure is increased against environmental loads. The formation of plastic hinge in a single girder in an indeterminate system will only cause failure local to that girder. Other girders will be able to distribute moments thereby offering combined resistance and avoiding failure. Hence, the greater the number of girders, the higher the degree of redundancy. A minimum of four girders is permitted by most states.

   Similarly, a single span system will theoretically require only one plastic hinge to form at the location of maximum positive moment (midspan) for failure. A continuous girder system will require two or more plastic hinges to be formed at the locations of maximum positive and negative moments. Hence, there is a need to adopt redundant systems.

2. Composite and non-composite sections: Use of shear connectors makes it possible for composite action between the deck slab and the top flange of the girder. This provides a tremendous advantage in preventing global buckling of the top flange. During construction stages and until the full strength of concrete is achieved, the girder section will behave as non-composite and will be designed as such.

3. Compact and non-compact sections: To prevent local buckling, the top flange needs to be braced laterally by the composite slab. Flange shapes and sizes which meet long column effect and local buckling requirements of slenderness and lateral support are called *compact sections*.

   The advantages of compact sections are that they are capable of generating full plastic behavior leading to improved strength, with local buckling resistance capability compared to the non-compact sections. Non-compact sections can develop yield moment, but only partially develop a plastic response.

Non-compact sections are generally safe but are uneconomical compared to compact sections. Hence, it is important to try and achieve a compact section while sizing a plate girder. One method of reducing local buckling in compression flange of a non-compact section is to provide transverse diaphragms. Long column effect still needs to be checked for global buckling for total compression flange length between bearings.

4.9.10 Selection of Cross Section/Shape for Steel and Prestressed Girders

For beam bending, an I-shape (wide flange section) or box section has maximum moment of inertia and gives minimum deflection and bending stress compared to other open shapes, such as a channel. An I-shape conforms better to conventional steel beam theory, in which the two flanges resist bending moment and the web resists shear force.

Design of an I-shaped girder consists of the design of its individual components, i.e., compression flange, tension flange, and web. Length, depth, width, and thickness and the relative proportions of each component of an I-girder determine local stress distribution and buckling. Sizing of girders plays an important role in both the elastic and plastic behavior of girders.

In steel beam theory, bending moment can be replaced by two axial forces, one compression and one tension, each acting at the other side of the neutral axis. Axial compressive stress in flange will cause local buckling if the flange area is small and also if the flange is long and slender (having high L/r ratio).

For a compact steel section, flange thickness needs to exceed a certain minimum, so that the width to thickness ratio is less than 16. Typical values of flange thickness are greater than 1 inch. Stress requirements must also be satisfied. At supports when bottom flanges are in compression and there is no lateral bracing available, the thickness of the bottom flange needs to be controlled by the slenderness ratio of the flange width to thickness. The compression part of web will also help prevent buckling by making a T-section with the flange.

In addition, torsion may occur at girder ends when girder ends are restrained at bearing supports. Torsional moment occurs due to eccentricity of load application when the result of vertical loads is not located at the center line of the web, especially at fascia girders.

4.9.11 Lateral Buckling of Girders

Failure mode of a steel girder is either by yielding or by buckling. If adequate lateral support is provided, failure will be by yielding. If the member does not satisfy slenderness (kL/r) requirement, failure will occur by out-of-plane buckling due to compression.

Failure can also occur by torsion due to lateral deflection. This mode of failure is known as "lateral buckling" or "lateral torsional buckling." Twisting is resisted by a combination of St. Venant (pure) torsion or warping torsion. In closed sections such as boxes and tubes, two webs are available; torsional stiffness is very high in resisting warping torsion, and lateral buckling is not important.

All I-shaped girders with thin webs are subjected to additional stress from lateral buckling and need to be designed for combined stresses. Warping torsion is higher for cantilever beams which have high deflection at the free end.

Formulae to calculate lateral bucking stress are provided by AASHTO LRFD specifications.

4.10 NONLINEAR ANALYSIS IN STEEL AND CONCRETE

4.10.1 Relationships between Load and Deflection, Stress, and Strain

Relationships between load and deflection, stress, and strain are initially linear. After a certain magnitude of applied load, the behavior of the beam or column becomes nonlinear.

For example, dead load or self weight analysis is usually linear. When response of the member under a higher load, as observed in the laboratory, shows second or third degree nonlinearity of load deflection or stress-strain relationships, the analysis is no longer straight forward. Also, significant changes in geometry of the structure can result in overstress, cracking or yielding of material.

Geometric nonlinearity: Examples are load deflection characteristics of suspension cables, long span trusses, and slender columns (P-Delta effect). Tied arches and trusses are not designed to undergo large deflections. In other structures long spans and heavy loads may lead to nonlinear load-deflection behavior of members. Hence, analysis should correspond to a series of unique loads under which the deformed structure is in equilibrium and has geometric compatibility.

Material non-linearity: When analysis is based on plastic deformation of certain materials, material nonlinearity is considered. Nonlinear analysis is required when studying post-elastic behavior of steel and concrete structures. An approximate upper bound linear analysis can first be carried out and then compared to results from nonlinear analysis to estimate the degree of nonlinearity and its effects. An example of nonlinear material behavior is shown by the C. S. Whitney stress-strain curve for reinforced concrete sections.

At higher loads as cracking develops, most materials exhibit a nonlinear behavior. Stiffness properties of concrete and composite members are based on uncracked sections or on cracked sections consistent with anticipated behavior. Progressive cracking based on experimental observations is included in the mathematical model.

A load deflection behavior is usually more accurately defined than stress-strain behavior since in the latter case physical changes such as crack formations would occur.

Principle of superposition: Large deflection analysis is inherently nonlinear, and loads are not proportional to displacements. Superposition of deflections and stresses cannot be used. From an analytical point of view, deflections, moments, and forces from dead load analysis cannot be directly added to those from large deflection live load analysis. In other words, live load analysis results cannot be "superposed" on dead load analysis results, and a unique solution will be required for the combined dead and live loads acting together.

4.10.2 Simplified Nonlinear Analysis

P-Delta method: Under concentric compressive axial forces, a slender column is subjected to out-of-plane deflection "Delta" giving rise to additional moment of P-Delta. Additional moment causes an equivalent eccentricity of axial load.

Synergism may be defined as the interaction of moments and forces that when combined produces a total effect which is greater than the sum of individual moment or force. Hence, synergistic effect of interaction results in apparent softening of the column, which can be expressed as a loss of stiffness. Quantitatively this can be classified as a second order effect. When axial compressive stress is high and reaches Euler bucking stress, failure by buckling takes place.

Approximate method can be used by selecting a moment correction factor. LRFD load factors are used for analysis rather than applying the load factors after analysis is complete. Changes in lateral deflection for each incremental load are incorporated in equilibrium equations. Iterative adjustment of deflections is carried out until convergence is reached.

Moment magnification of long columns (AASHTO 4.5.3.2.2b):

$$M_c = \Delta_b M_{2b} + \Delta_s M_{2s} \tag{4.21}$$

where M_{2b} = Moment on compression member, due to factored gravity loads (K-ft)

M_{2s} = Moment on compression member, due to factored gravity or lateral loads that result in side sway $> l_u/1500$

$$f_c = \Delta_b f_{2b} + \Delta_s f_{2s} \tag{4.22}$$

where f_{2b} = Stress corresponding to M_{2b} (ksi), f_{2s} = Stress corresponding to M_{2s} (ksi)

$$\Delta_b = C_m/(1 - X_1) > 1 \tag{4.23}$$

where $X_1 = P_u/(\phi_k P_e)$ where P_u = Factored axial load (Kips), ϕ_k = Stiffness reduction factor (1 for steel and 0.75 for concrete)

$$\text{Euler buckling load } P_e = \pi^2 EI/(K\, l_u)^2 \tag{4.24}$$

where K = Effective length factor,

l_u = Unsupported length of compression member

$\Delta_s = 1/(1 - X_2)$;

where $X_2 = \Sigma P_u/(\phi_k \Sigma P_e)$

Euler method of incremental loads: This method deals with first order or mild nonlinear problems. Select small increments in loads in the post-linear load deflection stage until convergence is reached.

Newton-Raphson (N-R) Method: For a higher degree of nonlinearity, the N-R Method is used. Derivatives of nonlinear terms are solved with linear terms. It is more accurate than the incremental load method since second order method is used.

4.10.3 Procedures for Analysis

1. Non-composite analysis cannot be combined with composite analysis. The two conditions need to be evaluated separately.
2. For LRFD construction load combination cases, construction processes cannot be superposed, and separate analysis for each construction stage needs to be carried out.

   Temporary construction loads of machinery and materials need to be considered, with the bridge performing only at its partial strength. In the final stage, multiple longitudinal joints need to be filled up with cast in place closure pours, which changes the strength of the superstructure.
3. The order of load application shall be consistent with that on an actual bridge, i.e., dead load stages followed by live loads, etc. This procedure is particularly important for slab and beam bridges using accelerated bridge construction and precast construction techniques.
4. Various stages of an unfinished bridge need to be modeled mathematically, with boundary conditions representing partial composite action. AASHTO code currently does not address the sequence, construction loads, or design methods of joints.
5. Mathematical modeling also applies to timber and metal formwork sheeting and temporary column supports to girders and deck slabs during construction. Strength of three-dimensional pipe assemblies or individual props needs to be evaluated. Overstress from materials and machinery loads, local buckling of pipe columns, temporary footing settlement, and unexpected lack of performance of temporary connections have led to many collapses and construction deaths in the past.

4.11 SINGLE SPAN LIVE LOAD ANALYSIS

4.11.1 Moving HL-93 Truckloads on a Single Span

The author has developed simplified expressions for maximum moments, their location, maximum shears, and reactions for design based on moving HS-20 truckload.

Combined HL-93 truck and lane loads:

1. Weight of HS-20 Truck = 72 kips used with additional lane load of 0.64 kips/ft.
2. Weight of alternate tandem truck = 50 kips used with additional lane load of 0.64 kips/ft.

A comparison between an HS-20 truck and an alternate tandem truck shows that up to a 40-foot span, tandem truck moments are highest.

For spans > 40 ft but < 150 ft span, HS-20 truckloads are highest.

For spans > 150 ft, lane load is higher. Lane load is combined with tandem truck for spans up to 40 ft and with HS-20 truckloads for spans > 40 ft.

Maximum shear/support reactions are, in all cases, highest for HS-20 trucks for spans up to 210 ft.

For spans > 210 ft, lane load shear is highest.

Theorem: Maximum live load BM occurs when the mid-span section divides the distance between the center of gravity of three axle loads and the adjacent wheel load equally.

Proof: Calculate the c.g. of loads where resultant is acting.

Assuming maximum BM occurs under middle heavy wheel at B.

$$M_2 = R1 \times (L/2 - x) - 4k \times 14 \text{ ft}$$

$$R1 = \text{Resultant} \times (L/2 + x - \bar{x})/L$$

$$R1 = 36 \times (L/2 + x - \bar{x})/L \tag{4.25}$$

For maximum BM rate of change of moment = 0

$$dM_2/dx = \text{Resultant} \times (\bar{x}/L - 2x/L) = 0$$

$$\bar{x} = x/2$$

Procedure for HL-93 truck:

Replace 3 wheel loads by a single resultant of wheel loads = Σ (16 + 16 + 4) = 36 kip

Locate the resultant force from nearest wheel by taking moments about that wheel.

Σ M = 0 at wheel B

(36 · X) + 16 · 14 ft − 4 · 14 ft = 0;

Distance (to c. g. O) X = (12 · 14)/36 = 14/3 = 4.667 ft from wheel B.

Calculate reactions at supports: For a given live load, reactions will vary with span lengths. Taking moment at about support 2,

R1 × L − 36 (L/2 − 2.333 ft) = 0;

Reaction R1 = 18L – 60;

$$\text{Mmax} = 18(L\text{-}1.55)\,(L\text{-}14)/L \text{ (Kip-ft)} \tag{4.26}$$

$$\text{Vmax} = 24(3L\text{-}28)/L \text{ (Kips)} \tag{4.27}$$

Comparisons of numerical values are shown in Table 4.4.

4.11.2 Procedure for HL-93 Alternate Tandem Truckloads (Table 4.5)

$$\text{Author's BM formula for tandem loads} = 12.5\,(L\text{-}4) \text{ (Kip-Ft)} \tag{4.28}$$

$$\text{Author's shear formula for tandem loads} = 50\,(1\text{-}2/L) \text{ (Kips)} \tag{4.29}$$

Single lane moments and forces are given. For multiple lanes, live load will be increased linearly. A multiple-lane reduction factor is allowed. A 6-foot transverse distance between tires is assumed.

HL-93 maximum lane loads

Author's use of simplified formula = $w\,L^2/8$

Table 4.6 shows total HL-93 truck and lane load moment and shear Σ (Lane loads of 0.64 kips/ft + HS-20) and Σ (Lane loads of 0.64 kips/ft + Alternate tandem trucks)

4.11.3 Envelope of Design Moments and Shears

Design Moment = Total HL-93 moment × Impact factor × Applicable distribution factor for single lane.

Dynamic load allowance (impact factors) = 1.33 applicable to truckloads only. For reduced posted speed, a lower factor may be used.

For multiple lane live loads, a reduction in total live load is applicable.

Fatigue dynamic load allowance (for fatigue analysis) = 1.15 applicable to truckloads only.

Comparison of maximum live load moments for internal beam. For truck load versus tandem load a comparative study is shown in Table 4.5.

Table 4.4 Comparative study of single span HS-20 moments and shear forces.AASHTO and simplified formula developed by the author.

	HS-20 LL Moments		HS-20 LL Shear	
Span Length (Feet)	AASHTO LL Moment (Kip-Ft)	Mmax = 18(L-1.55) (L-14)/L (Kip-Ft)	AASHTO Reaction/Shear Force (Kips)	Vmax = 24(3L-28)/L (Kips)
1. Small spans				
20	160	28′ distance between outer axles > 20′ formula N/A	41.6 > 36 k vehicle, not correct	
30	282.1	273.12	49.6	49.60
40	449.8	449.87	55.2	55.20
2. Medium spans				
50	627.9	627.91	58.5	58.56
60	806.5	806.61	60.8	60.80
70	985.6	985.68	62.4	62.40
80	1164.9	1164.98	63.6	63.60
90	1344.4	1344.44	64.5	64.53
100	1524.0	1524.01	65.3	65.28
110	1703.6	1703.65	65.9	65.89
120	1883.3	1883.36	66.4	66.40
3. Long spans				
130	2063.1	2063.10	67.6	66.83
140	2242.8	2242.89	70.8	67.20
150	↑	2422.704	↑	67.52
160		2602.54		67.8
170		2782.40		68.06
180	Lane load with 18 kips concentrated load for bending governs	2962.27	Lane load with 26 kips concentrated load for shear governs	68.27
190		3142.16		68.46
200		3322.05		68.64
210		3501.96		68.8
220		3681.88		68.95
230		3861.80		69.08
240	↓	4041.73	↓	69.2

4.12 SELECTION OF STEEL GIRDERS

4.12.1 Selection Based on Shapes

In practice, the following three types of steel girders are used:

1. Rolled sections.

AISC Manual of Steel Construction lists a wide range of grades 36, 50, and 50W (weathering steel) sections. Sections for grade 70W are not currently available. Since available depth of steel form is restricted to 42 inches, the application has been to smaller bridge spans only. LRFD increased live load, and deflection requirements are the constraints. Older bridges have used 36-inch depth rolled sections whenever possible.

Table 4.5 Selection of HS-20 and alternate HL-93 tandem truck moments and forces for smaller increments in span.

		Max. Truck Moments				Max. Truck Shear/Reactions			
Span Length (Feet)		Author's Formula* HS-20 (Kip-Ft)	Tandem† Two 25 Kip axles (Kip-Ft)	Governing Truck Moments (Kip-Ft)		Author's Formula** HS-20 (Kips)	Tandem Two 25 Kip Axles (Kips)	Governing Truck Shear (Kips)	
20	Small spans	28'distance between outer axles> 20' formula N/A	200 moments	200	Tandem governs	38.4	45	45	Tandem governs
22		133.85	225	225		41.45	45.45	45.45	
24		168.38	250	250		44	45.83	45.83	
26		203.12	275	275		46.15	46.15	46.15	
28		238.05	300	300		48	46.43	48	HS-20 governs
30		273.12	325	325		49.6	46.67	49.6	
32		308.31	350	350		51	46.88	51	
34		343.59	375	375		52.24	47.06	52.24	
36		378.95	400	400		53.33	47.22	53.33	
38		414.38	425	425		54.32	47.37	54.32	
40		449.87	450	450		55.2	47.5	55.2	
42	Medium spans	485.40	475	485.4	HS-20 governs	56	47.62	56	
45		538.78	512.5	538.78		57.07	47.78	57.07	
50		627.91	575	627.91		58.56	48	58.56	
55		717.20	637.5	717.20		59.78	48.18	59.78	
60		806.61	700	806.61		60.8	48.33	60.8	
65		896.11	762.5	896.11		61.66	48.46	61.66	
70		985.68	825	985.68		62.4	48.57	62.4	
75		1075.31	887.5	1075.31		63.04	48.67	63.04	

* $M_{max} = 18(L - 1.55)(L - 14)/L$ kip-ft

** $V_{max} = 24(3L - 28)/L$ kips

Advantages of rolled sections are thicker webs, and stocky, compact sections. Ready availability due to mass production is the reason for their widespread use for smaller spans. Due to thicker webs, the use of transverse web stiffeners is avoided. Weathering steel helps in maintenance costs through minimal painting.

The disadvantages are using a uniform flange thickness along the span which is uneconomical since bending moment is maximum at midspan but reduces along the length.

2. Rolled sections with cover plates.

With single or two layers of plates, the application of rolled sections can be extended to spans of about 80 feet. It has been a practice in the past to weld cover plates in regions of peak positive and negative bending moments. This resulted in economical design compared to using rolled sections only. However, tension welds at midspan and at supports are subjected to fatigue and need to be checked. Inspection of welds using NDT techniques and maintenance of fatigue prone details for older bridges are expensive. Current practice is to avoid this approach in favor of fabricated girders.

Span Length (Feet)		Max. Truck Moments Author's Formula* HS-20 (Kip-Ft)	Tandem† Two 25 Kip axles (Kip-Ft)	Governing Truck Moments (Kip-Ft)		Max. Truck Shear/Reactions Author's Formula** HS-20 (Kips)	Tandem Two 25 Kip Axles (Kips)	Governing Truck Shear (Kips)	
80	Medium spans (80–120)	1164.98	950	1164.98	HS-20 moments govern (80–240)	63.6	48.75	63.6	HS-20 shear force govern (80–240)
85		1254.70	1012.5	1254.70		64.09	48.82	64.09	
90		1344.44	1075	1344.44		64.53	48.89	64.53	
95		1434.21	1137.5	1434.21		64.93	48.95	64.93	
100		1524.01	1200	1524.01		65.28	49	65.28	
105		1613.82	1262.5	1613.82		65.6	49.05	65.6	
110		1703.65	1325	1703.65		65.89	49.09	65.89	
115		1793.50	1387.5	1793.50		66.16	49.13	66.16	
120		1883.36	1450	1883.36		66.4	49.17	66.4	
130	Long spans (130–240)	2063.10	1575	2063.10			49.23	66.83	
140		Lane load with 18 kips concentrated load for bending governs (140–240)	1700	2242.89		Lane load with 26 kips concentrated load for shear governs (130–240)	49.29	67.20	
150			1825	2422.704			49.33	67.52	
160			1950	2602.54			49.38	67.8	
170			2075	2782.40			49.41	68.06	
180			2200	2962.27			49.44	68.27	
190			2325	3142.16			49.47	68.46	
200			2450	3322.05			49.5	68.64	
210			2575	3501.96			49.52	68.8	
220			2700	3681.88			49.55	68.95	
230			2825	3861.80			49.57	69.08	
240			2950	4041.73			49.58	69.2	

† Tandem load moments exceed 20 moments for long spans.

3. Fabricated girders.

Due to improvements in fabrication technology, lighter girders are possible for a wide range of spans. It is more economical to use variable thickness flanges and a thinner web with transverse stiffeners at variable spacing. Welds in tension zones can be avoided.

Max. HL-93 moments are combined truck and lane moments. While maximum lane moments occur at midspan, maximum truck moments occur under the wheel load. An approximate method is to combine the two values. The lane moments at the location of wheel will be smaller due to parabolic shape of bending moment diagram.

With the availability of plates in a wide range, hybrid girders with 70W steel flanges and 50W webs are now possible. Some states have successfully used 100W steel for longer spans.

Use of Rolled Sections Versus Plate Girders

When designing a girder, a distinction is found between selections of rolled sections compared to fabricated plate girders. Due to rolling tolerance requirement of thinner webs, rolled

Table 4.6 HL-93 combined lane and truck design moments and forces.

Span Length (Feet)	Max. Lane Midspan Moment (Kip-Ft)	Max. BM Under Second Wheel (Kip-Ft)	Total Moment* Including Governing Truck (Kip-Ft)	Max. Lane Shear/Reaction (Kips)	Total Shear** Including Governing Truck (Kips)
1. Small Spans (Tandem Truck Governs)					
20	32	30.26	230.26	6.4	44.8
22	38.72	36.98	261.98	7.04	48.49
24	46.08	44.34	296.08	7.68	51.68
26	54.08	52.34	327.34	8.32	54.47
28	62.72	60.98	360.98	8.96	56.96
30	72	70.26	395.26	9.6	59.2
32	81.92	80.18	430.18	10.24	61.24
34	92.48	90.74	465.74	10.88	63.11
36	103.68	101.94	501.94	11.52	64.85
38	115.52	113.78	538.78	12.16	66.47
40	128	126.26	576.26	12.8	68
2. Medium Spans (HS-20) Truck Governs					
42	141.12	139.38	614.38	13.44	69.44
45	162	160.26	672.76	14.4	71.46
50	200	198.26	773.26	16	74.56
55	242	240.26	877.76	17.6	77.38
60	288	286.26	986.26	19.2	80
65	338	336.26	1098.76	20.8	82.46

sections are made stocky by using thicker webs. This resulted in no web stiffener design which is sometimes uneconomical. For example, unit weight of a well-proportioned plate girder will be less than that of a rolled section.

Traditionally, rolled sections are available in a large number of sizes. Their mass production keeps their unit costs low. They cater to the needs of building beams with small spans and small live loads. Due to their ready availability, they have been frequently used for smaller bridge spans. They lead to a compact design which is always desirable.

The limitation of rolled sections is that they are not commercially available in depths greater than 36 inches and seldom in depths up to 42 inches. Also, they are not manufactured for higher grade steel such as 70W.

Span Length (Feet)	Max. Lane Midspan Moment (Kip-Ft)	Max. BM Under Second Wheel (Kip-Ft)	Total Moment* Including Governing Truck (Kip-Ft)	Max. Lane Shear/Reaction (Kips)	Total Shear** Including Governing Truck (Kips)
70	392	390.26	1215.26	22.4	84.8
75	450	448.26	1335.76	24	87.04
80	512	510.26	1460.26	25.6	89.2
85	578	576.26	1588.76	27.2	91.29
90	648	646.26	1721.26	28.8	93.33
95	722	720.26	1857.76	30.4	95.32
100	800	798.26	1998.26	32	97.28
105	882	880.26	2142.76	33.6	99.2
110	968	966.26	2291.26	35.2	101.09
115	1058	1056.26	2443.76	36.8	102.95
120	1152	1150.26	2600.26	38.4	104.8
3. Long Spans (HS-20 Truck and Lane Loads Govern)					
130	1352	1350.26	2925.26	41.6	108.43
140	1568	1566.26	3266.26	44.8	112
150	1800	1798.26	3623.26	48	115.52
160	2048	2046.26	3996.26	51.2	119
170	2312	2310.26	4385.26	54.4	122.44
180	2592	2590.26	4790.26	57.6	125.86
190	2888	2886.26	5211.26	60.8	129.26
200	3200	3198.26	5648.26	64	132.64
210	3528	3526.26	6101.26	67.2	136
220	3872	3870.26	6570.26	70.4	139.34
230	4232	4230.26	7055.26	73.6	142.67
240	4608	4606.26	7556.26	76.8	146

* Includes governing truck moments from Table 4.5.

** Includes governing shear forces from Table 4.5.

Table 4.7 Comparison of permit loads in three states.

Comparisons of permit vehicles	NJ DOT permit vehicle	Caltran 9 axle (P 9)	PennDOT P-82
Distance between axles	16.0 m	22.64 m	16.65 m
Total weight	890 KN	973 KN	910 KN

Table 4.8 Comparison of maximum live loads with NJDOT permit vehicles.

Short spans	Factored AASHTO ML governs
Medium spans	NJ DOT Permit vehicle governs
Long spans	Factored AASHTO HL-93 governs

Steel, due to its superior strength and long-term performance, is best suited for medium and long spans. The old practice for medium span lengths for bridges was to weld cover plates on rolled sections in high-tension areas, such as at the midspan. Due to repetitive loads, fatigue of tension welds has been a problem, since it reduces the working life of expensive steel bridges. Also, regular monitoring of fatigue-prone details increases maintenance costs.

4.12.2 Selection of Girder Depth for Deflection Control

AASHTO requirements based on limiting maximum live load deflection:
AASHTO Table 25263-1 provides guidelines for minimum depth as follows:

Minimum overall depth of composite I − girder = 0.04 L for simple spans
= 0.032 L for continuous spans.

Minimum depth of steel truss = 0.1 L for simple or continuous spans.

Minimum depth of prestressed concrete I − girder and CIP box beams
= 0.045 L for simple span and
= 0.04 L for continuous spans.

Minimum depth of adjacent box beams = 0.03 L for simple spans and
= 0.025 L for continuous spans.

Any deviation for use of a shallower depth needs to be justified by detailed calculations.
Other structural requirements:

1. Repetitive deflections cause fatigue, debonding of reinforcing bars, cracking in deck slabs and wearing surfaces, and affects durability adversely.
2. High instantaneous deflections cause discomfort for motorists.
3. Bearing rotations during construction result in large deflections. Camber needs to be provided in girders to minimize dead load deflection.

Limiting live load deflections on long spans:

1. All design lanes shall be loaded. Lane reduction factors are applicable for two or more lanes.
2. For straight girders, distribution factor for deflection = Number of lanes/Number of girders
3. Dynamic load allowance (impact factor) shall be applied.
4. For vehicular loads, maximum computed live load deflection < L/800.
5. For vehicular and pedestrian loads, maximum computed deflection is 25 percent lower, i.e.,

$< L/1000$. $L \geq \Delta$ pedestrian/No. of beams + Δ vehicle $\dfrac{\text{x No. of lanes}}{\text{No. of beams}}$

6. For vehicular loads on cantilever arms, maximum computed deflection $< L/300$.
7. For vehicular and pedestrian loads on cantilever spans, maximum computed deflection is 25 percent lower, i.e., $< L/375$.
8. Deflection check is optional in current AASHTO LRFD code.

4.13 REVIEW OF COMMON FAILURE THEORIES OF MATERIALS

4.13.1 Failures Related to Construction Materials

In Chapter 3, external reasons for failure were based on:

1. Design defects such as incorrect assumptions, error in data, incorrect analysis, non-compliance with code guidelines, incorrect connection details, and mistakes in drawings.
2. Construction defects such as poor workmanship, substandard materials, inadequate concrete curing, imperfections in steel, lack of fit, and lack of quality control.
3. Investigating probable modes of failure: The response of materials to external forces like steel and concrete is discussed here.

Postmortem of collapse reveals details of sudden or progressive collapse. Large displacements result in combined shear and bending type overstress in members.

Critical sections for plastic hinges to form are located at midspan, under the concentrated load where deflection or positive bending moment is highest, or at a support where shear force, reaction, or negative bending moment is the highest. Tension yielding occurs in the flange.

4.13.2 Modes of Failure for Steel Bridges:

1. Bending tension stress in a member is exceeded due to long-term fatigue.
2. Shear stress or principal tensile stress is exceeded at girder supports.
3. Failure of bolts or welds at joints.
4. Local buckling of compression members.
5. Increased thermal stress in members due to malfunction of bearings.
6. Foundation movement due to flood scour during floods leading to settlement of pier or abutment.
7. Settlement of pier or abutment due to liquefaction during earthquake.
8. Lack of adequate support width at abutment during earthquake.

4.13.3 Review of Leading Theories of Yielding of Steel

1. When a specimen of steel is subjected to increasing axial load, a point is reached when axial stress is no longer proportional to strain. Hooke's Law is no longer applicable and the material is said to be *yielding*. Ductile metals like steel exhibit yielding and subsequent plastic deformation. Theories of yielding based on principal stresses and strains are summarized here.
2. Von Mises shear strain theory which is based on principal stress difference is most accurate and correlates best to experimental behavior.
3. Tresca maximum shear stress theory gives reasonable predictions and has a simpler mathematical form.
4. Rankine criteria for maximum principal stress: When a point in a material is subjected to principal stresses in three directions, yield of material will occur under the maximum of the three principal stresses for applied tension.

   As an alternative, yield of material will occur under the minimum of the three principal stresses for applied compression.

If $\sigma 1$, $\sigma 2$, and $\sigma 3$ are principal stresses, and if $\sigma 1$ is maximum of the three,

i.e., $\sigma 1 > \sigma 2 > \sigma 3$

Yield stress in tension = σyt

Yield stress in compression = σyc

Maximum principal stress $\sigma 1 = \sigma yt$ if tension force is applied

Minimum principal stress $\sigma 3 = \sigma yc$ if compression force is applied.

5. St. Venant Criteria for Maximum Principal Strain: When a point in a material is subjected to principal strain in three directions, yield of material will occur under the maximum of the three principal strains for applied tension or applied compression.

As an alternative, yield of material will occur under the minimum of the three principal strains for applied compression.

If $\sigma 1$, $\sigma 2$, and $\sigma 3$ are principal stresses, and if $\sigma 1$ is maximum of the three,

i.e., $\sigma 1 > \sigma 2 > \sigma 3$

$\sigma 1 = \sigma yt$ if tension force is applied

$\sigma 1 = \sigma 1/E - \nu (\sigma 2 + \sigma 3)/E = \sigma yt/E$, where ν = Poisson's Ratio (4.30)

$\sigma 1 - \nu (\sigma 2 + \sigma 3) = \sigma yt$

Similarly, $\sigma 1 = \sigma yc$ if compression force is applied.

$\sigma 3 - \nu (\sigma 1 + \sigma 2) = \sigma yc$.

6. Tresca-Guest criteria for maximum shear stress: Yield will occur when maximum shear stress T in the material reaches half the yield stress σyt.

$\tau = (\sigma 1 - \sigma 3)/2 = \sigma yt/2$

$(\sigma 1 - \sigma 3) = \sigma yt$

For two-dimensional stress, $\sigma 3 = 0$

$\sigma 1 = \sigma yt$.

7. Huber-von Mises shear strain energy theory: Huber proposed that Shear Strain Energy Theory = Total Strain Energy, U_t − Volumetric Strain Energy, U_v.

$$U_s = (U_t - U_v)$$

Based on Huber's theory, von Mises developed the following relationship between principal stresses:

$$U_s = [(\sigma 1^2 + \sigma 2^2 + \sigma 3^2) - 2\gamma (\sigma 1\, \sigma 2 + \sigma 2\, \sigma 3 + \sigma 1\, \sigma 3)]/2E \quad (4.31)$$

$$Uv = (\sigma 1 + \sigma 2 + \sigma 3)^2 (1 - 2\nu)/6E \quad (4.32)$$

$$(\sigma 1 - \sigma 2)^2 + (\sigma 2 - \sigma 3)^2 + (\sigma 3 - \sigma 1)^2 = 2\, \sigma yt^2$$

For a two dimensional system, $\sigma 3 = 0$; general equation can be simplified as

$$(\sigma 1)^2 + (\sigma 2)^2 - (\sigma 1\, \sigma 2) = \sigma yt^2$$

8. Beltrami-Haigh theory for total strain energy: Yielding will occur when total strain energy = Yield stress in tension.

$$(\sigma 1)^2 + (\sigma 2)^2 + (\sigma 3)^2 - 2\nu (\sigma 1\, \sigma 2 + \sigma 2\, \sigma 3 + \sigma 1\, \sigma 3)] = \sigma yt^2$$

9. Recommendations based on fracture mechanics for steel:

- Ductile metals like steel exhibit yielding and subsequent plastic deformation.
- Von Mises's shear strain theory based on principal stress difference is most accurate and

correlates best to experimental behavior. Apply Tresca and Von Mises principal strain and shear strain yield criteria for different types of steel.

- Tresca's maximum shear stress theory gives reasonable predictions and has a simpler mathematical form.
- St. Venant's and Beltrami's theories do not compare closely to experimental results and are of academic interest only.

4.13.4 Application to Failure of Steel Sign Structures from Yielding

1. The number of sign structures in use on U.S. highways is far greater than the number of bridges. In recent years, failures of cantilever types have been reported. Wind gusts from fast moving trucks have been responsible for causing torsion and shear stresses and thereby failures of sign structures. A method of calculation using yield criteria is presented here to evaluate combined stresses due to bending and torsion.

Minimum vertical under clearance for overhead and cantilever sign support structures may be assumed as 17 feet 9 inches.

2. Solved example for a sign structure: A cantilever sign structure beam 20 feet long is subjected to bending moment from the weight of the sign panel and torsion from wind. Beam is tube shaped with an external diameter of 8.5 inches and an internal diameter of 8.0 inches. Calculate maximum torsion due to wind to cause yielding.

M max = 12 kip-ft; Es = 30,000 ksi, Fy = 36 ksi

Maximum bending stress $\sigma x = 32M / (\pi\ d^3)$ from equation 4.37

$\sigma x = 32D\ M / \pi\ (D^4 - d^4) = 32 \times 8.5 \times 12.0 \times 12/\pi\ (8.5^4 - 8.0^4)$

$= 39168 / 3531.35 = 11.09$ ksi

$\tau xy = 16\ DT / \pi\ (D^4 - d^4)$

$= 16\ T \times 8.5 \times 12/3531.35 = 0.462\ T$ ksi

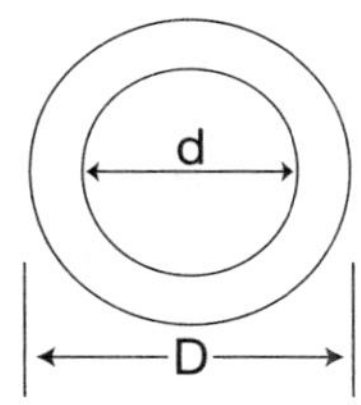

3. Using the Tresca's maximum shear stress theory:

$$\tau = (\sigma 1 - \sigma 2)/2 = Fy/2 \qquad (4.33)$$

$$1/2\ \sqrt{(\sigma x^2 + 4\tau xy^2)} = Fy$$

$$\sqrt{(123 + 4x\ (0.462T)^2} = 72;$$

$$T = 77 \text{ kip-ft}$$

4. Using the Rankine maximum principal stress theory:

$$\sigma 1 = \sigma x/2 + ½\ \sqrt{(\sigma x^2 + 4\tau xy^2)} = Fy \qquad (4.34)$$

$$\sigma 1 = 11.09/2 + 0.5 \times \sqrt{(11.09^2 + 4T^2 \times 0.462^2)} = 36 \text{ ksi}$$

$$11.09 + \sqrt{123 + 0.924^2\ T^2} = 36x^2$$

$$T = 64.8 \text{ kip-ft}$$

4.13.5 Fracture of Concrete

AASHTO LRED specifications are gradual, incorporating guidelines for a fracture mechanics approach. In the light of research carried out by Bazant and S. P. Shah at Northwestern University, it is expected that future codes will have recommendations.

4.14 PLASTIC BEHAVIOR OF STEEL SECTIONS

1. The first two diagrams show two types of steel tested (Figure 4.14). The stress-strain curve is linear up to yield point beyond which it is nonlinear. The third diagram shows an idealized stress-strain curve.

2. In the elastic range, stresses are proportional to strains. Also, both stresses and strains are proportional to applied loads. In plastic range, stresses are not proportional to strains. However, strains are always small and unlike stresses are proportional to applied loads in both elastic and plastic ranges. The magnitude of strains in the elastic range is considerably smaller than in the plastic range.
3. The plastic stress block across the depth of the beam is assumed rectangular for mathematical reasons.
4. If the beam is unloaded, the curve will not return to its origin, and there will be an offset with built-in strain showing that deformation has taken place.
5. In a simply supported beam, plastic moment will occur at midspan where elastic bending moment is also maximum. Near the supports, sections of beam will be elastic. Away from the supports closer to midspan, sections will gradually turn into elasto-plastic before becoming fully plastic at midspan (Figure 4.15).

   For fully plastic moment, $h = d/2$

   Elastic stress $\sigma y = M_{elastic}/S$ where S is section modulus; $S = b\,d^2/6$, see equation 4.2

$$M_{elastic} = \sigma y \cdot (b\,d^2/6) \tag{4.35}$$

   Due to ductile behavior of steel, moment capacity increases by 50 percent over maximum elastic capacity. This can be regarded as reserve of strength available beyond yield point of the additional strength available.
6. The LRFD method takes advantage of the additional available capacity resulting in a more accurate moment of resistance than developed for elastic range only in the allowable stress design method. In brittle materials there is no elasto-plastic or plastic behavior observed. Plastic hinge does not develop.
7. Shape factor = Plastic moment/Yield moment = 3/2.

   In steel

$$M_{plastic} = (3/2)\, M_{yield} \text{ for rectangular section}$$

   where b_1 = flange width: web side

   d_1 = clear web depth

   d = depth of beam

$$\text{For I-section } M_{plastic}/M_{yield} = 3/2\,(1 - b_1\,d_1^2/b\,d^2)/(1 - b_1\,d_1^3/b\,d^3) \tag{4.36}$$

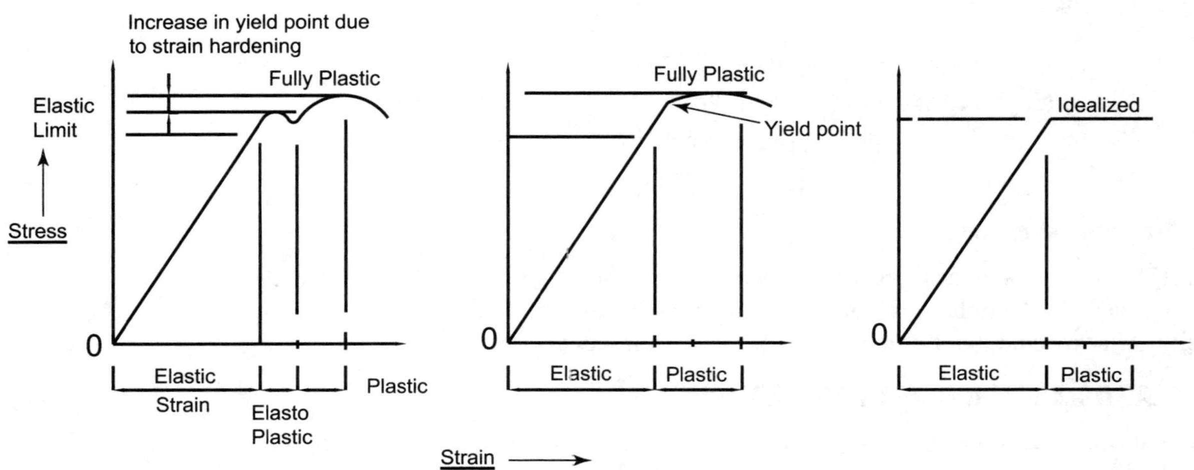

Figure 4.14 Nonlinear stress-strain plots for specimens tested in tension.

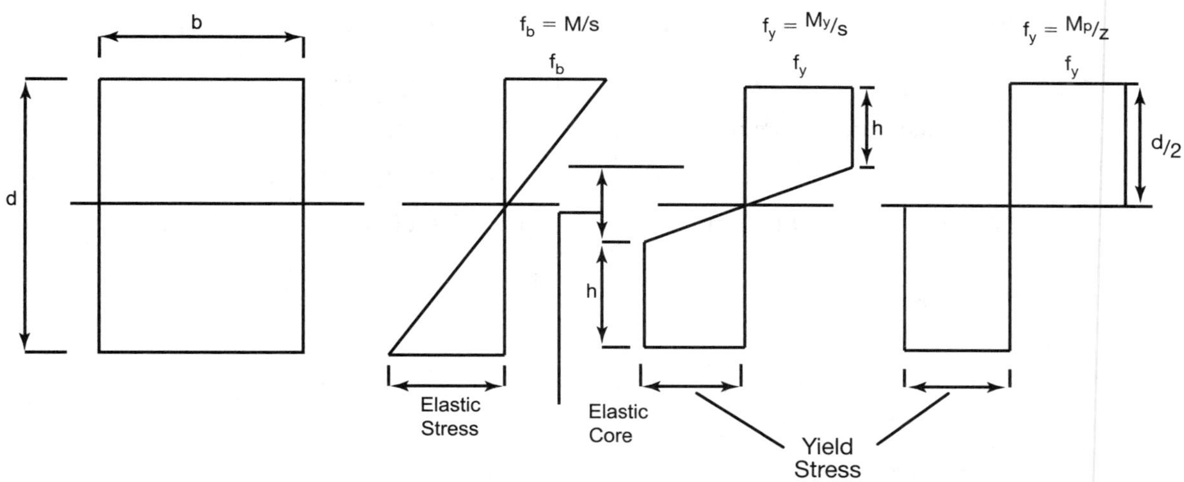

Figure 4.15 Elastic, elasto-plastic and plastic stages at location of peak moment in rectangular beam.

Circular	1.7
Rectangle	1.5
Wide flange weak axis	1.5
Round tube	1.27
Rectangular tube (width = depth/2)	1.20
Wide flange strong axis	1.14

4.15 PLASTIC BEHAVIOR OF STEEL NON-COMPOSITE SECTION

4.15.1 Plastic Moment of a Non-Composite Section

Part elevation cross section plastic stress distribution:
In Figure 4.16, the sum of all horizontal forces = 0

$$Fc + Fwc = Fwt + Ft$$

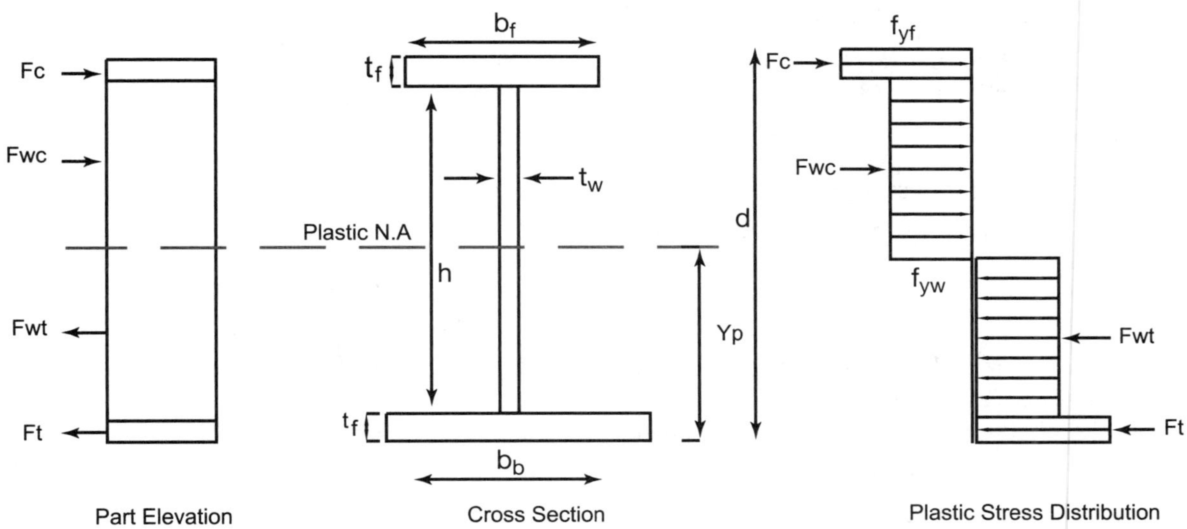

Figure 4.16 Plastic forces acting on non-composite section.

Yield stress in steel = Fy, taking moments about plastic N. A., calculate Yp

$Fy \times [b_f \times t_f + t_w \times (d - Yp - t_f)] = Fy\ [b_b \times b_f + t_w \times (Yp - t_f)]$

Plastic moment = Sum of all moments about N. A.

$$= Fc \times Yc1 + Fwc \times Yc2 + Fwt \times Yc3 + Ft \times Yc4 \tag{4.37}$$

From the equation (4.39), the value of Yp (location of N. A.), compression, and tension forces are known and require the following for solution:

1. Applicable I section.
2. Strength limit states I to V.
3. Construction limit state and uncured slab.
4. Reference to AASHTO Section 6.10.5.2.3.

Non-composite non-compact sections may require:

1. Strength limit states I to V.
2. Construction limit state and uncured slab.
3. Reference to Section 6.10.5.3.2 and Section 6.10.5.3.3.

4.15.2 Plastic Moment in Hybrid Girders

1. With the availability of 70W and 100W grade steel, the use of the higher-grade steel (such as 70W for flanges and 50W for web) is becoming popular.

   Moment causing initial web yielding = f_{yw} S (d/h)

   Moment causing initial flange yielding = f_{yf} S (d/h)

$$\text{Plastic bending moment } M_p = b_f\, t_f\, (d + h)\, f_{yf}/2 + t_w\, h^2\, f_{yw}/4 \tag{4.38}$$

2. Moment resisted by flanges and shear by web:
   - Strength limit states I to V
   - For 70W steel and construction limit state, assume non-compact
   - Redistribution of moments in continuous girders permitted.

4.15.3 Plastic Moment of Composite Section

When plastic N. A. is in web:
In Figure 4.18, the sum of all horizontal forces = 0

$$Fs + Fc + Fwc = Fwt + Ft \tag{4.39}$$

Yield stress in steel = Fy, for positive moment concrete is in compression

Ultimate stress in concrete = 0.85 fc′

$$0.85\, fc' \times be \times ts + Fy \times [bt \times tf + tw \times (D - Yp - tf)] = Fy\ [bb \times bf + tw \times (Yp - bf)]$$

From the above equation the value of Yp (location of N. A.), compression, and tension forces are known.

Plastic Moment = Sum of all moments about N. A.

1. Strength limit states I to V.
2. Redistribution of moments in continuous girders permitted.

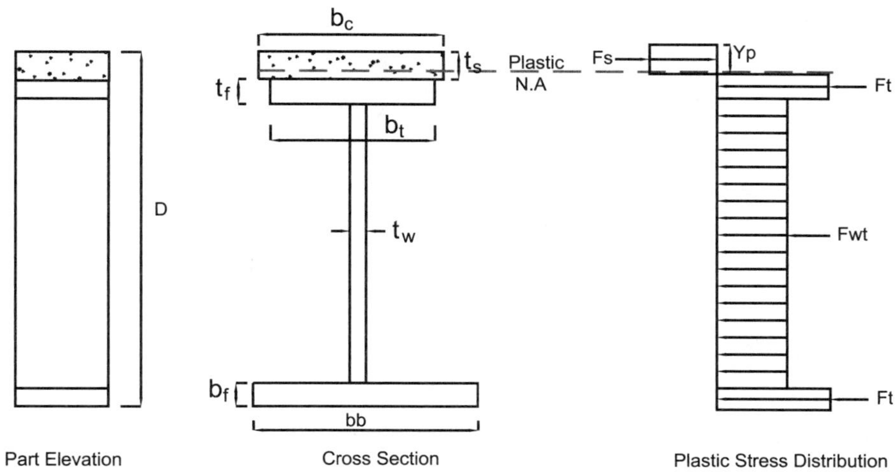

Figure 4.17 Plastic forces acting on composite section when plastic N.A. is located in concrete slab.

4.16 PLASTIC BEHAVIOR OF COMPOSITE SECTIONS

Concrete displays a small magnitude of tension compared to ductile material like steel. It has a very low modulus of elasticity (about one-tenth of steel) resulting in very small deformations (maximum strain of 0.3 percent, i.e., 0.003).

Since post-elastic behavior and the formation of cracks in tension are not clearly defined for different types of concrete, extra tensile capacity is not reliable, and any resistance to tension is conservatively neglected. Compressive stress distribution originally proposed by C. S. Whitney is parabolic but is currently assumed as an equivalent rectangle for design.

For negative moment, concrete is in tension and is neglected.

$$M_p = F_{st} \times Y_{st} + F_{sb} \times Y_{sb} + F_c \times Y_{c1} + F_{wc} \times Y_{c2} + F_{wt} \times Y_{c3} + F_t \times Y_c \quad (4.40)$$

where Y_{st} etc. are distances from plastic N.A.

4.16.1 Ultimate Load Behavior in Reinforced Concrete Sections

Concrete weakness in tension is made up by the use of reinforcing steel in the tension zone.

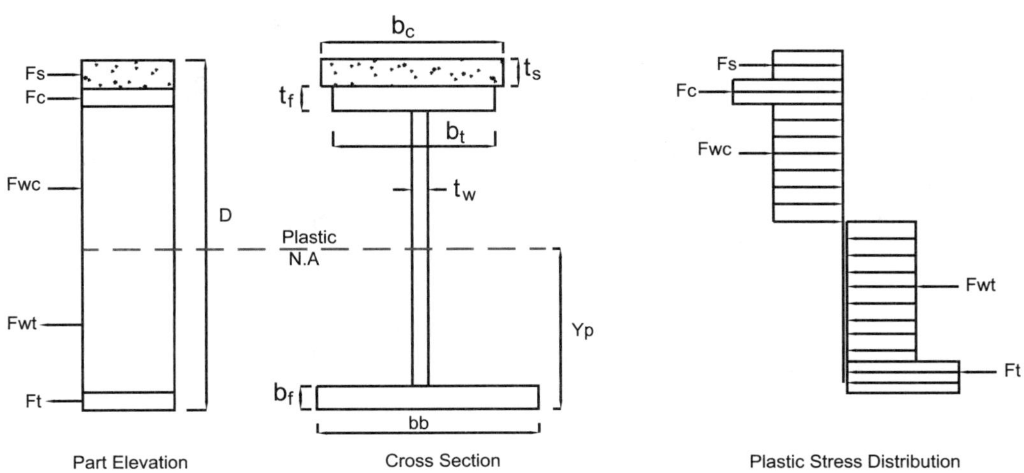

Figure 4.18 Plastic forces acting on composite section when plastic N.A. is located in web.

1. Ultimate moment $M_u = \phi M_n$

$$= A_s f_y (d - a/2) \quad (4.41)$$

where $a = A_s f_y / 0.85 f_c' b$ and $A_s = (M_u/\phi)/f_y jd)$ (4.42)

Parabola is replaced by an equivalent rectangle (0.85 fc′ × a) in area

2. Plastic hinge zone for substructure: Assume a pattern of plastic hinges at locations of peak bending moments. At ultimate loads, failure is likely to occur at these locations.

Equilibrium equations can be expressed in terms of factored moments and moments of resistance and solved on the computer as a matrix.

4.16.2 Shear Behavior of Reinforced Concrete Beams

1. Ritter-Morsch model:

The original truss model was proposed by Ritter in 1899 and was developed by Morsch in 1906 for bent-up bars in place of vertical stirrups. Basic assumptions are:

- Shear forces are resisted by shear reinforcement only
- Angle of inclination of diagonal struts is constant at 45°.

2. ACI 318-95 model:

ACI 318 Model is developed for beams with normal concrete strength i.e. $f_c' < 41.5$ MPa (6 ksi).

f_c' value > 41.5 MPa (6 ksi) (but < 69.2 MPa, i.e.,10 Ksi) can be used, provided the minimum web reinforcement is increased.

3. Plasticity model:

Lampert and Thurlimann in 1968 proposed a modification to the Ritter-Morsch truss by using a variable value of angle of cracking (Figure 4.19).

4. Modified compression field theory:

One of the many significant changes in design approach is the use of modified compression field theory (MCFT) for shear design of reinforced concrete and prestressed concrete girders. The Ritter-Morsch truss method was used for most of the twentieth century. Research at Northwestern University by the author, Edwin Rossow, and S.P. Shah has verified the validity of the method proposed by M. P. Collins. Diagonals represent concrete struts and verticals represent steel stirrups. Modifications to the Ritter-Morsch constant angle truss need to be introduced.

The compatibility model (MCFT) was recently proposed by M. P. Collins and further developed by T. C. Hsu. It proposes a variable angle of cracking (θ). It considers the important aspect of compatibility between concrete and steel stirrups. It explains the biaxial state of stress at a diagonal crack better than the plasticity model. The compatibility model is based on the modified compression field theory for concrete beams. Canadian and U.S. contributions to the ultimate load design method use Mohr's circle approach for principal stress and principal planes.

In prestressed concrete sections, both axial compressive stress and flexural compressive stresses are introduced. The law of superposition is applicable, and tensile stress from applied load is reduced or cancelled by axial compressive stress from prestressed tendons.

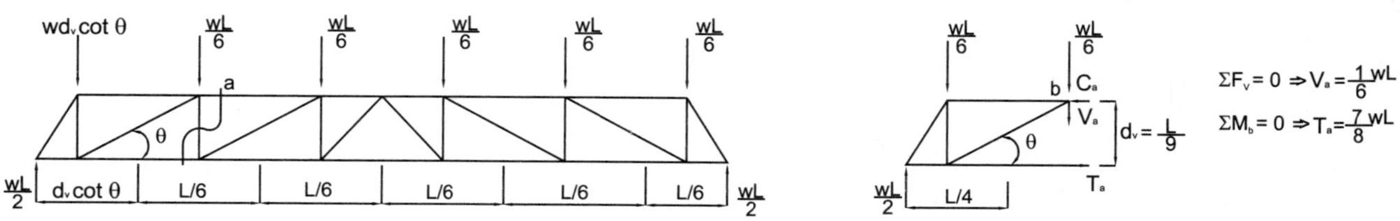

Figure 4.19 Reinforced or prestressed concrete beam idealized as a varying angle hybrid truss.

4.16.3 Compatibility Model (Strips and Elements)

1. In a research project carried out at Northwestern University as a follow-up of recommendations of the ACI Shear Committee, use of high strength concrete in the design of reinforced and prestressed girders was investigated. The author investigated the validity and application of MCFT to high strength concrete beams for both buildings and bridges. Many of AASHTO formulae for shear design have developed from widely used ACI codes including ACI 318. The method and results are briefly described here.

   A finite element approach was used. The beam is idealized into strips or multi-layered elements.

   Typical elements lie in the vertical plane of beams, with the depth of element equal to the depth of each strip (Figure 4.20a).

2. Two types of elements are considered:
   - Elements containing longitudinal rebars and transverse stirrups only (Figure 4.20b)
   - Elements containing transverse stirrups only (Figure 4.20c).

   The assumptions used are:

1. Steel and concrete have plasticity.
2. No bond exists between concrete and web reinforcement.
3. The angle of cracking varies and is dependent on the ratio of percentage of longitudinal steel to transverse steel.

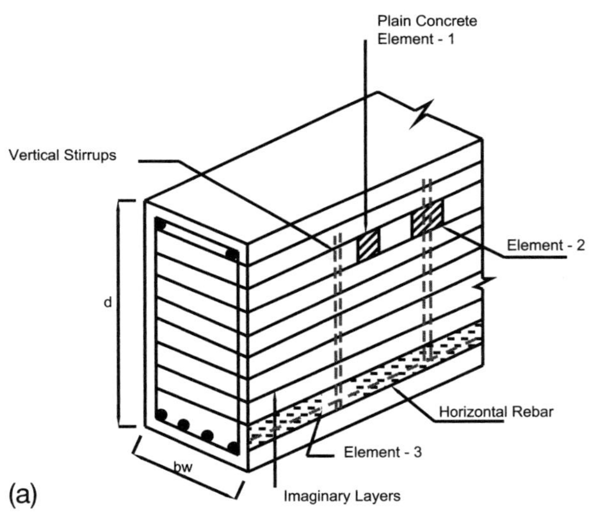

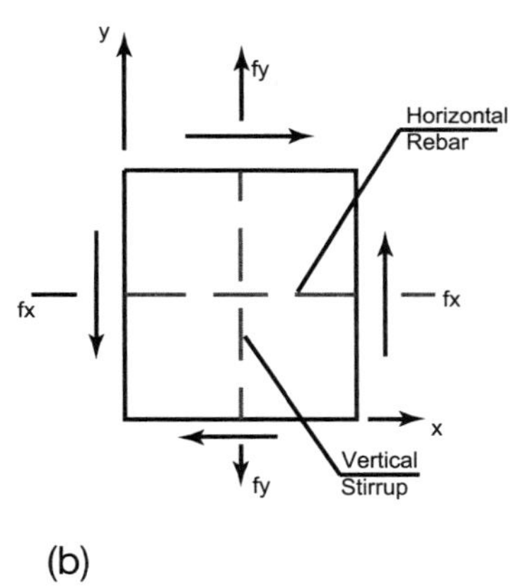

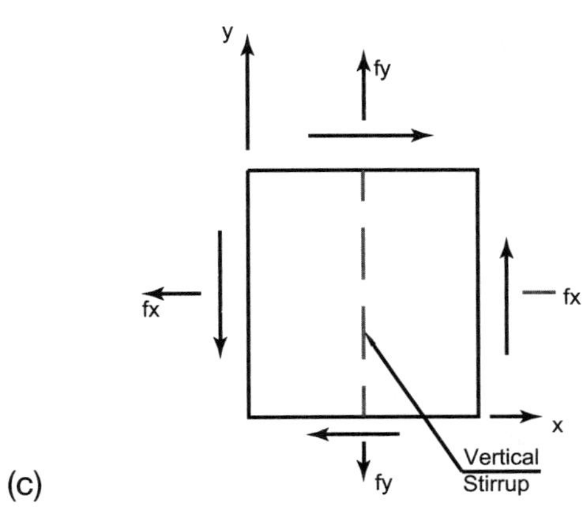

Figure 4.20 (a) Idealization of beam, (b) element-2, (c) element-3.

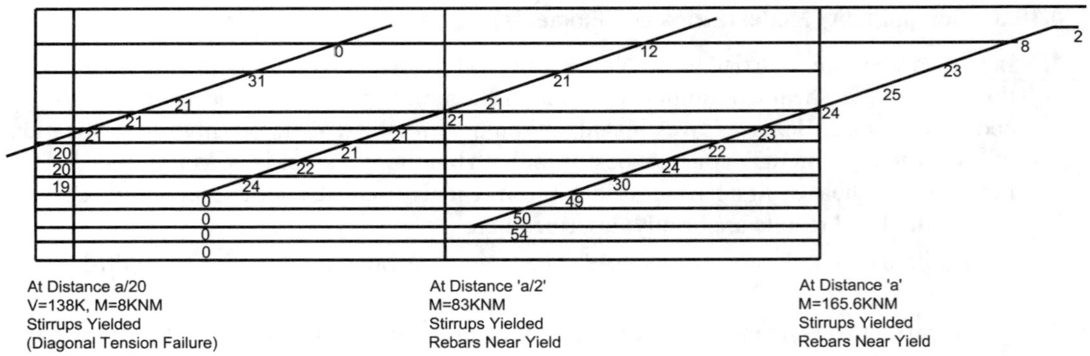

Figure 4.21 Distribution of angle of inclination θ at failure.

Computer results for shear failure from the above model were compared with experimental results from high strength beams tested by four universities, Cornell, North Carolina, Toronto and Imperial College London. The major outcome was that the values of θ range from 20° to 48° in the beams and unlike ACI model do not have a constant angle of 45° (Figure 4.21).

Strength comparisons: The ultimate shear capacity of the beam is considered to have been reached when one of the following conditions is met:

Capacity of stirrups in tension is exceeded, i.e., $\epsilon_{sy}/\epsilon_{yld\geq 1}$

Capacity of steel in tension is exceeded, i.e., $\epsilon_{sx}/\epsilon_{yld\geq 1}$

Various notations are described in papers by Khan et al, listed in the bibliography.

High strength lightweight concrete beams with stirrups failed in shear in diagonal tension, as would be expected for beams without stirrups.

Influence of parameters on beam behavior: It is observed that the ratio of $V_{ntheory}/V_{nACI}$ capacity of two of the beams tested at Imperial College is below 1.0. The a/d ratios of such beams are between 2 and 4.

Effect of ϵ_o:

ACI equation for stress-strain relation is $\epsilon_o = 0.00195 + 1.03 \times 10^{-4}$

$\epsilon_0 = 0.002 + 0.00012\ (f_c' - 3),\ f_c' \geq 3$ ksi (20.8 Mpa)

The major parameters are:

ϵ_o the compressive strain at the peak stress

f_{cr} the stress at which tensile fracture of concrete occurs.

The ACI equation gives a higher value of ϵ_0 and thus overestimates the failure strain.
Conclusions of theoretical and test beam studies of shear failure:

1. The results confirm the variable angle truss model theory for beams failing in shear.
2. The effect of concretes having different tensile stresses f_{cr} can be significant on shear capacity of beam, concrete stresses, and steel strains. Like f_c', f_{cr} is an important parameter.
3. Load capacities using the MCFT for high strength lightweight aggregate concrete beams show lesser agreement with experimental values. By using actual stress-strain curves in compression and tension for the lightweight aggregate concrete, it is expected that differences will be smaller.
4. As f_c' values increase, ACI code shear capacities get more conservative.
5. It is seen that cracking planes may be located at angles as low as 20 degrees. Due to changes in the orientation of cracking angles, stress patterns are modified.
6. Effect of shear deflection also needs to be included at cracking stage in future studies.
7. Accurate evaluation of modulus of elasticity of concrete is important since Ec values are

Table 4.9 Specifications of selected beams failing in shear.

Beam ID	a/d ratio	f_c' Mpa	ρSx	ρw	$\epsilon_0 \times 10^{e-3}$	$f_{cr} \times f_c'$ Mpa
Cornell G5	4.0	40.1 (5.8)	0.025	0.0017	2	13.8 (2)
Cornell G6	4.0	20.8 (3.0)	0.025	0.0017	2	13.8 (2)
Toronto SK3	1.8	28.4 (4.1)	0.046	0.0046	2	27.6 (4)
Imperial College IC2	3.4	38.2 (5.51)	0.030	0.0062	2	17.3 (2.5)
Imperial College IC4	2.27	38.2 (5.51)	0.030	0.0062	2	17.3 (2.5)
Cornell G4	4.0	63.0 (9.1)	0.033	0.0017	2	13.8 (2)
N. Carolina LRNS2.59	2.59	53.9 (7.79)	0.0145	0	2	15.9 (2.3)
N. Carolina LRNS3.63	3.63	52.3 (7.56)	0.0145	0	2	15.9 (2.3)

Note: Figures in brackets are ksi units.

Table 4.10 Shear strength of beams compared with test results and ACI code.

Beam ID	V_n Theory KN	V_n ACI KN	V_n Test KN	V_n Test/V_n ACI	V_n Theory/V_n ACI	Mode of Failure
G4	122.8 (27.5)	97.6 (21.86)	150.4 (33.7)	1.54	1.26	Diagonal tension
G5	96.0 (21.5)	84.8 (18.99)	113.4 (25.4)	1.34	1.13	Diagonal tension
G6	75.4 (16.9)	71.7 (16.06)	79.5 (17.8)	1.11	1.05	Diagonal tension
SK3	600 (134.4)	459.4 (102.9)	600 (134.4)	1.31	1.31	Diagonal tension
LRNS2.59	26.8 (6.0)	18.3 (4.1)	26.8 (6.0)	1.46	1.46	Diagonal tension
LRNS3.63	21.9 (4.9)	18.3 (4.1)	21.9 (4.9)	1.19	1.19	Diagonal tension
IC2	28.1 (6.3)	34.2 (7.67)	28.1 (6.29)*	0.82	0.82	Diagonal tension

used for calculating bending moments, shear forces, and deflections. Comparative studies for evaluating peak concrete strains and tensile strength of concrete have been made.

8. Critical section analysis can be used for evaluating reserve capacities and ultimate strengths of existing reinforced concrete and prestressed beams, which are being subjected to heavier loads than at the time of original design. Thus, failure of existing bridges on heavy traffic routes can be prevented.

ACI Code Sec. 11.5.5.3 recommends minimum shear reinforcement,

$$A_v = 50\, b_w \cdot s/f_y \text{ for } \sqrt{f_c'} < 0.69 \text{ MPa (100 psi)} \tag{4.43}$$

An earlier ACI Code Sec. 11.1.2.1 states that the full value of $\sqrt{f_c'}$ can be used for concrete strengths greater than 69.2 MPa (10 ksi) if the minimum shear reinforcement

= (f_c'/5000) × minimum web reinforcement (but no more than three times the amounts required by Section 11.5.5 is provided).

For $f_c' = 15$ ksi, the stirrups required would be three times higher, compared to stirrups area required for the range $f_c' < 10$ ksi.

It is recommended that the above formulae should be based on the important parameters a/d, f_c', f_{cr}, and longitudinal and transverse reinforcement percentages, and they are therefore approximate.

9. The modified compression field theory can be used to predict the ultimate load behavior of concrete beams for shear design. The MCFT with variable angle truss model is applicable to both normal and HSC and comparisons with FEM analysis and laboratory test results for beams (with stirrups) show good agreement.

The following formula may be used:

$$\theta = \tan^{-1}\sqrt{\frac{f_d - f_L}{f_d - f_t}} = \tan^{-1}\sqrt{\frac{f_{dt} + f_t}{f_{dt} + f_L}} \tag{4.44}$$

4.16.4 Fracturing Truss Model as an Alternate to MCFT

Recent research results reported by Bazant and Shah utilize a two-parameter fracture model, crack band, and fictitious crack models. Size effects and bond effects can be studied from nonlinear elastic fracture mechanics models using fracture-based finite element codes.

In the FTM, shear strength is based on fracture toughness, and it is assumed cracks are distributed throughout the concrete web. This assumption is more consistent with actual beam failure than the MCFT assumption that cracks are confined to critical sections only.

In addition to common parameters in the two methods, such as f_c' and E_c', the strength of beam in FTM depends on:

1. Fracture energy resulting from aggregate size.
2. Proportion of specimen dimensions.
3. Geometric similarity of beam shapes.

Also, shear strength is a function of bond effects, size effects, and the compression failure in inclined compression struts. Size effects and bond effects have practical importance in the design of concrete beam sections. FTM has the potential to be a powerful tool in future design codes.

4.17 SHEAR DESIGN FOR REINFORCED CONCRETE AND PRESTRESSED BEAMS

4.17.1 Shear Design for the Lifting and Erection of Prestressed Concrete Beams

A dynamic allowance factor for wind gust is applied for estimating the crane capacity. Two cranes may be required for lifting. The capacity of beam cross section, hooks, and anchorage length inside concrete needs to be checked using MCFT. A program may be developed in

Figure 4.22 Photo by author of the single crane lifting of a heavy prestressed concrete box beam.

Mathcad or Excel® spreadsheets. For precast concrete slab and beam bridges manufactured in factory conditions, the lifting and erection weights can be extremely high, and biaxial bending of beams and overhangs need consideration using finite element modeling.

For lifting and erection, the beam is non-composite and in some cases concrete has not achieved its 28-day strength. Shear and bending strength would be much lower than for composite action with the deck slab. Usually four lifting points are required. The criteria is to balance the vertical shear forces at lifting points, and also the centroid of dead load and the lifting points must coincide for stability.

4.17.2 Shear Design of Concrete Beams

A major revision to shear design deviating from the old Ritter-Morsch approach was recommended by AASHTO LRFD code. The method has been incorporated in several computer software programs such as PSLRFD prepared by PennDOT and will not be repeated here in detail.

The seven steps required as given below:

For values of parameters θ versus v/fc′ (using plots of distribution of ϵx) and β vs. v/fc′ (using plots of distribution of ϵx), refer to AASHTO Figure 5.8.3.4.2-1.

The LRFD code method is summarized as follows:

Step 1

Calculate M and V from elastic analysis

Calculate factored M_u and V_u

Use load factors for strength I limit state

Dead load factor = 1.25

Use with DC1 (deck slab DL) and DC2 (parapet DL)

Dead load factor for wearing surface = 1.5 use with DW

Live load + Impact factor = 1.75

Calculate V_u and M_u for 1.25(DC1 + DC2) + 1.5 DW + 1.75 (LL+I)

$a = A_s f_y / 0.85 f'_c b$

Check if $a < h_f$

Calculate maximum value of (lever arm) d_v

$d_v = d_e - a/2$, d_e is depth of beam (from top of deck to center line of rebar)

$d_v = 0.9 d_e$ or $d_v = 0.72$ h (AASHTO Section 5.8.2.7)

Step 2

$\phi_v = 0.9, V_p = 0$

$v = V_u / \phi_v b_v d_v$ (AASHTO Section 5.8.3.4.2-1)

calculate v/f'_c, $V_p = 0$

Step 3

Assume θ = 40 deg. (AASHTO Section 5.8.3.4.2-2)

$$\epsilon_x = \frac{M_u/d_v + 0.5\, V_u \cot\theta}{E_s A_s} \leq 0.002$$

Step 4

Use iterative process to calculate θ.

Using v/f'_c and ϵ_x, determine β and θ from AASHTO Fig. 5.8.3.4.2-1. If θ is different from assumed value, use the new value of θ and repeat.

Step 5

$V_n = V_c + V_s + V_p$ (AASHTO Section 5.8.3.3-1)

$V_c + V_s = 0.25\, f'_c\, b_v\, d_v + V_p$ (AASHTO Section 5.8.3.3-2)

$V_c = 0.083\, \beta \sqrt{f'_c}\, b_v\, d_v$ (AASHTO Section 5.8.3.3-3)

$V_s = A_v\, f_y\, d_v (\cot\theta + \cot\alpha) \sin\alpha / s$ (AASHTO Section 5.8.3.3-4)

$V_s = V_u/\phi_v - 0.083\, \beta \sqrt{f'_c}\, b_v\, d_v$

Step 6

If $A_v = 200\ \text{mm}^2$

Calculate spacing from three conditions and select minimum

1. $s \le A_v\, f_y\, d_v \cot\theta / V_s$

Check for minimum spacing required by AASHTO

$s \le A_v\, f_y / 0.083 \sqrt{f'_c}\, b_v$ (AASHTO Section 5.8.2.5)

2. $s \le 0.8\, d_v \le 600$ mm if $V_u < 0.1\, f'_c\, b_v\, d_v$

$s \le 0.4\, d_v \le 300$ mm if $V_u \ge 0.1\, f'_c\, b_v\, d_v$

Step 7

Check adequacy of longitudinal reinforcement. If reinforcement given by AASHTO Section 5.8.3.5 is not satisfied, increase longitudinal or transverse reinforcement (stirrups).

$A_s\, f_y \ge M_u/d_v\, \phi_f + (V_u/\phi_v - 0.5\, V_s) \cot\theta$

Increasing V_s to satisfy

$V_s \ge 2\, (V_u/\phi_v - [A_s\, f_y - M_u/d_v\, \phi_f] \tan\theta)$

$s \le A_v\, f_y\, d_v \cot\theta / V_s$

$$\text{Increase } A_v \ge \frac{M_u/d_v\, \phi_f + (V_u/\phi_v - 0.5\, V_s) \cot\theta}{f_y}$$

BIBLIOGRAPHY

Analysis and Design

ACI 318-89

Barker and Puckett, *Design of Highway Bridges: An LRFD Approach*, 2nd Edition, J. Wiley & Sons, 2006.

Benham, P. P. and F. V. Warnock, *Mechanics of Solids and Structures*, Pitman Publishing, 1973.

Boresi, A. P. and R. J. Schmidt, *Advanced Mechanics of Materials*, 6th Edition, J. Wiley & Sons, 2002.

Bridge Design Guidelines, Florida Department of Transportation, 2005.

Bridge Design Manual, New Jersey Department of Transportation (Sections 45 and 46 developed by Mohiuddin Ali Khan).

Bridge Design Manual, 2005, New York State Department of Transportation.

Bridge Design Manual, 2005, Ohio Department of Transportation.

Brockenbrough, R. L. and B. G. Johnston, *Steel Design Manual*, U.S. Steel Corporation.

Design Manual, Part 4, Volume 1, Structures, Pennsylvania Department of Transportation.

International Handbook of Earthquake Engineering: Codes, Programs, and Examples (IHEE), Kluwer Academic Publishers, 1995.

Khan, Mohiuddin Ali, "Elastic Composite Action in Slab-Frame Systems," *The Structural Engineer*, London, 1970.

Khan, Mohiuddin Ali and K. O. Kemp, "Elastic Full Composite Sction in Slab-Beam Systems," *The Structural Engineer*, London, 1969.

Khan, Mohiuddin Ali, Edwin Rossow and S. P. Shah, "Behavior of Reinforced and Prestressed High Strength Concrete Beams," NED University, Karachi, *Journal of Engineering Research*, Vol. 1, No. 1, pp. 1–29, January 1994.

Khan Mohiuddin Ali, Edwin Rossow and S. P. Shah, "Shear Design of High Strength Concrete Beams," *Proceedings of ASCE/SEI Structures Congress*, Philadelphia, PA, 2000.

Park, Sung H., *Bridge Superstructure Design and Rehabilitation*, Trenton, NJ, 2005.

Structural Standards Manual, 2006, Michigan Department of Transportation.

Timoshenko, S. P. and S. Woinowsky-Kreiger, *Theory of Plates and Shells*, McGraw-Hill, Inc., 1959.

Wasti, S. T. and A. C. Scordelis, "Comparative Structural Behavior of Straight, Curved, and Skew Reinforced Concrete Box Girder Bridge Models," *Analysis and Design of Bridges*, NATO Advanced Study Institute, Izmir and Middle East Technical University, Ankara, Turkey.

Xanthakos, Petros P., "Theory and Design of Bridges," John Wiley & Sons, New York, NY, 1993.

5

Load and Resistance Factor Rating and Redesign

Section 2
Strengthening and Repair Work

5.1 RATING AND REDESIGN METHODS

5.1.1 Rehabilitation, Replacement, and Widening of Structural Components

For rehabilitation of bridge components, rating is followed by redesign or replacement design. Dead load analysis is straight forward. However, live load analysis is complicated due to:

1. Moving loads
2. Dynamic impact factor
3. Vibrations and variable deflections
4. Live loads combined with environmental loads.

Both rating and design methods have undergone fundamental changes with the application of more refined ultimate load theories. In this chapter, an overview of the latest AASHTO recommended design methods are reviewed and presented to assist inspection and design engineers. Salient features such as many load combinations are explained in detail.

Limit state methods for rating (LRFR) and design (LRFD) are applied for structural evaluation of capacity and for redesign of members. AASHTO formulae are summarized for computation. Detailed load combinations for strength, serviceability, and extreme load conditions are given in tabular forms. In addition, construction load combinations and effects of non-symmetric pouring sequence on girder stress are developed and presented. Live load deflection criteria as opposed to stress criteria are reviewed. Suggested computer software for use is listed, as well.

There are several methods of structural evaluation of a bridge (rating). Both theoretical and physical methods are being used:

1. Live load rating based on LRFR methods.
2. Non-destructive static tests (such as placing sand bags with known loads).
3. Moving load tests and field observations (using calibrated vehicles): Structural capacities and loadings are used to analyze the critical members and determine the appropriate load rating. Load rating by load testing may be feasible in special cases such as:
   - When analytical results provide a posting or operating rating factor <1, but the bridge is otherwise showing no visual signs of distress
   - When construction plan records for the bridge are not available
   - When the bridge is of a special type that cannot be analytically rated.

   Load testing is performed by driving a truck of known axle weights over a bridge. Stresses are then measured in the load-carrying members with strain gauges and specially designed data analysis equipment. These axle weights and actual measured stresses are used to calibrate the computer input data.
4. Live load vehicles vary in intensity (Table 5.3), such as:
   - First level standard vehicles (HL-93).
   - Second level legal loads: AASHTO or state specified legal trucks are used. Weigh stations are common on important routes in the U.S. to ensure that no posted weight is exceeded.
   - Third level permit loads: Permit vehicles are two to three times heavier than HL-93. Permits are issued for a limited number of crossings or for the whole year. Heavier vehicles need to be escorted.
   - For complex bridges or those with non-structural members, evaluation of combined performance may not be accurate. The performance of existing components such as girders needs to be checked against different levels of live loads and rated using a common rating factor.
5. Using engineering judgment: For bridges with no engineering plans available, it is not easy to field measure and evaluate every structural detail. Stone masonry and concrete bridges provide problems since reinforcing, prestressing, or encased steel details are not obvious. A balance needs to be maintained so that load estimates are not rated too high (for safety) or too low (to disallow use by permit vehicles). A criteria needs to be developed based on engineering judgment and experience. Alternatively, expensive NDT techniques can be applied. Computation methods using theoretical vehicles are the least expensive.

5.1.2 Rating Factor

In the past, a rating factor was based on the LFD method as follows:

$$RF = (C - DA_1) / A_2 (L + I)$$

C = Capacity of member

D = Dead weight effects

L = Live load effects

I = Impact factor

A_1 = Factor for dead load

A_2 = Factor for live load.

RF for LFD is expanded as the general LRFR equation given below (refer to AASHTO Manual for Condition Evaluation and Load and Resistance Factor Rating of Highway Bridges, October 2003):

$$\text{Rating factor RF} = [C - (\gamma_{DC})\,DC - (\gamma_{DW})\,DW + (\gamma_P)\,P] / (\gamma_L) / (LL + IM) > 1$$

Capacity $C = \varphi_c \varphi_s \varphi R_n$ where $\varphi_c \varphi_s > 0.85$

φ_c = Condition factor

φ_s = System factor

φ = Resistance factor.

For service limit state $C = f_R$ (allowable stress specified in LRFD specifications):

DC = Dead loads

DW = Dead load of wearing surface and utilities

γ_{DC} and γ_{DW} = LRFD load factors, replacing A_1 factor for dead load in LFD rating equation

γ_L = Evaluation live load factor, replacing A_1 factor for live load in LFD rating equation

P = Permanent load other than dead load

LL = Live load effect

IM = Dynamic load allowance.

Rating is calculated in tons and standard truck weight is multiplied by rating factor to evaluate capacity.

$$RT = W \times RF$$

If the rating factor is < 1 the following measures are required:

1. Load posting using AASHTO Manual on Uniform Traffic Control Devices (MUTCD): When RF is between 0.3 and 1, safe posting load = (RF − 0.3) W/0.7. RF < 0.3, a lower legal load vehicle is posted.
2. Strengthening: When RF < 0.3 for the three AASHTO specified legal trucks, the bridge is due for strengthening or partial replacement.
   - Girder replacement: If it is not feasible to strengthen, girders need to be replaced with deck slab and parapets.
   - Bridge replacement: If substructure requires major repairs, the entire bridge needs to be replaced.
   - Strengthening and replacement of members requires detailed design for all possible environmental loads and hence many more design load combinations are applicable than for the load rating.
   - Rating is followed by design. LRFD computer software computes and prints out ratings in addition to design results. Hence, rating and design usually go hand in hand.
   - Live load deflection control such as L/800 is not be applicable for rating.
   - Wind loads, temperature effects, and earthquakes are not considered for load ratings.
   - Loads on substructures are both vertical and horizontal. Any redesign of substructures may require applying all AASHTO LRFD load combinations.

5.1.3 Types of Ratings

Load rating analysis of bridges is performed to determine the live load that structures can safely carry. Bridges are rated at three different stress levels, referred to as:

1. Inventory rating: Inventory rating is the capacity rating for the vehicle type used in the rating that will result in a load level which can safely utilize an existing structure for an indefinite period of time. Inventory load level approximates the design load level for normal service conditions.

2. This value is typically used when evaluating overweight permit vehicle moves. Operating rating: Operating rating will result in the absolute maximum permissible load level to which the structure may be subjected for the vehicle type used in the rating. This rating determines the capacity of the bridge for occasional use. Allowing unlimited numbers of vehicles to subject the bridge to the operating level will compromise the bridge life.
3. Posting rating: The posting rating is the capacity rating for the vehicle type used in the rating. It is a load level which may safely utilize an existing structure on a routine basis for a limited period of time. The posting rating for a bridge is based on inventory level plus a fraction of the difference between inventory and operating.

   This may lead to load restrictions of the bridge or identification of components that require rehabilitation or other modification to avoid posting of the bridge.
4. Preparing a load rating report: A load rating report shall include the following:
   - Material properties
   - Loading assumptions
   - Plans or sketches showing all properties and assumptions
   - Printout of computer data file(s) where appropriate
   - Documentation of structural model used in analysis
   - Inventory, operating, and posting summary for HS-20 and all legal loads
   - Electronic copies of data file(s).

5.1.4 Reduced Rating—Allowable Strengths of Materials

For old bridges, where possible coupons obtained from the bridges may be tested in a laboratory:

1. In the absence of coupon tests, for wrought iron girders, strength in tension and bending for inventory rating is 14,600 psi and operating rating is 20,000 psi.
2. For concrete, maximum allowable bending stress may be reduced as per Section D.6.6.2.4. and D.6.6.3.2 of the Manual of Condition Evaluation and LRFR.

Upon completion of replacement or rehabilitation of a posted structure, the need for load restriction no longer exists. The Bridge Management Section would then prepare a "Removal of Load Restriction Resolution."

5.1.5 Rating Live Loads

Bridge capacity depends upon bridge geometry, material strength, condition, structure type, etc. As related to trucks, a bridge's capacity depends not only upon the gross weight, but also upon the number and spacing of axles and distribution of load between the axles.

Since it is not practical to rate a bridge for the countless axle configurations, bridges are rated for six standard vehicles (including HS-20) which are representative of actual vehicles on the highways.

5.1.6 Analytical Steps in Load Rating

The following analytical steps are required to determine load rating:

1. Determine section properties.
2. Determine allowable and/or yield stresses.
3. Calculate section capacities.
4. Determine dead load effects.
5. Calculate dead load portion for section capacity and live load effect.
6. Calculate live load impact and distribution, and allowable factor.

The stress levels used to analyze critical members and determine an appropriate inventory and operating rating are outlined in AASHTO's Manual for Condition Evaluation of Bridges.

5.1.7 Selection of Members for Rating

1. It is transportation policy to rate only the primary load-carrying members in a bridge. For superstructure, this is normally the slabs, girders, trusses, or arch ring.
2. Concrete box culverts are rated as rigid frames.
3. Not included in the load rating are the piers, abutments, and foundations. The condition of these elements shall be considered, and they shall be assumed to safely carry the loads transmitted to them unless there is evidence of serious deterioration.

5.1.8 Restricted Live Load Posting

The Bridge Management Section implements load restrictions by preparing a "Load Restriction Resolution." The Bridge Management Section then informs public officials about the intended load restrictions. Those receiving information include local fire companies, school transportation directors, the local authority for regional transport, senators, and representatives.

When a bridge is not able to safely carry the loads allowed by state statute, it is posted for its reduced capacity. The policy of most agencies is to restrict loads on bridges when the posting-rating factor drops below 1.0 for any of the state legal truck loads. The minimum posting is 3 tons.

5.1.9 Analytical Differences between Design and Rating

1. Rating is based mainly on live loads. Available live load capacity of a bridge can be compared with various levels of truck weights such as:
   - AASHTO defined HL-93 and legal loads and state defined permit loads or
   - County defined reduced truck loads and maximum legal loads.

   For selecting live load trucks for rating, several lighter truck loads such as H-20 and ML-80 are also considered. A port authority may use special heavy intensity truck loads on its local bridges for high density loading on freight trains.
2. Rehabilitation covers both rating and redesign aspects. In fact, redesign of an existing member is dependent upon the rating. Dead load analysis is common to both design and rating. In both cases basic consideration is the effective use and application of construction materials, namely steel, reinforced concrete, prestressed concrete, or timber. Replacement or rehabilitation of bridges is based on detailed analysis and design which includes checking the strengths of members and connections for a variety of primary and secondary effects including extreme events.
3. Deflection control is not an essential part of the rating but may be required for rehabilitation and redesign.
4. For important bridges, seismic vulnerability and scour sufficiency may be evaluated on a case-by-case basis, and bearings, piers, or foundation retrofit may be implemented. Service load combinations are for deflection control and crack control.
5. Due to the small probability of occurrence, bridge girders are not rated for wind, temperature, or extreme events, although they are retrofitted in detail for seismic forces, accidents, and scour.
6. Only steel girders are rated for fatigue.
7. Timber girders are only rated for strength I and II but not for service load.
8. Ultimate loads and service load combinations will be based on the following:

   Strength I: Relating to normal use without wind

   Strength II: Permit vehicles without wind

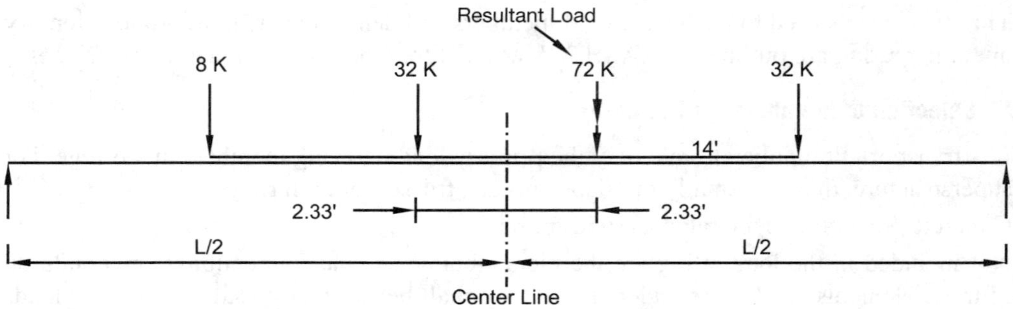

Figure 5.1 Position of 3-axle deflection truck for maximum deflection.

Strength III: Wind velocity exceeding 90 KM/h (no significant LL)

Strength IV: Very high DL/LL ratio for large span bridges

Strength V: Normal use with wind of 90 KM/h velocity

Extreme Events (major earthquake, flood, collision of a vessel or vehicle, and ice flow)

Extreme Event I: Load combination relating to earthquake is determined on a project-specific basis.

Extreme Event II: Ice load, collision by vessels and vehicles with reduced live load.

9. Service limit state:

Service I: Normal use with wind 90 KM/h – deflection and crack control

Service II: Related only to steel - control yielding and slip in connections

Service III: Related only to tension in prestressed concrete.

10. Fatigue limit state:

Stress range, f = 1.0

Fatigue loads for steel girders: A load factor of 0.75 and trucks at lower weight than the HL-93 design truck will be used.

11. User defined construction loads: Construction loads with appropriate load factors shall be considered for short-term live loads.

12. Use of a deflection design truck to limit maximum deflections: A standard deflection rating vehicle is used. AASHTO LRFD code has prescribed a moving 72 kips deflection vehicle (a variation of the basic HS-20 truck) to compute maximum live load deflection (Figure 5.1).

5.2 UTILIZING ULTIMATE LOAD BEHAVIOR OF MATERIALS

5.2.1 Introduction to Load and Resistance Factors in Rating and Design

1. While allowable stress design (ASD) is based on use of a safety factor, LRFD uses both load factors and resistance factors. It captures much of the post elastic behavior for computing member sizes. Load factors are based on probability theory and also resistance factors from material behavior observed during laboratory testing.

Primary analysis remains unfactored and is modified by applying appropriate load and resistance factors. Also, the history of stress development from no load to full load is maintained, as small strains are assumed during both elastic and post-elastic ranges.

2. The load and resistance factors are calibrated from actual bridge statistics to ensure a uniform level of safety. Hence, transitioning to load and resistance factor is to produce consistent levels of reliability across a broad range of designs. It ensures that by applying appropriate conservatism where needed, significant cost savings can be realized. AASHTO LRFD Specifications 2007, Section 4 recommends the use of the LRFD method based on plastic theory and ultimate loads. LRFD incorporates state-of-the-art analysis and design methodologies with load and resistance factors based on the known variability of applied loads and material properties.

3. The plastic theory is a direct approach since the failure of a beam or column will occur by the yielding of steel or crack formations in concrete. Compared to the yield point limit for applied load used in the elastic theory, most materials are capable of resisting even higher loads beyond yield and prior to failure. To a great extent, the true material behavior can be represented by applying the plastic theory. Redistribution of stresses in the member and formation of plastic hinges prior to failure lead to an accurate determination of reserve capacity of the member and ultimate loads.

5.2.2 Applicable AASHTO Codes

All construction activities associated with rehabilitation require planning, analysis, and compliance with design codes. In the U.S., the entire highway inspection, monitoring, repair, and rehabilitation of bridges and highway structures is governed by the provisions laid down in the following codes:

1. Manual for Condition Evaluation of Bridges, 2nd Edition, AASHTO, 2000 using load factor design (LFD) *(for rating analysis)*.
2. Manual for Condition Evaluation and Load and Resistance Factor Rating (LRFR) of Highway Bridges, 1st Edition, AASHTO 2003, Washington D.C., ISBN: 1-56051-283-0 using load resistance factor design *(for rating analysis; supersedes LFD)*.
3. AASHTO LRFD Bridge Design Specifications, 2007 *(for redesign)*.
4. AASHTO LRFD Bridge Construction Specifications, 2nd Edition, 2006 Interim Revisions *(for reconstruction)*.
5. The state or agency technical specifications covering the type of materials, the method of measurements and payments, etc.

5.2.3 History of Design and Rating Specifications

In the U.S., there has been a progressive evolution of structural design codes. Existing bridges have been constructed based on one of the following codes:

- Original AASHO code.
- AASHTO specifications.
- AASHTO Manual for Condition Evaluation.
- AASHTO English standard specifications.
- AASHTO metric standard specifications.
- AASHTO LRFD specifications.
- AASHTO LRFR Condition Evaluation Manual.

1. Additional rules are usually supplemented by a state bridge design manual.

AASHTO first introduced the Standard Specifications for Highway Bridges (Standard) in 1931, and since then these specifications have been updated through 19 editions. The last edition was published in 2007. The standard specifications were based on the allowable stress design (ASD) philosophy until 1970, after which the load factor design (LFD) philosophy was incorporated into the specifications.

Earlier methodologies provide a desirable level of safety for bridge designs, but do not ensure a uniform level of safety. AASHTO introduced the LRFD specifications in 1994 to bring consistency to the design safety level. AASHTO also adopted the Guide Manual for Condition Evaluation and Load and Resistance Factor Rating (LRFR) of Highway Bridges in 2002, which incorporates the same philosophies into the rating of bridge conditions.

2. The uniform service levels and bridge reliability resulting from using LRFD should ensure superior serviceability and long-term maintainability. These specifications were calibrated using structural reliability techniques that employ the probability theory. New specifications are probability-based and incorporate serviceability design while accounting for "extreme events" such as ship collisions and earthquakes. They can help produce more reliable structures.
3. Significant differences between the standard and LRFD specifications include the live load model, the dynamic load (impact) factor, live load distribution factors (DF), and the load combinations. The standard specifications use the greater of the HS 20-44 truck or the design lane loading for live load. The LRFD specifications use the HL-93 model truck.
4. The standard specifications provide a simple expression for DF for girder bridges in S/D format, where, for example, D = 5.5 for a bridge constructed with a concrete deck on prestressed concrete girders carrying two or more lanes of traffic, and S is the girder spacing in feet. The effects of various parameters such as skew, continuity, and deck stiffness are ignored. There was a need to develop DF formulas that apply to a broad range of beam and slab bridges. The live-load DF formulas in the LRFD specifications consider the effects of different parameters, including skew, deck stiffness, and span length.
5. The fatigue and fracture limit state, which applies primarily to steel bridges, is not applicable to the design of wood components under current design practices. Provisions of the LRFD specifications that apply are:

    - Use of load and resistance factors.
    - Inclusion of a dynamic load allowance for static truck loads.
    - New live load deflection criteria.
    - Revised load combinations.
    - Live load distribution requirements.
    - New values for material strength (base resistance).
    - The extreme event limit state, which is intended to ensure structural survival in major earthquakes, floods, and vehicle collisions, will generally not control the design of most wooden bridges. For wood bridge design, the most applicable of these limit states are the strength limit state, which is intended to ensure that the structure will provide the required strength and stability over the design life, and the service limit state, which restricts stress and deformation under regular service conditions.

6. Concrete bridges: The majority of U.S. bridges in the past were designed using AASHTO standard specifications. Several differences exist between the standard and LRFD specifications with respect to ultimate strength design in prestressed concrete beams:

    - The LRFD load factors for the ultimate flexural strength design load combinations, strength I, are less than those provided by the standard specifications.
    - Overall, the LRFD shear design provisions were found to yield conservative results as compared with the standard specs. LRFD shear design provisions overestimate the shear strength of girders with span-to-depth ratios of 1.5 and below and underestimate the shear strength of girders with span-to-depth ratios of 2.0 and above.
    - The design of an AASHTO Type III girder bridge design was similar in most respects for both the specifications. The most significant changes observed were in the shear design, where the LRFD skew factor and reinforcement requirements increased the required concrete strength and reinforcement.
    - In a study presented at the 2006 Concrete Bridge Conference, Adil, Hueste, and Keating investigated that the LRFD designs tend to have a slight reduction in the maximum span

length for AASHTO Type IV girders. LRFD specifications required a larger number of strands for the Type IV girders. The differences in the required number of strands for LRFD versus standard design increase with an increase in span length and decrease with an increase in girder spacing and skew angle.

7. For steel bridges, the behavior of homogeneous and isotropic types of steel is well defined compared to that of timber or concrete. The fabricated structural steel industry is more comfortable with the use of LRFD as the preferred specification. The reasons are the ongoing advancements in strength-design-specific areas such as composite systems design, design of systems using partially restrained connections, and seismic design. The advances in steel design have simply been applied entirely within the context of the standard specifications. The LRFD specifications help to eliminate some of these limits.

   The theory of plasticity can be expressed in terms of a limit state for material behavior. The limit state specified in AASHTO LRFD was originally introduced in British standards following the development of the plastic hinge theory for steel frames by J. F. Baker at the University of Cambridge and for reinforced concrete frames by A. L. L. Baker at Imperial College London.
8. Composite beam design: A simple comparison between the LRFD and ASD methods illustrates the logical and rational basis of the LRFD strength model. Because concrete inherently behaves inelastically, a composite section does the same. Although the ASD approach provides a safe design for common loadings, it can be seen that the assumed elastic neutral axis is incorrect when the failure occurs in the inelastic range. Inelastic behavior makes it inconsistent or inefficient to use a model that is based upon elastic assumptions.
9. Wood or timber bridges are still being designed and built, especially in the nation's federal lands and parks.

5.2.4 Limit State Applied to Design and Rating

It may be defined as a state or condition in which a partial collapse or total collapse may occur due to flexure or shear failure (LRFD/LRFR strength limit state). The component material then ceases to meet the provisions for which it was designed.

For general bridge design, four limit states are defined:

1. Strength.
2. Serviceability.
3. Fatigue and fracture.
4. Extreme events.

An important goal of a designer is to prevent a limit state from being reached.

5.2.5 Salient Features of the New LRFD/LRFR Method

1. Changes in basic vehicular loads such as the use of HL-93 and tandem vehicles.
2. Permit vehicles shall be used for long-term live loads.
3. Use of the probability theory for equations that predict strength, workmanship, quality control, and consequence of failure to account for uncertainties in load combination.
4. A variety of distribution factors to simplify the theoretical approach.
5. Resistance calculations for strength I to V and serviceability I to III will be based on correction factors such as $f < 1.0$ to account for uncertainties in material properties.
6. Fatigue and fracture: Applications of fracture mechanics theories in design are being introduced. A composite reinforced concrete section may behave inelastically even at small loads.

7. Scour design (local or contraction scour): For new bridges, scour depth as calculated may require deep foundations. For existing foundations subject to scour, adequate countermeasures such as sheet piling or stone riprap protection of the riverbed and foundations are required.
8. Seismic design.
9. Modified compression field theory for shear design of concrete.
10. Use of segmental concrete bridges for longer spans.

5.2.6 Experimental Verification of Plastic Behavior

1. Experimental verification of plastic behavior is desirable since the number of parameters affecting failure is large. Both the yield line theory for slabs and the plastic hinge theory for beams and columns are based on experimental verification.
2. For more complex structural systems, the application of the plastic theory is not fully developed, and approximate applied factors are assumed.
3. When the plastic theory is used, it is important to anticipate the failure mechanism, which should be based on observed response of extreme force effects. The location of plastic hinges in a theoretical model should coincide with those observed in well-instrumented experimental laboratory models or observed in the field after seismic events.
4. If flexural failure is identified, failure should not occur due to deficient shear design, buckling, or bond between concrete and steel.

5.2.7 Use of in-Built Ductility in Materials

The response of material beyond the elastic limit can be either ductile or brittle. Ductile behavior can be defined as members displaying significant inelastic deformation and energy dissipation, without any loss of load carrying capacity. Failure is not sudden, and there is warning in terms of excessive deformations prior to collapse.

Brittle behavior implies sudden loss of load carrying capacity after elastic limit is reached. To prevent failure, the system should have more ductility as opposed to more brittleness. Greater ductility will lead to greater economy in design.

5.2.8 Comparisons of LRFD Design Method with ASD and LFD Methods

Using appropriate load factors, strength load combinations for typical dead and live loads may only be expressed as:

$$\text{ASD} - 1.0\ \text{DL} + 1.0\ (\text{LL} + \text{I}) \leq \text{Resistance safety factor}$$

$$\text{LFD} - 1.3\ \text{DL} + 2.17\ (\text{LL} + \text{I}) \leq \varphi\ R_n$$

$$\text{LRFD} - 1.25\ \text{DC}_1 + 1.25\ \text{DC}_2 + 1.50\ \text{DW} + 1.75\ (\text{LL} + \text{I}) \leq \varphi\ R_n$$

5.3 LRFD SERVICE LOAD REQUIREMENTS

5.3.1 Deflection Control

1. AASHTO LRFD live load deflection criteria is based on arbitrary limits:
   Minimum single span length = 20 feet. Maximum single span length = 300 feet.
2. For the design or rehabilitation of bridge girders, the two criteria are:
   - Providing adequate strength
   - Deflection control—Live load deflection should be less than:
     L/800 for bridges without sidewalks
     L/1000 for bridges with sidewalks

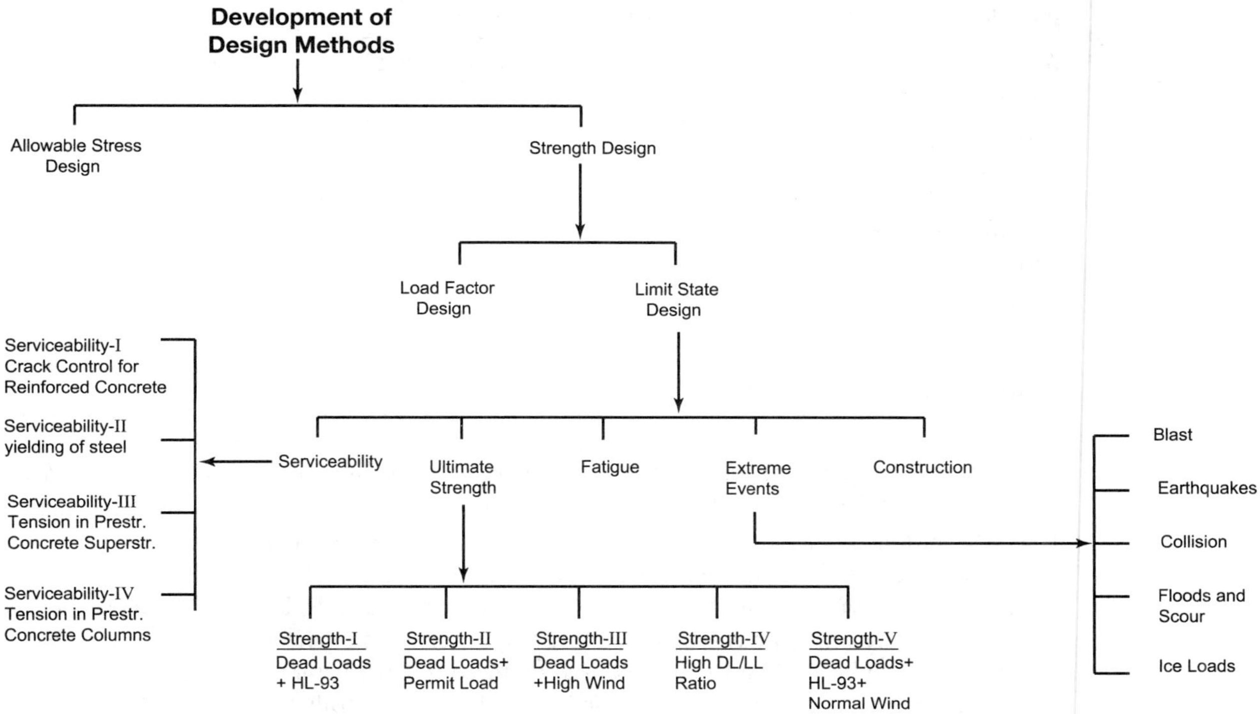

Figure 5.2 Development of LRFD load combinations tree.

L/300 for vehicular load on cantilever arms
L/375 for pedestrian loads on cantilever arms
L/425 for timber girders.

3. A broad-based comparison with a building floor design shows that allowable live load deflection is nearly three times higher for buildings, e.g., maximum live load deflection limit = L/325. However, span lengths in bridges are five to ten times higher. The magnitude and frequency of live load is also much higher. Hence, allowable deflections in bridge girders can be as low as L/1000, although some states like New Hampshire have used an L/1600 limit.

4. A committee of the American Society of Civil Engineering (ASCE) in 1958 reviewed the history of bridge deflection limits, along with a survey of data on bridge vibration, field measurements, and human perception of vibration. The committee questioned the applicability of the original deflection limits to those used at the time. Their limited survey showed no evidence of serious structural damage attributable to excessive live load deflection. It found no clear structural basis for the deflection limits. Amid all the advances in manufacturing, design, and construction, there is a need to reexamine the deflection criteria.

5. Existing deflection limits vary in their application from state to state. A comparative study has shown that the difference between the most and least restrictive approaches can be as high as 1000 percent. This is due to many factors such as: variation in the actual limit (e.g., one state uses L/1600), variation in load magnitude and pattern, and application of load and distribution factors. Such a wide variation highlights the need for further study and demonstrates that in their present form it clearly cannot address serviceability and durability issues as they are intended to do.

6. Controlling L/D ratios: The origin of L/D limits is traced to 1905 American Railway Engineering Association (AREA) specifications used for railroad bridges. Table 5.1 shows the L/D limits that were incorporated in previous AREA and AASHTO specs. The 1935 AASHTO specifications stated: *"If depths less than these are used, the sections shall be so increased*

Table 5.1 Span-to-depth (L/D) limits in original AREA and AASHO

Year(s)	Trusses	Plate Girders	Rolled Beams
A.A.S.H.O.			
1913, 1924	1/10	1/12	1/20
1931	1/10	1/15	1/20
1935, 1941, 1949, 1953	1/10	1/25	1/25
A.R.E.A.			
1905	1/10	1/10	1/12
1907, 1911, 1915	1/10	1/12	1/12
1919, 1921, 1950, 1953	1/10	1/12	1/15

that the maximum deflection will be not greater than if these ratios had not been exceeded." Since then AASHO has changed to AASHTO and AREA has changed to AREMA.

According to Saadeghvaziri of New Jersey Institute of Technology, explicit deflection limits were introduced in the early 1930s after Bureau of Public Roads conducted a study on the impact of vibration on humans. Bridges used wood plank decks, and the superstructure samples were either pony trusses, simple beams, or pin-connected through trusses. Nevertheless, the limit of an L/800 set based on these basic structures and materials almost a century ago is still used. The deflection limit of L/1000 for bridges used by pedestrians was set in 1960, reportedly after a prominent mother complained about her baby awakening as she drove over a bridge.

5.3.2 AASHTO Optional LRFD Deflection Criteria

Optional criteria uses:

1. Span/girder depth ratio.
2. Span/superstructure depth ratio.
    - Superstructure depth = Total depth of structural slab + depth of haunch + depth of girder.
    - Thickness of deck pan plate connected to top of top flange is neglected.
    - Composite action between slab and beam is assumed due to shear connectors.

5.3.3 Parametric Study of Deflection

1. The following live loads need to be considered in sequence:
    - AASHTO LRFD defined deflection truck.
    - HL-93 truck and lane load.
    - Alternate truck live load.
    - Maximum legal load.
    - Permit load.
2. For a three dimensional model, only the vertical component of deformation is considered. Under heavy nonsymmetric truck loads, horizontal displacements take place at bearings but may be neglected compared to vertical deflections.

Maximum dead load deflection of slab and beam system = Dead load deflection at center of symmetric system due to:

Weight of (deck slab + girders + median barrier + parapets) −
(initial upwards deflection of girders due to camber)

3. Maximum live load deflection of slab and beam system = deflection of composite beam due to flexure + shear deflection due to impact − deflection due to arching action.

Maximum live load deflection would occur under the heaviest axle of a permit load.

4. The National Cooperative Highway Research Program (NCHRP) published a comprehensive report, "Improved Live Load Deflection Criteria for Steel Bridges." An important task under this study was a review of existing literature and the state of practice for steel bridge deflection control. It included:

- Overview and historical perspective of the subject.
- Review of effect of bridge deflections on structural performance.
- Review of effect of bridge deflection on superstructure bridge vibration (human response, field studies, analytical studies).
- Review of alternate live load deflection serviceability criteria.
- Survey of state bridge engineers to determine how AASHTO deflection limits are actually applied in bridge design.

5. Sause and Fisher, in "Application of High Performance Steel in Bridges" (Figure 5.3) have shown that an increase of yield strength between 50 ksi to 70 ksi decreases the composite girder weight by 14 to 19 percent. However, the AASHTO deflection limit of L/800 can be reached earlier for the lighter HPS beam. Hence, there is a need to revisit the deflection limit for HPS 70W and higher yield strength steel girders.

Deflection limitation for 100W steel: Some states have initiated the use of 100W flange plates resulting in shallower girders. In addition to live load deflections, vibration studies need to be carried out since it is not always possible to reduce girder spacing due to the need for multiple utility pipes to be located between the girders.

From the in-depth studies above, it is envisaged that graphs and tabulated results of span versus live load deflections can be plotted for a variety of girder depths, high performance materials, and skew boundary conditions. These easy and ready to use results will enable the

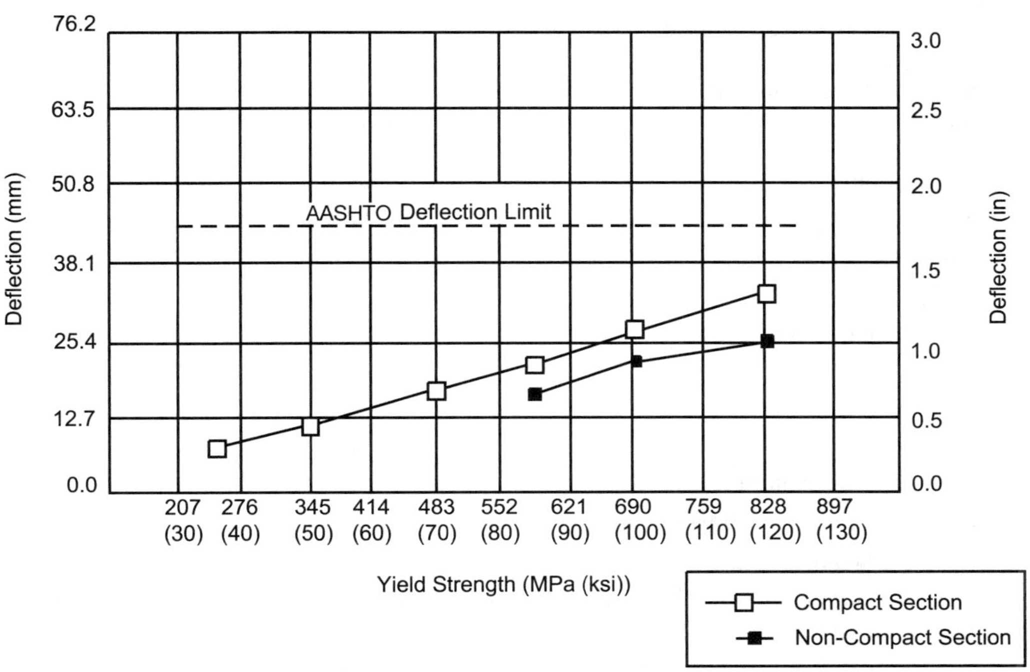

Figure 5.3 Deflection versus yield strength for the Lehigh Street Bridge in PA (Homma and Sauce 1995).

designer to select safe and economical girder sizes in the future. Travel comfort, durability, and vibration-free bridges will result.

6. Use of shallow girders case study for Magnolia Avenue Bridge in City of Elizabeth, New Jersey:

Preliminary design by the author shows that the AASHTO/NJDOT vertical underclearance requirement on the heavily used bridge over Route 1 and 9 was 16 ft 6 in while a higher girder depth using 50 ksi yield strength would only provide 14 ft 6 in. By using hybrid 50W and 70W girders it was possible to restrict girder depth to 3 ft 6 in over a span of 129.5 feet. Hence, the vertical underclearance increased by 1 ft 3 in to 15 ft 9 in. This was accepted as a design exception by NJDOT considering that other existing bridges in the corridor had lower than a 15-foot vertical underclearance.

The bridge had two 15-foot lanes with two sidewalks. Both AASHTO LRFD and NJDOT Bridge Manual required L/1000 live load deflection criteria (i.e. 1.5 inch maximum under HL-93 live load). With only 3 ft 6 in girder depth, maximum deflection under live load plus impact exceeded well over 1.5 in. However, even with reduced girder spacing of 6 ft 6 in, girder deflection exceeded well over 1.5 in.

The AASHTO requirement of L/1000 is conservative, especially for HPS 70W steel, and can only be met at the expense of reducing vertical underclearance. The L/D for the girders is on the higher side amounting to 37. It was a source of concern for the QA/QC Department, which finally understood the need for a design exception for hybrid closely spaced girders.

7. On other HPS 70W projects in New Jersey, where vertical underclearance is not an issue, current practice is to use a deeper 50W web with flanges using HPS 70W steel. Also, when adjacent bridges or those on local roads have a vertical underclearance of 14 ft 6 in, a deeper girder web can be used and no design is needed.

Composite sections with HPS 70W and HPC decks with 5 or higher strengths: Since deflection is inversely proportional to the EI value of the composite section, high performance concrete is also playing an indirect role in the sizing of girder depths. Deflection control needs to be investigated for a variety of HPC strengths and deck slab thicknesses, with the view to enhance performance of such bridges.

5.3.4 Mathematical Approach

The following parameters need to be considered in analysis:

1. Plan aspect ratio.
2. Skew angles and curved slabs.
3. Number of girders.
4. Spacing of girders.
5. Boundary conditions:
   - Simply supported
   - Continuous
   - Fixed
   - Girders framing into integral abutments
   - Girders framing into semi-integral abutments.

Fascia girder deflections are also affected by the thickness of the sidewalk on the cantilever side. Hence, variations in the deck slab thickness need to be considered for fascia girders for bridge decks with sidewalks.

For serviceability and durability, the maximum live load deflection calculated from any of the above methods must be less than the AASHTO criteria.

6. Method of computing deflections: Deflections affect bending moment and shear force distribution. Live load deflections are usually computed by one of the following methods:
   - Stiffness matrices.
   - Strain energy.
   - Double integration method.
   - Harmonic analysis.
   - Finite difference method.
   - Finite element method.

   The current method prescribed by AASHTO is using a line girder and applying multiple lane reductions. There are approximations in the method of applying live load distribution from each lane for load sharing on the single girder under consideration. Code deflection calculations are based on the use of distribution coefficients (DF), which may not give a true deflection value in all cases due to the complexity of bridge geometry. It may be desirable to calibrate the line girder code method against a three-dimensional model using the stiffness method. Software such as SAP2000 or ADINA may be used.

5.3.5 Large Deflections in Single Spans

Large deflections in a single span truss (Figure 5.5) require nonlinear analysis since the load deflection curve becomes nonlinear when compared to continuous span deflections.

1. Large deflection causes fracture and debonding of the wearing surface due to excessive work done. In timber construction, fasteners loosen.

2. Stiffness of entire deck width considered:

   Deflection DF = Number of lanes/number of lanes

3. Additional shear deflection occurs under heavy truck load, especially for cantilever overhangs.

4. Arching action at the supports in composite deck slabs may reduce deflection.

5.3.6 Primary Effects of Deflections

1. The two design criteria for strength and deflection are linked together. When there is no deflection, there is no stress. The higher the deflection, the higher the slope and resulting curvature (Figure 5.5). An increase in curvature gives rise to an increase in bending moment

Figure 5.4 Use of three column hammerhead piers in place of wall piers.

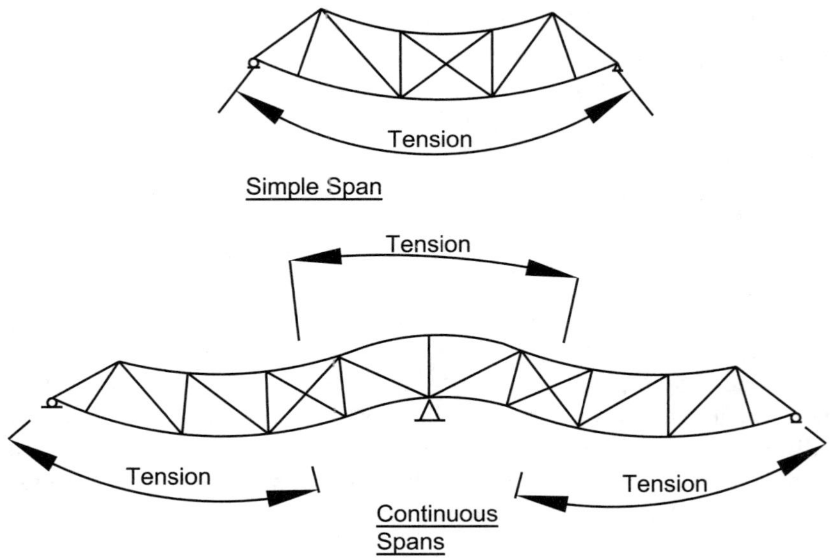

Figure 5.5 Positive and negative curvature in trusses.

and bending stress. Strength is based on ultimate load conditions, while deflection is based on service conditions. Total deflection due to dead and live load plus impact will not exceed yield stress.

2. Sophistications in today's bridge designs combined with advances in development of high performance materials of various grades demands an equally advanced and sophisticated approach to considering serviceability and durability requirements such that it will not negate the economic benefits of advances made in material development. A simple example of a simply supported beam is used to demonstrate why present serviceability requirements (e.g., deflection limits) can have such a significant impact on the design of HPS girder bridges. The maximum moment, M_{max}, which is equal to PL/4, is used in strength-based design to size the member cross section.
3. According to Saadeghvaziri of NJIT, the flexural equation stress-load relation can be represented as follows:

$$\sigma = Mc/I = PLc/4I \tag{5.1}$$

In this equation, ***c*** is distance to extreme bending fiber and ***I*** is moment of inertia. In typical designs the above equation is solved for required moment of inertia to determine the section geometry. Subsequently, deflection is determined based on the following equation and checked against codes limits.

4. Maximum deflection,

$$\Delta\ max = PL^3/48EI \tag{5.2}$$

For most cases the deflection limits are easily satisfied, often with a large margin. However, existing deflection limits negate the economical use of high performance materials because the original basis for these limits were not well established, and they did not consider existing bridge systems and the range of materials currently available.

5. Developing a deflection-strength relation: The required moment of inertia, I, is determined based on material strength = PLc/4 σ.

As can be seen, the higher the material strength the lower the required moment of inertia. The required moment of inertia considering deflection limit:

$$I = PL^3/48E\ \Delta\ lim \tag{5.3}$$

The smaller the deflection limit, the higher the required moment of inertia:

$$\Delta \text{ lim} = L^2 \sigma / 12 \text{ Ec} \tag{5.4}$$

The higher the material strength, the lower the deflection limit, i.e., a more stringent requirement. This limitation in existing design specifications penalizes the use of high strength material. Rational design methods ensure that higher performance materials are used, while structural serviceability and durability are achieved.

Equation 5.4 also shows that the ratios of span to depth and span to deflection are not independent, as this equation can be rewritten in the following form (noting c = d/2):

$$\Delta / L = L \sigma / 24 d \text{ E c} \tag{5.5}$$

$$\text{If } k = 1/24E, \Delta / L = k \sigma / 24 \text{ E c} = k \sigma L/d \tag{5.6}$$

Stress, σ, increases for stronger steel and a shorter span. Therefore, the span-to-deflection ratio (L/Δ) tends to control for shorter spans while the span-to-depth (L/d) limit controls for longer spans.

Figure 5.6 shows deflection as a function of span-to-depth ratio. L/Δ, limit has significant impact on the use of high strength steel. For shallow HPS 70W steel, girders, L/800 deflection limit is exceeded while strength is acceptable. Design specifications need to revise the limit for HPS 70W and above.

5.3.7 Factors Affecting Deflections

These factors may be summarized as:

1. Effective span length (between center lines of bearings).
2. Varying beam depth (with and without cover plates).
3. Beam width (assumed constant).
4. Depth of haunch (minimum 1 in to maximum 4 in).
5. Depth of slab (assumed constant).
6. Effective width of flange (Ell beam width for fascia girder and T-beam width for interior girder as defined by AASHTO code).
7. Beam stiffness or EI value (including top and bottom longitudinal reinforcement in deck slab) for longitudinal bending.

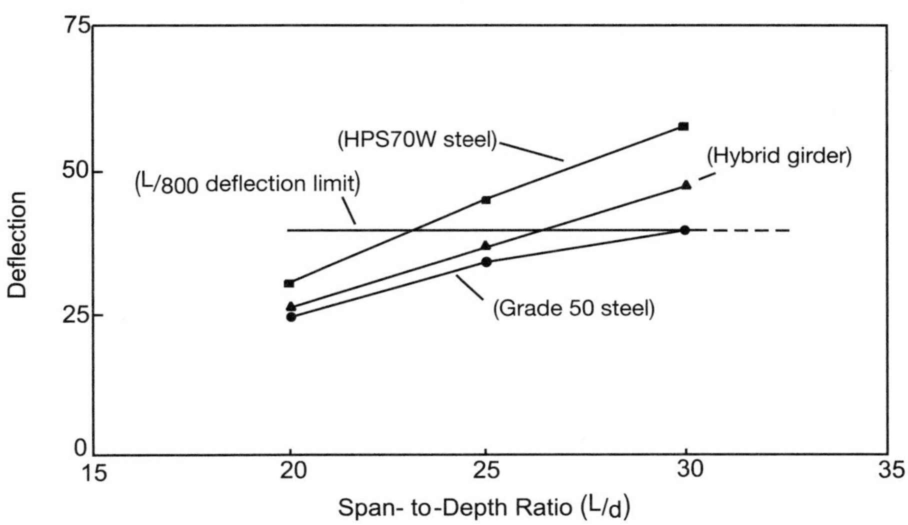

Figure 5.6 Comparison of deflection curves versus L/d for different material strength.

8. Transverse stiffness of diaphragms and spacing.
9. Intensity of load (AASHTO 72 kips HS 20 deflection vehicle with dynamic allowance of 30 percecnt).
10. L/d ratio where d is beam depth (for shallow, medium depth, and deep beams).
11. L/D ratio where D is beam depth + haunch depth + deck slab depth (for composite section).
    - Shallow beams with high L/D ratios are likely to result in higher deflections and cause vibrations as compared to the following two types.
    - Medium depth beams are commonly used in practice and are usually based on AASHTO LRFD optional deflection criteria.
    - Deep beams (with small L/D ratios) have lower deflections. The depth of beam approaches a small height wall. Stress strain behavior is nonlinear and the beam cannot be modeled as a line girder. Deep beam/wall effects need to be considered using the finite element method.

In design, dead load deflection is practically eliminated by providing initial camber in the beam. Live load deflection can be reduced to some extent, for example, by providing high tensile strands in prestressed concrete beams and pre-tensioning or post-tensioning the strands. The practice of prestressing a steel beam is less common.

5.3.8 Effect of Beam Materials on Deflections

AASHTO criteria of L/800 or L/1000 is generally applicable to beams that are made of composite with deck slabs using shear connectors. Modulus of elasticity of the material affects deflection. For deflection control, the response of a variety of concrete, steel, and timber beams needs to be considered such as:

1. Reinforced concrete beam.
2. Prestressed concrete I girders and box beams with normal strength.
3. Prestressed concrete I girders and box beams with HPC. Composite beams with HPC decks (fc′ values ranging from 4000 psi to 8000 psi are permitted by most states).
4. Prestressed concrete I girders and box beams with ultra high performance concrete.
5. Grade 50W steel beams (both rolled sections and fabricated welded sections).
6. HPS 70W steel beam (fabricated welded sections).
7. Fabricated welded sections using hybrid HPS 70W/50W and hybrid HPS 100W/70W.
8. Sawn timber beams.
9. Glue-laminated timber beams.

5.3.9 Secondary Effects of Deflections

Vibrations can result from live load deflections on highway, pedestrian, and equestrian bridges. It is assumed that vibrations will depend upon differential (varying) live load deflections due to fast moving vehicles. There may be harmful effects of rapid variations in live load deflections giving rise to vibrations and fatigue.

1. Deflection resulting in bridge vibration: Control of undesirable psychological effects on passengers and pedestrians is apparently one of the primary reasons for AASHTO deflection limits (Table 5.2). However, prior research indicates that it is not just live load deflection but vertical acceleration and bridge dynamic characteristics (e.g., frequency) that control vibration and human perception. Acceleration is the most important parameter affecting psychological discomfort.
2. Use of the finite element model representing deck and girder stiffness in two directions: Analysis of bridge systems with applications of 2-D and 3-D finite element models is required to study statics and dynamics response of highway bridges. The finite element analysis

Table 5.2 Human effect versus acceleration (as reported in literature).

Effect	Acceleration(m/sec2)
Humans cannot perceive motions	<0.05
Sensitive people can perceive motion, hanging objects may move slightly	0.05–0.10
Majority of people will perceive motion or desk work becomes difficult or almost impossible	0.1–0.4
People strongly perceive motion, difficult to walk naturally, standing people may lose balance; most people cannot tolerate motion and are unable to walk naturally	0.4–0.6
People cannot walk or tolerate motion	0.6–0.7
Objects begin to fall and people may be injured	>0.85

employs both 2-D and 3-D models. 2-D models are more efficient computationally and will be used to better identify important parameters that will be further studied with the 3-D models. Typical 2-D models include beam elements to represent both the superstructure and substructure. The superstructure composite properties are typically used in modeling the deck since the shear studs are headed and it is well established that there is perfect bonding between the concrete deck and the steel girder.

5.3.10 Accelerations from Moving Loads

The acceleration that affects human reaction is due to overall superstructure response, which a 2-D model can easily simulate. It is the spatial effects (such as load distribution or relative girder vibration) that is not represented with a 2-D model and requires application of a 3-D model. Use of compatible eccentric beam elements will ensure proper modeling of the cross-sectional geometry while enhancing computational efficiency. A vibration analysis will be needed, especially for longer spans of pony truss-through girders. There are many possible varying locations of moving loads and load patterns. One such method, which considers the vertical acceleration as the most important parameter affecting human comfort, was proposed by Wright and Walker for steel bridges as published by AISI:

1. Determine static deflection, Δ_s, due to live load.
   Ensure the acceleration, a, does not exceed 100 in/sec^2 or 0.25g.
2. Among limitations of the above method are: The frequency equation is suitable only for simply supported bridges, and there is a need for estimating the frequency of practical bridges; the DI factor might not necessarily be accurate for existing system and loading; and lack of consideration to the relationship between the bridge natural mode and dynamic load. Relaxing L/D limits or not considering deflection and L/D limits will not necessarily result in designs with large deflections.
3. For flexible girders with higher L/D ratios:
   - Loads are better distributed to other girders (decreased distribution factor and load per girder).
   - Reduced live load moment due to a decrease in distribution factor.
   - Reduced peak negative live load moment that can aggravate deck cracking.
   - Better accommodation of shrinkage strains due to higher deck-to-girder stiffness.

   The latter advantage can minimize transverse deck cracking potential as recommended by Saadeghvaziri and Hadidi.
4. Vibrations may lead to overstress. Analytical grid modeling of deck, girders, and diaphragms for vibration shall consider:

- Alternate single longitudinal T-girder analysis: A simplified approach will be used to compare/check results from three-dimensional model.
- Field study of bridges damaged by large deflection: Explanation of damage in the light of vibration and fatigue studies may be required.
- Correlation between deflection and vibration: Correlation between vibration model results and field measurements may be carried out.

5. Boundary conditions may be fixed, pinned, or free.

6. For pedestrian bridges, moving load patterns of varying numbers of people on a bridge can be applied. Alternate ways to model them will offer different results. For example, one person walking across the bridge will cause minimal vibration. However, several individuals jumping on the deck can create higher oscillations.

Following the failure of Tacoma Narrows Bridge under wind, bridges with vibration effects have been retrofitted with stiffening decks. Although wind analysis is seldom a primary consideration, it can cause an increase in vibrations when combined with live load.

Member sizes allow designers to predict the dynamic behavior of a footbridge to a certain extent. Computational analyses for live loads and lateral forces for pedestrian bridges needs to be carried out.

5.3.11 Parameters Affecting Vibrations

1. Structural system: Through girders or multiple composite girders.

2. Truck load, lane load, or combination with dynamic load allowance (impact).

- All design lanes loaded.
- Increase in vibrations and fatigue may result for certain types of load patterns.

All beams may be assumed to deflect equally with CG of live loads to coincide with CG of dead loads.

3. More advanced finite element responses (deflections and stresses) will be compared to design values. This will help with better identifying the effect of spatial stiffness and load distribution on bridge response.

Furthermore, dynamic analysis will be performed to investigate vibration characteristics.
Damping effects: These depend on a number of parameters such as:

- Random live loads and modeling.
- Materials used.
- Complexity of the structure.
- Type of surfacing.
- Bearing conditions.
- Weight of railings and parapets.

Bridges designed for equestrian loads are subject to galloping horse loads and impact and need to be checked for vibration. For damping, use of a timber deck supported on lightweight aluminum or steel floor beams of pony or through truss is preferred.

5.4 FATIGUE AND FRACTURE

5.4.1 Fatigue Analysis

Old steel bridges to be rehabilitated require a fatigue analysis of all existing steel members to provide an estimate of the remaining fatigue life. The analysis shall be performed in accordance with the:

1. Latest edition of the AASHTO Guide Specifications for Fatigue Evaluation of Existing Steel Bridges.

Figure 5.7 Construction of a three-column bent pier in New Jersey (designed by the author).

2. Current Standard Specifications for Highway Bridges. Results serve as useful indicators of the severity of the fatigue detail.

   The fatigue analysis shall include the following:

3. Tables showing:
   - Remaining safe and mean fatigue life
   - Moments and stress ranges at each detail and the location being evaluated.
4. A list of assumptions and input values used for each detail and the location:
   - Live load distribution factor
   - Wheel and axle spacing of the fatigue truck used.
5. Location and composite or non-composite section properties of the detail.
6. ADTT and present age of structure in years.
7. Impact percentage.
8. Calculated reliability factors.

5.4.2 Fatigue Retrofit

Strict attention needs to be paid to practicality as well as strength and fatigue requirements.

1. End bolted cover plates: Bolted splice plates shall be provided. When the fatigue category of the welded cover plate ends or welded flange and web splices are to be upgraded. The bolted plates shall be designed to carry the fatigue live load truck.

   Members and connections subjected to repeated flucutations of stress shall be designed for fatigue vehicle. The number of stress cycles, the magnitude of stress range, and the type and location of construction detail need to be considered in fatigue design.

2. Dynamic analysis will include a determination of the effect of moving load on load reversal (cyclic) and consequently durability (fatigue). That is, to perform fatigue analysis, high amplitude dynamic cycles will be added to the highest ADTT over 75 years recommended AASHTO life.

- Fatigue analysis will be based on S-N curves. Maximum N value of 2 million cycles is assumed. For steel, maximum S value will depend upon fatigue category of B or C as per AISC Manual of Steel Construction.
- AASHTO LRFD fatigue vehicle is considered: According to a study carried out by WSDOT, the fatigue load vehicle specified in the steel structures section of LRFD specifications is more reflective of the fatigue loads experienced by highway bridges and produces a lower calculated stress range than that of AASHTO standard specifications.
- Existing experimental and theoretical results from available fatigue studies on bridges can be utilized to evaluate the effect of live load deflection on long-term fatigue. Figure 5.7 and 5.8 shows crack location.
- AASHTO recommended 75-year life is considered for number of cycles. The highest projected ADTT prevalent on the bridge will be used.

3. Small web gaps for fatigue prone details may result in stress concentration and subsequent cracking, intersecting welds, lateral connection plates, longitudinal stiffeners, and cracks.

4. Notch effects, such as rivet holes and non-radius cuts, cause increase in stress.

5. All poor details, fatigue sensitive details, and stress risers of all types need to be removed.

- Rivet holes should be made round by reaming to eliminate crack initiation sites.
- Lateral connection plates should not be welded to tension flanges.
- The stiffness of the new members should be considered and how the existing members should be strengthened in order to carry the new loadings should be determinded.
- Major repair issues: For the alternate selected, the impact on traffic disruption needs to be a minimum. Items to be evaluated include crack sealing.
- Fatigue is the lower than expected performance of the member or a joint under repeated cyclic loads when it fails at a stress level below yield stress (Figure 5.7).
- Analytical grid modeling of deck, girders, and diaphragms for fatigue.

The flow diagram of Figure 5.9 demonstrates the procedure to underline the importance of fatigue in railway bridge girders under high fluctuating wheel loads.

5.4.3 Fracture Critical Member (FCM) and Redundancy

1. FCM is a steel member in tension whose failure is likely to cause the entire bridge to collapse. These members are highly vulnerable to a fatigue type failure. The three conditions for identifying fracture critical members are steel material, tensile stress, and vulnerability to collapse due to simple configuration.

2. Redundancy helps to delay fracture and increase fracture toughness. Redundancy is a structural condition resulting when there is more than the minimum number of structural elements for stability.

The framing or configuration of the bridge such as number of girders, number of continuous supports, or multiple numbers of members forming a connection would increase redundancy.

- Load path redundancy: When three or more main load carrying members are present between substructure units and if one member is close to failure, load can be safely distributed to other members.
- Structural redundancy: A configuration which provides continuity of load path to adjacent spans due to redistribution of moments. Only internal spans, but not the end spans, will provide structural redundancy.

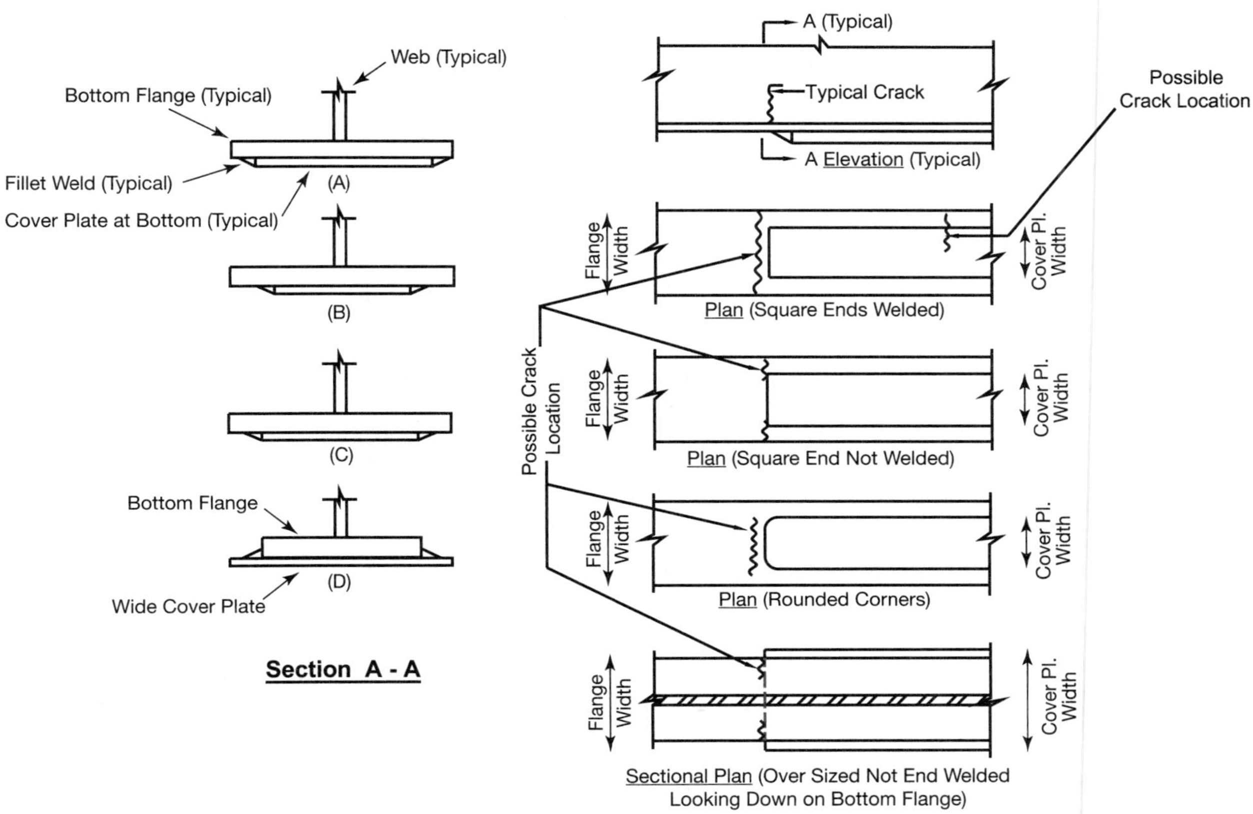

Figure 5.8 Possible crack propagation at cover plate ends.

- Internal or member redundancy: A configuration that contains three or more elements that are fastened mechanically such that multiple independent load paths are formed. Failure of one member element would be local and would not cause multiple member failure.
- Intersecting welds or welds which overlap or are closely located will have high stress concentration. Examples are welds located at the intersection of a transverse and longitudinal stiffener. Tack welds or weld repairs may also reduce the internal or member redundancy. Transverse welds that are perpendicular to the applied stress field exhibit a greater tendency to crack than continuous longitudinal welds that are parallel to the applied stress field.

The average volume of traffic per day shown on any route would vary according to rush hours, weather conditions, or for an important event. Some lanes may be closed occasionally due to accident or repair. Some states, such as Michigan, have the highest truck loads in the country. There is little control over intensity of out-of-state traffic using interstate highways. Also, projected traffic in 20 years may be much higher.

Traffic counts are needed for the planning of each bridge and for fatigue studies. It would be uneconomical to design bridges located on local roads for the same intensity of truck load as the heaviest interstate trucks. Also, the fatigue of girders will be proportional to the frequency and intensity of trucks.

Individual traffic counts may only be representative of the time when they were taken. Even an envelope of counts made over a long period may change due to detours implemented elsewhere. At best, these are approximate and are guidelines only; however, as a check a system of weigh stations has been installed that permits loads specified by different states.

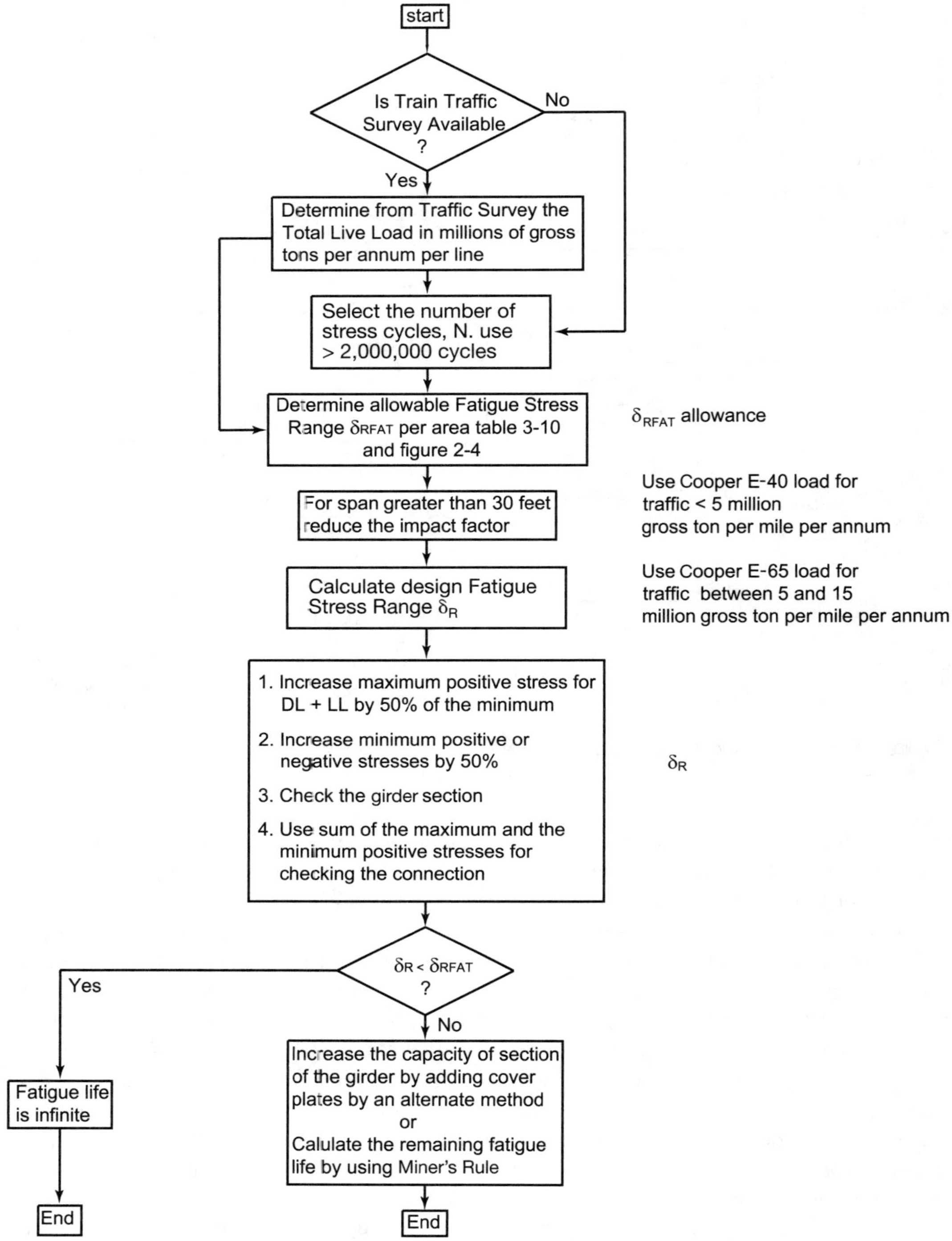

Figure 5.9 A flow diagram for fatigue evaluation of railway bridges based on AREMA code.

5.5 SELECTION OF TRUCK LIVE LOADS

5.5.1 The Role of the Highway Administration

In the U.S., truck loads such as HL-93 are based on the combined intensity of truck and lane loads. Three types of traffic volumes can be summarized as:

1. Recommended truck loads are HL-93. Legal and permit loads for the state in which the bridge is located are also applicable.
2. High through traffic on interstate and major highways between states or cities where the average number of trucks per lane per day exceeds 1000 and the average of all vehicles exceeds 4000.
3. Moderate traffic on arterial roads serving property and business access where the average number of trucks per lane per day is between 250 and 1000, and the average of all vehicles is between 1000 and 4000. Recommended truck loads are either HL-93 or other as specified by the state in this category.
4. Limited traffic on local roads serving small communities where the average number of trucks per lane per day is less than 250 and the average of all vehicles is less than 1000. Recommended truck loads are for the state or county in which roads are located.

Older bridges with occasional truck load are posted for the maximum weight of truck permissible to prevent overstress.

5.5.2 Historical Perspective of Live Load Vehicles Specified for Design

1. According to increasing traffic demands, truck loads have been upgraded in the U.S. Major changes occurred in 1944 and 1993. These have been supplemented by special truck loads for which permits are required. There are restrictions on use by permit trucks such as use during night time windows with very light traffic and police escort. Weigh bridges are installed to monitor the maximum weights of trucks.
2. HS-20 truck: An idealized truck load of 72 kips is applied using the HS-20 vehicle load configuration, which has not changed since 1944 when it was first introduced as a tractor-semi-trailer combination. In 1944 the idealization was based on a three-axle group of vehicles known as 3-S2's. Load length varied between 28 feet and 44 feet.
3. Interstate or tandem load: The drawback with the above idealization was the long length between axles that did not govern for the design of smaller span bridges. It was felt that a smaller distance was needed between axles, of minimum 4 feet spacing. During construction of the interstate system, the two-axle tandem load of 25 kips each was introduced. Those states that adopted HS-25 also increased the tandem load by 25 percent.
4. Some states have specified rating vehicles and legal vehicles.

5.5.3 Details of 2007 LRFD Specifications

1. Separate analyses are required to obtain an envelope of bending moments and shear forces for each set of vehicles. Basically, stress, fatigue, and deflection limits are applicable based on selected vehicle types causing maximum effects.
   - HS-20 truck + lane load (HL-93) as shown in Figure 5.10
   - 90 percent HS-20 truck pair + lane load (alternate to HL-93 tandem pair spaced at 50 ft min.; 14 ft typical spacing between 32 kips axle)
2. Tandem truck + lane load (alternate HL-93) as shown in Figure 5.11.
   - 90 percent tandem pair + lane load (alternate to HL-93 tandem pair spaced at 50 ft min.; 14 ft typical spacing between 32 kips axle)

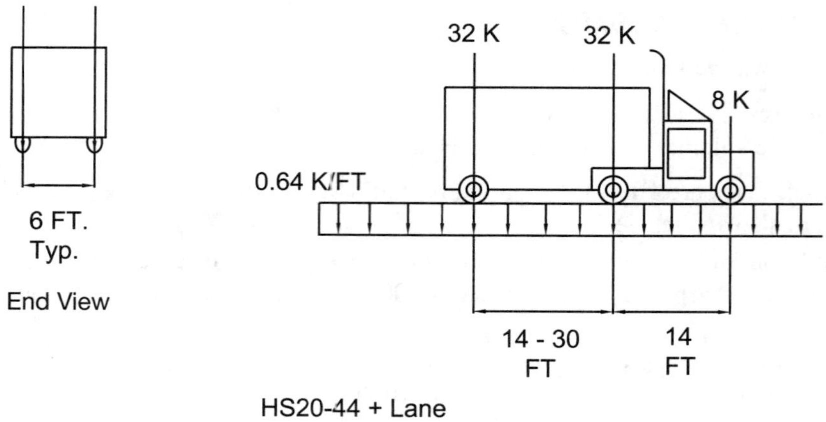

Figure 5.10 Standard HL-93 live load.

- Truck pair for negative moments with partial lane load (HL-93 for continuous spans) as shown in Figure 5.12.

3. Fatigue vehicle: Live load vehicle for fatigue load without lane load per lane:

- HS-20 truck alone with 14 ft fixed axle spacing for evaluating fatigue as shown in Figure 5.13.

4. Deflection vehicle and 25 percent HS-20 truck + lane load (alternate deflection). Deflection vehicles are as shown in Figures 5.14.

5.5.4 Increase in HL-93 and Tandem Loads

1. In view of the truck traffic, some states such as Pennsylvania have increased HS-20 truck to HS-25, which is a 25 percent increase. Lane load of 0.64 kips/ft is added to the truck load without any change.

2. Similarly, tandem loads were increased by 25 percent to two 31.25 kip axle loads spaced 4 ft apart.

5.5.5 Permit/Notional Loads Without Lane Load

1. In addition to HL-93 loads, following special loads need to be considered:

Legal, permit, and military loads: Both the superstructure and foundation need to be designed to sustain these loads. In earthquakes, all gravity loads including live load on the bridge contribute to seismic forces and moments. The length of moving vehicle is also considered to compute design moments and shear forces.

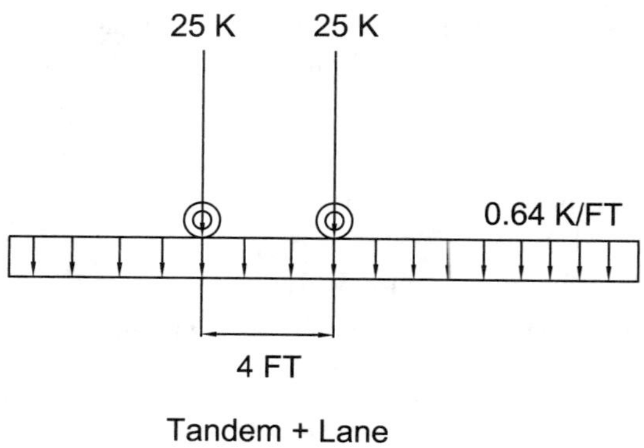

Figure 5.11 Alternate live load for maximum positive moment.

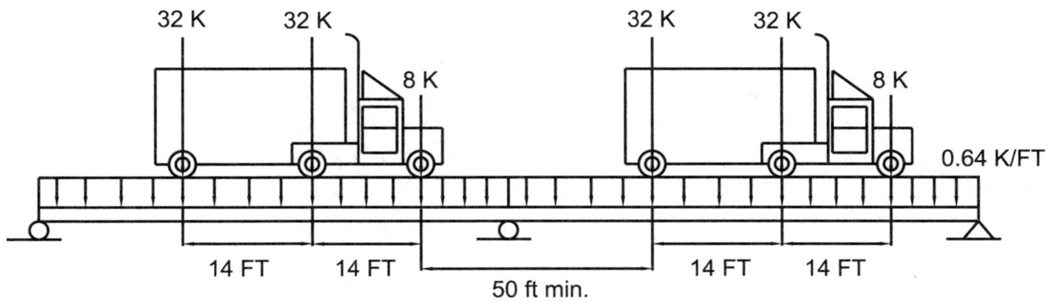

Figure 5.12 Continuous spans alternate live loads, reduce loads by 10 percent.

2. Permit load for Pennsylvania: minimum length of P-82 permit load is 55 ft with 204 kips weight (Figure 5.15).

   AASHTO lengths of permit (notional) loads are 51 ft between the first and last axles with 199 kips weight.

3. Permit load for New Jersey (Figure 5.16).

5.5.6 Comparison of Permit Loads with Cooper E-80 Train Loads.

1. Live loads for railway bridges are many times higher than even the highest of truck loads requiring a permit. A comparison of moving loads is shown in Figure 5.16 between the latest New Jersey permit load (790 KN or 180 kips total), cooper loads (Figure 5.17 and 5.18), 1800 kips or 400 kips for alternate load (Figure 5.19).
2. It is interesting to note that a bridge is normally designed for single truck occupancy per lane but with simultaneous lane load added. Correction factors are applicable to multiple lanes, with live load modified by applying a factor.

   Figure 5.20 shows a comparative study of moments with HS-20 and PA trucks and permit load P-82.
3. Application of Impact Factors and Multiple Lane Presence Factors
   - Refer to AASHTO Section 3.14.14: Impact factors are applicable to truck loads and not to lane loads. As required in AASHTO standard specifications, no concentrated load is required in LRFD lane load.
   - Unfactored live load analysis is first carried out without impact factors. An impact factor of 1.3 is applied on truck load elastic analysis. In addition, a multiple lane presence factor is applicable.

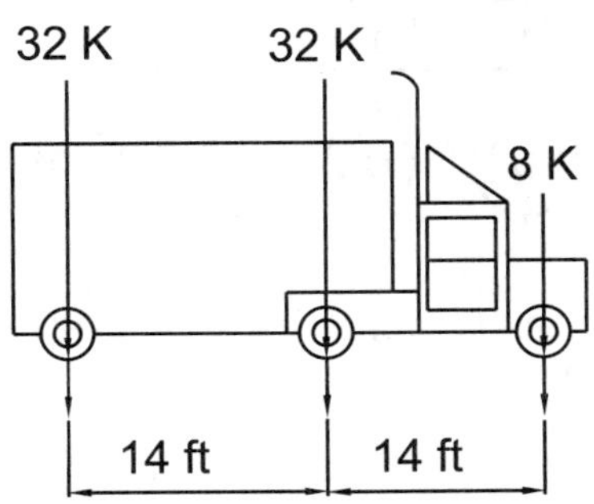

Fatigue Loads Truck

Figure 5.13 Fatigue load truck with constant 14 ft spacing.

Table 5.3 Classification of live load vehicles.

Limit State Load Case	LRFD Live Load Analysis	State Permit Live Load Analysis	State Maximum Legal Live Load Analysis
Strength I	HL-93, H-20, HS-20	N/A	ML-80
Strength II	H-20, HS-20	Permit Load	ML-80
Strength III	N/A	N/A	N/A
Strength IV	N/A	N/A	N/A
Strength V	HL-93, H-20, HS-20	N/A	ML-80
Service II	HL-93, H-20, HS-20	N/A	ML-80
Fatigue	Fatigue vehicle	Fatigue vehicle	Fatigue vehicle
Deflection	Deflection vehicle	N/A	N/A
Construction/Uncured Slab	User defined	User defined	User defined
Ratings	HL-93, H-20, HS-20	Permit load	ML-80

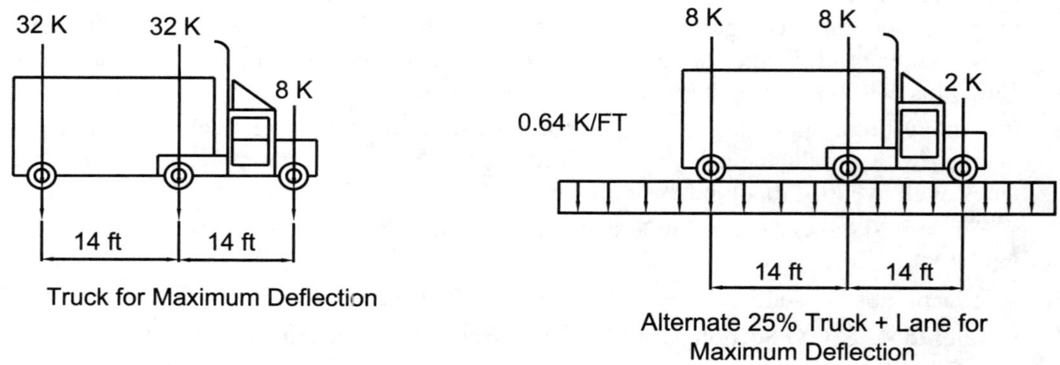

Figure 5.14 Truck and alternate loads for maximum deflection.

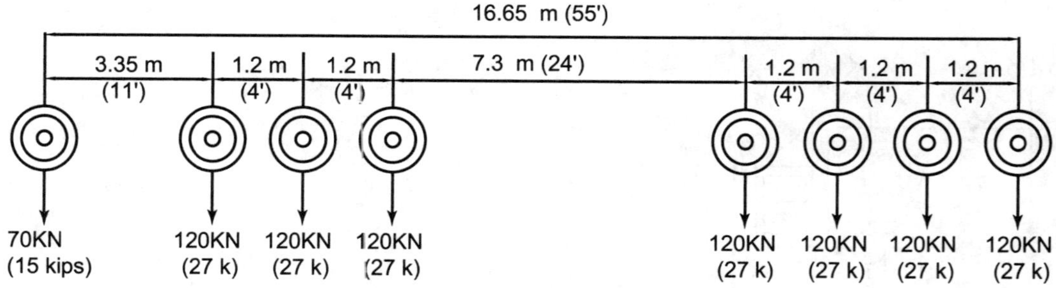

NOTE: P-82 width is the same as the Design Truck.
Transverse wheel location is the same as Design Truck.

Figure 5.15 Pennsylvania 8-axle permit load (P-82).

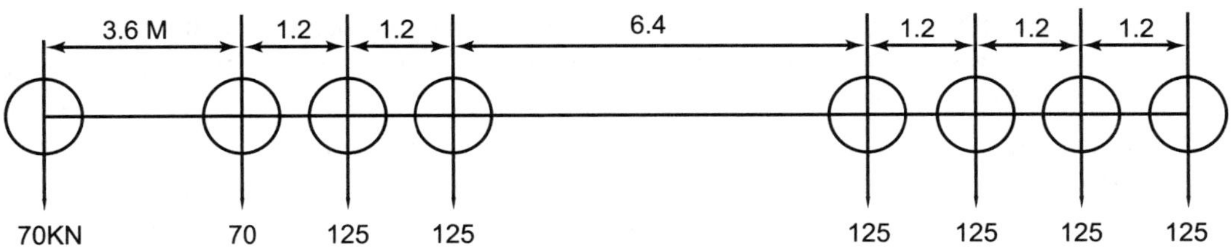

LRFD Permit Vehicle, NJDOT (KN)

Figure 5.16 New Jersey 8-axle permit load.

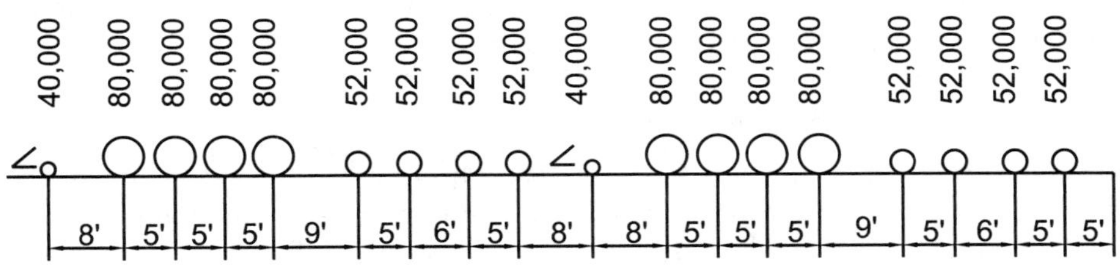

Figure 5.17 Cooper E-80 load (lbs) for comparison with permit loads.

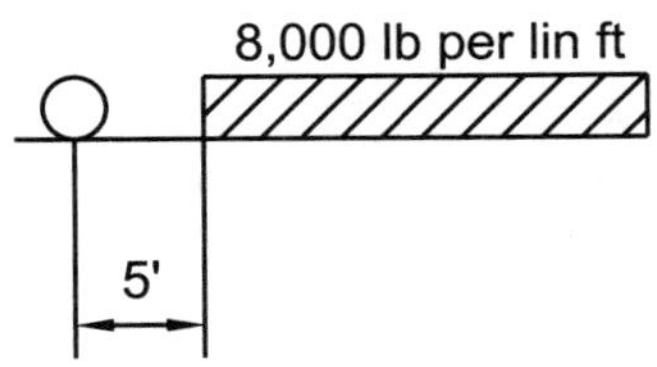

Figure 5.18 Additional Cooper E-80 distributed load of 8000 lb/linear foot.

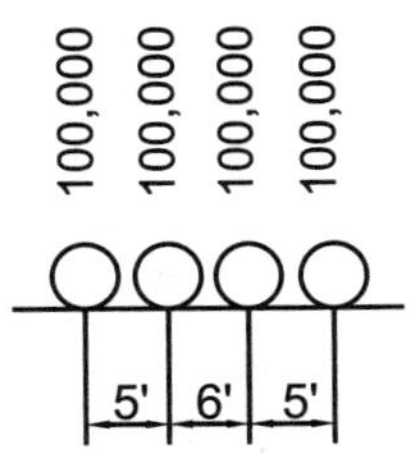

Alternate Train Load on 4 Axles

Figure 5.19 AREMA alternate train load.

5.5.7 Use of Influence Lines for Live Load Analysis

1. Unit load influence lines for peak moments have been generated at every beam location for a variety of span lengths.

   At any section of a simple span, peak bending moment = Wab/L (a and b are distances from left and right supports).

   If there was only one axle with full-lane load, peak moment for a single span = WL/4 due to axle and w $L^2/8$ due to lane load, both located at midspan. For three axle loads on a two-span bridge peak, positive moment will occur at approximately 0.4L.

   Maximum shear force = Maximum reaction at supports.

2. For multiple lanes, the multiple-lane reduction factor is applicable. Also, for skew slabs, the skew correction factor is applicable.

3. Usually, the center of gravity of truck load coincides with the peak moment at the section to give maximum positive bending moment. Similarly, maximum negative moment will occur at a support due to different positions of axle loads.
4. Using the law of superposition, any section of beam magnitude of bending moment may be added for each axle load to produce the combined bending moment from three or more axles. The first axle is placed at the unit peak value and multiplied by the axle load to give peak moment due to that load. Values of other axle loads, say spaced at 14 ft, are multiplied by the corresponding unit values and added. For certain positions of truck axles, absolute maximum moment may be obtained by comparison. Therefore, many positions of the truck need to be considered.
5. For convenience, influence lines may be divided into regions to extract maximum positive and negative moments. Maximum moments so generated will be multiplied by distribution factors and impact factors to give design moment.

5.5.8 The Need for Live Load Distribution Factors

1. As discussed in Chapter 4, Section 4.4, the theoretical approach for slab and beam systems should be considered. The true physical model is three-dimensional, with multiple beams located under multiple lanes. Truck loads, with two or more rows of wheels placed 6 feet apart transversely, will be distributed to more than one beam and will be shared by all the beams to varying degrees. Wheel load is one-half of axle load.
2. For analysis of the superstructure, the designer has the option to use:
    - A three-dimensional model including diaphragms and deck slab thickness using the finite element method, Fourier Series, and Finite difference method. The time spent in modeling with appropriate boundary conditions is usually beyond the scope of a regular bridge project. Complex bridges and bridges with unusual features such as very long spans and very tall piers may warrant a complete model.
    - A two-dimensional grillage model of beams in longitudinal direction and diaphragms in transverse direction using differential equations: The finite difference method or Fourier series approach is used to compute deflections, moments, and shear forces. Based on previous investigations, simplified methods are available, and a modified grillage analysis is used for saving time.
3. A significant contribution to the use of distribution factors was made in 1956 by Morice and Little in England when a distribution coefficient method similar to that proposed by Guyon and Massonnet in France was generalized. Harmonic analysis of orthotropic plates was used. Cusens and Pama refined the method by considering higher order terms in harmonic analysis.

    As a most general case of two-directional bending of orthotropic slab-beam systems, both flexural and torsional rigidity of longitudinal and transverse strips were considered. For a bridge deck of span L and width 2b,

    $\theta = b/L\,(D_x / D_y)^{0.25}$;

    $\alpha = (D_{xy} + D_{yx}) / 2\sqrt{D_x . D_y}$; where

    D_x = Longitudinal flexural rigidity per unit width corresponding to EI of longitudinal beam

    D_y = Transverse flexural rigidity per unit length corresponding to EI of transverse beam

    D_{xy} = Longitudinal torsional rigidity per unit width corresponding to GJ of longitudinal beam

    D_{yx} = Transverse torsional rigidity per unit length corresponding to EI of longitudinal beam.

Both Θ and α are plotted in chart form and coefficients are read based on longitudinal and transverse bending and torsion based on orthotropic plate theory. Applied loads are converted into equivalent concentrated loads at standard locations for which charts are given. For small values of Θ, the method gets approximate and is not applicable to skew bridges.

4. A simplified one-direction longitudinal beam model: Empirical procedures of distribution factors for transverse load distribution have been widely used in the U.S. Since deck slab spans are in a transverse direction, a concentrated load will mainly be distributed in a transverse direction.

Distribution coefficient formulae developed by AASHTO LRFD make use of transverse distribution. The simplified load distribution approach has been refined for accuracy. The accuracy of results has been calibrated against the first two methods and is found to be acceptable.

The reasons for the success of the simplified approach are as follows:

- The observed deflections of a bridge can be approximated to that of a single longitudinal beam.
- Direction of traffic flow coincides with the length of primary beams.
- Convenience factor—The effort required for preparing data and interpreting results is not time consuming for any design office.

5. Distribution factors for moments, shears, and deflections: In a multiple-girder system, it is assumed that load path is in the direction of slab bending. If beams are placed parallel to the direction of traffic, distribution from the deck slab is mainly in the transverse direction. The combined lane load from all lanes is shared by the total number of beams. Maximum distribution to the beam will be less than the full truck wheel load, due to Poisson's ratio effect and multiple-beam load sharing of the system.

An example of transverse load distribution from the LFD method: DF is a function of girder spacing (Figure 5.20). If beam spacing is 5.5 ft and distribution coefficient used is S/5.5, the coefficient = 1.0. Hence, one line of wheel load is assumed to be distributed to the beam for this spacing. Longitudinal distribution may be neglected since beam span is much longer than the spacing between adjacent beams. However, for short span lengths and wide decks, this method becomes approximate and is modified in the LRFD Method.

If beams are spaced at 11 ft, the distribution coefficient = 2.0, i.e., two lines of wheel loads from two adjacent trucks will be shared by the beam. The in-built conservatism has been corrected in the LRFD method, in which the longitudinal stiffness of the girder is considered. An empirical formula based on the longitudinal stiffness of the slab and beam spacing is used. It results in reduced distribution between 15 to 25 percent from the LFD method, and resulting beam design is more economical as seen by the example given below:

$DF_m = 0.075 + (S/9.5)^{0.6} (S/L)^{0.2} (Kg/12\ L\ ts^3)^{0.1}$ for two lanes loaded. Kg needs to be calculated separately.

Assume L = 100 ft, S= 12 ft, ts = 8 in, Kg = 1317,726

DF_m = 1.773 wheels per beam

LFD distribution factor = S/5.5 = 2.18 wheels per beam for moments.

Reduction in BM = 1.773/2.18 =0.81; reduction = 19 percent for interior beam.Hence beam design will be approximately 19% more economical. See Appendix for solved example.

6. In addition to live load, distribution factors for the following fatigue and deflection trucks need to be computed:

- Fatigue distribution factor
- Deflection distribution factor
- Sidewalk deflection distribution factor if applicable.

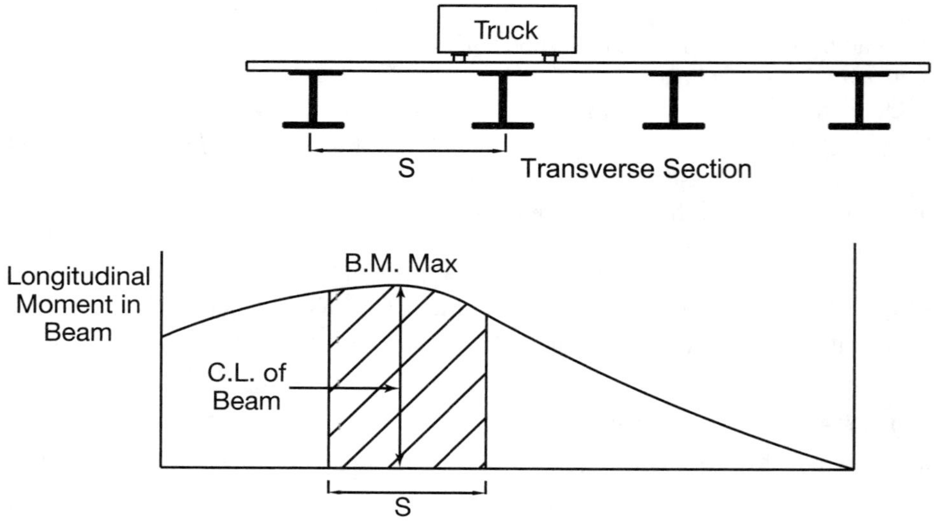

Figure 5.20 Maximum distribution of beam moment under wheel loads. (Resultant of wheel load coincides with center line of beam).

5.6 DETAILED AASHTO LOAD COMBINATIONS FOR RATING AND DESIGN

5.6.1 Load Combinations for Rating and Magnitude of Load Factors

1. For load rating, a simplified format is used as compared to LRFD. Only primary loads need to be considered. Bridges need not be rated for wind and extreme loads.
2. Load factors for legal loads in Tables 6-5 of AASHTO Manual of Condition Evaluation and LRFR are based on ADTT volume.
3. Load factors for permit loads in Tables 6-6 are based on permit truck weights and ADTT volume.
4. Service I for permit live loads is optional.
5. Service II for all live loads is applicable to the rating of steel bridges. In addition, optional fatigue loads without dead loads may be used.

Table 5.4 Load factors for rating of bridges.

Limit State	Materials	Dead Load DC	Dead Load DW	Level I Live Load Inventory Rating	Level I Live Load Operating Rating	Level II Legal Load	Level III Permit Load
Strength I (HL-93)	All	1.25	1.50	1.75	1.35	Table 6.5	N/A
Strength II (Permit)	All	1.25	1.50	N/A	N/A	N/A	Table 6.6
Service I (Permit)	Reinforced/ prestressed concrete	1.00	1.00	N/A	N/A	N/A	1.00
Service II (All live loads)	Steel	1.00	1.00	1.30	1.00	1.30	1.00
Service III (Legal)	Prestressed concrete	1.00	1.00	0.80	N/A	1.00	N/A

Note: All bridges will be rated for strength I (HL-93) and strength II permit loads.

6. Service III for legal live loads is optional.
7. Both legal and permit live loads vary for each state as given in the state bridge design manual.
8. Bridges need to be load posted if the requirements are not met.

5.6.2 Detailed Load Combinations

The author has detailed a large number of load combinations recommended in the LRFD method. AASHTO LRFD specifications address a total of 14 load combinations:

1. Total strength limit state (5).
2. Total serviceability limit state (4).
3. Total fatigue and fracture limit state (1).
4. Total extreme events limit state (2); one each for earthquake and water loads (from floods). Earthquake can be replaced by alternate extreme events such as ice, vehicle collision force, and vessel collision force. Water load, flood, and scour may be classified as three separate alternate load cases, earthquake as two cases, and construction as three separate cases as detailed below. Selection of an extreme event such as a bomb blast, earthquake, scour, vessel collision force, or ice load would primarily affect the substructure design. Piers and abutments would be most vulnerable. The importance of a bridge being on a military route and its proximity to a hospital or school also need to be considered.
5. Three independent alternate load combinations are added here for blast loads.
6. AASHTO Sections 3.4.2 and 3.6.1.3.2 recommend load combinations for
   - Total deflection limit state (1)
   - Total construction limit state (1).

Design moments and forces are based on an initial elastic analysis for each of the load combinations on which load factors are applied. Magnitude of load, positions of loads, and all possible combinations of loads that may occur in practice are covered. An envelope of maximum values for girder, bearing, or pier design can be generated, thus covering all possible environmental conditions that may be experienced during the life of the structure.

5.6.3 Limit State for Substructure (Abutment, Backwall, Bearing, Wingwalls, Pier)

Type of AASHTO primary and secondary loads and specified load factors, Table 5.5.

Permanent Loads

D Downdrag
C Dead load of structural components and nonstructural attachments
W Dead load of wearing surfaces and utilities
H Horizontal earth pressure load
S Earth surcharge load
V Vertical pressure from dead load of earth fill

Transient Loads Description

BR Vehicular braking force
CE Vehicular centrifugal force
CR Creep
CT Vehicular collision force
CV Vessel collision force
EQ Earthquake
FR Friction
IC Ice load
IM Vehicular dynamic load allowance
LL Vehicular live load
LS Live load surcharge
PL Pedestrian live load
SE Settlement
SH Shrinkage
TG Temperature gradient
TU Uniform, temperature
WA Water load and stream pressure
WL Wind on live load
WS Wind load on structure

Note: The above notations and abbreviations are adopted by AASHTO (see Table 5.5).

Table 5.5 LRFD primary and secondary load combinations for substructures.

	Primary Loads				Secondary Loads					
Load Combinations	Unfactored Substr. DL + DL Reactions from Superstr.	Unfactored LL Reactions from Superstr.	Earth Loads on Substr.	Wind on Substr. + Unfactored Wind React. from Superstr.	Water Loads on Substr.	Wind on Live Load from Superstr.	Erection Load from Superstr.	Concrete Properties	Temperature Gradient	Support Settlement
	DC + (DC1 + DC2 + DW)	(LL + IM + BR + PL) + (CE) + (LS)	(EH) & (EV + ES)	(WS)	(WA)	(WL)	(EL)	(TU) + (CR + SH)	(TG)	(SE)
DL + HL-93 LL (Strength I)	γ_p	1.75	γ_p	—	1.00	–	γ_p	—	γ_{TG}	γ_{SE}
DL + permit LL (Strength II)	γ_p	1.35	γ_p	—	1.00	—	γ_p	—	γ_{TG}	γ_{SE}
DL + WL (Strength III)	γ_p	—	γ_p	1.40	1.00	—	γ_p	1.40	γ_{TG}	γ_{SE}
DL of superstr. + substr. (Strength IV)	γ_p	—	γ_p	—	1.00	—	γ_p	—	—	—
DL of super + substr. + live load + reduced wind on superstr. + substr. + wind on live (Strength V)	γ_p	1.35	γ_p	0.40	1.00	—	γ_p	0.40	γ_{TG}	γ_{SE}

Table 5.5a

$\sum$ FX, FY, FZ & MOMENTS MX, MZ FOR EACH LOAD COMBINATION

ABUTMENT FOUNDATION LOADS: Factored Primary Forces & Moments From Computer Output STRENGTH

LOAD COMBINATIONS FOR STRENGTH I, STRENGTH II, STRENGTH III & STRENGTH V	Substr. DL + DL Reactions From Super Structure - FACTORED	LL Reactions From Superstructure - FACTORED	Earth Loads on Substructure FACTORED	Wind On Substr. + Wind React. From Superstructure - FACTORED	Wind On Live Load From Superstructure - FACTORED
USE ENVELOPES OF FACTORED LOADS	(DC + DW) MAX. (DC + DW) MIN.	(LL+IM+BR)+(CE)+(LS)	(EH) + (EV+ES) MAX. (EH)+(EV+ES) MIN.	(WS) (WL)	
DL-SUPER+DL-SUBSTR+HL-93 L L + EARTH PR. STRENGTH I	**1.25DC + 1.5 DW** **0.9DC + 0.65 DW**	1.75(LL+I+BR)	1.5EH+1.35EV+1.5ES 0.9EH+1.0EV+0.75ES	—	
DL-SUPER+DL-SUB+PERMIT LL + EARTH PR. STRENGTH II	**1.25DC + 1.5 DW** **0.9DC + 0.65 DW**	1.35(LL+I+BR)	1.5EH+1.35 EV+1.5ES 0.9 EH+1.0EV+0.75ES	—	
DL + WS + EARTH PR. STRENGTH III	**1.25DC + 1.5 DW** **0.9DC + 0.65 DW**	–	1.5EH+1.35EV+1.5ES 0.9 EH+1.0EV+0.75ES	1.4 WS	–
DL-SUPER+DL-SUBSTR.+ LIVELOAD+REDUCEDWIND ON SUPERSTR. + WIND ON LIVE STRENGTH V	**1.25DC + 1.5 DW** **0.9DC + 0.65 DW**	1.35(LL+I+BR)	1.5EH+1.35EV+1.5ES 0.9 EH+1.0EV+0.75ES	0.4 WS1.0 WL	

FOR FOOTING DESIGN
STRENGTH IV = 0MY = 0LS = 0ES = 0

CE = 0
IM = 0

ES = EARTH SURCHARGE
LS = LIVE LOAD SURCHARGE

Table 5.5b

ABUTMENT FOUNDATION LOADS: Factored Primary Forces & Moments From Computer Output					
SERVICE LOADS					
LOAD COMBINATIONS FOR SERVICE I, SERVICE II & SERVICE III	Substr. DL + DL Reactions From Superstructure (DC + DW)	LL Reactions From Superstructure (LL+IM+BR)+(CE)+(LS) FACTORED	Earth Loads on Substructure (EH) + (EV+ES)	Windon Substr. + Wind React. From Superstructure (WS) FACTORED	Wind on Live Load From Superstructure (WL)
DL + HL-93 L L + EARTH PR. SERVICE I	1.0DC + 1.0 DW	1.0(LL+I+BR)	1.0EH+1 .0EV+1.0ES	0.3 WS	1.0 WL
DL+PERMIT LL+EARTH PR. SERVICE II	1.0DC + 1.0 DW	1.3(LL+I+BR)	1.0EH+1.0EV+1.0 ES	–	–
DL + WS + EARTH PR. SERVICE III	1.0DC + 1.0 DW	0.8(LL+I+BR)	1.0EH+1.0EV+1.0ES	–	–

Table 5.5c

ABUTMENT FOUNDATION LOADS: Factored Primary Forces & Moments From Computer Output					
EXTREME LOADS					
LOAD COMBINATIONS FOR EXTREME EVENT I	Substr. DL + DL Reactions From Super Structure - FACTORED	LL Reactions From Superstructure	Earth Loads on Substructure FACTORED	Wind On Substr. + Wind React. From Superstructure - FACTORED	Wind On Live Load From Superstructure
	(DC + DW) MAX. (DC + DW) MIN.	(LL)	(EH) + (EV+ES) MAX. (EH)+(EV+ES) MIN.	(WS)	(WL)
DL + EARTHQUAKE EXTREME EVENT I	1.25DC + 1.5 DW 0.9DC + 0.65 DW	0.15(LL)	1.5EH+1.35EV+1.5ES+ 1.0ED 0.9EH+1.0EV+0.75ES+ 1.0ED	0.3 WS	1.0 WL

ED - MONONOBE-OKABE DYNAMIC FORCE
ES = EARTH SURCHARGE
LS = LIVE LOAD SURCHARGE

5.6.4 Primary Load Combined with Extreme Events

Table 5.6a LRFD primary and extreme load combinations for substructures.

	Primary Loads					Extreme Loads							
	(DC1 + DC2) + DW	(LL+ FR + PL) + (LS)	(EH) + (EV) + (ES)	FR	WA	IC	CT	CV	AS	PS	EQ	SE	BL
Earthquake	γ_p	1.75	γ_p	1.00	1.00						1.00		
Ice loads	γ_p	1.35	γ_p	1.00	1.00	1.00							
Vehicle collision	γ_p	—	γ_p	1.00	1.00		1.00						
Vessel collision	γ_p	—	γ_p	1.00	1.00			1.00					
Contraction and local scour at abutments	γ_p	1.35	γ_p	1.00	1.00				1.00				
Contraction and local scour at piers	γ_p	1.75	γ_p	1.00	1.00					1.00			
Superstructure collapse due to blast load	γ_p	1.75	γ_p	1.00	1.00								1.00
Settlement of foundation due to design or construction error	γ_p	1.35	γ_p	1.00	1.00							1.00	

5.6.5 Substructure Design—Service Loads

Table 5.6b LRFD service load combinations for substructures.

DL of super + DL of substr. + live load + reduced wind on super + substr. + wind on live (Service I)	1.0	1.0	1.0	0.30	1.0	1.0	1.0	1.00/1.20	γ_{TG}	γ_{SE}
DL of super + DL of substr. + with max. live load (Service II)	1.0	1.30	1.0	—	1.0	—	1.0	1.00/1.20	—	—
DL of super + DL of substr. + with min. live load (Service III)	1.0	0.80	1.0	—	1.0	—	1.0	1.00/1.20	γ_{TG}	γ_{SE}
DL of super + DL of substr. + with reduced wind on superstr. + substr. (Service IV)	1.0	—	1.0	0.70	1.0	—	1.0	1.00/1.20	—	1.00

5.6.6 Limit State Design for Superstructure Components

A. Deck slab, haunch, girders, parapet, median barrier, sidewalk, railing, sign panel supports
Four extreme events may be described as follows:

- Three load combinations for flood conditions and scour
  - Water impact on pier and superstructure
  - Settlement of pier or abutment from soil erosion under footings
  - Increased axial stress in piles due to reduced embedment length from soil erosion.
- Two seismic load combinations in transverse and longitudinal directions
- Two vehicle collision impacts on superstructure and on pier
- Vessel collision impact on piers located in navigational channel.

Table 5.7 LRFD primary and secondary load combinations for superstructures.

Primary Loads				Secondary Loads			
Load Combinations	Dead Loads (Components + Wearing Surface)	Live Loads and Friction	Wind on Superstr. + LL	Erection Load	Concrete Properties	Temperature Gradient	Support Settlement
	(DC1 + DC2) + DW	(LL+ IM + PL) + (CE) + (LS) + (FR)	(WS) + (WL)	(EL)	(TU) + (CR + SH)	(TG)	(SE)
(DL + LL) + secondary loads (Strength I)	γ_p	1.75	—	—	0.5/1.20	γ_{TG}	γ_{SE}
(DL + permit load) + secondary loads (Strength II)	γ_p	1.35	—	—	0.5/1.20	γ_{TG}	γ_{SE}
(DL + wind) + secondary loads (Strength III)	γ_p	—	1.40	—	0.5/1.20	γ_{TG}	γ_{SE}
(DL) + secondary loads (Strength IV)	γ_p	—	—	—	0.5/1.20	—	—
DL + LL + wind + EL (Strength V)	γ_p	1.35	0.40	1.0	0.5/1.20	γ_{TG}	γ_{SE}

B. Superstructure Design for Extreme Events

Table 5.8 LRFD primary and secondary load combinations for extreme events.

	Primary Loads			Extreme Loads						
	(DC1 + DC2) + DW	(LL+ IM + PL) + (CE) + (LS)	FR	BR	CT	CV	BL	DER	CER	EQ
Earthquake	γ_p	γ_{EQ}	1.00							1.00
Braking force				1.00						
Vehicle collision with superstructure/fire					1.00					
Vessel collision with superstructure						1.00				
Connections failure due to design error								1.00		
Connections failure due to construction error									1.00	
Superstructure collapse due to blast load							1.00			

Note: 1. Besides earthquakes, there are additional extreme events such as vessel loads.

2. The live load and dynamic impact factor for equestrian and elephant loads are different from pedestrian loads.

5.6.7 Superstructure Design for Service Loads

Table 5.9a LRFD primary and secondary load factors for superstructures.

	PRIMARY LOADS			SECONDARY LOADS				
	(DC1 + DC2) + DW	(LL + IM + BR + PL) + (CE) + (LS)	(WS)	(WL)	(EL)	(TU) + (CR + SH)	(TG)	(SE)
(DL + LL + wind) + secondary loads Service I	1.0	1.0	0.30	1.0	1.0	1.00/1.20	γ_{TG}	γ_{SE}
(DL + max. LL) + secondary loads Service II	1.0	1.30	—	—	1.0	1.00/1.20	—	—
(DL + min. LL) + secondary loads Service III	1.0	0.80	—	—	1.0	1.00/1.20	γ_{TG}	γ_{SE}
(DL + wind) Service IV	1.0	—	0.70	—	1.0	1.00/1.20	—	1.00

5.6.8 Superstructure Design for Fatigue Stress (Applicable to Steel Girders)

1. Fatigue limit state.
2. Category C fatigue details and at sections:
   - Refer to Section 6.6.1.2.5
   - Refer to Section 6.10.4.3 for web fatigue stress limit for moment
   - Refer to Section 6.10.4.4 for web fatigue stress limit for shear
   - Refer to Section 6.10.7.4 for stud and channel shear connectors fatigue.

Table 5.9b LRFD primary and secondary load combinations for fatigue.

Fatigue vehicle only	—	0.75	—	—	—	—	—	—

5.6.9 Superstructure Design for Live Load Deflection

Two load combinations for LL deflection, one for HS-20 truck and other with 25 percent HS-20 truck + Lane load.

1. Service limit states I to IV.
2. Refer to Section 6.10.3.2.
3. Control of permanent deflection.

Table 5.9c LRFD primary and secondary load factors for deflection.

Deflection vehicle only	—	1.0	—	—	—	—	—	—

5.6.10 Design Checks

Service limit state checks:

1. L/D ratio < 20.
2. Optional live load deflection check.
3. Flexural stress check.
4. Optional moment redistribution in continuous beams.

Fatigue and fracture limit state: Stress check from single fatigue truck (0.75 × HS-20 for life or 1.5 × HS-20 for infinite life).

Shear stress: Web shear < Buckling shear.

5.6.11 Compact Sections

1. For steel girders, member proportions and compactness will be checked.

   For compact sections at supports of continuous beams, positive and negative moment redistribution is permitted by LRFD specifications which results in economically designed moments (Equations 6.10.5.2.3b1 to 3d1).
2. For redistribution options, negative moment at support is reduced, and positive moment is increased (Section 6.10.2.2).

5.6.12 Non-Compact Sections

For non-compact sections, lateral torsional buckling applies. Moment redistribution facility is not applicable. The following equations are applicable:

Braced compression flange – Yielding limit: R_h is hybrid girder reduction factor for web yielding

1. $f_{bu} + f_l < 6_f R_h F_{yc}$

   Web buckling limit

2. $f_{bu} < 6_f F_{ctw}$

   Strength limit: 1/3 based on inelastic analysis

3. $f_{bu} + f_l/3 < 6_f F_{nc}$

Braced tension flange – Yielding limit: No instability

1. $f_{bu} + f_l < 6_f R_h F_{yf}$

   No lateral bending: Continuously braced flange

2. $f_{bu} < 6_f R_h F_{yf}$

5.7 CONSTRUCTION LOADS AND LOAD COMBINATIONS

5.7.1 Design of Temporary Works

1. Design of temporary works during construction: Design criteria for falsework and formwork will conform to the AASHTO Guide Design Specifications for Bridge Temporary Works.
2. The following construction load combinations based on strength I, III, and V conditions may be used. Construction loads include:
   - Weight of equipment
   - Weight of formwork
   - Weight of materials.
3. When deck pours are accomplished in stages on different days, stresses are induced in concrete due to curing, chemical action, temperature, and shrinkage. Due to compatibility between composite steel flanges and concrete, stresses are locked in steel girders. Such stresses are generally neglected in design. Figure 5.21 shows linear stress distribution for shored beam conditions.

$$\nu = VQ/I_{tran}$$

Shear failure can result during unshored construction.

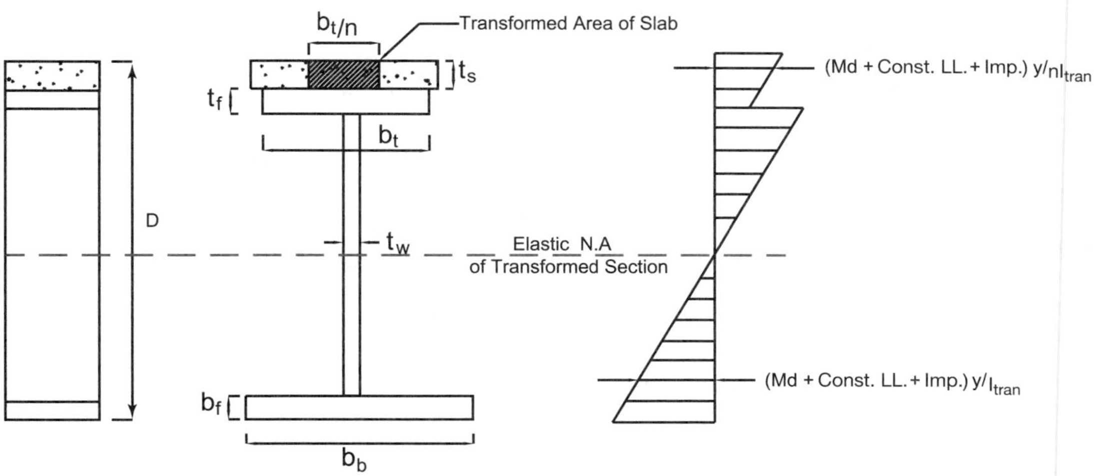

Figure 5.21 Elastic stress distribution for shored beam construction condition.

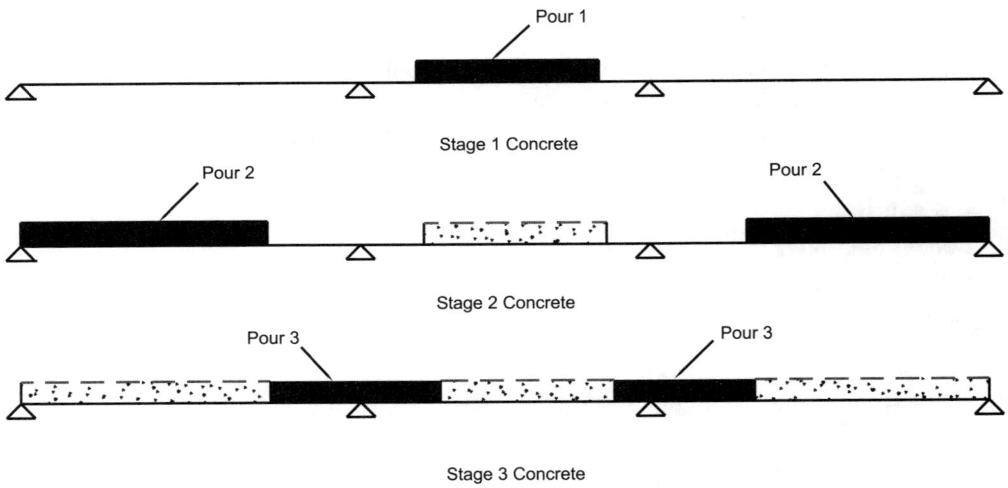

Figure 5.22 Stage concrete deck construction for long spans (density of wet concrete = 160 pcf).

4. The following combinations are added to AASHTO LRFD bridge design code:
For strength I condition use
- All dead load of bridge components + wet concrete symmetric or unsymmetric pours (DC1)
- Utilities (DC2)
- Construction equipment (CE1)
- Construction material such as screed (CE2)
- Construction live load, such as moving trolleys and deck finishing machines (CLL).
  $\gamma_p = 0.9$ to 1.25

5. General dtrength 1 construction live load combination is:
$\gamma_p(DC1 + DC2) + 1.5\,(CE1 + CE2) + 1.5\,(CLL)$.
- Stage 1 symmetric pattern of deck pour is assumed.

$\gamma_p(DC1_{\text{pour-1}} + DC2) + 1.5\,(CE1 + CE2) + 1.5\,(CLL)$
- Stage 2 symmetric pattern of deck pour is assumed.

$\gamma_p(DC1_{\text{pour-2}} + DC2) + 1.5\,(CE1 + CE2) + 1.5\,(CLL)$
- Stage 3 symmetric pattern of deck pour is assumed as shown in Figure 5.22.

$\gamma_p(DC1_{\text{pour-3}} + DC2) + 1.5\,(CE1 + CE2) + 1.5\,(CLL)$

6. Alternate non-symmetric deck pour:
Nonsymmetric pattern of deck pour is assumed. Pattern to be provided by contractor and will vary for each project.
$\gamma_p(DC1_{\text{nonsym pour}} + DC2) + 1.5\,(CE1 + CE2) + 1.5\,(CLL)$

7. For strength III condition, use wind on superstructure including forming:
General construction wind load combination: Wind on superstructure (WS)
$\gamma_p(DC1 + DC2) + 1.5\,(CE1 + CE2) + 1.4\,(WS)$
- Stage 1 symmetric pattern of deck pour is assumed.

$\gamma_p(DC1_{\text{pour-1}} + DC2) + 1.5\,(CE1 + CE2) + 1.4\,(WS)$
- Stage 2 symmetric pattern of deck pour is assumed.

$\gamma_p(DC1_{\text{pour-2}} + DC2) + 1.5\,(CE1 + CE2) + 1.4\,(WS)$
- Stage 3 symmetric pattern of deck pour is assumed.

$\gamma_p(DC1_{\text{pour-3}} + DC2) + 1.5\,(CE1 + CE2) + 1.4\,(WS)$

5.7.2 Symmetric and Non-Symmetric Patterns of Deck Pour

1. Alternate non-symmetric deck pour: Nonsymmetric pattern of deck pour is assumed. Pattern to be provided by contractor

   $\gamma_p(DC1_{nonsym\ pour} + DC2) + 1.5\,(CE1 + CE2) + 1.4\,(WS)$

   For strength V condition, use construction wind load on construction equipment (WCEL), the load combination is

   $\gamma_p(DC1 + DC2) + 1.5\,(CE1 + CE2) + 1.35\,(CLL) + 0.4\,(WS) + 1.0\,(WCEL)$;

   $\gamma_p = 0.9$ to 1.25

   - Stage 1 symmetric pattern of deck pour is assumed.

   $\gamma_p(DC1_{pour\text{-}1} + DC2) + 1.5\,(CE1 + CE2) + 1.35\,(CLL) + 0.4\,(WS) + 1.0\,(WCEL)$

   - Stage 2 symmetric pattern of deck pour is assumed.

   $\gamma_p(DC1_{pour\text{-}2} + DC2) + 1.5\,(CE1 + CE2) + 1.35\,(CLL) + 0.4\,(WS) + 1.0\,(WCEL)$

   - Stage 3 symmetric pattern of deck pour is assumed.

   $\gamma_p(DC1_{pour\text{-}3} + DC2) + 1.5\,(CE1 + CE2) + 1.35\,(CLL) + 0.4\,(WS) + 1.0\,(WCEL)$

2. Alternate nonsymmetric deck pour:

   Nonsymmetric pattern of deck pour is assumed. Pattern to be provided by contractor.

   $\gamma_p(DC1_{nonsym\ pour} + DC2) + 1.5\,(CE1 + CE2) + 1.35\,(CLL) + 0.4\,(WS) + 1.0\,(WCEL)$

   Specification developments and special provisions will ensure quality assurance.

5.7.3 Summary of AASHTO LRFD Load Combinations

Older bridges were originally designed as per ASD or LFD. For new bridges, almost all structural designs in the U.S. follow AASHTO LRFD code.

There are 14 load combinations based on probability for evaluating strength, serviceability, and fatigue in AASHTO code. External actions need to be clearly defined in short-term and long-term behavior. Excessive deformation such as deflection, rotation, curvature, vibration, side sway, and settlement needs to be kept to a minimum.

Principal stresses due to combined bending and shear resulting from settlement of supports may exceed what is allowable and lead to cracking, formation of plastic hinges, and disintegration of elements. When designing for maximum shear stress, principal stress along diagonal planes also needs to be checked.

5.7.4 AASHTO LRFD Shear Capacity Evaluation of Steel Girders

All section numbers are based on AASHTO 2007 specifications.

1. Compact and non-compact sections:
   - Strength limit states I to V
   - Construction limit state and uncured slab
   - For stiffened webs refer to Section 6.10.7.3
   - For unstiffened webs refer to Section 6.10.7.2.
2. Interior panels of compact sections:
   - Strength limit states I to V
   - Uncured slab
   - Refer to Section 6.10.7.3.3a.
3. Interior panels of non-compact sections:
   - Strength limit states I to V
   - Construction limit state and uncured slab
   - Refer to Section 6.10.7.3.3b.

4. Non-compact section compression flange buckling:
   - Strength limit states I to V
   - Construction limit state and uncured slab
   - Refer to Section 6.10.5.3.3c and LRFD Equation 6.10.5.3.3c.
5. Non-composite non-compact section lateral torsional buckling:
   - Positive and negative flexure
   - Strength limit states I to V
   - Construction limit state and uncured slab
   - Refer to Sections 6.10.5.5 and 6.10.6.4.
     - $L_b < 1.76\ r_1\ (E/F_{yc})^{1/2}$

5.7.5 Transverse and Bearing Stiffeners

1. Strength limit states I to V.
2. Refer to Section 6.10.8.
3. Refer to Section 6.10.8.2.1 for bearing stiffener location.
4. Refer to Section 6.10.8.2.4 for bearing stiffener geometry.

5.7.6 Shear Connectors

1. Strength limit states I to V.
2. Refer to Section 6.10.7.4.

5.7.7 Evaluation of Wind Effects Mw During Construction

1. Strength limit states III and V only.
2. Construction limit state and uncured slab.
   Refer to Sections 3.8.1, 4.6.7.2.1 and 6.10.5.7.
3. Construction limit state.
   Refer to Sections 6.10.5.3.2 and 6.10.5.3.3.
4. Section 6.10.5.3.2b for non-compact web slenderness.

5.8 LRFD LOAD COMBINATIONS FOR STRENGTH, SERVICEABILITY, AND EXTREME CONDITIONS

5.8.1 AASHTO Load Combinations

In addition to old non-composite bridges, both composite and hybrid bridges are now being built as a replacement.

The behavior of hybrid bridges is not fully known. They comprise:

1. Precast post-tensioned beams and cast-in-place deck slabs
2. Precast post-tensioned beams and precast deck panels
3. Steel girders with precast deck panels.
4. Cast-in-place reinforced concrete T-beams
5. HPS 70W steel girders with HPC
6. Pedestrian bridges.

Applicable load factors for temperature, shrinkage and creep etc. will vary for each type. Dead load (DC1) load factors may not be identical in all cases.

Fatigue behavior for HPS beams may require a refined analysis. Similarly deflection trucks need to be of different lengths for small and long spans.

For Extreme load conditions the bridges located in different Seismic performance category need to be rated on a different scale. Scour categories also need to be identified.

Future AASHTO Specifications for design and rating are expected to consider the variations in practice and are likely to come up with detailed load factors and resistance factors. The

Figure 5.23 Use of additional pier adjacent to abutment minimizing sharp skew effects.

following load combinations for hybrid (cast-in-place deck and precast beams or precast panel decks and cast-in-place beams) cases are expressed as examples using different load factors for each condition.

AASHTO LRFD Table 3.4.1-1 does not differentiate between primary and secondary loads/effects. Superstructure and substructure load combinations need to be separated. Earth pressure, water loads, and most extreme events need to be tabulated separately. In the following load combinations, the most commonly used loads such as dead load, live load, and wind are used. For extreme events such as floods, AASHTO Extreme Event II is revised.

Unit weights of cast in place (CIP) wet concrete and precast concrete will be different and are differentiated. Appropriate γ_i factors need to be used.

Factors for temperature, creep, and shrinkage will be different for precast construction than those shown in AASHTO Table 3.4.1-1. For a composite or unified behavior between precast units and CIP units, AASHTO factors will not be applicable. The probability of vehicle impact (IM) to happen at the same time as extreme events is very low and is neglected. Construction load combinations not fully covered by AASHTO Table has been added.

5.8.2 Strength (see Table 5.7)

For strength I load combination with HL-93 live load:

$$\varphi R_n > [\eta_{(CIP\ DC1)} \cdot \gamma_{(CIP\ DC1)} Q_{CIP} + \eta_{(Precast\ DC1)} \cdot \gamma_{(Precast\ DC1)} \cdot Q_{Precast} + \eta_{DC2} \cdot \gamma_{DC2} Q_{DC2} + \eta_{WS}\, \gamma_{WS}\, Q_{WS})] + [\eta_{LL} \cdot \gamma_{HL93} (1 + I) Q_{HL93}]$$

Where I is dynamic load allowance or impact factor.

For strength II load combination with permit live load:

$$\varphi R_n > [\eta_{(CIP\ DC1)} \cdot \gamma_{(CIP\ DC1)} Q_{CIP} + \eta_{(Precast\ DC1)} \cdot \gamma_{(Precast\ DC1)} \cdot Q_{Precast} + \eta_{DC2} \cdot \gamma_{DC2} Q_{DC2} + \eta_{WS}\, \gamma_{WS}\, Q_{WS})] + [\eta_{LL} \cdot \gamma_{PERMITL} \cdot Q_{PERMITL}]$$

For strength V load combination with live load:

$$\varphi R_n > [\eta_{(CIP\ DC1)} \cdot \gamma_{(CIP\ DC1)} Q_{CIP} + \eta_{(Precast\ DC1)} \cdot \gamma_{(Precast\ DC1)} \cdot Q_{Precast} + \eta_{DC2} \cdot \gamma_{DC2} Q_{DC2} + \eta_{WS}\, \gamma_{WS}\, Q_{WS})] + [\eta_{LL} \cdot \gamma_{LL} (1 + I) Q_{LL}]$$

For other load combinations:

For strength III load combination with no live load, but wind forces:

$$\varphi R_n > [\eta_{(CIP\ DC1)} \cdot \gamma_{(CIP\ DC1)} Q_{CIP} + \eta_{(Precast\ DC1)} \cdot \gamma_{(Precast\ DC1)} \cdot Q_{Precast} + \eta_{DC2} \cdot \gamma_{DC2} Q_{DC2} + \eta_{WS}\, \gamma_{WS}\, Q_{WS})] + [\eta_{wind} \cdot \gamma_{wind}\, Q_{wind}]$$

5.8.3 Extreme Conditions (see Tables 5.6a and 5.8)

For earthquake load combination with live load:

$$\varphi R_n > [\eta_{(CIP\ DC1)} \cdot \gamma_{(CIP\ DC1)} Q_{CIP} + \eta_{(Precast\ DC1)} \cdot \gamma_{(Precast\ DC1)} \cdot Q_{Precast} + \eta_{DC2} \cdot \gamma_{DC2} Q_{DC2} + \eta_{WS}\, \gamma_{WS}\, Q_{WS})] + [\eta_{LL} \cdot \gamma_{EQ}\, Q_{LL} + \eta_{EQ} \cdot Q_{EQ}]$$

For collision load combination with partial live load:

$$\varphi R_n > [\eta_{(CIP\ DC1)} \cdot \gamma_{(CIP\ DC1)} Q_{CIP} + \eta_{(Precast\ DC1)} \cdot \gamma_{(Precast\ DC1)} \cdot Q_{Precast} + \eta_{DC2} \cdot \gamma_{DC2} Q_{DC2} + \eta_{WS}\, \gamma_{WS}\, Q_{WS})] + [\eta_{CV} \cdot 0.5\, \gamma_{CV}\, Q_{CV}]$$

For collision load combination with partial live load:

$$\varphi R_n > [\eta_{(CIP\ DC1)} \cdot \gamma_{(CIP\ DC1)} Q_{CIP} + \eta_{(Precast\ DC1)} \cdot \gamma_{(Precast\ DC1)} \cdot Q_{Precast} + \eta_{DC2} \cdot \gamma_{DC2} Q_{DC2} + \eta_{WS} \gamma_{WS} Q_{WS})] + [\eta_{CV} \cdot 0.5\, \gamma_{CV} Q_{CV}]$$

5.8.4 Service (see Tables 5.6b and 5.9a)

For service I with live load and partial wind:

$$R_S > [Q_{CIPDC1} + Q_{Precast\ DC1} + Q_{DC2} + Q_{WS})] + Q_{LL} + [\gamma_{wind} Q_{wind}]$$

For service II and III, with factored live load and no wind:

$$R_S > [Q_{CIPDC1} + Q_{Precast\ DC1} + Q_{DC2} + Q_{WS})] + \gamma_{LL} Q_{LL}]$$

For service IV, with no live load and reduced wind:

$$R_S > [Q_{CIPDC1} + Q_{Precast\ DC1} + Q_{DC2} + Q_{WS})] + \gamma_{LL} Q_{LL} + [\gamma_{wind} Q_{wind}]$$

For fatigue with partial live load and impact:

$$\varphi R_n > [\eta_{LL} \cdot \gamma_{HL93} (1 + I) Q_{HL93}$$

$$\varphi R_n > (\gamma_{CIP\ DC1-} + \gamma_{Precast\ DC1-} + \gamma_{DC2} + \gamma_{WS}) Q_i / \eta_i$$

DC is dead weight of component; CIP is cast in place construction; precast is factory made component.

5.8.5 Fatigue (see Table 5.9b)

5.8.6 Deflection (see Table 5.9c)

5.9 LRFD T-BEAM BRIDGE DESIGN

5.9.1 Summary of Design Method

A general method of solution based on LRFD sections of 2007 specifications is presented here. Steps shown are for both primary and secondary loads. An excel Spreadsheet or Mathcad program may be developed for ready use of large number of equations given in AASHTO Specifications. For small spans < 40 feet, reinforced concrete T-beam design is suitable. Interior and exterior beams need to be designed separately.

5.9.2 Structural Planning and Girder Design Data

1. Data required:
- Width of bridge
- Span lengths-(continuous spans are small depth)
- Skew angle
- Girder spacing
- f_c', f_y
- Live load.

2. Develop general plan, elevation, and section, units in mm or inches.

3. Develop typical section and design basis on a trial basis:
- Use top flange thickness from deck design
- Check maximum slab span $< 20 \times t_s$
- Check web thickness

  $b_{min} = 2 \times (\text{conc. cover}) + 3\, d_b + 2\,(1.5\, d_b)$ (AASHTO Sec. 5.10.3.1.1)

4. Check minimum depth of beam:

$h_{min} = 0.065L$ for continuous spans (Table 2.5.2.6.3-1)

5. Reinforcement limits:
- Minimum reinforcement: Use the lesser from

  $\varphi M_n > 1.2\, M_{cr}$

  $\varphi M_n \geq 1.33 \times$ Factored moment required for strength I limit state (5.7.3.3.2)
- Minimum reinforcement: $A_s \geq 0.03\, f_c' / f_y \times$ Area of x-section
- Crack control: $f_s \leq Z / (d_c A)^{0.33} \leq 0.6\, f_y$ (5.7.3.4)

- Longitudinal skin reinforcement required if web depth > 900 mm (5.7.3.4)
- Shrinkage and temperature reinforcement
  $A_s \geq 0.75\ A_g / f_y$ (5.10.8.2)

6. Effective flange widths: (4.6.2.6.1)
- Effective span length = distance between points of permanent load inflexions
- Interior beams $b_i \leq$ ¼ effective span
  $12\ t_s + b_w$
  Average spacing of adjacent beams
- Exterior beams b_e ½ $b_i \leq$ ⅛ effective span
  $6\ t_s +$ ½ b_w
  width of overhang.

7. Non-composite and composite section properties

5.9.3 Resistance Factors

1. Select resistance factors: Table 7.10 (5.5.4.2)

2. Select load modifiers: (1.3.2.1)
$\eta_i = \eta_D\ \eta_R\ \eta_I$ (0.95 1.0 1.0)

3. Select applicable load combinations: (Table 3.4.1-1)
- Strength I limit state
  $U = \eta_i\ [1.25\ DC + 1.5\ DW + 1.75\ (LL + IM) + 1.0\ (WA + FR)]$
- Service I limit state
  $U = 1.0\ (DC + DW) + 1.0\ (LL + IM) + 1.0\ WA + 0.3\ (WS + WL)$
- Fatigue limit state $U = 0.75\ (LL + IM)$

4. Calculate live load force effects:
- Select number of lanes (3.6.1.1.1)
- Select multiple presence factors (3.6.1.1.2)

5. Dynamic load allowance—IM (3.6.2.1)

5.9.4 Live Load Force Effects

1. Distribution factors for moment (4.6.2.2.2)

2. Cross section type (Table 4.6.2.2.1-1)

3. Interior beams with concrete decks (4.6.2.2.2b) and (Table 4.6.2.2.2b-1)

4. Exterior beams (4.6.2.2.2d) and (Table 4.6.2.2.2d-1)

5. Skewed bridges (4.6.2.2.2e) Reduction of live load distribution factors for moment in longitudinal beam are permitted.
$\gamma_{skew} = 1 - c_1\ (\tan \Theta)^{1.5}$
$c_1 = 0.25\ (K_g / Lt_s^3)^{0.25}\ (S/L)^{0.5}$ (Table 4.6.2.2.2e-1)

5.9.5 Load Combination Dead Load Moments and Shears

1. Dead loads for interior girders − DC, DW:
- Calculate interior girder unfactored moments and shears
- Dead loads for exterior girders − DC, DW
- Calculate exterior girder unfactored moments and shears.

2. Investigate service limit state:
- Investigate durability (C.5.12.1)
- Crack control (5.7.3.4).

5.9.6 Plastic Distributed Live Load Moments and Shear

$M = mgr\,[(M_{Tr} \text{ or } M_{Ta})\,(1 + IM/100) + M_{Ln}]$
Tr – Truck load
Ta – Tandem load
LN – Lane load

1. Distribution factors for shear: (4.6.2.2.3)
Cross section type (e) (Table 4.6.2.2.1-1)
- Interior beams (4.6.2.2.3a and Table 4.6.2.2.3a-1)
- Exterior beams (4.6.2.2.3b and Table 4.6.2.2.3b-1)
- Skewed bridges (4.6.2.2.3c and Table 4.6.2.2.3c-1)
- Distributed live load shears
  $V_{LL+IM} = mgr\,[(V_{Tr} \text{ or } V_{Ta})\,1.33 + V_{Ln}]$

2. Reactions to substructure: (3.6.1.3.1)
Reactions are per design lane without any distribution factors. The lanes shall be positioned transversely to produce extreme force effects.
$R_{sup} = V_{sup} = 1.33\,V_{Tr} + V_{LN}$
Calculate force effects from other loads.

5.9.7 Investigate Strength Limit State

1. $U = \eta\,[1.25DC + 1.25DC2 + 1.5DW + 1.75\,(mgr)\,LL\,(1 + IM)]$
- Flexure
  $M = \eta\,\Sigma\gamma_i M_i = 0.95\,(1.25\,M_{DC} + 1.50\,M_{DW} + 1.75\,M_{LL+IM})$
  Check $\varphi\,M_n > M_u$
  Check $\rho = A_s / A_g > \rho_{min}$
- Analysis
  Main reinforcement perpendicular to traffic
- Deck thickness for deflection control
  $h\,min = (s + 3000) / 30 < 175$ (Table A2.5.2.6.3-1)

2. Dead loads:
- beams
  $M\,max = + wL^2 / 12$ and $- wL^2 / 8$
- 4 beams
  $M\,max = + wL^2 / 10$ and $- wL^2 / 10$
- 5 or more beams
- Use influence lines at 0.4 L of first span
  $M = 0.0744\,wL^2$
- Adjust for overhang moment $- w\,l^2 / 2$
  2nd span $M = + 0.0471\,wL^2$ and $- 0.1141\,wL^2$ plot B.M. and S.F. diagrams.

3. Live loads:
Overhang 1140 + 0.833 X
- IM = 33 percent of LL (3.6.2.1)
- Number of lanes
  N = Int (Roadway width / 3600) (3.6.1.1.1)
- Multiple presence factor

M = 1.2 for one loaded lane
= 1.0 two
= 0.85 three

- Tire contact area

Wheel load is applied as a distributed load. For design live load both HL-93 and state permit HL-93 single span loads are applicable (for truck and lane live load moments see Chapter 4, Sections 4.11 and 4.12. Tables 4.5 and 4.6 refer to Section 5.8 for strength and serviceability conditions). For continuous beam positive moments Tables 4.5 and 4.6 may be used for preliminary design and checked by a continuous beam analysis software.

5. Design:

$M_u = \varphi M_n = A_s f_y (d - a/2)$

$a = A_s f_y / 0.85 f_c' b$

$A_s = (M_u / \varphi) / f_y jd$

6. Check ductility: $a \leq 0.35$ d

Minimum reinforcement $\rho \geq 0.03 f_c' / f_y$ (5.7.3.3.2)

Maximum spacing smax = 450 mm or $1.5 \cdot h$ (5.10.3.2)

Distribution reinforcement = $3840/\sqrt{S_e}$ % (9.7.3.2)

where S_e is the effective span length. (9.7.2.3)

Shrinkage and temperature reinforcement:

Temp $A_s \geq 0.75 A_g / f_y$ (5.10.8.2)

Crack control: $f_s \leq Z / (d_c A)^{0.33} \leq 0.6 f_y$ (5.7.3.4)

Z = 23000 N/mm

$d \leq 50$ mm

5.9.8 Calculate Deflection and Camber (Table 3.4.1-1)

Service I limit state = 1, gravity load factors = 1

- U = DC + DCE + DW + LL + IM
- LL deflection criteria
- Distribution factor for deflection.

Δ_{allow} = Span / 1000 for bridge with sidewalks

Section properties $E_c = 4800 \sqrt{f_c'}$, $f_r = 0.63 \sqrt{f_c'}$

Calculate I_g,

Calculate I_e

$M_{cr} = f_r I_g / y_t$

Calculate live load deflection at location where moment is maximum.

$I_e = (M_{cr} / M_a)^3 I_g + [1 - (M_{cr} / M_a)^3] I_{cr} \leq I_g$

Dead load camber w DL = w DC + w DC2 + w DW

Perform unit load analysis and calculate deflections.

5.9.9 Investigate Fatigue for Steel

Fatigue limit state (Table 3.4.1-1)

$U_f = 0.75$ (LL + IM)

- Determine need to consider fatigue

- Allowable fatigue stress range (5.5.3.2)

$f_f = 145 - 0.33f_{min} + 55\ (r/h)$ MPa

f_{min} = Algebraic minimum stress level from fatigue load; $r/h = 0.3$

5.10 SOFTWARE FOR SUPERSTRUCTURE

Tables 5.10a and 5.10b list some of the software that may be used. Over time, more sophisticated software is being developed.

Table 5.10a Frequently used computer programs for superstructures.

Program I.D.	Authors	Web Site/E-Mail**
AASHTO Opis	AASHTOWare®	Vigesa@email.msn.com
LRFD Live Load Generator LRFD Prestressed Beam	Florida DOT	www.dot.state.fl.us
CONSYS 2000 CONSPAN LA	LEAP Software	www.leapsoft.com
MDX Curved and Straight Steel Girder	MDX Software, Inc.	www.mdxsoftware.com
Merlin-DASH	OPTI-MATE, Inc.	www.opti-mate.com
STLRFD—LRFD Steel Girder Design and Rating Composite Non-compact Section PSLRFD—LRFD Prestressed Concrete Girder Design and Rating FBLRFD—LRFD Floorbeam Analysis and Rating	PennDOT	www.dot. state.pa.us
BPLRFD—LRFD Bearing Pad Design and Analysis SPLRFD—LRFD Steel Girder Splice Design and Analysis		
Sam	Bestech Systems Ltd.	www.bestech.co.uk/Bridges@ bestech.co.uk
BRASS-GIRDER (LRFD)	Wyoming DOT	www.wydotweb.state.wy.us
SAP2000	Computers & Structures, Inc.	www.csiberkeley.com
WinStrudl Pro	CAST	www.civilstructure.com
ROBOT Millenium	Integrated Structural Software, Inc.	www.robot-structures.com
RAM Structural System	Bently Systems, Inc.	www.bentley.com
Dynamics & Vibration Analysis	Universal Technical Systems, Inc.	www.uts.us.com
RISA-3D Integrated Software	RISA Technologies, LLC	www.risatech.com

Table 5.10a Frequently used computer programs for superstructures *(continued).*

Program I.D.	Authors	Web Site/E-Mail**
DESCUS Merlin-DASH	OPTI-MATE, Inc.	www.opti-mate.com
GT STRUDL	Georgia Tech CASE Center	www.gtstrudl.gatech.edu
STAAD.Pro MicroStation	Bentley Systems Inc.	www.bentley.com
AASHTO Virtis (Bridge Load Rating)	AASHTOWare®	Vigesa@email.msn.com

**E-mail address is subject to change.

Table 5.10b Frequently used computer programs for substructures.

Program I.D.	Authors	Web Site/E-Mail
RC-PIER®	Leap Software	www.leapsoft.com
Merlin-RCWall	OPTI-MATE, Inc.	www.opti-mate.com
Sam	Bestech Systems Ltd.	www.bestech.co.uk/Bridges@bestech.co.uk
ABLRFD—LRFD Abutment and Retaining Wall Analysis and Design PAPIER Pier Design BXLRFD—LRFD Box Culvert Design and Rating	PennDOT	www.dot.state.pa.us

**E-mail address is subject to change.

5.11 SOFTWARE FOR SUBSTRUCTURE

Seismic analysis and design software is listed in Chapter 12.

Engineers are advised to review and evaluate program defaults and make the necessary modifications to ensure that bridge components are designed in accordance with the AASHTO design criteria as modified by this section.

- Construction loads and procedures—Many failures seem to happen during construction. Site organization is based on selection of one general contractor, who in turn selects several subcontractors, who have specialized in a particular trade such as concreting, formwork, steel fabrication, bearings, reinforcing steel, etc. One of the difficulties is that construction practice varies from state to state and from job site to job site.
- Current design specifications do not seem to cover in detail construction related design for temporary loads. Future construction codes should address issues created by use of the latest technology. Technical specifications may also be made comprehensive to give minute details about construction procedures.
- Accelerated bridge construction—Modern construction technology seems to be pulling the train on design methods. Precast technology is a world apart from traditional wet construction methods. Self-propelled modular transportation (SPMT) has enabled the transportation of long span girders without the need for splices.
- Connections used for precast methods are different from those used in traditional construction.

- Stresses at demolition or at failure: Many of the failures can be avoided by paying attention to changes in design technology and details. For example, new concrete materials such as lightweight and heavyweight aggregates, structural plastics, and glass composites have different unit weights to that of conventional wet concrete.

Such materials display nonhomogeneous and non-isotropic behavior. Current load factors and resistance factors need to be modified in the light of experimental results. AASHTO load combinations given in Table 3.4.1-1 do not differentiate between primary and secondary effects. Also, superstructure and substructure load combinations need to be separated. Earth pressure, water loads, and most extreme events need to be tabulated separately.

BIBLIOGRAPHY

AASHTOWARE "OPIS Software for Bridge Design."

AASHTOWARE "VIRTIS Software for Bridge Rating."

AASHTOWARE "PONTIS Software for Bridge Management."

American Iron and Steel Institute, "Useful Information on the Design of Plate Structures," 1979 AREMA Manual for Railway Engineering, Vol. 2.

Barker R. M. and J. A. Puckett, *Design of Highway Bridges*, John Wiley & Sons Inc., 1997.

Brockenbrough, R. L. and Johnston, B. G., Steel Design Manual vs Steel Corporation.

Deflection Limitations of a Bridge: Proceedings of the American Society of Civil Engineers, American Society of Civil Engineers, J. Struct. Div., 84, Rep. No. ST 3, 1958.

Evaluating Scour at Bridges, Federal Highway Administration, HEC-18, Washington, DC, 2001.

Fountain, R. S. and C. E. Thunman, "Deflection Criteria for Steel Highway Bridges," *Proceedings of the AISC National Engineering Conference*, New Orleans, LA, 1987.

Guide Specifications for Design of FRP Pedestrian Bridges, American Association of State Highway and Transportation Officials, 2008.

Guide Specifications—Thermal Effects in Concrete Bridge Superstructures, American Association of State Highway and Transportation Officials, Washington, DC, 1989.

Inspection of Fracture Critical Bridge Members, Federal Highway Administration, Washington, DC, 1999.

Khan, Mohiuddin, Ali, Elastic Interaction of Floors and Frames, Master of Philosophy thesis, University of London, 1968.

LRFD Bridge Design Specifications, 4TH Edition, American Association of State Highway and Transportation Officials, Washington, DC, 2007.

Manual for Condition Evaluation of Bridges, 2nd Edition, American Association of State Highway and Transportation Officials, Washington, DC, 2000.

Movable Bridge Inspection, Evaluation, and Maintenance Manual, American Association of State Highway and Transportation Officials, Washington, DC, 1998.

NBIS Code of Federal Regulations, 23 Highways Part 650 Subpart C.

NHI Bridge Coatings Inspection Course, National Highway Institute.

NHI Fracture Critical Inspection Techniques for Steel Bridges Course, National Highway Institute.

NHI Stream Stability and Scour at Highway Bridges Course, National Highway Institute.

Oehler, L. T., "Bridge Vibration—Summary of Questionnaire to State Highway Departments," *Highway Research Circular, Highway Research Board*, No. 107, National Research Council, Washington, DC, 1970.

Park, Sung H., *Bridge Inspection and Structural Analysis: Handbook of Bridge Inspection*, S. H. Park, Trenton, NJ, 1980.

Park, Sung H., *Bridge Rehabilitation and Replacement (Bridge Repair Practice)*, S. H. Park, Trenton, NJ, 1984.

PCI Bridge Design Manual, Pennsylvania Department of Transportation, Chapter 8.

Proceedings of the 12th U.S.–Japan Bridge Engineering Workshop, October 1996, Buffalo, New York.

"Recording and Coding Guide for the Inventory and Appraisal of the Nation's Bridges," Report No. FHWA-PD-96-001, Washington, DC, 1995.

Rehm, K.C., "Future of Bridge Design," *Bridges*, January/February 2007.

Roeder, C. W., K. E. Barth, A. Bergman and R. A. Christopher, "Improved Live-Load Deflection Criteria for Steel Bridges," National Cooperative Highway Research Program, NCHRP Interim Rep. 20-07/133, National Research Council, Washington, DC, 2001.

Saadeghvaziri, M. Ala, "Finite Element Analysis of Highway Bridges Subjected to Moving Load," *Journal of Computers and Structures*, Vol. 49, No. 5, 1993.

Saadeghvaziri, M. Ala, "Response of Struve-Slough Bridge Under the Loma Prieta Earthquake," *Second Workshop on Bridge Engineering Research in Progress*, Reno, NV, October 1990.

Saadeghvaziri, M. Ala and R. Hadidi, "Cause and Control of Transverse Cracking in Concrete Bridge Decks," Final Report, FHWA-NJ-2002-19, December 2002.

Saadeghvaziri, M. Ala and Feizi Bakhtiar, "Beneficial Aspects of a Multi-Hazard Approach to Design for Highway Bridges," *Proceedings, Sixth National Conference on Bridges and Highways*, Charleston, SC, July 2008.

Saadeghvaziri, M. Ala and R. Hadidi, "Transverse Cracking of Concrete Bridge Decks: Effects of Design Factors," *Journal of Bridge Engineering*, American Society of Civil Engineers, Vol. 10, No. 5, 2005.

Seismic Retrofitting Manual for Highway Bridges, Federal Highway Administration, Publication No. FHWA-RD-94-052, Washington, DC, 1995.

Standard Specifications for Highway Bridges, 17th Edition, American Association of State Highway and Transportation Officials, Interim Revisions, 1st Edition, Washington, DC, 2002.

Standard Specifications for Structural Supports for Highway Signs, Luminaires, and Traffic Signals, 6th Edition, American Association of State Highway and Transportation Officials, 2008.

Timoshenko, S. P. and Woinowsky-Krieger. S., "Theory of Plates and Shells." McGraw-Hill International Edition, 2nd Edition, 1959.

Walker, W. H. and R. N. Wright, "Criteria for the Deflection of Steel Bridges," *Bulletin for the American Iron and Steel Institute*, No. 19, AISI, Washington, DC, 1971.

6

Applications of Bridge Design and Rating Methods

Section 2

Strengthening and Repair Work

6.1 INTRODUCTION

In earlier chapters, LRFD and LRFR methods based on ultimate strength design for bridge members were introduced for the purpose of rehabilitation design and replacement of bridges. This chapter deals with an extension of the methods covered earlier. Prime equations and formulae are developed for flexural capacity under moving H-15 and H-20 truck loads for single spans using LRFD and LRFR methods, including application of shear design using MCFT.

Examples for deck slab, reinforced concrete, prestressed concrete and steel beams, and connections design are presented. It is possible to program the extended formulae, such as for construction load combinations for the routine design of members:

1. Using Excel spreadsheets.
2. Using Mathcad.
3. Developing commercial software.

AASHTO bridge design methods are expressed in detail.

6.2 LIMIT STATES DESIGN EQUATION

6.2.1 Design Methods and Details

The three criteria for choosing a design method are:

1. Strength control or stress criteria.
2. Deflection control.
3. Crack control.

Stress criteria: There are several considerations in bridge design, of which avoiding overstress is the most important. The magnitude of stress depends upon response of the material and its age.

The response of material is based on physical properties such as elastic modulus and modulus of rupture and on the size and shape of the member. It is usually expressed as:

1. Elastic stress.

2. Yield stress.

3. Plastic stress.

4. Failure or collapse stress.

Generally, the extreme limit of elastic stress is yield stress, the extreme limit of yield stress is plastic stress, and the extreme limit of plastic stress is failure or collapse stress.

- Glass, ceramic, or plastics display fragile behavior
- Concrete displays brittle behavior
- Steel displays ductile behavior.

Hence, glass needs to be reinforced with wire, plastics reinforced with fiber, and concrete made composite reinforced with rods.

5. Stress history is based on applied loads. Evaluation of various stages in the life of bridge components and chronological assessment of their performance are required so that limits can be placed either in design or in practice to prevent failure. Sources of stress include:

- Fabrication stress
- Transportation stress
- Erection stress
- Stresses resulting from maintenance loads.

6. Stresses at demolition or at failure: Many bridge failures can be avoided by paying attention to changes in design technology and details. For example, new concrete materials such as lightweight and heavyweight aggregates, structural plastics, and glass composites have different unit weights than conventional wet concrete. Such materials display nonhomogeneous and non-isotropic behavior. Current load factors and resistance factors need to be modified in the light of experimental results.

6.2.2 Internal and External Effects

Factored resistance at any location of structure > Σ factored load effects acting at that location.

Factored resistance = ϕR_n

$\phi R_n > \eta_i \Sigma \gamma_i Q_i$ when γ_i selected is maximum.

$\phi R_n > \Sigma \gamma_i Q_i / \eta_i$ when γ_i selected is minimum.

Factored resistance = ϕR_n, ϕ = Resistance factor

γ_i = Load factor (AASHTO Table 3.4.1-1 and Table 3.4.1-2)

Q_i = Nominal force effect

R_n = Nominal resistance—it is the mean or an identified level of strength.

η_i = Load modifier

Table 6.1 AASHTO recommended values for load modifier η.

		Strength	Service	Fatigue	Deflection	Construction (Wet Concrete)
Ductility	η_D	0.95	1.0	1.0	1.0	1.0
Redundancy	η_R	0.95	1.0	1.0	1.0	1.0
Importance	η_I	1.05	N/A	N/A	1.0	1.0
η_i =	($\eta_D \eta_R \eta_I$)	0.95	1.0	1.0	1.0	1.0

LRFD maximum value of η_i = 1.0, Table 5.5.

Minimum value of $\eta_i = 1/(\eta_D \eta_R \eta) = 0.95$. Some states like Pennsylvania use a higher maximum value of η_i.

When $\eta_i = 1$, both equations are the same.

Example of load modifier η value used for reinforced concrete bridge design: (1.3.2.1)

6.2.3 Resistance Factors

Minimum value is 0.85, and maximum is 1.0 (Table 5.6).

Table 6.2 AASHTO recommended values for resistance factors.

Material	Limit State	$\phi = \gamma_R(1 - 2V_R)$	ϕ_{LFD}*
Noncomposite steel	Moment	0.90	0.90
Noncomposite steel	Shear	0.90	0.85
Composite steel	Moment	0.90	1.00
Composite steel	Shear	0.90	1.00
Reinforced concrete	Moment	0.85	0.90
Reinforced concrete	Shear	0.85	0.85
Prestressed concrete	Moment	0.90	0.90
Prestressed concrete	Shear	0.85	0.85

* Strength reduction factor

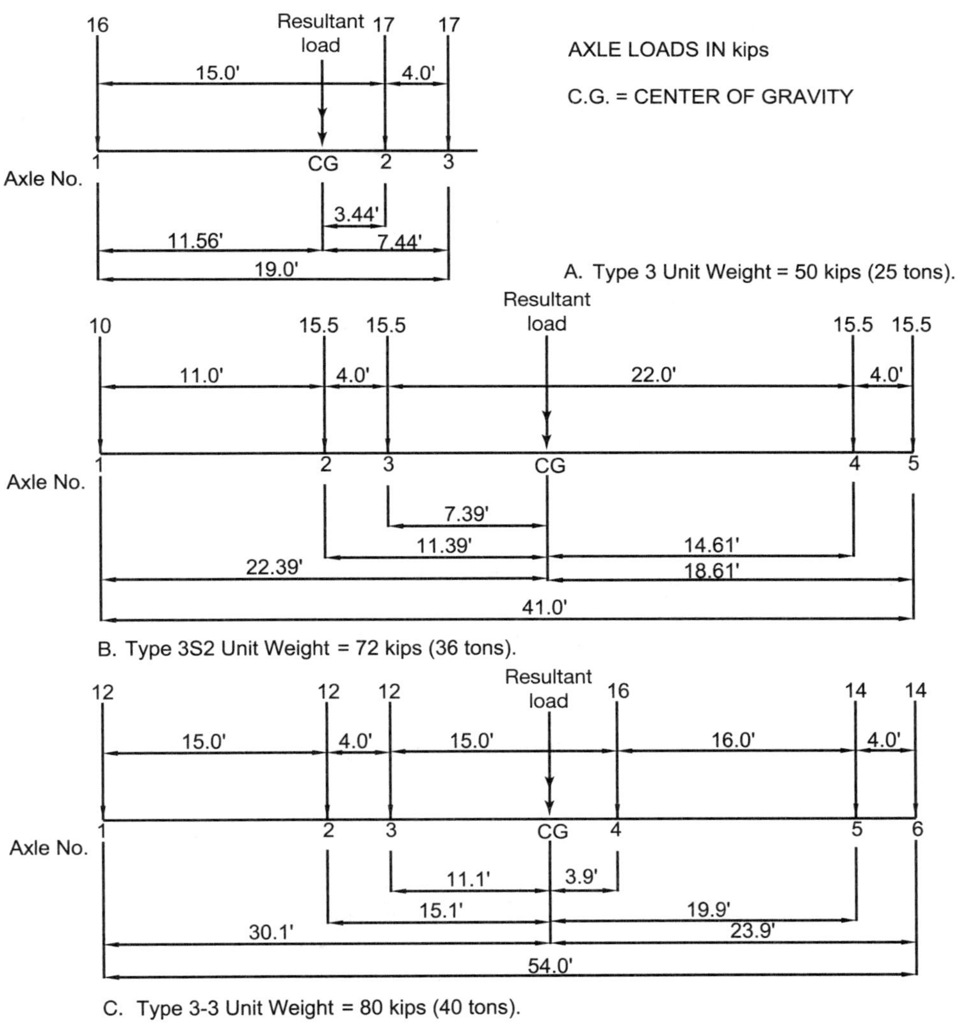

Figure 6.1 AASHTO recommended legal loads.

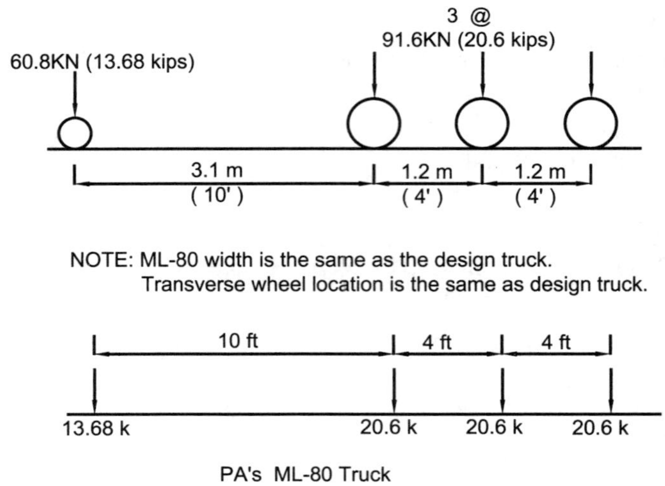

Figure 6.2 Pennsylvania's 4-axle (ML-80) truck.

6.3 LEGAL LOADS

6.3.1 AASHTO Legal Loads for Rating

6.3.2 Examples of Modifications to AASHTO Legal Loads

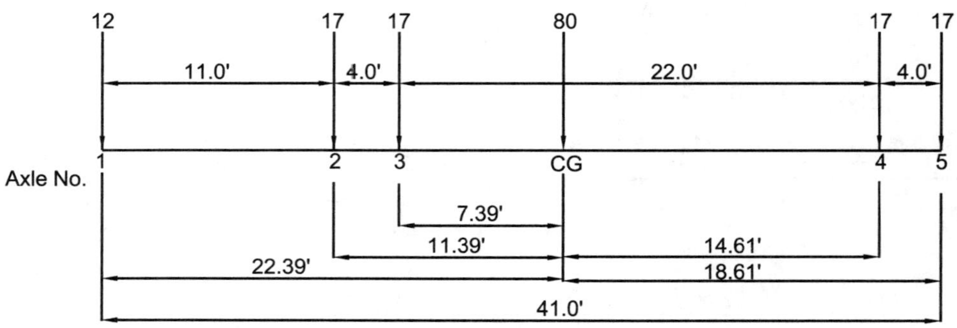

Figure 6.3 New Jersey's modifications to AASHTO vehicle rating (3S2).

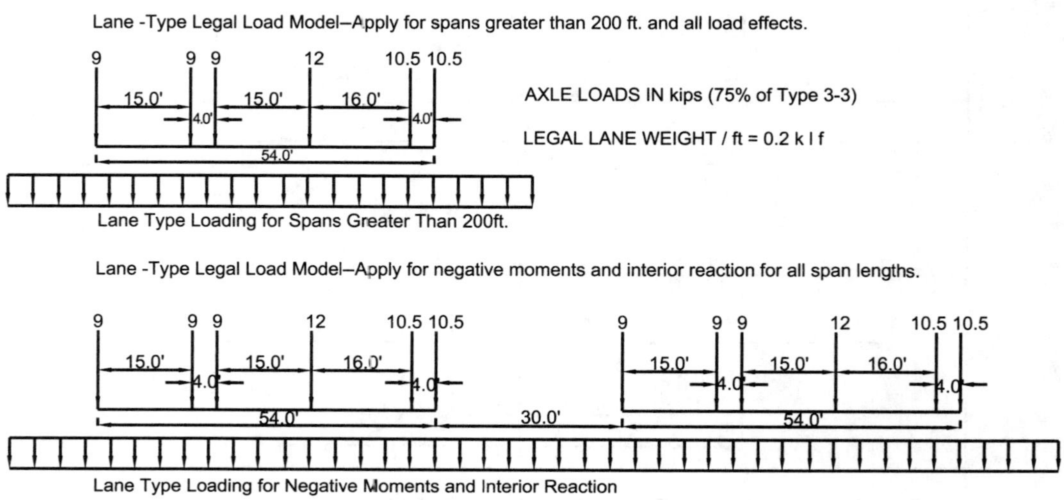

Figure 6.4 Legal loads for rating.

Some states such as Pennsylvania and New Jersey have modifications to AASHTO legal loads. Pennsylvania has adopted a four-axle 18 ft long vehicle known as ML-80. New Jersey has adopted legal vehicles types 3 and 3-3 but modified vehicle type 3S2.

6.3.3 Deficient Bridge

Detailed structural evaluation is required to establish deficiency. The steps include:

1. Frequent inspection
2. Material testing to assess strength of structural components and condition of materials
3. Load rating
4. Nondestructive load testing
5. Remaining fatigue life evaluation.

For conventional bridges, if there is a functional or structural deficiency the following options are available:

- Repair
- Rehabilitate
- Replace
- Widen
- Load posting
- Abandon or no action.

6.3.4 Bridge Design and Rating Engineering

A designer has to face nitty-gritty of developing design calculations; CADD drawings and the related documentation. It appears that in the recent years this multi-disciplinary task has become specialized and requires the following:

- Public outreach to educate about staging; temporary lane closure and detour
- Obtaining as-built drawings
- Field visit and reconnaissance
- Generating geometric data
- Performing deck drainage calculations and providing scuppers.
- Coordinating with the following disciplines:
    1. Highway engineer to obtain alignment; baseline and profile grade line
    2. Surveyor to obtain field survey results for topography; grade elevation;
    3. Hydraulics engineer for river characteristics such as flood elevation; size of opening
    4. Environmental engineer for maintaining air quality or water quality
    5. Traffic engineer for number of lanes; acceleration and deceleration lanes; location of sign structures
    6. Geotechnical engineer for soil information; shallow and deep foundations
    7. Utilities engineer for bridge mounted utilities
    8. Right of way engineer to allow access to site during construction and materials storage
    9. Electrical engineer for bridge lighting

10. Railway engineer for bridges locate over railroad for span length and construction issues
11. Vendors for availability of products in time. Examples are: Precasting; welded and bolted connections; bearings; deck joints; water proofing post-tensioning; repair materials and any special products etc.
12. Ensuring availability of high performance steel; high performance concrete
13. Obtaining approval from client for using a commercial software
14. Project management: Client coordination; In-house progress meetings; setting milestones and submission dead lines.

6.3.5 Design Requirements

All calculations are code based. LRFD Specifications and LRFR Manual for Condition Evaluation have provided design and rating guidelines. There are dozens of formulae in each chapter which need to be understood.

Rehabilitating designing and constructing may cost multi-billion dollars for highway and transit structures; the cost of NJ Turnpike widening project on which author worked recently is alone worth 3 billion dollars while cost for Oakland Bay Bridge in San Francisco is over twice as much.

In this book emphasis is laid down on explaining and extending AASHTO codes for design and rating. LRFD numerical examples such as for deck slab replacement and the necessary steps for concrete and steel bridges are given. Focus is on concepts rather than details.

The difficulties in giving too many examples are that they become specific for each revision of AASHTO code. With changing regulations solved examples cannot be used with latest revisions and it may sometimes even be misleading when using a superseded procedure.

Computer programs are also being revised and the output format also changes.

Advantage of hand calculations seems to be in preliminary sizing of members or checking the results of a computer output.

Most highway agencies no longer request formal design calculations as part of the submission and others only require computer solutions of approved software.

Use of computer has the advantage that any errors in arithmetic or misuse of units are avoided.

6.3.6 Rating Aspects for Existing Bridges

Rating is dependent upon:

- Magnitude of live load
- Intensity and frequency of traffic
- Number of lanes
- Existing conditions of structure
- Bridge age
- Material properties

6.3.7 Knowledge Database for Rating and Design

Excellent LRFD solved examples are provided on various aspects in the following text books. These numerical examples may be referred to for use in detailed rating and design calculations as they are useful in understanding the subject fundamentals.

The following references are available on the subject of rating and design:

1. AASHTO Manual for Condition Evaluation for LRFR—Rating formula generally refer to AASHTO LRFD Specifications.

2. Sung H. Park, "Bridge Superstructure Design and Rehabilitation, Trenton, NJ - Examples of rating and design using both WSD LFD and LRFR methods.

- LRFD design aspects for new and replacement bridges:
  1. Tonias, D. and Zhao, J., Bridge Engineering - Design of composite concrete and steel bridges using both LFD and LRFD methods.
  2. Barker, R. and Puckett, J., Design of Highway Bridges - Design of single and multiple span composite concrete; steel and wood bridges using LRFD methods.
  3. FHWA, "LRFD for Highway Bridge Substructures and Earth Retaining Structures", Reference Manual, Publication No. FHWA-NHI-05-095.
- LFD design aspects:
  1. Xanthakos, Petros P., "Theory and Design of Bridges," John Wiley & Sons, New York, 1993.
  2. PennDOT Bridge Design Training Manual, PDT- Pub. No. 302, 1994.
  3. Mufti, Aftab A., Bakht, Baidar, Jaeger Leslie G., "Bridge Superstructures New Developments", National Book Foundation, Islamabad, 1996.
  4. Heins, Conrad and Lawrie, Richard, "Design of Modern Concrete Highway Bridges", J. Wiley and Sons, 1984.
- FHWA and NCHRP publications
- ASCE Journal of Bridge Engineering
- International conference proceedings - Bridge conferences in Pittsburgh and New York held regularly and worldwide conferences.
- Foreign publications in English languages - Canadian codes British codes Indian Australian and New Zealand codes.
- Short courses:

  Bridge engineers are referred to specialist courses in rating or design being offered by National Highway Institute (NHI) and some universities.

  One or two day courses offered by the author on LRFD design of bridges; seismic analysis and design; long span steel cable structures; scour analysis and design of scour countermeasures.

6.3.8 Details of Rating Methods

The condition of a bridge is evaluated by structural assessment of its components. As described in AASHTO Rating Manual, ratings types are:

1. Live load rating: This is used to determine the usable live load capacity by inspection and by rating. Each component is evaluated and the lowest component rating is the most critical. It is expressed in tons.

The load factor method used the following approach for a long time:

Rating Factor $= (C - A1D) / A2\ (L + I)$ AASHTO LRFR Eq. (D.6-1)

$C =$ Capacity of member

$D =$ Dead load effect on member

$L =$ Live load effect on member

$I =$ Impact factor

$A1 =$ Dead load factor (AASHTO LRFR D 6.5.2)

$A2 =$ Live load factor (AASHTO LRFR D 6.5.3)

The following formula is recommended:

Live load capacity rating =

$$\frac{\text{(Allowable load} - \text{Dead load)}}{\text{(Rating vehicle live load} + \text{Impact)}} \times \text{Vehicle Weight (tons)}$$

Rating vehicles include:

- HS vehicles
- H vehicles
- Legal load vehicles
- Military loads
- Permit loads
- Type 3 unit; Type 3-S2 unit; Type 3-3 unit.

LRFR Factor = $(C - \gamma\ DC - \gamma\ DW + -\gamma\ P) / \gamma\ (LL + I)$

$C = \phi c\ \phi s\ \phi\ Rn$

= LRFD Resistance Factor. $\phi s = 1$ for redundancy

$\phi c\ \phi s > 0.85$

C = Allowable stress in the Service Limit State.

Live load rating is for two load levels

- Inventory rating: A load level which can safely utilize a bridge for an indefinite period of time. Inventory level reflects the existing bridge and material conditions with regard to corrosion, loss of section and other deficiencies. It does not exceed the design stresses.

  For steel girders the allowable inventory stress is 55% of yield stress:

  Grade A 36 – 19.8 ksi

  Grade 50 or 50W – 27.5 ksi

  Grade 70W – 38.5 ksi

- Operating rating: The absolute maximum permissible load to which the bridge can be subjected. The use of bridge by unlimited number of heavier vehicles would exceed the capacity and is not permitted.

For example the load produced by a permit vehicle must be lower than the structural capacity determined by operating rating.

2. Sufficiency Rating: It is defined as a *calculated rating indicating the bridge's sufficiency or capability.*

Ratings range from 100 (entirely sufficient) to 0 (entirely insufficient or deficient).

Sufficiency Rating is considered by the federal government when a state requests federal bridge dollars to improve the condition of the bridge. Bridges with low sufficiency ratings are eligible for more funds.

Sufficiency Rating vs.	Funding Eligibility
0–49	Eligible for costs to replace bridge
50–79	Eligible for costs to rehabilitate or refurbish bridge
80–100	not available

Factors included in the calculation are:

- Structure's adequacy and safety - accounting for 55% and based on inspection data,

- Structure's serviceability and functional obsolescence—accounting for 30% and based on ability of bridge to meet current traffic conditions, and
- How essential the bridge is for public use - accounting for 15%.

3. Vulnerability rating

If rating is based on extreme conditions to which the bridge may be subjected in its life time and is vulnerable to fail as a result it is called vulnerability rating.

Examples are seismic vulnerability or scour vulnerability etc. The behavior of the substructure or superstructure components under such cases is studied so that a failure can be avoided by a retrofit etc.

4. Rating of historic bridges

A record of bridges of historic importance is maintained by National Register of Historic Places.

Survey information is available on a database kept by the state. General requirements are:

1. Over 50 years age of bridge
2. Stone arch, metal truss and bridges with technological importance.
3. Bridges located in on historic routes or districts.

Instead of replacement of original components, repair is preferred. Strengthening is carried out at a higher cost than nonhistoric bridges. Nondestructive testing methods to verify component performance are applied. State guidelines or Secretary of Interior's Standards for the treatment of Historic Properties are a good guide for maintenance criteria.

6.3.9 Posting Analysis

As described in AASHTO Rating Manual:

1. When RF > 1.0 Posting is not required.
2. When RF > 0.3 but < 1.0 the safe posting load = W (RF −0.3)/0.7 AASHTO LRFR Eq. (6-7).
3. When lane load governs, W = 40 Tons.
4. For any vehicle when RF < 0.3 the vehicle type should not be allowed on the span.
5. When RF for all three legal loads (Type 3; Type 3-S2 and Type 3-3 units) < 0.3 the bridge shall be closed for vehicular traffic.
6. Speed limit may be lowered to reduce Impact live load.
7. All Regulatory or warning signs shall conform to AASHTO Manual on Uniform Traffic Control Devices (MUTCD).

6.3.10 Diagnostic Load Testing as an Alternative or a Supplement to Load Rating

A nondestructive load test is useful for important or complex bridges. It provides in-depth structural data. In most cases a theoretical analysis or rating is carried out. In some cases as built drawings may not be available to perform theoretical analysis. After analysis of field test results the live load rating may be revised. In Section 3, methods of nondestructive testing, repair and retrofit are discussed. The test vehicle weight is selected.

Cost of load test and condition of bridge determines the need for load test.

Load test is performed by installing instrumentation without closing the bridge.

1. Both linear and nonlinear behavior is studied.
2. Strains, deformations and rotations are measured by field observation. Strain sensors, electrical resistance gages, acoustic strain gages or strain transducers are attached to critical locations. Measurements are taken by remote sensors both at the start of test and at selected increments. Displacements are monitored by electrical, mechanical and optical measurements.

3. The measurements of end rotations are by mechanical tiltmeters.
4. Optical methods use surveying tools or laser methods.
5. Electrical methods use linear variable differential transducers that transform displacements to voltage.
6. Dynamic characteristics of the bridge, longitudinal and torsional mode shapes, frequency of vibrations and damping effects can be effectively studied.
7. For bridges located in severe earthquake zones, earthquake response can be studied by measuring bridge frequency, vibration and damping.
8. Vibration tests are usually performed by portable sinusoidal shakers, impulsive devices such as hammers, sudden release of applied deflections and sudden braking of vehicles. For modal frequency, mode shapes and damping ratios, accelerometers are used.
9. Weigh in motion testing (WIM) is commonly used to survey the truck volume and weight spectra. By using axle sensors, WIM tests provide data on vehicle arrivals, speed and axle loads.

6.3.11 Increasing Remaining Fatigue Life

Fatigue life shall be evaluated for stress reversal (as per flow diagram developed by the author). Governing truck weight, effective stress range, traffic count for correct value of ADTT and number of cycles per truck passage shall be considered for an accurate fatigue analysis.

If remaining fatigue life of a steel bridge is not acceptable, the following steps are applicable:

1. Load restriction
2. Identifying fatigue prone details, retrofitting critical details to change AASHTO detail category.
3. Accepting greater risk due to redundancy and increased inspection.

6.4 SIMPLIFIED FORMULA

6.4.1 Use of Influence Lines

1. Reciprocal theorem and Muller-Breslau Method for elastic deflections and forces are applicable. Effect of all unit moving loads at specific sections of beams is computed. At that section the envelope of all axle loads is generated. Location and magnitude of absolute maximum positive bending moment and shear force or reactions is determined.
2. This method is particularly useful for redundant or continuous beams with equal spans, where absolute negative bending moment and shear maximum values for trucks with 3 or more axles are required. The raw values need to be multiplied by load factors, distribution factors, multiple lane reduction factor and impact factor etc to obtain design moment and shear force.
3. However, it gets cumbersome when one or more spans are unequal. With the advent of fast computers and stiffness matrix software it is much quicker to resort to computer solution.
4. Majority of bridges in practice is of single span for which the author has developed approximate simplified formulae for maximum bending moment and shear forces listed below for HS, H and S trucks. The formula can be used for preliminary design. For HS-20 formula see Chapter 4.

Table 6.3 Maximum live load moments and shear forces for HS-15 truck single, span only (alternate lane loads apply).

	HS-15 Truck Moments		HS-15 Truck Shear	
Span Length (Feet)	AASHTO LL Moment (Kip-Ft)	M max = 13.5 (L-1.55) (L-14)/L* (Kip-Ft)	AASHTO Reaction/Shear Force (Kips)	Vmax = 18 (3L-28)/L (Kips)
Small spans				
30	211.6	204.84	37.2	37.2
40	337.4	337.40	41.4	41.4
Medium spans				
50	470.9	470.93	43.9	43.92
60	604.9	604.96	45.6	45.6
70	739.2	739.26	46.8	46.8
80	873.7	873.74	47.7	47.7
90	1008.3	1008.33	48.4	48.4
100	1143.0	1143.00	49.0	48.96
110	1277.7	1277.74	49.4	49.42
120	1412.5	1412.52	49.8	49.8
Long spans				
130	1547.3	↑	50.7	↑
140	1682.1		53.1	
150	1856.3		55.5	
160	2076.0	Lane load governs	57.9	Lane load governs
170	2307.8		60.3	
180	2551.5		62.7	
190	2807.3		65.1	
200	3075	↓	67.5	↓

* Developed by the author.

Table 6.4 HS-15 truck loads (introduced 1944).

Span Length (Feet)	Lane Load 0.48 Kips/ft + Conc. Load of 18 Kips (BM)	Lane Load 0.48 Kips/ft + Conc. Load of 26 Kips (SF)	Max. Moment HS-15 Truck 13.5 (L-1.55) (L-14)/L* (Kip-Ft)	Max. Shear/Reaction HS-15 Truck 18 (3L-28)/L* (Kips)
Small spans				
22	103.29	24.78	100.390	31.090
24	115.56	25.26	126.281	33.000
26	128.31	25.74	152.342	34.615
28	141.54	26.22	178.537	36.000
30	155.25	26.7	204.840	37.200
32	169.44	27.18	231.229	38.250
34	184.11	27.66	257.691	39.176
36	199.26	28.14	284.212	40.000
38	214.89	28.62	310.784	40.736
40	231	29.1	337.398	41.400
Medium spans				
42	247.59	29.58	364.050	42.000
45	273.375	30.3	404.085	42.800
50	318.75	31.5	470.934	43.920
55	367.125	32.7	537.901	44.836
60	418.5	33.9	604.957	45.600
65	472.875	35.1	672.081	46.246
70	530.25	36.3	739.260	46.800
75	590.625	37.5	806.481	47.280
80	654	38.7	873.736	47.700
85	720.375	39.9	941.021	48.070
90	789.75	41.1	1008.330	48.400
95	862.125	42.3	1075.659	48.694
100	937.5	43.5	1143.005	48.960
105	1015.875	44.7	1210.365	49.200
110	1097.25	45.9	1277.738	49.418
115	1181.625	47.1	1345.122	49.617
120	1269	48.3	1412.516	49.800

(continued on next page)

Table 6.4 HS-15 truck loads (introduced 1944) *(continued)*.

Span Length (Feet)	Lane Load 0.48 Kips/ft + Conc. Load of 18 Kips (BM)	Lane Load 0.48 Kips/ft + Conc. Load of 26 Kips (SF)
Long spans		
130	1452.75	50.7
140	1648.5	53.1
150	1856.25	55.5
160	2076	57.9
170	2307.75	60.3
180	2551.5	62.7
190	2807.25	65.1
200	3075	67.5

Table 6.5 Maximum live load moments and shear forces for H-20 truck (alternate lane loads apply).

Comparative Study of AASHTO and Simplified Formula			
	H-20 Truck Moments		**H-20 Truck Shear**
Span Length (Feet)	**AASHTO LL Moment (Kip-Ft)**	**$M_{max} = 10(L-2.8)^2/L$ (Kip-Ft)**	**AASHTO Reaction/Shear Force (Kips)**
Small spans			
20	160	147.92	34.4
30	246.6	246.61	36.3
40	346.0	345.96	36.8
Medium spans			
50	445.6	445.57	58.5
60	558.0	545.31	60.8
70	707.0	645.12	62.4
80	872.0	744.98	63.6
90	1053.0	844.87	64.5
100	1250.0	944.78	65.3
110	1463.0	1044.72	65.9

* Developed by the author (lane load may govern).

Table 6.6 H-20 and H-15 truck loads (introduced 1944).

Span Length (Feet)	Max. Moment H-20 Truck Using 10(L-2.8) (L-2.8)/L* (Kip-Ft)	Max. Shear/ Reaction H-20 Truck Using 8(5L-14)/L* (Kips)	Max. Moment H-15 Truck 7.5(L-2.8) (L-2.8)/L* (Kip-Ft)
Small spans			
20	147.92	34.4	110.94
22	167.56	34.91	125.67
24	187.27	35.33	140.45
26	207.02	35.69	155.26
28	226.8	36	170.1
30	246.61	36.27	184.96
32	266.45	36.5	199.84
34	286.31	36.71	214.73
36	306.18	36.89	229.63
38	326.06	37.06	244.55
40	345.96	37.2	259.47
Medium spans			
42	365.87	37.33	274.4
45	395.74	37.51	296.81
50	445.57	37.76	334.18
55	495.43	37.96	371.57
60	545.31	38.13	408.98
65	595.21	38.28	446.40
70	645.12	38.4	483.84
75	695.05	38.51	521.28
80	744.98	38.6	558.74
85	794.92	38.69	596.19
90	844.87	38.76	633.65
95	894.83	38.82	671.12
100	944.78	38.88	708.59
105	994.75	38.93	746.06
110	1044.72	38.98	783.53
115	1094.68	39.03	821.01
120	1144.65	39.07	858.49

* Developed by the author (lane load may govern).

Table 6.7 Lane loads of 0.64 kips/ft (+18 kips concentrated load for moment and 26 kips for shear) in lieu of H-20 truck.

Span Length (Feet)	Max. Lane Mid-span Moment (Kip-Ft)	Governing Moment (Kip-Ft)	Governing Shear Force (Kips)
Small spans			
20	122	147.92	32.4
22	137.72	167.56	33.04
24	154.08	187.27	33.68
26	171.08	207.02	34.32
28	188.72	226.8	34.96
30	207	246.61	35.6
32	225.92	266.45	36.24
34	245.48	286.31	36.88
36	265.68	306.18	37.52
38	286.52	326.06	38.16
40	308	345.96	38.8
Medium spans			
42	330.12	365.87	39.44
45	364.5	395.74	40.4
50	425	445.57	42
55	489.5	495.43	43.6
60	558	558	45.2
65	630.5	630.5	46.8
70	707	707	48.4
75	787.5	787.5	50
80	872	872	51.6
85	960.5	960.5	53.2
90	1053	1053	54.8
95	1149.5	1149.5	56.4
100	1250	1250	58
105	1354.5	1354.5	59.6
110	1463	1463	61.2
115	1575.5	1575.5	62.8
120	1692	1692	64.4

Table 6.8 Typical format for computation of unfactored dead loads and wind loads.

Dead Load			*	
1	Slab			
		Thickness =	8.5	in
		Length =	70	ft
		Weight =	150	lb/ft3
		Width =	63.75	ft
2	Beams			
		Number of beams =	10	
		Length =	70	ft
		Weight =	670	lb/ft
3	Haunch			
		Thickness =	1.5	in
		Width =	18	in
		Length =	70	ft
		Weight =	150	lb/ft3
		No. of beams/haunches =	10	
4	Diaphragm			
		5% of girder wt =	23450	lbs
5	Parapet		*	
		Area of X-Section	513.13	in*2
		Length =	70	ft
		Weight =	586.91	lb/ft
6	SIP			
		Length =	70	ft
		Weight =	13	lb/ft2
		Width =	63.75	ft
7	Stem			
		Thickness =	2.5	ft

* Assumed data for a typical slab and beam bridge.

(continued on next page)

Table 6.8 Typical format for computation of unfactored dead loads and wind loads *(continued)*.

		Length =	70	ft
		Weight =	150	lb/ft3
		Depth	4	ft
8	Bearings			
		Weight =	300	lbs
		Number =	10	
9	Suspended Backwall			
		Width	2	ft
		Length =	70	ft
		Weight =	150	lb/ft3
		Depth	4	ft
10	Relief Slab			
		Width =	5	ft
		Thickness =	1.5	ft
		Length =	70	ft
		Weight =	150	lb/ft3
		Total DC1		
	FWS			
		Thickness =	2.5	in
		Length =	70	ft
		Weight =	150	lb/ft3
		Width =	63.75	ft
		With Stem		
		Without Stem		
		Load on Abutment		
	Pile Loads			
	Pile Cap Thk. =		2.5	ft

* Assumed data for a typical slab and beam bridge.

(continued on next page)

Table 6.8 Typical format for computation of unfactored dead loads and wind loads *(continued).*

		Length =	70	ft	
		Weight =	150	lb/ft3	
		Width =	7.25 *	ft	
		Total 5 22.1 + 2.72 5			
		Avg. 70 ft length			
		13 Piles	DL per pile		
	Geometry				
		Depth of beam =	45	in	
		Depth of deck =	8.5	in	
		Height of haunch =	1.5	in	
		Total height =	55	in	
		Total height =	4.5833	ft	
	Horizontal loads				
		Wind Pressure on beam =	50	ksf	
		Wind Load on Structure =	16042	lb	
		Wind on Live Load (A3.8.1.3)			
	Load =	0.1	k/ft		
	Wind load on LL =	7 kip			
		Wind on substructure (A3.8.1.2.3)			
	Load =	40			psf
	Abutment height =	5			ft
	Wind load on LL =	200			lb/ft
		Braking force (A3.6.4)			
	Load =	25%			
	Axle wt. of truck =	72			kip
	Lanes =	3			
	Braking force =	54			kip
		Temperature force			
	Load =				kip

* Assumed data for a typical slab and beam bridge.

Figure 6.5 Use of transverse steel box beam to eliminate pier permitting through traffic (similar concept used by the author on I-80 and Route 23 intersection in northern New Jersey). For curved steel girder design, refer to *AASHTO Guide Specifications for Horizontally Curved Steel Girder Highway Bridges.*

6.5 RATING PROCEDURES FOR CONCRETE AND STEEL BRIDGES*

6.5.1 Rating of Composite Single Span Steel Bridge

The following steps need to be computed:

1. Section properties of composite section.
2. Dead load moments for DC1, DC2, and DW.
3. Distribution factors for flexure for one lane loaded.
4. Distribution factors for shear for one lane loaded.
5. Distribution factors for flexure for two or more lanes loaded.
6. Distribution factors for shear for two or more lanes loaded.
7. Maximum design moments for HL-93 live load.
8. Maximum design shear for HL-93 live load.
9. Web slenderness.
10. Ductility requirements.
11. Plastic moment of resistance.
12. Yield moment of composite section.
13. Nominal flexural resistance.
14. Nominal shear resistance.
    - General load rating equation and evaluation factors.
    - Design load rating for Strength I (inventory and operating level).
    - Design load rating for Service II (inventory and operating level).
    - Rating factors for fatigue limit state.

*For detailed design of concrete and steel girders, refer to AASHTO LRFD Specifications Section 5, Appendix A5; Basic Steps for Concrete Bridges and Section 6, Appendix C6; Basic Steps for Steel Bridge Super structures.

- Infinite life fatigue check.
- Calculate remaining fatigue life.
- Legal load rating.
- Permit load rating.

6.5.2 Flexural Capacity Rating of Steel Girders for Load Combinations

(Summary of steps refers to LRFD sections)

In Chapter 5, steps for AASHTO design method was summarized.

Member proportions (I_{yc}/I_y) web slenderness (6.10.2.2):

For plate girder webs, $D/t_w \leq 150$

For compression flange composite with deck slab, $b_c > D/5$

$t_c > 1.2\ b_c$

$b_c/2\ t_c \leq 12$

For tension flange, $b_t > D/4$

$b_t > 1.2\ b_c$

$b_t/2\ t_t \leq 12$

1. Strength limit states I and II (6.5.4.1).
2. Service II limit state (6.10.4.1) and (6.10.4.3).
3. Fatigue limit state ((6.10.5).
4. Deflection limit state (6.10.3.2) and (6.10.4.2).
5. Refer to Section 6.10.1.1 (preliminary section calculations).
6. $0.1 \leq (I_{yc}/I_y) \leq 0.9$.

Flexural resistance M_n (6.10.8):

1. Strength limit states I and II (6.5.4.1).
2. Construction limit state.
3. Refer to Section 6.10.5.6.

Composite compact section:

1. Strength limit states I and II.
2. Service II limit state (C 6.10.4_1).
3. For 70W steel and construction limit state assume non-compact.
4. Redistribution of moments in continuous girders permitted.
5. Refer to Section 6.10.5.2.2c, 6.10.5.2.3, and 6.10.2.2.
6. Refer to Section 6.10.5 for I girder.
7. Section A6.2 for yield moment.

Composite non-compact section:

1. Strength limit states I to V.
2. Service II limit state.
3. Construction limit state.
4. Refer to Sections 6.10.5.3.2 and 6.10.5.3.3.
5. Refer to Section 6.10.5.3.2b for non-compact web slenderness.

Non-composite section (positive and negative flexure):

1. Applicable to I section.
2. Strength limit states I and II.
3. Construction limit state and uncured slab.
4. Refer to Section 6.10.5.2.3.

Non-composite non-compact section (positive and negative flexure):

1. Strength limit states I and II.
2. Construction limit state and uncured slab.
3. Refer to Sections 6.10.5.3.2 and 6.10.5.3.3.

6.5.3 Unfactored Dead and Live Load Moments with Impact

Multi-girder bridge span of 80 ft: Four steel plate girders @ 250 lbs/ft wt., spaced at 10.33 ft centers, slab thickness of 7 in and 3 in haunch, superimposed dl = 2.0 kips/ft,

DL moment of inertia, I = 24,416 in^4, y_b = 19 in

SDl moment of inertia, I = 46,613 in^4, y_b = 29.7 in

I = 67,576 in^4, y_b = 39.79 in

DL on interior girder:

1. Slab 0.15k/cft × 10.33 ft × 0.58 ft = 0.9 k/ft
2. Haunch 0.15 × 1.33 ft × 0.25 ft = 0.05 k/ft
3. Girder 0.25 k/ft + 0.03 k/ft (for stiff. and diaphragm) = 0.28 k/ft

Total DL = 1.23 k/ft

Mdl = 1.23 × $(80)^2/8$ = 984 k-ft

Msdl = (2.0/4) × $(80)^2/8$ = 418 k-ft

Live load of HS-20 acting on single 80 ft span, Ml = 582.4 kip-ft, impact factor = 0.244

M (L + I) = 1.244 × 582.4 × 1.88 = 1362 kip-ft

6.5.4 Buckling Capacity Rating of Steel Girders for Load Combinations

(Summary of steps refer to LRFD sections)
Non-compact section compression flange buckling:

1. Strength limit states I and II.
2. Construction limit state and uncured slab.
3. Refer to Section 6.10.5.3.3c and LRFD Eq. 6.10.5.3.3c

Non-composite non-compact section lateral torsional buckling:

1. Positive and negative flexure.
2. Strength limit states I and II.
3. Construction limit state and uncured slab.
4. Refer to Sections 6.10.5.5 and 6.10.6.4.
5. $L_b \leq 1.76\, r_l\, (E/F_{yc})^{1/2}$

6.5.5 Shear Capacity Rating of Steel Girders for Load Combinations

(Summary of steps refers to LRFD sections)
Compact and non-compact sections:

1. Strength limit states I and II.
2. Construction limit state and uncured slab.
3. For stiffened webs refer to Section 6.10.7.3.
4. For unstiffened webs refer to Section 6.10.7.2.

Interior panels of compact sections:

1. Refer to Section 6.10.7.3.3a.
2. Strength limit states I and II.
3. Uncured slab.

Interior panels of non-compact sections:

1. Strength limit states I and II.
2. Construction limit state and uncured slab.
3. Refer to Section 6.10.7.3.3b.

6.5.6 Evaluation of Deflection

(Summary of steps refer to LRFD sections)

1. Service limit State II.
2. Refer to Section 6.10.3.2.
3. Control of permanent deflection.

Compute C: D_W / t_w

Compute k: $k = 5 + 5/ (d_0 / D)^2$

$10.10\,(Ek/F_{yw})^{1/2} \leq D_W / t_w$

$1.38\,(Ek/F_{yw})^{1/2} \leq D_W / t_w$

$$C = \frac{1.52\,(Ek/F_{yw})}{(D_W / t_w)^2}$$

$V_p = 0.58\,F_{yw}\,D_W\,t_w$

$V_n = C\,V_p$

Both inventory rating and operating rating > 1.0 for HL-93 loads.

Nominal shear resistance for unstiffened web (LRFD Section 6.10.7.2):

If $2.46\,(E/F_{yw})^{1/2} > D/ t_w$

$V_n = 0.58\,F_{yw}\,Dt_w$ (Eq. 6.10.7.2-1)

Service II limit state. Section 6.6.4.1.

$$\text{Let rating factor RF} = \frac{f_R - (\gamma_D)(f_D)}{(\gamma_L)(f_{LL+IM})} \geq 1$$

$f_R = 0.80\,R_b\,R_h\,F_{yf}$ LRFD 6.10.5.2

For tension flange load shedding factor $R_b = 1.0$; LRFD 6.10.4.3.2 b

For compression flange, if $2Dc/t_w \leq \gamma_b\,(E/f_c)^{1/2}$, $R_b = 1.0$; LRFD 6.10.4.3.2a

For homogeneous sections, $R_b = 1.0$; LRFD 6.10.4.3.1

$(\gamma_{DC}) = (\gamma_{DW}) = 1.0$; Table 6-1

$(\gamma_L) = 1.3$ for inventory; $(\gamma_L) = 1.0$ for operating.

- Floor beam analysis:
  1. Floor beam moments
  2. Maximum SF
- Rating of main girder:
  1. For one loaded: Multiple presence factor m = 1.2
     Distribution factor DF = g1 LRFD 3.6.1.1.2
  2. For both lanes loaded: DF = g2
  3. Nominal flexural resistance of section LRFD C6.10.4 -1
  4. Classify section LRFD 6.10.3.3.2
     Non-composite symmetric section, $D_{cp} = 0.5 \times D_w$
  5. Check web slenderness for compact section LRFD 6.10.4.1.2
     $3.76\ (E/f_{yc})^{1/2} \leq 2D_{cp}/t_w$
  6. Check compression flange slenderness at midspan LRFD 6.10.4.1.4
     $b_f/2\ t_f \leq 1.38\ (E/f_{yc})^{1/2}$; where $f_{yc} = (2D_c/t_w)^{1/2}$
  7. Check compression flange bracing LRFD 6.10.4.1.9
     $L_b \leq 1.76\ r_t\ (E/F_{yc})^{1/2}$
  8. Compute I_{yc}

Compute A_c (compression flange area at midspan)

$r_t = (I_{yc}/A_c)^{1/2}$

- Check flexural resistance and non-compact section LRFD 6.10.4.2.4

Load shedding factor, R_b LRFD 6.10.4.3.2a

Flexural resistance:

$F_n = R_b\ R_h\ F_{yf}$, $R_h = 1.0$

Compute $2D_c / t_w$

$\lambda_b = 5.76$ when compression flange area $\geq$ tension flange area

Compute $\lambda_b\ (E/\ f_c)^{1/2}$

$R_b = 1 - [\ a_r / (1200 + 300\ a_r)]\ [2D_c / t_w - \lambda_b\ (E/f_c)^{1/2}]$

$a_r = 2D_c t_w / A_c$

- Using general load rating Eq. (6-1) Section 6.4.2
- Evaluation factors for strength limit states
  - Resistance factor ϕ
    $(\phi_f) = (\phi_v) = 1.0$ for flexure and shear
  - Condition factor $\phi_c = 1.0$ Section 6.4.2.3
  - System factor ϕ_s
    $\phi_s = 0.90$ for flexure of riveted two-girder system Section 6.4.2.4
    $\phi_s = 1.0$ for shear Section 6.4.3
- Strength I limit state Section 6.6.4.1
  Flexural stress at midspan (unfactored):
  $f_{DC+DW} = (M_{DC+DW})/S$
  $f_{LL+IM} = (M_{LL+IM})/S$
  Flexural resistance at midspan: F_n
  Inventory RF
  Operating RF
- Service II limit state

For non-composite and non-compact sections, service II limit state does not need to be checked.

Maximum positive moment evaluation method is applicable to both single span and continuous beams. However, the maximum negative moment evaluation method is only applicable to continuous beams. Maximum positive moments in equal span continuous bridges can be calculated using a similar approach as used for single spans.

Footprints of continuous span bridges show that they are likely to be of unequal span lengths. Hence, a simplified analysis approach giving a summary of results cannot be formulated. Instead, the influence line method or continuous beam analysis results using conventional three moments theorem are presented and compared to those from computer programs.

Maximum bending moments:

1. Positive moment at midspan for a variety of loads (non-symmetric axle + lane).
2. Positive moment at midspan for alternate tandem axle loads (symmetric axle + lane).
3. Negative moment at supports for non-symmetric axle loads + lane loads.

6.5.7 Shear Design

Strength I limit state:

SF at girder ends: Compute V_{DC+DW}

Compute V_{LL+IM}

Girder web D_W, t_w:

Required end panel transverse stiffener spacing for stiffened girders $\leq$ 1.5 D.

Shear resistance of end panel:

Maximum shear forces:

Maximum shear force at supports for variety of loads (non-symmetric axle + lane)

Results of a simplified analysis approach can be presented in a summary of tables. The comparative study would help in frequently used design and rating for moments and forces.

6.6 RATING OF SECONDARY STRUCTURAL MEMBERS

6.6.1 Summary of Rating Procedure for Transverse and Bearing Stiffeners

(Provided for ready reference)

1. Strength Limit States I and II
2. Refer to (6.10.8) Stiffeners
3. Refer to (6.10.8.2.1) for Bearing Stiffener Location
4. Refer to (6.10.8.2) for Axial Resistance and Bearing Stiffener Geometry

6.6.2 Shear Connectors

(Provided for ready reference)

1. Strength Limit States I and II
2. Refer to Sec. 6.10.7.4 for shear connectors.

The following steps are provided for ready reference. Prior to developing a software or for solving the equations using hand calculations, equations need to be checked against the latest version of applicable AASHTO LRFD Specifications or LRFR Manual.

6.7 EXAMPLE OF LOAD RESISTANCE FACTOR RATING (LRFR)

6.7.1 Analysis Procedure for a Two-Girder Steel Bridge

The following steps are provided for ready reference. Prior to developing a software or for solving the equations using hand calculations, equations need to be checked against the latest version of applicable AASHTO LRFD Specifications or LRFR Manual.

- Refer to AASHTO Manual for Condition Evaluation of Bridges Example A8.

  Geometry, materials, traffic data:

  LL = HL-93 (HS-25 truck and lane)

  Span = 100 ft

  fc′ = 2.5 ksi

  Fy = 30 ksi

  ADTT = 500

  Skew = 0 deg.

  Spacing of floor beams = 10 ft

  Overlay thickness = 1.5 in

- For equilibrium, Σ horizontal resisting forces above NA = Σ horizontal resisting forces below NA

  Force in top flange = $P_c = F_{yc}\, b_c\, t_c$ = Force in bottom flange P_t

  Force in web = $P_{wc} = P_{wt}$

  $\bar{a} = D/2$ (LRFD Appendix A 6.1, Case 1)

  $d_c = d_t$

  For an economical design, when shear and deflection criteria are met:

  Maximum positive moment = Maximum negative moment.

  For floor beams, cantilever overhangs at each end are provided.

  Plastic moment of resistance (M_p) = Force in top flange × Distance to NA + Force in bottom flange × Distance to NA + Force in web x Distance to NA

  $= (P_c \times d_c + P_t \times d_t) + P_w \times [(\bar{a})^2 + (D - \bar{a})^2] / 2D$

- Check for compactness:

  Positive moment compression flange is fully in contact with the deck and is adequately braced.

  Assume section as compact (LRFD Section 6.6.9.3).

  Negative moment section: Lateral bracing $L_b \leq (r_y\, E/F_{yc})\, [0.124 - 0.0759\, (M_1 / M_p)]$ (LRFD Section 6.10.4.1.7)

6.7.2 Sample of a Rating Worksheet

Based on AASHTOWare software an example of output ratings are presented below. In this case HL-93 loads were not applied as overstress was possible on some of the members. Instead HS 20 vehicle (36 Tons) results are compared with Type 3 (25 Tons), Type 3-S2 (40 Tons) and Type 3-3 (40 Tons) vehicles.

Structure No.:______________ Route:________________ Cycle No.: 16
Insp. Date: 11/07/2008

SUMMARY OF RATINGS

The Load Factor and Working Stress ratings, computed in the 9th Cycle report in accordance with the FHW directive dated November 1993 and AASHTO Manual for Condition Evaluation of Bridges, 1994, as modified by Division 4 of the New Jersey Department of Transportation Design Manual, Bridges and Structures, are as follows:

Computer Program Used: AASHTOWare Virtis (Version 6.1.0)

PERCENT (%) SECTION LOSSES: N/A

Material	Compressive Strength f'c	Allowable Stresses (Psi) Yield	Inventory	Operating
Concrete	3,000	–	1,200	1,650
Concrete (Beam)	–	40,000	20,000	28,000
Reinforcing Steel	–	33,000	18,000	24,500

Member	Truck Type (Tons)		Rating (Tons) Load Factor Inventory	Operating
	HS-20	(36T)	37*	61*
Span 1	HL-93	(100T)	–	–
Interior Stringers S4,	Type 3	(25T)	32*	53*
S5, S6, S7, and S8	Type 3S2	(40T)	46*	77*
	Type 3-3	(40T)	62*	104*
	HS-20	(36T)	48	81
Span 2	HL-93	(100T)	–	–
Interior Stringer S2	Type 3	(25T)	45	76
	Type 3S2	(40T)	61	102
	Type 3-3	(40T)	72	120
	HS-20	(36T)	50	84
	HL-93	(100T)	–	–
Span 3	Type 3	(25T)	47	79
Interior Stringer S2	Type 3S2	(40T)	65	108
	Type 3-3	(40T)	77	128

* Controlling Member

6.7.3 Sheikh Ibrahim's Method for Checking Gusset Plate Connections in Steel Truss

Following the collapse of I-35W Bridge, FHWA developed a simplified method for gusset plate design. (Refer to—Sheikh Ibrahim, F. I., *Load Rating Evaluation of Gusset Plates in Truss Bridges,* FHWA Bridge Design Guidance No. 1, February, 2008)

Due to the need for checking the failed gusset connections, it is important to apply Ibrahim's method to similar connection details.

Gusset Plate Resistance in accordance with the Load and Resistance Factor Rating Method (LRFR)

Summary Procedures: The evaluation of gusset connections shall include:

1. The evaluation of the connecting plates and
2. Evaluation of fasteners.
   a. The resistance of a gusset connection is determined at the strength limit state only and is the smaller resistance of the fasteners or gusset plates.
   b. For safety, owners may require that connections be checked at other limit states such as the service limit state to minimize serviceability problems.

The Resistance of Fasteners

For concentrically loaded bolted and riveted gusset connections, the axial load in each connected member may be assumed to be distributed equally to all fasteners, at the strength limit state.

1. The bolts in bolted gusset connections shall be evaluated to prevent:
   - Bolt shear
   - Plate bearing failures at the strength limit state.
   - At the strength limit state, the provisions of AASHTO LRFD Article 6.13.2.7 and 6.13.2.9 shall apply for determining the bolts' resistance to prevent bolt shear and plate bearing failures.
2. Riveted gusset connections

   The rivets shall be evaluated for:
   - Rivet shear
   - Plate bearing failures at the strength limit state.
   - The plate bearing resistance for riveted connections shall be in accordance with AASHTO LRFD Article 6.13.2.9 for bearing at bolt holes.

The Resistance of Gusset Plates

The resistance of a gusset plate shall be determined as the plate's least resistance in:

- Shear,
- Tension including block shear,
- Compression, and
- Combined flexural and axial loads.

Gusset Plates in Tension

Gusset plates subjected to axial tension shall be investigated for three conditions:

- Yield on the gross section,
- Fracture on the net section,
- Block shear rupture

The factored resistance, *Rr*, for gusset plates in tension shall be taken as the least of the values given by either:

1. Yielding,
2. Fracture, or
3. The block shear rupture resistance.
   - Gross Section Yielding Resistance
   - Block Shear Rupture Resistance

The resistance of block shear rupture is that of combination of parallel and perpendicular planes, one in axial tension and the remainder under shear. The factored resistance of the plate for block shear rupture shall be considered.

Gusset Plates In Shear

The factored shear resistance, *Rr*, for gusset plates in shear shall be taken as the least resistance against shear yielding and net section fracture.

Gusset Plates under Combined Flexural and Axial Loads

1. The maximum elastic stress from combined factored flexural and axial loads

   $< \phi f\, Fy$ based on the gross area of the plate.

   where:

   ϕf = resistance factor for flexure = 1.00

   Fy = specified minimum yield strength of the plate

The analysis of gusset plates for combined flexural and axial loads involves the evaluation of several sections to arrive at the critical section.

The large number of equations listed are based on FHWA's latest research on the subject and need to be developed in a computer program.

6.7.4 Connections Design Using High Strength Bolts (Example Only)

Case study of connection design for I-95 state road viaduct, north of Philadelphia: On a recent project for PennDOT, AASHTO LRFD specifications were used for the equations evaluating strength and service design of diagonal bracing members. Forces output was from STAAD-Pro frame analysis. Six or nine bolt connections were used.

Table 6.9 Sample spreadsheet for check bolts for strength I, III, and V load combinations.

Strength I / III Axial Force	Diagonal	Horizontal	SMSQ		Resultant
LOCATION 1					
$R = (H^2 + V^2)^{0.5}$	1.5	2.6	9.01		3.0017
LOCATION 2	10	15	325		18.0278
LOCATION 3	15	10	325		18.0278
NO. OF BOLTS					6
FORCE/BOLT = R/N					3.0046
Shear Resistance					
$R_t = (\phi)\ R_n$					

(continued on next page)

Table 6.9 Sample spreadsheet for check bolts for strength I, III, and V load combinations *(continued)*.

Strength I / III Axial Force	**Diagonal**	**Horizontal**	**SMSQ**		**Resultant**
AASHTO LRFD	A_b	F_{ub}	N_s		R_n
Sec. 6.13.2.2 Eq.2					
$R_n = 0.38\, A_b\, F_{ub}\, N_s$	0.6	120	1		72
Red. Factor = 0.8					57.6
				(ϕ)	
(ϕ) R_n				1	57.6
If (ϕ) $R_n/R_t > 1$ OK					19.1704
Slip Resistance	Slip-Critical				
$R_t = R/N$, (ϕ) = 1					
AASHTO LRFD	K_h	K_s	N_s	P_t	R_n
Sec. 6.13.2.8 Eq.1					
$R_n = K_h\, K_s\, N_s\, P_t$	0.6	0.33	1	39	7.722
Tables 6.13.2.8	Table 2	Table 3		Table 1	
Sec. 6.5.4.2 (ϕ)				(ϕ) shear	
(ϕ) R_n				1	7.722
If (ϕ) $R_n/R_t > 1$ OK					2.5700
Strength V Axial Force	**Diagonal**	**Horizontal**	**SMSQ**		**Resultant**
LOCATION 1					
$R = (H^2 + V^2)^{0.5}$	1.5	2.6	9.01		3.0017
LOCATION 2	10	15	325		18.0278
LOCATION 3	15	10	325		18.0278
N = NO. OF BOLTS					6
FORCE/BOLT = R/N					3.0046
Strength V Slip Resistance	Slip-Critical				
$R_t = R/N$, (ϕ) = 1					3.0046
AASHTO LRFD	K_h	K_s	N_s	P_t	R_n
Sec. 6.13.2.8 Eq.1					
$R_n = K_h\, K_s\, N_s\, P_t$	0.6	0.33	1	39	7.722
Tables 6.13.2.8	Table 2	Table 3		Table 1	

(continued on next page)

Table 6.9 Sample spreadsheet for check bolts for strength I, III, and V load combinations *(continued).*

Strength I / III Axial Force	Diagonal	Horizontal	SMSQ		Resultant
Sec. 6.5.4.2 (ϕ)				(ϕ_s)	
(ϕ) R_n				1	7.722
If $R_n/R_t > 1$ OK					2.5700

Service II Condition

Service II Axial Force	Horiz.Comp.		Vert.Comp.	SMSQ		Resultant
LOCATION 1						
$R = (H^2 + V^2)^{0.5}$	100.9		22.7	10696.1		103.422
LOCATION 2	10		15	325		18.0278
LOCATION 3	15		10	325		18.0278
NO. OF BOLTS						9
Force/Bolt = R/N	26.7	25.6	56			11.4913
Service II Slip Resistance		Slip-Critical				
$R_t = R/N$, (ϕ) = 1						11.4913
AASHTO LRFD	K_h		K_s	N_s	P_t	R_n
Sec. 6.13.2.8 Eq.1						
$R_n = K_h K_s N_s P_t$	0.6		0.33	2	39	15.444
Tables 6.13.2.8	Table 2		Table 3		Table 1	
If $R_n/R_t > 1$ OK						1.344

6.8 DESIGN OF A DECK SLAB

6.8.1 Design and Rating of a Reinforced Concrete Slab or a Slab Bridge

Of all bridge components, the deck is the most vulnerable to wear and tear and may be replaced after every 10 years when subjected to heavy daily traffic and deicing salts.

Slab bridges: Both in the U.S. and worldwide there are many existing bridges which are beamless or "slab bridges." In the following solved examples, only small and medium spans are considered for design and rating.

According to LRFD 5.14.4.1, concrete slabs and slab bridges designed in conformance with AASHTO specifications are considered satisfactory for shear. Also, as per LRFD 6.5.9, shear need not be checked for design load and legal load ratings of concrete members.

No service limit states apply to reinforced concrete bridge members. Only strength I for HL-93 or legal live load or strength II for permit loads will be considered.

Comparison between design and rating: Design requires balancing the external dead and live load moments with moment of resistance, using load factors and resistance factors. Design is done before rating. Rating requires evaluating the reserve strength or moment capacity available for live loads after deducting the dead load moments. Rating is expressed as a ratio of available live load moment capacity to peak external live load moment from HL-93 legal or permit loads and needs to be greater than one.

If RF $\leq$ 1 for the legal loads, the bridge needs to be posted for the lower live load. For a new bridge, rating is generally not a problem since the rating factor is > 1.

For rating, the load factors are different from those used for design. Load factors used are higher for rating, and the approach is more conservative.

However, for an existing deteriorated bridge the moment of resistance is lower. Older bridges were designed for lower truck loads and need to be evaluated against new truck types.

The following equations for slab bridge design or rating may be programmed using Excel spreadsheets or Mathcad:

Using standard definition of nominal moment of resistance, $M_n = A_s f_y (d_s - a/2)$

$a = c\,\beta_1$

$c = A_s f_y / (0.85 f_c' \beta_1 b)$

Minimum reinforcement required to develop M_r

$\leq$ lesser of 1.2 M_{cr} or 1.33 M_u (LRFD 5.7.3.3.2)

$M_r = \phi M_n$ Eq. (6-4)

$1.2 M_{cr} = 1.2 (f_r + f_{pb}) S_{bc} - M_{d,nc} (S_{bc} / S_b - 1)$ (LRFD 5.4.2.6)

$M_{d,nc}$ = Non-composite dead load moment

f_{pb} = Compressive stress in concrete due to effective prestress in the prestressed tensile zone

S_{bc} = Uncracked section modulus neglecting steel $S_{bc} = I/y_t$

I = Moment of inertia of uncracked section

y_t = Distance to extreme tension fiber from neutral axis of uncracked concrete

$f_r = 0.24 (f_c')^{0.5}$

Maximum reinforcement: For ductility, limit neutral axis depth. (LRFD 6.5.6)

$$c/d_e \leq 0.42$$

Rating of interior strip:

Calculation of equivalent interior strip width 'E' (LRFD 4.6.2.3)

$$L_1 \leq 60 \text{ ft}$$

Number of loaded lanes > 1

$$E = [84 + 1.44 (L_1 W1)^{0.5}] \leq 12 W/N_L \text{ LRFD Eq. (4-18)}$$

W_1 is lesser of 43 ft or 60 ft.

One lane loaded

$$E = 10.0 + 5.0 (L_1 W_1)^{0.5} \text{ LRFD Eq. (4-18)}$$

W_1 is lesser of 43 ft or 30 ft.

Longitudinal edge strip width (LRFD 4.6.2.1.4b)

which supports one line of wheels + a tributary portion of design lane load

$= \Sigma$ (Width of strip under the barrier/parapet + E/2 + 12) (LRFD 4.6.2.1.4a)

$\leq$ E or 6 ft.

6.8.2 Rating Method

The following steps are provided for ready reference. Prior to developing a software or for solving the equations using hand calculations, equations need to be checked against the latest version of applicable AASHTO LRFD Specifications.

1. General load rating equation (LRFD 6.4.2)

   $RF = [C - (\gamma_{DC})(DC) - (\gamma_{DW})(DW) \pm \gamma_P (P)] / \gamma_L (LL + IM)$ (EQ 6-1)

   Factors for inventory rating: $\gamma_{DC} = 1.25$; $\gamma_{DW} = 1.25$; $\gamma_{LI} = 1.75$

Factors for operating rating: $\gamma_{DC} = 1.25$; $\gamma_{DW} = 1.25$; $\gamma_{LO} = 1.35$

Operating rating = Inventory rating × (γ_{LI} / γ_{LO})

2. Design rating equation

Strength I limit state:

$C = (\phi_c)(\phi_s)(\phi) R_n$

$RF = [(\phi_c)(\phi_s)(\phi) R_n - (\gamma_{DC})(DC) - (\gamma_{DW})(DW)] / \gamma_L (LL + IM)$ (LRFD 6.5.4.1)

Condition factor (ϕ_c) = 1.0 for no deterioration (LRFD 6.4.2.3)

System factor (ϕ_s) = 1.0 for a slab bridge (LRFD 6.4.2.4)

Resistance factor (ϕ) = 0.9 for flexure (LRFD 5.5.4.2)

IM = 1.33

6.8.3 Beamless Reinforced Concrete Slab Bridge

Some of the older bridges constructed in the earlier part of twentieth century have smaller spans less than 30 feet and were constructed in the pre-prestressed concrete era. Nearly all of them were cast in place construction requiring heavy formwork. Both single and multiple spans have been used. Original design in reinforced concrete was based on AASHO code, which was popular at that time. The advantages were that design of beams and bearings was not required. Due to live load restrictions they are commonly used for pedestrian bridges or are posted for about 15 tons.

They are uneconomical since live load deflection requirements lead to a small span/depth ratio. Main reinforcement is placed parallel to direction of traffic. Distribution reinforcement is required. Due to continuity in transverse direction, shear design is not required. Although there are lesser local effects of shear distribution due to moving wheels than from vehicles which may be stationary in a traffic jam, the live load impact factor is higher. Fatigue stress needs to be considered since reversal of stress from moving vehicles will induce fatigue stress in bending.

6.8.4 Solved Example for Design of Single Span Slab Bridge:

Data: Single span = 30 ft with uneven surface of deck overlay

Clear distance between curbs = 20 ft including 5 ft wide, 5 in thick sidewalk on deck slab.

Out-to-out bridge width = 20 ft + 2 × 1.75 ft (New Jersey barrier width) = 23.5 ft

Slab thickness = 18 in with 2 in overlay; $f_c' = 0.30$ ksi, $f_y = 30$ ksi

$d = 18 - 1.5 - 0.5 = 16$ in

$L/D \leq 20$ i.e. $\leq 30 \times 12/18 = 20$, hence assumed thickness is okay. (AASHTO 8.24.1.2.1)

Effective span = Clear span + Depth of slab or distance to center lines of bearings

$\leq$ Center-to-center distance

Effective width of slab $E = 4 + 0.06\,S = 4 + 0.06 \times 30$ ft = 5.8 ft (AASHTO 3.24.3.2 Case B)

Dead load of slab and overlay = $0.15 \times (18\text{ in}/12 + 2\text{ in}/12) = 0.25$ ksf

Manual of Condition Evaluation of Bridges (3.3.2.1 to 3.3.2.3)

Dead load of parapet + Sidewalk to be distributed over 0.5 k/ft / (23.5/2) + (0.15 × 5)/12 / (23.5/2)

$= 0.043 + 0.005 = 0.048$ ksf; use 0.05 ksf

Total DL = $0.25 + 0.05 = 0.30$ ksf

Maximum DL negative moment at supports = $-wl^2/10 = -0.30 \times 30^2/10 = -27.0$ kip-ft

Per ft width = 27.0/5.8 ft = + 4.66 kip-ft/ft

Maximum positive moment at 0.4 L= + $wl^2/12 = +0.30 \times 30^2/12 = 22.5$ kip-ft

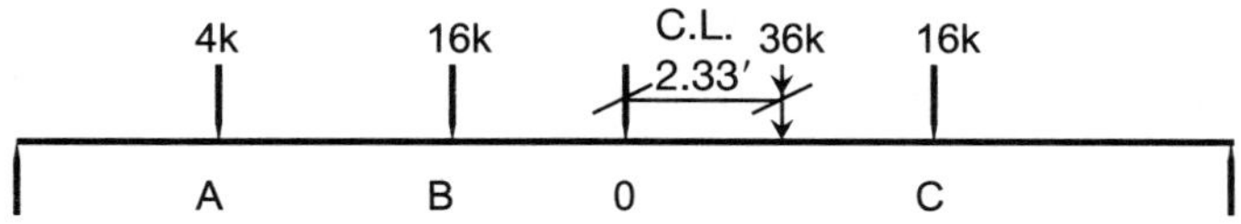

4.14 + 36.X - 16.14 = 0
X = 168/36 = 4.66: X' = 4.66/2 = 2.33'

Figure 6.6 HS-20 truck wheel loads.

Per ft width = 22.5/5.8 ft = + 3.88 k-ft/ft

Use HS-20 truck with two axles of 32 kips each and one axle of 8 kips.

Two wheels/axle = 16 kips/wheel; imp. factor = 1.10

Distance between back axles = 14 ft; distance between front axles varies.

Theorem: Max. live load BM occurs when midspan section divides the distance between the resultant and second wheel load equally.

Procedure:

1. Replace three wheel loads by a single resultant of wheel loads = Σ (16 + 16 + 4) = 36 kip
2. Locate the resultant force from the nearest wheel by taking moments about that wheel.
   Σ M = 0 at wheel B
   (36.X) + 16.14 ft − 4.14 ft = 0; X = (12.14)/36 = 14/3 = 4.667 ft from wheel B.
3. Calculate reactions at supports.
   R1 × L − 36 (L/2 − 2.333 ft) = 0; L = 30 ft; R1 = 15.20 kip.
4. Maximum moment occurs under the second wheel = 4 × (14 ft) − 15.20 × (15 ft − 2.333 ft) = 136.54 kip-ft at wheel B
5. Consider total vehicle load distributed over an effective width of 5.8 ft
6. Since every axle has two wheels, maximum BM = 2 × 136.54 = 273.08 kip-ft. Design as a beam of cross section 5.8 ft width × 1.5 ft depth.

6.8.5 Rating of Continuous Span Slab Bridge

Data: Three spans 25 ft each; clear distance between curbs = 30 ft including 6 ft wide, 8 in thick sidewalk.

Out-to-out bridge width = 30 ft + 2 × 1.75 ft (New Jersey barrier width) = 33.5 ft

Slab thickness = 18 in with 2 in overlay; $f_c' = 0.28$ ksi, $f_y = 33$ ksi

d = 18 in − 1.5 in − 0.5 in = 16 in

L/D $\leq$ 20, i.e., $\leq 25 \times 12/18 = 16 \leq 20$ (AASHTO 8.24.1.2.1)

Hence, assumed thickness is okay.

Effective span = Clear span + Depth of slab or distance to center lines of bearings

Deducting width of bridge seat at abutments, clear span = 30 ft − 2 × 1.5 ft = 27 ft

Effective span = 27 ft + 1.5 ft = 28.5 ft $\leq$ Center to center distance of 30 ft.

Using the greater of the two distances, L = 30 ft

Effective width of slab E = 4 + 0.06 S = 4 + 0.06 × 25 ft = 5.5 ft (AASHTO 3.24.3.2 Case B)

Dead load of slab and overlay = 0.15 × (18 in/12 + 2 in/12) = 0.25 ksf

Manual of Condition Evaluation of Bridges (3.3.2.1 to 3.3.2.3)

Dead load of parapet and sidewalk to be distributed over = 0.5 k/ft / (33.5/2) + 0.15 × 8/12/(33.5/2) = 0.030 + 0.006 = 0.036 ksf (Use 0.04 ksf.)

Total DL = 0.25 + 0.04 = 0.29 ksf

Maximum DL negative moment at supports = $-\,wl^2/10 = -\,0.29 \times 25^2/10 = -18.13$ kip-ft

Per ft. width = 18.13/5.5 ft = + 3.3 kip-ft/ft

Maximum positive moment at 0.4 L = $+wl^2/12 = +0.29 \times 25^2/12 = 15.1$ kip-ft

Per ft. width = 15.1/5.5 ft = +2.75 k-ft/ft

Use HS-20 Truck with two axles of 32 kips each and one axle of 8 kips.

Two wheels/axle = 16 kips/wheel; Imp. factor = 1.33

Distance between back axles = 14 ft; distance between front axles varies.

Using influence lines M_{L+I} at midspan or support = 1.33 × (16 kip × k1 + 16 kip × k2) span

Positive moment at midspan with one wheel in negative moment region,

k1 = 0.175, k2 = − 0.017 *(Check coefficients from influence lines method and STAAD. Pro output)*

$M_{L+I} = 1.33 \times 25 \times 16(0.175 - 0.017) = 532 \times 0.158 = 84.1$ kip-ft

Per ft. width = 84.1/5.5 ft = + 15.28 k-ft/ft

Maximum positive moment occurs at 0.4 L of end span,

k1 = 0.204, k2 = 0

$M_{L+I} = 1.33 \times 25 \times 16(0.204) = 532 \times 0.204 = 108.53$ kip-ft

Per ft. width = 108.53/5.5′ = + 19.73 k-ft/ft

Negative moment at penultimate support,

k1 = −0.102, k2 = −0.09

$M_{L+I} = 1.33 \times 25 \times 16(-0.102 - 0.09) = 532 \times -0.192 = -102.14$ kip-ft

Per ft width = −102.14/5.5 ft = −18.57 k-ft/ft

Load factors: $\gamma_{DL} = 1.2$; $\gamma_{LL} = 1.3$ (Section 3.3.4.1 and Table 2)

Maximum total moment = 1.2 × 3.3 +1.3 × 19.73 = 29.61 kip-ft/ft

$\phi = 0.85$ (Section 3.3.4.2 and Table 3a and 3b)

Reinforcement: ¾ in diameter bars at 8 in centers (0.66 in^2 top) and 6 in centers (0.89 in^2 bottom)

$a = A_s f_y / 0.85 f_c' b$

a = 0.89 × 33.0 ksi / 0.85 × 2.8 ksi × 12 in = 1.03

$M_u = \phi M_n = A_s f_y (d - a/2) = 0.85 \times 0.89 \times 33.0 \times (16 - 1.03/2) = 24.970 \times 14.48$ kip-inch

$M_u = 361.71$ kip-in = 30.14 kip-ft/ft > 29.61 kip-ft

Hence okay.

Existing slab bridge has the capacity to carry HS-20 live load.

6.8.6 Deck Replacement of Slab Composite with Repeated Beams

Empirical design: Alternate method to traditional method (9.7.2.4)

Flexure in longitudinal direction: Design as a series of longitudinal strips. Idealize width of strip = Width of wheel + 2 × Effective depth of slab

Max. BM due to dead load = $w_d (L^2/8)$

Max. BM due to lane load = $w_L (L^2/8)$

Max. BM due to truck load is calculated from influence lines. For deck slabs, Puncher's influence line diagrams were used.

AASHTO analytical method: Analyze as a strip of unit width continuous over beam flanges.

Many states in the U.S. have ready-made simplified structural solutions for deck thickness and rebars based on AASHTO empirical methods. These are frequently used in practice and were developed by examination of repeated drawing details and calculations performed for a large number of spans. Some of the approximations in current AASHTO or state code design methods are:

1. Boundary effects of skew and curved decks not considered.
2. Arching action at supports (9.7.2.1) arising from reverse bending curvature: Planar or membrane forces will be generated in addition to bending. Three dimensional modeling and analysis will be required.
3. Conventional methods do not consider additional thickness for transverse deck drainage, camber thickness, thickness of concrete haunches on top of flanges, or any groove formations. In addition,
    - Added stiffness due to stay-in-place folded steel or aluminum formwork
    - Secondary stresses such as resulting from creep and shrinkage stresses contributing to cracking
    - Daily thermal stress variation during summer and winter months
    - Composite behavior of wearing surface thickness, using special concrete (such as latex modified or corrosion inhibitor aggregates for forming defense against tire friction and braking forces).
4. Effects of shear deflection are neglected.
5. Applications of fracture mechanics formulae for deck cracking.
6. Approach slab analysis: Approach slab behind integral abutments is itemized as a structural member during construction. For analysis, it needs to be idealized as slab on grade and acts as a plate on elastic foundations. Geotechnical properties of subgrade material will be required.

    More accurate analytical method: Deck slab is idealized using FEM. Ultimate load behavior of RC elements needs nonlinear analysis. For concrete, stress-strain curve is nonlinear during cracking stage. The tangential stiffness method is used.

6.8.7 LRFD Design Methods for Deck Slab Design

Use Mathcad or spreadsheets to program equations, load, and resistance factors to comply with AASHTO LRFD code (refer to Chapter 7).

The following steps are provided for ready reference. Prior to developing a software or for solving the equations using hand calculations, equations need to be checked against the latest version of applicable AASHTO LRFD Specifications or LRFR Manual.

Analysis: Main reinforcement perpendicular to traffic. Analyze as a strip of unit width continuous over beam flanges.

Deck thickness: It is based on live load deflection control, which is a function of girder spacing, $s/h \leq 20$ where s is girder spacing, h min $= (s + 3000) / 30 \leq 175$ (Table A2.5.2.6.3-1)

Structural depth > 7 in (for thinner slabs and wide girder spacing, transverse prestressing may be used to prevent crack formation).

Dead load moments: Dead load bending moments can be evaluated by the moment distribution method, stiffness matrix method, or by coefficients given in handbook.

Highest positive and negative moments are in end spans of a continuous deck and less in middle spans. The greater the number of beams, the higher the redundancy.

Equal spacing of girders is assumed.

Transverse dead load bending moments, for repeated beams, M max $= +wL^2/12$ and $-wL^2/8$

For four beams, M max $= +wL^2/10$ and $-wL^2/10$

For five or more beams (use influence lines at 0.4 L of 1st span), $M = 0.0744\ wL^2$

Adjust for overhang moment $-wl^2/2$

For scond span, $M = +0.0471wL^2$ and $-0.1141wL^2$

For analysis, external actions due to load must balance internal actions: $\Sigma M = 0$

Hence $M_u = \phi M_n = A_s f_y (d - a/2)$; $a = A_s f_y / 0.85 f_c' b$

$A_s = (M_u/\phi) / f_y j.d)$

For Design, $\phi M_n \geq M_u$ Note: Sign ($\geq$) is to ensure minimum requirements and results in higher overall safety factor for the bridge.

Check for shear $\Sigma V = 0$ is not required.

- Check ductility: $a \leq 0.35\ d$
- Minimum reinforcement $\rho \geq 0.03\ f_c'/f_y$ (5.7.3.3.2)

$1.2\ M_{cr} \leq M_u$ or $1.33\ \Sigma\ M$ (factored moments)

$M_{cr} = f_r I/(y t)$

$M_u = M_{DC} + M_{DW} + M_{LL+I}$

Rupture stress $f_r = 2.4 \sqrt{f_c'}$

- Maximum spacing s max $= 450$ mm or 1.5 h (5.10.3.2)
- Distribution reinforcement $= 3840/\sqrt{S_e}$ % (9.7.3.2) and (9.7.2.3)

  where S_e is the effective span length

  $220/\sqrt{s} \leq 67$ percent (bottom distribution bars)

  $A_s > 0.11 A_g / f_y$ (top distribution reinforcement)

- Maximum reinforcement:

  $c / d_e \leq 0.42$ (5.7.3.3.1-1)

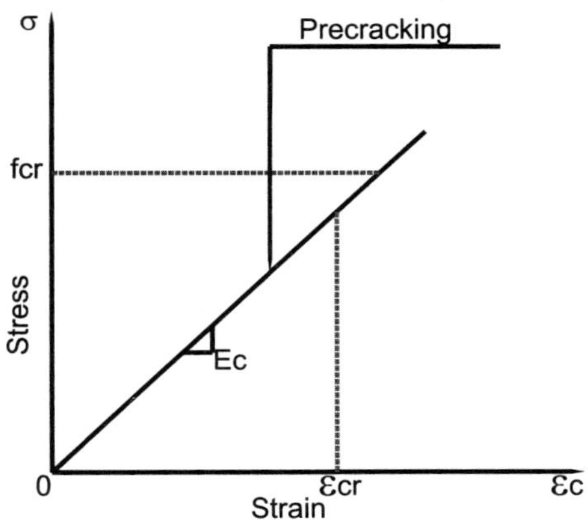

Figure 6.7 Stress-strain diagram for concrete.

- Shrinkage and temperature reinforcement:

  Temp $A_s \geq 0.75\ A_g / f_y$ (5.10.8.2)
- Minimum cover is $\geq 1.0''$ based on Table 5.12.3.1 or state code requirements.
- Crack control: $f_s \leq Z / (d_c A)^{0.33} \leq 0.6\ f_y$ (5.7.3.4)

  $Z = 23000$ N/mm, $d \leq 50$ mm

Service I limit state.

Unfactored moment to calculate tensile stress in reinforcement

$M = M_{DC} + M_{DW} + M_{LL+I}$

$n = E_s / E_c$, $Es = 200000$ Mpa (5.4.3.2)

$E_c = 0.043\ \gamma_c \sqrt{\ } f_c'$ (5.4.2.4)

Overhang design: Design as nonredundant member.

Cantilever span ≤ 6 ft (3.6.1.3.4) or ≤ 0.65 girder spacing.

Distance to face of barrier ≤ 3 ft (4.6.2.2.1)

Minimum edge depth = 8 in for overhang supporting parapets or deck mounted posts

= 12 in for side mounted posts.

Live loads:

Equivalent live load = 1.0 kip/ft located 1.0 ft from face of railing.

Strength I limit state for HL-93 loads

Extreme event II for collision from vehicles

Negative moments at fascia girders from parapet loads have beneficial effects of reducing positive moments in deck continuous spans.

Since extreme state collision moment is distributed over 5 to 10 ft width of deck, it is generally smaller than deck negative moment.

Due to small cantilever span, strength I overhang moment $\leq$ deck negative moments.

Cantilever design does not control the deck design.

Overhang slab is usually cast composite with the parapet by placing U- or ell-shaped rebars from overhang slab inside the barrier. Overhang thickness is increased for extreme cases of collision of parapet or railing. In such cases, overhang will not crack,and parapet or railing can be made sacrificial. (A13.4.2)

Overhang 1140 + 0.833 X

+ ve moment 660 + 0.55 S

− ve moment 1220 + 0.25 S

IM = 33 percent of LL (3.6.2.1)

- Number of lanes

  N = Int (roadway width / 3600) (3.6.1.1.1)

- Multiple presence factors

  m = 1.2 for one loaded lane

  = 1.0 for two

  = 0.85 for three

- Tire contact area = l × 510 mm

  Where l = 2.28 γ (1 + IM/100) P (3.6.1.2.5)

  Wheel load is applied as a distributed load.

- Fatigue limit state: Not required for multi-girder applications (9.5.3)
- Empirical design: Alternate method to traditional method (9.7.2.4)

6.9 RATING PROCEDURE FOR REINFORCED CONCRETE T-BEAM BRIDGE

Refer to AASHTO LRFR Specifications.

1. Perform dead load analysis for DC and DW.
2. Perform live load analysis for design live loads.
3. Compute live load distribution factors. (LRFD Table 4 – 11)
   - One lane loaded: DF for flexure

     $g_{m1} = 0.06 + (S/14)^{0.4} + (S/L)^{0.3} + (K_g/12Lt_s^3)^{0.1}$

   - Two lanes loaded: DF for flexure

     $g_{m2} = 0.075 + (S/9.5)^{0.6} + (S/L)^{0.2} + (K_g/12Lt_s^3)^{0.1}$

   - 6.9.1 Computation of Distribution Factors
   - Comparison with LFD method for steel girders:
   - DF = 0.15 + $(S/3)^{0.6}$ $(S/L)^{0.2}$ $(Kg/12\ L\ ts^3)^{0.1}$
   - Assume L = 1000 ft, S = 12 ft, ts = 8 in, Kg = 1317,726
   - DF = 1.773 wheels per beam
   - LFD distribution factor = S/5.5 = 2.18 wheels per beam
   - Reduction in BM or SF = 1.773/2.18 = 0.81
   - Reduction = 19 percent for interior beam
   - One lane loaded: DF for shear

     $g_{f1} = 0.36 + S/25.0$

   - Two lanes loaded: DF for shear

     $g_{f2} = 0.2 + (S/12) - (S/35)^{2.0}$

   Compute design live load moments for HL-93 (HS-20 truck, lane, or tandem) and multiply by DF.
4. Compute effective flange width (LRFD 4.6.2.6.1)

   Select minimum of L/4; Spacing or t_s + Greater of t_w or $b_f/2$
5. Compute distance to neutral axis c

   $c = A_s f_y / (0.85 fc' \beta_1 b)$; $a = c \beta_1$

6. $M_n = A_s f_y (d_s - a/2)$

7. Minimum reinforcement required to develop M_r

$\leq$ Lesser of 1.2 M_{cr} or 1.33 M_u (LRFD 5.7.3.3.2)

$M_r = \phi M_n$ Eq. (6-4)

$1.2\, M_{cr} = 1.2\, (f_r + f_{pb})\, S_{bc} - M_{d,nc}\, (S_{bc} / S_b - 1)$ (LRFD 5.4.2.6)

$S_{bc} = I / y_t$

$f_{pb} = 0$ (no prestress)

$M_{d,,nc} = 0$ (non-composite dead load moment)

8. Critical section for shear (LRFD 5.8.3.2)

Effective shear depth $d_v = M_n / (A_s f_y + f_{ps} A_{ps})$ (LRFD 5.8.2.9)

Maximum of 0.9 d_e and 0.72 h LRFD Eq. (C5-9)

or distance between resultants of tensile and compressive forces.

Compute dead load shear V_{DC} and V_{DW}

$V_n = V_c + Vs$

$V_c = 0.0316\, b_v\, d_v\, (fc')^{0.5}\, \beta$

$Vs = d_v \operatorname{Cot} \theta\, (A_s f_y)$

6.9.1 General Load Rating Equation (LRFD 6.4.2)

$RF = [C - (\gamma_{DC})(DC) - (\gamma_{DW})(DW) \pm \gamma_P (P)] / \gamma_L (LL + IM)$ (EQ 6-1)

1. Factors for inventory rating: $\gamma_{DC} = 1.25$; $\gamma_{DW} = 1.25$; $\gamma_{LI} = 1.75$

Factors for operating rating: $\gamma_{DC} = 1.25$; $\gamma_{DW} = 1.25$; $\gamma_{LO} = 1.35$

Operating rating = Inventory rating x $(\gamma_{LI}/\gamma_{LO})$

2. Design rating equation

Strength I limit state:

$C = (\phi_c)(\phi_s)(\phi) R_n$

$RF = [(\phi_c)(\phi_s)(\phi) R_n - (\gamma_{DC})(DC) - (\gamma_{DW})(DW)] / \gamma_L (LL + IM)$ (LRFD 6.5.4.1)

Condition factor $(\phi_c) = 1.0$ for no deterioration. (LRFD 6.4.2.3)

System factor $(\phi_s) = 1.0$ for a slab bridge. (LRFD 6.4.2.4)

Resistance factor $(\phi) = 0.9$ for flexure (LRFD 5.5.4.2)

IM = 1.33

3. Compute legal load rating

4. Compute permit load rating

Table 6.10 Rating summary table.

Limit State		Design Live Load Rating		Legal Load Rating			Permit Load Rating
		Inventory	Operating	T3	T3S2	T3-3	
Strength I	Flexure						
	Shear						
Strength II	Flexure						
	Shear						
Service II							
Fatigue							

6.10 RATING OF PRESTRESSED CONCRETE GIRDER

Simple Span I-Girder Bridge

The following steps are provided for ready reference. Prior to developing a software or for solving the equations using hand calculations, equations need to be checked against the latest version of applicable AASHTO LRFD Specifications or LRFR Manual.

Effective flange width (LRFD 4.6.2.6.1)

Select minimum of L/4; Spacing S or t_s + Greater of t_w or $b_f/2$

$Ec = 33000\,(Wc)^{1.5}\,(fc')^{0.5}$ (LRFD 5.4.2.4)

A. Dead load analysis for DC1, DC2, DW

$M\ max = wL^2/8$

B. Perform live load analysis

C. Calculate DF

1. One lane loaded: DF for flexure

$g_{m1} = 0.06 + (S/14)^{0.4} + (S/L)^{0.3} + (K_g/12Lt_s^{\ 3})^{0.1}$

Longitudinal stiffness parameter $K_g = n\,(I + A\,(e_g)^2)$

2. Two lanes loaded: DF for flexure

$g_{m2} = 0.075 + (S/9.5)^{0.6} + (S/L)^{0.2} + (K_g/12Lt_s^{\ 3})^{0.1}$

3. One lane loaded: DF for shear

$D_{f1} = 0.36 + S/25.0$

4. Two lanes loaded: DF for shear

$D_{f2} = 0.2 + (S/12) - (S/35)^{2.0}$

Compute design live load moments for HL-93 (HS-20 truck, lane or tandem) and multiply by DF.

Flexural resistance:

$f_{ps} = (1 - K\,c/\,d_p\,)\,(f_{pu})$ LRFD Eq. (5.16)

$K = 0.28$ for low relaxation strands

$f_{pu} = 270$ ksi

d_p = Distance from extreme compression fiber to the centroid of prestressing tendons.

5. Distance to neutral axis c

$c = (A_{ps}\,f_{pu})\,/\,(0.85\,fc'\,\beta_1\,b + K\,A_{ps}f_{pu}/d_p)$ LRFD Eq. (5-19)

$a = c\,\beta_1$

6. $M_n = A_{ps}\,f_{ps}\,(d_p - a/2)$

7. Minimum reinforcement required to develop M_r

$\leq$ Lesser of 1.2 M_{cr} or 1.33 M_u (LRFD 5.7.3.3.2)

$M_r = \phi\,M_n$ Eq. (6-4)

$M_{cr} = (f_r + f_{pb})\,S_{bc} - M_{d,nc}\,(S_{bc}/S_b - 1)$ (LRFD 5.4.2.6)

Modulus of rupture $f_r = 0.24\,f_c{}')^{0.5}$ (LRFD 5.4.2.6)

$f_{pb} = P_{pe}/A + P_{pe}\,e/S_b$

$S_{bc} = I/y_t$

$f_{pb} = 0$ (No prestress)

$M_{d,nc} = 0$ (non-composite dead load moment)

8. Determine effective prestress force P_{pe}

f_{pe} = Initial prestress − Total prestress losses.

$f_{pe}\,A_{ps} = P_{pe}$

Total prestress losses = $(\Delta f_{pES} + \Delta f_{pTL} - \Delta f_{pRI})$ (LRFD 5.9.5.1)

Loss due to elastic shortening $\Delta f_{pES} = E_p f_{cg\,p} / E_{ci}$ (LRFD 5.9.5.2.3a)

$f_{cg\,p} = (P_i/A + P_i e^2 / I - M_D e / I)$

For low relaxation prestressing strands, initial prestress = $0.75 f_{pu}$ (LRFD Table 5-7)

$P_{PR} = f_{py} A_{ps} / (f_{py} A_{ps} + fy\,As)$ (LRFD Table 5-12)

Time-dependent losses for I-girder:

$\Delta f_{pTL} = 33[1.0 - 0.15 (fc' - 6.0)/6.0] + 6.0 P_{PR} - 6.0$ (LRFD 5.9.5.3)

Relaxation at transfer $\Delta f_{pRI} = \log(24.0\,t) [(f_{pj}/f_{py}) - 0.55] f_{pj} / 40.0$

For As = 0, P_{PR} = 1 LRFD Eq. (5-8)

$f_{pb} = P_{pe}/A + P_{pe}\,e/S_b$

Maximum reinforcement: (LRFD 6.5.6)

$c/d_e \leq 0.42$ (LRFD 5.7.3.3.1)

9. Critical shear check (LRFD 5.11.4)

Shear check is not required if there is no visible sign of shear distress.

Critical location for shear occurs near the supports and it is greater of d_v or $0.5\,d_v\,(\text{Cot}\,\theta)$. (LRFD 5.8.3.2)

Effective shear depth d_v is maximum of:

$0.9\,d_e$; $0.72\,h$; distance between resultants of tensile and compressive forces.

Assume θ = 30 degrees; $0.5\,d_v\,(\text{Cot}\,\theta) = 0.87\,d_v \leq d_v$

Minimum transfer length = 60 × strand diameters (LRFD 5.11.4)

If section is outside transfer length, full value of f_{po} is used in calculating shear resistance.

Maximum shear at critical section near supports:

Total shear = $V_{Lane} + V_{Truck} \times IMP$

Nominal shear resistance = $V_n = V_s + V_c + V_p$ LRFD Eq (5-66)

For straight tendons, $V_p = 0$

Provide minimum transverse reinforcement:

10. Modified Compression Field Theory (MCFT)

$V_{Lane} = 0.0316 (fc')^{0.5} b_v S / f_y$ (LRFD 5.8.2.5)

$V_{Lane} = 0.0316\,\beta\,(fc')^{0.5} b_v d_v$ LRFD Eq (5-68)

$V_s = A_v f_y d_v \text{Cot}\,\theta / S$ LRFD Eq (5-69)

11. Simplified approach for shear design

θ = 45 degrees; β = 2

$V_c = 0.0316\,\beta\,(fc')^{0.5} b_v d_v$

Table 6.11 Sample summary table of rating factors.

Limit State		Design Load Rating		Permit Load Rating
		Inventory	Operating	
Strength I	Flexure			
	Shear			
Strength II	Flexure			
	Shear			
Service II				
Fatigue				Stress Ratio 5

Total nominal shear resistance $V_n = V_c + V_s$

$v = (V_u - \phi V_p) / \phi b_v d_v$ LRFD Eq (5-65)

$v/fc' \leq 0.25$

At first critical section for shear, calculate design (factored) dead load and live load moments.

Using the LRFD shear design flow chart and Table 1 (LRFD 5.8.3.4.2), LRFD Table 5-5, and Figure 5-19, calculate ε_x:

$\varepsilon_x = [M_u/d_v + 0.5 N_u + 0.5 (V_u - V_p) \cot\theta - A_{ps} f_{p0}] / 2 (Es A_s + E_p A_{ps})$ LRFD Eq. (5-70) ≤ 0.002 (LRFD 5.8.3.4.2)

Calculate $V_n = V_s + V_c$ and compare with the simplified method.

12. Check longitudinal reinforcement LRFD 5.8.3.5

Force $T = [M_u / \phi d_v + 0.5 N_u / \phi + (V_u / \phi - 0.5 V_s - V_p) \cot\theta]$ LRFD Eq. (5-74)

Development length $L_d = (f_{ps} - ⅔ f_{pc}) d_b$ LRFD Eq. (5-36)

13. Compute nominal shear resistance at stirrup change (at 20 ft. from center line of bearing) (LRFD 5.11.4)

Compute maximum shear at stirrup change.

Check longitudinal reinforcement.

Check tensile capacity on flexural tension side of member > Force T.

14. Load rating equation

$RF = [C - (\gamma_{DC})(DC) - (\gamma_{DW})(DW) \pm \gamma_P (P)] / \gamma_L (LL + IM)$ (EQ 6-1)

Condition factor $(\phi_c) = 1.0$ for no deterioration. (LRFD 6.4.2.3)

System factor $(\phi_s) = 1.0$ for a multi-girder bridge. (LRFD 6.4.2.4)

Resistance factor $(\phi) = 1.0$ for flexure (LRFD 5.5.4.2.1)

Design rating equation LRFD 6.4.3

Strength I limit state:

$C = (\phi_c)(\phi_s)(\phi) R_n$

$RF = [(\phi_c)(\phi_s)(\phi) R_n - (\gamma_{DC})(DC) - (\gamma_{DW})(DW)] / \gamma_L (LL + IM)$ (LRFD 6.5.4.1)

Condition factor $(\phi_c) = 1.0$ for no deterioration. (LRFD 6.4.2.3)

System factor $(\phi_s) = 1.0$ for a slab bridge (LRFD 6.4.2.4)

Factors for inventory rating: $\gamma_{DC} = 1.25$; $\gamma_{DW} = 1.5$; $\gamma_{LI} = 1.75$

Factors for operating rating: $\gamma_{DC} = 1.25$; $\gamma_{DW} = 1.5$; $\gamma_{LL} = 1.35$

Flexure at midspan: Operating rating = Inventory rating × $(\gamma_{LI}/\gamma_{LO})$

A systematic evaluation of the shear and longitudinal yield criteria based on shear-moment interaction should be performed along the length of the beam at L/20, in addition to points of special interest. Flexural ratings should be checked at maximum moment sections and at sections where there are changes in flexural resistance.

15. Service III limit state

Rating factor for inventory level

f_{pb} = Compressive stress due to effective prestress

Allowable tensile stress = $0.19 (fc')^{0.5}$ (LRFD 5.9.4.2.2)

Flexural resistance $f_R = f_{pb}$ + Allowable tensile stress

Compute dead load stresses f_D

Compute live load stresses f_{LL+IM}

$RF = [(f_R - (\gamma_D) f_D] / \gamma_L f_{LL+IM}$ (LRFD 6.5.4.1)

Legal load rating: When inventory design load rating factor > 1, legal load rating does not need to be performed and no posting is required. (LRFD 6.4.3.1)

Permit load rating—the following types of permit loads apply:

special, single trip and mix with traffic, no traffic control.

Compute undistributed maximum live load moment and shear.

Calculate distribution factors.

Compute flexure RF and shear RF.

BIBLIOGRAPHY

AASHTOWare catalog, www.aashtoware.org.

"AASHTO-LRFD Bridge Design Specifications, Background and Live Load Model Development and Applications," *ASCE Structures Congress*, NHI Course No. 13061, Philadelphia, PA, May 2000.

Agrawal, A., Y. Zhihua and Mohiuddin Ali Khan, *Handbook of Scour Countermeasures*, National Cooperative Highway Research Program, Trenton, NJ, 2007.

Brockenbrough, R. L. and F. S. Merritt, *Structural Steel Designer's Handbook*, 4th Edition, McGraw-Hill Inc., New York, NY, 2006.

"Detailed Planning Research on Accelerating the Renewal of America's Highways," National Cooperative Highway Research Program, 2003.

FHWA Manual on Use of Self-Propelled Modular Transporters to Remove and Replace Bridges, Federal Highway Administration.

FHWA Bridge Design Guidance No. 1, "Load Rating Evaluation of Gusset Plates in Truss Bridges," Firas Sheikh Ibrahim, 2008.

Guide Design Specifications for Bridge Temporary Works, American Association of State Highway and Transportation Officials,1995.

"Improved Live Load Deflection Criteria for Steel Bridges," National Cooperative Highway Research Program, Project 20-7, November 2002.

Jakovich, Gary and Vasant Mistry, "National Perspective on Accelerated Bridge Construction," *24th International Bridge Conference*, Pittsburgh, PA, 2007.

Khan, Mohiuddin Ali, "Simplified Seismic Design," (textbook under prep.), Elsevier (Butterworth) Publications.

LRFD Bridge Construction Specifications, 2nd Edition, American Association of State Highway and Transportation Officials, 2004.

LRFD Bridge Design Specifications, 3rd Edition, American Association of State Highway and Transportation Officials, 2004.

Manual for Condition Evaluation and Load and Resistance Factor Rating of Highway Bridges, American Association of State Highway and Transportation Officials, 2003.

Michael Baker Corporation, "Software for AASHTO LRFD Bridge Specifications," *International Bridge Conference*, Pittsburgh, PA, 1998.

Mufti, Aftab A., Baidar Bakht and Leslie G. Jaeger, *Bridge Superstructures: New Developments*, National Book Foundation, Islamabad, 1996.

Park, Sang H., *Bridge Inspection and Structural Analysis: Handbook of Bridge Inspection*, Park, Trenton, NJ, 1980.

Park, Sang H., *Bridge Superstructure Design and Rehabilitation*, Trenton, NJ.

Steel Bridge Design Handbook, "Design for Constructability," Chapter 13, National Steel Bridge Alliance, 2007.

7

Bridge Widening and Deck Replacement Strategy

Section 2
Strengthening and Repair Work

7.1 SELECTING BRIDGE WIDENING OR REPLACEMENT

7.1.1 Introduction

In Chapter 2 the five types of rehabilitation were discussed. Rehabilitation may require continued maintenance and high cost since the greater the number of bridges selected for repair, the greater the investment by the responsible agency. Besides rehabilitation, only widening, widening with repair and retrofit, and full or partial replacement are the other options. In this chapter, selection of either widening or widening with a retrofit or replacement method are discussed. The procedures for superstructure and substructure replacement are based on alternatives analysis and evaluation of life cycle costs.

Widening entails retaining and reconstructing an existing bridge. It gives the bridge a new function and would cost only a fraction of the cost of a new bridge. Over decades, demographic changes have increased the ADTT on many important routes. To meet the growing traffic needs, both approach highways and bridges need to be widened. One or more traffic lanes, shoulders and/or sidewalks may be needed. Symmetric widening is not always possible but preferred. Providing continuity over piers for the widened deck area and new girders may not be easy due to lack of continuity over supports in the original design. Resulting widening of abutments, piers, and their foundations are unavoidable.

Major widening is new construction work for an existing bridge facility, which may nearly double the deck area of the existing bridge and increases the number of traffic lanes. It may be widening an existing substandard lane width, or adding overhangs or sidewalks. In each case, the widened area to be constructed is the transverse out-to-out dimension of the proposed deck.

The salient features are selection of structural solutions, feasibility, methodology, planning, alternatives analysis, stress and deformation analysis, detailed design, constructability issues, economics, life cycle costs, and aesthetics. Case studies of successfully completed projects are also examined.

7.1.2 Widening Practice

Requirements for preparing widening plans are shown in Figure 7.1.

An example of deck slab design based on the LRFD method is included in this chapter. Guidelines for girder selection based on optimum considerations are also summarized in this chapter. After selecting the girder type, the steps required for preliminary or advanced design described in Chapters 5 and 6 should be followed. Although familiarity with hand calculations is required, the designer should be familiar with the listed approved computer software and its supporting manual, which will outline the applicable methods of analysis and design, in keeping with AASHTO requirements.

The current design method is the LRFD method, while the one used for original design is likely to be ASD or LFD. The truck live load intensity and deflection criteria would be different. Comprehensive load combinations and applications of AASHTO LRFD methods described in Chapters 5 and 6 are as important for widening as they are for replacement bridges.

1. In order to match the existing deck thickness and reinforcement area, live load capacity for new design may be different due to differences between old and new regulations. Following are some general guidelines:

   The transverse reinforcement in the new deck should be spaced to match the existing spacing. Different bar sizes may be used if necessary.
2. The new girders in the widened portion must match the existing span lengths.
3. Avoid mixing concrete and steel beams in the same span. The use of beams of the same type as those used in the existing structure is preferred.
4. If the existing beams are cast-in-place concrete, detail the widened deck supported on precast prestressed beams.

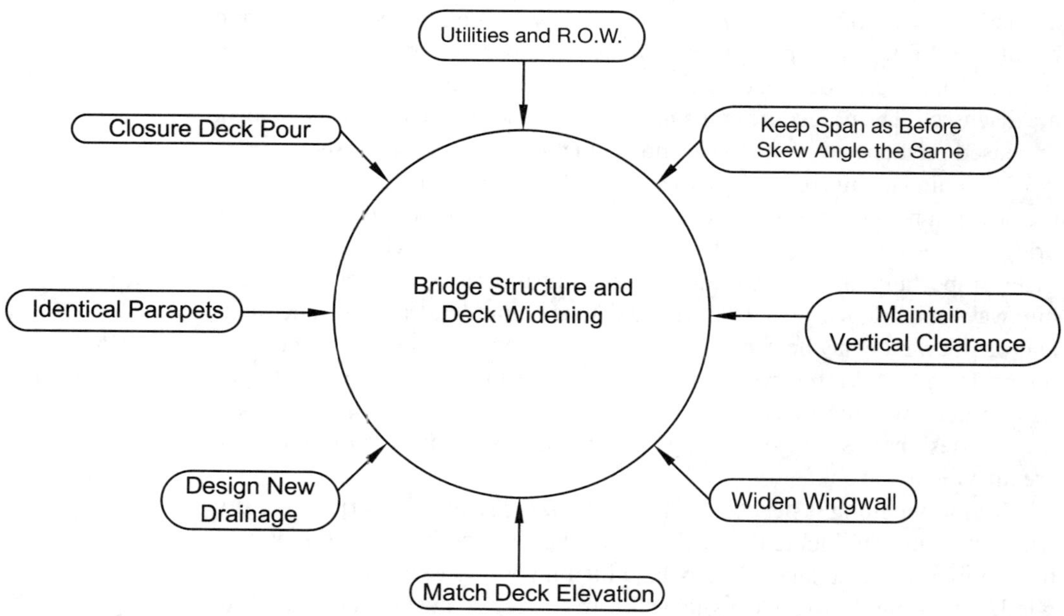

Figure 7.1 Steps for planning deck widening.

5. Provide the required vertical clearance on interstate structures unless an exception is granted. To assist in meeting the minimum vertical clearance requirements on a widening, the standard beam depth may be decreased. This is generally possible by using modern HPS 70W steel. Where the existing bridge does not satisfy current vertical clearance requirements and where the economics of doing so are justified, the superstructure may be elevated.

7.1.3 Widening Design Methods and Applicable Computer Software

Bridge widening that will exceed one-third of the deck slab area will be designed according to the AASHTO LRFD Bridge Design Specifications. Both non-composite and composite design may be required. LRFD steel beam design software such as STLRFD, Merlin-DASH, or alternate software will be used.

A list of commonly used computer software is provided in Chapter 6.

7.1.4 Widening of Voided Slab Bridges and Through Girder Bridges

Spacing and use of adjacent units require the widening width to be a multiple of the voided slab width. Transverse diaphragms require special attention. Recommendations and criteria for the widening of the particular structure are on a case-by-case basis. Other critical features include:

1. Attachments such as utilities, lighting poles, and sign structures to existing structure.
2. When performing stress checks of the existing structure, consider the construction sequence and the degree of interaction between widening and the existing structure for the completed structure.
3. Thermal properties of materials: Materials used in the construction of the widening should have the same thermal and elastic properties as those of the existing structure.
4. On through girder bridges, widening is provided with cantilever sidewalks or by adding a lane. To increase the width of a bridge, the bridge deck will be made wider. Transverse floor beams supported between bottom flange of the two through girders can be made longer with their lengths in the transverse direction increased or by utilizing maximum cantilever action of say 5 feet on each side of longitudinal through beams. For continuity, new cantilever beams can be connected to ends of existing floor beams located at the bottom flanges of through girders. Replacing the existing floor beams with maximum 10 feet longer floor beams (i.e., two cantilevers of 5 ft each) is feasible.
5. Maintaining symmetry and balanced cantilever action will not affect the stability of through girders. However, an asymmetric increase of dead and live load will introduce additional stress and torsion in the through truss members. A three-dimensional stress analysis for new loads will be necessary to check and prevent overstress.
6. The main exterior member of the existing structure should be checked for both construction conditions and the final condition, i.e., after attachment of the widened portion of the structure.
7. The alternate for a through truss, with the bottom chord post tensioned by rods, is replacement of the damaged/missing post tensioning rods along the bottom chord with post tensioning cables.
8. Drilling:
    - When drilling into heavily reinforced areas, main reinforcing bars need to be exposed by chipping.
    - Drilled holes should have a minimum edge distance of three times the metal anchor diameter (3d) from the free edges of concrete and 1-inch minimum clearance between the edges of the drilled holes and existing reinforcing bars.
9. Widening closure pour: Closure pours on bridge structures with integral or semi-integral abutments shall include the concrete for abutments' diaphragm. The purpose of the closure

pour is to accommodate the differences in deflection that can occur between the new and the old decks during construction.

7.2 SUPERSTRUCTURE WIDENING

7.2.1 General Guidelines for Design

Widening may require near-term repairs to the existing deck and girders. The objectives are to restore serviceability and the original functionality following distress from severe localized deterioration or from vehicle impact damage.

Consider the following physical parameters:

1. Bridge location, pier location, skew angle, and stationing are to be verified in field.
2. Span lengths will be maintained.
3. The number of beams will increase. The type of new beams will be shallower to maintain minimum vertical underclearance.
4. Similar types of parapets will be built for historic reasons.
5. The pier and abutment will be widened to accommodate the new widths of decks.
6. Utilities supported on the bridge may need to be relocated during construction.
7. New finished grade elevations will match existing elevations.
8. A design exception may be required in some cases if minimum vertical clearances cannot be maintained.
9. Drainage scuppers will be relocated and redesigned.
10. Closure deck pour between old and new concrete shall be as per technical specifications.
11. Existing wing walls need to be demolished, and new walls on new locations are required.

7.2.2 Additional Widening Requirements

Many of the following requirements may be applicable:

1. To ensure the safety of the bridge through increase of redundancy.
2. To keep traffic detour to a minimum. Repairs also require minimal impact to traffic.
3. To respond to public needs through outreach regarding bridge closure and the construction schedule. Coordinate with the local communities and adjacent roadway agencies to identify acceptable reconstruction solutions.
4. To minimize any environmental impact and prevent any river pollution during construction.
5. To maintain the historic appearance and shape of members.
6. To introduce speed restriction to minimize live load impact.
7. To post the bridge for a lower live load.

7.2.3 Typical Activities

On a bridge widening project, some of the typical activities will include:

1. A field inspection will be made to determine the condition of existing deficiencies.
2. For work on floor beams and on main trusses, the bridge will be closed for restricted duration after public outreach is favorable.
3. Some floor beams supporting the deck may have tilted. A minor twist (buckling) to their web mainly at supports will require web stiffening by bolting web plates. Bolts to stringer connections will be added. Damaged rivets and bolts will be replaced.
4. Repairing portions of the floor system and the sidewalk damaged by impact from flood: All damaged steel members will be repaired and painted.

5. Any pitting of top flange areas of floor beams will be repaired by grinding steel surfaces.
6. Structural painting may be required.

7.2.4 Staging Planning

To maintain minimum discomfort to the public, sensitive planning is required, including the following steps:

1. A traffic count needs to be performed to assess impact on traffic flow during construction. Warning signs must be placed weeks in advance so that the users may select an alternate route to avoid congestion.
2. Local authorities should be contacted to determine if they have any restrictions regarding lane closures.
3. Prior to developing staging plans, the agency's traffic operations department will provide the maximum allowable lane closure hours in each direction and the maximum number of lanes that can be closed at one time. An eight- to ten-hour night window is required for the contractor to properly complete his work. Extra hours will be permitted for weekend work.
4. Plans must comply with MUTCD and AASHTO LRFD regulations. All non-standard signs shall be sized according to the MUTCD with letter heights and alphabet size given for each line.
5. Construction staging plans shall include cross sections of the bridge for each stage of construction. Fewer stages will require less time for completion. Time required for the completion of concrete work for each stage is nearly that of a bridge. Two main stages are preferred over three or four, although there may be substages.
6. Structural drawings showing construction in each stage should conform to traffic control plans. A set of applicable standard traffic control plans are to be used as a basis for developing the final traffic control plans. These plans shall be customized to reflect site conditions and the ability of the shoulder to withstand traffic.

7.3 REHABILITATION/RETROFIT OR REPLACEMENT DECISIONS

7.3.1 Management Methods Associated with Design

Teamwork and organization are required for all public projects. Design is a step-by-step process. Hence, final design should be carried out only after the following necessary documentation is in place.

1. Permit approval: In addition to analysis, design will be completed by simplifying the permitting application and approval process using computer software and Internet media. Compliance with dozens of requirements causes delays in getting construction permits approved. Sometimes permit approvals may take longer than the bridge design effort. Changes requested by DEP may alter design.
2. A Web site needs to be developed. Both designers and government agencies will log into the secure Web site. Online meetings will take place there. Through electronic submission, online tracking of plans, documents review, and payment of approved fees will be possible and permits will be issued online. Through software applications the system will:
   - Simplify and accelerate the development review and permit issuance process
   - Improve uniformity and predictability of code applications
   - Use Internet-based technology and computer software to connect customers and regulators
   - Create a computer network linking the entire construction industry
   - Incorporate the best practices in construction, reviews, and field inspections.

3. Iterative planning: Today's right-of-way, utilities relocation, environmental protection ordinances, and construction codes are complex. As many as 15 agencies representing right-of-way, local government, citizen review committees, fire department, health department, police, utilities, zoning, land use, highway, historical preservation, environmental protection, etc. are likely to be involved.

7.3.2 Use of Application Modules

1. Employ a separate Web site for each project, where all documents will be posted and transactions will take place.
2. Employ electronic white boards to facilitate interactions and on-the-spot markups of plans by plan reviewers and design professionals.
3. Allow automatic updates of submissions.
4. Enable field inspectors to use wireless communications tools to access project drawings online and transmit their inspection results right away.
5. Automatically route all documents to appropriate review agencies.
6. Enable applicants to monitor the progress of reviews carried out by each agency.

The system will use application modules that can be downloaded on an applicant's personal computer. Modules will include the road map of the permitting process and resources needed for permit approval. An inventory and checklist will be maintained for status and any missing details in documents submitted. It will identify violations of air quality, water quality, fauna and flora and adverse environmental impacts. Based on technical information submitted, computer programs may suggest remedies and request an alternate design or construction method. The new system will:

An education and training process for the new computer software will be required for applicants and design professionals. Agencies will suggest ways and means to iron out any inherent procedural difficulties. It will save man-hours for both applicants and agency officials and will ensure an early start of construction.

7.3.3 Project Management Aspects for Widening or Replacement

All major repairs or retrofit will fall under general rehabilitation as distinct from replacement. These guidelines are intended to be used during project scoping. The following important factors need to be addressed for any superstructure and substructure rehabilitation issues:

1. Funding and cost.
2. Development of rehabilitation schemes.

Table 7.1 Feasibility and cost comparisons for alternative methods.

Component for REH or REP	Rehabilitation (REH)	Replacement (REP)	Remarks
Deck Slab	For minor repairs	For major repairs	Replacement of deck every 15 to 20 years is common
Fascia or interior girders	For minor repairs	For major repairs	Heat straightening may be done
Bearings	Not usually done	For all types of repairs	Jacking of ends of girders
Substructure	For minor repairs	Substructure cannot be replaced	Bridge is replaced for major repairs
Foundations	Retrofit with mini piles	Not usually done	Bridge is replaced for major repairs
Entire bridge	For minor repairs	Bridge is replaced for major repairs	Staging may be used

3. Value engineering.
4. Satisfactory functional requirements.
5. Aesthetics.
6. Public involvement.

7.3.4 Tips for Good Planning

Good planning includes the following:

1. Planning for geometry, minimum skew or curvature, adequate sight distance, sufficient horizontal or vertical clearances, and adequate opening over waterway.
2. Meeting the functional requirements such as providing an adequate number of lanes to prevent overload, and posting of warning signs and directions ahead of the bridge.
3. Using design aspects resulting in minimum deflection and vibration of girders, using jointless deck, keeping deck surfacing uncracked, unrestricted bearing movements, and ductility of joints.
4. Providing facilities for ease of maintenance such as a provision for inspection chambers, structural health monitoring by remote sensors, and nondestructive testing.
5. The construction industry has also benefited from the use of new machinery, cranes, and tractor trailer vehicles for freight. Precast concrete technology and pre-assembled replacement bridges offer quick and reliable solutions by minimizing delays and reducing construction time.
6. Using modern high strength and corrosion resistant materials.

Experience has shown that if any of these is lacking, indirect costs in terms of structural damage, accidents, or delays (which were not provided for in the original budget) will accrue.

7.3.5 Further Planning Considerations

1. Tables 7.1 to 7.3 list a variety of available options between minor and major repairs, retrofit, rehabilitation, and replacement. In general, replacement alternatives for a substructure need not be considered. In considering superstructure replacement, the substructures must first be evaluated. This evaluation may include in-depth inspection, performing NDT, and taking cores to verify their condition.
2. For historic bridges, replacement is generally not an option except for safety reasons, and rehabilitation is to be carried out.
3. The functional importance and how important a bridge is to the overall transportation system of the area need to be considered. In addition to structural adequacy there are other social, political, and capacity related considerations.
4. Accident history and potential must be examined.

   They must be examined for the bridge project. In terms of safety for the RH/RP decision, accident history is the most important element. The accident history can be determined by examining the accident reports on file. Although sometimes inconclusive, this review should look for trends in accident patterns that would point to whether the bridge caused or contributed to the accidents.
5. Expected useful life.

   Tables 7.2 and 7.3 show a comparative study of expected useful life.

Table 7.2 Useful service life for superstructure and substructure members.

No.	Location of Member	Rehabilitation/ Protective Measures	Useful Service Life* (Years)	Remarks
1A	Terminal decks	Minor patching and bituminous overlay	5	Maximum life of bituminous overlay is eight years
1B	Deck to remain in place	Membrane waterproofing and bituminous overlay	20	Replace membrane each time overlay is replaced
1C	-do-	Latex modified concrete overlays, cathodic protection	20	
1D	New concrete deck	With epoxy-coated rebars	40	Can be extended to 50 years with maintenance
2	Deck joints/expansion dams	Periodic replacement of glands or trough	40	-do-
3	Beams and connections	Repairs and/or rehabilitation	40	Painting required
4	Other types of superstructure and elements	-do-	40	Can be extended to 50 years with maintenance
4A	New superstructure	Replacement	75	Can be extended to 100 years with maintenance
5	Bearings	Replacement	75	Require inspections and maintenance
6	Existing substructure	Repairs and/or rehabilitation	75	Can be extended to 100 years with maintenance
6A	New substructure	Replacement	100	Can be extended to > 100 years with maintenance
7A	Existing R.C. retaining walls	Repairs and/or rehabilitation	50	Can be extended to > 50 years with maintenance
7B	New R.C. retaining walls	Replacement	75	Can be extended to > 75 years with maintenance
7C	New PMU/earth reinforced	Replacement	50	Can be extended > 50 years with maintenance

7.3.6 Compare REP/REH Ratios

The next step is to compare rehabilitation and replacement costs assuming both are viable possibilities. This relationship can be established in terms of the rehabilitation cost being a percentage of the replacement cost (REP/REH percentage).

1. REP/REH percentage > 120 percent.

The preliminary choice in this case is replacement. Other factors given below be examined for compatibility with replacement.

- Constructing a temporary structure may not be possible from a right-of-way point of view.
- Construction for replacement on a new alignment may not be possible due to right-of-way restrictions, even with stage construction.

Table 7.3 Useful service life for highway structures

No.	Location of Member	Rehabilitation/ Protective Measures	Useful Service Life* (Years)	Remarks
1A	Existing culverts	Repairs and/or rehabilitation	25	Can be extended to 50 years with maintenance
1B	Culverts with extension	Rehabilitation	75	Can be extended to 100 years with maintenance
1C	New culvert	Replacement	100	Can be extended to > 100 years with maintenance
2A	Existing framed sign structures	Repairs and/or rehabilitation	50	Maintenance required
2B	New framed sign structures	Replacement	75	Maintenance required
3A	Existing cantilever sign structures	Repairs and/or rehabilitation	25	Fatigue prone—Maintenance required
3B	New cantilever sign structures	Replacement	50	Fatigue prone—Maintenance required
4A	Ground-mounted sound barriers	Replacement	50	Maintenance required
4B	Structure-mounted sound barriers	Replacement	75	Maintenance required

* Varies from project to project and depending on ADT and no extreme conditions of earthquake or flood.

- It may be necessary to take disruption of traffic and user costs into account on replacement projects since there would be a change that would impact the traveling public on a permanent basis.
- Detouring traffic in highly urbanized areas may not be feasible from a capacity point of view.

2. REP/REH percentage is between 120 and 150 percent.

In this range, rehabilitation or replacement may be the preliminary choice. Other factors must be examined to establish the appropriate type of work.

3. REP/REH percentage > 150 percent.

The preliminary choice in this case is rehabilitation. Other factors, such as bridge type, must be examined to ensure compatibility with rehabilitation.

7.3.7 Cost-Benefit Analysis

Due to the huge investment involved, reconstruction and maintenance of bridges need to be considered as running and operating a modern industry. Planning and utilization of funds are sensitive but important issues. If repairs are not made in time, progressive retrofit or rehabilitation costs will be incurred. Providing repair costs by a highway agency is mandatory, however, investing in a new bridge or highway is optional.

If sufficient thought goes into the planning of a new bridge, fewer problems will be encountered down the road. Proper investment at the construction stage will minimize subsequent mainatenance, repair, and rehabilitation costs.

Total cost = Initial cost + Lifecycle cost (useful service life for old and new components)

Lifecycle cost = Σ (Cost of routine inspections + Maintenance and retrofits)
+ Σ (Repairs from extreme events + Cost of demolition)

Total cost is computed over the life of the bridge. Extreme events may or may not apply within the life of the bridge. They may be unforeseen events and include accidents such as vehicle and vessel collision, floods and scour, earthquakes, fires, bomb blasts, etc.

Life cycle costs are linked to the quality of planning. If the initial cost does not cover all the structural requirements, the life cycle costs for repair and rehabilitation will be much higher.

Screening criteria is based on several practical considerations, such as the owner's preference, the local community's preference, and using a merit-based point system.

A matrix based on a point system can be used (Table 7.4). The mark for each criterion will be based on engineering judgment, experience, and intuition.

The following grading is suggested for each abutment or pier alternate:

Condition multiplier

Poor = 1

Fair = 2

Good = 3

Very good = 4

Excellent = 5

A list of viable superstructure or substructure components is first prepared. Issues may be addressed in the following order of importance and given highest to lowest marks:

1. Constructability: Ease of construction.
2. Cost: Initial cost.
3. Maintainability and lifecycle cost.
4. Performance: Safety and durability.
5. Compatibility with environment: No adverse impact on environment.
6. Aesthetics: Pleasant appearance.
7. Construction schedule: Minimum period of construction.

An Excel® spreadsheet may be used for this process.

Other things being equal, alternate 2 with a total of 229 points would be the selection. The above method can be used, for example, to select a prestressed concrete girder compared to a

Table 7.4 Matrix point system for ranking hypothetical alternates using screening criterion.

	Constructability	Cost	Maintainability	Performance	Compatibility	Aesthetics	Schedule	Total
Grade point	10	9	9	8	7	6	5	—
Maximum multiplier	5	5	5	5	5	5	5	
Maximum points	50	45	45	40	35	30	25	270
Alternate #								
1	5 x 10 = 50	4 x 9 = 36	3 x 9 = 27	2 x 9 = 18	1 x 9 = 9	2 x 9 = 18	3 x 9 = 27	185
2	4 x 10 = 40	4 x 9 = 36	5 x 9 = 45	3 x 9 = 27	2 x 9 = 18	3 x 9 = 27	4 x 9 = 36	229
n	3 x 10 = 30	2 x 9 = 18	3 x 9 = 27	4 x 9 = 36	3 x 9 = 27	2 x 9 = 18	3 x 9 = 27	183

steel girder or type of abutment or pier. Fixing (repairing or rehabilitating) old bridges versus building new ones is an engineering decision guided by the condition of the old structure and available funding. A cost benefit analysis needs to be carried out prior to funding in millions of dollars. The engineer needs to determine the remaining useful life of the existing structure, its maintenance cost per lane per year, and compare that to the corresponding cost for a new one.

7.3.8 FHWA Deterministic Approach to LCCA (Refer to Chapter 2 for LCCA Application)

Structural engineering projects can cost hundreds of millions of dollars. LCCA is an asset management and economic analysis tool useful in selecting the preferred alternative by assigning values to relative merits. LCCA considers both short-term and long-term costs.

FHWA lists the following steps for a deterministic approach:

1. Establish design alternatives.
2. Determine activity timing.
3. Estimate costs (agency and user).
4. Compute life cycle costs.
5. Analyze the results.
6. Effects of inflation: Expenditures occur in the past or future and are therefore measured in different value units because of inflation.
7. Effects of discounting:

   Adjusting for the opportunity value of time is known as discounting.

   $(\text{Dollar})_{\text{base year}} = (\text{Dollar})_{\text{dats year}} \times (\text{Price index})_{\text{base year}} / (\text{Price index})_{\text{dats year}}$

   The consumer price index for each year is available.

   Real discount rate ranges from 3 to 5 percent.

   Discount future constant value costs to present value by using the formula:

   Present value = Future value $\times 1/$Discount factor

   $(1 + r)^n$ = Discount factor

   Where r = Real discount rate

   n = Number of years in the future when cost will be incurred

   Computer software is available to compute life cycle costs. (Reference FHWA, Lifecycle Cost Analysis Primer)

7.3.9 Evaluating the Relative Cost of Rehabilitation versus Replacement

A significant portion of the nation's inventory of bridges is rapidly approaching the end of its intended design life. Therefore, it is helpful to understand the processes which cause reduction in service life. Introducing innovative methods for extending the life of these structures through quick construction needs to be encouraged. The following factors need to be evaluated:

- Condition of deck, superstructure, and substructure
- Cost of rehabilitation of substructure and superstructure
- Cost of replacement of substructure and superstructure
- User cost of delays and detour during rehabilitation or replacement
- Life cycle cost analysis of rehabilitation and replacement.

All conclusions drawn in the replacement versus rehabilitation discussion process must be fully documented to discuss the differing viewpoints and gain the knowledge and experience of the team before making the final decision.

7.3.10 Performing a Comparative Study for Selection

1. Replacement costs: Major reconstruction or replacement should be the last resort for a variety of reasons. A new bridge is likely to cost millions of dollars. At any given time, a highway agency may be looking at thousands of bridges for reconstruction. Extended budgets are generally met by taxpayers, for example, through an increased tax on gasoline.
2. Accelerated schedule: A rehabilitation project takes less construction time than replacement.
3. Administrative and environmental impacts: In addition to the administrative efforts in resolving right-of-way legal issues, relocation of utilities and obtaining environmental permits can prevent the start of a new project for quite some time. Rehabilitation projects involve fewer social and environmental impacts than replacement projects. Hence, project delivery and procedural requirements are expedited with rehabilitation.
4. Redundancy: In a non-redundant structure, a failure of one principal load carrying member would result in probable collapse. The possibility of adding redundancy favors replacement.

   Two girder bridges with welded construction have a greater risk of failure than trusses. Concrete arches and concrete rigid frames are difficult and expensive to rehabilitate because of their monolithic type of construction.
5. Foundation costs: For scour critical bridges, deep foundations are preferred. The type of soil and scour depth will determine the type of foundation. If there are no serious scour problems, rehabilitation should be preferred.

Table 7.5 The feasibility of rehabilitation versus replacement.

Task	Purpose	Method
1. Collect detailed structure condition data.	To collect sufficient data to assess the viability of the work alternates. The data should be detailed enough to allow the completion of a level 1 load rating.	Perform an in-depth inspection in accordance with the requirements of the specifications for in-depth inspection. This activity could include taking cores of existing concrete elements.
2. Assess the condition of the structural deck.	To determine whether a deck can be rehabilitated or must be replaced.	Perform a deck evaluation in accordance with the current deck evaluation manual. The decision to rehabilitate or replace a deck can significantly impact associated rehabilitative work, design criteria, and the resulting costs. It is therefore imperative to accurately define the condition of the structural deck.
3. Assess the structural integrity.	To assure serviceability of the structure during construction and to define the extent of rehabilitative work required.	Perform a level 1 load rating. The level 1 load rating will provide a base structural capacity for the bridge from which the necessity and potential for improvement can be judged.
4. Assess the structure's vulnerabilities.	To identify impact to project scope and cost to address the structure's vulnerabilities prior to design approval.	Evaluate the structure and its details using the procedures provided in the Bridge Safety Assurance Policy.
5. Assess the feasibility of rehabilitation versus replacement.	Refine project cost and further assess the alternate's cost effectiveness and technical feasibility.	Update project costs and schedule based on more detailed information. Perform rehabilitation versus replacement evaluation. This evaluation provides direction concerning reasonable costs of various alternates and technical considerations that correspond to feasibility.

6. Accident potential: Geometrics, such as substandard sight distances, grades, and super elevations which contain clear potential for accidents, need to be improved. Those improvements may favor replacement. The review of geometrics should include bridge width, horizontal clearances, alignments, etc.
7. Live load restriction: When considering rehabilitation, the load rating will be checked, especially if it is posted.
8. Accelerated construction: The resulting rehabilitation or reconstruction activities would require temporary closures or detours, thereby causing significant reduction in the public's mobility. It would lead to traffic congestion, delays, and work zone accidents. Interruptions interfere with the ability to reliably plan travel time and adversely affect commerce and

Table 7.6 Comparative study of bridge rehabilitation (REH) versus replacement (REP).

Subject	Impact	Reconstruction Option
1. Historic Bridge?	Yes	REP not feasible
	No	REH or REP
2. Deficient geometry or road condition causing fatal accidents?	Number of fatal accidents < statewide average but no potential for accidents	Warning signs to be posted for deficiency
	Number of fatal accidents < statewide average but potential for accidents	REH or REP to correct safety problem
	Number of fatal accidents > statewide average	REP
	Impact of deficiency affects highway, navigable waterway, or railroad traffic below bridge	REP
3. Relative cost of rehabilitation/ replacement?	REH/REP > 1.5	REH
	REP/REH = 1.33 TO 1.5	REH or REP
	REP/REH > 1.33	REP
4. Redundancy of bridge?	Non-redundant	REH or REP by adding redundancy
	Redundant	REH or REP
5. Fatigue sensitive details?	Yes	REH or REP by removing critical details
6. Bridge conforms to AASHTO and state code standards?	Conforms to standards	REH or REP
	Deviation from standards are acceptable	-do-
	Can be brought up to standards by REH	REH
	Cannot be brought up to standards by REH	REP
7. Inadequacy of river opening size or poor stream alignment?	Hydraulic inadequacy	REP with span adjustment
	No hydraulic inadequacy	REH or REP
8. Maintenance and protection of traffic?	Construction staging is feasible	-do-
	Detour is feasible	-do-
	Bridge on new alignment is feasible	-do-
	Temporary structure is feasible	-do-

industry. To some extent the public seems to be losing patience with the prolonged duration of temporary construction work, which should provide comfort. Hence, focus should be placed on accelerating the bridge construction.

9. Experience with community outreach has shown that people care about sentimental values and preservation over construction of a new bridge or going through the hassles of lane closures, traffic jams, and detours during long construction periods.
10. Most bridges have a reserve strength which needs to be utilized. With the desired emphasis on effective repairs or rehabilitation, engineering efforts can be focused on developing and applying new technologies in this specialized discipline. It would be prudent to discourage outright replacement. The decision to replace may be postponed for as long as conditions permit, although the following considerations may point in favor of replacement:
    - For a new structure, there is unlimited choice in selecting aesthetics, location, span length, alignment, geometry, construction material, number of beams, member sizes, type of bearings, and types of substructures and footings. An alternative analysis needs to be carried out for finalizing the physical parameters.
    - Heavier truck loads such as permit loads may be applied.
    - A provision may be kept for future widening.
    - A rehabilitation project may be designed to conform to FHWA criteria of "highway for life" or "smart bridges."

7.3.11 Plan Preparation and Presentation

1. For rehabilitation projects, the type, size, and location (TS&L) plans shall have the normal TS&L plan details plus a complete scope and extent of work. When developing bridge rehabilitation plans, all pertinent details should be shown on the contract drawings.
2. In plan preparation, actual field measurements should be considered more reliable than the drawings.
3. The contract plans shall be sufficiently detailed to provide an overall view of the bridge indicating:
    - The existing and proposed geometric dimensions
    - Limitations and restrictions
    - Extent and type of work to be performed
    - Construction stages
    - Material information
    - All related information needed to rehabilitate the bridge.
4. Pay limits, quantities, and pay items should be adequately defined to eliminate ambiguity or confusion. All work shall be accounted for by specific pay items.
5. Where applicable, reasons for critical limitations and restrictions should be explained to assist the contractor and the field inspector in adjusting to the field conditions.

7.3.12 Duties for Reconstruction

1. Marketing, writing proposals, and winning a project.
2. Hiring engineers.
3. Engineering activities.
4. Political aspects (good public relations).
5. Social aspects (teamwork, awards, and recognition).
6. Educational aspects (training courses, conferences, engineering societies).

7.3.13 Engineering Activities

1. Analysis (moment distribution, stiffness matrix, and finite elements methods, computer software).
2. Design (ASD and LRFD methods).
3. Construction documents (contract drawings, technical specs, schedule).
4. Permits, right-of-way, utilities relocation.
5. Construction coordination (QA/QC, schedule).
6. As-built drawings (deviations from contract drawings).

7.4 REPLACEMENT PLANNING OR A NEW BRIDGE AT A DIFFERENT LOCATION

7.4.1 Replacement Issues

In earlier chapters, repairs and rehabilitation issues were discussed. Replacement is the last resort among reconstruction options. This is a comprehensive design subject, which allows the designer greater flexibility in decision making than repair or rehabilitation tasks.

Replacement is needed when an existing bridge has become dangerous or is functionally obsolete. A bridge becomes "functionally obsolete" due to:

1. Substandard travel lane width.
2. Lack of shoulder and median.
3. Inadequate stopping sight distance.
4. Horizontal alignment with a sharp radius.
5. Vertical profile/cross slope.
6. Low design speed.
7. Substandard guard rail.
8. Other deficiencies and problems not listed here.

The types of replacement options are:

1. Deck replacement only.
2. Superstructure replacement only, with or without substructure repairs.
3. Complete superstructure and substructure replacement is the most expensive of the options. It takes longer planning, design, and construction time and effort compared to the other options of rehabilitation, restoration, retrofit, and widening. Demolition of existing bridge also adds to the effort. The new structure can be constructed either at the very same location or at a new alignment.

7.4.2 Replacement Strategy

A study needs to be carried out to evaluate:

1. Initial cost and life cycle cost analysis. FHWA requires an alternate design for a bridge project over $10 million.
2. Level of service to be provided for traffic, i.e., number of lanes, shoulders, and sidewalks for pedestrian and bicycle use.
3. Provision for future widening.
4. Minimizing environmental impact and developing an in-kind replacement scheme.
5. Comparative study of schemes for desirable future service and a scheme for minimum service. Compare advantages of both schemes.
6. Planning a functional and attractive bridge.

7. Providing an adequate waterway opening.
8. Providing minimum clearances—horizontal and vertical.
9. Rapid construction by using prefabricated components.

7.4.3 Feasibility of Replacement Projects

1. If a decision is made in favor of replacement, several options need to be considered, such as replacing the bridge on:
    - The same alignment using staged construction.
    - A new alignment.
    - The same alignment with partial detour.
    - The same alignment with full detour.
    - The same alignment with a temporary bridge on a new alignment.
2. The feasibility study needs to consider traffic count, right-of-way, and utility relocation issues. Prepare a structure study plan using the following steps:
    - Ensure conformance to accepted standards and policies.
    - Highlight the need to consider exceptions.
    - Provide information to advance the proposed design.
    - Allow an initial constructability review.
    - Prepare a structure study plan which is a conceptual presentation of the proposed work.
3. Prepare a structure justification report using the following steps:
    - Provide a mechanism to achieve consensus on the appropriateness of the proposed structure.
    - Present logical decisions to select or discard the various design alternates for the project.

7.4.4 Replacement Methodology

In recent years, there have been notable changes in the design philosophy, environmental criteria, staged construction techniques, and computer methods involved in bridge replacement. New types of materials such as concrete, steel, aluminum, timber, bamboo, and composites are available. For example, the use of precast components and high early strength concrete would lead to quick reconstruction.

1. Safety requirements and precautions: The contractor must meet the safety requirements of the Occupational Safety and Health Administration (OSHA), in addition to the scaffolding requirements. The contractor must install and maintain suitable shields or enclosures to prevent damage to adjacent buildings, parked cars, trucks, boats, or vehicles traveling on, over, or under structures undergoing galvanized repairs.
2. Pollution control: The contractor must take all necessary precautions to comply with pollution control laws, and rules and regulations of federal, state, and local agencies.
3. Use of finger joints for bridge superstructures: Either a galvanized or metalized coating for the final joint is required. The finger joint system must be shop assembled with all components except for the drain trough. Where a finger joint is installed at an abutment, backwall concrete shall not be placed until the concrete deck pour is complete.
4. Girder replacement:
    - A limited vertical under clearance usually causes accidents and may damage girders. Replacement with a shallow girder depth is a possibility. The proposed depth of girders will be adequate to accommodate utilities under the bridge deck.

- For aesthetic reasons, if existing girders are steel, the proposed replacement girders preferably should be in steel.
- Fracture and fatigue cracks reduce the level of live load.
- Heat straightening of bent plates and welding of flange cover plates at midspan may be required.
- Girders need to be rated and the bridge posted for lower live loads if required.
- Hybrid system: Use of a slab and beam system with HPS 70W may be investigated. High performance girders have the advantage of weathering steel. Using hybrid high performance steel 70W, fabricated girders with 50W webs will lead to thinner flanges and fewer girders. Cost comparisons will be made to determine optimum depths. Live load deflection checks, although not a requirement in LRFD code, may be required to meet the L/800 or L/1000 requirement.

7.4.5 Selection of Type, Size, and Location of Replacement Bridge

Planning new and replacement bridge types: Nature seldom forgives a mistake in planning. Hence, some basic considerations are important in planning. Structural engineering has been to some extent an application of science, but continues to be an art and covers the following:

1. Intuition: Intuition is the insight that comes from both common sense and engineering sense.
2. Common sense involves aesthetics and choosing bridge designs that have stood the test of time and withstood environmental forces and vagaries of nature.
3. Engineering sense involves discarding the planning that was responsible for earlier failure of bridges.
4. Rule of thumb based on experience: There is no substitute for experience. For example, restricting maximum live load deflection by limiting girder span-to-depth ratio leads to travel comfort and minimizes long term fatigue.
5. The application of science covers the following:

    - Mathematics: Unfortunately, Tensor calculus and Riemannian algebra have found limited uses in computation. It has been a long time since Pythagoras theorem and other basic concepts came into effect. There is a need for renaissance in developing new mathematical theorems.
    - For convenience, closed form structural solutions rather than numerical methods are preferred.
    - The use of calculators and fast computers: Computer output helps in documentation and preparing bridge drawings. Use of color codes in printing and formatting eliminates the possibility of errors.

7.4.6 Planning Parameters

The following planning parameters need to be considered:

1. Ideal location of abutments and piers: It is desirable that a reconnaissance of the highway route be carried out on both sides of the abutments. Difficulties such as frequency of traffic accidents and inadequate sight distance will be studied.
2. A checklist will be prepared to document the present and future intensity of traffic, number of lanes, and any speed restriction. For replacement cases, change of alignment or refinements in bridge approaches needs to be considered. In some cases, a narrower bridge width than the width of approach may be desirable. This is usually achieved by eliminating shoulders on the bridge.

3. Deck geometry: Alternates for curved, skew, and straight geometry should be considered.
4. Number of spans: A single long span using a special method of construction, such as suspension cable, cable-stayed, arch, or segmental bridge, may be the answer. Aesthetic considerations and pleasing geometry are overriding factors generally influenced by public outreach.

   Cost comparisons are necessary to meet budgetary constraints and funding availability. If a slab and beam bridge is selected, a feasibility study is usually carried out to determine span configuration and location of piers. Symmetric and equal span lengths are preferable to non-symmetric and unequal spans.
5. Top of concrete elevation (bridge seat elevation): In many cases the existing substructure is retained and the bridge seat elevation or top of concrete elevation remains unchanged. Since the top of deck elevation for a replacement bridge is unlikely to change, the sum of girder and bearing depths also needs to fit into the existing dimensions. Table 7.7 may be used to ensure a good fit.

7.4.7 Procedure of Computing Bearing Seat Elevations

Consider the following items prior to making a replacement decision:

1. Condition of the deck, superstructure, and substructure.
2. Rehabilitation costs of the substructure and superstructure.
3. Replacement costs of the substructure, and superstructure.
4. User costs of delays and detours from rehabilitation and replacement.
5. Life cycle cost of rehabilitation and replacement.

The geometric design policy of the state must be considered as well as the design report, site data package, and projected traffic requirements 20 years ahead. The profiles and sections of the features being crossed as well as the crossing feature create two geometric reference planes. The relationship of these planes to each other can be determined by vertical underclearance. Software such as InRoads, COGO, or CADD routines can also be used to determine the location of the minimum critical vertical clearance point and the maximum available beam depth. Alternate span lengths, size of the bridge opening, and configurations and construction staging need to be considered and a feasibility study carried out.

7.4.8 Replacement Span Lengths—A Major Factor

1. Span lengths less than 40 feet:
   - The various types of units and materials available for this span range include: deep fills, culverts, underpasses, tunnels, aluminum and steel plate pipes, masonry and concrete arches, and crossings due to corrosion concerns. Environmental and size constraints normally dictate the type of bridge to be selected. The latest specifications and manufacturer's catalogues need to be consulted.
   - Precast or cast-in-place reinforced concrete structures: Reinforced concrete structures for culverts and short span bridges consist of four-sided boxes, three-sided frames, and arch shapes. These structures are usually precast in segments and assembled in the field. The precast segments are usually designed by a professional engineer employed by the contractor after award of the contract.
   - Four-sided boxes have a maximum practical single-cell clear span of approximately 20 feet.
   - Three-sided structures have a maximum practical clear span of approximately 50 feet.

     These units are supported on strip footings founded on rock or piles. A precast or cast-in-place, full-invert slab/footing unit can also be used. Both three-sided structures and precast arches can be used for many of the same situations identified for the larger pipes. In order

Table 7.7 A practical example of computing bridge bearing seating elevations.

Superstructure Depth (Inches)		Bearing Depth Plate Thickness	Expansion End West Abutment	Fixed End East Abutment	Remarks
Deck Slab	8.5″	Sole Pl. (T3 + T4)/2	1.0″	1.0″	
Haunch	2.5″	Steel Load Pl.T2	1.0″	1.0″	
Girder Depth	47″	Elast. Pad	5.5″	3.375″	
Bearing Depth	8.5″/6.5″	14″ x 16″			
		Masonry Pl.T1	1.0″	1.0″	
			8.5″	6.375″	Use 6.5″ at Fixed End
Total Thickness					
Total West/East Abutment	66.5″/64.5″				
	5.54′/5.38′				

	Line A	Distance From A (Ft.)	Difference (Ft.)	Deck Elevation	Elevation	Superstructure Depth (Ft.)	Top of Conc. El.
Girder G1							
C.L. East Abut. Brgs.	48.37	1.25	0.025	48.345	48.35	5.38	42.97
C.L. West Abut. Brgs.	45.83	1.25	0.025	45.805	45.81	5.54	40.27
Girder G2							
C.L. East Abut. Brgs.	48.37	9.25	0.185	48.185	48.19	5.38	42.81
C.L. West Abut. Brgs.	45.83	9.25	0.185	45.645	45.65	5.54	40.11
	Line B	**Distance From B**	**Difference**	**Deck Elevation**	**Elevation**		
Girder G3							
C.L. East Abut. Brgs.	48.06	2	0.04	48.02	48.02	5.38	42.64
C.L. West Abut. Brgs.	45.53	1.25	0.025	45.49	45.49	5.54	39.95
Girder G4							
C.L. East Abut. Brgs.	48.06	10	0.2	47.86	47.86	5.38	42.48
C.L. West Abut. Brgs.	45.53	10	0.2	45.33	45.33	5.54	39.79
Girder G5							
C.L. East Abut. Brgs.	48.06	18	0.36	47.7	47.7	5.38	42.32
C.L. West Abut. Brgs.	45.53	18	0.36	45.77	45.17	5.54	39.63
	Line C	**Distance From C**	**Difference**	**Deck Elev.**	**Rounded Elev.**		
Girder G6							
C.L. East Abut. Brgs.	47.52	1.25	0.025	47.545	47.55	5.38	42.17
C.L. West Abut. Brgs.	44.98	1.25	0.025	45.005	45.01	5.54	39.47

to obtain the necessary headroom for some cases, the units may be raised by supporting them on a pedestal wall.

- Deck slabs or deck/girder designs: Prestressed slab units, stress-laminated timber decks and concrete or timber decks with steel or timber girders cover this entire span range. Conventional reinforced concrete slabs, however, are inefficient for spans greater than 25 feet due to their excessive depth and heavy reinforcement.

2. Composite deck systems utilizing concrete with built-up steel girders or rolled sections can also be considered for spans in this range.

Spans between 40 and 100 feet:

- Adjacent prestressed concrete slab units can be used to a maximum span of about 60 feet. Prestressed concrete box units, concrete I-beams, bulb-tee sections, etc., are used for the 60 to 100 feet span range.
- Bulb tees are usually preferred over concrete I-beams. Deck/girder systems using laminated timber beams have a maximum span of about 80 feet. Conventional composite design systems utilizing concrete decks and steel stringers can be used for the entire span range. At the lower end of the span range, rolled beam sections would be used. Fabricated, welded plate girders would more likely be used at the upper end.
- Special prefabricated bridge panels with concrete decks and steel beams can reach spans approaching 100 feet. They have the advantage of reduced field construction time.

3. Span lengths between 100 and 200 feet:

- Special modified prestressed concrete box beam units up to 60 inches deep can span up to 120 feet. Prestressed concrete I-beams and bulb tee beams can span up to approximately 140 feet. The designer should investigate the feasibility of transporting and erecting the beams, especially those with a span longer than 140 feet.
- Composite steel plate girder systems can easily and economically span this range. Single spans up to 200 feet have been used. Once the single span exceeds 200 feet, alternate multiple span arrangements should be considered. The cost of additional substructures must be compared to the greater superstructure cost.

4. Span lengths between 200 and 300 feet:

For the majority of cases only a thru or deck truss should be considered. Plate girders or spliced concrete girders can be used at the lower end of this span range. Special designs utilizing arches, slant leg rigid frames, and concrete or steel box girders are also viable options. These types of special structures are used to address limited member depths, aesthetics, and compatibility with site conditions. Constructability concerns and possible alternatives should be discussed in detail due to higher cost considerations.

5. Multiple-span arrangements:

For multiple-span bridges, a continuous design should be used whenever possible to eliminate deck joints. In the case of multiple simple-span prestressed unit bridges, the deck slab should be made continuous for live load over the intermediate supports. Span arrangements ranging from equal span viaduct type structures to proportionally increasing relative span lengths should be evaluated. Continuous design using steel rolled beams or built-up plate girders takes into account the continuity over the interior support points. Poor continuous span ratios may result in uplift. For longer spans, live load deflection requirements become important.

6. Spans over 300 feet:

- Multiple-span arrangements in this range will involve balancing superstructure and substructure costs to achieve an optimum design. Site restrictions will often impact efficient substructure placement. Long multiple span structures can utilize a variety of construction types and materials.

- Steel:
  - Through or deck trusses with girder approach spans
  - Trapezoidal box beams
  - Variable depth girders (I shaped beams and box girders)
  - Hybrid girders utilizing conventional steel for the web and high-performance steel for the flanges
  - Cable-stayed girders or box beams
  - Deck or through arches
  - Cable-stayed bridges
  - Suspension bridges.
- Concrete:
  - Segmental box designs
  - Cable-stayed trapezoidal boxes
  - Deck arches
  - Floating bridges/pontoons
  - Post-tensioned, spliced bulb tees
  - Segmental viaducts with variable depth units.

7.4.9 Selection Guidelines for Replacement

Selecting preferred alternatives based on parameters discussed below:

1. Superstructure alternatives.
2. Substructure alternatives.
3. Preparing an evaluation matrix based on:
   - Initial construction cost
   - Life cycle maintenance cost
   - Environmental and social impact
   - Constructability
   - Future maintenance/inspection.

7.4.10 Practical Considerations Based on Experience

1. A vast majority of bridges in practice are small single-span structures. When skew angles over 45 degrees are involved, adjacent prestressed concrete beam design should be chosen only after careful review, since conventional joint details and reinforcement become quite complicated, as do the size of the bearings and bridge seats. Bulb tees or I-beams would be preferred.
2. For curved spans with prestressed concrete, adjacent box beams or slab units are seldom chosen because of the increased cost of the wider chord alignment and the complications that arise with regard to bridge railing anchorage and end transitions. Prestressed concrete bulb tees, I-beams, or spread boxes are alternates worth considering.
3. Concrete bulb tees, I-beams, or spread box beams should be considered if vertical clearance requirements can be satisfied.
4. At locations where either long piles or poor bearing capacity is anticipated, prestressed adjacent box or adjacent slab design has the disadvantage of having a heavier superstructure. Under these conditions, a spread box, bulb-tee, or concrete I-beam with deck slab configuration might be considered to reduce the loads.
5. Prestressed concrete adjacent beam design is often chosen over steel beams when a structure must be opened to traffic quickly. This type of construction eliminates the need for deck

slab forming. It can also accommodate a temporary asphalt wearing surface if the time of the year prohibits placement of the concrete deck.

6. Where significant space must be provided for utilities, a spread system using steel girders, concrete I-beams, or bulb tees is the preferred choice. Spread concrete box units can also accommodate some utilities.
7. Vertical curves are better handled with multi-girder systems, since camber can be fabricated and controlled with greater accuracy. Adjacent prestressed units must accommodate any curve correction by placing a variable depth deck slab. This can result in considerable additional dead load, necessitating a deeper beam.
8. Adjacent prestressed concrete boxes or slabs are preferred over streams where ice and/or debris are a problem. The smooth underside of adjacent units reduces the snagging potential.

7.4.11 Related Economic Considerations

1. It is not economical for the bridge to be "fixed" when it is beyond repair or:
   - When the remaining useful life of the bridge is just a few years, either due to fatigue or deficiencies, and investing money in rehabilitation may not provide a cost-benefit solution.
   - When the bridge requires continuous maintenance and occasional closure of traffic lanes.
   - When there are structural difficulties for widening and providing additional lanes, replacement may be the way to go. Right-of-way issues and utilities relocation may govern the feasibility of replacing an existing bridge.
2. Design techniques for economy:
   - Optimization of the substructure design.
   - Use of the LRFD method.
   - Use of HPC and HPS.
   - Prestressing deck slab and longitudinal beams.
3. Economy of construction:
   - Accelerated bridge construction.
   - Quick erection using subassemblies.
   - Increased shipping lengths.
4. Additional details for economy:
   - Reduce superstructure depth.
   - Wider girder spacing.
   - Use of jointless (integral abutment) bridges.
   - Eliminating bearings.
   - Reducing field splices.

7.4.12 Meeting the Available Construction Budget

Most traffic congestions are in urban areas where transportation needs are intense. The capacity to build new roads in urban areas is restricted due to the cost of land, utility relocation, and right-of-way issues. With construction budgets being limited, more funding is being allocated to maintaining existing bridges rather than going for new construction. An approximate unit cost estimate of different new bridge types is provided in Table 7.8 in dollars per square foot (SF). The cost will vary with the span length, skew angle, and for the state or location of the bridge.

Hence with time, as a bridge gets older, maintenance costs increase and cost per vehicle usage is likely to increase. The higher the occupancy of vehicles, the lower the cost per person.

Table 7.8 Approximate unit cost of normal bridges.

Bridge Type	Unit Cost	Range of Span Length
Steel girder	$250 /SF	Small
Steel girder	$275 /SF	Medium
Steel box girder	$300/SF	Medium
Steel arch	$450/SF	Long
Concrete box culvert	$200/SF	Small
Concrete PC girder	$225/SF	Medium
Concrete PC girder	$220/SF	Small
Concrete box girder	$230/SF	Small
Segmental concrete: span-by-span	$400/SF	Long
Segmental concrete: balanced	$375/SF	Long
Cable-stayed bridge	$550/SF	Long
Suspension bridge	$600/SF	Very long

* Unit costs are for 2007 and vary for each location and project.

From records of past expenditures, a computer program can be developed for estimating daily cost per vehicle usage.

For selection of major repairs, retrofit, rehabilitation, or replacement, a discount rate may be applied. Only broad concepts are presented to help in selecting the type of reconstruction, whether rehabilitation or replacement. For more accurate calculations, with variable interest and inflation rates, software can calculate present worth and annual worth.

Activities for rehabilitation and repair are absorbing an increasing share of public funding. The bulk of highway funds are being allocated to maintain existing bridges rather than funneling the money to construct new ones, so with limited funding many existing bridges may be kept open. To reduce the demands on already strained construction and maintenance budgets, it may be beneficial to pursue the option of preservation rather than going for complete replacement.

7.4.13 A Comparative Study of Steel and Concrete Bridges for Selection Purposes

Two examples of major items for steel and concrete bridges are presented here. Minor items are not listed and are covered by the 20 percent contingencies. Unit costs may differ for each state, but are usually based on average bid tabs of projects completed in the past two or three years. These costs are readily available and can be downloaded from Web sites. It may be noted that the cost of a concrete bridge is about 13 percent less than that for a steel bridge, the main difference being the use of prestressed concrete box girders rather than steel girders.

7.5 CONSTRUCTABILITY ISSUES

7.5.1 Construction Scheme

Structural analysis and design are based on theoretical assumptions. It is important that the constructed model matches as closely as possible to the theoretical model. Also, designing for construction loads should ensure trouble-free fabrication, erection, and construction.

Failure during construction is the biggest single source of bridge collapse. Even though bridge members are designed for long-term loads, short-term design of temporary works is required.

1. There is a need to define a feasible method of construction in the contract documents. However, an alternate proposal from a contractor for a detailed workable procedure needs to be considered. It will be reviewed for safety and quality assurance.

2. The selection of a contractor shall be based on his available resources, ingenuity, and experience.
3. Fabrication procedures shall be described in technical specifications. Vertical camber of girder can be achieved by cutting the web to the required contour and by heat curving the flanges to the required contour prior to welding with the web.
4. An erection plan showing the location and capacity of cranes and the sizes and elevations of temporary supports shall be developed. It will be unique for each project for access to site, construction easement, and utilities relocation.
5. Temporary supports and false work shall be designed for horizontal and vertical reactions from construction loads.
6. Precautions during construction: To prevent debris from falling below during deck slab, containment methods, such as installing wire nets below the deck, will be evaluated.
7. Anticipated construction loads shall be specified for each stage of sequence construction.
8. Construction specifications: The above provisions may be included in special provisions.

7.5.2 Constructability Review

The constructability issues for proposed widening should be carefully considered. The subassemblies fabricated should be able to fit properly in the field. Constraints and field verification of conditions will help in planning. The major constraint is that the widened bridge shall conform to existing conditions, including traffic count of truck live loads. For all widening, the available existing bridge plans must depict the actual field conditions.

1. The constructability review should include a simulation of construction sequencing requirements. An on-site review may be necessary to ensure that all project elements are considered in the constructability review. Designers should become familiar with current local bridge construction practices so most designs can be built with conventional equipment or local materials.
2. Overseeing construction and quality control: Identification of constructability issues at the design stage, preparation of detailed contract documents, a realistic construction schedule, application of the critical path method (CPM), effective shop drawings review, request for information (RFI) and design change notices (DCN) procedures, and development of quality assurance (QA) and quality control (QC) procedures for the project.
3. Common issues are:
   - Utilities that may cause interference need to be relocated.
   - Shop and field welded joints should be designed and detailed from constructability considerations making easy access to all joints possible. When designing connections, interference with other members should be considered. It may be easier to construct bolted connections.

     Only fillet welds and complete joint penetration welds should be shown. The existing steel to be welded may require special preheat because of its chemistry, especially when making spans continuous for live load or full dead load. Lateral gusset plates may have to be moved since welding of stiffeners is not allowed to the splice plates.
   - When designing additional girders for a structure, widening the deflected profile of the new girders should match the deflected profile of the existing girders. The new girders should not include camber for superimposed dead load if the existing girders were not cambered for that load. The designer must provide for any differential deflection between existing and new girders; otherwise, due to unequal deflections, future deck cracking in the longitudinal direction may result.
   - When detailing connections and selecting construction methods, consider the differential camber present prior to placing the new deck.

- Avoid open or sealed longitudinal joints in the riding surface which may become a safety hazard.
- Specify that live load vibrations from the existing structure be minimized during deck pour and curing.
- Provide adequate clearance between proposed driven piles and existing piles, especially for battered piles.
- Bearing fixity and expansion devices should be similar in both widened and existing bridges.

4. Summary: The following constructability issues need to be evaluated:

- **Duration of construction.**
- **Maintenance and protection of traffic.**
- **Underwater work:** The health and safety of construction personnel may be of concern if the water depth is high. Trained divers will be required.
- **Access to site:** A temporary road for transportation of materials and equipment adjacent to the channel bank may be difficult to construct.
- **Temporary works:** Temporary construction works may be required. More economical alternatives implementing quick construction and safety need to be carefully evaluated.
- **Safety of personnel:** OSHA safety standards must be followed.
- **Environmental risks:** Pollution of rivers from construction material may occur. The channel needs to be cleaned. Approvals for stream encroachment permits would be necessary.
- **Impact on existing utilities:** The effect of driving sheeting or bed armoring on existing utilities needs to be evaluated. Utilities may be relocated in such cases. Coordination and approval from utility companies would be required.
- **Impact on right-of-way:** A construction easement needs to be determined and permits obtained.
- **Specialized work:** The contractor performing such tasks needs to train his construction crew for such techniques.
- **Availability of labor and plant:** Some types of bridge construction, such as gabions, interlocking blocks, and stone pitching, require experienced labor. Since local labor may not be familiar with the work, bringing labor from long distances may be expensive.
- **Limited vertical clearance under the bridge:** It may be difficult to construct a cofferdam or drive sheeting under a bridge if a restricted vertical clearance exists.
- Usually, there is more than one method of construction because of variations in the contractor's resources. However, in bridge repairs not many alternate solutions may be feasible and a single solution may be the most appropriate. If design takes into account the appropriate method of construction, there will be fewer constructability issues. However, changes in structural details during the long construction process may be necessary due to unforeseen field conditions such as the discovery of different soils. Coordination between design and construction teams is essential for answering RFI's and issuing agreed DCN's as revisions to contract drawings to finish work within the construction schedule. For delicate repairs a more stringent QC procedure may be necessary.

5. Designs should be reviewed to identify any of the following potential construction problems:

- Is the sequence of construction practical?
- How will each component be constructed?
- Can the structure actually be constructed in accordance with the plans?
- Is the design economical or can it be constructed with conventional equipment by experienced contractors?

- Transportation and storage of materials at the site.
- MPT: How will traffic be maintained at each stage of construction?
- ROW during construction: How will the contractor access the site?
- MSF: Maintaining stream flow during construction for a bridge over a stream.
- Dewatering the work area.
- Removal of forms and temporary sheet piles.
- Utility relocations.
- Delays in acquiring construction permits, affecting the schedule.

7.5.3 Erection Stresses and Maintaining Small Deflection during Erection

The following steps are required:

1. All fit up shall be at no load condition. During fabrication and erection, lack of fit among components may result from imperfect workmanship and thermal changes, etc. It would lead to stress concentration and locking in of stress and should be prevented.
2. Temporary bracing or support will be pre-designed for use during construction to prevent any failure. Reactions at temporary supports and forces in bracing will be computed using standard software.
3. Welded and bolted connections may cause brittle failures. They will be designed to withstand construction stresses.
4. Staged construction may present additional issues. When staging is done in sequence, it is important to match the existing camber with the camber of the adjacent girder being placed.
5. For skewed deck construction, any lift-off of the girder end at bearings during deck placement will not be permitted.
6. Girder length will be based on 68°F temperature. Holes in girders should match anchor bolt holes which have a clearance of as little as $\frac{1}{16}$ in around the anchor bolt. Correction in length needs to be applied before drilling holes in the field.
7. Bearing rotation during erection due to girder dead weight should not exceed theoretically computed values. Additional bearing rotations may cause malfunction of the bearing. Expansion bearings have slots, the midpoint of which should coincide with the center line of support at 68°F. Adjustment will be required at other temperatures.

7.5.4 Allowances to Be Made in Design and Construction

The following conditions may require construction adjustments:

1. Subzero and freezing temperatures.
2. Hot weather.
3. Low and high wind.
4. Construction during all traffic lanes open.
5. Construction during partial lanes closure.
6. Nighttime construction.

7.5.5 Construction over Rivers

Before starting a design, it is important to know the ground realities, river realities, and construction factors, which are likely to affect the selection of countermeasures. The following constructability issues need to be evaluated:

1. Duration of construction: The available flow width may be reduced due to construction of cofferdams, embankment, and cams.

2. Flow velocities through the reduced channel opening will increase, and thereby scour in the channel and around the structure is increased. Hence, construction of the above items shall be done during flood off-season.
3. Maintenance and protection of traffic: During installation, small cranes or pile driving equipment may be parked on a lane or shoulders. A lane closure would then be required.
4. Coordination with traffic police and local officials would be necessary.
5. Underwater work: Health and safety of construction personnel may be of concern if the depth of water is high. Trained divers will be required.
6. Access to site: A temporary road for transportation of materials and equipment adjacent to the channel bank may be difficult to construct. Wooden mats should be used when lane width is restricted.
7. Temporary works: Temporary construction works may be required. More economical alternatives implementing quick construction and safety need to be carefully evaluated.
8. Safety of personnel: Due to the instability of banks because of recent floods (for banks with slopes steeper than 1:1), sudden collapse of a bank may occur. OSHA safety standards must be followed.
9. Environmental risks: Pollution of a river from construction material may occur. The channel needs to be cleaned. Approvals for stream encroachment permits would be necessary.
10. Impact on existing utilities: The effect of driving sheeting or bed armoring on existing utilities needs to be evaluated. Utilities may be relocated in such cases. Coordination and approval from utility companies would be required.
11. Impact on right-of-way: Countermeasures may extend into adjacent property limits. Right-of-way needs to be purchased in such cases. Similarly, encroachment of adjacent property during construction may occur. A construction easement needs to be determined and permits obtained.
12. Specialized work: Modern countermeasures require new construction techniques. The contractor performing such tasks needs to train his construction crew for such techniques.
13. Availability of labor and plant: Some types such as gabions, interlocking blocks, and stone pitching require experienced labor. Since local labor may not be familiar with the work, bringing labor from long distances may be required and expensive.
14. Limited vertical clearance under the bridge: It may be difficult to construct a cofferdam or drive sheeting under a bridge if restricted vertical clearance exists. Placement of countermeasures will also be difficult.

7.5.6 Safety Considerations

1. Working adjacent to fast, unpredictable currents and rapidly rising water levels can be extremely dangerous. The safety of construction workers is a very important aspect of emergency work.
2. Floating (or subsurface) debris and woody materials contribute to hazards during emergency work.
3. Weather conditions (rain, snow, or darkness) may further endanger safety.
4. OSHA recommendations for slopes of excavations in soils.

The following maximum values of slopes shall be used for excavation of sloping structures:

- Solid rock—90 degrees
- Compacted angular gravels—0.5:1 (63 degrees, 26 ft)

- Average soil 1:1— (45 degrees)
- Compacted sand 1.5:1— (33 degrees, 41 ft)
- Loose sand 2:1— (26 degrees, 34 ft)

The angle of repose shall be flattened when an excavation has water conditions.

7.5.7 Emergency Protection Measures

1. Any design and installation of bridge protection measures during high water can be difficult, if not impossible. A planned response for bridge scour is much preferred over a reactive response.
2. An emergency installation is typically much more costly. There is usually an increased cost of mitigation since damage during an emergency project can be greater and equipment remobilization may be required for post-project mitigation. Project impacts (i.e., damage to trees and vegetation) in carrying out emergency work must be mitigated in the same way as for projects with normal timing.
3. Impacts of carrying out emergency work should be minimized. Under emergency scenarios, the tendency is to take actions to protect a bridge at the expense of existing trees and other vegetations. However, these trees and vegetations may be providing protection or may eventually protect bridge abutment or approach. The trees and vegetation also provide important riparian habitat and should be protected even if they don't offer any direct stabilization of bridge countermeasures.

7.5.8 Traffic and Utilities Issues

The following steps are required:

1. Site access: Adequate access to the site shall be provided for trucks to deliver countermeasures material.
2. Right-of-way: Construction easements and right-of-way may have to be purchased for the duration of construction.
3. Possible detours: Detour, lane closures, or nighttime work may be necessary. Coordination with traffic control would be required.
4. Emergency vehicles and school bus services shall not be affected by lane closures.
5. Utilities: Relocation of utilities at the sides of abutments or piers may be necessary for the duration of construction. Coordination with utility companies would be required.
6. Four weeks before the start of construction, traffic police should to be informed of shutdown or detour of one or more lanes.
7. Warning signs showing dates and times of shutdown are required to be posted well in advance for the information of users.
8. Construction time must be kept to a minimum or performed at nighttime.

7.6 SUPERSTRUCTURE REPLACEMENT PROCESS

7.6.1 Introduction

The collapse of the I-35 Bridge over the Mississippi River on August 1, 2007 and many others have confirmed the need for regular maintenance. Otherwise, replacement is inevitable. The replacement process includes:

1. Decision making to replace the bridge based on inspection and structural evaluation.
2. Securing funding approval.
3. Advertising or solicitation for design.

4. Selection of consultants.
5. Feasibility studies.
6. Holding public meetings.
7. Resolving right-of-way and utilities relocation issues.
8. Approving preliminary and detailed design, contracts, and documents.
9. Selection of contractors.
10. Monitoring construction and implementing a construction schedule.

From start to finish the planning and design duration may extend well over one year. Even construction of a bridge smaller than 20 feet requires the same legal and financial approval procedures as those required for a much longer bridge.

7.6.2 Deck Replacement Projects

1. Benefits of prefabrication: Precast panels, such as the Fort Miller Co.™ Effideck, may be used. A precast deck can be cast in panels of required sizes, usually 5 to 6 feet wide or as dictated by existing supporting girder spacing, with the longer side placed transversely. The construction is repeated for each stage. Deck panels are designed for main reinforcement either perpendicular or parallel to the direction of traffic flow, as applicable. Shear studs on top of the beams are arranged in groups and are grouted inside pockets provided in the precast panels. Composite construction results in savings of time and better quality control.

   The Federal Highway Administration cites a number of advantages of prefabricated bridges/decks, because of the quality obtainable under controlled production conditions and the speed with which the bridge can be erected. The process minimizes traffic impacts, improves construction zone safety, creates less environmental disruption, makes bridge designs more constructible, and improves quality and life cycle costs. The cost of stay-in-place forms required for cast-in-place deck construction is avoided by using prefabricated panels.
2. Fiber reinforced polymer (FRP) bridge deck: A recent, innovative use of an FRP reinforcement cage in concrete bridge decking will be investigated. University researchers investigated an innovative, modular, three-dimensional, FRP-pultruded grid reinforcement system to construct a concrete bridge deck on a major bridge structure. The FRP systems can be utilized to replace concrete decks on steel girders, or to serve as self-supported short span bridge superstructures. The two concepts rely on using cellular components to form the core of the deck system and an outer shell to wrap around those cells to form the integral unit of the deck. Carbon fiber reinforced polymer is also being used.
3. Precast panels using high-performance concrete (HPC): HPC has become a conventional bridge construction material partly due to the Strategic Highway Research Program research by the Federal Highway Administration. HPC is a set of specialized concrete mixes which provide added durability for concrete structures. Their benefits include ease of placement and consolidation without affecting strength, long-term mechanical properties, early high strength, and longer life in severe environments. They also conserve material, require less maintenance, deliver extended life cycles, and, if designed well, enhance aesthetics. The use of HPC with galvanized reinforcement steel enhances durability.
4. Deck redesign: A one-course deck slab with a corrosion inhibitor admixture can be planned. A minimum top reinforcement cover of 2¾ inches is usually required. Two-course construction with the overlay of LMC or silica fume requires an additional one to two weeks construction time.
5. Use of Exodermic deck slabs: For rapid deployment in deck replacement projects, the use of adequate span lengths of a lighter Exodermic deck system offers benefits. A cost comparison can be made.

6. Selection of rolled steel girders versus fabricated plate girders: Rolled sections with welded or bolted cover plates at the bottom flange of the midspan were popular at one time for the small and medium span ranges. However, due to the limited depth of 36 inches and fatigue at welds of tension connections, they were not economical and had maintenance problems.

Bridge designers frequently use cold-formed plate girders with variable sizes, shapes, and strengths. Typically, 80- to 100-foot girder lengths are easier to galvanize, transport, hoist in the air, and erect in position. Splice plates are used for longer lengths.

7.6.3 Deck Slab Replacement Design

1. When complete deck replacement is anticipated and where feasible, consider elimination of deck joints first. The decision to eliminate existing joints should be based on length of the structure, the type of bearings used, and substructure/foundation compatibility.

Deck slab replacements shall be designed in accordance with the AASHTO LRFD Bridge Design Specifications and state code manual.

2. Special measures such as requiring the use of removable deck forms, retrofitting stringers with shear connectors, design criteria exceptions, etc. may be required.

3. Precast members have better quality control and should be preferred.

Simplified LRFD Calculations for Replacement of Deck (cast-in-place)*

A. Data − 6 HPS 70W girders; Spacing = 8 ft centers;
Corresponding slab depth D = 8.5 in (refer to state bridge design manual)
Effective depth d = 8.5 in − (1 in + 0.75 in/2) assume rebar diameter = 3/4 in;

$d = 7.125$ in; b_f (top) = 16 in; $t_f = 1$ in;
Clear span = 8 ft − 1.333 ft = 6.667 ft; fc′ = 4.0 ksi
Effective span = Clear span + d = 6.667 ft + 7.125 in/12 = 7.26 ft

B. Dead load moment due to self weight, overlay and stay-in-place form (SIP).
1. Compute self weight for B = 12 in, D = 8.5 in
2. Assume 2 in thick future wearing surface weight = 25 PSF
3. SIP weight = 5 PSF

Spacing of SIP form ribs shall match the spacing of main bottom reinforcement steel.

Load factors: (DL Max. = 1.25; DL Min. = 0.90); refer to Table 5.7

Factored moment $M = 1.25\ M_{DC} + 1.5\ M_{DW} + 1.75\ M_{(LL+I)}$

Unfactored moments: For exterior span use, $M_{DC} = +\ w\ (L)^2/10$

$M_{DC} = 150$ pcf × (12 in × 8.5 in)/144 × (7.26 ft)²/10 = 0.56 kip-ft/ft
$M_{DW} = (25 + 5)$ psf × (7.26 ft)²/10 = 0.158 kip-ft/ft
Total dead moment = 0.56 + 0.158 = 0.718 kip-ft/ft

Live load moment: Refer to AASHTO LRFD Specifications 2004, Section A4, Table A4-1;

Strength I − (LL Max. = 1.75; Impact factor IMP = 33 percent; S = 8.0 ft)

AASHTO Table A4-1 moments were calculated using equivalent strip method and are based on:
HL-93 moving loads of HS-20 truck and lane load of 220 psf acting over 10 ft lane width:
Max. − ve $M_{(LL+I)} = -4.81$ kip-ft/ft @ 6 in from center line of beam.

* For precast construction load combinations, refer to Sec. 5.8.2).

Max. + ve $M_{(LL+I)}$ = +5.69 kip-ft/ft (impact included)
Factored + ve moment $M_u = (1.25 \times 0.56 + 1.5 \times 0.158 + 1.75 \times 5.69) =$ +10.89 kip-ft/ft
Factored − ve moment $M_u = -(1.25 \times 0.56 + 1.5 \times 0.158 + 1.75 \times 4.81) =$ −9.35 kip-ft/ft

C. Assume bar diameter and spacing, compute Mn and compare with Mu,

Ensure that $M_n > M_u$ or calculate as $= M_u /\phi$ /fy jd

Resistance factors: Strength = 0.9; Serviceability = 1.0 (refer to Sec. 6.2.3)

D. For + ve moment, assume 5/8 in diameter bars at 9 in centers at bottom (approximate solutions are given in design manuals):

$d = 8.5 - 1.0 - 0.625/2 = 7.19$ in

$As = \pi/4\,(0.625)^2 \times (12 \text{ in}/9 \text{ in}) = 0.41 \text{ inch}^2$

$a = As\, fy /0.85\, fc'\, b = As\, 60/\,(0.85 \times 4.0 \times 12) = 1.47\, As$

$M_n = M_u$

As fy (d − a/2) = Mu results in a non-linear equation for As

$0.9 \times As \times 60\,(7.19 - 1.47/2\, As) = 10.89 \times 12$
$7.19 \times 54\, As - 0.735 \times 54\, (As)^2 = 130.68$

$(As)^2 - 9.782\,(As) + 3.293 = 0;$

Add and subtract $(4.891)^2$

$(As - 4.891)^2 - 23.922 + 3.293 = 0$

$(As) = +v20.628 + 4.891 = 0.349 \text{ in}^2$

Bottom reinforcement 5/8 in diameter @ 10 in centers. ($0.362 \text{ in}^2 > 0.349 \text{ in}^2$) Okay.

Check: Equation of type $ax^2 + bx + c = 0$

$x = \{-b \pm \sqrt{(b^2 - 4ac)}\}/2a$

$a = 1, b = -9.782, c = 3.293$

$x = \{9.782 \pm \sqrt{9.7822 - 4 \times 1 \times 3.293}\}/2 \times 1$

$x = \{9.782 \pm \sqrt{82.52}\}/2 = \{9.782 + 9.08\}/2 = 0.349 \text{ or } -9.431$

Neglecting imaginary root, x = 0.349. Calculations are Okay.

E. For − ve moment, assume 5/8 in diameter bars at 9 in centers at top (approximate solutions are given in design manuals).

$d_e = 8.5 - 1.5 - 1.25 - 0.625/2 = 5.44$ in; much lower than for bottom section.

As fy $(d_e - a/2) = M_u$ results in a nonlinear equation for As

$0.9 \times As \times 60\,(5.44 - 1.47/2As) = 9.35 \times 12$

$293.76\, As - 39.69\,(As)^2 = 112.2$

$(As)^2 - 7.4\,(As) + 2.83 = 0$, Add and subtract $(3.7)^2$

$(As - 3.7)^2 + 2.83 - (3.7)^2 = 0$

$(As - 3.7) = +3.295; As = 0.40 \text{ in}^2$

Top reinforcement 5/8 in diameter @ 9 in centers. ($0.41 \text{ in}^2 > 0.40 \text{ in}^2$) Okay.

Haunch thickness is neglected. Hence conservative.

Check maximum reinforcement: d = 5.44 in, $\beta = 0.85$

$a = T/\beta\, fc' \times$ beam spacing; $T = As\, fy = 0.41 \times 60 = 24.6$ kips

$a = 24.6/0.85 \times 4 \times 9 = 0.803;$

$c = a/\beta = 0.803/0.85 = 0.94;$

$c/de = 0.94/5.44 = 0.174 < 0.42$ Okay. (AASHTO 5.7.3.3.1)

F. Check for flexural cracking under service limit state. (AASHTO 8.16.8)

Check for fs < fsc

$fsc = z / (dc\ Ac)^{1/3} < 0.6\ Fy$

For severe condition, z = 130 kip/inch

Minimum required dc = 1 + 0.625/2 = 1.312 in < 2 in Okay.

$Ac = 2\ dc \times \text{Bar spacing} = 2 \times 1.312\ \text{in} \times 9\ \text{in} = 23.62\ \text{in}^2$

$fsc = 130/(1.312 \times 23.62)^{1/3} = 41.4\ \text{ksi} > 0.6 \times 60 = 36\ \text{ksi}$

Use fsc = 36 ksi.

Calculate fs

$fs = n\ M_u\ y / It$ where It is transformed moment of inertia.

$n = Es/Ec = 29{,}000/3{,}640 = 7.97$ (Use 8.0)

Calculate y and It:

$It = b\ (k\ de)\ 3 + n \times As\ (de - k\ de)^2$

Calculate ρ, k:

$k = \{(n\rho)^2 + 2n\rho\}^{1/2} - n\rho$

$\rho = As/bde = 0.41/12 \times 7.19 = 0.0048$ (d for + ve moment region)

$n\rho = 8 \times 0.0048 = 0.0384$

$k = \{(0.0384)^2 + 2 \times 0.0384\}^{1/2} - 0.0384 = 0.28 - 0.0384 = 0.242$

$k\ de = 0.242 \times 7.19 = 1.735$ in

$It = 12\ (1.735)\ 3/3 + 8 \times 0.41(7.19 - 1.735)^2 = 63.22 / 3 + 97.60 =$ 118.67 in 4/ft width

$y = de - k\ de = 7.19 - 1.735 = 5.455$ in

From Section B, total dead moment = 0.56 + 0.158 = 0.718 kip-ft/ft;

Max. + ve $M_{(LL+I)} = +\ 5.69$ kip-ft/ft

Mservice = 0.718 + 5.69 = 6.408 kip-ft/ft

$fs = n\ Mservice\ y / It = 8 \times 6.408 \times 5.455 \times 12/118.67 = 28.28$ ksi

< fsc = 36 ksi (No cracking under service limit state in + ve moment region.)
Hence section rebars are okay.

Similarly, it can be shown that there is no cracking under service limit state in − ve moment region. Bottom reinforcement 5/8 in diameter @ 10 in centers and top reinforcement 5/8 in diameter @ 9 in centers meet service limit state.

Overhang Design

A. Data

New Jersey barrier (1.75 ft base width and 2.67 ft height) is considered as parapet.

NJB unit weight = 0.5 kips/ft

Effective design cantilever span = 3.33 − 1/4 of flange width = 3.0 ft from edge of deck.

Depth of overhang = 8.5 in + 1.5 in = 10 in

Loads

Overhang wt. = 0.15 × 10/12 = 0.125 ksf

F.W.S = 0.025 ksf

Total DL = 0.15 ksf

Approx. LL = 1.0 kip/ft acting at 1.0 ft from face of barrier.

B. Strength 1 limit state: Total $M_u = (1.25\ M_{DC} + 1.5\ M_{DW} + 1.75\ M_{(LL+I)} \times \rho$

$M_{DC} = 0.125 \times 3.02 / 2 + 0.5\ (3.0\ \text{ft} - 0.67\ \text{ft}) = 0.563$ kip-ft/ft,

where CG of barrier is 0.67 ft from edge of deck/barrier.

$M_{DC} = 0.563 + 1.165 = 1.728$ kip-ft/ft

$M_{DW} = 0.025 \times (3.0\ \text{ft} - 1.75)^2 / 2 = 0.02$ kip-ft/ft, neglecting the barrier width.

$M_{LL} = (1 + I_{MP}) \times W_{LL}\ (3.0\ \text{ft} - 1.75\ \text{ft} - 1.0\ \text{ft}) = 1.33 \times 1.0 \times 0.25 = 0.333$ kip-ft/ft

Modification Factor $\eta = \eta_D\ \eta_R\ \eta_I$; (refer to Sec. 6.2.2)

Ductility Factor $\eta_D = 1.0$

Redundancy Factor $\eta_R = 1.05$

Importance Factor $\eta_I = 1.0$

$\eta = 1.05$

Overhang negative moment $M_u = (1.25 \times 1.728 + 1.5 \times 0.020 + 1.75 \times 0.333) \times 1.05$

= 2.911 kip-ft/ft < 9.35 kip-ft/ft (deck moment).

Hence extend 5/8 in diameter top bars to edge of overhang.

C. Extreme Event II - Collision with parapet

Forces are acting in transverse and longitudinal directions

Collision moment = 22 kip-ft/ft

Conservatively assume 5 ft width of deck slab to resist the moment.

Transverse negative moment = 22/5 = 4.4 < 9.35 kip-ft/ft

Hence extend 5/8 in diameter bars at 9 in centers at top into overhang slab. (For construciton load combinations, refer to Sec. 5.7.2, Figure 5.22).

7.6.4 Deck Joint Reconstruction/Replacement

Conventional deck joints have been a source of local concrete cracks.

1. Elastomeric deck joints that cannot be repaired should be scheduled for replacement. Several manufacturers have developed strip seal joints that are embedded in elastomeric concrete which can be used to bond the total joint to the concrete blockout.
2. Elastomeric concretes are furnished by various manufacturers, i.e., D. S. Brown Co., Watson-Bowman Association, and Epoxy Industries, Inc. These proprietary systems include trade names such as Delcrete, Wabocrete, and Cevacrete, respectively. They are particularly well-suited as replacement systems for existing elastomeric expansion dams because the required blockout depth is between 2 and 2½ inches, and required blockout widths are comparable to existing blockouts that accommodate elastomeric seals.
3. The joints should be a combination of the elastomeric concretes, appropriate extrusions, and compatible waterproof neoprene strip seals (Figure 7.2).

Strip seals are available for movements up to 5 inches. However, they should not be used for more than 4-inch movement classification due to concern for load carrying capacity of wider

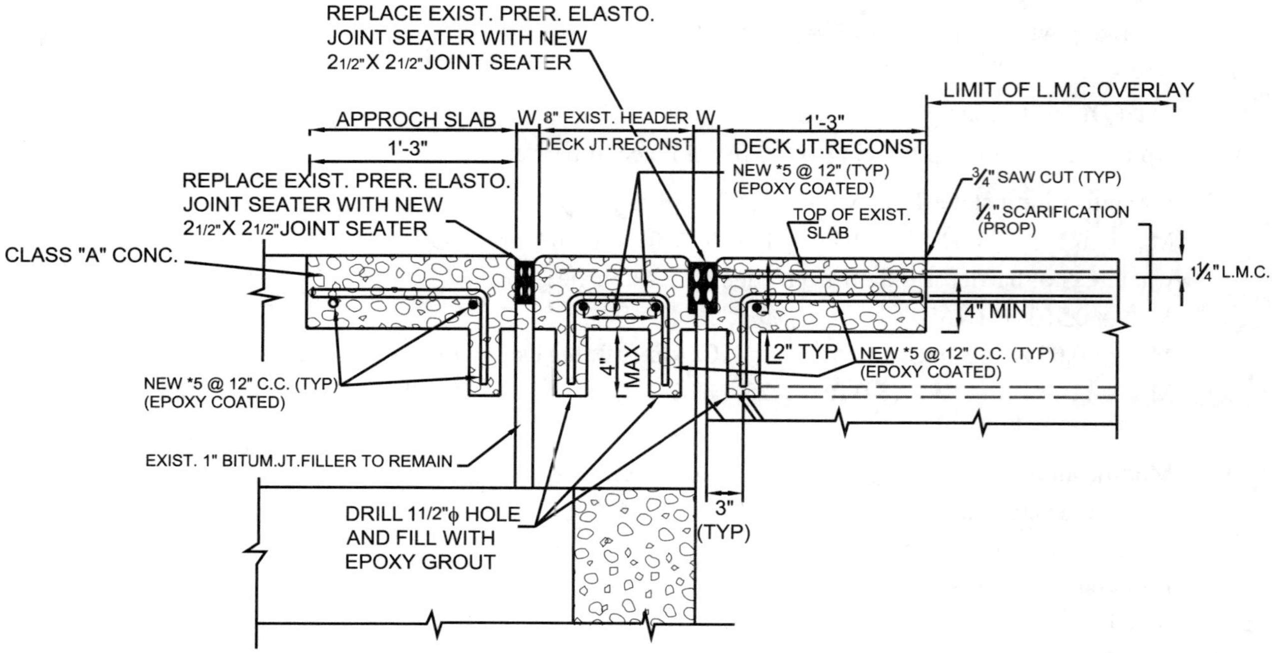

Figure 7.2 Example of a deck joint reconstruction at an abutment. (W is from deck joint manufacturers table).

openings. Some of these systems use heat fusion bonding, which heat cures in about two hours, bonds to irregularities such as spalls in the blockout area, and is suitable for stage construction.

Deck joint reconstruction shall include:

- Saw cut ¾ in wide.
- Concrete removal and disposal of deck slab and sidewalk to limit shown.
- Concrete removal and disposal of abutment header (backwall).
- Traffic shielding.
- Removal of preformed joint filler.
- Placement of new epoxy-coated reinforcement.
- Reconstruction with Class A concrete.
- Hot pour mastic 2 in × 1 in.

7.6.5 Designing Haunches

The purpose of the haunch is:

1. To provide a means for final adjustment of the deck slab elevation to match the designed roadway profile and cross slope.
2. It allows adjustment to correct construction and fabrication variations without having the top flange of the girder project into the structural depth of deck.
3. Haunch tables shall always be computed considering stage construction and the assumed pouring sequence for continuous span structures.
4. In addition, the computations shall consider differing dead loads applied to fascia and interior girders, especially with construction loads such as barriers.

The calculated depth of the haunch shall have a 2-inch minimum concrete thickness as measured at the centerline of the beam from the top of the beam to the bottom of the slab. A deeper minimum is required when the top flange equals or exceeds 16 inches in width to allow for roadway cross slope. The total haunch depth shown on the plans shall include the thickness of the top flange for fabricated steel girders.

At all splice locations for steel girders, the top flange splice plates will reduce the haunch depth. It is not necessary to provide the full 2-inch minimum haunch at the splice location.

5. For simple span bridges, the calculated depth of the haunch at the centerline of bearings shall be the minimum depth, plus the difference in thickness between the maximum and minimum top flange plates plus increases to account for cross slope and horizontal curvature when straight girders are used.
6. The haunch shall be reinforced when the depth of the concrete portion of the haunch exceeds 4 inches. Haunches on fascia beams of multispan bridges shall be set so that the top of the webs of fascia beams in adjacent spans line up.
7. A haunch table shall be shown on the plans to assist in construction. For spans of 60 feet and under, the haunch table should be done for span quarter points. For spans over 60 feet, the haunch table shall give elevations at span tenth points, but not to exceed a spacing of 20 feet.

   Bridges with curved girders should have a haunch table with the elevations given at the diaphragm lines. The predicted concrete slab and superimposed dead load deflections are shown at these points.

7.6.6 Concrete Placement Method for Decks

Deck placement sequence: For large deck areas, a symmetric pattern of pouring needs to be followed.

1. Maximum deflections during construction shall be computed. Undue deflection may adversely affect deck drainage.
2. Longitudinal reinforcement in concrete will be adequate in area to limit tensile stress in concrete. This may happen when fresh concrete is poured adjacent to hardened concrete, which may already be subjected to tensile stress. Concrete tension shall be restricted to less than 0.9 times modulus of rupture.
3. f_c' value will be based on the actual age of concrete at the time of construction.
4. During deck replacement operation in one lane, with live load on adjacent lanes, an increase in deflection and stress needs to be checked for asymmetry.
5. Deck placement sequence shall preferably follow a checkered pattern. Sufficient time shall elapse between adjacent casts to prevent damage to new concrete.
6. Formwork may be made up of trough or folded metal deck, plywood, or precast panel. Stay-in-place formwork need not be removed except at overhangs.
7. Overhang brackets: Triangular frame brackets to support formwork for overhangs are commonly used. The formwork in turn supports rails for moving trolleys filled with concrete. Typical spacing for frames is 2 to 3 feet centers.
   - A space frame analysis is required to evaluate forces being transferred to the fascia web.
   - The fascia girder shall be checked for lateral flange bending stress, rotation, and deflection from weight of a moving trolley, wet concrete, and construction equipment.
   - The connection of bracket diagonal members to the fascia girder web shall be avoided to prevent web damage. Instead, a connection to the lower flange shall be made.

7.6.7 Evaluating the Substructure's Load-Carrying Capacity for Superstructure Replacement

Substructures should be analyzed for adequacy for the following conditions:

1. Evidence of substructure distress.
2. The minimum load-carrying capacity is HS-20.
3. AASHTO criteria supplemented by state design code shall be used for the analysis.
4. Foundation bearing resistance should be determined based on available test boring data. The need for a detailed foundation investigation including the drilling of borings should be determined.
5. If no boring data is available and no borings are planned for the project, an assessment of the adequacy of the foundations to sustain the bridge in its rehabilitated condition should be made based on known substructure information and local geologic data.
6. Where end or point bearing piles are present, pile resistance may be based on current design criteria. For friction piles, static analysis should be performed using available soil and pile data.
7. When deciding where to locate substructures, the designer should identify all appropriate horizontal offsets, standards, and requirements to select either a single or multiple span arrangement, whichever is most appropriate. The available beam depth is factored in along with any special concerns such as:
   - Sheeting requirements for staging and substructure construction. Cantilever sheeting design versus tied-back sheeting versus pile and lagging wall costs, and deep water cofferdam construction versus shallower depths or causeway construction.
   - Treatments such as high abutments with large reveal heights for form liner, masonry, or brick treatments.
   - Wetland encroachments: Longer spans to avoid wetlands will require additional beam depth. This can raise a profile and move the toe of slope out or require a retaining wall.
   - Shorter spans may disturb more of the area and require additional wetland mitigation.
   - Staging problems: Includes interference between the existing and new features, (e.g., substructures, beams, pier caps, and pile driving—especially battered piles) as well as utilities that must remain in service.
   - Utility conflicts: Avoidance of utilities that would require costly relocations can further restrict the location of substructures.
   - Pile driving and sheeting placement may be limited by overhead or underground interference.
   - Integral abutments: Deck joints are avoided, and improved seismic performance results.

7.6.8 Foundation Assessment for Superstructure Replacement

1. The *site data* package includes the substructure boring logs for the bridge. These logs should be evaluated with regard to location of the new bridge and consistency of the soil with respect to each log. There should be sufficient information to estimate pile lengths and sheeting depth.
2. Water crossings: The following criteria shall be applied to all structures crossing water,
   - Unless founded on rock, all structures crossing water shall be supported on piles or have other positive protection to prevent scour of the substructure.
   - Cofferdams should be evaluated with regard to need, type, size, constructability, and cost. Alternative types of construction such as causeways, caissons, or drilled shafts should be considered and compared to conventional cofferdam costs.

- The estimated maximum depth of scour should be used to determine overall structure stability. Piles should be socketed into rock if scour can affect their stability. Recommendations for details will be contained in the foundation design report (FDR).

7.7 CASE STUDIES OF ALTERNATES FOR REPLACEMENT OR NEW BRIDGES

7.7.1 Planning Study of a 240-Foot Span Length on the New Jersey Turnpike

The following considerations were adopted by the author:

1. For the required distance between abutments of 240 feet, the possibilities are:
   - One single span of 240 feet
   - Two equal spans of 120 feet
   - Two unequal spans of perhaps 100 feet and 140 feet length
   - Three equal spans of 80 feet each
   - Three unequal spans.

   The use of unequal spans in urban areas is more frequent when required horizontal clearance from the new pier to a skew railway track is considered mandatory. However, by providing a guide rail in place of the 30-foot minimum horizontal clearance, abutments and piers can be protected and the span length reduced.
2. The number of girders and spacing: Equal spacing (preferably between 6 and 12 feet) is most desirable. Fewer but deeper girders may reduce the minimum vertical clearance to16 feet 6 inches.

   It is desirable to use an odd number of girders (with the middle girder logistically placed to carry a median barrier). For future staging, a longitudinal construction joint at the top flange of the middle girder location helps.
3. The fascia girder needs to be designed independently of the interior girder. However, the fascia girder is generally lighter than the interior girder and the latter is used in place of the fascia. For future widening it is important to have equal strength girders.

   Through girders are not generally used due to a lack of redundancy. The risk of failure is greater. The minimum number of repeated girders is four.
4. Effect of deck overhang: Overhang deadweight negative moment at the fascia girder reduces deck positive midspan moment in the adjacent panel. Without an overhang, fascia girder web is subjected to out-of-plane bending and torsion.

   For arriving at an optimum cantilever span length:

   Maximum negative dead load moment at fascia = Maximum positive dead load moment at midspan of deck panel.

   It can be calculated as: $wd_1\,(L1)^2/2 = wd_2\,(L2)^2/10$; $L1 = L2/(5)^{0.5}$ if $wd_1 = wd_2$
5. Generally overhang is made thicker than the deck. The additional weight of the parapet at the edge of the overhang compensates for one truck wheel on the deck panel. The other wheel is assumed to act at or near fascia girder. Maximum overhang span is restricted to 5 feet due to expected large deflection at the free end of the cantilever which supports a parapet weight of 600 lbs/ft.
6. Full composite action between slab and beam is assumed due to adequate use of shear connectors. Composite action helps in distributing dead loads from parapets, median barrier, and sidewalks to longitudinal girders equally.
7. Resolving difficult issues such as right-of-way, environmental permits, and utilities relocation may increase design or construction time.
8. Planning for minimum dead weight: In seismic zones it is important to minimize inertia forces.

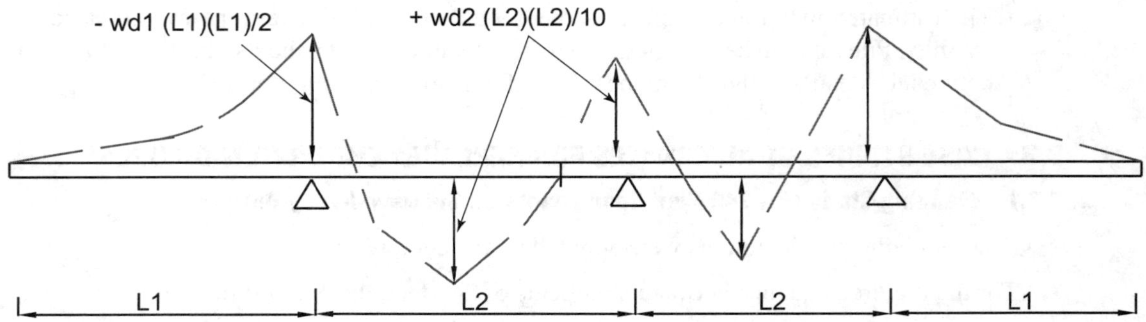

Figure 7.3 Balancing peak positive and negative moments.

Deck slab weight contributes the most to total dead weight. It can be made lighter by:

- Using lightweight aggregate concrete
- Using an orthotropic slab.

9. Maintenance and protection of traffic: A detour during construction may not always be feasible. Staged construction or an accelerated construction schedule may be necessary.

There will be no impacts on traffic or on wetlands by using top-down construction and a temporary bridge.

10. Height of abutments such as full height, medium height, and stub abutment must fit into the existing topography. Extensive cut and fill should be avoided. Types of piers are column bent, pile bent, wall, or hammerhead.

11. Public involvement: Feedback from drivers and taxpayers helps in selection of an appropriate bridge type and in planning for minimum traffic disruptions.

12. Aesthetics

- Open appearance
- Avoid abrupt changes
- Pier geometry can be made aesthetically pleasing
- Use of MSE walls with patterns.

13. Context sensitive design (CSD)

- Community or social group involvement
- Historic or sensitive features.

7.7.2 Replacement of Deck Joints with Integral Abutment Construction

The author was responsible for the design of two integral abutment type bridges in New Jersey. Alternate superstructure to pile cap connection details are presented in Figures 7.4. This technology has improved bridge performance, and its use is on the increase. Alternative details are shown in Figures 7.5 and 7.6.

7.8 PRACTICAL CASE STUDIES

7.8.1 Author's Alternate Analysis Case Study for New Jersey Turnpike Ramp TN over I-195

Due to high traffic volume on the New Jersey Turnpike, serving three airports, accelerated construction and economical maintenance-free ramp design was required. An alternate analysis was desirable. Selection of a viable structure for a curved steel bridge was required at Interchange 7-A.

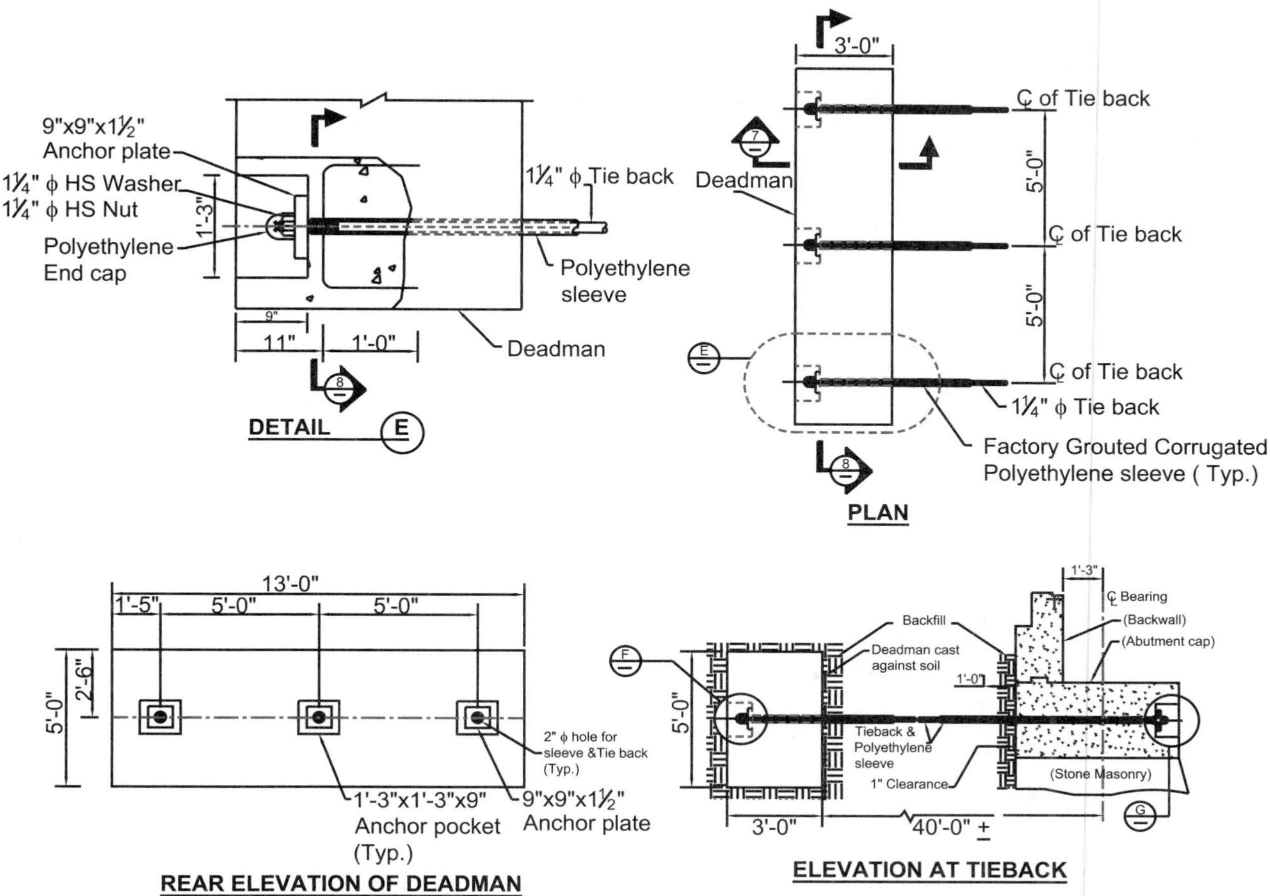

Figure 7.4 Example pf strengthening of existing abutment by use of tieback.

1. Ramp TN has both sharp curvature and a skew. From the user's point of view, the ramp is important since it has one of the highest ADTT's in the country.
2. Effect of combined curvature and skew edges:

   This type of complex structure is sometimes unavoidable due to site topography, vertical profile, curved alignment of base line, and horizontal clearance. Curvature produces torsional moments. The problems are lift off of fascia girder ends, unanticipated twisting, and lateral deformations of girders and bearings. The role of end diaphragms during erection is important in keeping webs vertical. They act as primary rather than secondary members in transverse load distribution.
3. Analysis method: A three dimensional finite element method study is required, which would need to model girder depth and details of cross frames. A correct analysis is of paramount importance since camber computations as dead load deflection for each curved girder depend upon cross frame configuration and transverse load distribution. Among available finite element software are DESCUS or BSDI.

7.8.2 Parametric Study

1. For selection of number of spans: Alternates were single span, two equal spans, and two unequal spans. Additional screening criteria which have an impact on the number of spans, such as environment, fatigue, and splices, were included.

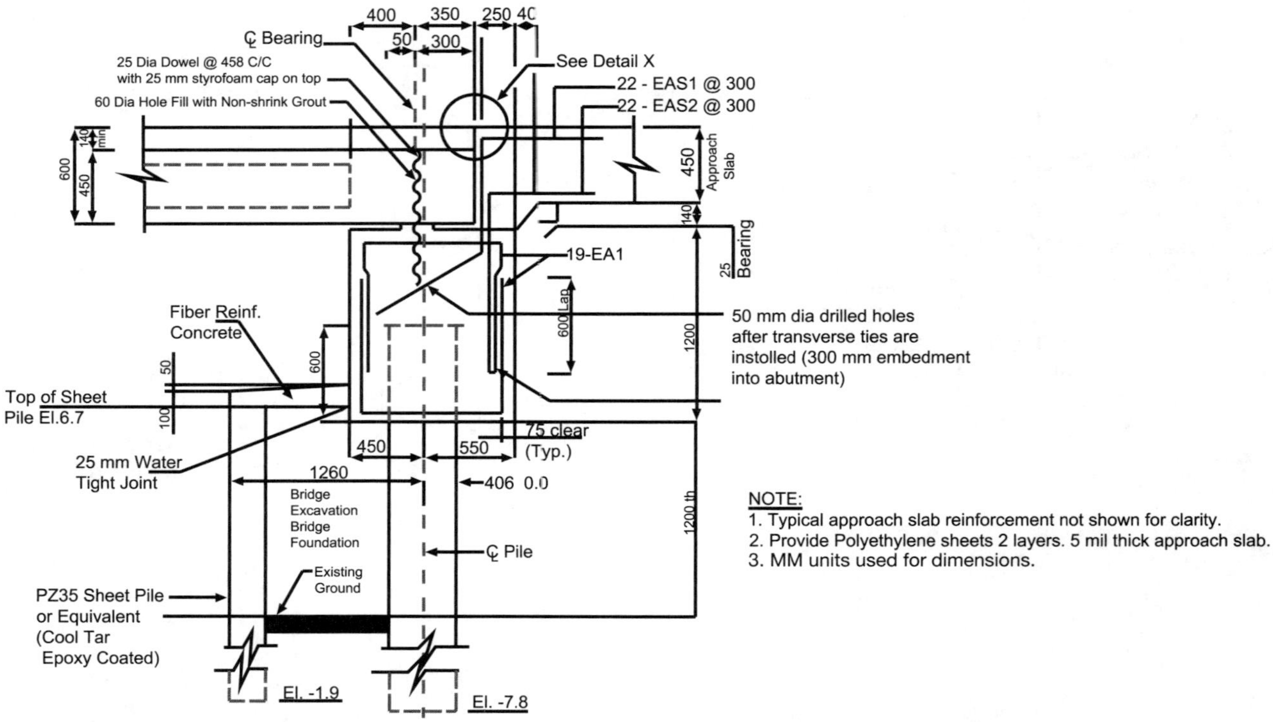

ROUTE U.S. 322 / N.J. 50 TYPICAL SECTION OF INTEGRAL ABUTMENT

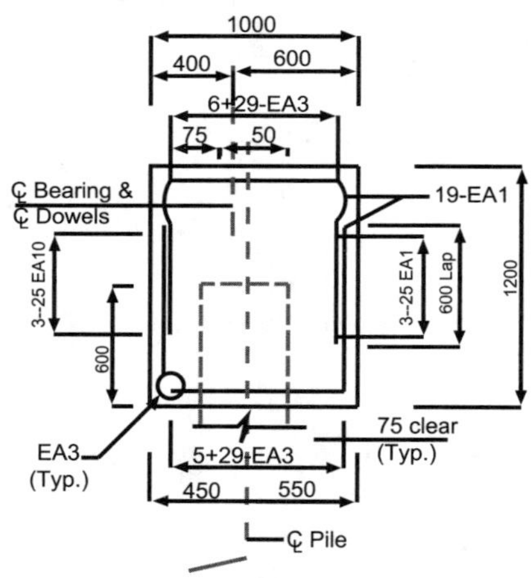

ABUTMENT CAP REINFORCEMENT

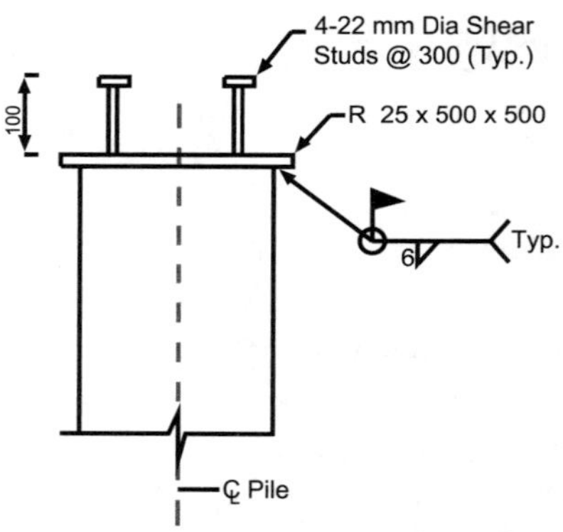

ABUTMENT CAP ANCHORAGE DETAIL

Figure 7.5 Alternate details of integral abutment bridge.

2. For selection of number of girders: Alternates were five, six, and seven girders. Additional screening criteria which have an impact on the number of spans, such as provision for future staging, were included.
3. For selection of abutment type: Alternates were full height, medium height, stub and spill through types. Additional screening criteria which have an impact on the abutment behavior, such as seismic resistance and soil pressure, were included.
4. For selection of pier type: Alternates were keeping center line of bearings normal or skew to base line, frame bents, or hammerhead types. Additional screening criteria which have an impact on the pier behavior, such as seismic resistance, were included.

Figure 7.6 Precast pier construction.

7.8.3 Planning Recommendations

1. There is a need for substantial member sizes for end diaphragm, web thickness, and moment connections between cross frames and girder webs, through bolted or welded connections.
2. Cross frames need to be spaced closely for maintaining equilibrium between torsional moments arising out of curvature and forces in cross frames.
3. For large skew angles with curvature, box girders are preferred to I shaped girders to resist torsion.
4. To cater for lateral deformation, elastomeric pads are preferred.
5. Future jacking of ends of curved girders may be required for bearing replacement or repairs. It is important that the end diaphragm has level (not sloping) bottom members. Channel section is preferred.

7.8.4 Using Screening Criteria for Selecting Number of Spans

1. Performance.
2. Cost.
3. Schedule.
4. Constructability.
5. Environmental impact.
6. Maintenance.
7. Aesthetics.
8. Site compatibility.
9. Fatigue.
10. Splices.

Several alternates which have a bearing on existing conditions at the ramp are considered. Only major types are considered important and are likely to include: performance, cost, schedule, constructability, environmental impact, maintenance, aesthetics, site compatibility, fatigue, and need for splices.

Points are allocated under each category. Selection of a preferred alternate is made based on comparative studies and the point system summarized in Tables 7.9 and 7.10.

Table 7.9 Comparison of span alternates using screening criteria.

NJ Ramp Over I-195			Span Option (Rank/Score)					
S. No.	Screening Criteria	Weight	Single Span		Two Equal Spans		Two Unequal Spans	
1	Performance	100	3	70	1	90	2	80
2	Cost	90	3	70	2	80	1	85
3	Schedule	80	3	70	1	80	2	75
4	Constructibility	70	3	60	1	70	2	65
5	Environmental	60	1	60	2	50	3	45
6	Maintenance	50	1	50	2	45	3	40
7	Aesthetics	40	1	40	2	35	3	30
8	Site compatibility	30	3	20	2	25	1	30
9	Fatigue	20	3	10	1	20	2	15
10	Splices	10	3	5	1	10	1	10
Total score		550		455		505		475
Overall ranking			3		1		2	

Final selection for the preferred alternate is made by elimination of those that do not fully conform to the above criteria and display greater difficulties.

Of the three alternates for span arrangements, the alternate for two equal spans of 110 feet each with the highest points is recommended.

7.8.5 Curved Girder Framing Alternates

Three alternate girder arrangements were considered:

1. Five girders with four spacing of 9 feet 9 inches.
2. Six girders with five spacing of 7 feet 9 inches.
3. Seven girders with six spacing of 6 feet 6 inches.

Overhang was kept constant at 2 feet 6 inches.

Table 7.10 Comparison of framing alternates.

Ramp NJ Over I-195			Number of Girders Option (Rank/Score)					
S. No.	Screening Criteria	Weight	Five		Six		Seven	
1	Performance	100	3	70	2	90	1	100
2	Cost	90	1	90	2	85	3	80
3	Schedule	80	1	80	2	75	3	70
4	Constructibility	70	1	70	1	70	1	70
5	Environmental	60	1	55	2	50	2	50
6	Maintenance	50	1	50	2	45	2	45
7	Aesthetics	40	1	40	2	35	2	35
8	Site compatibility	30	3	20	2	25	1	30
9	Future staging	20	2	5	2	5	1	20
10	Fatigue	10	3	5	1	10	1	10
Total score		550		485		490		510
Overall ranking			3		2		1	

Design was based on an alternative analysis of continuous girders using LRFD software. Merlin-DASH and STLRFD software consider straight continuous spans. Since girder curvature is not sharp, any approximation in initially neglecting curvature is applicable equally to all alternates. Doing independent finite element studies using DESCUS software for three types of framing is time consuming and is not warranted. A comparative study of deflections and maximum stresses, using both STLRFD and DESCUS software for the LRFD method, was made. The preliminary study was carried out using a single continuous girder analysis.

The preferred alternate was also designed in detail using the University of Maryland DESCUS Program for Curved Bridges.

7.8.6 Girder Selection Based on a Comparative Study

1. Using an odd number of girders is better since the middle girder reduces vibrations and is conveniently located for future widening.
2. Construction joints for deck replacement are located in the middle of the girder flange. This method prevents the possible opening of joints under repeated live loads.
3. For long curved spans with a skew of 17.6 degrees, the possibility of end of fascia girder lifting over the bearing is eliminated due to closer spacing.
4. The degree of redundancy is increased by using more girders.
5. Future widening is easier.
6. The five-girder option results in design of deeper girders and thicker flanges, which more than offsets the economy resulting from the use of fewer girders.
7. The disadvantage with five girders is that for future deck replacement only one lane will be available. Also, the location of construction joints falls in the middle region of slab spans, which requires cantilever slabs (the thickness or existing reinforcement of which is not designed to support live loads during construction) and is not feasible.
8. The use of six girders allows only one lane to be available for future deck replacement. Without a middle girder, higher deck deflection results, and there is increased difficulty in future widening.

7.8.7 Screening Criteria for Selecting Number of Girders

Choosing the appropriate number of girders requires analysis of the following criteria:

1. Performance.
2. Cost.
3. Schedule.
4. Constructability.
5. Environmental impact.
6. Maintenance.
7. Aesthetics.
8. Site compatibility.
9. Future staging.
10. Fatigue.

For future widening or deck replacement with two lanes open, only the seven-girder option seems to work. The seven-girder arrangement secures the highest points with 6 feet 6 inch spacing resulting in lighter girders (16-inch wide flange and $\frac{9}{16}$-inch web) and is recommended.

7.8.8 Alternates for Bearings

Since girders have large curvature, centrifugal forces are small. It appears that multi-rotational pot bearings are not required. Elastomeric pads are recommended. Bearings may be designed using PennDOT BRLRFD software or other approved software.

7.8.9 Case Study of Future Deck Repairs/Replacement Construction Staging

1. Seven girders alternate: As required by the Turnpike Design Manual, a provision is made for MPT including safe demolition and deck replacement, in the future. The number of girders and spacing selected would make possible two lanes of traffic. Deck cutting (construction joint) will be allowed only at the middle of girder flanges. Night windows for construction trucks and temporary materials may be required.

   Stage 1—Place temporary barrier on deck over girder G3 location to protect the side of the work zone with the precast concrete construction barrier, type 4.

   Shift traffic to the southern side of the road:

   - Remove existing striping and place temporary traffic striping
   - Create two adjacent 11-foot lanes in both directions
   - Cut the existing northern deck and parapet as shown on the construction sections
   - Remove the existing deck
   - Construct the proposed northern side of the bridge including the new parapet.

   Stage 2—Move temporary barrier on the new deck by resetting the precast concrete construction barrier and creating one 11-foot lane.

   - Restriping the roadway within the limits of the work
   - Shift westbound (WB) traffic to the newly constructed northern side of the road
   - Cut the existing deck between girders G3 and G5 as shown on the construction sections
   - Remove the existing deck
   - Construct the proposed middle side of the bridge.

   Stage 3—Place temporary barriers on the deck over girder G5 location to protect the side of the work zone with the precast concrete construction barrier.

   - Shift eastbound (EB) traffic to the middle side of the road by removing existing striping and placing temporary traffic striping
   - Create two adjacent 11-foot lanes in both directions
   - Cut the existing southern deck and parapet as shown on the construction sections
   - Remove the existing deck and parapet
   - Construct the proposed southern side of the bridge with the new parapet.

   Stage 3A—Remove the temporary precast concrete construction barrier. Install the new traffic markings and striping. Restore traffic to its original position with two lanes and shoulders.

2. Bridge opening height: The deck elevation is on a vertical curve, and a 4.8 percent transverse slope, maintaining the superelevation and roadway profile, is required.

   Width: The proposed abutments will be placed approximately 9.5 ft ± from the existing abutments by increasing the bridge span by 19 feet.

3. Traffic control: The bridge will be constructed over virgin land. Traffic control or detour will not be applicable except for future widening or deck replacement.

4. Design speed: Maximum speed limit of 40 MPH is assumed for calculations for centrifugal forces acting on the deck superelevation.

5. Controlling design elements:

   - Cross slopes: On ramp TN, the proposed cross slopes have been set at 4.8 percent at each lane. The recommended adjacent slope of 5 percent over 3-foot shoulder width is not doable. The difference of 0.2 percent at break of slope is applicable to 3-foot shoulder only and is ⅛ of an inch. Permitted construction tolerance is higher. Hence, uniform 4.8 percent cross slope will be used for deck construction.

- Minimum radius and superelevation: The minimum radius proposed for baseline is 1024 feet.
- Grades: The gradient for the crest vertical curve on which the bridge lies will be held below the maximum of 15 percent allowed by AASHTO.
- Sight distance: The minimum horizontal sight distance will be provided at all points along the ramp within the project limits. All vertical curves have been designed to provide desirable stopping distance.
- The width of the proposed acceleration and deceleration lanes of future I-195: Varying widths are required. Bridge span is based on maximum width.

Bridge width and structural capacity: The total deck width including sidewalks and parapets will be 44 feet. The structure has been designed for a capacity of HL-93 loading.

- Vertical clearance: The proposed bridge will attain the minimum vertical clearance of 16 feet, 6 inches.

6. Public involvement: Public outreach issues will be performed.
7. Cost estimate of preferred type.

7.8.10 Superstructure Alternatives

The following design criteria will be followed:

1. Deck slab construction: One course deck construction is proposed. Design will include a provision for a future 2-inch overlay.
2. Deck joints: Deck joints will be provided only at the abutments. It is anticipated that strip seal expansion joints will be used.
3. Deck drainage: Longitudinal and cross slopes are provided on the deck slab. Scuppers will be provided only if required, otherwise drainage inlets will be provided outside the deck area. The total length of 220 feet is less than the maximum spacing of scuppers of 300 feet. Deck drainage calculations will be performed using rainfall intensity for Mercer County.
4. Truck live loads: Since projected ADTT exceeds 500, the bridge will be designed for a live load of HL-93 (LRFD version of HS-25 truck load). Fatigue analysis will be based on projected ADTT.
5. Utilities: Except for bridge lighting, no bridge-mounted utilities are planned.

Span length: Three alternate span arrangements were considered:

Single curved span of 220 feet—There are many disadvantages to using a single curved span, namely:

1. Bending moment is increased four times for a single span. It results in design of deep girders (7 feet 6 inches depth for single span compared to 5 feet for two spans) and reducing minimum vertical underclearance to 14 feet 3 inches.
2. Construction will be more expensive since loads on bearings, abutment walls, and foundations will be nearly doubled, and heavier sizes will result.
3. A relative increase in live load deflections would cause vibrations in the bridge.
4. The architecture of a single span with no pier will not match with the architecture of adjacent NJDOT bridges on the I-195 corridor.

Two equal curved spans of 110 feet each—Advantages are reductions in maximum bending moment and economical design of superstructure. Also, 16 feet 6 inch vertical underclearance is achieved.

Two unequal curved spans with one span of 110 feet and the other of 100 feet—Due to unequal requirements for future acceleration and deceleration lanes of I-195, unequal ramp spans are required. However, the pier must be located at the middle of the 60 foot median barrier.

There are many disadvantages to using a single span, namely:

- Non-symmetry in construction leads to an increase in costs.
- An unequal span bridge is not aesthetically pleasing. Architecture does not match that of adjacent bridges.

7.8.11 Substructure Alternatives

Abutment types: Several alternates, including three alternate abutment heights, were considered:

1. Full height—Placing the footing below the existing grade requires abutment and retaining wall heights of 38 feet. It leads to an uneconomical design of abutment and wingwalls.
2. Medium height—Desirable height is 25 feet, which includes frost depth over top of spread footing. Footing can be placed on compacted fill of embankment.
3. Stub abutment—It is possible to consider 12-foot high stub abutment. The disadvantage is that a 15-foot longer span is required on each side, due to a 2:1 slope of fill.
4. Integral abutments, MSE walls, and spill-through abutments—Performance of integral abutments supporting curved girders and without piles is not fully known. Since spans are long, it is highly probable for the MSE wall to act as an abutment. Spill-through abutment supported on fill may lead to foundation movements.

7.8.12 Screening Criteria for Selecting Abutment Type

Soil pressure and seismic resistance play an important role in selecting an abutment type. The following criteria are examined when considering the type of abutment to use.

1. Performance.
2. Cost.
3. Schedule.
4. Constructability.
5. Environmental impact.

Table 7.11 Comparison of abutment types.

Ramp TN Over I-195			Abutment Option (Rank/Score)							
S. No.	Screening Criteria	Weight	Full Height		Medium Height		Stub		Spill-Through	
1	Performance	100	2	9	2	9	1	10	3	8
2	Cost	90	3	7	2	8	1	9	4	6
3	Schedule	80	2	8	1	9	1	9	3	7
4	Constructibility	70	1	9	1	9	2	8	2	8
5	Environmental	60	2	7	1	8	1	8	3	6
6	Maintenance	50	2	7	1	8	1	8	2	7
7	Aesthetics	40	2	9	1	10	2	9	3	8
8	Site compatibility	30	3	5	1	10	2	6	3	5
9	Soil pressure	20	4	7	3	8	2	9	1	10
10	Seismic resistance	10	3	6	1	8	2	7	1	8
Total score		100		75		87		83		73
Overall ranking			3		1		2		4	

6. Maintenance.
7. Aesthetics.
8. Site compatibility.
9. Soil pressure.
10. Seismic resistance.

Medium height abutment of ± 20 feet securing highest points is recommended.

7.8.13 Screening Criteria for Selecting Pier Types

Several alternates were considered:

1. Pier center line skew to bearing center line—This option requires the pier face to be parallel to the direction of traffuc flow. It matches with the existing architecture of the middle pier on adjacent bridges.
2. Pier center line normal to bearing center line—This option is more economical for framing of girders and in placing the diaphragms. However, the deviation in architecture will be a distraction for vehicle drivers.
3. Frame bent type—Three round columns and a curved pier cap are being widely used for NJDOT and Turnpike bridges. PennDOT's PAPIER will be used.
4. Hammerhead or solid wall type—These shapes are less economical than pier bents. Also, architecture does not match that of adjacent bridges.

Frame bent type with pier face parallel to the direction of traffic flow (pier center line skew to bearing center line) is recommended for cost and aesthetic reasons.

Alternate 1—Two span HPS 70W hybrid steel girders with new central pier

Advantages:

- Lower superstructure depth (girder depth 5–33 in)
- Lighter abutments
- Economical spread footing foundations
- Longer superstructure life due to smaller deflections, vibrations, and fatigue.

Alternate 2—Single span HPS 100W hybrid steel beam without pier

Advantages:

- Minimum disruption to traffic
- Increased sight distance and minimum traffic accidents.

Disadvantages:

- Costs of both the superstructure and substructure is higher compared to two spans.
- High performance 100W steel is a new material. Its long term fatigue behavior cannot be field verified.

Recommendations: Alternate 1 is the preferred alternate. It is more economical and meets the vertical clearance criteria better than other alternates. The impact of temporary traffic lane closure can be minimized by night work construction.

7.8.14 Retaining Walls and Wingwalls

Due to large variations in grade elevation and undulations of topography close to abutments, retaining walls will be required along I-195. Exact locations, lengths, and heights will be investigated. It is proposed to construct varying heights of walls in incremental steps.

Both MSE walls and cast-in-place reinforced concrete alternates will be considered, including two cast-in-place reinforced concrete alternates similar to abutment heights:

1. Full height—Placing the footing below the existing grade requires wingwall heights of approximately 38 feet. It leads to an uneconomical design of wingwalls.
2. Medium height—Desirable height is 25 feet, which includes frost depth over top of spread footing. Footing can be placed on compacted fill of embankment.

   A 2:1 slope of embankment will be maintained, and sloping face will be protected by concrete slab.

   Embankment grading of 2:1 will be required at the face of wingwalls. The height of the wingwall will conform to vertical curve.

   Since medium height walls are recommended, additional alternate of prefabricated modular wall construction will be investigated.

   Medium height wall of ±20 feet is recommended.

7.8.15 Approach Slabs

Since projected ADTT exceeds five percent of total traffic volume per day, it is proposed that approach slabs be provided. Approach slabs will be provided on vertical curve. The length of approach slabs will be calculated from geometry of curve.

Approach slabs will be supported on compacted fill material.

7.8.16 Foundations

Spread footings are anticipated on compacted fill material pending the results of the subsurface investigation for placing footings on newly compacted fill approximately 15 feet above grade. At this elevation, the water table will not be a problem. It will not be feasible to place footings on rock due to very high depth. Usually, differential settlement is not a concern for abutment wall spread footings.

7.9 SELECTED CASE STUDIES

7.9.1 C and D Canal Equestrian Bridge in Delaware

1. The author recently designed a 42-foot span steel pony truss bridge located at the intersection of fast flowing Guthrie Run and C and D Canal (Chesapeake Bay and Delaware River Canal). Abutments and wingwalls were supported on piles. During floods the bridge receives a high level of flow from two directions at right angles. Abutments were constructed behind existing stone walls, which increased span, but scour of piles was eliminated.

   A timber deck with timber stringers and decking was provided for horse rides over the bridge.

Figure 7.7 Proposed site for equestrian bridge over Guthrie Run discharging in C and D Canal, DE.

Figure 7.8 Proposed equestrian pony truss bridge with timber deck and approaches.

2. Rock Creek, Maryland— Trail and Trestle Bridge project in Rock Creek, Maryland. The trail serves as a vital link in the county's bicycle and pedestrian trail system and at the center of the project is an historic railroad trestle that the team converted to a pedestrian bridge.
3. Siebenthaler Avenue Bridge project in Dayton, Ohio, completed in 2004. The structure features four precast, post-tensioned arches,at an overall length of 263 feet and a deck area of 15,780 square feet. The main arches frame into massive thrust blocks using post-tensioned bars. The PTI awarded the bridge a merit award for superior post-tensioning projects for design and construction focusing on creativity, innovation, ingenuity, cost-effectiveness, functionality, constructability, and aesthetics.
4. The LeJeune Flyover was designed and built to ease traffic around the Miami International Airport. During initial construction of the three-span, twin steel box girder bridge, crews noticed cracks in two piers. To complicate matters, the bridge is located over a canal, so the structure and the repair solution had to be resistant to corrosion from natural elements. The project engineers formulated an innovative repair solution using post-tensioning for the pier caps and columns, and completed the work in just four weeks. Post-tensioning was an economical solution to rehabilitate the structure with the least amount of disruption to the public.
5. Lake Villa, Illinois Bridge: The Illinois Department of Transportation (IDOT) recently announced completion of a new bridge in Lake Villa, Lake County, Illinois, constructed with high-performance steel developed by engineering researchers at Northwestern University. About 500 tons of the copper alloy steel, known as ASTM A710 Grade B high-performance structural steel, was used in constructing the 430-foot span that carries Illinois Route 83 over the Canadian National Railroad tracks. Not only is this steel strong, tough, and easy to fabricate, but it withstands the elements better than typical steel, meaning it doesn't require painting. This makes construction easier and significantly reduces long-term maintenance costs.

The only previous use was in the rehabilitation of the Poplar Street Bridge over the Mississippi River in the Metro East area. The steel has a strength of 70,000 pounds per square inch (psi) compared with 50,000 psi in commonly used structural steel. It is also easy to weld, and tests have shown it has high-impact toughness at low temperatures. In addition, the high copper content gives the alloy much better resistance to atmospheric corrosion than other high-performance steels.

7.9.2 Summary of Planning and Design Criteria

Based on case studies by the author, planning aspects should address the following:

1. Meeting constructability and minimum cost requirements.
2. Provision of future widening: The distance between abutments to compensate for varying widths of shoulder and future acceleration and deceleration lanes.

3. Providing minimum lateral distance of 30 feet from the edge of the travel lane to the abutment and 16 ft 6 in minimum vertical clearance.
4. Maintenance and protection of two lane traffic on the ramp during maintenance or repairs.
5. Providing 2:1 slope for embankment height: Total length of bridge is governed by optimum abutment locations on fill.
6. Bridge aesthetics will be in conformance with adjacent bridges. Pier will be located in the middle of median.

7.9.3 Case Study of 32 Sign Structures for New Jersey Turnpike

The author was project manager for a recent project that included a variety of sign structures such as overhead, cantilever, butterfly types, VMS and CMS structures. The variety of foundations included drilled shaft, spread footings, bridge mounted, and retaining wall mounted types.

In addition to regular meetings, modern management technology was used. A software known as ProjectSolve was developed. Electronic files for each of the disciplines were uploaded to the ProjectSolve Web site on Fridays. Relevant files were downloaded on Mondays by the concerned receivers. This fast Internet communication reduced the need for countless interdisciplinary meetings and expedited progress on the job. Use of electronic files also reduced recordkeeping of hundreds of files required for each discipline.

Some of the procedures were as follows:

1. Drawings preparation: The two types of electronic CAD files were *sheet files* and multiple *reference files*. The sheet files were structural plans, while reference files were plans of highways, right-of-ways, utility locations, drainage, geotechnical boring locations, soil information, lighting, fiber optics, power supply, etc.
2. A master reference folder was created for the selected reference. Usually structural drawings had 1:10 scales with details shown as ¼ in = 1 foot. Individual reference files were of 1:30 or smaller scales. Converting each reference file to the scale required on the sheet files required additional work.
3. For uniformity in the presentation of thousands of construction drawings for the entire project, the owner developed standards for CAD, which had a specialized approach to font size, scales, text style, line weight, etc.
4. To access and make use of the reference files, a large number of engineers were required to be trained in ProjectSolve.
5. An "Alert" option to inform the concerned team members was also used at the time of uploading the file. It was addressed to an individual from whom feedback was required. The alert key sent a message to the project manager as to the status of a certain activity. The PM could then pursue the matter to resolve any outstanding issues. For example, if boring information was not available within 100 feet of the foundation location, the geotechnical engineer would receive an alert. He could then initiate additional borings and provide the soil properties for design. Electronic coordination was quick and decisive.

The above management approach was very useful for a large size project and the computer maintained a chronology of events for meeting the required milestones. It helped in quality control, cost control, and in monitoring the progress by meeting individual deadlines on time.

7.9.4 Need for Further Research and Investigation

Due to numerous rehabilitation issues, it should be understood that both design and construction procedures need to be further refined. Research in the following areas is needed:

1. Implementation of advanced materials and continuation of materials research, e.g., high-performance materials, materials durability, lightweight concrete to provide lower self-weight for larger components, etc.

2. Identification of technical and cultural barriers, both real and perceived.
3. Implementation and further development of rapidly assembled connection details and joints that are constructible, durable, and repairable.
4. Implementation and further development of cost analysis and risk assessment.
5. Development of maintenance needs, accessibility, reparability, and inspection criteria.
6. Identification of transportation and erection issues including loads and equipment.

Accelerated bridge construction:

1. Establishment of a database to track accelerated construction of bridge and highway structures and substructures to demonstrate and document successes, including costs.
2. Implementation and further development of innovative construction methods, including total bridge movement systems such as self propelled modular transporter (SPMT), launching, etc.
3. Development of prefabricated seismically resistant systems, including substructures.
4. Development of more efficient modular sections.
5. Development of quality assurance measures for accelerated techniques for superstructure and substructure construction.
6. Implementation and further development of design considerations for hardening of existing structures and rapid recovery after disasters (natural and man-made).
7. Implementation of and further development of contracting strategies that encourage speed and quality.
8. Active and structured dissemination of information on available technologies and successful accelerated bridge construction projects to both decision makers and designers.
9. Identification of methods to accelerate construction of bridge foundations and earthwork, and identification of demonstrated sources of construction delays.

Possible topics for a master's thesis:

1. Comparative study of integral abutments, with or without staged construction.
2. Comparative study of integral abutments on rivers and crossings.
3. Comparative study of structural connections between deck and sleeper slab.
4. Effect of skew angles.
5. Effect of thickness and length of relief slab on seismic behavior of piles.

Possible topics for a Ph.D. thesis: In addition, the following topics are possible:

1. Effect of beam spacing on seismic forces/pile top movements for prestressed adjacent box beam versus spread box beams and I girders.
2. Study of semi-integral abutments.
3. Study of integral piers.

The purpose of research is to study the behavior of bridge structural systems—different spans lengths, ratios, and skew conditions—when they are subject to seismic forces as well as to normal vertical forces.

BIBLIOGRAPHY

Agrawal, A., Y. Zhihua and Mohiuddin Ali Khan, *Handbook of Scour Countermeasures*, National Cooperative Highway Research Program, Trenton, NJ, 2007.

American Iron and Steel Institute Steel Bridge Forum, *Structures Congress*, American Society of Civil Engineers, Seattle, WA, 2003.

Benham, P. P. and F. V. Warnock, *Mechanics of Solids and Structures*, Pitman Publishing, London, 1973.

Bell, Brian, *Advanced Theory of Structures*, McDonald and Evans, Plymouth, 1978.

British Standards Institute, British Standard 5400, Parts 1 through 10, London.

Brockenbrough, R. L and B.G. Johnston, *Steel Design Manual*, U.S. Steel Corporation, Pittsburgh, 1968.

Brockenbrough, R. L. and F. S. Merritt, *Structural Steel Designer's Handbook*, 4th Edition, McGraw-Hill, 2006.

Connor, Robert J., "Extending the Service Life of Steel Bridges Through Field Instrumentation and Preemptive Retrofits," ATLSS Engineering Research Center, Lehigh University, Bethlehem, PA.

Construction Handbook for Bridge Temporary Works, American Association of State Highway and Transportation Officials, 1995.

"Detailed Planning Research on Accelerating the Renewal of America's Highways," National Cooperative Highway Research Program, 2003.

Guide Design Specifications for Bridge Temporary Works, American Association of State Highway and Guide Design Specifications for Bridge Temporary Works, American Association of State Highway and Transportation Officials, 1995.

Guide Specification and Commentary for Vessel Collision Design of Highway Bridges, American Association of State Highway and Transportation Officials, 1991, Washington, DC.

High Performance Steel Bridge Concepts, American Iron and Steel Institute, J. Muller International, November 1996.

Jakovich, Gary and Vasant Mistry, "National Perspective on Accelerated Bridge Construction," *24th International Bridge Conference*, Pittsburgh, PA, 2007.

Khan, Mohiuddin A., Edwin Rossow and Surendra Shah, "Shear Design of High Strength Concrete Beams," *ASCE-SEI Congress Proceedings*, Philadelphia, PA, 2000.

Khan, Mohiuddin Ali, "Elastic Composite Action in Slab-Beam Systems," M. Phil. Thesis, University of London, 1968.

Khan, Mohiuddin Ali, "Long Span Prestressed Cable Nets," Ph.D. Thesis, University of Southampton, London, 1972.

LRFD Bridge Construction Specifications, 2nd Edition, American Association of State Highway and Transportation Officials, 2004.

LRFD Bridge Design Specifications, 3rd Edition, American Association of State Highway and Transportation Officials, 2004.

LRFD Design Manual, New Jersey Department of Transportation, 2001.

Manual for Condition Evaluation and Load and Resistance Factor Rating of Highway Bridges, American Association of State Highway and Transportation Officials, 2003.

Manual on Use of Self-Propelled Modular Transporters to Remove and Replace Bridges, United States Department of Transportation, Federal Highway Administration, June 2007.

Morice, P. B. and G. Little, *The Analysis of Right Bridge Decks Subjected to Abnormal Loading*, Cement and Concrete Association, Wrexham Springs, UK.

Mufti, Aftab, Baidar Bakht and Leslie Jaeger, *Bridge Superstructures New Developments*, National Book Foundation, Islamabad, 1997.

National Highway Institute, Participant Notebook, Course No. 13061, United States Department of Transportation, 1999.

NYSDOT Bridge Manual, New York State Department of Transportation, April 2006.

NYSDOT LRFD Bridge Design Specifications or the NYSDOT Standard Specifications for Highway Bridges.

Ontario Highway Bridge Design Code, Ontario Ministry of Transportation and Communication, Toronto, Canada.

PCI Bridge Design Manual, Precast/Prestressed Concrete Institute, www.pci.org.

Rahrig, Philip, "Galvanizing Plate Girders," American Galvanizing Association, *Modern Steel Construction*, August 2002.

Standard Specifications for Roads and Bridge Construction, as modified by special provisions, New Jersey Department of Transportation, 1996.

Steel Bridge Design Handbook, "Design for Constructability," Chapter 13, National Steel Bridge Alliance, 2007.

Steel Bridge Design Handbook, National Steel Bridge Alliance, Chapter 23.

Timoshenko, S. P. and S. Woinowsky-Kreiger, *Theory of Plates and Shells*, 2nd Edition, McGraw-Hill International Edition, 1959.

"Transverse Stiffener Requirements in Straight and Horizontal Curved Steel I-Girders," *Journal of Bridge Engineering*, American Society of Civil Engineers, March/April 2007.

Wasserman, E. P. and J. H. Walker, *Integral Abutment for Steel Bridges*, Tennessee Department of Transportation, Knoxville, TN.

AASHTO Publications

10-43 Inspection, Evaluation, and Maintenance Manual for Movable Bridges. Adopted by AASHTO in 1997.

NCHRP Products for AASHTO Committees

Performance TESTS of modular bridge deck joints. Completed in August 2001. Published as NCHRP Report 467.

Establish the fatigue stress category of existing, retrofitted, and new cost-effective connection details for high-level lighting and for cantilever structures, 2010.

Develop GUIDANCE for the design and construction of durable cast-in-place reinforced concrete connections for precast deck systems that emulate monolithic construction, 2010.

PROCEDURES for testing bridge decks. 2009.

NCHRP Products for AASHTO Committees

Project Product Status

Procedure and software for estimating bridge life cycle costs, 2002. Published as NCHRP Report 483.

Methodology for considering redundancy in highway bridge substructures. Published as NCHRP Report 458.

Procedure for validating bridge design and rating software, 2002. Published as NCHRP Report 485.

Design Specifications and Details for steel bridge integral connections with substructures, 2004. Published as NCHRP Report 527.

Rational design and construction guidelines for geosynthetic reinforced soil bridge abutments and approaches with flexible facing elements, December 2004. Published as NCHRP Report 556.

NCHRP Products for AASHTO Committees

Project Product Status

Methodologies and software for network and project-level optimization of multiple, user specified performance criteria for bridge management, December 2006. Published as NCHRP Report 590.

Design guidance for improving the structural performance and resistance to explosive effects of new and existing bridges, 2008.

A design methodology, connection details, and design and construction specifications for precast bent cap systems under seismic loading, 2009.

Design methods and specifications for fiber reinforced polymer systems used for strengthening concrete girders in shear, 2009.

NCHRP Products for AASHTO Committees

A manual on service life of corrosion-damaged reinforced concrete bridge superstructure elements, 2006. Published as NCHRP Report 558.

Guidelines for the use of self-consolidating concrete in precast, prestressed concrete bridge elements for possible incorporation into the AASHTO LRFD Design and Construction Specifications, 2008. Published as NCHRP Report 628.

Guide specifications for the use of lightweight concrete in high-strength prestressed concrete girders and in high-performance bridge decks and related additions/changes to AASHTO LRFD Bridge Design and Construction Specifications, 2010.

Performance measures for bridges, 2000.

Strategic plan for bridge engineering, 2000.

Recommendations on live load deflection limits for steel bridges, 2002.

Summary of international practice on bridge girder and deck strengthening with fiber-reinforced polymer composite.

Seismic design guide specifications that reflect a consensus among state bridge engineers on how earthquake forces and ground motions should be considered when designing a new bridge, 2006.

A research problem statement identifying research needed to make all pertinent AASHTO documents applicable to design concrete strengths up to 15.0 ksi or higher, 2007.

Recommended AASHTO Design-Build Procurement Guide, 2005.

8

Inspection, Rating and Health Monitoring Techniques

Section Overview

- Structural evaluation for diagnostic design.
- Routine health monitoring and inspection.
- Rating systems used for the selection of repair method.
- Emerging technology for bridge protection against extreme events.

8.1 BRIDGE MANAGEMENT SYSTEM (BMS)

8.1.1 Introduction

1. Engineering management is the most important aspect of bridge maintenance. Bridge management is applicable to all existing bridges, old or new. Issues are technical as well as financial and social. The purpose of management is maintenance of a bridge by identifying deficiencies and ensuring the continued safety of traffic through rehabilitation.

A project manager and his team should have training in management aspects. Both structural expertise and major funding are necessary to meet the objectives. The design and construction of a new bridge, widening, and replacement are additional tasks outside the scope of the management system and are not included in rehabilitation or retrofit. Not long ago, the Transportation Equity Act of 1998 allocated over $200 billion for resurfacing, restoring, and rehabilitating various programs every six years.

Bridge management is usually handled by the structural evaluation department of various state agencies. Since it would be a herculean task to inspect thousands of bridges by a single department, the work is awarded to inspection teams of consultants. It consists primarily of:

- Inspection
- Structural evaluation
- Rehabilitation.

2. BMS is an integrated collection of the following elements:
- Organizational roles
- Procedures
- Data
- Analytical tools and computer programs
- Support services.

3. BMS's using a dynamic bridge substructure evaluation and monitoring system can be applied to:
- Monitoring
- Integration into a BMS of problem-specific inspection and condition data (for example, RETAIN, an expert system for retaining wall rehabilitation, may provide opportunities to integrate dynamic bridge substructure evaluations and monitoring systems into BMS's.)
- Determination of optimal inspection frequencies
- Measuring vulnerability to scour, seismic events, and impacts such as vehicles, boats, ice, and flood debris.

4. In general, visual inspection is the widely used means for including condition data in a BMS. This has several disadvantages, including:
- It is qualitative, and ratings generally do not exhibit a high degree of consistency or repeatability. The rating accuracy is unknown.
- The rating reflects an aggregate measure of condition. For example, microscopic flaws can be catastrophic, but will never be reflected in the rating.
- The ratings are not closely related to the cause of the problem or the response.

There is clearly a role for improving and enhancing the status of bridges through better NDE and its integration with BMS's.

Dynamic bridge substructure evaluation and monitoring system: It is just one aspect of nondestructive testing appropriate for bridges. NDE methods may be more cost effective.

BMS's currently include only visual condition data. The value of multiple sources of data needs to be compared with NDE results. NDE data can complement, reinforce, or support visual condition rating data. It is not a substitute for visual condition ratings.

Dynamic bridge evaluation searches a specific type of defect. The following needs to be investigated:
- Dynamic testing data in terms of an increase or decrease on bridge life cycle costs (the influence of low probability, high impact catastrophic events on bridge life cycle costs may be considered.)
- The effectiveness of remedial strategies.

Publications by AASHTO Planning Subcommittee on Asset Management

1. *Transportation Asset Management Guide,* 2002. Prepared for NCHRP Program 20–24(11).
2. *Analytical Tools for Asset Management,* 2005. Published as NCHRP Report 545.
3. *Performance Measures and Targets for Transportation Asset Management,* 2006. Published as NCHRP Report 551.

8.1.2 Objectives of Monitoring

1. There are numerous qualitative and quantitative tools that aid in the management of bridges. Bridge management's main goal is ensuring safety while minimizing costs.

Some of the procedures used are inspection, maintenance, and repair. Structural health

monitoring can aid in several aspects of bridge management, such as:
- Reducing inspection costs while improving quality.
- Prioritizing repair/maintenance schedules.
- Increasing the accuracy of deterioration estimations and improving the decision-making process.

2. The main goal of monitoring is to detect accurately and efficiently structural damage either due to long-term deterioration processes or due to extreme events (e.g., earthquakes, blasts).
3. The right decision is required at the right time. There is a need to dedicate more resources to inspecting for diagnostic purposes and to make correct decisions about:
   - Rating analysis.
   - Rehabilitation of those structures that may have outlived their cost-effectiveness.
   - Maintaining older structures that may benefit from low-cost maintenance.
   - Constructability.
   - Safety.
   - Aesthetics.
   - Environmental impact and permits.
   - Future maintenance, inspection access, and remote health monitoring.

8.1.3 Monitoring as an Effective Countermeasure

A multi-hazard design of a highway bridge may improve the reliability of the entire transportation network of the region that the highway bridge serves. A risk assessment is used that analyzes seismic retrofit and bridge security improvements. A common approach is:

$$\text{Risk to the bridge} = \text{Probability of collapse} \times \text{The consequences of collapse}$$

After prioritization of a major bridge, a risk assessment is performed to determine the vulnerability of its fracture critical and failure critical elements. The AASHTO *Guide to Highway Vulnerability Assessment for Critical Asset Identification and Protection* shall be used as the standard for conducting the risk assessment:

1. The consequences of partial or full collapse are based on vulnerability (expected damage, and loss of life and functional use) and the importance of the bridge.
2. Importance is based on:
   - Retrofit/replacement value
   - Emergency evacuation
   - Military importance
   - Importance to the regional infrastructure network.
3. Planning and execution of emergency plans at the local, state, and national levels.
4. Engineering and economic (such as toll revenue value) considerations. The severity of these hazards can significantly increase the costs of construction and maintenance, especially if they are for existing infrastructure requiring rehabilitation.
5. Monitoring structural health such as identification of failure critical and fracture critical members. Engineers routinely deal with natural hazards:
   - Earthquakes
   - Floods
   - Wind
   - Ice
   - Blast loads
   - Accident impact loads.

8.1.4 Multi-Hazard Consideration

A general theory of multi-hazards in infrastructure needs to be applied to structural analysis, design, life cycle costs, risk assessment, and structural health monitoring. While maintaining the needed safety levels, multi-hazards offer anticipated overall cost reduction. This approach considers the increasing complexity of the structural systems to meet the demands of the current environment. It takes advantage of the recent developments and innovations in computing, analytical, and sensing technologies.

Benefits of multi-hazard considerations include the following:

- A more accurate estimation of the inherent resiliency of the system
- A more accurate treatment/estimation of life cycle costs of the system
- Optimization of the structural health monitoring (SHM) to increase experimental efficiency.

Recent improvements in computational tools, database population (that include behavior of systems under several severe hazards), and structural health monitoring (SHM) all contribute to the increased consideration of the multi-hazard approach to systems.

The preliminary risk-based assessment of major bridges shall include the following tasks:

1. Perform cursory site inspections (which do not require special inspection equipment or access methods) to become familiar with the following:
   - The physical layout
   - Below deck and above deck features
   - Available access routes for each bridge.
2. Assemble and review the following documents:
   - Existing and available plans and documents for each bridge
   - Original as-built plans and shop drawings
   - Subsequent repair and rehabilitation plans
   - Annual, biennial, and special bridge inspection reports
   - Structure inventory and appraisal (SI&A) form updates
   - Load ratings
   - Fatigue evaluations
   - Structural models
   - All applicable reports and materials.
3. Compile and review traffic volume:
   - Current average annual daily traffic (AADT) data
   - Average annual daily truck traffic (AADTT) data
   - Associated peak occupancies for each bridge.
4. Compile and review current bridge replacement costs for each bridge.
5. Compile and review available overload analyses performed for each bridge.
6. Prior to a vulnerability assessment, perform a prioritization of the fracture critical and failure critical members and member components on an individual bridge basis.

8.1.5 Multi-Hazard Resiliency

1. For a given system that is exposed to multi-hazards, there exists an inherent multi-hazard resiliency within the system. This multi-hazard resiliency implies an interrelationship between the manners in which the system responds to different hazards.
2. Weakness in rating system: A bridge is as strong as the weakest component. Greater attention needs to be paid to design and maintenance for items that show a deficiency.

For example, in a Minneapolis bridge failure, less expensive items—the gusset plates—were under designed. Recording and coding guides should give a more stringent evaluation of failure prone and sensitive items.

3. Sufficiency rating calculations may show that a bridge superstructure is rated at 6 (satisfactory condition) and a substructure is rated at 7 (fair condition with sound structural members and minor deterioration). However, other items in National Bridge Inspection (NBI) can make it structurally deficient. To some extent there is an inherent contradiction in the rating system being used.
4. A reliability-based methodology for management of aging bridges needs to be developed.

   Optimum use of types of bridges or types of decks should be evaluated to streamline the type of traffic. Segregation of traffic consists of:

   - Pedestrian use (H-5 truck)
   - Passenger cars only (H-10 truck design)
   - Full service (HS-20 truck).

8.1.6 Prioritization of Bridges for Widening or Replacement

1. Prioritization shall be based on NBI data, as well as specific operational and logistical features of each bridge. These may include such features as:
   - Potential for loss of life based on traffic volumes
   - Importance to emergency response and evacuation
   - Military route designations
   - Availability of alternate routes
   - Bridge replacement values
   - Loss of use and associated toll revenue.
2. Summary report of findings and recommendations: Based on the prioritization and risk assessment of all major bridges, including estimates of capital and operating costs where applicable, a summary report shall be compiled with an executive summary, cost-benefit analysis, and implementation timeline over a minimum 10-year time frame.
3. Recommendations for future engineering phases in priority order include:
   - Additional special inspections
   - Detailed structural modeling
   - Redundancy and progressive failure analyses
   - Fatigue analyses
   - Conceptual retrofit design.

8.1.7 Risk Assessment Study of Major Bridges

Failure critical members are generally defined as members or member components whose failure would be expected to result in a partial or full collapse of the bridge.

Failure critical members are not restricted to being steel members in tension, and therefore may be constructed from any material and be designed to withstand non-tensile loads. Some examples of failure critical members include a concrete compression member such as a single column pier or a compression member of a truss chord.

The AASHTO *Guide to Highway Vulnerability Assessment for Critical Asset Identification and Protection* shall be used as the standard for conducting the risk assessment.

The risk-based assessment of major bridges shall include the following:

1. Perform cursory site inspections to determine the physical layout, below deck and above deck features, and available access routes for each bridge. The cursory site inspections shall be for visual inspections only, for which no special inspection equipment or access methods are required.
2. Research and identify current best practices for risk management of critical infrastructure, particularly as they pertain to the potential failure of fracture critical and failure critical bridge elements. Also, evaluate the current best practice policies and procedures of other transportation agencies.
3. Assemble and review existing and available plans and documents for each bridge:
   - Original as-built plans and shop drawings
   - Subsequent repair and rehabilitation plans
   - Annual, biennial, and special bridge inspection reports
   - Structure inventory and appraisal (SI&A) form updates
   - Load ratings
   - Fatigue evaluations
   - Structural models
   - All other applicable reports and materials.
4. Compile and review the following:
   - Current average annual daily traffic (AADT) data
   - Average annual daily truck traffic (AADTT) data
   - Associated peak occupancies for each bridge
   - Current bridge replacement costs for each bridge
   - Available overload analyses performed for each bridge
   - Identify all fracture critical and failure critical members
   - Identify connections and details for each bridge as part of a comprehensive assessment
   - Identify non-redundant members
   - Members and connections with fatigue sensitive details
   - Welded connections
   - Bearings and other special details as applicable.
5. Following prioritization of the major bridges, perform a risk assessment to determine the vulnerability of their fracture critical and failure critical elements.
6. Prepare recommendations for:
   - Additional special inspections
   - Detailed structural modeling
   - Redundancy and progressive failure analyses
   - Fatigue analyses and conceptual retrofit design.

The summary report shall include an executive summary, cost-benefit analysis, and implementation timeline over an estimated time frame. Follow a risk assessment similar to those previously developed for seismic retrofit and bridge security improvements, where

Risk to the bridge = Product of the probability of collapse × Consequences of collapse

The consequences of partial or full collapse may be based on:

- Vulnerability (expected damage, loss of life and functional use, etc.)
- Importance (retrofit/replacement value)

- Emergency evacuation and military importance
- Importance to regional infrastructure network and economy, toll revenue value, etc.).

Utilize structural health monitoring systems to detect and monitor deficiencies on the major bridges. The structural health monitoring systems typically entail the installation of sensors at key locations on the structure being studied. These may include loads, stresses, strains, and differential movements, as well as chemical composition of the concrete and steel structural components.

Review and compare proprietary structural health monitoring systems for their applicability, effectiveness, and installation and maintenance costs, and make recommendations for specific applications.

8.1.8 Identify and Repair Fracture Critical Members

The identification of all fracture critical members shall be based on the following definitions: Fracture critical members or member components (FCM's) are steel tension members or steel tension components of members, whose failure would be expected to result in a partial or full collapse of the bridge.

For a bridge member to be classified as fracture critical, it must satisfy two criteria:

1. The bridge member must be in tension or have a tension element.
2. The failure must cause a partial or full collapse of the structure.

Therefore, recognition and identification of the following physical conditions are required:

1. The types of forces carried by the bridge member.
2. Its degree of load path.
3. Structural and internal redundancy.

As part of a comprehensive assessment, identify:

1. All fracture critical members.
2. Connections and details for each bridge.
3. Non-redundant members.
4. Members and connections with fatigue sensitive details.

Examples of failure critical members include a concrete compression member, such as a single column pier or a compression member of a truss chord, welded connections, bearings, and all special applicable details.

8.1.9 Literature Review on Superstructure Condition Evaluation

Structural parameter identification systems consist of both mathematical (static and dynamic) and non-mathematical activities. Mathematical activities generally include strain measurements, displacement and rotation measurements, time domain and frequency domain dynamic measurements. Non-mathematical activities include pattern recognition, signal processing, and expert system. Recent research on the role of NDE in BMS's suggests the desirability of integrating dynamic testing results with visual ratings data.

Ewins listed references on dynamic testing for modal vibration measurement and analysis. For the ultimate load tests, the bridges selected for removal from service were tested to failure. These studies generally provided some insight on the ultimate load capacity and mechanisms of failure that could be used in the future.

Bridge superstructure condition evaluation research programs focused on two primary areas: ultimate load tests and dynamic tests. Laboratory and field studies to evaluate dynamic properties of bridges and relate them to condition assessments have been reported extensively in past years.

1. Baumgartner and Waubke showed that frequency measurements in tension hangers under traffic loading can be related to the end fixity of the hangers.
2. Biswas et al. reported a component evaluation technique using a hammer impact using dynamic responses. Results were confirmed with laboratory models, but field verification was limited.
3. Cawley and Adams related changes of successive mode frequencies to the existence and location of structural deterioration in beams.
4. DeWolf et al. and Gregory et al. demonstrated a relationship between dynamic testing results and structural deterioration. Sensitivity of dynamic characteristics to deterioration was shown to depend on the particular modes being observed.
5. Huston et al. reported various full-scale bridge dynamic tests, showing that dynamic characteristics may be revealed using vibratory shakers, impact hammers, and traffic and wind loads.
6. Manning suggested that a localized dynamic analysis might be advantageous because serious loss of strength of a single member may occur before it can be observed on the entire structure.
7. Mazurek and DeWolf showed in field tests that ambient traffic loads could be used as a basis for an automated monitoring scheme based on changes in vibration signatures. Their laboratory results encouraged further field investigations. Changes in support condition and crack development affect natural frequencies and modal amplitudes. Changes in modal frequency were up to 30 percent for changes in support condition and up to 10 percent for cracking.
8. New York State DOT used continuous monitoring of bridge dynamic characteristics. A remote bridge monitoring system is based on measuring dynamic motion (using accelerometers) as well as strain and rotation (using inclinometers). A warning alarm is detected when significant changes in modal frequencies occur. Dynamic response data collected show up to 10 percent scatter in the modal frequency measurement.
9. Salane et al. reported dynamic tests of a bridge for detecting structural deterioration caused by girder fatigue cracks. A concrete deck on steel girders was loaded with an electrohydraulic actuator system up to 465,000 load cycles.

   Accelerometers were used to determine damping ratios, frequency contents, and impedance at various stages during the loading. The test results indicated increases in damping ratios with cycles of loading caused by cracking and a decrease in amplitude at resonant frequencies, as well as a 20 to 40 percent change in computed stiffness coefficients.
10. Salawu and Williams reported a study of the forced vibrations of a bridge before and after repair. The test results demonstrated small changes in natural frequency induced by the repair.
11. Woodward et al. conducted dynamic tests for a full-scale bridge subject to artificially induced fatigue cracking (vertical cuts) in a main girder. Preliminary field test results showed changes in dynamic characteristics due to a maximum amount of damage.

It is possible to conclude from the above studies that simpler interpretations for vibration measurements have not been reported in the literature.

Dynamic measurements can be used to evaluate the distribution of loading in axially loaded members such as cables and truss diagonal braces. The natural frequency of such axially loaded members is highly sensitive to the magnitude of the axial load.

Further work is needed to relate dynamic properties to component deterioration.

8.1.10 Literature Review on Substructure Condition Evaluation

Many state DOT's such as Connecticut, Florida, and New York have used instrumentation in their bridge inspection programs. During the North American Workshop on Instrumentation and Vibration Analysis of Highway Bridges in 1995, researchers and practitioners agreed that instrumentation is a viable tool for bridge inspection.

Huang et al. recently developed the HHT, a technique for applying time domain data that makes it possible to analyze vibration data and determine the resonant frequencies of systems instantaneously by location throughout a time domain record for nonlinear, non-stationary systems. In other words, the HHT method helps determine short-duration changes in the system response frequencies that indicate the lower frequency resonance associated with damage to a structural member. This technique promises to be more sensitive to short-term changes through lower frequency, nonlinear responses when a moving or varied excitation force is most actively exciting and closest to a damaged member. Thus, the masking of lower frequency responses associated with damage to a single member is better analyzed using the HHT approach.

A bridge abutment generally is designed to resist backfill soil pressures; however, for a rigid frame abutment, the thermal deck expansion causes backfill pressures that are far in excess of the active soil pressures used in design. In addition, bridge skew results in a large horizontal gradient of the backfill pressures, producing local backfill pressures that could exceed the capacity of the abutment walls.

1. Raghavendrachar and Aktan demonstrated that multi-reference impact testing could serve as the main experimental component for comprehensive structural identification of large constructed facilities. Demanding standards are required from modal testing designs for the accurate experimental measure of flexibility, which is the inverse of stiffness (also known as displacement divided by force as a function of frequency).
2. Aktan and Helmicki performed a study in instrumented monitoring of a full-scale bridge. For structures subjected to lateral loads, impact testing may not be the appropriate method; forced excitation modal testing using larger vibrators is desirable.
3. California DOT (Caltrans) has an instrumentation program for bridge inspection. It involves monitoring seismic excitations and foundation systems. Practicing bridge engineers recognized the need to evaluate and formalize the use of structural identification and instrumentation.
4. Chen and Kim developed the "bending wave" method for investigating pier conditions and local defects by measuring the velocity dispersion curve of the transverse waves propagated down from the top of a pier. The method proved suitable for short piles in softer soils.
    - The dispersion of the reflected waves from the pier bottom was used to assess overall pier conditions.
    - The dispersion of the directly arrived wave was used to assess local damages.
5. Finno and Prommer studied the impulse response (IR) method for inaccessible drilled shafts under pile caps. Several drilled shafts connected together with concrete grade beams were tested using the nondestructive IR method. Based on the field data, it was found that shaft heads that were more rigid because of larger or several grade-beam connections exhibited greater signal attenuation.
6. Hussein et al. reported the use of compression waves for investigating single pile length and integrity, settlement, and scour.
7. On the Interstate 15 bridges in Salt Lake City, UT, research on dynamic testing for the condition evaluation of bridge bents was performed using vibration tests with horizontal excitation. Modeling and experimental modal vibration test results were compared in terms of mode shapes and frequencies. The estimated location and intensity of the damage or retrofit was identified. Both damaged and repaired substructure states were used to identify the condition of the structure.
8. Pierce's and Dowding's method focused on the determination of internal cracking and large local deformations caused by earthquakes for concrete bridge piers using time domain reflectometry (TDR).

9. Warren and Malvar used a falling weight deflectometer to assess structural conditions of reinforced concrete piers. Based on matching dynamic responses, deflected shapes from the FEM and testing results were compared and local stiffness and soft areas of the piers were determined. Automatic Dynamic Incremental Nonlinear Analysis (ADINA) system software was used for systematic changing of the stiffness parameters. This method was tested on a real bridge with timber piles in New Jersey. The geometry data was deemed sufficient to identify the effective stiffness of the piers and damaged areas.
10. Washington State DOT studied lateral load responses of a full-scale reinforced concrete bridge to investigate the seismic vulnerability of bridge piers. This study points out that the seismic design of pier columns in the 1950s and 1960s was not adequate to sustain displacements during earthquakes. This study concluded that the latest pier design was more suitable for earthquakes.
11. University of California, Berkeley developed a software program in 1996. BASSIN was developed for dynamic analysis of a bridge-abutment-backfill system that is subjected to traveling seismic waves. It can compute 3-D dynamic responses of an arbitrarily configured bridge-abutment-backfill system induced by compression, vertical shear, horizontal shear, and surface waves with arbitrary wavelength, amplitude, and direction of incidence.

8.1.11 Unknown Foundations and Non-Availability of As-Built Plans

Out of approximately 580,000 highway bridges in the National Bridge Inventory, a large number of older bridges have no design or as-built bridge plans. Consequently, little or no information is available to document the type, depth, geometry, or materials incorporated in their foundations. NCHRP 21-5 research study for determination of unknown subsurface bridge foundations evaluated and developed NDE methods and equipment. It allowed the determination of subsurface bridge foundation depths and other characteristics where such information is unavailable, unknown, or uncertain.

Another study evaluated many existing and new NDE methods including five acoustic methods (sonic echo and impulse response, bending wave, ultra seismic, parallel seismic, and borehole sonic), one modal vibration method (dynamic foundation response), and one electromagnetic (borehole and surface ground penetrating radar) method.

A follow-up phase II study focused on researching and developing equipment, field techniques, and analysis methods for the surface-based ultra seismic and borehole-based parallel seismic methods. These two methods showed the most promise for immediate application to the determination of unknown foundation depths for the most substructures.

8.2 BRIDGE INSPECTION LEADING TO DIAGNOSTIC DESIGN

8.2.1 Field Inspection and Structural Evaluation

Bridge mechanics is a subject that can help an inspector understand how a bridge functions and how certain defects affect the load carrying capacity. Hence, it is desirable that an inspector should possess a high degree of knowledge and skill in structure condition evaluation, analytical techniques, design requirements, bridge components, and the variety of load application and materials' response.

An inspector is a highly trained structural engineer whose skill and expertise is expected to perform the duties of recognizing and identifying:

1. Bridge components and load path in each bridge structural system.
2. Bridge live loads.
3. Properties, performance, and response of the materials.
4. Basic concepts of elasticity, plasticity, and failure modes from corrosion and cracking, etc.

5. The monitor of sensitive aspects such as truss joints, deck joints, bearings, scuppers, and railings.
6. Inventory of items and requirements laid down by NBIS.
7. Use of sensors and SHM measurement methods.
8. Costs of repairs, rehabilitation, and replacement.

In this chapter, methods of inspection, frequency, structural health monitoring, the Pontis System, rating methods, and qualifications and training needs of inspectors are presented.

Monitoring is different from inspection. Monitoring detects any noted deficiencies or existing conditions that are long-term or slow developing—deficiencies which do not require remedial action at the present time but should continue to be monitored so appropriate remedial action can be taken as it becomes essential or economically feasible.

This category covers conditions related to the age and use level of the structure, or slow long-term deterioration not yet at the threshold of regular repairs. Items may be deferred for efficient and economical inclusion in a planned or future contract. Defects in this category may include:

1. Concrete shrinkage or settlement cracks.
2. Surface scaling.
3. Isolated or slightly abnormal bearing devices.
4. Steelwork corrosion and paint deterioration.
5. Superstructure, substructure, or approach misalignment or subsidence, etc.

There is an old proverb, "A stitch in time saves nine." It holds very true for bridges which are in need of repair or rehabilitation. The longer inspection and repairs are postponed, the more the cost increases. Further, too much delay may make the bridge beyond repair, leading to failure and eventual replacement. Many bridges were constructed when the interstate highway system was built and have served well in the past. But problems such as fatigue, over stress from heavier trucks, corrosion, vibration, and soil erosion require immediate attention. Bridges are expensive structures in terms of initial and recurring costs when compared to other structures, and that is all the more reason for vigilant action.

It is estimated that over half a million bridges in the U.S. require regular inspection. Old bridges are candidates for some form of rehabilitation. The primary reason for inspections is to ensure safety and a longer service life for bridges. Inspection reports point out deficiencies and a suggested scope of work. Each subject is covered by codes of practice either from FHWA or the state agency.

More than one discipline is applicable in bridge engineering and asset management. These include design, construction, maintenance, and rehabilitation. An efficient design should address issues such as:

1. Life cycle costs.
2. Constructability.
3. Inspectability.
4. Maintainability.

The above commitments may not always be reflected in the design process. New designs should provide details to facilitate adequate inspection and maintenance.

The inspection process should recognize and document critical deficiencies in keeping with National Bridge Inspection Standards (NBIS), such as cracking in concrete, decay in timber, and corrosion and fatigue in steel, and bring them to the timely attention of the owners. The procedure is to:

1. Recognize critical findings.
2. Reevaluate bridge capacity.

3. Take appropriate corrective action.
4. Review bridge data to ensure that condition ratings are accurate.
5. Implement corrective measures.

8.2.2 Engineering Inspection and Structural Evaluation

Federal law requires states to periodically inspect public road bridges (normally every two years) and to report these findings to FHWA. This information characterizes the existing condition of a bridge (compared with one newly built) and identifies deficiencies, either structural or functional.

The federal government sets the standards for inspection through the National Bridge Inspection Standards (NBIS; 23 CFR 650 subpart C), which is the basis for the *Bridge Inspectors Reference Manual* used by federal, state, and contractor personnel for guidance in inspection. The manual sets forth how, with what frequency, and by whom inspection is to be completed.

The federal government has not established the same inspection requirements for signs or luminaries as for bridges. As a result, some states do not inspect signs or luminaries.

Evaluating physical conditions:

1. Structurally deficient bridge: It is considered structurally deficient if significant load-carrying elements are found to be in poor condition, to the point of causing intolerable traffic interruptions. A structurally deficient bridge is one that has major deterioration, cracks, or other flaws that reduce its ability to support vehicles. A structurally deficient bridge can suffer partial failures that further decrease its capacity and can pose a risk to public safety. It is not necessarily unsafe, but may require the posting of a vehicle weight restriction.
2. The reasons for substandard conditions are:
   - Deterioration
   - Damage
   - Inadequate waterway opening.
3. Functionally obsolete bridge: It is one where its current geometric characteristics are deficient compared with current design standards and traffic demands:
   - Deck geometry (such as the number and width of lanes)
   - Roadway approach alignment
   - Vertical underclearance.
4. A bridge can be both structurally deficient and functionally obsolete. Structural deficiencies take precedence. About half of structurally deficient bridges are also functionally obsolete.
5. In 2006, about 26 percent of bridges were classified as either structurally deficient, functionally obsolete, or both. Only about 12 percent (approximately 74,000) were classified as structurally deficient. Theoretically, the flow of funding should keep pace with the increase in deficiencies as time passes. However, it appears that the deficiencies are increasing exponentially in many cases.

8.2.3 Evaluation of Existing Structures

An accurate description of a critical deficiency is important in determining appropriate responses, designing repair procedures that would eliminate the cause of deficiency, developing risk assessment, and arriving at a baseline for monitoring programs. Some bridges are weakened to the point that signs must be posted to bar vehicles heavier than the calculated maximum load.

The following deficiencies may be found as a result of inspection:

1. Structural deficiencies.
2. Poor condition of deck, superstructure, and substructure.

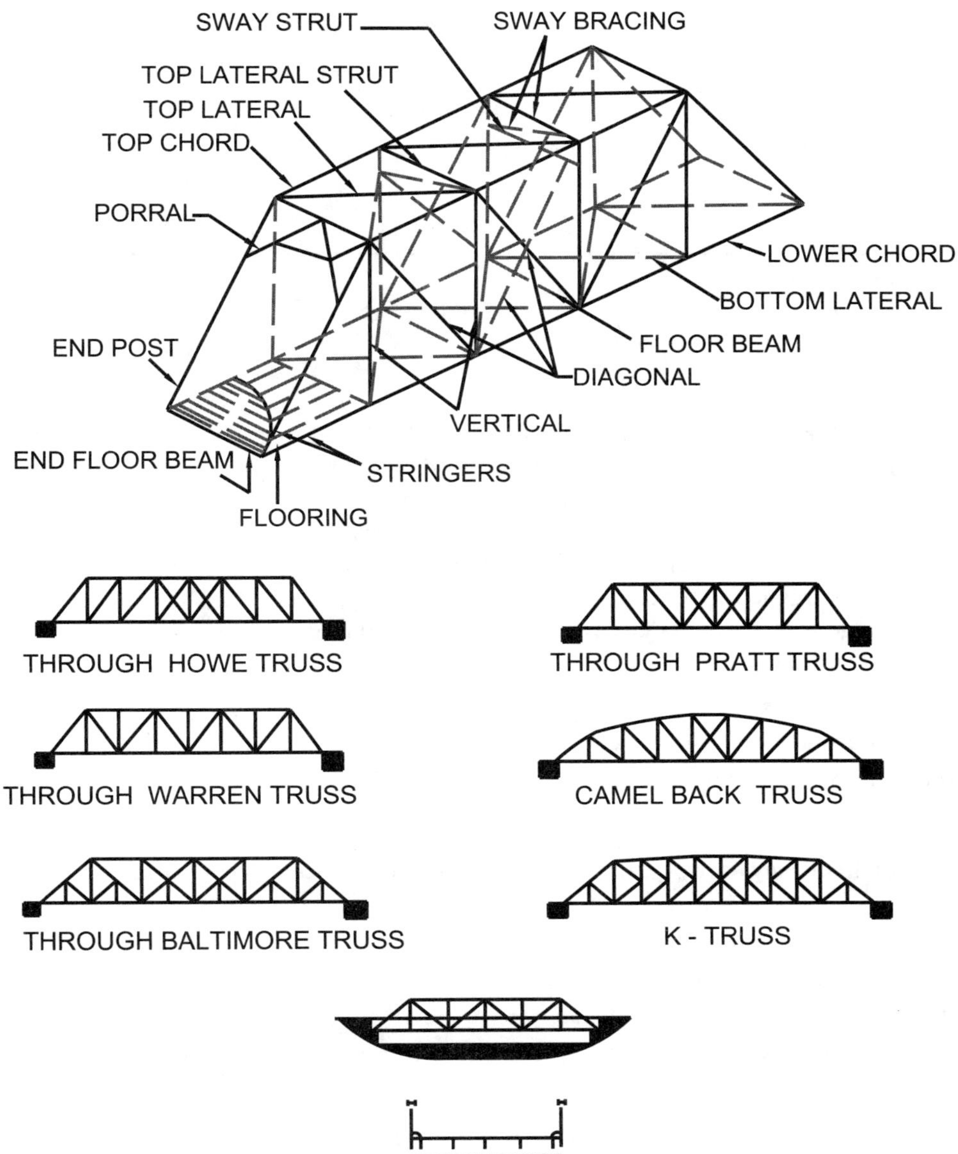

Figure 8.1 Configuration of existing steel trusses that require rehabilitation.

3. Low structural capacities of girders.
4. Substandard deck geometry and lateral and vertical clearance.
5. Scour deficiencies.
6. Seismic deficiencies.

8.2.4 History of Bridge Inspections

In 1998, the FHWA estimated that almost 30 percent of nearly 600,000 bridges in the U.S. were considered structurally deficient—over a quarter of a million bridges built between 1956 and 1975 would require major repairs or deck replacement in the near future. To ensure the best use of limited funds, wise engineering management is required.

Bridge failures need to be avoided at a predetermined cost. Load posting requirements are a step in the right direction.

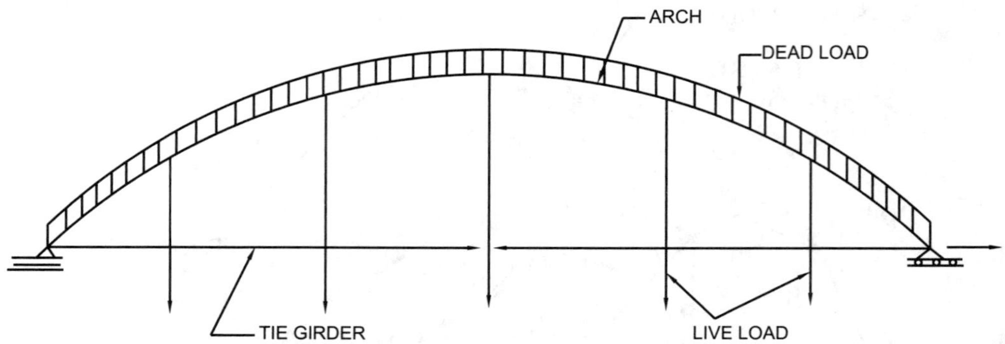

Figure 8.2 Illustration of applied forces causing tension in tie.

History of the National Bridge Inventory database: The U.S. Congress passed the Federal Highway Act of 1968 after the collapse of the Silver Bridge in West Virginia in 1967 which had resulted in the death of nearly 50 people. Following subsequent bridge failures, such as those of the I-95 Mianus River Bridge in 1983, the Schoharie Creek Bridge in 1987 in NY, and the Cypress Viaduct due to the California Loma Prieta earthquake in 1989, the inspection and rating procedures have undergone further changes.

National Bridge Inspection Standards (NBIS) were introduced in 1971, which established national policy regarding:

1. Inspection procedures.
2. Frequency of inspection.
3. Qualifications of personnel.

The following revisions were made to NBIS:

- Inspection frequency of two-year cycle must be maintained for bridges exceeding 20 feet in length
- Fracture critical members must be identified
- Underwater inspection procedures are required
- Team leader certification requirements were revised
- Inventory data for state bridges must be updated within 90 days of any change in load restrictions.

Responsible agencies for developing the Bridge Management System (BMS): In the U.S., the BMS is developed at the highest echelon by:

1. AASHTO.
2. FHWA.

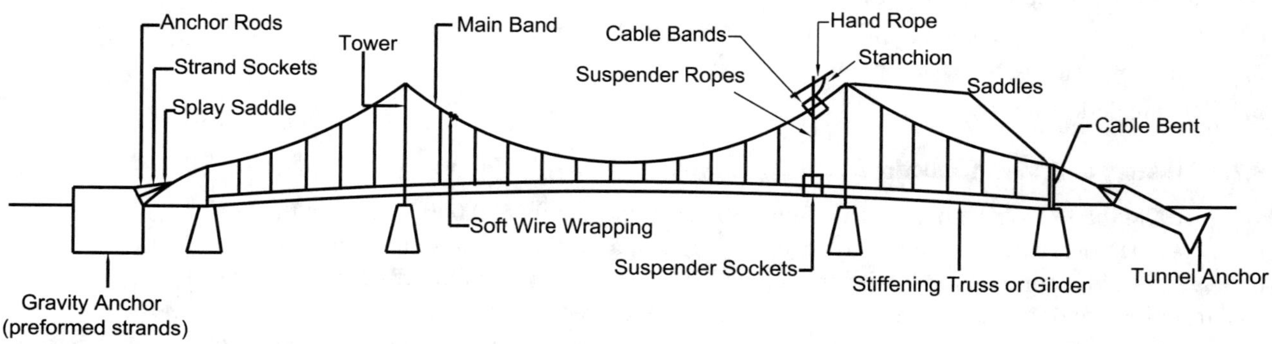

Figure 8.3 Identification of primary structural parts of a cable suspension bridge.

3. States.
4. Counties.

The organization, inspection, and diagnostic design teams: According to the National Bridge Inspection Program (NBIP), each state is responsible for the inspection of all public highway bridges within the state (except for those owned by the federal government or that are tribally owned).

8.2.5 FHWA and AASHTO Procedures

Each subject is covered by codes of practice from FHWA or the state agency. The following steps are generally needed:

1. Inspection.
2. Diagnostic testing.
3. Applying maintenance policy principles.
4. Evaluating types of distress and the NBIS rating.
5. Preparing an inspection report recommending types of repairs and cost estimates.
6. Performing emergency repairs.
7. Structural evaluation and load capacity evaluation, diagnostic design, and preparing construction drawings.
8. Using advanced inspection techniques.
9. Securing or obtaining the highway agency's approval of funds for rehabilitation work.
10. Based on evaluations, carrying out substructure and superstructure repairs being carried out as required.

The vast majority of inspections are done by state employees or by certified inspectors employed by consultants under contract to a state DOT. FHWA inspectors do, at times, conduct audit inspections to assure that states are complying with bridge inspection requirements.

FHWA also provides onsite engineering expertise in the post examination of reasons for a catastrophic bridge failure. The most common onsite inspection is a visual inspection by trained inspectors. One of them must meet the requirements of a team leader. Damage and special inspections do not require the presence of a team leader.

References on bridge inspection:

1. *Bridge Inspector's Training Manual,* 1970
2. *Manual for Maintenance Inspection of Bridges,* 1970
3. *Recording and Coding Guide for the Structure Inventory and Appraisal of the Nation's Bridges,* Report No. FHWA-PD-96-001, USDOT, FHWA, 1995, Washington DC.

8.3 TYPES OF INSPECTIONS

8.3.1 Routine Inspections

Usually, visual inspection, nondestructive testing, and underwater inspection are carried out. Traffic should be routed through work areas by trained personnel with traffic control devices. Approaching motorists should be guided in a clear and positive manner.

The Federal Manual on Uniform Traffic Control Devices for Streets and Highways is generally used. OSHA regulations should be consulted. Access equipment and access vehicles including bucket trucks, cables, platforms, scaffolds, man lifts, and ladders may be required.

Special equipment for underwater inspection, nondestructive testing, and surveying equipment also may be required.

Personal and public safety during inspection: Field attire should be properly sized, include reflecting safety vests, and be suitable for the climate. Hard hats, climbing boots, leather gloves, safety goggles, and tool pouches, as well as life jackets over waterways are required for safety. Established safety procedures shall be followed.

8.3.2 Additional Inspections

Examples of inspections include:

1. Periodic/routine.
2. Interim/special to monitor a deficiency.
3. Damage caused by accidents or environment (Figure 8.4).
4. In-depth using nondestructive testing, etc.
5. Inventory/initial with load capacity ratings from the structural inventory and an appraisal sheet using the AASHTO sufficiency rating formula evaluating structural deficiency.

Engineering judgment alone shall not be used to determine the live load capacity of a bridge component where sufficient structural information is known to utilize a rational method of analysis and rating.

A bridge rating based upon engineering judgment should consider, but not be limited to, the following factors:

- Condition of the load carrying components
- Material properties of members
- Redundancy of load path
- Traffic characteristics: Number and size of trucks, loading, and projected traffic
- Performance of bridge under current traffic: Evidence of distress and evidence of excessive movement under load
- Bridge restrictions (past, current, and proposed).

When using engineering judgment for the main live-load-carrying members/components, ratings are to be determined for all of the bridge posting vehicles at the inventory and operating rating levels for the live load carrying component of the bridge. These vehicle ratings are needed for the posting evaluation.

However, for many older concrete or masonry structures (including slabs, beams, and arches), the structural components either cannot be measured (arches) or critical details (e.g., reinforcement details) are not known with sufficient confidence to evaluate through computations. For these structures, a rating based on engineering judgment by a qualified engineer familiar with the bridge and the factors listed above may be appropriate.

8.3.3 Inspection Data

Using inspection data, an engineer will be able to:

1. Predict deterioration.
2. Improve safety or serviceability.
3. Estimate savings and costs.
4. Optimize a program with limited funds.
5. Generate reports for state legislation.

Further, the inspection system can be used to:

1. Identify maintenance needs
2. Develop repair and rehabilitation strategies.
3. Plan functional improvements.
4. Suggest replacement.

8.4 BRIDGE COMPONENTS FOR INSPECTION AND STRUCTURAL EVALUATION

8.4.1 Superstructure Components

Inspection and evaluation procedures cover each of the multiple superstructure components:

Deck slabs:

1. Types of deck repairs.
2. Timber, concrete, steel, joints, drainage, lighting, and signs.

Timber bridges: Solid sawn, glulam

Concrete bridges:

1. Slab bridge, T-beam, reinforced and prestressed concrete girders, box girders.
2. Precast segmental, box culverts.

Steel bridges:

1. Fatigue and fracture.
2. Nondestructive testing.
3. Rolled steel and fabricated plate girders.
4. Multi-girder and through bridges.
5. Box girders, trusses, arches, and rigid frames.

Special bridges:

1. Suspension cable and cable-stayed.
2. Movable.
3. Pipe culverts.

Inspection and evaluation procedures will cover the bearings and multiple substructure components.

8.4.2 Superstructure Inspection

Structural deck ratings:

1. Sidewalk rating.
2. Rail/parapets rating.
3. Curbs rating.
4. Wearing surface rating.
5. Monolithic surface rating.

Scupper ratings:

1. Grate rating.
2. Inlet rating.
3. Primary and secondary girders rating.
4. Superstructure joint rating.
5. Superstructure paint rating.
6. Median rating.
7. Lighting rating.
8. Sign rating.
9. Utility rating.

Table 8.1 Action required for the inventory of superstructure members.

Superstructure Component	Activity
Deck	Clean/flush
Scupper/downspout	Clean/flush
Bearing seat	Clean/flush
Structural steel	Clean/flush
Deck joints	Reseal/repair
Compression seal deck joints	Repair/rehabilitate
Modular dam deck joint	Repair/rehabilitate
Steel dam deck joints	Repair/rehabilitate
Expansion dams	Repair/rehabilitate
Parapet mounted railing	Repair/replace
Deck mounted railing	Repair/replace
Pedestrian railing	Repair/replace
Median barrier	Repair/replace
Bearings	Lubricate/rehabilitate/retrofit
Bearing pedestal/seats	Reconstruct
Approach slab	Repair/replace
Bituminous wearing surface	Repair/replace
Timber deck	Repair/replace
Steel open grid deck	Repair/replace
Concrete deck	Repair/replace
Concrete sidewalk	Repair/replace
Concrete curb	Repair/replace
Parapet	Repair/replace
Scupper grate	Repair/replace
Drain scupper	Install
Timber stringer	Repair/replace
Other timber members	Repair/replace
Steel truss (FCM)	Repair/replace
Steel girders	Repair
Steel floor beams	Repair/replace
Steel diaphragm/bracing	Repair/replace
Prestressed concrete beam	Repair/replace
Reinforced concrete beam	Repair/replace
Reinforced concrete diaphragm	Repair/replace
Other reinforced concrete members	Repair/replace
Superstructure members	Spot painting
Superstructure members	Apply protective coating
Temporary bridge	Modular/construct

8.4.3 The Pontis System for Bridges

The earlier used SI&A sheet for field data input about each element of a bridge has now been replaced by a Pontis-based data input sheet. In many states, PONTIS data sheets are being used to help with input in a computer program. Otherwise the objectives of the method are the same. Old structural inventory and appraisal (SI&A) sheets and new PONTIS sheets:

1. SI&A sheets outline the inspected condition of a bridge, the ratings, and proposed improvements.
2. Based on these sheets, a rehabilitation report is prepared.
3. Many of the existing inspection reports are based on SI&A sheets, which had extensive data in a condensed format on a single page.

SI&A sheets include the following inspection related data:

1. Bridge name and location.
2. Identification.
3. Classification.
4. Structure data for load posting, type of service, structure type for main and approach spans, minimum vertical and lateral clearances, etc.
5. Proposed improvement costs in thousands and future ADT.
6. Inspection data and other inspection dates.
7. Condition ratings for deck, superstructure and substructure, channel and channel protection, culvert and approach condition, critical features such as fracture critical, underwater, or special inspections.
8. Inventory and operating rating.
9. Appraisal ratings, structural evaluation, deck geometry, vertical and lateral underclearance, bridge posting, and waterway adequacy.
10. Railroad items.
11. Programming.
12. Remarks.

8.4.4 PONTIS and BRIDGIT

Following the National Bridge Inventory rating, Pontis provides bridge substructure as a default element category definition. Specific elements can constitute a condition.

Elements can be user defined or drawn from a list of default element definitions. Piers, pier walls, and abutments of various materials (concrete, reinforced concrete, masonry, and timber) are included.

Condition states and actions for each element are provided. A reinforced concrete submerged pile (Element 227) in Condition State 3 (exposed steel) would require cleaning and patching (action item 41).

Pontis allows users to define condition units that are in different environments as parts of the same element and may have different conditions.

BRIDGIT is similar to Pontis in inventory and condition rating. BRIDGIT appears to offer some additional flexibility in bridge-specific condition data such as load rating.

User defined data items can be added to the inventory and in a given condition state, condition information is included as a percentage of a particular element.

Pontis consists primarily of commonly recognized (CoRe) elements and smart flags. In addition, it consists of any sub-elements, smart flags and special non-CoRe elements.

1. CoRe elements are commonly recognized nationwide. A CoRe element will have nationwide consistency regarding condition, units of measure, and relationship to NBI.

Examples of sub-elements are exterior girders, girder ends, hinges, and other paint systems.

2. Smart flags are necessary to track distresses not included in the CoRe condition that would lead to an accurate NBI translation. Examples of smart flags are fatigue, rust, deck cracking, soffit, settlement, scour, and traffic impact.
3. Non-CoRe elements are unique items that states can add to their Pontis database. Examples are rigid frames, diaphragms, lateral bracing, splice plates, slope protection, and tunnels.
4. Data changes include rating of up to 160 bridge elements in quantitative units, so that a major bridge element can be rated and subdivided into various condition states.

   Conventional rating software such as BAR7 and the latest Virtis® software using LRFR methods developed by AASHTO are reviewed. In the interim, BAR7 (developed by PennDOT) is still being used from SIA sheet field data by some states.
5. Advanced inspection techniques such as remote sensor monitoring and underwater inspection are reviewed.
6. It leads to rating scales for a deficient bridge: Sufficiency rating, seismic, and scour rating for over topping floods.
7. Each state must maintain an accurate and current inventory of the status and condition of all bridges. The inventory data is used by FHWA for its National Bridge Inventory which is submitted to the U.S. Congress for funding purposes. It reflects the condition of bridges by assigning condition codes. It includes a section on appraisal. A copy of a Pontis Data and Rehabilitation Report is included in this book's appendix.
8. In 1985, FHWA initiated a two-phase demonstration project:

   Phase 1 called for a review of existing state BMS practices.

   Phase 2 required a microcomputer tool that any state could use to manage its bridge inventory. The phase 2 tool was named Pontis.

   The two highway bridge management systems deal with the management of bridges on state highways and interstate highways.

An analytical tool is needed to utilize the data. Through a systematic procedure, a network level analysis and optimization of bridge inspection data can result. The network level must be examined rather than only the project level. The following revisions were made to NBIS:

- A two-year inspection frequency cycle for bridges exceeding 20 feet in length must be maintained.
- Fracture critical members must be identified.
- Underwater inspection procedures are required.
- Team leader certification requirements were revised.
- Inventory data for state bridges must be updated within 90 days of any changes in load restriction.

The Pontis System

Recently, the Pontis method using a network-level bridge management system was introduced. It is basically bridge inspection software developed by FHWA. Pontis, a Latin word meaning bridge, is a tool for storing and analyzing bridge condition data. The bridge engineer inputs the data into a computerized database system.

The database contains bridge condition data, traffic needs, accident data, maintenance, improvement and replacement costs, available funding, etc., from which a prioritized list of bridge needs can be produced that optimizes the limited funds available.

It incorporates dynamic, probabilistic models and a detailed bridge database to predict maintenance and improvement needs, recommends optimal policies, and schedules projects within budget and policy restraints. Pontis field inspection data can be input directly in a computerized database. With the Pontis system, more precise condition inspections will be required such as:

- Element level data inspection
- Environmental code (benign, low, moderate, and severe).

The NBI Translator converts Pontis data elements (using an FHWA computer program) to NBI. The new system can identify:

1. Maintenance needs.
2. Repairs and rehabilitation strategies.
3. Functional improvements.
4. Replacement options.

8.4.5 Inspection Frequency

1. With charging tolls on thousands of bridges, regular inspection has become an industry and is a requirement for asset management.
2. Maintenance principles are to some extent similar to those for vehicle maintenance. As automobile technology is required for vehicle maintenance, competence in structural engineering is required for bridge maintenance.
3. Inspection and monitoring play an important role in estimating the capacity of a bridge to carry maximum live load safely. NBIS and FHWA coding guides are used. A bridge is as strong as its weakest component. To prevent any unexpected failure, bridge performance needs to be rated, for example, for fatigue resulting from live load reversals.
4. Inspection frequency depends upon design details, materials, age, intensity of loading, importance of the bridge, and its fitness to resist extreme events. The Inspection Procedures Review Committee of ASCE-AASHTO recommends a tiered approach that alternates between in-depth inspections and less comprehensive inspections.

   States must identify bridges that require less than a 24-month inspection frequency.
5. The following list is intended as a guide for identifying classes of bridges that, in general, would not be considered for routine inspection at intervals longer than two years:
   - Bridges with any condition rating of 5 or less
   - Bridges that have inventory ratings less than the state's legal load
   - Structures with spans greater than 100 ft in length

Table 8.2 Frequency of required inspections.

Type of Inspection	Frequency of Inspections
Bridges with weight restrictions	Once every year
Most bridges	Once every two years
Frequency of underwater inspection	Generally 60 months, but may be increased to 72 months
Underwater inspections for river bridges	More frequent due to flood conditions
Fracture critical (non-redundant) bridges	Less than two years
Bridges subjected to earthquake, major flood, or any other potentially damaging event	Should immediately receive a damage inspection
With spans less than 20 feet and culverts	Every four years
European practice	Almost every six years

- Structures without load path redundancy
- Structures that are very susceptible to vehicular damage, e.g., structures with vertical over clearances or underclearances less than 14 ft, narrow through or pony trusses
- Uncommon or unusual designs or designs where there is little performance history, such as segmental, cable-stayed, etc.

6. Fracture critical members, distressed members, and underwater members may require different types and frequencies of inspections.
7. Currently, very old and deteriorating bridges with fracture critical members are considered at par with new bridges for frequency of inspections. An optimum frequency approach must be adopted.

8.4.6 In-Depth Inspections

1. The goal of a bridge inspection is to collect enough data to make an informed decision about the scope of the project. Other criteria, such as physical characteristics, capacity demands, hydraulic adequacy, and required maintenance of traffic could dictate decisions on the scope of work for a bridge project, regardless of what the inspection data provides.
2. Due to the variation in the types of problems encountered, the designer shall perform a subsequent in-depth inspection of the structure to identify and confirm the defects that exist, and develop a solution that is unique to the problems found. This field inspection should include color photographs and sketches showing pertinent details and field verified dimensions.
3. The purpose of the inspections is to identify, closely examine, and record levels and areas of deterioration of all structural and non-structural substructure elements in order to develop repair recommendations and details. This effort also includes correlating probing measurements taken near the pier edges with the previous substructure inspections.
4. In-depth bridge inspections are done to assist in making rehabilitation versus replacement decisions and to assist designers in progressing bridge rehabilitation projects. In general, an in-depth bridge inspection is a detailed inspection of an entire bridge that can include both destructive and non-destructive testing. It is more complete than a general inspection and results can be used to satisfy Uniform Code of Bridge Inspection requirements for general inspection.
5. The code requires that in-depth inspections be done in accordance with the specifications for in-depth bridge inspection. A professional engineer should review the specifications for applicability to a particular project and, if necessary, develop modifications in the form of an addendum.
6. The designer should consider all factors when determining which bridge elements do not need to be inspected. As an example, there is no need to inspect the girders and deck on a bridge with concrete T-beams that cannot be retrofitted or rate them when the project objective is to increase the capacity of the structure. In this case, the scope of the project should be at least a superstructure replacement because the concrete T-beams cannot be sufficiently strengthened to accommodate the project objectives. However, it is important to get expert interpretation of the existing conditions and their influence on each element of the structure before eliminating them from the rehabilitation inspection.
7. It is imperative that an in-depth, hands-on inspection is done by the design agency preparing the repair or rehabilitation plans to determine the extent of structural steel and concrete repairs. Large quantity and cost overruns result when this inspection is not adequately performed, resulting in substantial delays to completion of the project.
8. Every attempt shall be made to prepare plans that reflect the actual conditions in the field. All pertinent dimensions shall be field verified or field measured by the designer and incorporated into the plans. It is not permissible to take dimensions directly from old plans without checking them in the field because deviations from plans are common.

9. The engineer will furnish all equipment and inspection rigging necessary to perform the inspections and will perform a thorough, detailed NBIS in-depth above and below water inspection of the substructure elements. Inspectors will also utilize sounding and probing instruments, along with other standard inspector's equipment during the inspection.
10. Prior to the NBIS substructure inspection, the engineer will review available bridge plans and previous inspection reports of the structures included in the contract. Existing substructure deterioration and critical locations of section loss and possible areas of priority repairs will be highlighted. Boat access locations, current river conditions, and precipitation forecasts will also be reviewed in order to finalize the type of equipment needed and establish the sequence of inspection.
11. The engineer will resolve sensitive MPT situations, such as full or partial detour, to minimize the disruption to bridge traffic. Access may be made to the majority of substructure units from below deck, using amphibious equipment launched from the river banks.
12. All inspection will be in compliance with OSHA and DEP regulations, and FHWA and NBIS inspections. Inspectors will perform in-depth NBIS inspections of all above-water substructure components, and record their findings on field note templates that incorporate all possible relevant element conditions. In order to perform efficiently and meet the schedule, inspection team leaders will scan and e-mail PDF files of field notes, sketches, and photos to the office so that work may proceed on the inspection report concurrently with the inspection.
13. A complete report will be prepared and will include field notes, photographs of general conditions, observed deterioration of substructure elements, bridge substructure sketches with highlighted details and locations of noted deterioration, and a conclusion with repair recommendations and cost estimates.

8.5 SUBSTRUCTURE COMPONENTS

8.5.1 Bridge Bearings

1. Common types of transverse restraint include:
   - Anchor bolts
   - Keeper bars
   - Shear keys.
2. The following types of bearings may offer transverse restraint:
   - Substructure with concrete shear key
   - Elastomeric pad with bearing plates and anchor bolts
   - Elastomeric pad with center pin
   - Sliding or multi-rotational bearings with guide bars and four or more girders.
3. Bearing rating:
   - Pedestal rating
   - Top of cap rating
   - Stem rating
   - Backwall rating
   - Cap beam rating
   - Column rating
   - Footing rating
   - Pile rating.
4. High seismic forces are generated in the restraints during a seismic event. The least transverse restraint is offered by:
   - Roller bearings
   - Rocker bearings: Rocker bearings are most vulnerable.

- Elastomeric bearings
- Sliding
- Sliding or multi-rotational bearings without guide bars and three or less girders.

8.5.2 Substructure Items

Substructure items include:

1. Abutments
2. Wingwalls
3. Piers
4. Noise walls
5. Earth retaining walls, modular and cast-in-place.

Table 8.3 Action required for the inventory of substructure members.

Substructure Component	Activity: Remove/Construct/ Repair/Replace/Protective Coating/Painting, etc.)
Abutment headwall/head block	Repair/replace
Stem	Repair
Pier	Repair
Wingwalls	Repair/replace
Footings	Underpin
Abutment slope protection	Repair/replace
Masonry	Repoint
Piles	Apply protective coating
Substructure members	Spot painting
Substructure members	Apply protective coating
Scour countermeasures	
Substructure Component	**Activity (Repair/Remove/Construct, etc.)**
Streambed paving	Repair/construct
Rock protection	Repair/construct
Stream deflectors	Repair/construct
Scour hole	Repair
Debris	Remove
Vegetation	Remove

Table 8.4 Action required for the inventory of culvert members.

Culvert Culvert Component	Activity (Repair/Replace)
Culvert headwall	Repair/replace
Culvert barrel	Repair
Culvert apron	Repair/replace
Culvert wingwalls	Repair/replace
Slab box/culvert	Replace

8.6 FHWA CONDITION AND SUFFICIENCY RATINGS

8.6.1 Objectives

The objectives of bridge ratings are different from the objectives of new bridge design and these differences result in different limit states and reliability requirements.

FHWA uses inventory data to rate the bridge as:

1. Non-deficient (safely carries design traffic loads).
2. Structurally deficient (lighter load posting is required).
3. Functionally obsolete (does not meet the highway criteria).

As defined earlier, a structurally deficient bridge is one:

1. Whose design or existing condition has impacted its ability to adequately carry its intended traffic loads.
2. That has a span > 20 ft and has not had major reconstruction in the past 10 years?

A functionally obsolete bridge is one in which there is reduced ability to adequately meet the traffic needs and is below the accepted design standards for:

1. The deck geometry.
2. Load carrying capacity.
3. Clearance.
4. Approach roadway alignment.

The LRFR manual proposes a system for bridge load rating and permit evaluation based on a uniform reliability index (where possible) that uses site-specific information to narrow the uncertainty in some of the evaluation variables. Condition rating assessment and analysis is the first step in structural evaluation. It needs to be followed by other steps such as scour rating or seismic rating if applicable.

The National Bridge Inventory (NBI) is used for selection of bridges for federal aid highways.

Sufficiency rating (SR) is generated automatically using various bridge management software programs or is hand calculated using SR formulae.

- Bridges with SR < 80 are considered for selection
- Bridges with SR < 50 are considered for rehabilitation or replacement

8.6.2 Condition Ratings

1. Condition ratings are important to automated decisions on actions for maintenance, repair, or replacement of bridges. Bridge management systems use condition ratings to identify bridges that need repair now or are expected to need repair within a selected planning period.

   Condition states, the definitions of ratings, determine what is reported and what kinds of decisions can be supported. Damage, reported as poor condition ratings, is one factor that triggers repair or replacement actions for bridges. There are other factors such as functional inadequacies and vulnerability to sudden failure.

   Typically, a bridge abutment is designed to resist lateral movement and overturning created by soil pressure and settlement resulting from dead and live loads. The bridge abutment and its connection to the footing must resist moments and shear forces, and the footing must provide resistance to vertical, lateral, and overturning forces.

   Live loads slightly add to the vertical dead loads, but they also add to the resistance to overturning and sliding. Therefore, the bridge superstructure usually controls the load ratings.

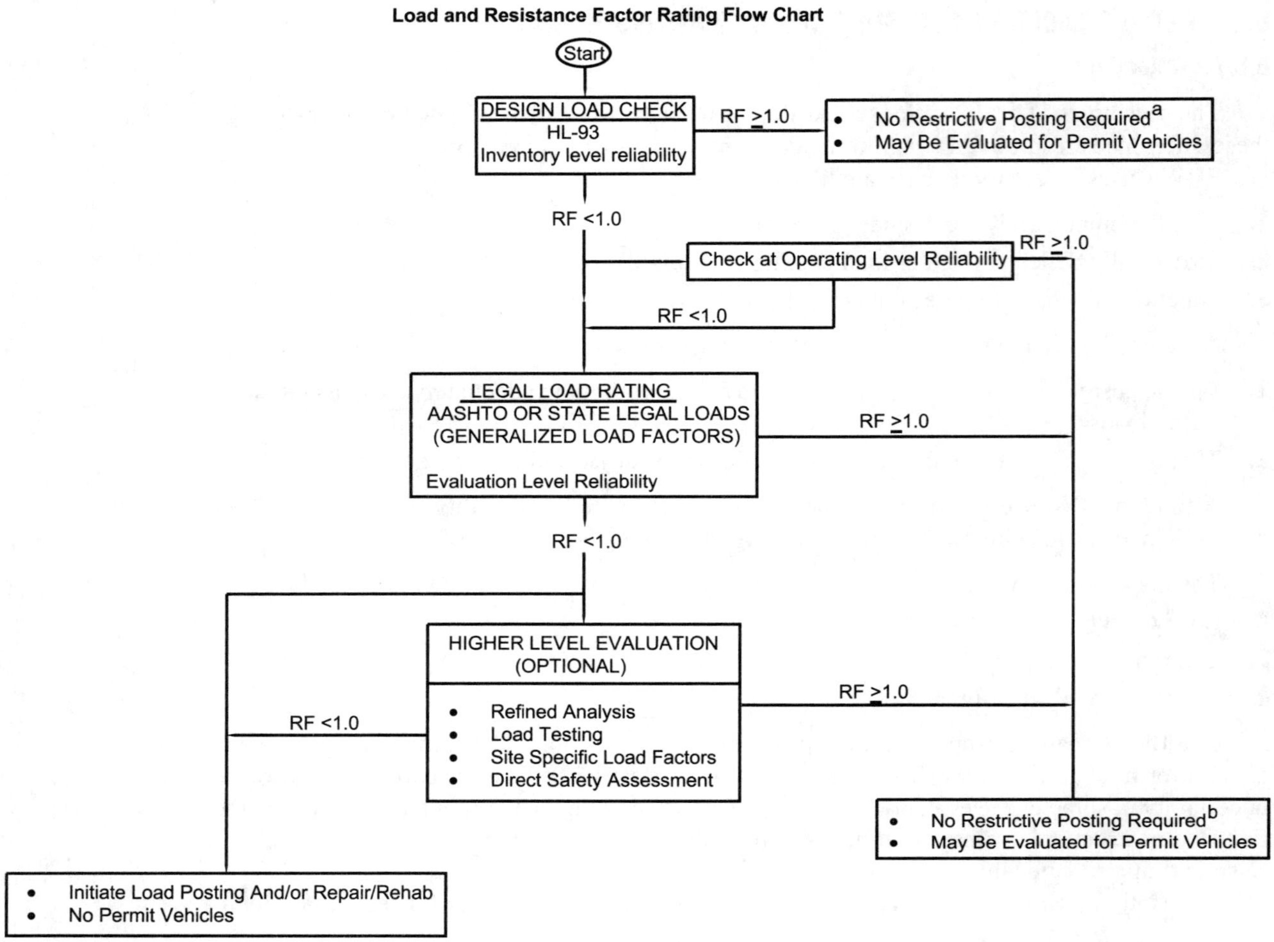

a. For AASHTO legal loads and state legal loads within the LRFD exclusion limits.
b. For AASHTO legal loads and state legal loads having minor variations from the AASHTO legal loads.

Figure 8.4 Flow diagram for evaluating load resistance factor rating (refer to AASHTO LRFR Manual).

2. A bridge abutment condition rating is governed by three factors:
   - The presence of excessive soil pressure caused by poor drainage
   - The condition of the abutment structure
   - The dimension and type of foundation (shallow, deep, or combined such as footing, pile, or pile cap on pile).
3. A bridge pier is designed to resist vertical settlement resulting from dead and live loads, and lateral movement and rotation caused by temperature change, friction, wind, water, and seismic loads. The bridge pier and its connection to the footing must resist moments, shear, and compressive forces. The footing must resist lateral, vertical, and rotational movements. The bridge pier condition rating is governed by the condition of the pier structure and the dimension and type of footing (shallow, deep, or combined).
4. Actions are needed for deficient bridges; that is for bridges where damage, function, or vulnerability is unacceptable. Each factor, damage, function, and vulnerability, can be expressed as a condition and reported as a condition rating. Bridge condition is simply the need for action.

5. These sufficiency conditions become definitions of condition states. The number of feasible actions determines the number of condition states. A system is achieved for:
   - Definition of condition states.
   - Selection of inspection and testing operations.
   - Coordinated examination of deficiencies.
6. The inspector will notify the appropriate agency immediately from the site if he discovers a potentially major finding during the condition assessment and subsequent inspections that could:
   - Reduce the load rating capacity of the bridge as determined by previous inspections.
   - Require Priority 1 (emergency response) repairs.
   - Require vehicular or pedestrian traffic restrictions to be imposed.

   This will enable the agency staff to observe the condition as soon as practical after receiving notification regarding the damage or deterioration encountered while the necessary traffic control and special access equipment or rigging is available, Upon viewing and discussing the area(s) in question, the inspector and the agency engineer will jointly determine if any immediate corrective and/or remedial measures are warranted and the nature of such measures.
7. A condition rating is a value that represents an overall assessment of the condition of a bridge. It is a numerical value from 1.000 (poor) to 7.000 (excellent). The condition rating is computed to three decimal places using the ratings of the elements with whole number values assigned. The three decimal point accuracy is significant only for the purpose of "breaking ties" when listing bridges by rank order of condition rating.

The computation uses bridge elements considered most important for an overall condition appraisal. Each element is weighted in proportion to its relative importance. The condition of each element is multiplied by the assigned weight for that element, with the result divided by the sum of the weighted values, resulting in the condition rating for the bridge.

The following general condition ratings are used as a guide. The overall condition of items 58 (deck), 59 (superstructure), and 60 (substructure) are of main concern in the sufficiency calculation. They will be assigned an NBIS condition rating given in Table 8.5.

Table 8.5 Condition rating for deck, superstructure, and substructure.

Number	Condition Rating Description
9	Excellent condition
8	Very good condition—no problems noted
7	Good condition—some minor problems
6	Satisfactory condition—structural elements show some minor deterioration
5	Fair condition—all primary elements are sound, but may have minor section loss, cracking, or spalling
4	Poor condition—advanced section loss, deterioration, or spalling of primary structural elements.
3	Serious condition—loss of section, deterioration, spalling, or scour have seriously affected primary components, and local failures are possible
2	Critical condition—advanced deterioration of primary structural elements—fatigue cracks in steel or shear cracks in concrete
1	Imminent failure condition—major deterioration or section loss present in critical structural components; bridge is closed to traffic
0	Failed condition—out of service, beyond corrective action

If any of these items are coded 4 or less, the structure is classified as structurally deficient.

Subpart G—Discretionary Bridge Candidate Rating Factor

8. In 1983, the FHWA described a rating factor used as part of a selection process of allocation of discretionary bridge funds made available to the Secretary of Transportation.

According to some of the provisions of § 650.703 Eligible Projects:

- Deficient highway bridges on federal aid highway system roads may be eligible for allocation of discretionary bridge funds to the same extent as they are for bridge funds apportioned under 23 U.S.C. 144, provided that the total project cost for a discretionary bridge candidate is at least $10 million.
- After November 14, 2002, only candidate bridges not previously selected with a computed rating factor of 100 or less and ready to begin construction in the fiscal year in which funds are available for obligation are eligible for consideration.

FHWA § 650.707 Rating factor: The following formula is to be used in the selection process for ranking discretionary bridge candidates.

$$\text{Rating Factor (RF)} = \frac{\text{SR TPC}}{\text{N ADT}'} \times \left[1 + \frac{\text{Unobligated HBRRP Balance}}{\text{Total HBRRP Funds Received}}\right]$$

The lower the rating factor, the higher the priority for selection and funding.

The terms in the rating factor are defined as follows:

SR is sufficiency rating computed as illustrated in Appendix A of the *Recording and Coding Guide for the Structure Inventory and Appraisal of the Nation's Bridges,* USDOT/FHWA (latest edition). (If SR is less than 1.0, use SR = 1.0).

ADT is average daily traffic in thousands, taking the most current value from the national bridge inventory data.

ADTT is average daily truck traffic in thousands (pickup trucks and light delivery trucks not included).

For load posted bridges, the ADTT furnished should be that which would use the bridge if traffic were not restricted. The ADTT should be the annual average volume, not peak or seasonal;

ADT′ is ADT plus ADTT.

N is National Highway System status. N = 1 if it is not on the National Highway System; N = 1.5 if bridge carries a National Highway System road;

The last term of the rating factor expression includes the state's unobligated balance of funds received under 23 U.S.C. 144.

TPC is total project cost in millions of dollars.

HBRRP is Highway Bridge Replacement and Rehabilitation Program.

Minimum sufficiency rating used will be 1.0. If the computed sufficiency rating for a candidate bridge is less than 1.0, use 1.0 in the rating factor formula.

If the unobligated balance of HBRRP funds for the state is less than $10 million, the HBRRP modifier is 1.0.

Special considerations: The selection process for new discretionary bridge projects will be based upon the rating factor priority ranking. Special consideration will be given to bridges that are closed to all traffic or that have a load restriction of less than 10 tons.

8.6.3 Sufficiency Rating (SR)

FHWA uses a sufficiency rating to provide an overall assessment of a bridge's condition. It is a quantitative measure of the degree of "sufficiency" of a bridge, a numeric value which indicates a bridge's relative ability to serve its intended function.

SR is calculated from an AASHTO formula, which evaluates the factors indicative of bridge competence. A sufficient bridge has a 100 rating, and an insufficient bridge has zero. A number is assigned from 0.0 (poor) to 100.0 (excellent) to represent both structural and functional adequacies. A fairly complex formula is used and is described in FHWA's *Recording and Coding Guide for Structure Inventory and Appraisal of the Nation's Bridges,* December 1995.

1. Structural adequacy and safety—S1 (bridge superstructure and substructure, 55 percent maximum rating).
2. Serviceability and functional obsolescence—S2 (deck condition, drainage, under clearances, narrow lane width, 30 percent maximum rating)
3. Essentiality for public use—S3 (loss of access to business due to closure or posting)
4. Special reductions—S4 (loss of essential service—fire, police etc.)
5. When SR > 50 but < 80, rehabilitation is required.
   SR < 50, replacement is required.

Sufficiency rating (S1 + S2 + S3 − S4) = 100 maximum.

8.6.4 Vulnerability Evaluation Codes

The following codes may be used for vulnerability evaluation:

Table 8.6 Vulnerability evaluation codes.

Cause	Code	Rating Category	Classification	Failure Vulnerability
Hydraulic	HYD	Safety priority - 1	High - H	Structural damage - 1
Overload	OVL	Safety program - 2	Low - L	Partial collapse - 3
Collision	COL	Capital program - 3	Medium - M	Catastrophic - 5
Seismic	SMC	Inspection program - 4	Not vulnerable - N	
Steel	STL	No action - 5		

If the inventory LRFR rating (LRFD design strength and service combinations) is less than 1.0, a more refined analysis must be performed to accurately load rate the structure. Report the LRFR inventory and operational rating factors for the HL-93 design loading. Report the LRFR operational rating factors for both the permit and HS-32 vehicles. Use no posting avoidance techniques for these calculations and report the data for review.

Before a variance or exception is approved for a bridge widening, use LRFR Appendix D.6.1 to determine the LFD inventory rating factor. If any rating for design and legal vehicles is below 1.0, replacement is preferred, and a variance or exception is required for rehabilitation.

The state maintenance office should be contacted regarding any calculated operating ratings, service, or strength less than 1.67 for the HL-93 design truck on any state road to assure appropriate permitting operations.

Although in the original design the bridge is designed for certain theoretical live loads, it needs to be rated for the actual live load that the bridge can support. This evaluation may show a reduction in strength due to fatigue, corrosion, and an increase in the magnitude of truck load.

8.7 DUTIES, QUALIFICATIONS, AND TRAINING OF INSPECTORS

8.7.1 Duties and Qualifications of the Inspection Team

The NBIS set minimum requirements for program managers, team leaders, and various inspection personnel. The following requirements are for routine inspections and do not consider the complexity of a bridge. Requirements for complex bridges are developed at the state level.

1. The team must have a program manager and a team leader, whose qualifications and experience are laid down in the AASHTO *Manual for Condition Evaluation of Bridges.* Inspectors must be well-versed in structural behavior and must be provided with an array of assessment techniques so they can accurately evaluate the structure. Their primary duties are:
   - Planning and preparing for inspection, such as determining the type of inspection, selecting the team, defining activities, developing a sequence and schedule, arranging for traffic control and method of access, organizing equipment and tools, subcontracting, and reviewing safety precautions
   - Reviewing bridge files such as previous inspection reports, as-built plans, retrofit plans, geotechnical, utility and right-of-way plans, and preparing forms and sketches
   - Performing inspections, such as visual and physical examination and evaluation of bridge components by following procedures
   - Preparing reports such as completion of forms, objective written documentation, photos, references, evaluations, recommendations, and cost estimates
   - Identifying items for repair and maintenance.
2. The program manager shall be a registered professional engineer or have a minimum of 10 years experience in bridge inspection assignments and have completed a training course based on the Bridge Inspector's Reference Manual (BIRM). The team leader in charge of an inspection team shall have the qualifications specified for a program manager or have a minimum of five years experience in bridge inspection assignments and have completed a training course based on BIRM or possess a current certification as a Level III or Level IV bridge safety inspector under the NSPE program.
3. The qualification requirements would not change under the Pontis System.
4. An inspector's qualifications should match the complexity of the inspection. Hence, training may need to be upgraded to match the requirements. Retraining would ensure promoting uniform inspection practices and consistency in condition ratings for thousands of bridges throughout the U.S.

8.7.2 The Inspector's Responsibilities

The inspector's responsibilities are linked to the designer's role and supplement the designer's responsibilities of providing adequate safety factors and cost-effective designs. In particular, an inspector's role is to:

1. Maintain public safety by identifying defects.
2. Protect public investment by initiating timely action.
3. Participate in the bridge inspection program by following inspection procedures and frequency, ensuring qualifications, reporting, and preparing inventory.
4. Provide accurate bridge records through high-quality inspection.
5. Fulfill legal responsibilities in preparing legal documents.

The man-hours required for an inspection usually consist of the following.

1. Office preparation and gathering tools and equipment.
2. Acquiring plans, access permits, and previous reports.

3. Travel.
4. Field duration.
5. Report preparation.
6. Maintenance of state bridge inventory.

The need exists for continuous coordination and close collaboration between the bridge inspection and condition team and those responsible for maintenance and repair.

8.7.3 Inspection Access

Items included in access:

- Walking
- Step ladder
- Extension ladder

40 foot under bridge inspection unit
60 foot under bridge inspection unit

- Lightweight under bridge inspection unit

30 ft lift
30 − 90 ft lift
> 90 ft lift

- Rowboat
- Barge
- Diving
- Railroad flagman
- Railroad electrical
- Scaffolding
- Lane closure
- Shadow vehicle.

Items included for field repair: The contractor must furnish, erect, and move scaffolding and other appropriate equipment to permit the inspector the opportunity to closely observe all affected surfaces. This opportunity must be provided to the inspector during all phases of the work and continue for a period of at least 10 working days after any repair work has been completed.

8.7.4 Inspector Preparation of Report

Format of report

Project #______________________________ Date: ____________

Inspector: ____________________________________

1. Review of plans and specifications. ____________________

   Comments:

   ____

   ____

2. Field review of site, including surrounding area. ________________

   Comments:

   ____

   ____

3. Inspection of structure and existing coating. ______________________

   Comments:

   ____

   ____

4. Inventory and calibration of equipment. ______________________

   Comments:

   ____

   ____

5. Scheduling of pre-painting conference. ______________________

   Comments:

   ____

   ____

6. Erosion ratings:
   - Stream alignment rating.
   - Channel erosion rating.
   - Waterway opening rating.
   - . Bank protection rating.
   - Embankment rating.
7. Coding: Each span element is given a single digit numeric rating using the following rating scale:
   - Totally deteriorated, or in failed condition: Ratings between 1 and 3.
   - Serious deterioration, or not functioning as originally designed: Ratings between 3 and 5.
   - Minor deterioration, but functioning as originally designed: Ratings between 5 and 7.
   - New condition: No deterioration.
   - Condition and/or existence unknown.

8.8 OVERLOAD PREVENTION

8.8.1 Introduction

Load rating and strength analysis can be carried out based on field data. A load rating is a calculation of the weight-carrying capacity of a bridge and is critical to the bridge's safety.

A load rating is performed separately from the bridge inspection, but is based upon design capacities supplemented with data and observations of the bridge's physical condition provided by a bridge inspector.

Proper and regularly scheduled reviews of the calculations of all bridges' maximum safe load ratings are important because as a bridge ages, corrosion and decay can decrease its capacity to support vehicles. The current rating method is the LRFR strength method, while the one used for original design is likely to be based on ASD or LFD. Rating vehicles, load combinations, and applications of AASHTO LRFR methods described in Chapter 6 are important.

The preventive method consists of:

1. Identifying bridges at risk for overloading.
2. Data collection of overweight trucks approaching a bridge.

3. Installing weigh-in-motion technology.
4. Advising overweight trucks to follow a detour route.
5. Installing flashing static signs or variable message signs so that a truck driver has ample time to follow an alternate route. This situation may arise if a bridge has recently been posted for a lower capacity after routine structural evaluation and the occasional driver of that route is unaware of the new restriction.

Automated weigh-in scales generate truck weight data, which leads to an accurate evaluation of maximum legal and permit load in design. Equipment being used for this purpose includes videos, license plate readers, and weigh-in scales for axle weight and gross weight of trucks.

This process, employed at a selected location prior to a bridge, has prevented overload, large deflection, and overstress of bridges.

8.8.2 Audit of Weight Posting Procedures

Greater use of computerized bridge management systems is needed to improve states' bridge inspection programs and enhance the accuracy of bridge load ratings. The load rating, expressed in tons, serves as the basis for posting signs noting the vehicle weight limit restriction, which can be referred to more simply as the bridge's maximum weight limit.

Load rating is calculated at two design strength capacity levels: the operating, which is the higher level, and the inventory, which is the lower level.

In spite of the earlier audit outcome, the more recent collapse of the I-35 West Bridge in Minneapolis, Minnesota due to inadequate design of gusset plate connections, has resulted in public hue and cry for fixing widespread deficient bridges.

8.8.3 Quality Assurance and Quality Control

The following issues need to be addressed to ensure adequate quality assurance and quality control of bridge structures.

1. Following systematic quality control (QC) and quality assurance (QA) procedures (NBIS 23 CFR 650.313g).
2. Compiling data on deterioration rates of bridges and of specific components should be compiled.
3. Developing public relations guidelines for funding needs.
4. Researching and developing a model inspection manual that would make use of detailed photographs, illustrations, and detailed drawings. Such a manual can be incorporated into state and local inspection manuals.
5. Developing a central database of bridge deterioration information to indicate signs and symptoms of distress.
6. Incorporating inspectability guidelines (for example, providing easier access for inspection, chambers for bearing inspection, etc.) into design specifications for new bridges so that inspection tasks are made easier.
7. Increasing or reducing frequency of inspections based on risk assessment.
8. Increasing use of NDE technologies: Educating inspectors and owners on readily available NDE technologies without extensive funding.

8.9 SUPERSTRUCTURE LOAD CARRYING CAPACITY

8.9.1 LRFR and LRFD Application

Minimum load carrying capacity for all rehabilitated bridges shall be the same as for a new design using the LRFD method. Analysis should include a minimum 30 psf for future wearing surface. If existing beams do not adequately rate for shear using the current LRFD shear criteria, the beams should be rated using the criteria used for the original design. A note on the rating table should indicate which criteria were used in determining the shear rating.

All superstructure components must be checked for the remaining fatigue life. The remaining fatigue life must be at least as long as the expected life of the type of rehabilitation being considered. A procedure for evaluating the remaining fatigue life of in-service bridges is included in the *Manual for Condition Evaluation and Load and Resistance Factor Rating of Highway Bridges,* which was developed in NCHRP Project 12-46.

An inventory rating (Item 66) will result in a load level which can safely utilize an existing structure for an indefinite period of time.

Table 8.7 Rating analysis.

Code	Description
1	Load factor
2	Allowable stress
3	Load and resistance factor
4	Load testing
5	No rating analysis performed

For compact sections, $M_u = fy \times Z$
For non-compact braced sections, $M_u = fy \times S$
For non-compact unbraced sections, $M_u = M_r \times Rb$

Magnitude of rating vehicles: Refer to Chapter 6, Sections 6.3 and 6.4 for bending moments and shear forces. Examples of classified truck loads are:

H-20 (20 tons)
HS-20 (36 tons)
Type 3 (25 tons)
Type 3-S2 (36 tons)
Type 3-3 (40 tons)
Usually axle loads are considered.

The load resistance factor method (LRFD): The LRFD method provides only one rating factor corresponding to the strength limit case, whereas the load factor method has an inventory and operating rating for both a serviceability capacity and maximum strength capacity.

$$\Phi\, Rn = \gamma_d\, D + (RF)\, \gamma_l\, (L+I)$$

where Φ = Resistance factor 0.90 for concrete and 0.95 for steel.
γ_d = Dead load factor of 1.20
γ_l = Live load factor which varies between 1.3 to 1.8

8.10 CONDITION RATING FOR EXTREME EVENTS

8.10.1 Bridges on Rivers

Evaluation of the overall condition of substructures will be assigned the following NBIS condition rating in the absence of an in-depth scour evaluation:

Very good condition—no problems noted.
Good condition—some minor problems.
Satisfactory condition—structural elements show some minor deterioration.
Fair condition—all primary elements are sound, but may have minor section loss, cracking or spalling.

Poor condition—advance section loss, deterioration, or spalling of primary structural elements.

8.10.2 Scour Vulnerability Assessment (SVA)

1. The purpose of an SVA is to identify which bridges are stable for current conditions and which bridges are potentially scour critical. Scour assessment using existing bridge data and field observations will be performed based on procedures set forth in USGS Procedures for Scour Assessments to assess each bridge's vulnerability to scour and risk of failure due to scour.

A scour study will be based on scour vulnerability analysis, which is a 1988 mandatory requirement of FHWA for bridges on waterways.

2. Both boring information and grain size analysis would be needed for accurate determination of scour. Collection and processing of geomorphic, hydrologic, and hydraulic data for assessment of scour at bridges require borehole information for soil characteristics.

3. The scour sufficiency rating needs to be evaluated. A scour analysis is carried out to evaluate scour depths based on FHWA Publication HEC-18. Scour depths for long-term scour, contraction scour, and local scour are assumed to develop independently. A scour assessment rating is based on the type of bed material. For example, rock can be classified as erodible, highly erodible, and non-erodible.

4. A scour rating can be computed using a computer program. An approved software such as Scour Calculator Manual will be used for computing scour assessment rating.

8.10.3 Seismic Rating for Earthquakes

Seismic vulnerability rating; Screening inventory data: AASHTO Specifications are limited to regular bridges. Their susceptibility grouping may be classified as follows,

Group 1: High seismic vulnerability

Group 2: Moderate-high vulnerability

Group 3: Moderate-low vulnerability

Group 4: Low seismic vulnerability

There are six basic steps to the group assignment process.

Step 1—For a single span bridge, vulnerability is limited to bearings and connections at abutments, as described in steps 2 and 3.

For multi-span bridges, pier and foundation types affect vulnerability, as described in steps 4, 5, and 6.

Step 2—Integral abutment bridge is assigned to Group 4.

Step 3—Bridges with steel rocker bearings have a tendency to overturn and are assigned to Group 2. For abutment skew $>$ 30 degrees, a bridge is assigned to Group 2 regardless of bearing type.

Step 4—Continuous girder bridges have greater stability than single spans since vulnerability to unseating mode of collapse is less. Continuous girders may be examined according to step 5 below.

Step 5—Continuous girders supported on steel rocker bearings are assigned Group 2. For two or three continuous girders or trusses, a bridge has poor redundancy and little resistance to collapse if lateral resistance is lost at an edge. Such bridges are assigned to Group 2.

If piers are not reinforced, the bridge is assigned to Group 2.

If each pier is a single column, the bridge is assigned to Group 2.

If each pier is timber or steel pile bent, the bridge is assigned to Group 3.

For spread footings or pile footings, the bridge is assigned to Group 3. For spread footings on rock, the bridge is assigned to Group 4.

Step 6—If simply supported girders are supported on steel rocker bearings, the bridge is assigned to Group 1.

If skew $>$ 30 degree, the bridge is assigned to Group 1, regardless of bearing type.

For two or three continuous girders or trusses, the bridge has poor redundancy and little resistance to collapse if lateral resistance is lost at an edge. Such bridges are assigned to Group 1.

8.10.4 Seismic Ranking and Prioritizing

1. Retrofit priorities are defined in the FHWA Seismic Retrofitting Manual for Highway Bridges and will depend upon:

- Structural vulnerability (some components are more vulnerable than others, such as girder connections, bearings, seat width, piers, abutments, and soils).
- Seismic and geotechnical hazards.
- Importance factor.

2. Rating system consists of a quantitative seismic rating (bridge ranking) and is based on:

- Structural vulnerability
- Seismic hazard.

3. The qualitative part consists of an overall priority index consisting of importance factor, remaining useful life, non-seismic deficiencies, and redundancy.

Priority index P = f (R, importance, non-seismic, and other issues)

R = Rank based on structural vulnerability and seismicity

Ranking factor R is based on other parameters such as:

- Soil type
- Seat width
- Expansion joint
- Pier details.

Rank (R) = Vulnerability V (0 to 10) × Earthquake/seismic hazard rating E (0 to 10)

R = 0 to 100

The higher the score, the greater the need for a bridge to be retrofitted.

V = V1 or V2

V1 = Vulnerability based on connections, bearings, and seat widths

V2 = Vulnerability based on column, abutment, and soil liquefaction

= CVR + AVR + LVR $<$ 10

If V $>$ 5, the bridge has high vulnerability to collapse and needs retrofit.

Column vulnerability due to shear failure (CVR between 0 and 10):

CVR = Q − R (derived empirically during the San Fernando earthquake of 1971 and does not apply to tall and slender columns)

Q = 13 − 6(Lc /Ps Fb max)

Lc = Effective column length

Ps = % of column longitudinal steel area

F = Framing factor 1, 1.25, 1.5, or 2 according to number of columns and end conditions

bmax = max. transverse column dimension

R = Factor to reduce susceptibility to shear failure

= 1, 2, or 3 depending on acceleration coefficient, skew, redundancy, and steel grade.

4. Pier vulnerability due to flexure failure at column reinforcement splices:
(CVR between 0 and 10)

CVR = 7 for A < 0.4

CVR = 10 for A > 0.4 (when micro zoning is considered)

For piers with unknown reinforcement:

Q = 7 for pier height < 23 ft.

Q = 10 for pier height > 23 ft.

5. Column vulnerability due to foundation deficiencies (for single column bents on piles):

CVR = 5 for 0.4 < A

CVR = 10 for 0.5 < A

Abutment vulnerability, AVR:

Abutment failure is linked indirectly with settlement of approach fill such as for spill through abutments and those without wingwalls.

AVR = 0 if bridge is classified as SPC B.

AVR = 5 if fill > 6 inch.

AVR = 5 if bridge is classified as SPC D.

6. Liquefaction vulnerability rating, LVR:

Bridges with discontinuous superstructure and non-ductile supporting members have a higher vulnerability than continuous superstructures.

LVR = 10 for bridges subjected to severe liquefaction.

LVR = 5 for bridges subjected to moderate liquefaction.

LVR = 0 for bridges subjected to low liquefaction.

7. Bridge types which are exempt from ranking process (no remedial action is required):

- Bridge designed to seismic standards.
- Bridge has SPC A and is not a critical facility.
- For bridge with SPC B, only vulnerability rating for restrainers, support lengths, and LVR for certain sites is required.
- For bridge with SPC C or D, in addition to vulnerability rating for restrainers, support lengths, and LVR for certain sites, a vulnerability rating for column, abutment, and foundation is required.

If the V1 value is determined from a high bearing vulnerability, only the bearings need to be retrofitted.

If V1 < V2, other components would also need retrofit.

8. Seismic hazard rating, E:

E = 1.5 A S < 10

S = Site coefficient

N. Minimum seat width, N

For SPC A & B, N = (8 + 0.02L + 0.08H) (1 + 0.000125 S2) inch

For SPC C & D, N = (12 + 0.03L + 0.12H) (1 + 0.000125 S2) inch

H = Height of column, pier or abutment in feet

S = Skew angle at support

Simply supported spans with high skew and inadequate seat width are most vulnerable to overturning.

Continuous superstructures with integral abutments are most stable.

Flow diagrams for calculating vulnerability in transverse direction (VT) and in longitudinal direction (VL) need to be utilized.

8.10.5 Failure Rating

Substructure failures are caused by tilting or settlement of a pier or abutment. Shear and flexure are other modes of failure. Failure types are defined as:

Catastrophic failure rating, FR = 5

Partial collapse, FR = 3

Structural damage, FR = 1

8.11 STRUCTURAL HEALTH MONITORING (SHM) AS A RATING AND DIAGNOSTIC TOOL

8.11.1 Utilize Structural Health Monitoring Systems

1. Structural health monitoring systems are used to detect and monitor deficiencies on the major bridges. Structural health monitoring systems typically entail the installation of sensors at key locations on the structure being studied. The technologies vary among the systems; however, they possess the common feature of measuring desired physical properties.

These may include loads, stresses, strains, and differential movements, as well as chemical composition of the concrete and steel structural components.

Review and compare proprietary structural health monitoring systems offered by the specialized firms for their applicability, effectiveness, and installation and maintenance costs, and make recommendations for specific applications.

SHM is a practical tool to detect, measure, and record the field performance of a bridge. It is particularly useful for structurally deficient and historic bridges. The structure is an intelligent or smart system that is capable of sending information and providing warnings before any major failure. The dependence on bridge inspection teams and their numbers will be reduced in the future.

With the emergence of new technologies, structures can now be monitored remotely from a central monitoring station located several miles away from the field. The trend is to place sensors at several critical locations along the structure, and send structural information to a central station. Technologies cover smart sensors, wireless networks, data acquisition, damage identification and localization, model updating, safety evaluation and reliability forecasting, damage control, life cycle performance-based design, and codes of SHM.

SHM is a multi-disciplinary evaluation of the condition of a bridge. It serves as additional eyes and ears of the engineer. It involves:

- Practical rating techniques.
- Diagnostic health evaluation technology.

- Risk evaluation.
- Remote sensor technology.
- Data communication and processing.
- Influence of soil-structure interaction.

2. SHM uses one or more in situ sensing systems placed in or around a structure, providing real-time evaluation of its performance and ultimately preventing structural failure. When dealing with extreme load conditions such as earthquakes, floods, or hurricanes, SHM gives a sense of comfort against overload. Any changes in physical conditions can be monitored and timely repairs applied.

Dynamic performance can be studied by vibration data recorded on bridges from fast moving traffic, wind, flood, and earthquake. Changes in physical properties, including mass, stiffness, and damping of a bridge resulting in changes in dynamic modal characteristics such as frequency and mode shapes. During numerous seismic events, these changes are detected by recording measurements of vibration sensors distributed over girders.

3. SHM basic operation procedures consist of:

- Loading tests on a bridge using calibration trucks.
- Installation of sensors.
- Installation of modern surveillance equipment to continuously monitor effects of wind and temperature.
- Using a remote monitoring station to analyze the results. Average values and standard deviations can be displayed automatically.
- A field inspection can then focus on visual assessment of the identified spots which may otherwise be ignored. Any structural deficiency can be rectified through repair, retrofit, or strengthening.
- Results from field measurements based on mass, stiffness, and damping of a bridge member can also be correlated to those obtained by computation or by using rating software. Such a comparative study would reveal differences based on assumptions and limitations of theoretical approach.
- Health monitoring through instrumentation, ultrasonic scanning, infrared cameras, piezoelectric sensors, etc. are possible alternatives since they display a higher degree of reliability when observations are not possible on hidden locations.

8.11.2 Monitoring Plan

1. A monitoring plan is likely to include:

- Taking photos.
- Measuring bank and channel cross sections.
- Bed elevations.
- Measuring lateral migration.
- Measuring plant densities and species composition.
- Estimating fish use.
- Developing a database of photographic records of one or more constant points above and below flow depths for scour critical bridges.
- Measuring scour depths regularly.
- Identifying eroded areas around the footings after major floods.

2. Monitoring may involve developing a photographic record from upstream and downstream of scour critical bridges.

A monitoring and inspection program, which includes taking scour measurements and planning for the inspection, traffic closures, etc. should be developed. Scour measurements may be recorded at regular intervals.

3. The following methods may be planned in developing the monitoring and inspection program:

- The normal two-year inspection cycle with soundings for all bridges, where required.
- For bridges whose substructure foundations cannot be visually inspected, underwater inspections may be necessary. Underwater inspections should be planned once every four years. Refer to "Underwater Inspection and Evaluation of Bridges."
- In the absence of a scour analysis, inventory item coding, as determined by state procedures or NBIS criteria, shall be used to classify the scour critical nature of the bridge.
- Periodic inspections, especially after major floods or coastal storm surges.
- Adequate measures to restore the eroded areas around the footings after major floods should be identified.

4. Innovations in management methods: The following are new techniques that can be utilized in asset management:

- Application of new management methods and management software.
- In-depth inspection and diagnosis using SHM.
- LRFR computer-based interpretation of field data using Pontis.
- Use of GIS and imaging technologies and as-built modeling by laser scanning: This technology for CAD allows accurate as-built conditions to be scanned and converted to drawings. Tedious field measurements are avoided.
- Remote monitoring sensors: Users can maintain an asset inventory database, collect inspection data, keep maintenance records, generate inspection reports, and provide decision support.

8.11.3 Remote Health Monitoring (RHM)

1. *Smart bridge* technology can improve the way structures are designed, evaluated, and maintained. Ultimately, service levels are maintained or improved. The economic gap between the cost of effective preservation decisions and available financial resources is reduced or eliminated. The new methods help to repair bridge decks effectively. They check corrosion in reinforcing bars embedded in concrete, look for cracks in welded joints in steel connections, and provide data on sizes and depths of unknown foundations.

SHM uses one or more in situ sensing systems placed on the bridge and provides real-time evaluation of performance, ultimately preventing its failure. In situ sensing and monitoring utilizing electrical conductivity measurements are suggested as effective ways to measure the properties of the in-place concrete. The different types of sensors include point sensors (traditional strain gauges), recently developed optical fiber sensors, and non-contact sensors (remote sensing, photographic inspection) that measure fundamentally different material response. Processing and interpretation of data can be carried out from a number of sensors placed in parallel. Depending on the complexity of bridge behavior, a tailor-made SHM system can be designed. To establish accuracy, a benchmark comparison of various SHM systems is preferred. Health monitoring is followed by prognosis or condition evaluation leading to hazard mitigation.

2. Examples of measurements are:

- Stress/strain.
- Acoustic/ultrasonic.
- Electrical.
- Temperature.

3. The in situ measurement of properties includes:
   - Porosity.
   - Pore connectivity.
   - Water permeability.
   - Ion diffusivity.
   - Life cycle monitoring of moisture movement.
   - Ion (e.g., chloride) penetration inside the material.

The applications of the sensing system and the role of each sensor, the parameters measured, and principles of how each sensor operates need to be studied.

8.11.4 Types of Instrumentation

1. Types of fixed instrumentation include:
   - Sonar devices.
   - Sounding devices.
   - Buried electromechanical devices.
   - Tethered sensors.

2. The latest instrumentation techniques include:
   - Mounting optical extensometer on bearings during jacking operation for prestressing.
   - Optical strands on decks or prestressed girders to measure average strain.
   - Installing digital cameras, clinometers, and fiber optic sensors for recording changes in displacements and strains.
   - Point sensors or conventional strain gauges to measure behavioral trends of dynamic deflection, strain, and vibration of structural components. They can be installed on superstructure members where the highest deformations are noted. Modern sensors are optical fiber sensors.

3. The following data processing and interpretation methods need to be studied:
   - Non-contact systems include visual or photographic inspection, remote sensing, and NDE techniques.
   - Base isolation bearings technology: It leads to effective decoupling of a superstructure from a substructure. Bearing friction in non-isolation systems can be avoided. Non-isolation bearings affect the expected performance of bridges at low levels of seismic excitations. Isolation systems separating the superstructure from the substructure such as laminated rubber bearings, lead rubber bearings, high damping rubber bearings, and Teflon sliding bearings can be used.
   - Instrumentations for seismic resistance evaluation, such as an accelerograph, can be mounted on piers, pier caps, foundation tops, and pile tips.

8.11.5 Monitoring Bridges through Remote Sensors

1. The system is intended to allow agencies to monitor the safety of bridges continuously to detect long-term deterioration that may lead to structural collapse. Current inspection requirements are intermittent and sporadic because they require bridges only to be inspected and rated every couple of years. For the future, develop a system that can continuously monitor piers and warn of structural weaknesses.

2. The use of wireless and remote sensors enables the movements of bridges to be monitored around the clock. This is most desirable in flood situations. Motiev Blog Sensors, when installed on scour critical bridges, would minimize the possibility of sudden collapse and would serve as a warning for the bridge to be closed.

Most of the system identification process thus far is concentrated in three phases—data acquisition, structural identification, and damage detection. The decision-making phase is left to the bridge engineer.

3. In a recent study, Farshad Rajabipour and Jason Weiss linked health monitoring in concrete structures with durability performance simulations. Service life prediction models provide a useful and practical approach to estimating the durability that may be expected in concrete structures. For reliable performance predictions, these life cycle models require accurate material property inputs that describe the quality of the placed concrete.

Remote sensors are installed both on bridge components and under the foundations. This may supplement condition evaluation of two-year cycle inspections to some extent. For timely action of repairs and retrofit, the monitoring information from electronic devices needs to be accurate. The cost of equipment and additional engineering manpower would add to the cost of the project. Also, engineers need to be trained to interpret the results and implement any corrective action procedures.

4. An integrated system for monitoring the condition of a bridge utilizes the following:

- Advanced sensing.
- Microprocessing.
- Wireless communication.
- Damage diagnosis methods.

5. Specifically, the following issues need to be addressed:

- Development of modular wireless vibration sensing, data acquisition, and processing units.
- Development of advanced structural damage assessment procedures.
- Environmental effects on experimentally obtained modal parameters.

6. For sensor development in wireless technology, Micro Strain, Inc. has produced a micro data logging transceiver. The high-speed system enables wireless data collection from up to eight channels of sensor input. The sensors can be triggered to initiate data collection remotely or by any specified sensor exceeding a programmable threshold. Users can transmit data from 1000 unique sensors to one Web-based receiver, enabling massive amounts of data to be shared globally in real time. Such a system can save owners money when compared to installing and protecting electrical cable-based monitoring systems.

7. Germany's Federal Institute for Material Research and Testing introduced the "RoboSense" project to produce a robotic tool. The system includes equipment such as impulse radar, impact-echo, cover meter, and cameras for visual inspection. Two climbing bigfoot-type robots that use vacuum feet for adhesion have been developed for the experimental approach.

Working conditions include a temperature range of 0° to 50° Centigrade, 90 percent humidity, and wind speeds of up to 20 km/hour.

Fiber optic smart sensors are both surface-mounted to the concrete and bonded to the deck reinforcing bars during the construction phase of a bridge. Static and dynamic testing of the bridge is performed using loaded trucks. A three-dimensional analytical finite element model of the bridge is developed, and its results are compared to the field data. Various live load combinations, dynamic effects, and secondary loading effects are applied to the model. After some adjustments of the global boundary (restraint) conditions, the FEM model and the RHM system can be compared for correlation. The study helps in estimating the bridge behavior under heavy loads.

The technology developed under this work will enable practical, cost-effective, and reliable maintenance of bridge structures.

8.11.6 Nondestructive Evaluation (NDE) Methods for Bridge Decks

1. Structural damage will result in permanent changes in structural stiffness, distribution of stiffness, and relevant material properties. These changes may be detected by monitoring dynamic behavior of the structure. Because of the direct relationship of mass, damping, and stiffness of a multi-degree-of-freedom system to the natural frequencies, mode shapes, and modal damping values, these properties can be used for structural health monitoring. For non-destructive evaluation and damage detection, there is increased interest nowadays in the application of vibration techniques.
2. There are several methods in use for SHM of deck slabs. Use of ground penetrating radar (GPR) for investigating conditions of bridge deck deterioration is popular. In addition, overlay thickness, depth of rebar, debonding, and delamination can be estimated by GPR.

   Ground penetrating radar (GPR) is a geophysical imaging technique used for subsurface exploration and monitoring. GPR provides an ideal technique for concrete evaluation in that it has the highest resolution of any subsurface imaging and is far safer than other methods such as x-ray technology. It is widely used within the forensic, engineering, geological, mining, and archeological communities. Concrete evaluation studies utilizing GPR include the inspection of structural deck slabs, post-tensioned or conventionally reinforced slab-on-grade foundation systems, and retaining walls.
3. GPR is a widely used geophysical method that uses high frequency pulsed electromagnetic waves to acquire subsurface information (Figure 8.5). GPR maps geologic conditions such as depth to bedrock, depth to water table, and thickness of soil layer sediment strata on land and under freshwater bodies. It can also locate fractures and cavities in bedrock.
4. Monitoring bridge foundation soil: Geotechnical engineers can measure lateral movement, vertical movement, and pore pressures by applying instrumentation technology. During construction, they can better predict the factor of safety against slope instability. Soft, saturated clays and silts settle and can drag down piles or drilled shafts, creating damaging stresses and movements. The soils can also move horizontally, creating stress on structures and even failing completely. Sands and gravels provide more reliable soil conditions.

   When water moves slowly through soil, pore pressures (the water pressure within the soil) can increase rapidly, reducing the strength of the soil. Time allows pore pressures to drop, soils to strengthen, and construction of projects to occur in areas that would otherwise not be possible.
5. It allows engineers to track the vital signs of the underlying soils. By keeping contractors apprised of the condition of the soil, project schedules can be optimized.
6. Instrumentation can be used to ensure that enough time is given to allow soils to settle, pore pressures to drop, and lateral movements to be managed without unnecessary delays.
7. Trained engineers should carefully select the locations of each instrument to ensure data will be collected from the most critical locations and interpret data from the geotechnical instruments.

8.11.7 Other Applications

1. Other applications include detection of flaws in the bridge deck during inspections. Other methods are half-cell to estimate corrosion activity in rebar mesh:

   - Chain drag and hammer sounding are used to estimate delamination between concrete and rebar mesh.
   - A portable seismic pavement analyzer relies on pulse propagation to accurately determine delamination between concrete and rebar mesh.

- Pacometers are used to determine location, depth, diameter, and spacing of rebar in the bridge deck.
- Infrared thermography is used to obtain thermal images of the bridge surface, identifying damp areas that are not visible to the naked eye.

2. NDE inspection using thermography is based on imaging surface temperatures in order to infer subsurface defects (Figure 8.8). With no internal defect, heat flow through the concrete deck will be uniform. Heat induction through a material will be altered if a delamination is present.

- Current advancements in radiographic and computer technologies have greatly improved performance characteristics for industrial computed tomography (CT). Non-film radiography methods necessary for CT have improved significantly and personal computers and software are advancing at an unparalleled rate. By combining these advancing technologies, fully functional CT systems with high data acquisition rates can be achieved.
- A consequence of collecting image plane views of a specimen, as opposed to line scans, is that the total data acquisition time can be reduced by up to a factor of one hundred. The next generation of technology is based on flat panel x-ray detectors that have a number of important advantages over the scintillator-optics-CCD approach. The rapid enhancements of personal computers (PC) and PC hardware and software have also contributed to advancements in CT technology. Hardware improvements result in efficient data acquisition, fast data transfer, precision staging and real-time high-resolution displays. Software improvements result in better hardware control, efficient data processing, and functional graphical user interfaces.
- Rotary percussion: This NDT method allows for quicker inspection on vertical surfaces and overhead surfaces. Advantages are low equipment cost and quick testing. It is similar to chain drag testing used for horizontal surfaces.
- Though several types of instruments are available, the three primary tools are inclinometers, settlement cells, and piezometers. Inclinometers measure horizontal movement of soil. Settlement cells measure the settlement of the soil supporting the embankment. These instruments are installed near the original ground surface prior to placement of embankment fill. As embankment fill is placed over the instruments and the supporting soils settle, the settlement cells measure the vertical movement of the underlying soils. Piezometers measure the pore pressure within the soil. The reaction of the pore pressure to fill placement provides an indication of stability and expected foundation soil movement. Piezometers can show when pore pressures are dangerously high, which may halt construction. When pressures are low enough, construction can continue or even speed up. Piezometers and settlement cells can also be used to double check expected settlement.

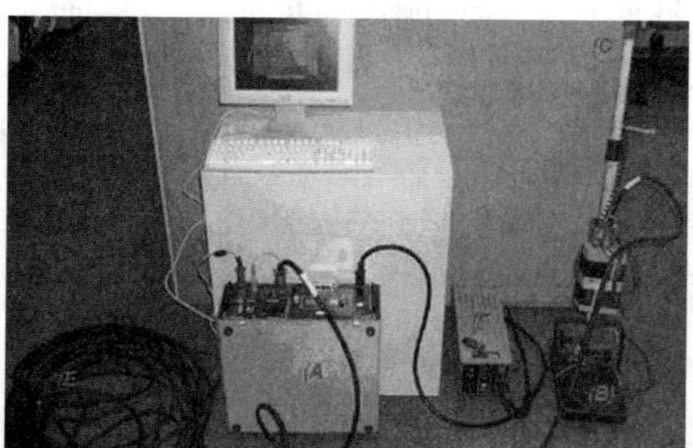

Figure 8.5 GPR equipment.

8.11.8 Acoustic Emissions

1. According to ASTM Standard E610-82 "acoustic emissions are transient elastic waves generated by rapid release of energy from localized sources within a material." Structural materials are known to release energy in the form of transient elastic waves when subjected to distress. Their detection and interpretation can be made by using the sound-print technique. This monitoring process can detect flaws and imperfections such as initiation and growth of fatigue cracks in steel, joints, connections, and welds. The real structural response of the bridge can be measured and studied, rather than assumed and accepted as modeled theoretical response.

   In addition, bond failure and fiber failures in composite materials can be detected. Acoustic emissions listen for sound from active defects. It is sensitive to detect defects relayed as acoustic emissions such as crack formations under overload condition.
2. Advanced signal processing and correlation to parametric measurements are used to separate noises generated by dynamic loading or crack growth.

   Developments of very sensitive acoustic sensors and availability of computing capacity now enable engineers to investigate the behavior of material under real loads rather than theoretical loads.
3. Sensors are now so sensitive that phenomena such as corrosion inside concrete, fatigue cracking, phase changes, and surface fretting can be detected and monitored through the science of acoustic emission.
4. The impact-echo method: The impact-echo method is a technique for flaw detection in concrete. The method overcomes many of the barriers associated with flaw detection in concrete based on ultrasonic methods. It is based on monitoring the surface motion resulting from a short-duration mechanical impact. One of the key features of the method is the transformation of the recorded time domain waveform of the surface motion into the frequency domain. The impact gives rise to modes of vibration and the frequency of these modes is related to the geometry of the test object and the presence of flaws. The ASTM standard governing the use of the impact-echo method for measuring the thickness of plate-like structures may be used.
5. The past few years have seen advancements in electrical engineering fields, such as microelectromechanical systems (MEMS) and micro sensors. Together with more mature technologies like wireless data communication, these advancements have begun to make embedded micro devices for use in a concrete structure a serious and cost-effective reality, and detection is possible with distributed, embedded micro devices.
6. In the area of local strength testing, where neither current NDE methods nor existing micro sensors have proven effective, novel MEMS testing devices are proposed. A concept is being developed for introducing a *smart aggregate* micro device for distributed, embedded concrete infrastructure monitoring.

8.11.9 NDT versus Acoustic Emissions

1. Conventional NDT measuring techniques used in the past:
   - Dye penetration
   - Visual inspection
   - Magnetic particle testing.
2. With acoustic emissions technology, structures can be monitored without the need to remove paint, gain access to difficult areas, or disrupt traffic.

   A network of microphones is used to detect acoustic waves. Signals are transmitted and stored on a data logger for pre-processing and post-processing. However, unwanted noise signals affect the accuracy of signals and need to be minimized or eliminated by pre-processing.

3. Stress concentrations in welded connections may lead to cracking, and acoustic emissions are being used to locate and identify defect.s
4. Through the vibration monitoring method, fatigue stress and active corrosion areas can be detected.
5. For timely action for repairs and retrofit, the monitoring information from electronic devices needs to be accurate. The costs of equipment and additional engineering manpower would add to the cost of the project. Also, engineers need to be trained to interpret the results and implement any corrective action procedures.

8.12 MONITORING USING NONDESTRUCTIVE EVALUATION (NDE) AND MODERN NONDESTRUCTIVE TESTING (NDT) METHODS

8.12.1 Introduction

Nondestructive testing methods use ultrasonic pulse velocity and full-scale NDT methods by replacing inspection and provide an accurate assessment for analysis.

Laser-based instrumentation for highway bridge applications: The FHWA Nondestructive Evaluation Validation Center (NDEVC) enhances the types of tools available to bridge inspectors. Systems are currently being developed that greatly increase the amount of data that can be gathered with conventional instrumentation.

One of these systems is a laser-based instrument that has been successfully used in a number of applications related to highway bridges. The applications include laboratory testing on a full-sized steel curved girder bridge, measurement of the as-built state of bridge abutments in a full-sized test bridge, load testing of in-service bridges, and tests related to fabrication of steel bridges.

Both field and laboratory inspection services for steel and concrete prestressed bridges, as well as NDT inspections and bridge paint inspections are possible. A complete sign structure inspection program provides weld procedure qualifications, bridge bearing pad testing, and numerous other testing services.

For strain sensing coating on conventional concrete, carbon fiber reinforced cement can be used. Strain sensing is similar to that of strain gauges. By attaching electrical contacts, resistance can be measured. The method is more reliable since strain gauges become detached from concrete surface.

1. The new methods help to repair bridge decks effectively. They check corrosion in reinforcing bars embedded in concrete, look for cracks in welded joints in steel connections, and provide data on the sizes and depths of unknown foundations.
2. Magnetostrictive sensor technology involves placing a coil around the cable or wire strand to be inspected. This locates wire breaks and corrosion. A pulse of electric current is sent through the coil. The coil generates an ultrasonic wave that propagates along the cable or strand. If the wave encounters a break or defect, part of the signal is reflected and picked up by a receiver.

The technique should prove valuable to bridge inspectors by replacing a process of unwrapping cables and removing suspenders during visual inspections, which can be costly and time consuming.

Dynamic load response: A parallel discovery in the performance evaluation is the impact of thermal effects on member response. The objective measure of structure performance relative to various live load combinations, dynamic load response, and thermal effects is a primary consideration in asset management.

8.12.2 Integration of Dynamic Testing Results with Bridge Management Systems (BMSs)

Commonly used bridge management systems such as PONTIS rely on subjective visual ratings to determine bridge element conditions. Recent research suggests that integrating NDE with visual ratings provides more consistent and richer bridge condition data.

A bridge management system must have a comprehensive bridge inventory that contains the number, type, size, and condition of each of the elements, the cost of maintenance and repair activities, and predictions for the future bridge conditions.

An FHWA research project investigated the possibility that, by measuring the dynamic response characteristics of a bridge substructure, the condition and safety of the substructure and its foundation type (shallow or deep) may be determined.

Determining bridge foundation conditions using dynamic response characteristics may be applied to quantify losses in foundation stiffness caused by earthquakes, scour, and impact events. Identifying bridge foundation type may be used to estimate bridge stability and vulnerability under dead and live load ratings, particularly for unknown bridge foundations.

Of several protocols evaluated, Hilbert-Huang Transforms (HHT) showed the most promise for structural damage diagnosis. Further work using the HHT method is recommended. The results of this study will be of interest to those who are involved in nondestructive bridge condition assessment.

Exploration of dynamic bridge substructure evaluation and monitoring systems by Olson (FHWA, 2005 Report # FHWA-RD-03-089) has shown that bridge foundation vertical stiffness is an appropriate indicator for the bridge condition evaluation, and it can be used to support BMSs in three ways:

- Inventory: Identification of a bridge foundation as either pile, pile with cap, or spread footing is possibly based on the bridge foundation vertical stiffness.
- Condition evaluation and monitoring: Changes in bridge foundation vertical stiffness over time and after major events such as earthquakes, floods, and ship impacts can be tied to the need for corrective action or closing or posting the bridge to protect users.
- Deterioration modeling: Historical data for a variety of bridge types help to assess the future costs and estimate the remaining service life of a bridge. According to the FHWA, dynamic bridge substructure evaluation and monitoring provide opportunities for improving bridge management systems. Measures of dynamic bridge foundation vertical stiffness or HHT results, or both, that identify downward frequency shifts indicating damage show:
- Monitoring bridge substructure conditions and assessing the remaining life of a bridge
- Assessing the effect of major events such as barge collisions, floods, and earthquakes on bridge substructure integrity
- Aiding the development of deterioration models for bridge substructures
- The role of NDE in BMSs suggests the desire to integrate dynamic testing results, including HHT results, with visual ratings data.

8.12.3 Case Studies Showing Improved Asset Management through RHM of Steel, Timber, and Concrete Materials

Installation of instrumentation to monitor erosion due to floods is discussed in the publication by Richardson E.V. & Lagasse, P.F., "Instrumentation for Measuring Scour at Bridge Piers and Abutments," Final Report, Phase III, NCHRP Project No. 21-3, Transportation Research Board, Washington, DC, 1994.

Instrumentation methods have been used successfully on a number of complex and newly constructed bridges:

1. Woodrow Wilson Bridge near Washington D.C. (for soil erosion monitoring).
2. I-95 Scot Road, NJ, integral abutment bridge (for live load and thermal stresses monitoring).

3. Hawk Falls Bridge, Beaver County, Pennsylvania (for monitoring stresses in truss members).
4. George Washington Bridge, NJ, managed by Port Authority of New York and New Jersey (use of magnetostrictive sensor technology).

The integration of an RHM system with a finite element model defines it as a "smart" bridge and empowers the client to make objective asset management decisions based on actual bridge performance.

The case study of a smart bridge is presented:

1. Case Study of Hawk Falls Bridge located in Beaver County, PA: This is a continuous, 680-foot-long, riveted steel deck truss structure built in the 1950s is owned by the Pennsylvania Turnpike Commission. To compliment this data, a 3-D finite element model (FEM) of the structure using MIDAS/Civil (Modeling, Integrated Design and Analysis Software for Civil Structures), a general purpose finite element software developed by MIDAS Information Technology Co., was used.

   Validation of the FEM model using the data from these sensors ensures accuracy in the analytical assessment of other bridge members. The performance of any component of the bridge can be modeled in detail and evaluated objectively. Coupling of the RHM system data stream with the FEM model resulted in Pennsylvania's first smart bridge. Smart bridge technology can improve the way bridges are designed, evaluated, and maintained.

   Stress wave timer: This NDT method offers the ability to determine the presence of internal decay in bridge members. It is especially useful for glulam or thick timbers where hammer sounding is not effective.

   Timber testing resistograph: This instrument drills a small 1-mm diameter hole in wood and it measures resistance output on wax paper. Decayed wood offers less resistance. A more expensive model offers PDF file output.

   Sound wave reflector: Impact generated sound waves are applicable to concrete and masonry. Impact waves that propagate through the structure are reflected by internal flaws and external surfaces. A piezoelectric crystal in a transducer produces a voltage proportional to displacement.

   This method can be used to determine the location and extent of cracks, delaminations, voids, honey combing, and debonding in reinforced and prestressed concrete.
2. Case study of smart bridge technology on I-35W replacement bridge, Minneapolis:

   The failure of the bridge (presumably due to failure of gusset plates and from additional dead load of storage material for resurfacing) cost almost $250 million. But the silver lining is the use of redundant steel girders and smart bridge technology to detect small problems before they turn into big problems in the future.

   A system of over 300 sensors and cameras providing traffic data including speeds, accidents, stalls, and traffic jams are embedded in concrete to generate an extensive record of heavy moving traffic and extreme Minnesota climatic conditions. In addition, some sensors will be used to prevent icing of the bridge deck. Additional sensors act as security sensors. During construction, sensors were placed in fresh concrete to monitor the curing process. Additional technology used on the new bridge includes:

   - Strain gauges to detect stress changes
   - Accelerometers for vibration study
   - Linear potentiometers at expansion joints and bearings to measure how the bridge expands and contracts
   - Chloride penetration sensors in the bridge deck for corrosion study from deicing salts. Sacrificial steel bars are placed at different depths to see the degree of corrosion and the need for deck replacement.
   - The data will feed into a bank of computers in a control room near the bridge and will be recorded and downloaded for analysis.

- The success of smart bridge technology on the I-35 bridge will promote the use of such monitoring methods on other bridges.

The advantage of using a smart bridge is to detect small structural problems through continued monitoring using remote sensors before they turn into major issues. The new bridge uses four concrete box girders and is built with redundant systems so that if one part fails, the bridge will not collapse. The old bridge was fracture critical, which means that the failure of a few structural elements would bring down the entire bridge. According to a report published by the Minneapolis DOT, the following innovative methods will be used:

- A record number of 323 embedded sensors will generate an extensive record of stresses and strains under live loads. Strain gauges and accelerometers measure strains and vibrations. Linear potentiometers will be located at the expansion joints and bearings to record response of extreme weather.
- Chloride penetration sensors in the deck slab will monitor corrosion of concrete due to deicing salt penetration. Sacrificial steel bars are implanted at various depths to indicate rate of corrosion from deicing salts.
- A system of cameras and sensors will feed data on traffic flow, speed, and accidents to a traffic management center. Other senors will automatically activate the anti-icing system in cold weather. Security sensors will detect intruders trying to access doors of box girders.

The data will be fed into a bank of computers in a control room located near the bridge. Such data will be downloaded for analysis by the highway agency. The long-term behavior of such bridge types will be understood much better than ever before. The calibration of sensors was possible by using moving sand trucks of known weights. Eight 25-ton trucks were used. Deep box girders would enable upgrading of the monitoring system from inside the box girders if required in the future.

8.12.4 Use of Optic Sensors

1. Long gage-length interferometric optical sensors for condition assessment in bridge structures: Optical sensor technology is a potentially important element in the development of a bridge monitoring system. Long gage-length sensors are expected to be more effective in detecting changes in global dynamic response, effectively integrating the response along predetermined lengths of the structure.

For detecting changes in bridge structure, dynamic response is used in conjunction with pattern recognition techniques. Research has utilized the inherent characteristics of fiber-optic cable to develop a variable gage-length interferometric sensor. Sensor location is critical when using point sensors.

2. Fiber optic distributed crack sensors for concrete structures: A distributed fiber optic sensor is developed for embedment in concrete structures. The sensor consists of a number of individual segments on one line with gauge lengths designed according to the structural and materials requirements. An optical time domain reflectometer (OTDR) is employed for interrogation of the sensor signal.

8.13 ADAPTIVE MANAGEMENT AND MAINTENANCE FOR BRIDGES OVER WATERWAYS

8.13.1 Higher Degree of Maintenance

1. Scour critical bridges typically require a higher degree of maintenance than those over intersections. Since streams are dynamic, and since many bridge protection measures include living plants and biodegradable material, the potential is high for stabilization measures to change in some way over time and through flood events. If monitoring indicates that a bridge protection countermeasure is no longer meeting design criteria, then adjustments can be made to ensure the continued long-term function of the technique. Such maintenance is called *adaptive management* because it is geared to identify over time what countermeasures are best for bridge scour, while minimizing impacts to flora and fauna.

2. Gradual changes that can be monitored include:
   - Migrating meander forms.
   - Sediment supply from upstream.
   - Impacts to vegetation.

   When a high flow event occurs, changes may be sudden and unexpected.

8.13.2 Use of Special Access Equipment for Above Water Inspection

1. A complete report will be prepared for each substructure. It will include:
   - Field notes.
   - Photographs of general conditions.
   - Locations of deterioration of substructure elements.
   - Bridge substructure sketches with highlighted details.
   - Conclusion with recommendations for repairs and cost estimates.
2. For bridges located over rivers, the engineer will utilize special access equipment, such as:
   - Bridge tracker (a rubber-tracked bucket truck that can operate in water depths of up to 8 ft and has a 42 ft working height).
   - Bucket boat (a medium boat-sized water vessel equipped with a hydraulic bow thruster that uses outrigger pontoons).
   - Other mechanical systems for stability that have a working height of 66 ft and need 12 inches of draft.
   - Spider or hi-lo system (allows vertical access for piers, walls, and abutments in excess of 1000 ft) for full-depth inspection of all bridge substructure elements. This special access equipment will not impact bridge vehicular and pedestrian traffic.
   - An Aspen Aerials UB truck with MPT will be utilized in limited situations. The Aspen Aerial UB truck will permit the inspector access to bridge component locations that are inaccessible from below.
3. The inspection sequence chosen by the engineer should be designed to:
   - Maximize the effective capabilities of the under-deck access equipment.
   - Take advantage of the average seasonal climate and low river depth conditions.
   - Minimize the impact to vehicular traffic.
   - Allow the completion of all inspection work within the proposed time frame.
4. Inspection of the bridge substructures will utilize the bucket boat before the river's drop in water depth precludes the use of a boat. The bridge tracker will be utilized on areas of the canal that are less than 8 ft deep and do not demand more than 40 ft working height.

   The bucket boat will be used in the main channel, and the bridge tracker will be used closer to the banks where the water depth is much lower. If site conditions permit, both pieces of access equipment will be used concurrently while inspecting a bridge.

8.13.3 Substructure Underwater Inspection

To ensure public safety and to protect the capital investment in bridges over water, underwater structural members must be inspected to the extent necessary to determine their structural condition with certainty. Underwater inspections must include the streambed for evaluating erosion.

The following three methods are commonly used:

1. A wading inspection in shallow waters: In shallow water, inspections may be made visually from above the water surface. Use of cofferdams will be avoided due to the expense involved.
2. Scuba diving in deep water: SCUBA stands for self-contained underwater breathing apparatus." Air is inhaled from portable high-pressure steel or aluminum supply cylinders.

   In deep water, inspections require diving or other appropriate techniques to determine conditions. Duration varies according to depth but will not exceed 30 minutes at a maximum 40 ft depth. Two types of exposure suits are required—the dry suit and wet suit for cold and hot temperatures.
3. Hardhat and lightweight diving for rivers with high velocities: This involves surface supplied air by a high volume low pressure compressor (150 to 200 psi). Diver's equipment weight is over 150 lbs.

8.13.4 FHWA Level III Diving Inspections

1. This work will be performed for developing design and construction details for underwater substructure repair and scour remediation. Inspections will be required to follow the guidelines and standards as stated in "Underwater Inspection of Bridges," FHWA-DP-80-1, 1989.
   - Inspection of sensitive substructure elements will be in accordance with FHWA criteria.
   - The subaqueous portions of the structures receive a visual and tactile inspection by the engineer/diver inspector.
   - All inspections will be in accordance with FHWA Guidelines for Level III Underwater Inspection.
   - The type, approximate size, location, and extent of deterioration will be documented and set forth in sketches and in the inspection report.
   - If necessary, cleaning tools, from brushes to water blasting equipment, are utilized to clean excessive marine growth and expose surface areas for inspections.
   - All structures are sounded with a hammer to locate and determine the extent of deterioration.
   - Probing rods are utilized to locate filled scour holes around substructure units.
   - Significant deterioration and damage to substructures may be documented with underwater photography utilizing a 35 mm camera, with close-up attachments and an underwater strobe. In addition, a channel bottom description is provided for each pier along with any noted scour.
   - The inspections will be performed using surface supplied air with hard wire communication and a safety line.
2. The following requirements need to be addressed:
   - Reasons for underwater inspection.
   - Types of condition surveys.
   - NBIS underwater requirements: All dive teams will be in compliance with OSHA regulations regarding underwater inspection of bridges by divers, 29 CFR PART 1910, SUBPART T.
   - Training of inspector divers: Trained underwater inspectors are required. Due to the sensitive nature of this work, professional engineer (PE) divers uniquely qualified to perform NBIS and underwater inspections are required. Rehearsals and guidelines for inspection may be provided.
   - Diving equipment: Underwater diving, inspection, and documentation equipment has improved in quality in the recent years. Visibility underwater may be limited.

3. Dive safety: Acquaint those responsible for bridge safety with underwater inspection techniques and equipment so the safety of divers is not jeopardized. Safety considerations include:
   - Working adjacent to fast, unpredictable currents, rapidly rising water levels can be extremely dangerous.
   - Floating (or subsurface) debris and woody materials contribute to hazard during emergency work.
   - Weather conditions (rain, snow, or darkness) may further endanger safety.
4. Identifying levels of inspection.
5. Conducting diving inspection: Standard procedures for underwater inspection shall be followed. Observations shall be accurate.
6. Flow conditions in the river may make it difficult to place filter layers. Precautionary measures are required.
7. Presenting methods of repair for commonly found defects:
   - An accurate and consistent classification of damage, deterioration, and substandard conditions encountered during inspections
   - The appropriate level of communication of such conditions to the agency for implementation
   - Recommendations for repair
   - Selection of repair methods
   - Estimate of cost of selected alternate.
8. All inspectors will use bridge repair priority codes to evaluate observed conditions and prioritize needed repairs to the facilities. Priority codes will identify descriptions and conditions for low priority and routine maintenance, which may be performed separately.
9. Preparing the condition report for repair: Refer to FHWA Technical Advisory No. TA 5140.20, Sep. 1988.

8.13.5 Fathometer Survey

1. Preparation prior to underwater inspection: Conditions in the field are adverse and ever changing. Proper planning, equipment, and personnel are vital for safety.

   Review available plans and previous underwater inspection reports to determine:
   - The configuration of the substructure units and channel.
   - Previous conditions.
   - Areas of concern to perform preliminary dive planning.
   - The most appropriate method of access, type of equipment, and personnel required.
   - A suitable dive plan.
   - A schedule and sequence to perform the work in the most efficient and safe manner.
2. A fathometer survey will also be conducted along fascias, 100 ft upstream and downstream, and at the bridge centerline. The fathometer survey will be tied into an existing benchmark.
3. A complete report for each abutment and pier will be prepared. This report will include:
   - General information.
   - Field conditions during the inspection.
   - Findings.
   - Recommendations for repair.
   - Cost estimates.
   - Detailed bridge drawings, including sounding profiles.

- A bathymetric plan of the channel bottom.
- Structural defects.
- Bridge plan view.

4. In order to maximize efficiency, it is appropriate to inspect during low tide. This decreases the amount of underwater inspection work and permits tide zone inspections "in the dry."

- Currents and visibility have a considerable effect on the quantity and quality of underwater inspections.
- Short, slack tide periods reduce the time in which critical inspections can be performed.
- Coordinate dive activity with neap tide, when the tidal fluctuations and currents are minimal.

5. Underwater photography and video equipment: Waterproof 35 mm cameras designed for underwater use equipped with a variety of lenses and flash unit will be used. Visibility in most rivers is less than 2 feet. Modern 8 mm camcorders are equipped with voice recording equipment ,as well.

Underwater inspection techniques: These include site reconnaissance and data collection, as-built drawings and previous reports, inspection of pier, abutments, piles, and fenders using ultrasonic measuring devices for concrete.

6. Structural defects:

- Examples for concrete are cracking, scaling, spalling, chemical attacks, and abrasion from moving ice or vessel impact. Deterioration can be prevented by appropriate scour countermeasures.
- Examples for steel are corrosion of steel piles and columns. Corrosion can be prevented by applying coatings and by cathodic protection.
- Examples for timber are decay from bacteria, fungus, marine borers, abrasion, collision, and damaged connections. Preservatives may be used.

8.13.6 Scour Safety Evaluation

A scour report consists of a detailed field survey, substructure information, scour analysis results, hydraulics related findings, and countermeasure recommendations. Some of the common types of countermeasures are riprap, gabion baskets, concrete blocks, and sheet piling. Harnessing the river to reduce flood velocities may also be used. A sample copy of a scour report is attached in the Appendix.

A flow diagram (Figure 8.6) can be used for procedures. It is desirable to inspect scour critical bridges after every flood and after installation. Typical issues that require monitoring include:

1. Erosion to bridge footings.
2. Performance assessment of installed countermeasures.
3. Migrating meander forms.
4. Adjustments to water and/or sediment supply from upstream.
5. Impacts to vegetation survival from onsite land use.

8.14 CONCEPT STUDY REPORT, PLANS, AND RECOMMENDATIONS

8.14.1 Purpose of Report

1. The intent of the draft concept study report is to provide a concise aggregation of the important elements of the condition evaluation and an overall synthesis of conclusions and recommendations. The draft concept study report will provide a summary of construction costs for the overall project and a recommendation for packaging of the construction documents.

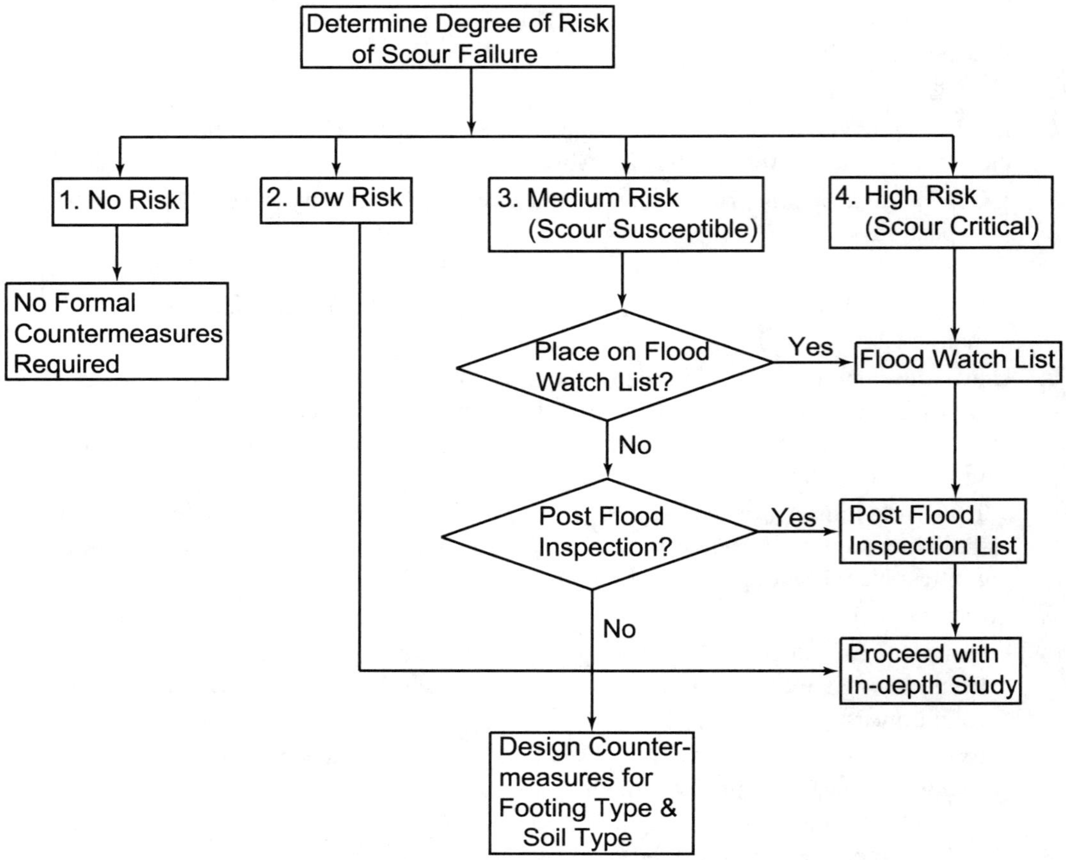

Figure 8.6 Typical flow diagram for preparing flood watch list.

At the conclusion of the task, the draft concept study report, plans, and recommendations (report) will be prepared for use at the value engineering/constructability workshop. The report will include a certification statement by the licensed professional engineer under whose direction the inspection was conducted.

2. Contents of report: The draft concept study report shall include:

- Condition assessment
- Scour vulnerability assessment
- Discussion of alternatives, including a discussion of potential impacts for each alternative
- Recommended repairs/remediation, including supporting justification
- Requirements for design, including required survey, geotechnical evaluation, and hydraulic analysis
- A list of applicable permits, including the permit cost and the anticipated duration to obtain the permit
- Right-of-way requirements, including a discussion of required temporary construction easements, if any
- MPT requirements, including sketches and/or conceptual plans for construction staging and detours
- Construction costs
- Anticipated construction schedule

- 11 in × 17 in plans: Plans include detail to a level sufficient to clearly demonstrate the repair/remediation type and location, along with construction staging and construction access.

8.14.2 Preparing a Rehabilitation Report

All inspection observations and results will be summarized in a report using a standard format. A sample copy of a rehabilitation report is attached in the Appendix. It addresses the following issues:

1. Description of project area and structure.
2. Condition of structure and ratings.
3. General considerations and constraints.
4. Roadway and safety improvements.
5. Structural rehabilitation and reconstruction scheme.
6. Widening and/or vertical clearance improvement.
7. Maintenance of traffic.
8. Cost estimate for alternates and cost analysis.
9. Recommendation.

Rehabilitation report for emergency, interim, and priority repairs: This report is required when a bridge needs to be repaired without delay as a result of fire, accident, terrorism, or extreme conditions such as earthquake or floods.

Major defects, if not repaired immediately, may require closing the bridge or a portion thereof and could even lead to a total collapse of the structure. Most repairs should start within 3 to 14 days of notification by the field staff.

Much depends upon the judgment and training of the bridge inspector in the field in eliminating safety hazards to the traveling public. For example, the FHWA *Recording and Coding Guide for Structure Inventory* identification of old SI&A coded as 2 or less on an inventory sheet is important.

Item #58—Deck

Item #59—Superstructure

Item #60—Substructure

Code 2 = Critical condition—Advance deterioration of primary structural elements.

Fatigue cracks in steel or shear cracks in concrete may be present. Scour may have removed substructure support. Unless closely monitored, it may be necessary to close the bridge until corrective action is taken.

Code 1 = Imminent failure condition—Major deterioration or section loss present in critical structural components or obvious vertical or horizontal movement affecting structure stability. The bridge is closed to traffic, but corrective action may put it back into light service.

Rapid design may require ad hoc procedures using engineering judgment, a higher factor of safety, and little consideration of cost. If emergency funds are not readily available, funds may be transferred from other account heads. Rapid design and construction procedures may be used and a contractor may be appointed without a bid process.

8.14.3 Emergency (Highest Priority) Category Repairs

1. Examples of damage requiring emergency repair include:
   - Crack in a non-redundant primary load carrying steel member.
   - Localized failure of bridge deck.

- Deterioration which causes a main load carrying member to become unstable.
- More than 50 percent undermining of bearing area of a non-redundant member.
- Damaged or missing sections of bridge railing.

2. In many instances, it may be necessary to:
- Block off the affected area partially.
- Block off the affected area completely.
- Post load restrictions.
- Post speed restrictions.

Interim repairs: The need for interim repair measures shall be considered when permanent repairs may not be constructible in a timely manner. Also, when the importance factor of the bridge is high, emergency funds may readily be made available on a limited basis. In which case emergency design can be split into an interim design plus a final design to follow.

8.15 VALUATION OF UNKNOWN FOUNDATIONS

Potential of buried footing to erosion: Many of the older substructures do not have as-built drawings. However, for scour vulnerability assessment, the physical size and type of foundation needs to be investigated. Using FHWA's Geotechnical Engineering Notebook Issuance GT-16 "Determination of Unknown Subsurface Bridge Foundations," 1998, the following approach can be employed to determine unknown foundations of bridges on timber cribbing that require a scour study:

The engineer will determine footing depths using borings.

8.16 DISASTER MANAGEMENT

The author was a member of a UNESCO sponsored team, which visited Erzurum in Eastern Turkey in 1987 after a devastating earthquake. As a member of a National Academy of Sciences team, the author visited North Pakistan in 2005 to inspect seismic damage and prepared a report for rehabilitation and reconstruction measures on behalf of USAID.

The relief and rehabilitation report and other details are provided in an earlier chapter.

Based on findings of the condition assessment, potential repair/remediation recommendations will be developed. Fluctuating river elevations will be taken into account when developing repair recommendations and reviewing the construction feasibility.

Evaluation shall include, but is not limited to:

1. Evaluation of constructability.

2. No construction staging would be required for substructure rehabilitation work.

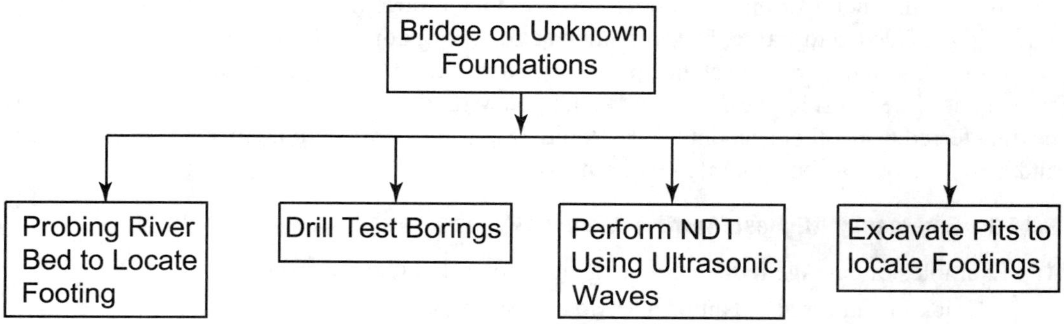

Figure 8.7 Procedures for locating unknown foundations.

Figure 8.8 Author (left) and members of U.S. National Academy of Sciences seismic team visiting Islamabad in December 2005 for inspection of structural damages.

3. Community impacts during construction: Impacts to emergency vehicle response, tourist industry, traffic delays, impacts to pedestrians and bicyclists, impacts to local business, and noise shall be considered.
4. Construction cost estimates shall be developed for each feasible alternative along with the anticipated construction schedule.
5. Maintenance protection of traffic schemes may be required for certain alternatives.
6. Right-of-way requirements and environmental impact to the waterway and endangered species shall be evaluated. Permit applications with supporting technical documentation need to be prepared.

8.17 TRAINING AND REGISTRATION NEEDS

FHWA has developed a three-week training program based on BIRM. The program consists of a one-week course on Engineering Concepts for Bridge Inspectors and a two-week course on Safety Inspection of In-Service Bridges. The following courses extending between one to three weeks are usually required for registration as a bridge inspector:

- NHI Bridge Inspector's Training Courses, Part I and Part II
- NHI Stream Stability and Scour at Highway Bridges
- NHI Bridge Coatings Inspection
- NHI Fracture Critical Inspection Techniques for Steel Bridges
- NHI Bridge Inspector Refresher Training.

BIBLIOGRAPHY

21st Century Asset Management, American Association of State Highway and Transportation Officials, Download Item Code 10—AM-2-OL.

Asset Management Primer, Federal Highway Administration, 1999.

Aktan, A.E., "The Integrated Asset Management Paradigm for Highway Transportation Infrastructure," *Structures 2004*, American Society of Civil Engineers.

Bridge Inspectors' Reference Manual, Federal Highway Administration, FHWA NHI 03-001, 2002.

Bridge Design Manual No. 4., Pennsylvania Department of Transportation.

Bureau of Design, Bridge Management System (BMS) Coding Manual, office version: Pennsylvania Department of Transportation Publication 100A, 1993.

Chaudhary, M.I.A. and Y. Fujino, "Structural Health Monitoring and Performance Evaluation of Bridges Using Recorded Seismic Data," *Proceedings of International Conference on Earthquake Engineering*, Lahore, 2006.

Cioara, Titus Gh., "Particular Methods in Signal Processing for Remote Bridge Monitoring Systems," Vibration Testing and Research Laboratory, University Politehnica, Timisoara, Romania.

Collins, Thomas J. and Ryan P. Breen, "Ireland's Bridge Management System," CEI Collins Engineers, Ltd, Dublin, Ireland.

Cross, Brad "Structural Engineering and Public Safety," *Proceedings of the 2006 Structures Congress*, John Finke, Editors.

Dewberry, S., "Easing the Way for E-Permitting," *Civil Engineering*, September 2000.

"Dynamic Bridge Substructure Evaluation and Monitoring," Larry D. Olson, Report No RD-03-089, Federal Highway Administration, September 2005.

Elias, Victor, "Strategies for Managing Unknown Bridge Foundations," FHWA-RD-92-030, Federal Highway Administration, U.S. Department of Transportation, Washington, DC, February 1994.

Ettouney, Mohammed M. "Engineering Structural Health Principal," Weidlinger Associates, Inc., Sreenivas Alampalli, Head, *Structures Research*, New York State DOT.

Frangopol, Dan M. and Jung S. Kong, "Life-Cycle Safety and Costing for Maintenance of Aging Bridges," University of Colorado, Boulder.

Guide for Commonly Recognized Core (CORE) Structural Elements, American Association of State Highway and Transportation Officials, Item Code 10-CORE-1-M.

Guidelines for Bridge Management Systems, American Association of State Highway and Transportation Officials Item Code 10-GBMS-1.

Hajdin, Rade, "BMS Development in Switzerland," *Structures 2000*, American Society of Civil Engineers, April 3, 2000.

Hearn, George, "Condition States for Highway Bridges," University of Colorado, Boulder.

Henneberg, Mark F. and Jeffrey L. Strause, "Rating for Roadway Bridges," by Open-File Report 01-446, (in cooperation with PENNDOT), New Cumberland, PA, 2002.

KUBA-MS: The Swiss Bridge Management System, Rade Hajdin, M., *Proceedings of American Society of Structural Engineers*, 2004.

Lengyel, T. and Y. Fu "Remote Monitoring of Bridges in Connecticut," University of Connecticut.

Manual for Condition Evaluation of Bridges, American Association of State Highway and Transportation Officials, 1994.

Manual for Condition Evaluation and Load and Resistance Factor Rating (LRFR) of Highway Bridges, American Association of State Highway and Transportation Officials, October 2003.

Mehrani, Ebrahim,"Remote Health Monitoring of the First Smart Bridge in Florida," University of South Florida, Tampa, Ashraf Ayoub, University of Missouri-Rolla, Rolla, MO.

Minervino, Charles M. and William M. Edberg, "Proposed AASHTO Manual for Bridge Evaluation Using LRFD Philosophy," A.G. Lichtenstein & Associates, Inc., Paramus, New Jersey.

Oh, Byung Hwan, Young Lew and Young Cheol Choi, "Realistic Assessment for Safety and Service Life of Reinforced Concrete Decks in Girder Bridges," *ASCE Journal of Bridge Engineering*, American Society of Civil Engineers, July/August 2007.

"Procedures for Bridge Scour Assessments at Bridges in Pennsylvania," U.S. Department of the Interior, U.S. Geological Survey, Open File Report 00-64, edited by Peter J. Cinotto and Kirk E. White (in cooperation with PENNDOT), Lemoyne, PA 2000.

Recording and Coding Guide for the Structure Inventory and Appraisal of the Nation's Bridges, Federal Highway Administration Report No. FHWA—PD-96-001, December 1995.

Recording and Coding Guide for the Structure Inventory and Appraisal of the Nation's Bridges, Federal Highway Administration, Report FHWA-ED-89-044, p. 115, 1989.

Revisions to the National Bridge Inspection Standards (NBIS), Federal Highway Administration,1988.

Richardson, E. V. and P. F. Lagasse, "Instrumentation for Measuring Scour at Bridge Piers and Abutments," Final Report, Phase III, Transportation Research Board, Washington, DC, 1994.

Smith, S., "Data Collection, an Advanced Art," *CE News*, April 2002.

"Software User's Guide for Determining the Pennsylvania Scour Critical Indicator Code and Streambed Scour Assessment, U.S. Department of Interior, Guidelines for Bridge Maintenance and Rehabilitation," U.S. Department of the Interior, U.S. Geological Survey, edited by Secretary, United States Department of the Interior, Washington, DC.

Stallings, J. Michael and Stephen B. Franklin, "Remaining Fatigue Life Estimates Using LRFD," *Proceedings of ASCE Conference*, 2000.

Steel Bridge Design Handbook, "Selecting the Right Bridge Type," Chapter 7, National Steel Bridge Alliance.

Thompson, Paul D., "Decision Support Analysis in Ontario's New Bridge Management System," Castle Rock, CO.

Thompson, Paul D and Reed M. Ellis, "Advanced Decision Support in the Ontario Bridge Management System."

Thurgood, Travis, Marvin W. Haling and Paul J. Barr, "Structural Health Monitoring of the Cherry Hill Bridge," Utah State University.

Underwater Evaluation and Repairs of Bridge Components, Federal Highway Administration, Report No. FHWA-DP-98-1 Final Report, November 1995.

Underwater Inspection of Bridges, Federal Highway Administration, Report No. FHWA-DP-80-1 Final Report, November 1989.

Wang, Ming L., Fan L. Xu and George M. Lloyd, "Health Assessment of a Post-Tensioned Concrete Bridge, "University of Illinois at Chicago.

Wells, Andrew, "Asset Management—Listen and Learn," *Bridge Design and Engineering,* Third Quarter 2001.

Weston, Daniel F., Timothy J. Stuffle, Julia West, Harry W. Shenton III and Michael J. Chajes, "Plan for Structural Health Monitoring of the Indian River Inlet Bridge," University of Delaware, Newark, DE.

Wroclaw, Jan Bieñ, "Neural Networks in Bridge Management Systems," University of Technology, Institute of Civil Engineering, Wyspianskiego, Poland, ASCE 2005.

Life Cycle Cost Analysis

2006 Status of the Nation's Highways, Bridges, and Transit: Report to Congress, Federal Highway Administration.

Audit of Oversight of Load Ratings and Postings on Structurally Deficient Bridges on the National Highway System, Federal Highway Administration Report No. MH-2006-043, March 21, 2006.

Bridge Inspection Terminology, Pennsylvania Department of Transportation.

Das, Parag C., *Management of Highway Structures*, Thomas Telford, 1999, Federal Highway Administration, 23 CFR Chapter 1, April 1, 2004 Edition.

Economic Analysis Primer, Federal Highway Administration, 2003.

Fiber-Reinforced Polymer, Federal Highway Administration, 2002.

Field Inspection of Reinforcing Bars, Concrete Reinforcing Steel Institute, Engineering Data Report No. 49.

Frangopol, D. and L. Neves, *Probabilistic Maintenance and Optimization Strategies for Deteriorating Civil Infrastructures*, Chapter 14 in Progress in Computational Structures Technology (Edited by B.H.V. Topping and C.A. Mota Soares), Saxe-Coburg Publications, Stirling, Scotland, 2004, 353–357.

Frangopol, D. and M. Liu, *LCCA for Highway Bridges: Accomplishments and Challenges.*

Horvath, A., *Life Cycle Environmental Accounting and Decision Making for Bridge Structures.*

Huang K., Y. C. Shen and L. J. Feeser, "An Information System for Bridge Life-cycle Assessment," *Proceedings of Second New York City Bridge Conference*, 2003.

Keating, P. B. and J. W. Fisher, "Review of Fatigue Tests and Design Criteria on Welded Details," Fritz Engineering Laboratory, Report No. 488-1 85, Table 4.

Life Cycle Cost Analysis Primer, Federal Highway Administration, 2002.

Life Cycle Cost Analysis, Utah Department of Transporation, Engineering Services Division.

Oshea, Dennis M. and George R. Deering, *Bridge Maintenance Manual*, Delaware Department of Transportation.

Priority Repair Procedure, New Jersey Department of Transportation Bureau of Structural Evaluation, Trenton, NJ.

Ryall, M. J., *Bridge Management*, Butterworth-Heinemann, Oxford, 2002.

Sivakumar, Bala and William M. Edberg, "Trial Rating of Bridges Using Load and Resistance Factor Procedures," Lichtenstein Engineering Associates, New York, NY.

Structures Manual 2007, Florida Department of Transportation Modifications to LRFR.

Underwater Evaluation and Repair of Bridge Components, Federal Highway Administration, Demonstration Projects Division, 1995.

Underwater Inspection of Bridges, Federal Highway Administration, Demonstration Project Division, 1989.

Modern Condition Evaluation Methods

Aktan, A. E., C. Chuntavan, K. L. Lee and T. Toksoy, "Structural Identification of a Steel-Stringer Bridge for Nondestructive Evaluation," Transportation Research Record, Number 1393, 1993, 1–36.

Aktan, A. E. and A. J. Helmicki, "Instrumentation, Testing, and Monitoring of Reinforced Deck-on-Steel Girder Bridges: Mid-Term Project Report," Report No. UCCII94, FHWA/ODOT Contract No. 7665, College of Engineering, University of Cincinnati, OH, 1995.

Aktan, A. E., V. Dalal, D. Farhey, A. Helmicki, V. Hunt and S. Shelley, "Condition and Reliability Evaluation for Bridge Management," *Journal of Infrastructure Systems*, 2(3), 1996, 108–117.

Aktan, A. E., R. Miller, B. Shahrooz, M. Zwick, M. Heckenmueller, I. Ho, W. Hrinko and T. Toksoy, "Nondestructive and Destructive Testing of a Reinforced Concrete Slab Bridge and Associated Analytical Studies," Report No. FHWA/OH-93/017 and UCCII 92/02. ODOT Contract No.14482, College of Engineering, University of Cincinnati, Cincinnati, OH, 1992.

Alampalli, S. and G. Fu, "Instrumentation of Remote and Continuous Monitoring of Structure Conditions," Transportation Research Board preprint No. 94-0261, 1994.

Bakht, B. and L. Jaeger, "Bridge Testing—A Surprise Every Time," *Journal of Structural Engineering*, American Society of Civil Engineers, 116(5), 1990, 1370–1383.

Baumgartner, W. and H. Waubke, "Service Life Estimation of Bridges," *Proceedings of the Bridge Management 2nd International Conference*, Guildford, UK, 1993.

Bay, J. A., K. H. Stokoe II and J. D. Jackson, "Development of a Rolling Dynamic Deflectometer," Transportation Research Record, Number 1473, 1995, 43–54.

Ben-Akiva, M. and R. Ramaswamy "Approach for Predicting Latent Infrastructure Facility Deterioration," *Transportation Science*, Vol. 27(2), May 1993, 174–193.

Biswas, M., A. K. Pandex and M. M. Samman, "Diagnostic Experimental Spectral/Model Analysis of a Highway Bridge," *International Journal of Analytical and Experimental Modal Analysis*, 1990, 33–42.

BRIDGIT: Bridge Management System Software, Technical Manual, NCHRP Project 1994 12-28 (2) A.

Chase, S. B., "FHWA NDE Contract Research, an Overview," *Presentation at North American Workshop on Instrumentation and Vibration Analysis of Highway Bridges*, Cincinnati, OH, 1995.

Cawley, P. and R.D. Adams, "The Location of Defects in Structures from Measurements of Natural Frequencies," *Journal of Strain Analysis*, 4(2), 1979, 49–57.

Chen, S. and Y. R. Kim, "Condition Assessment of Marine Timber Piles Using Stress Wave Method," Building an International Community of Structural Engineers, *Proceedings of the Structures Congress XIV*, 2, 1996, 853–860.

Clough, R. W. and J. Penzien, *Dynamics of Structures*, McGraw-Hill, 1993.

DeWolf J., D. A. Bagdasarian and P. N. O'Leary, "Bridge Condition Assessment Using Signatures, Nondestructive Evaluation of Civil Structures and Materials," *Conference sponsored by the FHWA and the National Science Foundation*, Boulder, CO, 1992, 171–179.

Eberhard, M. O., G. Hjartarson, J. A. MacLardy and M. L. Marsh, *Lateral-Load Response of a Reinforced Concrete Bridge*, WA-RD 305.2, Final Report, U.S. Department of Transportation, Washington, DC, 1993.

Egri, R., G. Henderson, A. Lichtenstein, S. Madanat and K. Maser, *Development of a Wireless Global Bridge Evaluation and Monitoring System: Literature Review*, Florida Highway Administration (FHWA) Contract No. DTFH61-94-C-00217, 1995.

Ewins, D. J., *Modal Testing: Theory and Practice*, John Wiley, New York, NY, 1985.

Farrar, C. R., W. B. Baker, T. M. Bell, K. M. Cone, T. W. Darling, T. A. Duffey, A. Eklund and A. Migliori, "Dynamic Characterization and Damage Detection in the I-40 Bridge over the Rio Grande," Los Alamos Report LA-12767-MS UC-906, Los Alamos, NM, 1994.

Finno, J. R. and J. P. Prommer, "Evaluation of Inaccessible Drilled Shafts Using the Impulse Response Method," ITI Technical Report #3, Northwestern University, Evanston, IL, 1994.

Gregory, A. E., N. C. Yang and R. C. Y. Young, "Bridge Inspection by Dynamic Loading Technique," *Proceedings of the ASCE National Convention*, Denver, CO, 1985.

Halling, M. W., I. Muhammad and K. C. Womack, "Dynamic Field Testing for Condition Assessment of Bridge Bents," *Journal of Structural Engineering*, American Society of Civil Engineers, Vol. 127(2), 2001, 161–167.

Hearn, B. and H. Shim, "Integration of Bridge Management Systems and Nondestructive Evaluation," *Journal of Infrastructure Systems*, 5(2), June 1998, 49–55.

Huang, K., "A Nondestructive Instrument Bridge Safety Inspection System Using a Transient Load," U.S. Patent Application Number 09/210.693; patent allowed April 2001.

Huang, N. E., R. S. Long and Z. Shen, "A New View of Nonlinear Water Waves-Hilbert Spectrum," Annual Review of Fluid Mechanics, 31, 1999, 417–457.

Huang, N. E., Z. Shen, S. R. Long, M. C. Wu, H. S. Shih, Q. Zheng, N. C. Yen, C. C. Tung, and H. H. Liu, "The Empirical Mode Decomposition and Hilbert Spectrum for Nonlinear and Non-Stationary Time Series Analysis," *Proceedings of the Royal Society of London*, A 454, 1998, 903–995.

Hussein, M., G. Likins and G. G. Goble, "Determination of Pile Lengths under Existing Structures," *Proceedings of Deep Foundations Institute 17th Annual Members Conference*, New Orleans, LA, 1993.

Huston, D. and M. G. Gardner-Morse, "Modal Identification of Cable-Stayed Pedestrian Bridge," *Journal of Structural Engineering*, American Society of Civil Engineers, 119(11), 3384-04, 1993.

Lichtenstein, A., "Bridge Rating Through Nondestructive Load Testing," Transportation Research Board, 1993.

Liu, S. C., "Natural Hazard Mitigation: Exploring the Technological Frontiers," *Structural Health Monitoring 2000* (F. K. Chang, ed.), Technomic Publishing, 36–55.

Manual For Bridge Rating through Load Testing, Final Draft, National Cooperative Highway Research Program (NCHRP), 12-28(13).

Manning, D. G., "Detecting Defects and Deterioration in Highway Structures," NCHRP Synthesis of Practice No. 118, 1985.

Mazurek, D. and J. DeWolf, "Experimental Study of Bridge Monitoring Technique," *Journal of Structural Engineering*, American Society of Civil Engineers, 116(9), 2532-49, 1990.

Olson, L. D. and M. F. Aouad, "Unknown Subsurface Bridge Foundation Testing," NCHRP 21-5(2), Final Report, 2001.

Olson, L. D., F. Jalinoos and M. F. Aouad, "Determination of Unknown Subsurface Bridge Foundations," Research Report NCHRP 21-5, 1995.

Performance and Accomplishments Report, Federal Highway Administration Pub. No. FHWA-HPL-05-009, Washington, DC.

Pierce, C. E. and C. H. Dowding, "Long-Term Monitoring of Bridge Pier Integrity with Time Domain Reflectometry Cables," Northwestern University, Evanston, IL, 1995.

Prine, D. W., "Problems Associated with Nondestructive Evaluation of Bridges," Northwestern University, Infrastructure Technology Institute, http://www.iti.northwestern.edu/publications/technical_reports/tr13.pdf, December 1998.

Raghavendrachar, M. and A. E. Aktan, "Flexibility by Multi-Reference Impact Testing for Bridge Diagnostics," *Journal of Structural Engineering*, American Society of Civil Engineers, 1992, 2186–2203.

Salawu, O. S. and C. Williams, "Bridge Assessment Using Forced-Vibration Testing," *Journal of Structural Engineering*, American Society of Civil Engineers, 121(2), 1995, 161–173.

Sandford, T. C. and M. Elgaaly, "Skew Effects on Backfill Pressures at Frame Bridge Abutments," Transportation Research Record, Number 1415, Transportation Research Board, Washington, DC, 1993, 1–11.

Salane, H. J., J. W. Baldwin and R. Duffield, "Dynamics Approach for Monitoring Bridge Deterioration," Transportation Research Record, Number 832, 1981, 21–28.

Sanford, K., "Improving Condition Assessment; Data Requirement for Bridge Management," Ph.D. Thesis, Department of Civil and Environmental Engineering, Carnegie Mellon University, Pittsburgh, PA, 1997.

Shelley, S. J., K. L. Lee, T. Aksel and A. E. Aktan, "Active Control and Forced Vibration Studies on a Highway Bridge," *Journal of Structural Engineering*, American Society of Civil Engineers, 121(9), 1995, 1306–1312.

VanDiver, J. K., "Detection of Structural Failure on Fixed Platforms by Measurement of Dynamic Response," Technical Report 78-10, Department of Ocean Engineering, Massachusetts Institute of Technology, Cambridge, MA, 1978.

Warren, G. E. and L. J. Malvar, "Structural Assessment of Reinforced Concrete Piers Using the Impact Load Method," *Proceedings of the International Conference on Nondestructive Testing of Concrete in the Infrastructure*, Dearborn, MI, 1993, 378–390.

Woodward, C., J. Minor, K. White and R. Idriss, "Nondestructive Evaluation of Fracture Critical Bridge," Structural Materials Technology, *Proceedings of the Nondestructive Testing Conference*, Atlantic City, NJ, 1994.

Wipf, T., "Use of Tilt Sensing Equipment for Monitoring Long-Term Bridge Movement," *Canadian Journal of Civil Engineering*, 18, 1991, 1033–1046.

Recommended field procedures, evaluation guidelines, and reporting standards for periodic inspection of in-service FRP bridge decks, December 2005. Published as NCHRP Report 564.

9

Conventional Repair Methods

9.1 SCOPE OF CONCRETE AND STEEL REPAIRS AND RELATED WORK

9.1.1 Introduction

In the U.S., the transportation departments of each state have been responsible for selecting applicable repair methods, developing procedures, and providing guidelines for implementation. The selection is based on effectiveness of the method, its feasibility in terms of relative cost in replacement with new bridges, detour and staged construction options, MPT during repair or reconstruction and duration of construction schedule. In this chapter, conventional repair methods currently in use are described. The objective is to present repair methods for use by the practicing engineer. Use of alternate methods is described in Chapter 8.

1. A report commissioned by the Transportation Construction Coalition (which represents trade groups and unions with a vested interest in funding for road construction) suggests that highway and bridge related conditions were a major factor in 22,000 fatalities and cost \$219.5 billion each year. By comparison, similar crashes where alcohol was a factor cost \$130 billion, speeding cost \$97 billion, and failure to wear a seat belt caused losses of \$60 billion. Almost 42,000 people die in traffic accidents per year.

   It recommends several improvements such as adding and widening shoulders, widening or replacing narrow bridges, realigning crooked roads, requiring breakaway signposts and light poles, using more brightly colored pavement markings, installing signs that are easier to read and decipher, and adding rumble strips and guardrails.

   A lot of this is a problem of old roads built in horse-and-buggy days. They had lots of trees for shade but didn't need wide bridges, so narrow bridges were built.

   With fast moving traffic and undesirable environmental effects, the type of material deficiencies and repair issues are diverse. Many types of repairs may require a cookbook approach, giving recipes for new materials and details of construction.

2. The objectives of identifying defects in members are to meet FHWA goals and ensure the safety and comfort of the public by making use of experience and guidelines provided by others, so that the latest and the best repair materials, methods, and design and field procedures can be recommended and made available for future use. Rehabilitation includes steel girder repairs, bearing repairs, repairs to welded, bolted connections, and gusset plates.
3. The following detailed recommendations are based on current practices and the literature review of a large number of publications, including FHWA, AASHTO, ACI, ASTM, and NCHRP. The popularity of concrete as a building material is universal. One of the reasons is concrete's ability to accommodate continued repairs. Structural repairs represent the most common activity in rehabilitation of any structure. The importance of repairs cannot be underestimated.
4. Maintenance includes, besides routine bridge work, accident cleanup, and repair work that includes: roadway pothole repairs; bridge lighting facilities; navigation lighting; emergency deck failure repairs; concrete median collision repairs; earth, stone slope, and ditch repairs; overlay crack sealing; delineator/milepost/signing repairs; bridge fencing repairs; drainage inlet cleaning; main line and ramp guide rail repairs; line striping; sign panel/sign structure repairs and right-of-way maintenance, bridge sweeping, and security items.

   Rehabilitation is a means to an end and is not an end in itself. Other civil, environmental, and management activities are indispensable and are complimentary to rehabilitation.

   Of all the bridge components a bridge deck which supports moving traffic constantly, is liable to be replaced most often. Increase in weight of truck is another reason. Bearings may need to be replaced for seismic upgrade or corrosion. Substructure which is partly buried is less susceptible.

   Foundation erosion after floods would require upgrading scour countermeasures. Post earthquake emergency repairs to piers, abutments and their foundations may be necessary.
5. In Chapter 3 the structural deficiencies listed in the documented history of inspection reports were addressed. They are site specific and time dependent. They may include:
   - Inability to define loads accurately, such as from foundation settlement.
   - Inability to include creep and shrinkage strain distribution in the deck slab.
   - Unpredictable behavior of connections and joints, splices, gusset plates, bolts, and welds.
   - Delamination and reduction in strength of concrete deck due to deicing salts.
   - Malfunction and locking of old bearing assemblies due to lack of maintenance.
   - Large thermal forces causing compression and local buckling of truss members and flanges.
   - Scour at pile top.
   - Lack of drainage behind abutments and pressure build-up behind abutments.

9.1.2 Key Issues and Critical Problems

On a medium or large size rehabilitation project, both structural and non-structural activities are required. The following are typical project activities for a steel bridge:

Condition Assessment of Approach Roadways

Pavement testing and evaluation will typically be required. The objectives of the pavement evaluation include identification of the underlying causes of the existing distresses observed in the pavement and assessment of the condition of the pavement system. Such findings will then enable development of recommendations concerning innovative and cost effective rehabilitative measures that could be used to restore the integrity of the pavement system to enable it to withstand the traffic loadings for at least a 15-year service life.

The pavement evaluation will be based on data collected with the following:

- The Falling Weight Deflectometer,
- Ground Penetrating Radar,
- Dynamic Cone Penetrometer, and
- Other relevant field testing techniques.

Generally, the governing factor for selecting the appropriate alternatives is the design requirements and life-cycle costs (determined via a life-cycle costs analysis).

Condition Assessment of the Bridge

In-depth bridge deck condition evaluation: As discussed in Section 9.12, the objective of bridge deck condition evaluation is to evaluate the concrete deck in order to identify the necessary rehabilitation schemes. Experience in testing and evaluation of bridge decks is required. The main evaluation tool would be a visual condition survey along with the half-cell and chloride content, under proper lane closure. Any deck treatment should consider the effect on the bridge load rating, the effect on the portion of the substructure below the deck, and the effect on bridge traffic during construction. A report on the condition evaluation and an overall synthesis of conclusions and recommendations is required.

In selecting the preferred alternative, an inspection of bridges/facilities will address

- The current condition of the superstructure and substructure
- Environmental restrictions
- Estimated short-term costs
- Life-cycle cost of major items (capital cost, interest rate, salvage value, rehabilitation, and maintenance).

The evaluation will include the type of deck overlay, rehabilitation, and replacement alternatives and will recommend deck sealing, rehabilitation, or replacement, based on the results of the deck analysis.

Structural Evaluation and Condition Survey

1. Defects and the need for repairs are observed on existing bridges by condition survey and standard inspection or monitoring procedures.

   A diagnostic survey of conditions is initially required to:

   - Identify the need for repair.
   - Identify the type of repair needed.
   - Perform load tests.
   - Perform field and laboratory tests, such as visual observation, impact-rebound, petrography and chemical tests, and compression tests on cores.

2. A field inspection and an underwater inspection, if applicable, will be carried out for verification of the latest conditions. An estimate of cost and repair quantities will be prepared for performing substructure repairs, above and below the water line.

   - Deteriorated backwall elements need to be fixed.
   - If concrete aprons at the piers exhibit wide cracks, repairs are required.
   - Deck joints and tooth dam if damaged needs to be replaced.
   - Removing any buildup of sand debris at piers.
   - Removing any tree trunks or tree roots between piers.

3. A condition survey may include:

- Hand investigation with a chipping hammer.
- Testing concrete cores.
- Drilling into unsound concrete to determine the depth of deterioration.

4. Degradation of concrete is a universal problem. The causes, type, and history of deterioration need to be evaluated in detail. They may include poor practices during the original construction. Improper joint spacing and any load imbalances also contribute to cracking and spalling. Typical causes of concrete cracking include:

- Load induced cracks: Cracks result when tensile stress due to repeated heavy wheel loads exceeds the tensile capacity of concrete.
- Shrinkage: Water reducers, though beneficial for workability, do not fully reduce the amount of water and water evaporation still causes shrinkage. During the chemical process of hydration there is reduction in volume.
- Resistance to consolidation and shrinkage from closely spaced reinforcement: Rebars may prevent free movement of concrete during the chemical process.

5. Additional causes for concrete cracking include:

- Drying shrinkage.
- Thermal contraction or expansion.
- Lack of appropriate control joints.
- Overload conditions that produce flexural, tensile, or shear cracks in concrete.
- Restraint of movement.
- Excessive deflection.

NCHRP and Other Studies

1. High temperature, wind velocity, and low humidity during placement and ineffective curing appear to accelerate cracking. Literature shows little evidence of cracking due to large deflections. Corrosion of reinforcing steel within concrete is recognized as a significant problem facing present-day owners and engineers.

2. A recent study by NCHRP on causes of bridge deck deterioration on the nation's bridges has shown typical deck deterioration and damages that may have resulted from vibration, excessive deflection, or fatigue:

- Transverse cracking—The most common form of deck damage is due to plastic shrinkage of concrete, drying shrinkage of hardened concrete combined with deck restraint, settlement of finished concrete around top mat of reinforcement, and traffic induced vibrations.
- Longitudinal cracking—This occurs due to poor mix design, temperature changes, shrinkage or live load impact. Multiple cracks occur due to pounding caused from wheel impact on rough surface and at deck joints.
- Spalling—Caused by corrosion of reinforcement, freeze/thaw cycles of concrete.
- Surface scaling—Improper finishing and curing.
- Effects of deicing salts: Exposure to deicing chemicals and marine-sourced chloride is a significant cause of corrosion, playing a more detrimental role than originally anticipated.

3. Krauss and Regale in 1996, after an extensive survey of U.S. and Canadian bridges, concluded:

- Cracking is more common among multi-span continuous steel girder structures due to restraint.
- Longer spans are more susceptible than shorter spans.
- Design variables that affect cracking are size, placement, and protective coating of reinforcement bars. Closely spaced small diameter bars with adequate cover of 1 to 3 inches would minimize cracking.

- Issa and Issa in 2000 reported that the sequence of deck casting contributes to deck deterioration at early ages of concrete.
- Bridge Deck Patching Survey by FHWA:
- Bridge deck rehabilitation projects are frequently subject to large variations between the amount of deck repair shown on the plans and the actual deck repair work completed. The primary purpose of this review is to evaluate the effectiveness of current policies, practices, and procedures that the (STA) Department ofTransportation (DOT) implements in estimating and tracking the quantities for bridge deck patching.
- In 2001 the DOT issued new mixture requirements for Portland cement concrete patching. The secondary purpose of the review is to evaluate the usage of these new mixtures and the possible need for additional high early strength mixtures.

4. STA introduced the use of four new concrete mix designs for bridge deck patching. The mix designs are PP-1, PP-2, PP-3, and PP-4. Contractors may minimize patch cure times before opening lanes to traffic. The procedures and practices followed often vary from agency to agency. The survey will also cover curing methods and time allowed for curing before opening to traffic. Methods used by agencies to estimate deck repair quantities include:

- Deck survey
- NBIS/PONTIS reports
- Delamination survey
- Ground penetrating radar
- Infrared thermography
- Other methods.

Load Rating and Fatigue Evaluation

The approach should include an independent load rating and fatigue evaluation report to determine the remaining useful life of the bridge.

- This will be based on an in-depth inspection and any documented section loss.
- Use reduced sectional areas of truss chords due to corrosion or damage.
- Fatigue evaluation will be based on information from existing inspection reports, the existing load rating, and existing plans.
- Load ratings will be based on load histories in accordance with latest AASHTO Manual for Condition Evaluation of Bridges, including all interims as modified by Design Manual. the Latest version of Bar 7 for load ratings for the vehicles: H, HS, ML, TK, P-82, Type 3, Type 3-3, and Type 3S2 will be used.
- Use ASD for older trusses and LRFD for deck stringers and floor beams.
- Full sidewalk live loading, acting with full live loads, shall be considered only for the operating rating levels.
- All load rating and structural modeling work will be preferablyconducted under the direction of a professional engineer licensed in the state.

Fixing Structural Items

Typical examples are as follows:

- Repairs to composite deck: The function of expansion joints will be investigated. Recommend deck sealing, rehabilitation, or replacement based on the results of the deck analysis.
- Repairs to steel members.
- Repair damaged portions of floor system and sidewalk: Any pitting of flange areas of floor beams will be investigated.
- Grinding in patches and structural painting may be required.

- For bracings, joints with loss of section will be repaired and will be checked for any defects.
- Any gusset plates showing loss of section will be replaced.
- Cracks in parapets will be repaired by pressure grouting.

Constructability Review

Addressing constructability issues early in the design process and performing constructability reviews throughout the design process will prove valuable in serving to eliminate changes during construction. A constructability review will include the following to avoid any surprises during construction.

- Site issues, such as transport to site and site accessibility
- Equipment locations
- MPT
- Construction work area restriction
- ROW encroachment, geometrics
- Access to adjacent properties
- Drive up costs
- Project duration
- Material availability
- Fabrication/erection feasibility
- Construction sequencing.

A review will be done during design and performed by construction managers. At the conclusion of the preliminary design, submit plans along with the construction cost estimate, outline of specifications, and the construction schedule for a constructability review by an independent consultant. The objective is to keep traffic moving, while providing a safe project work zone for all. Explore and evaluate methods to minimize construction costs and time, as well as the aforementioned risks.

A workshop will be conducted to evaluate constructability. Topics may include, but are not limited to, construction duration, access/ROW, and material availability.

Potential construction restrictions are identified as:

- The location of longitudinal construction joints with respect to the top flange locations.
- Staging widths and work area widths have been developed to avoid construction joint location away from steel flanges. Otherwise, under repeated heavy truck loads, unsupported construction joints can open at a future stage.
- Since shear connectors require field welds, small angles or channels bolted to top flanges will be considered. A storage area close to the bridge will be restricted. Staged construction versus detour shall be considered.
- Staged construction shown on the attached sketch will maintain uninterrupted traffic flow and eliminate potential traffic congestion associated with a detour. Some stages can be eliminated if the access is available from under the superstructure. Any impacts to daily commuters will be minimized by maintaining two lanes of traffic.

Coordination with any adjacent contracts is essential to the success of this project.

The primary goal for the project will be to design improvements that provide a higher level of safety and comfort to vehicular and pedestrian traffic, while having the least possible impact on existing surroundings.

The initially preferred alternate (IPA) should be able to be constructed in one season, as shown on the schedule. The fewer stages the lesser will be the completion duration. The structure, width from curb-to-curb, can be reconstructed in three stages by maintaining two 11-foot lanes of traffic at all times.

Evaluate the integrity of the existing shoulders to determine if any improvements will be required prior to shifting traffic onto them.

Review existing accident information to determine if any specific type of vehicular accidents may affect the proposed staging plans.

Traffic control plans will be prepared in accordance with the Manual on Uniform Traffic Control Devices (MUTCD) so that they will be comprehensive and clear in delineating how the traffic control is to be accomplished.

Maintenance of pedestrian access will be included in the traffic control plans. The use of various multi-media outlets should be considered for this project.

Develop construction schedule to account for work hour restrictions during Christmas and other holidays.

Maintenance and Protection of Traffic (MPT)

1. Traffic control plans must comply with the state, MUTCD and AASHTO LRFD regulations. Prior to developing staging plans, the project manager shall contact traffic operations to determine what are maximum allowable lane closures hours in each direction and the maximum number of lanes that can be closed at one time.
2. All traffic control schemes and detour plans on local roads must be approved by local authorities. It is important that early in the design a set of applicable traffic control and staging plans be sent to them for their approval.
3. A minimum ten hour night window may be required for the contractor to properly complete his work. Weekend work may be considered in addition.
4. Staging plans shall show a cross-section of the bridge for each stage of construction (two stages are preferred). In each of the planned stages, repair areas shall be distinguished from travel lane areas.

Plans, Preliminary Estimates; Final Report Preparation and Specifications

Plans and specifications will be developed in accordance with policy/procedures of the agency and the applicable AASHTO design standards for load resistance factor design (LRFD).

- Members will be checked for effective stress range for fatigue (load induced and displacement induced) and seismic and thermal analyses
- State or AASHTO recommended permit loads will be applied
- Impact on traffic from staged construction
- Structural stability of trusses
- Constructability issues
- Superstructure jacking
- Cracking of FCM and diaphragms will be investigated
- Expansion dams: If field visit has shown non-functioning deck expansion joints, adequate expansion joints need to be provided
- Bearing retrofit: If existing bearings need to be replaced, the type will be investigated.

Consideration of seismic retrofit for bearings will be coordinated with the system-wide seismic study. The width of bearing seat will meet current AASHTO LRFD requirements.

Based on detailed field verification, identify the repair requirements for the steel bearings to preclude the need for major repairs. It is anticipated that these repairs are likely to be designed as "replacements" in kind for the keeper angles, shoulder bolts, anchor bolts, etc.

The scope of the project is to complete the work within the existing footprint. Therefore, the first task for our design team will be to identify all substandard geometric features. This will establish the baseline conditions as a checklist that will be used to help minimize significant impacts that could slow down the implementation of the proposed improvements.

A design exception report will be prepared by utilizing the latest design exception manual for substandard controlling design elements that cannot be easily corrected. The existing deck cross slopes will be evaluated, and a spread analysis will be performed and compared with the allowable spread. If it is found to be substandard, we will try to correct this deficiency, in the design of the replacement deck and reconstruct/ add additional scuppers, as necessary. Otherwise, a design exception will be required. Existing lighting will be evaluated and upgraded as necessary.

Deck/Superstructure Replacement and Superstructure/Substructure Repair

Superstructure repairs will include -

- Structural steel painting
- Drainage system repairs/retrofits
- Specialty repairs
- Reinforced concrete slab beam
- Sign structures; bridge mounted signs
- Bridge railing/barrier repairs
- Roadway approaches: resurfacing, curb addition/repair, GR

1. Superstructure repairs will be made to the bridge to restore serviceability, original functionality, and extend the bridge life. A field inspection will be made to determine if conditions to the bridge have changed since the last inspection.
2. The deck will be evaluated using non-destructive testing such as ground penetrating radar to determine the need for repairs prior to milling and resurfacing. Recommend future work if required, including preventative maintenance.
3. Review the latest bridge inspection report and visit the site. Investigate replacing the deck and repairing the superstructure versus complete/partial superstructure replacement within the existing bridge footprint.
4. Environmental restrictions, estimated short-term costs, and life-cycle costs of major items (capital cost, interest rate, salvage value, rehabilitation, and maintenance cost) will be considered in selecting the repair methodology. Minimum service life of 25 years is targeted.
5. Examples of near-term repairs: These will include repair to deck joints at piers and abutments, patching spalls on deck underside and sidewalk, cleaning and painting the railing, rebuilding approach curbs at all four corners with epoxy concrete, removing deteriorated encasement, clean and paint exposed steel, and waterproofing. Cracks in deck and parapet concrete will be repaired by pressure grouting.
6. Repair corroded sectional areas of flanges, deteriorated asphalt overlay, curbs, sidewalks, under deck, exposed rebars, encasement deterioration, severe spalls, severe scaling, and spread of cracks.
7. Replacement deck design: One course deck construction is proposed. Design a minimum 8 inch thick composite deck. The new deck slab will extend over multiple approach spans at each end and uncased steel spans.
8. Increasing strength and service life: Making the new deck composite will increase section properties, thereby increasing the service life. Accordingly, bolt small angles or channels on top flange in place of welding shear studs. The increase in cost is small. Upgrading existing simply supported beams to continuous beams will eliminate deck joints over piers. This will also increase both the rating (load carrying capacity of beams) and the life of the deck slab. Continuity can be established by bolting flange plates at the piers.
9. Methodology: Decks will be designed using LRFD Method and LRFD Loading. As-built plans and other records of bridge history will be considered.
10. Fender system repairs/reconstructions.
11. Seismic hazard criteria—The earthquake locations and magnitudes shown in the database for national seismic hazard maps developed by the United States Geological Survey (USGS)

in 1996 shall be used for acceleration coefficient. The seismic hazard maps provide spectral accelerations having probabilities of exceedance of 10 percent, 5 percent, and 2 percent in 50 years, corresponding to return periods of approximately 475, 1000, and 2500 years, respectively.

12. Soils Criteria—Due to difficulty in obtaining borehole information at each site the national seismic hazard maps developed by the USGS can used as a basis for considering soil vulnerability characteristics, in the selection process.
13. Bearing vulnerability ratings will be based on the following characteristics:
    Bearing type—all major bearing types
    Bearing framing—especially for steel box-beam on bearing superstructures
    Expansion rocker bearings with varying heights
    Number of bearings
    Anchor bolt arrangements (various diameter, number per bearing, strength, condition)
14. Seat Length Vulnerability—Seat length vulnerability is based directly upon AASHTO specifications. Tall piers and/or long continuous spans coupled with high skews are typically to blame for inadequate seat lengths. With this in mind, the following characteristics are considered in seat length vulnerability ratings.
15. Pier height and span length between joints.
    - Variations in seat length capacity at abutments/piers/drop-in spans.
    - Substructure vulnerability.
    - Pier material types—steel, reinforced concrete, masonry.
    - Pier structure types—single columns/multi-column bents/wall piers/with and without cap beams etc.
    - Pier heights—range of tall and short piers.
16. Foundation types—pile foundations/spread footings.
17. Varying abutment heights/types.

9.1.3 Need for Accelerated Bridge Repair and Construction

Accelerated construction typically provides the shortest construction duration and the least traffic impacts.

1. Precast panels: Precast concrete deck panels are formed and poured in a precasting yard prior to construction, placed in the field, made composite with the girders via shear studs, post-tensioned and typically overlaid. The advantages of using precast concrete deck panels include the following:
    - Shorter on-site construction duration.
    - Minimized traffic impacts.
    - Improved quality control and less cracking.
2. Prefabricated superstructure systems.
3. Accelerated cure cast-in-place concrete.
    The disadvantages of using precast concrete deck panels include the following:
    - Maintenance and long term performance concerns (susceptible to leakage) due to the number of joints.
    - Involves post-tensioning, typically in both the longitudinal and transverse directions.
    - Difficulty in accommodating anything but simple geometric conditions (horizontal curves, skews, cross slopes, profiles).
    - Difficulties in constructing future replacement/repairs due to joints and post-tensioning.
    - Initial construction cost is greater than cast-in-place deck replacements.

4. Localized deck replacement/repair with overlay.

The localized deck replacement and deck repair approach has been used for nearly 40 years on bridges. This approach typically involves annual contracts that replace the highest priority localized areas of decks, repair deck spalls, repair/replace deck joints, and replace surfacing. The benefits of performing this type of work include:

- Targeting the areas which are in the worst condition and to avoid emergency repairs.
- Maintains structures in functional condition with a limited budget.
- Allows for work on numerous structures by focusing on the highest priority areas not just a single bridge.
- Cost effective when considering the limited construction budget and age of the bridge.

However, there are a few downfalls to this method of repair, including:

- Difficult to maintain traffic depending on number of lanes, shoulder widths, and superstructure configuration.
- Very expensive due to MPT costs and limited duration of lane closings.
- Repairs are made on the same bridges on a periodic basis.
- Limited number of contractors available who are capable of performing this type of work.
- Long term quality is compromised due to limited duration of construction stages.
- Increased unit cost because of reduced quantity.
- Eventually no longer economically feasible as the deterioration rate increases due to the age of the bridge.

5. Program schedule and funding:

The bridges in this study were evaluated to determine the preferred reconstruction/rehabilitation alternative at each location, as previously summarized. Then, considering condition, life expectancy and MPT impacts, the structures were prioritized to assist in developing a reconstruction/rehabilitation program for implementation by the Authority. The following page is a suggested schedule for design and construction of the preferred alternative at each bridge location. The schedule also provides an approximate duration for preparation of the environmental permits and for environmental agency review.

- Construction includes removal of existing concrete deck, balustrades, setting of the precast deck units, grouting longitudinal joints, grinding and grooving the final riding surface and general clean-up.
- A comparison of costs, durability, long term behavior, and reduction in construction schedule will be made. Meet the optimum conditions of fastest construction techniques with minimum cost.
- Some deck alternatives for consideration are:
  - Precast panels use of Effideck or a similar proprietary panel system will be investigated: Precast deck can be cast in panels of required sizes.
  - FRP (Fiber Reinforced Polymer) bridge deck: FRP systems can be utilized to replace decks on steel beams.
  - Exodermic deck slabs: For rapid deployment in deck replacement projects, a lighter exodermic deck alternative will be considered.

However, the disadvantages are increase in cost, lack of continuity, making composite connection of precast panel with shear studs on top of beams and grouting inside pockets provided in the precast panels.

6. Temporary shielding: To prevent debris from falling below during deck slab, containment methods such as installing wire nets below the deck will be evaluated.

9.2 ASSOCIATED TASKS AND TEAMWORK

9.2.1 Interdisciplinary Activities

Rehabilitation projects are dependent upon a host of other disciplines for completion. The project manager will be responsible to set up a team with relevant expertise for an early and economical completion. Activities on the critical path shall be identified and allocated adequate resources for timely completion. A concept study report shall be developed and approved by the authority.

9.2.2 Project Triple Constraint Management (Scope/Schedule/Budget)

The triple constraint to project management involves making tradeoffs between scope, schedule, and budget. It is inevitable in a project life cycle that there will be changes to the scope, schedule, or budget. Most projects fail when one of the areas changes and appropriate adjustments are not made to the other areas. We are thoroughly familiar with and routinely prepare projects per the project management body of knowledge (PMBOK).

1. Scope

   Our approach to project scope management includes the processes required to ensure that the project includes all the work required to complete the project successfully:
   - Project initiation—We are committed to begin the final design phase upon NTP.
   - Scope planning—A written scope statement will be prepared as the basis of project decisions.
   - Scope definition—Project deliverables such as highway designs, structure designs, and environmental permitting will be broken into smaller manageable components.
   - Scope verification—This is the formal acceptance of the scope deliverables by NJTA.
   - Scope change control—This will include controlling changes to the project scope. If there is a need to perform work not originally scoped, our staff and full service capabilities will provide the resources to complete any unanticipated tasks without impact to schedule. In addition to the identified services, we also offer traffic data collection; traffic modeling and forecasting; ITS; Web page design; landscape design; and ROW acquisition, negotiation, and relocation services. One important service we provide is a prompt re-evaluation of previously scoped improvements for appropriateness and adequacy as conditions change.
2. Schedule: Approach to schedule management includes processes required to ensure timely project completion:
   - Activity definition—Identifying specific tasks.
   - Activity sequencing—Identifying inter-task dependencies; Activity Duration Estimating – number of work periods per task.
   - Schedule development—Creating the schedule.
   - Schedule control—Controlling schedule changes. We will prepare a critical path model using Primavera Software for each task assigned under this term agreement. Regular NJTA schedule updates do not require that our PM use the program, but by duplicating the schedule in-house we can run the various reports to monitor the work being completed, identify items falling behind, and develop a recovery plan. We also issue reports to the various task managers identifying the tasks, corresponding budgets, and actual costs charged to their tasks. Budgets are directly impacted by schedules. When scheduled tasks are late, the amounts budgeted for the activities are usually exceeded. Therefore, our goal is to beat the schedule so that staying within budget happens almost automatically.
3. Budget approach to budget management includes processes that ensure the project is completed within budget:

- Resource planning—Determines what resources and quantities are needed for tasks.
- Cost estimating—Estimate of cost to complete tasks.
- Cost budgeting—Allocating the cost estimate to individual tasks.
- Cost control—Controlling budget changes. Be in concert with authority's "fix it first," "right sizing," and "smart solutions" concepts. With this understanding, the team will be able to fully focus on the project scope so as to minimize the potential for scope creep and accelerate the design process. Fully recognize current financial constraints. The approach to the project will be to design improvements while giving the authority the most cost-effective designs.
- Meet/contact the PM at least weekly to discuss project status, potential project challenges, and the risk associated with project decisions relative to tasks related to schedule and budget. Ensure uninterrupted communication with the NJTA and continuity of management throughout the project.

9.2.3 Preliminary Design Submission

The findings of the existing pavement slab condition evaluation, including drainage improvement recommendations as approved in the concept study report will be developed during preliminary design.

1. This submission will include 60 percent complete contract drawings of the preferred alternative, showing the scope of work, items of repair/modification, and details for new design or new features.
2. Preliminary design will be based on available data of member sizes from inspection reports.
3. Truss and floor beam members: Section loss due to corrosion will be compensated by designing and bolting steel plates.
4. Joints and connections: All riveted and bolted connections will be checked for strength and replaced if necessary.
5. A preliminary cost estimate and outline specifications will be furnished.

Submission will be accompanied by a QA form that indicates that PSQAP procedures were implemented.

9.2.4 Final Design Submission

Pre-final design submission: The pre-final design submission will be 100 percent complete, including plans, specifications, engineer's construction cost estimates, and construction schedules. The submission will include a quality assurance form that indicates the PSQAP procedures were implemented in the development of the submission.

Final design submission: This submission will include Mylar, along with a CD containing: CADD files, specifications, and the engineer's estimate. This submission will incorporate the commission's pre-final submission comments.

The final report will include recommendations/special provisions for modifying existing specifications

1. Worker safety during painting will be considered.
2. Nighttime construction will be guided by NCHRP Report # 475 and NCHRP Report # 476.
3. Structural steel replacement and/or strengthening: Review comments on the preliminary submission will be addressed.

9.2.5 Post-Design Services

Post-design/pre-award services will include:

1. Preparation for and attendance at the pre-bid meeting, and preparation of minutes.
2. Support and assistance in answering questions, and preparation of addendums.
3. Review and analysis of bids, and recommendation for award.

Upon completion of the bidding phase, incorporate addenda and re-issue original documents. post design/post-award services will include:

- Review of all shop drawings.
- Attendance at meetings.
- Responses to requests for information
- Provision of plan changes, responses to any design related question and/or request for clarification
- Preparation of as-built drawings.

9.2.6 Public Involvement and Community Outreach Program

The authority's project development process and the important role that public involvement plays in this process shall be considered. Providing the public with an understanding of the project and establishing trust with the communities is critical to the project's success. The public participation effort will provide an open and responsive forum where public input is considered during the decision making process.

1. The public participation effort will reach out to local residents, businesses, and elected officials. Stakeholders may include the merchants, inns, bed and breakfasts, county park, restaurants, and local commercial and retail businesses.
2. Inform the public of the authority's commitment to community improvements throughout the public participation process.
3. Plan to organize an approach that suits the project's needs, complies with federal and state requirements, and fits in the project's budget.
4. Public involvement will be completed with input from the project team, particularly the authority. The lines of communication will be kept open throughout the project.
5. A successful community involvement action plan (CIAP) identifies stakeholders early in the process and determines their sensitive issues. The team will implement authority's policy to conceive, scope, design, and build projects that incorporate design standards, safety measures, environmental stewardship, aesthetics, and community sensitive planning and design.
6. Coordinate with the PM in identifying key staff in local and county authorities, fire departments, rescue units, and other agencies or individuals required by the state to ensure that a proper level of attention will continue during the design process. The CIAP will effectively communicate the rationale for the project need, seek stakeholder input/feedback early in the project process, obtain stakeholder consensus, and maintain continuous lines of communication. Effective, early interaction with the stakeholders and formation of a partnership will result in the project flowing more smoothly and contribute to the project being completed ahead of schedule. Designs will be in accordance with the context sensitive design (CSD) philosophy.
7. Community outreach services will establish the identity of the project, disseminate information to the stakeholders about the anticipated impacts of the project schedule and scope, and respond quickly to questions from the community and residents pertaining to construction and changes to traffic patterns.

8. Coordinate closely with the community on traffic control planning, both business and residential accesses; utility relocations; and/or potential disruptions.
9. Develop and maintain a database of state, county, and local government officials; community groups; tenant organizations; local newspapers; block associations; local businesses; neighborhood improvement districts; schools; religious institutions; major employers; and state and local police, fire departments, and rescue units. The database will be updated regularly and used to obtain community input; distribute invitations to the public meetings, to be held early on during the design process; circulate press releases describing project progress; and distribute periodic community newsletters as necessary.
10. Also consider outreach methods such as flyers, newsletters, local community newspapers, notices to local community organizations, and developing/updating a project summary in the news release section of the authority's Website. This will keep the public informed of conditions, including traffic and traffic control.
11. Work with all project stakeholders, some of which include local officials, local residents, and businesses, to foster a plan to provide the least inconvenience to area residents and businesses, while providing sufficient time and area for construction. Public information meetings will be hosted and presentation graphics will be provided. The graphic information will identify land use, property ownership, maintenance of traffic, and work zone as directed by the PM.
12. Coordination and communication with the affected parties, such as state and local police, fire departments, rescue units, school districts, public transit agencies, and businesses, will include notice for lane closures, pattern changes, and traffic stops, especially during traffic control set-up and takedown, as necessary.
13. Consider CSD including aesthetic treatments (landscaping, railing treatments, etc.) pedestrian access, bicycle compatibility, transit friendly amenities, and preservation of environmental features to strengthen consistency with the state development and redevelopment plan. Depending upon budget constraints, transit enhancements could include any requested new bus stops

9.2.7 Utility Relocation and Coordination

1. Due to the critical nature of this task and its impact on schedule, the project team will start this activity immediately after receiving the notice to proceed.
   - Develop alternatives to minimize or eliminate the impact on utility facilities.
   - Based on proposed improvements, utility poles will be relocated, and the existing manholes and valves will be reset to the proposed elevations. Advance utility relocation will be considered to minimize construction duration. Work with the utility companies to satisfactorily resolve the possible conflicts in the most economical way and in a timely manner. Pole relocations with fiber optics will require relocation of the entire run and require longer time for construction.
   - Deck replacement may require relocation of the water main with fire hydrants. Determining the location of all project utilities is critical to avoid unwanted impacts during construction. Through the use of as-built plans obtained from the utility companies through our contacts that will be established for this project, approximate locations of underground utilities will be determined.
2. Overhead utilities will be documented in detail by a thorough field check of all electric, transformers, telephone, and cable lines. Completing this step will speed the utility documentation process by making it easier for utility companies to confirm their data. In order to avoid delays due to slow responses from the utilities, it is essential to stay in close communication with these companies. Accordingly, meet with each contact person identified in our initial contact letters to each utility.

3. The overall project schedule will be presented to the utilities, along with a detailed breakdown of the utility activity time line and the proposed relocations required for their facilities. Schemes of accommodation (SOA) will be developed for the proposed work. We find it effective to color code the SOA plans, enabling easy identification of the utilities and the proposed relocation scheme.
4. Working meetings will be held until an agreed upon SOA and timeline is reached. This will enable the utility companies to plan their work ahead of time so that the authority's schedule can be met or exceeded. Accordingly, the utilities can contribute as a member of the project team, rather than as an outside force that inhibits the schedule of this critical path item.
5. Rather than base utility impact costs on a percentage of construction that can be very inaccurate, use an existing database of utility costs from previous projects to accurately estimate the cost of utility work.

9.2.8 Drainage Design

The existing roadway drainage facilities within the project limits normally consist of primarily the storm sewer system.

1. Drainage design will be performed per the current drainage design manual. It is anticipated that a roadway drainage system primarily consisting of storm sewers and inlets will be provided, which will be tied into the existing storm sewer system at the project limits.
2. As with any highway located within a developed area, the presence of various businesses in close proximity to the roadway will require the runoff from the roadway to be completely contained within the roadway drainage system. Since the project will result in a newly created pavement area, storm water management (SWM) design will be provided.
3. We will determine the number of scuppers; replace existing short downspouts with downspouts carried down to ground level. Spray from the existing short downspouts usually contributes to beam section loss.

9.2.9 Lighting Design

A lack of lighting or poor lighting designs make the roadways unsafe and inefficient. A good walkway lighting system is crucial for safe pedestrian movement. Existing lights within the project limits will be impacted by replacement.

1. Lighting warrants will be studied for mainline, deceleration/acceleration lanes, and sign lighting to determine the need for continuous lighting.
2. Proposed lighting can use cutoff luminaires for better quality of illumination and driver comfort. An approved lighting analysis program such as CALAPro/Visual or an approved equivalent will be utilized to perform the photometric analysis.
3. Sign lighting will be provided using 250 watt mercury vapor luminaires. Lighting plans will be prepared in accordance with authority's procedures manual, CADD standards, and sample plans.
4. The clearances from the overhead utilities will be coordinated with the utility company and vertical and horizontal clearances will be provided in accordance with N.J.A.C. Chapter 25. The conflicts with other underground utilities will be identified and resolved.
5. Conflicts between proposed electrical facilities and other utilities (overhead & underground utilities), if not resolved during the design process, become major issues during construction. These conflicts can be eliminated with proper planning and coordination between the different disciplines during the design phase of the project by bringing in utilities and lighting task leaders for regular meetings on this issue.

6. Electrical service if required will be coordinated with the utility company and confirmation will be obtained prior to the final design. Safe and easy access for electrical maintenance will be considered during the design process.
7. A new lighting system will be evaluated to improve pedestrian and motorist safety during nighttime and bad weather. Electrical design will be performed in accordance with NEC requirements and NJDOT Roadway Design Manual.
   - Bridge lighting will be designed in accordance with ANS standard RP-8 Roadway Lighting. The existing offset type non-cutoff light fixtures will be replaced, if required, with full cutoff conventional type light fixtures.
   - High pressure sodium cutoff light fixtures are recommended for reducing light spillover, pollution, and glare and for increasing efficiency.
   - Approach sidewalks and bridge walkways will be illuminated in accordance with IES DG-5-94 Recommended Lighting for Walkway and Class I Bikeways. Light fixtures with decorative design will be considered. Fixture finish will be specified to match the structure paint color for bridge aesthetics.
   - An aesthetically pleasing lighting design will improve bridge aesthetics. Environmental specifications will be considered during the selection of the proposed lighting fixtures. Light fixtures should be suitable for installation on bridge structures.
   - The electrical installations will be designed in accordance with latest standards of National Electrical Code and NEMA and UL requirements.

9.2.10 Project Delivery on an Accelerated Schedule

For an early finish, try to accelerate the design process related to fast-track tasks and comprehensive assignments of rehabilitation and repair. A well-equipped matrix-formatted organization with a large, diverse staff, which permits the flexibility and capability to work on numerous simultaneous assignments, is desirable. A proactive, "fix-it-first" approach, allowing full focus on the project scope, will minimize the potential for scope creep. It is important to communicate with the authority to discuss project status, potential "bumps in the road," and the risk associated with project decisions relative to tasks related to schedule and budget. The design of the project can be accelerated by judiciously shaving the calendar time needed for specific work activities. Depending on complexity, most bridge, roadway, traffic signal, and drainage improvement projects completion should not exceed one year, including reviews and permitting, following notice to proceed. An estimated time savings on individual tasks of one to three months can add to commensurate cost savings of up to 10 percent of the final design cost.

Electronic submissions: Maximize the use of electronic submission of reports/deliverables, via an FTP site, to save on reproduction/shipping costs and time. Documents will be received by SME reviewers more quickly so that they can be reviewed expeditiously, thereby saving time and electronically archived.

On-board reviews: On-board reviews with the project team may be held throughout design. Prints may be hand-delivered to the project manager and SMEs at key development points and a team meeting held within two weeks. At this meeting, appropriate team members would be present and agree on the resolution of outstanding items. If any items are left unresolved at the end of this review meeting, dates for resolution would be agreed to, with remaining outstanding items bumped up to the next level for final resolution. Individual discussions will be held with key personnel prior to scheduling these meetings and all key personnel will have ample time to preview the plans in order to make the meetings most productive.

9.2.11 Environmental Issues/Permits

1. Environmental documentation: The scope of this project will qualify it for a categorical exclusion (CE) under NEPA regulations. Documentation will be prepared to confirm the

eligibility of the project per CE.

2. Cultural resources: By preliminary research ensure that there are no resources presently listed which make the bridge eligible for the National Register. Should eligibility be determined, consult with the HPO on appropriate treatments and include the results of that consultation.

9.2.12 Unanticipated Studies, Evaluation, and Assessment

If there is a need to perform work not originally scoped, add staff to the project, if necessary, so as to complete any additional tasks within schedule. In addition to the identified services, additional work for SUE, geotechnical, GIS processing of data; Web page design; landscape design; and ROW acquisition, negotiation, and relocation services may be required. Also provide prompt reevaluation of previously scoped improvements for appropriateness and adequacy as conditions may change.

9.2.13 Progress Reports

Monthly progress reports, generated using in-house software, should analyze each project task by comparing the percent complete to the percent of man-hours expended and the percent of budget spent. Both consultants and sub-consultants can use a similar monthly progress report format for the project. This will enable the PM to track actual costs versus budgeted costs both in-house and for the sub-consultants.

9.2.14 Quality Assurance/Quality Control

Quality assurance is the verification of the effectiveness of quality control measures. QA is a formalized system that documents the structure, responsibilities, and procedures required to achieve and deliver a quality product. Prepare a project specific QA plan (PSQAP) including the authority's capital project delivery procedures: design submission procedure, contract meeting checklist, CPM audit procedure, CPM quality management policy, department certification, designer certification, interactive communications procedures (prepare/submit a design communications report (DCR), quality management plan procedure, quality checklists, design services certification, and capital project delivery responsibilities matrices. The checklist items are the major design items that are important to assure accurate development of the contract documents. The authority's subject matter experts (SMEs) will continually interact with these units during the design process. This proactive approach serves to minimize comments and shorten comment resolution. In concert with our QA/QC team leader, both our PM and DPM will make sure that the team adheres to an approved QA plan. Corrective measures—an informal review of work on an ongoing basis with the appropriate SMEs prior to submissions—are required so that actual submissions generate a minimal number of comments, or preferably none. Expedite the resolution of any authority comments. Perform construction inspection as an integral QA/QC tool during the entire design process. Experienced construction engineers will be asked to make reviews periodically during the design process to ensure that the designs are constructible, practical, safe, and economical.

Regularly review the updates of the baseline document change, corrective action notices, and quality improvement advisory announcements on the NJTA Website to incorporate into the design. Consider including value engineering as part of normal QA/QC practices.

9.3 SHORT-TERM REPAIRS IN LIEU OF REPLACEMENT

9.3.1 Introduction

Anytime there are major and extensive repairs being proposed, an in-depth and thorough investigation of the condition of the concrete will be required:

The deck replacement decision shall be made based on whether the bridge structure geometry is substandard or is functionally obsolete, or if other major work is to be undertaken.

Repairing concrete materials, which are more than superficially damaged, is both expensive and problematic. Salvaging concrete containing corroding reinforcing steel or critically saturated aggregate may not result in a long-lasting component. When components can be completely replaced for less than the cost of extensive repair, replacement of deteriorated members should be pursued.

Bridge replacement: There are benefits for replacing an entire structure instead of rehabilitating the deteriorated areas. Complete bridge replacement is initially the most expensive solution, but can be necessary based on the overall extent of deterioration of the structure or inability to maintain traffic due to existing bridge width.

If the substructure of a particular structure is severely deteriorated, it is not cost effective to construct a new deck or superstructure when major substructure rehabilitation will be required in a few years. This approach also provides opportunities for geometric, traffic capacity, and live load capacity improvements.

When considering life cycle costs, bridge replacement can be the preferred alternative. It is also worthwhile to consider the advantages of constructing a new structure off-line from maintenance of traffic perspective. Traffic can move unimpeded on an original structure while the new structure is built along side of it.

The major disadvantage of a complete bridge replacement is the cost and construction duration. Costs for right-of-way, permitting, utilities, design, and construction can be significant and will limit funding to be spent on other needs.

9.3.2 Assigning Priority Repair Categories

A highway agency is required to maintain and fix several hundred bridges at any given time. With limited funding, degree of deficiency and relative risk factor need to be evaluated and selection of a series of bridges needs to be made. Priority repairs are also done in-house by maintenance forces and usually comprise of:

1. Deck joint repairs/reconstruction
2. Deck repair/reconstruction and resurfacing
3. Repairs to bridge and approach safety walk/sidewalk/curb/parapet/median
4. Bearing repairs/retrofits.

9.3.3 Functionally Obsolete Bridges

The causes of functionally obsoleteness include:

1. Substandard travel lane width.
2. Lack of shoulder and median.
3. Inadequate stopping sight distance.
4. Sharp horizontal alignment.
5. Sharp vertical profile/cross slope.
6. Low design speed.
7. Substandard guardrail.

Such bridges need to be replaced irrespective of their excellent structural condition.

9.3.4 Special Provisions

Special provisions that can be required in the plans and specifications for the following:

1. Construction staging.
2. Traffic controls and diversions.
3. Authorized detours.
4. Restricted working hours or days.

5. Load restrictions for construction equipment.
6. Posting for reduced speeds, substandard vertical under clearances, and/or load capacities.

9.3.5 List of Abbreviations with Descriptions

RSG	rolled steel girder
WPG	welded plate girder
RPG	riveted plate girder
MST	modular space truss
TWT	twin warren truss
CSBG	composite steel box girder
TBG	twin steel box girder
PCG	prestressed concrete girder
SRF	steel rigid frame
RCRF	reinforced concrete rigid frame
CSA	composite steel arch
RCBC	reinforced concrete box culvert
RCB	reinforced concrete beam
PCBB	prestressed concrete box beam

9.4 PRIORITIZATION OF BRIDGE DECKS

9.4.1 Recommended Deck Reconstruction Summary

The following steps may be followed:

1. Selection of sample population of bridges: Each bridge selected for the prioritization program must be representative on average of not less than three bridges. The sample is intended to be as representative as possible. However, the sample population should also contain bridges which exhibit a wide range of characteristics that are used in the vulnerability and importance algorithms, which covers all possible variations anticipated in the bridge population.
2. Bridge selection criteria— priority table format is given below.
3. Availability of information—the following information is required for preparing a study report: 7. Bridge importance criteria: The criticality of its role to the overall function of a network is considered important:
   - Bridge function (feature carried/feature crossed)
   - Traffic volume
   - Detour length considerations
   - Bridge size
   - Utilities.
4. Proposed list of bridges.
5. Format of priority. Table 9.1 is an example of bridge selection by tabulating the available information.

9.4.2 Highest Repair Category (Emergency)

Significant safety hazards to the traveling public need to be eliminated. The NBIS Recording and Coding Guide is used for condition evaluation and identifying deck defects during inspections.

Major defects are those that, if not repaired immediately, would require closing some lanes or the bridge itself. Such defects would affect the structural integrity of the bridge and could lead to a total collapse of the structure.

Table 9.1 Priority selection for rehabilitation.

Priority Structure	Method of Rehabilitation/ Reconstruction	Year Built	Rehabilitation/ Reconstruction Cost (2009 $)	FHWA Condition	NBIS/SI&A Sufficiency Rating (0–100)	BMS Deck Condition Rating
Bridge #1	Widen structure and overlay	1973	$7 million	Satisfactory	80.0	1.5

1. Deck (old SI&A Item #58) — coded 2.

   When from inspection, the following items are found to be in critical condition, emergency repairs are needed:
2. Superstructure (old SI&A Item #59) — coded 2.
3. Substructure (old SI&A Item #60) — coded 2.
4. Culvert (old SI&A Item #62) — coded 2.

Repairs under the emergency category must be done without delay. When delays are unavoidable, temporary repairs must be utilized. Examples are:

1. Crack in a non-redundant primary load carrying steel member.
2. More than 50 percent undermining of the bearing area of a non-redundant member.
3. Deterioration which causes a main load carrying member to become unstable.
4. Loose sections of concrete encasements located above the traveled roadway or sidewalks.
5. Missing sections of bridge railings.
6. Localized failure of the bridge deck or sidewalk.

9.4.3 Priority #1 (High)

Priority #1 defects are major defects such as those affecting the stability of the structure, which if not repaired may cause significant load restriction or partial collapse. Load restriction of affected area may be necessary. Major defects in the superstructure or substructure may cause a significant load restriction or partial collapse of the structure. Included in this category are defects affecting the stability of the structure.

Repairs under the Priority #1 category must be done as soon as possible when, from inspection, the following items are found to be in poor condition:

1. Superstructure (old SI&A Item #59) — coded 3.
2. Substructure (old SI&A Item #60) — coded 3.

Figure 9.1 Spalling of reinforced concrete diaphragm and rusting of exposed rebars.

3. Culvert (old SI&A Item #62) — coded 3.FHWA publication on Recording and Coding Guide is applicable. Generally, one of the inventory items (59 to 60 or 62) is coded 3 or less (poor condition).
4. Major repairs must start within one month. Examples are:
   - Longitudinal crack in a primary load carrying steel member in a redundant structure.
   - Collision damage causing major section loss to main load carrying member and requiring load posting.
   - Substantial (more than 50 percent) bearing area undermining of redundant load carrying member.
   - Major scour problem with footings not on piles.
   - Pier cap of a column bent in distress.
   - Damaged or missing sections of bridge railing.

   It is necessary to post load or speed restrictions to vehicular traffic until repairs are made.

9.4.4 Priority #2 (Medium)

Major defects in the superstructure, substructure, or deck, which if not repaired may cause a load restriction or partial collapse of the structure. When, from inspection, one of the following items is found to be in poor condition or coded 4 or less, repairs are required within three months:

1. Superstructure (old SI&A Item #59) — coded 4.
2. Substructure (old SI&A Item #60) — coded 4.
3. Culvert (old SI&A Item #62) — coded 4.

Examples are:

1. Short crack in a primary load carrying steel member in a redundant structure.
2. Substantial undermining of bearing area of a redundant load carrying member.
3. Major section loss or collision damage.
4. Major problems with the bridge railing.
5. Major scours problems with footings on piles.

9.4.5 Priority #3 (Low)

All other repair recommendations not falling into the previous three categories, emergency, high, and medium, including maintenance items, shall be classified under the low category.

Necessary repairs will be recommended in the bridge inspection report for further action.

9.5 SCOPE OF REPAIRS

9.5.1 Introduction

The scope of repairs is varied and is based on a host of conditions:

- Repairs cover a wide range of issues such as strength and durability of repair materials.
- Various procedures of maintenance such as strengthening and retrofit.
- Alternate repair option needs to be addressed.

Scope includes causes of cracking, testing for crack detection, control of cracking and practices for preventing cracks. Ways of minimizing strains and stresses that cause cracking are discussed:

- Common repairs or replacement is selected on the basis of each component

- Concrete repairs for deck (recurring/most common), curb, sidewalk, and parapet
- Deck overlay replacement; sealants
- Deck joints/expansion dam replacement at abutment backwall and pier caps and stems
- Concrete repairs for prestressed beams
- Concrete repairs for substructure
- Concrete repairs for foundations.

1. The scope of rehabilitation shall include all work required to assure satisfactory performance of the concrete deck, as well as supporting superstructure and substructure units. This may include items such as:
   - The removal of existing overlays
   - Removal and replacement of all deteriorated components
   - Complete removal and replacement of the entire bridge deck if necessary.

2. This work may also include repair or removal and replacement of deteriorated concrete curbs, sidewalks, parapets, as well as rail, deck joints, bearings, or similar incidental items which are associated with proper functional restoration of the structure.

3. Safety improvements should be undertaken with the above described work when such improvements eliminate an established hazardous condition. Such safety improvements may include widening, elimination of hazardous walks and substandard safety hardware, removal of hazardous fixed objects or the installation of an energy absorbing barrier system, and any other features that are consistent with current safety standards.

9.5.2 Major Repairs

Major repairs are expensive to fix and are often issues for older bridges, except where there may be accidental damage to a new bridge. For the alternative selected, the impact on disruption to traffic needs to be a minimum. Typical examples of repairs to steel and concrete are presented.

Major repair procedures:

1. As-built plans and shop drawings should be reviewed followed by a thorough site inspection making note of the material condition, fatigue prone details, utilities, geometry, girder alignment, and possible paint removal and containment considerations.

2. Details of particular importance to check are butt welded splices, partial length cover plate ends, welded lateral gusset plate connections, connection plate/stiffener welds, and shear connector welds in tension or reversal zones. Non-destructive testing should be performed on butt welded top flange splices to ensure weld soundness.

Figure 9.2 Deterioration of reinforced concrete beams due to lack of maintenance.

Fatigue sensitive details and stress risers of all types may be removed. All fatigue sensitive details must be analyzed.

1. Rivet holes and non-radius cuts cause stress increase.
2. Lateral connection plates should not be welded to tension flanges.
3. Rivet holes should be made round by reaming to eliminate crack initiation sites.
4. Riveted girders should not be retrofitted for continuity due to their uncertain fatigue performance and difficult splice detail requirements.
5. To carry the new live loads when widening, new load paths may be created. The stiffness of the new members and how the older adjacent members are to be strengthened should be considered.
6. Short term crack sealing and joint repair in steel.
7. Webs are fracture critical members (FCM). Fracture of thin webs is a dangerous scenario and needs to be fixed immediately.
8. Steels used in main members should be ordered to the correct level of strength and toughness. For main members, the material should specify Charpy V Notch (CVN) requirements for the FCM zone and reference the direction of rolling. FCM tension members on the drawings should be marked with (T).
9. Corrosion prevention: Maintenance painting of steel bridges is required.

9.6 ALTERNATIVE METHODOLOGY AND TECHNIQUES OF REPAIRS

9.6.1 General Repair Methods

The study involves review of:

1. As-built plans.
2. Reports.
3. Rating calculations.
4. A field/visual assessment at each of the seven structures using lane closings for top deck inspection and for bridges on rivers by use of boat for the underdeck inspection.

9.6.2 Alternate Analysis

Following the evaluation phase, an alternative analysis is prepared for bridge deck rehabilitation and reconstruction. Potential deck reconstruction methods to be considered for each bridge shall account for:

1. The deck condition.
2. Overall bridge condition.
3. MPT/staging.
4. Bridge age, structure, or component life expectancy.
5. Construction schedule.
6. Constructability issues.
7. Outstanding design issues.

9.6.3 Deck Rehabilitation Methods

The deck rehabilitation and reconstruction methods include:

1. Staged cast-in-place high performance concrete bridge deck reconstruction.
2. Superstructure replacement.
3. Accelerated construction methods such as prefabricated deck panels and superstructure systems, and use of accelerated cure cast-in-place concrete.

4. Localized deck replacements and repairs with new overlays.
5. As a last resort, bridge replacement.

Additional rehabilitation work items may be required with deck reconstruction projects, such as:

1. Bearing or superstructure repairs.
2. Bridge widening.
3. Superstructure painting.
4. Lighting improvements.
5. Drainage improvements.
6. Elimination of deck joints.
7. Seismic retrofit.
8. Approach roadway work.

9.6.4 Rehabilitation Study Report

A rehabilitation study report is prepared and recommendations based on the above criteria are made. It is better to prioritize the group of bridges which are due for repairs since it is difficult to work on all bridges at the same time. The following detailed assessments are needed:

1. Maintenance and protection of traffic.
2. Study of full or partial detour.
3. Cost estimates for each alternative.
4. Life cycle cost.
5. Environmental permit assessment.
6. Right-of-way acquisition.

9.6.5 Comparative Study of Rehabilitation Alternatives

The following rehabilitation methods need to be considered:

1. Cast-in-place bridge deck reconstruction such as a single course cast-in-place high performance concrete (HPC), normally with staged construction. The advantages of conventional staged cast-in-place HPC bridge deck reconstruction include:
    - Composite action can be achieved improving the live load capacity.
    - C.I.P. construction more readily accommodates geometric conditions.
    - High performance concrete provides an extended service life.
    - The entire deck replacement at the same time leads to uniform performance for shrinkage and creep in concrete.
    - Greater contractor familiarity.
2. The disadvantages of conventional staged cast-in-place HPC bridge deck reconstruction include:
    - High traffic impacts
    - Long construction duration
    - Weather restrictions
    - Non-uniform stresses in deck.
3. Superstructure replacement: It is particularly advantageous for rapid construction and in addressing deterioration or deficiencies in the superstructure. This method has the same advantages as conventional deck replacement:
    - Increased life expectancy reduces future temporary traffic impacts that would otherwise be experienced under a repair approach (for example bearing retrofit or bridge painting).

- Design issues such as span length, cost, and importance of bridge may influence the decision for replacement.

9.6.6 Accelerated Construction

Accelerated construction typically provides the shortest construction duration and the least traffic impacts.

1. Precast panels: Precast concrete deck panels are formed and poured in a precasting yard prior to construction, placed in the field, made composite with the girders via shear studs, post-tensioned and typically overlaid. The advantages of using precast concrete deck panels include the following:
   - Shorter on-site construction duration
   - Minimized traffic impacts
   - Improved quality control and less cracking.
2. Prefabricated superstructure systems.
3. Accelerated cure cast-in-place concrete: The disadvantages of using precast concrete deck panels include the following:
   - Maintenance and long term performance concerns (susceptible to leakage) due to the number of joints
   - Involves post-tensioning, typically in both the longitudinal and transverse directions
   - Difficulty in accommodating anything but simple geometric conditions (horizontal curves, skews, cross slopes, profiles)
   - Difficulties in constructing future replacement/repairs due to joints and post-tensioning
   - Initial construction cost is greater than cast-in-place deck replacements.
4. Localized deck replacement/repair with overlay: The localized deck replacement and deck repair approach has been used for nearly 40 years on bridges. This approach typically involves annual contracts that replace the highest priority localized areas of decks, repair deck spalls, repair/replace deck joints, and replace surfacing. The benefits of performing this type of work include:
   - Targeting the areas which are in the worst condition and avoiding emergency repairs
   - Maintains structures in functional condition with a limited budget
   - Allows for work on numerous structures by focusing on the highest priority areas, not just a single bridge
   - Cost effective when considering the limited construction budget and age of the bridge.

   However, there are a few downfalls to this method of repair, including:
   - Difficult to maintain traffic depending on number of lanes, shoulder widths, and superstructure configuration
   - Very expensive due to MPT costs and limited duration of lane closings
   - Repairs are made on the same bridges on a periodic basis
   - Limited number of contractors available who are capable of performing this type of work
   - Long term quality is compromised due to limited duration of construction stages
   - Increased unit cost because of reduced quantity
   - Eventually no longer economically feasible as the deterioration rate increases due to the age of the bridge.
5. Program schedule and funding: The bridges in this study were evaluated to determine the preferred reconstruction/rehabilitation alternative at each location, as previously summarized. Then, considering condition, life expectancy, and MPT impacts, the structures were

prioritized to assist in developing a reconstruction/rehabilitation program for implementation by the authority. The following page is a suggested schedule for design and construction of the preferred alternative at each bridge location. The schedule also provides an approximate duration for preparation of the environmental permits and for environmental agency review.

A comparative study of alternative methods of repairing cracks and joints is required. Case studies of successful repairs for past projects need to be made.

Figure 9.3 summarizes some of the common approaches to repair of concrete, including full or partial replacement or crack sealing. Other methods are partial casting, packing, overlaying, and grouting. Repairs are based on condition of defects and can be classified as:

1. No repairs needed.
2. Minor repairs needed.
3. Major repairs needed.
4. Partial replacement needed.
5. Full replacement needed.

Minor repairs are commonly needed. Major repairs are issues for older bridges except where there may be accidental damage to a new bridge. Even newer bridges may sometimes require repairs to decks, parapet, etc. resulting from lack of quality control during construction. The cost of new repairs is usually covered under a warranty.

9.6.7 Typical Repair Items and Approximate Unit Repair Costs

In preparing cost estimates for inspection (Items 94 to 96 of SI&A Sheet) and rehabilitation reports, unit repair costs for various items are required. Unit costs vary from state to state and an inflation index needs to be used.

An alternative is to use bid tabs or records of latest contractor's bid prices, preferably in the same county and for a similar bridge. An average value for a number of similar bridges may be used.

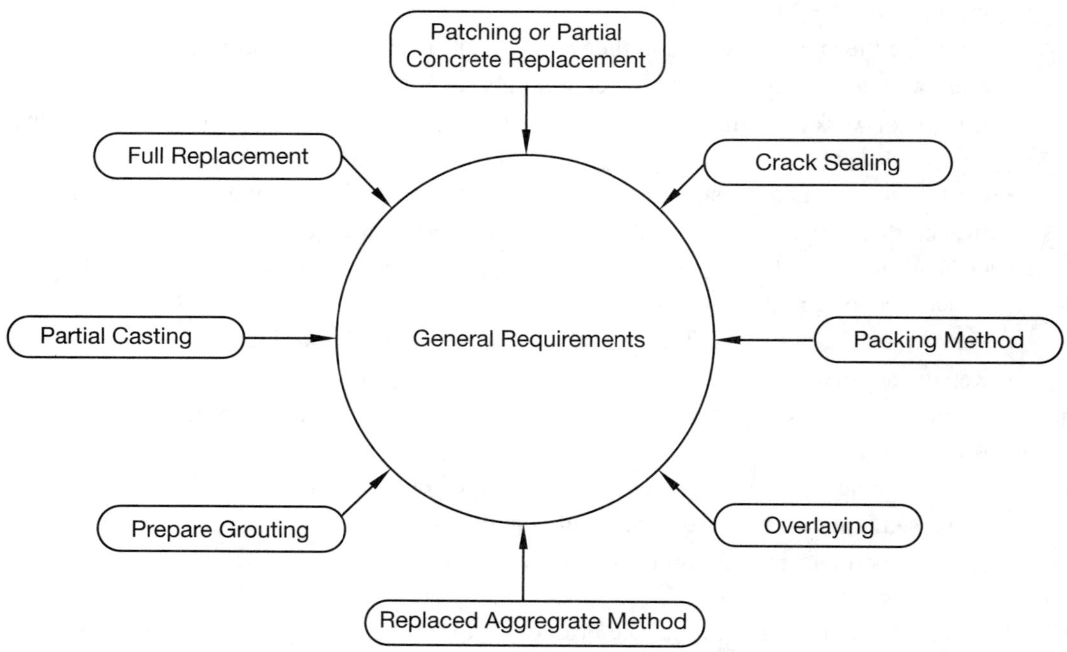

Figure 9.3 General repair methods.

1. Unit concrete repair costs (costs include: material, labor, equipment, and high reach rental).

	Item	Estimated Cost
Minor Repairs		
Type 1	Routing and sealing cracks	$10 per ft.
Type 2	Epoxy injection	$50 per ft.
Major Repairs		
Type 3	Demolition, cleaning rebars, putting repair material area over 50% of beam length.	$80 per sq. ft.
Type 4:	For smaller areas less than 50% of beam length	$55 per sq. ft.
	Corrosion Inhibitor for rebars	$2.5 per sq. ft.
	Painting and coating of concrete surfaces	$6.0 per sq. ft.

2. General bridge repair/replacement approximate costs (for relative differences only).

Table 9.2 Relative costs of rehabilitation.

Repair Item I. D.	Description	Estimated Cost
1	Mobilization, field office, providing site access, etc.	5–10% of project cost
2	Clearing site and demolition (depending on size of project and what can be salvaged)	10–15% of project cost
3	Staged construction (depending on number of stages)	10–20% of project cost
4	Preliminary engineering—small projects	10–15% of project cost
5	Preliminary engineering—large projects	8–10% of project cost
6	Bridge replacement—small simple bridges	$250 per sq. ft.
7	Bridge deck replacement—8½" to 9" thick (includes removal, traffic control and safety)	$175 per sq. ft.
8	Temporary shielding during deck replacement	$15 per sq. ft.
9	Bridge replacement—medium simple bridges	$300 per sq. ft.
10	Bridge Replacement—large simple bridges	$350 per sq. ft.
11	Bridge Replacement—complex/movable bridges	$390 per sq. ft.
Raising Superstructure to Increase Clearance—Two Lanes		
1	Jacking and support	$200,000/span
2	Modify abutment	$95,000 each
3	Modify pier	$75,000 each
4	Approaches reconstruction	$130,000
5	Traffic control	$80,000
Raising Superstructure to Increase Clearance—Four Lanes		
1	Jacking and support	$350,000/span
2	Modify abutment	$175,000 each
3	Modify pier	$95,000 each
4	Approaches reconstruction	$195,000
5	Traffic control	$130,000

Repair Item I. D.	Description	Estimated Cost
Lowering Roadway up to 3 Inches to Increase Clearance (including traffic control)—Two Lanes		
1	Arterial highway (1000 feet improvement length)	$110,000/span
2	Other highway (500 feet improvement length)	$90,000 each
Lowering Roadway up to 12 Inches to Increase Clearance (including traffic control)—Two Lanes		
1	Arterial highway (1000 feet improvement length)	$700,000/span
2	Other highway (500 feet improvement length)	$450,000 each
Lowering Roadway up to 3 Inches to Increase Clearance (including traffic control)—Four Lanes		
1	Arterial highway (1000 feet improvement length)	$240,000/span
2	Other highway (500 feet improvement length)	$180,000 each
Lowering Roadway up to 12 Inches to Increase Clearance (including traffic control) —Four Lanes		
1	Arterial highway (1000 feet improvement length)	$1,200,000/span
2	Other highway (500 feet improvement length)	$800,000 each
Re-striping Lanes to Increase Lateral Clearance (including traffic control)		
1	Re-striping lanes	$7,000 each
Repair Items		
1	Milling and removing bituminous pavement surface	$18 per sq. yd.
2	Scarify concrete decks (up to 1 ¼" thick)	$45 per sq. yd.
3	Waterproof membrane (includes traffic control & safety)	$95 per sq. yd.
4	Bituminous concrete (includes tack coat) 1½" thick	$80 per sq. yd.
5	Bituminous concrete (includes tack coat) 2" thick	$90 per sq. yd.
6	Bituminous concrete (includes tack coat) 3" thick	$100 per sq. yd.
7	Latex modified concrete 1¼" thick (includes traffic control & safety)	$130 per sq. ft.
8	Partial depth concrete deck repairs to 1" below top rebar mat (includes removal, traffic control & safety)	$120 per sq. ft.
9	Full depth concrete deck repairs to 1" below top rebar mat (includes removal, traffic control & safety)	$175 per sq. ft.
10	Concrete approach slabs (18" thick)	$240 per sq. yd.
11	Rebuild settled concrete approach curb	$70 per LF
12	Structural steel (large quantity)	$2.5 per LB
13	Replace structural steel (small quantity)	$25 per LB
14	Heat straighten structural steel (large beam)	$25,000 each
15	Temporary shoring of girder/beam	$6,000 per beam
16	Remove concrete encasement/ton of steel	$250 per ton
17	Prestressed beam end repair	$950 each
18	Resetting tilted bearings	$6,000 each
19	Concrete replacement (class B with rebars)	$1600 per CY
20	Epoxy waterproofing seal coat	$12 per sq. ft.
21	Pressure inject concrete crack with epoxy	$90 per LF

Repair Item I. D.	Description	Estimated Cost
22	Seal concrete crack with epoxy	$15 per LF
23	Soil anchors	$1,800 each
24	Soil anchors (prestressed)	$6,500 each
25	Rock anchors	$1,900 each
26	Rock anchors (prestressed)	$7,500 each
27	Steel sheet piling	$55 per LF
28	Timber piles (12" diameter, treated)	$70 per LF
29	Erosion fill material	$75 per CY
30	Channel excavation (small quantity)	$75 per CY
31	Channel protection (rip rap)	$85 per CY
32	Channel protection (gabion)	$95 per CY
33	Hot poured deck joint sealer	$15 per LF
34	Repack and seal pourable deck joint	$40 per LF
35	Replace deteriorated elastomeric compression joint sealer	$50 per LF
36	Replace deteriorated elastomeric strip joint sealer	$50 per LF
37	Replace deteriorated modular deck Joint	$60 per LF
Slope Protection		
1	Concrete slab	$160 per sq. yd.
2	Concrete bag	$250 per sq. yd.
3	Rip rap	$90 per sq. yd.
4	Bituminous treated	$100 per sq. yd.
Concrete Parapet and NJ Barrier Curb (Class A Concrete)		
1	2' 8" high rectangular	$180 per LF
2	NJ barrier type	$230 per LF
3	Parapet repair (patch with epoxy concrete)	$120 per LF
4	2' 8" x 2' 0" (double)	$275 per LF
5	2' 8" x 1' 3" (single)	$220 per LF
Metal Bridge Railing		
1	One rail	$90 per LF
2	Two rail	$130 per LF
3	Three rail	$140 per LF
4	Four rail	$160 per LF
Replace Deck Joints: Elastomeric Compression Joint Sealer/No Armoring		
1	1¾" x 2"	$60 per LF
2	2½" x ¾"	$75 per LF
3	3" x 3"	$85 per LF
4	5" x 5⁵⁄₁₆"	$95 per LF

Repair Item I. D.	Description	Estimated Cost
Replace Deck Joints: Elastomeric Compression Joint Sealer/With Armoring		
1	1¾" x 2"	$175 per LF
2	2½" x 2 ¾"	$190 per LF
3	3" x 3"	$210 per LF
4	5" x 5⁵⁄₁₆"	$225 per LF
Replace Deck Joints: Elastomeric Compression Joint Sealer/With Armoring (With Deck Replacement)		
1	1¾" x 2"	$160 per LF
2	2½" x 2¾"	$165 per LF
3	3" x 3"	$175 per LF
4	5" x 5⁵⁄₁₆"	$190 per LF
Guide Rail		
1	Steel beam rail	$80 per LF
2	Steel beam rail – bridge mounted	$130 per LF
3	Attachment to parapet	$1800 each
4	Stiffening in transition zone	$650 each
5	Rubrail	$30 per LF
6	Additional posts	$60 each
7	Slotted rail end terminal	$2500 each
8	Spacer blocks	$25 each
Paint Superstructure Steel		
1	No sand blasting—wire brush clean	$250 per ton
2	Sand blasting needed	$380 per ton
3	Sand blasting and containment needed	$580 per ton
4	Bearings (cleaning and painting)	$250 each

9.7 REHABILITATION OF STEEL AND PREVENTION OF CORROSION

9.7.1 Steel Trusses and Floor Beams

1. Structural evaluation for the condition of existing deficiencies will be based on FHWA publication on Recording and Coding Guide.
2. Section loss due to corrosion will be compensated by designing and bolting steel plates.
3. All damaged steel members will be repaired in the floor system and the sidewalk damaged by impact from flood.
4. Any tilt of floor beams on supporting deck needs to be rectified.
5. Structural steel painting: Working within standard technical specifications, recommend a paint system including surface preparation, primer, and protective layers.

   Any pitting of top flange areas of floor beams will be repaired by grinding steel surfaces. Structural painting will be required.
6. Damaged rivets and bolts will be replaced.

7. Post-tensioning of trusses: Existing post-tensioning rods serve an important structural function of maintaining the stability of deep trusses, both for near-term and long term. They help to increase the redundancy of fracture critical members and their strength. However, rods may need replacement due to elongation and sagging with higher tensile strength steel cables.
8. Staging: For work on floor beams and main trusses, the bridge will be closed for a restricted duration, after public outreach.
9. Accelerated construction: As an alternate, use of prefabricated units (former Inverset System) will be considered to minimize construction duration. With a precast deck, there may be difficulties in relocating and re-installing utility cables under the deck and attaching them to new girders since the deck slab needs to be composite with the girders.
10. Gusset plates and connections: All riveted and bolted connections will be checked for strength and replaced if necessary. Gusset plates showing loss of section will be replaced. Cracked welds for flange cover plates due to reversal of stress and fatigue need attention. Any cracked gusset plates for truss compression members may lead to failure of the bridge and need to be replaced. Failure of the Minneapolis Bridge is a case in point.
11. Sidewalks, if present, will have bicycle/pedestrian compatibility.

9.7.2 Heat Straightening

Heat straightening is a very old technique to restore deformed steel member by gradual heating and cooling. The procedure is more of an art than a science and requires experienced craftsmen. Beams or girders that have been struck by trucks or are bent by other causes can often be repaired by heat straightening only, or in combination with field welding to install new sections for the damaged steel member portions. Steel can be bent from overload, collision, earthquake or fire. If heat straightening is deemed to be practical, a detail showing the location of the repair and procedures needs to be prepared in the form of a report.

A repair procedure is generally used to straighten plastically deformed regions of damaged steel by applying repetitive heating and cooling cycles. Each cycle leads to a gradual straightening trend. Maximum temperature is controlled so that thermal stress from heat shall not increase the yield stress of steel. The damaged bridge girder is not removed while the heating operation is in progress. Only the regions local to the damaged area need to be heated.

Steel has the capacity to restore to its original condition through heating. Performance of repaired steel does not change. The alternate method of hot mechanical straightening uses an external force by which properties of steel are affected and early fracture can take place.

9.7.3 Use of Weathering Steel

It is recommended to replace any existing A7 steel with ASTM A709 Grade 250 or Grade 345 whenever possible. FCM Zone 2 steel should be used for FCM members.

Painting structural steel: Only 3 to 5 foot lengths at the end of beams need to be painted. Recommend a paint system including surface preparation, primer, and protective coatings. This task will include: a paint condition assessment (including document review and preliminary project planning); field survey (including visual quantification of degree of rusting or coating defects, dry film thickness, adhesion, substrate condition, sample collection, photographs, and limited environmental/worker risk assessment); preliminary design, quantity take-off, and preliminary estimates; and final report preparation and specification recommendations (the final report will include recommendations/special provisions for modifying specifications).

9.7.4 Structural Steel Painting

1. Coating system assessment: Any loss of section will be investigated. The following information will be collected: visual quantification of degree of rusting or coating defects; dry film

thickness; adhesion; substrate condition; sample collection; and photographs. Following the completion of the field survey and laboratory testing, alternative maintenance strategies will be assessed. This will include evaluating the potential maintenance painting strategies, coating systems, and options for surface preparation.

Working within standard technical specifications, only paint experts will recommend a paint system including surface preparation, primer, and protective coatings. This task will include:

- Paint condition assessment (including document review and preliminary project planning)
- Field survey (Including visual quantification of degree of rusting or coating fefects, dry film thickness, adhesion, substrate condition)
- Sample collection, photographs
- Environmental/worker risk assessment
- Following the completion of the field survey and laboratory testing, assess alternative maintenance strategies which will include evaluating the potential maintenance painting strategies, coating systems, and options for surface preparation.
- Drawings, field measurements, and photographs will be used to quantify the amount of steel painting. This will likely involve establishing zones of the structure (e.g., the splash zone above the roadway, floor system below the roadway) that may require different surface preparation or coating materials than other zones.
- Preliminary cost estimates will be developed based upon a painting program with due consideration for the potential worker, environmental, and containment issues.
- This information will be presented in a draft report and submitted for comments. A final report will be prepared and submitted following the review of all comments. Appendices will be included to contain all new data, laboratory results, calculations, or other technical data developed during the course of this work.

In 1999, FHWA carried out a review program for the bridge coating process. The purpose of the review was to evaluate the overall quality of the bridge coatings program and perform an assessment of the adequacy of the current bridge painting specifications as they relate to lead paint removal and field coating. Also, statewide policies for determining full paint removal versus spot painting and overcoating were examined to determine if cost-effective decisions for the preservation of steel structures were being made. The outcomes of this review included:

- Modifying and simplifying the current bridge coating specifications
- Revising statewide programming and scoping policies
- Forming a conclusion as to whether the shift from painting during rehabilitation projects to paint-alone contracts is beneficial to the state.

9.7.5 Scope of Review for Metal Corrosion

Visual inspection was used to assess coating performance. Cost data for these projects was collected whenever possible. The review included an assessment of some experimental coating systems that were recently applied. The goal of the review was to:

1. Determine the condition of existing steel substrate for the purpose of developing surface preparation recommendations.
2. Evaluate methods for surface cleaning and paint removal.
3. Recommend a final coating system.
4. Test existing paint for heavy metals.

These are more fully elaborated in the following:

- Identify chloride contaminated steel surfaces and determine chloride concentration on steel surfaces. Determine the extent of contamination on the existing steel surface using appropriate field measurement techniques with sensitivity approaching achievable in the laboratory. SSPC SP-12, "Surface Preparation and Cleaning of Steel and Other Hard Materials by High-and Ultrahigh-Pressure Water Jetting Prior to Recoating," contains definitions for surface cleanliness that include measurement of water soluble chlorides, iron-soluble salts, and sulfates.
- Cleaning methods for typical bridge components: Evaluate alternative methods for surface cleaning and paint removal. For each method, discuss the merits, environmental considerations, containment requirements, costs, and other issues. Evaluate areas of high chloride contamination, complex geometry, horizontal and vertical areas, and thick, adherent paint. The evaluation should be consistent with the objective of the project.
- Recommendations for the final coating system: Projects requiring a new paint system utilize a proprietary organic zinc rich self-curing primer coat, moisture cured aliphatic urethane intermediate coat containing a micaceous iron oxide, and moisture cured aliphatic urethane finish coat.
- Provide recommendations and necessary backup documentation for the surface preparation and high performance coating system to be used on the structure.
- Lead exposure: During cleanup operations, make sure that no lead-contaminated debris is left on the ground nor allowed to enter any nearby waterways. Lead is a very common element in our environment and has been used in materials such as paints and car batteries for many years. Lead can be hazardous to humans, particularly children, under certain conditions. Lead was a common component of industrial paints until the 1980s, and many of the steel bridges in the highway system are still coated with paint that contains up to 50 percent lead by weight.

High lead-containing primers can often be identified by their red or bright orange color. However, not all red and orange paints contain lead, and some paints of different colors can also contain a significant amount of lead. Lead hazards should be taken seriously for a relatively small amount of ingested or inhaled lead dust can elevate a person's blood lead level. Protection from lead hazards is not difficult to achieve. Proper respiratory protection should be worn. "Proper" protection consists of either air-fed, positive pressure respirator hoods (as worn by abrasive blasters), or negative pressure, filter-cartridge respirators. The required level of respiratory protection depends on the concentration of lead in the breathing air, and on the amount of time you are exposed.

Studies have shown a direct correlation between elevated blood lead levels in workers and "hand-to-mouth" lead ingestion. Washing your hands and face prior to eating or smoking is essential to avoid ingestion of lead particles.

9.7.6 Coating Systems Using Inorganic Zinc-Rich Primer

Standardization of design, fabrication, and erection processes is possible by applying Standard Guide Specifications developed by the AASHTO and NSBA. The painting document establishes and defines the functions, operations, requirements, and activities needed to achieve consistent quality in steel bridge painting.

To simplify the application parameters for similar zinc-rich primer on new steel bridges, a series of charts have been developed. These charts provide a convenient summary listing the detailed requirements for surface preparation, environmental conditions, coating application, curing, and verification testing.

9.7.7 Coating Systems Using Fluoro-Polymer Resins (FEVE)

Fluoroethylene vinyl Ether resins were developed in Japan. They can be applied in the field or in the shop with ambient curing. They offer excellent weatherability against corrosion under field conditions. The longest suspension bridge in the world, Akashi Kaikyo Bridge in Japan, was painted using FEVE and the coating is expected to have a life of at least 60 years.

In the U.S., the historic Shelby Street Bridge in Nashville TN was repainted using Fluoroethylene coating in 2003.

9.7.8 Primers, Intermediate, and Top Coats

The specifications cover three coating systems as described in Table 9.3.

Table 9.3 System identification.

System	Primer	Intermediate	Topcoat
System 1	Shop	Shop	Shop
System 2	Shop	Shop	Field
System 3	Shop	Field	Field

Table 9.4 Top coat inspection.

Requirement	Basis for Acceptance
1. Current painter qualification verified	Applicator QC plan
2. Ambient temperature	Product data sheet
3. Dew point and humidity	Product data sheet
4. Surface temperature	Product data sheet
5. Top coat component batch number	Owner approved batch numbers
6. Intermediate coat evaluation and repair	SSPC PA 1 and approximate procedure
7. Intermediate coat recoat time	Product data sheet
8. Verification of int. coat surface cleanliness	SSPC-SP 1
9. Date and time	N/A
10. Piece mark or bundle	N/A
11. Temperature of mixed top coat	Product data sheet
12. Top coat mixing and/or straining	Product data sheet
13. Top coat induction time	Product data sheet
14. Top coat pot life	Product data sheet
15. Stripe coat	Product data sheet
16. Topcoat dry time	Product data sheet
17. Top coat DFT	Table 3.1
18. Visual inspection	SSPC-PA 1
19. Adhesion	ASTM D 3359
20. Paint system final evaluation and repair	SSPC procedure

9.7.9 Protective Coating Type

Record the type of protective coating that has been applied to the span being inventoried.

1. Painted, lead based—The superstructure has a lead-based topcoat or non lead-based topcoat applied over existing lead based paint.

2. Painted, not lead-based—The superstructure is completely painted with non-lead based paint.
3. Painted, unknown—The superstructure is painted, but it is not known if lead is present in paint.
4. Unpainted (no coating) —The superstructure is not painted and has no other coating (e.g., weathering steel).
5. Galvanized or metalized—The primary members are hot-dip or mechanically galvanized, have flame sprayed coating system or an aluminized surface.
6. Bituminous based coating.
7. Concrete coated—A concrete coating has been sprayed on the primary members, or the primary members have been encased in concrete.
8. Coating containing asbestos—Coating contains asbestos.
9. Other coating—For any coating that does not fall in any of the above categories.
10. Localized painting, lead based—The superstructure has been painted in specific areas only (under joints, in splash zones, etc.).

9.7.10 Painting Inspection

Project #___________________

Date:______________

Inspector:___________________

Verification that the paint has not exceeded its shelf life. ___

Verification that paint is not stored in areas subject to temperatures beyond recommended limits. ___

Verification that the paint has not exceeded its pot life. ___

Documentation of any use of thinner. Do not exceed recommended maximum. ___

Verification of curing time. After painting, inform the contractor of the estimated time that should be allowed for the paint to cure. ____

Do not allow another coat to be put on until the appropriate amount of time has elapsed per the existing weather conditions. ___

Comments:

9.8 GALVANIC TECHNOLOGY

9.8.1 Introduction

During concrete condition inspections, areas of potentially active corrosion of the reinforcing steel can be seen in sound concrete. Galvanic methods of corrosion protection are used to combat the underlying corrosion. For repairs in either chloride-contaminated or carbonated concrete, embedded galvanic anodes minimize corrosion of the reinforcing steel adjacent to the repair. Successful methods in practice are shown in Figure 9.4. For details of the specialized method, see a paper by Glass, et al and others (Chapter 7 Bibliography).

Examples of uses include:

- Bridge deck widening

- Replacement of deck joint nosing
- Concrete pile jacketing.

Corrosion of steel can be slowed down by supplying a small electrical current to the reinforcing steel. The anode corrodes to galvanically protect the reinforcing steel. Anodes are installed in chloride-contaminated or carbonated concrete surrounding a patch repair or along a joint between new and existing concrete (Figure 9.4a). Embedded galvanic anodes are installed in a grid pattern by burying them within concrete. Anodes are installed in areas of the concrete where there is a high likelihood of corrosion occurring or recurring (Figure 9.4b). When mortar is placed around the anode, it begins to sacrificially protect the adjacent reinforcement.

All deteriorated concrete should be removed from around and behind the reinforcing steel inside the repair area (Figure 9.5a). Prior to installation, electrical continuity of the reinforcing bar within the repair area should be confirmed with the use of a DC ohm meter (Figure 9.6b).

Anodes are used in sound chloride-contaminated or carbonated concrete to prevent the onset of delamination or spalling of the concrete (Figure 9.4c). Embedded galvanic anodes reduce the corrosion activity of the reinforcing steel in the vicinity of the installed anode.

9.8.2 Repair Procedures for Galvanic Technology

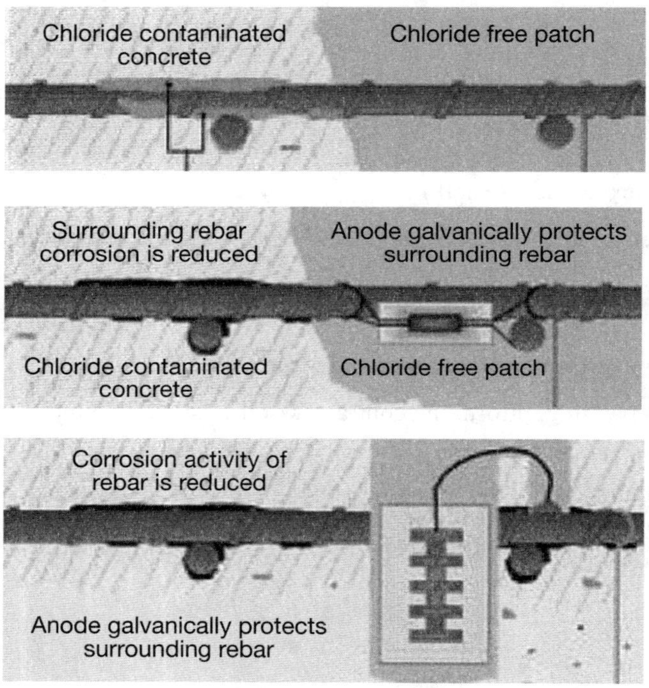

(a) accelerated corrosion

(b) ring anode corrosion

(c) chloride accelerated corrosion

Figure 9.4 Accelerated corrosion due to potential difference between patch and chloride contaminated concrete.

9.8.3 Use of High Performance Corrosion Protection Reinforcing Bars

1. The dual-coating process for rebar uses a thermally applied zinc coating for cathodic protection and durability. The Electro-statically applied powder outer coating acts as first line of defense against corrosion caused by water and chlorides.
2. Steel reinforcing bars are dual coated with zinc and epoxy coatings. Such rebar conform to ASTM A1055 Specifications. This process involves thermal bonding to a conventional reinforcing steel bar, a zinc inner layer under a polymer outer layer. These dual layers serve as a redundant and powerful system against corrosion.

3. Saving rebar splice lengths: Opposing thread bars can be spliced together by using couplers with hexagonal nuts and lock nuts (supplied for example by DYWIDAG Threadbar Reinforcing Systems).

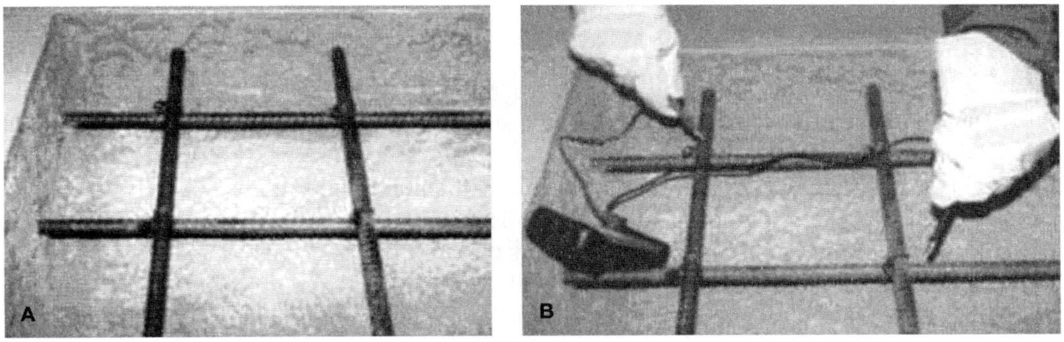

Figure 9.5 Reducing corrosion in steel rebars.

9.9 CORROSION IN THE POST-TENSIONING REGIONS OF BEAMS

1. Post-tensioning with high tensile strands of 270 ksi yield resists applied loads via increased compression. However, corrosion of prestressing strands is a common problem and has been a cause of failure.
2. For example, corrosion in post-tensioning in Florida bridges can be traced to several contributing factors.
    - Florida's humid saltwater environment
    - Insufficient grouting procedures produced air voids in the tendons that were grouted
    - Shrinkage cracks and leaks can compromise anchor protection by ordinary concrete pour backs
    - Flawed sealing of epoxy joints in precast segmental bridges can compromise the corrosion protection of internal tendons. Furthermore, discontinuous ducts at precast segment joints, along with imperfect epoxy joint seals, allow direct access of water to tendons not fully grouted.
    - High-density polyethylene ducts of some external tendons suffer splits, allowing moisture direct access to grout or strands.

9.10 DEVELOPING NEW CONSTRUCTION PRODUCTS

9.10.1 Use of HPS

In recent years, new stronger type of steels have revolutionized the load carrying capacity of steel girders beyond the capacity for Grade 36 used for older bridges and Grade 50 for the newer bridges. It is now possible to use shallower girders, if they meet maximum deflection criteria. Further, HP weathering steel minimizes repainting requirements, which reduces maintenance costs significantly. Shallower girders may be best suited for deck widening projects, where vertical under clearance is limited.

HPS may be used as stay-in-place formwork, open and closed wire rope and strand sockets, strand assemblies, open and closed bridge sockets, anchor sockets, turnbuckles, and specialized cable castings and forgings.

9.10.2 Waterproofing

Waterproofing for bridge decks uses a prefabricated membrane consisting of a reinforcement of synthetic, non-woven material thoroughly impregnated and coated with modified bitumen. The reinforcement provides remarkable physical and mechanical properties. It is protected by a thermo-fusible plastic film on the underside and has mineral surface protection in the upper face.

9.11 DRAINAGE

9.11.1 Superstructure

Ensure that the existing scuppers and down spouting are repaired, cleaned, or replaced and splash blocks are provided if none exist.

Lightweight drainage system: A bridge drain, which uses lightweight fiberglass scuppers, can be installed in the bridge deck surface to collect storm water. The scuppers are connected into a socket and spigot pipe system that transfers the runoff to catch basins under the bridge. A shallow, invert design accommodates bridge deck depth restrictions. A collection bell pipe fits onto the scupper outlet (allowing for bridge movement) and connects the scupper to the pipe system.

The FRP pipe system is resistant to weathering, including freeze/thaw cycles and temperature differences. The frame for the scupper is offered in stainless steel or coated steel, depending on whether it is used in corrosive or non-corrosive environments.

Class E ductile iron grates on the scupper can be specified as slotted.

9.11.2 Substructure

For MSE wall abutments, provide drainage according to standard drawings. For other abutment types, provide drainage as necessary. For off structure drainage provide an appropriate roadway inlet to eliminate shoulder washouts in accordance with standard drawings.

BIBLIOGRAPHY

AASHTO/NSBA Steel Bridge Collaboration.

Guide to Formwork for Concrete (ACI 347-01), American Concrete Institute, Committee 347, Farmington Hills, MI, 2001.

Concrete Repair Guide, (ACI 546R-96 [Reapproved 2001]), American Concrete Institute, Committee 546, Farmington Hills, MI, 1996.

Use of Epoxy Compounds with Concrete, (ACI 503R-93 [Reapproved 1998])," American Concrete Institute, Committee 503, Farmington Hills, MI, 1998.

Aramraks, T., "Highway Bridge Vibration Studies," Joint Highway Research Project No. JHRP-75-2, Purdue University, 1975.

Aspire, Summer 2007.

Austin, Deron and Ariel Soriano, "Geotextiles for Transportation Projects," *CE News*, April 2005.

Biggs, J. M., H. S. Suer and J. M. Louw, "The Vibration of Simple Span Highway Bridges," *Journal of Structural Division,* American Society of Civil Engineers, Vol. 96, No. EM5, October, pp. 593–620.

"Field Guide to Concrete Repair, Repair Application Procedures," Bulletins 1–9, American Concrete Institute.

Gaidis, J. M. and A. M. Rosenberg, "Estimating Life Cycle Cost of Concrete Using Life-365," *Structure*, March 2004.

"Guide for Surface Preparation for the Repair of Deteriorated Concrete Resulting from Reinforcing Steel Corrosion," No. 03730, International Concrete Repair Institute, 1995.

"Guide for Selecting and Specifying Concrete Repair Materials," No. 03733, International Concrete Repair Institute, 1996.

Issa, M. A., A. A. Yousif, and M. A. Issa, "Investigation of Cracking in Concrete Bridge Decks at Early Ages," *Journal of Bridge Engineering*, May 1999.

Krauss, P. D. and E. A. Rogalla, "Transverse Cracking in Newly Constructed Bridge Decks," National

Cooperative Highway Research Program, Report No. 380, TRB, 1996, Washington, DC.

Nowak, et al, "Development of a Guide for Evaluation of Existing Bridges," University of Michigan, Ann Arbor, 2000.

Oehler, L. T., "Vibration Susceptibilities of Various Highway Bridge Types," Michigan State Highway Department, 1957.

Park, Sung H., *Bridge Rehabilitation and Replacement (Bridge Repair Practice)*, S. H. Park, P.O. Box 7474, Trenton, NJ, 08628-0474, 1984.

Park, Sung H., "Bridge Topics," A Power Point Presentation, Hardesty & Hanover, New Jersey.

Roeder, C. W., K. Barth and A. Bergman, "Improved Live Load Deflection Criteria for Steel Bridges," National Cooperative Highway Research Program Report Project 20, November 2002.

"Seismic Prioritization Program—Seismic Coding Guide," New Jersey Turnpike Authority, Garden State Parkway.

Corrosion Control and Galvanic Technology

Bertolini, et al., "Cathodic Protection and Cathodic Prevention in Concrete: Principles and Applications," *Journal of Applied Electrochemistry*, V. 28, 1998.

Concrete Repair Guide, (ACI 546R-04), American Concrete Institute, Committee 546, 2004, Farmington Hills, MI.

Glass, G., J. Taylor, A. Roberts and N. Davison, "The Protective Effects of Electrochemical Treatment in Reinforced Concrete," NACE Paper No. 03291, *Corrosion*, 2003.

Glass, G. K., A. M. Hassenein and N. R. Buenfeld, *Corrosion Science*, V. 43, No. 6, 2001, pp. 1111–1131.

Hausman, D. A., "Steel Corrosion in Concrete," *Material Protection*, 1967, pp. 19–23.

Healer for Sika Corporation, Report No. 831298, American Concrete Institute, Committee E 706.

Guide for Surface Preparation for the Repair of Deteriorated Concrete Resulting from Reinforcing Steel Corrosion, International Concrete Repair Institute, Technical Guideline No. 03730, 1995.

Jerzak, H. P., "High Molecular Weight Methacrylate Resins in California," State of California Department of Transportation, January 1991.

Krauss, P. D., "Bridge Deck Repair Using Polymers," paper presented at International Congress on Polymers in Concrete, San Francisco, CA, Sept. 24–27, 1991.

Monomers, P. "Polymers in Concrete: Advances and Applications," SP-116.

Murray, M. A. and W. O. Eisenhut, "A Concrete Healer and Sealer," *Concrete Repair Digest*, V. 6, No. 3, June/July 1995.

Pfeifer, D. W. and W. F. Perenchio, "Coating, Penetrants, and Specialty Concrete Overlays for Concrete Surfaces," paper presented at NACE seminar, Chicago, IL, Sept. 27–29, 1982.

Protection of Metals in Concrete Against Corrosion, (222R-01), American Concrete Institute, Committee 222, 2001, Farmington Hills, MI.

Rodler, D. J., D. P. Whitney, D. W. Fowler and D. L. Wheat, "Repair of Cracked Concrete with High Molecular Weight Methacrylate," 1989.

Specifications and Special Provisions for Use of High Molecular Weight Methacrylate Monomers to Seal Bridge Decks, California Department of Transportation, Contract No. 04-001754.

Sprinkel, M. M., *Use of High Molecular Weight Methacrylate Monomers to Seal Cracks in Bridge Decks, Retard Alkali-Silica Aggregate Reactions, and Prime Bridge Surfaces for Overlays*, Virginia Department of Transportation Research Council.

Structural Crack Repairs Using Gravity Feed with Resin, (ACI RAP-2), American Concrete Institute.

Wiss Janney Elstner Associates, "Corrosion Protection Tests on Reinforced Concrete Treated with Sikaguard 70 and SikaPronto 19 Crack," 1986.

Epoxy Injection (ACI RAP-1)

Causes, Evaluation, and Repairs of Cracks in Concrete Structures, (224.1R-93), American Concrete Institute, Committee 224, 1993, Farmington Hills, MI.

Concrete Repair Guide, (546R-96), American Concrete Institute, Committee 546, 1996, Farmington Hills, MI.

Emmons, P. H., *Concrete Repair and Maintenance Illustrated*, R. S. Means Co., Inc., 1994, Kingston, Massachusetts.

Guide for Evaluation of Concrete Structures Prior to Rehabilitation, (364.1R-94), American Concrete Institute, Committee 364, 1994, Farmington Hills, MI.

Guide for Repair of Concrete Bridge Structures, (546.1R-80 [Reapproved 1988]), American Concrete Institute, Committee 546, 1988, Farmington Hills, MI.

Guide for Verifying Performance of Epoxy Injection of Concrete Cracks, 1998, ICRI Technical Guideline No. 03734.

Murray, M. A., 1987, "Epoxy Injection Welds Cracks Back Together," *Concrete Repair*, V. 3.

Promboon, Y., L. D. Olsen and J. Lund, "Nondestructive Evaluation (NDE) Methods for Quality Assurance," *ICRI Bulletin*, V. 15, No. 1, January/Febrary 2002, pp. 12–16.

Standard Specification for Epoxy-Resin Based Bonding Systems for Concrete, ASTM C 881-90, 1990, American Society for Testing and Materials, West Conshohocken, PA.

"State-of-the-Art Adhesives for Concrete Construction," *Construction Canada*, May/June 1998.

Trout, J. F., "Epoxy Injection in Construction," 1998, The Aberdeen Group.

Use of Epoxy Compounds with Concrete, (ACI 503R-93 [Reapproved 1998])," American Concrete Institute, Committee 503, 1998, Farmington Hills, MI.

A. Spall Repairs Using Preplaced Aggregate Method (ACI RAP-9)

Guide for the Use of Preplaced Aggregate Concrete for Structural and Mass Concrete Applications, (ACI 304.1R- 92), American Concrete Institute, Committee 304, 1992, Farmington Hills, MI.

Guide to Curing Concrete, (ACI 308R-01), American Concrete Institute, Committee 308, 2001, Farmington Hills, MI.

Guide to Surface Preparation for the Repair of Deteriorated Concrete Resulting from Reinforcing Steel Corrosion, International Concrete Repair Institute, Guideline No. 03730.

Guide to Using In-Situ Tensile Pull-Off Test to Evaluate Bond of concrete Surface Materials, International Concrete Repair Institute, Technical Guideline 03739.

King, J. and E. Bush, "Grouting of Granular Materials," ASCE Convention, 1960.

King, J. and A. Wilson, "If It's Still Standing, It Can Be Repaired," *Concrete Construction*, July 1988.

Selecting and Specifying Concrete Surface Preparation for Sealers, Coatings, and Membranes, International Concrete Repair Institute, Guideline No. 03732.

Standard Test Method for Flow of Grout for Preplaced Aggregate Concrete, American Society for Testing and Materials, C 939, 2002, West Conshohocken, PA.

Standard Test Method for the Compressive Strength of Grouts for Preplaced-Aggregate Concrete in the Laboratory, American Society for Testing and Materials, C 942, 1999, West Conshohocken, PA.

Standard Practice for Making Test Cylinders and Prisms for Determining Strength and Density of Preplaced-Aggregate Concrete in the Laboratory, American Society for Testing and Materials, C 943, 2002, West Conshohocken, PA.

Standard Test Method for Potential Alkali Reactivity of Aggregates (Mortar-Bar Method), American Society for Testing and Materials, C 1260, 2001, West Conshohocken, PA.

B. Vertical and Overhead Spall Repair by Hand Application (ACI RAP-6)

Dorsch, D. F., "A Guide to Placement Methods for Repair Materials," *Concrete Repair Digest*, December 1995, pp. 320–325.

Emmons, P., *Concrete Repair and Maintenance Illustrated*, RS Means, 1993.

Guide for Evaluation of Concrete Structures Prior to Evaluation, (ACI 364.1R-94), American Concrete Institute, Committee 364, 1994, Farmington Hills, MI.

Guide for Making a Condition Survey of Concrete in Service, (ACI 201.R-92), American Concrete Institute, Committee 201, 1992, Farmington Hills, MI.

Guide for Selecting and Concrete Repair Materials, International Concrete Repair Institute, Technical Guideline No. 03733, 1997.

Guide for Surface Preparation for the Repair of Deteriorated Concrete Resulting from Reinforcing Steel Corrosion, International Concrete Repair Institute, Technical Guideline No. 03730, 1995.

Guide for Using In-Situ Tensile Pull-Off Tests to Evaluate Bond of Concrete Surface Materials, International Concrete Repair Institute, Technical Guideline No. 03739, 1997.

Guide to Curing Concrete, (ACI 308R-01), American Concrete Institute, Committee 308, 2001, Farmington Hills, MI.

Guide to the Use of Waterproofing, Dampproofing Protective, and Decorative Barriers for Concrete, American Concrete Institute, Committee 515, 1985, Farmington Hills, MI.

Selecting and Specifying Concrete Surface Preparation for Sealers, Coatings, and Polymer Overlays, International Concrete Repair Institute, Technical Guideline No. 03732, 1997.

Smoak, W. G., *Guide to Concrete Repair*, U.S. Department of the Interior, Bureau of Reclamation Technical Service Center, April 1997.

Standard Specification for Liquid Membrane-Forming Compounds for Curing Concrete, American Society of Civil Engineers, C 309, 1998, West Conshohocken, PA.

C. Low-Pressure Spraying (ACI RAP-3)

Guide for Selecting Application Methods for the Repair of Concrete Surfaces, International Concrete Repair Institute, Guideline No. 03731, 1996.

Guide to Curing Concrete (ACI 308R-01), American Concrete Institute, Committee 308, 2001, Farmington Hills, MI.

Guide to Shotcrete, American Concrete Institute, Committee 506, 1995, Farmington Hills, Michigan.

Selecting and Specifying Concrete Surface Preparation for Sealers, Coatings, and Membranes, International Concrete Repair Institute, Guideline No. 03732, 1997.

Standard Specification for Curing Concrete, (ACI 308.1-98), American Concrete Institute, Committee 308, 1998, Farmington Hills, MI.

Surface Preparation for Repair of Deteriorated Concrete Resulting from Reinforcing Steel Corrosion, International Concrete Repair Institute, Guideline No. 03730, 1995.

D. Horizontal Concrete Surfaces (ACI RAP-7)

Guide for Evaluation of Concrete Structures Prior to Evaluation, American Concrete Institute, Committee 364, 1994, Farmington Hills, MI.

Guide for Making a Condition Survey of Concrete in Service, American Concrete Institute, Committee 201, 1992, Farmington Hills, MI.

Guide to the Use of Waterproofing, Dampproofing, Protective, and Decorative Barriers for Concrete, American Concrete Institute, Committee 515, 1985, Farmington Hills, MI.

Guide to Using In-Situ Tensile Pull-Off Tests to Evaluate Bond of Concrete Surface Materials, International Concrete Repair Institute, Technical Guideline No. 03739.

Selecting and Specifying Concrete Surface Preparation for Coatings, Sealers, and Polymer Overlays, International Concrete Repair Institute, Technical Guideline No. 03732.

Standard Specification for Curing Concrete, American Concrete Institute, Committee 308, 1992, Farmington Hills, MI.

Surface Repair Using Form-and-Pour Techniques (ACI RAP-4)

Concrete Repair Guide, American Concrete Institute, Committee 546, 1996 (Reapproved 2001), Farmington Hills, MI.

Guide for Selecting and Specifying Concrete Repair Materials, No. 03733, International Concrete Repair Institute, 1996.

Guide for Surface Preparation for the Repair of Deteriorated Concrete Resulting from Reinforcing Steel Corrosion, No. 03730, International Concrete Repair Institute, 1995.

Guide to Formwork for Concrete, American Concrete Institute, Committee 347, 2001, Farmington Hills, MI.

Use of Epoxy Compounds with Concrete, American Concrete Institute, Committee 503, (Reapproved 1998), Farmington Hills, MI.

American Association of State Highway and Transportation Officials (AASHTO)

ASHTO M 160, Standard Specification for General Requirements for Rolled Steel Plates, Shapes, Steel Pilings, and Bars for Structural Use.

AASHTO M 300, Standard Specification for Inorganic Zinc-Rich Primer.

American Society for Testing and Materials (ASTM).

ASTM A 6, Specification for General Requirements for Rolled Structural Steel Bars, Plates, Shapes, and Sheet Piling.

ASTM D 3359, Standard Test Methods for Measuring Adhesion by Tape Test.

ASTM D 4138, Standard Test Method for Measurement of Dry Paint Thickness of Protective Coating Systems by Destructive Means.

ASTM D 4285, Standard Test Method for Indicating Oil or Water in Compressed Air.

ASTM D 4414, Standard Practice for Measurement of Wet Film Thickness by Notch Gages.

ASTM D 4417, Standard Test Methods for Field Measurement of Surface Profile of Blast Cleaned Steel.

ASTM D 4541, Standard Test Method for Pull-Off Strength of Coatings Using Portable Adhesion Testers.

ASTM D 4752, Standard Test Method for Measuring MEK Resistance of Ethyl Silicate Zinc-Rich Primers by Solvent Rub.

Related Reference Documents

Applicable Ordinances and Regulations.

Equipment and Coating Manufacturer's Published Instructions and Product Data Sheets.

AASHTO Standard Specifications for Transportation Materials and Methods of Sampling and Testing, 28th Edition.

AASHTO LRFD Bridge Construction Specifications, 2nd edition, with 2008 Interim Revisions, Item Code 10-LRFDCONS-2-13.

AASHTO Guide Specifications for Highway Bridge Fabrication with HPS70W Steel, 2003.

AASHTO Guide Design Specifications for Temporary Works, with 2008 Interim Revisions, Item Code 10-GSTBW-1-M.

AASHTO Construction Handbook for Bridge Temporary Works, 2008 Interim, Item Code 10-CHBTW-1-11-UL.

AASHTO Inspector's Guide for Shotcrete Repair of Bridges.

AASHTO Guide Specifications for Applications of Coating Systems with Zinc Rich Primers to Steel Bridges, S 8.1- 2006, Item Code 10 – NSBASBCS-2-OL.

AASHTO Steel Bridge Fabrication QC/QA Guide Specification, S 4.1-2002.

AASHTO Shop Detail Drawing Review/Approval Guidelines, G 1.1-2000.

AASHTO Steel Bridge Erection Guide Specification, S 10.1-2007.

AASHTO Best Practices in Context Sensitive Solutions, 2006 Competition, Item Code 10-BPCSS-2-OL.

FHWA-RD-78-133, Extending the Service Life of Existing Bridges by Increasing Their Load Carrying Capacity, 1978.

NCHRP Report 206, Detection and Repair of Fatigue Damage in Welded Highway Bridges, 1978.

NCHRP Report 222, Bridges on Secondary Highways and Local Roads—Rehabilitation and Replacement, 1980.

NCHRP Project 12-17 Final Report, Evaluation of Repair Techniques for Damaged Steel Bridge Members: Phase l, 1981.

NCHRP Report 243, Rehabilitation and Replacement of Bridges on Secondary Highways and Local Roads, 1981.

NCHRP Report 271, Guidelines for Evaluation and Repair of Damaged Steel Bridge Members, 1984.

NCHRP Report 293, Methods of Strengthening Existing Highway Bridges, 1987.

NCHRP Report 333, Guidelines for Evaluating Corrosion Effects in Existing Steel Bridges, 1990.

10

Concrete Repair Methods

10.1 INTRODUCTION TO COMMON REPAIR PROCEDURES

Development of concrete as an every day construction material has been a benefit of paramount importance to the mankind. Over the years, conventional lime and Portland cement concrete evolved into:

- High early strength concrete
- High performance concrete (HPC)
- Fiber reinforced concrete (FRC)
- Fiber reinforced polymer concrete (FRPC)
- Carbon fiber reinforced polymer concrete (CFRPC)
- Other composites and use of additives.

Society's dependence on its applications has created continued demands for maintenance and repairs.

In this chapter, procedures and practices for repairs and related issues are discussed. The ever growing list of publications on specialized topics is listed at the end of chapter.

10.1.1 Need for Near-Term Repairs

In Section 9.1, the scope of concrete repairs and causes of concrete deterioration were discussed. Repairs will be accomplished with minimal impact to traffic. The objectives are to restore serviceability and original functionality.

1. Based on field verification, the repair requirements for the concrete and steel items need to be identified if possible to preclude the need for major repairs.

2. To identify acceptable reconstruction solutions, coordination with the local communities and other agencies is required.

3. Preliminary design will be based on available data of member sizes from inspection reports.

4. Any drainage improvement recommendations will be developed during preliminary design.

5. Due to distress from severe localized deterioration, vehicle impact damage and observed scour needs to be fixed.

10.1.2 Long-Term Repair Procedures for Concrete

Issues can be summarized as:

1. Assessing damage and deterioration.
2. Identifying the causes.
3. Load testing.
4. NDT techniques.
5. Developing reports.
6. Cementitious materials selection process.
7. Surface preparation.
8. Placement methods.
9. Crack and joint repairs.
10. Use of protective systems such as membranes and waterproofing, sealers, and coatings.

10.1.3 Deck Repairs

A one-course deck slab with a corrosion inhibitor admixture may be preferred. Minimum top reinforcement cover is higher (e.g., 2¾ inches).

1. Two-course construction with the overlay of LMC or silica fume requires an additional one to two weeks construction time.
2. Deck joints: Expansion joints may need replacement. Alternates such as compression or strip seal deck joints will be considered, and the joints should be repaired (by installing strip seal joints or eliminating the joint by providing a continuous deck as part of a rehabilitation project).
3. Precautions during demolition: To prevent debris from falling below, shielding (such as installing wire nets) below the deck will be provided.
4. The sidewalk decking will be evaluated and upgraded, if necessary, to improve safety.
5. Expansion dams: If a field visit has shown non-functioning deck expansion joints, adequate expansion joints need to be provided.

10.1.4 Fender System Reconstruction

An inspection report may cite fixing of damaged or missing planks in timber fenders around piers. This may result from ship collision or floating ice blocks. Damaged planks will be replaced and re-bolted. Check the fender system's ability to resist vessel collision and ice loading. Analysis will be by using a three-dimensional model. The STAAD Pro program or alternate computer software will be used.

10.2 DETAILS OF REPAIRS

10.2.1 Flow Diagram for Repair of Cracks and Defects

Essential activities of repairs are shown in the flow diagram in Figure 10.1. Details of repair methods are given in Table 10.1, mostly using concrete materials, although steel wire and steel plate may be required for jacketing.

For repairs based on functional requirements, basically two functions are served.

1. Architectural repairs to maintain aesthetics: Cosmetic and surface cracks are generally repaired by sealants, grouts, or other protection system.

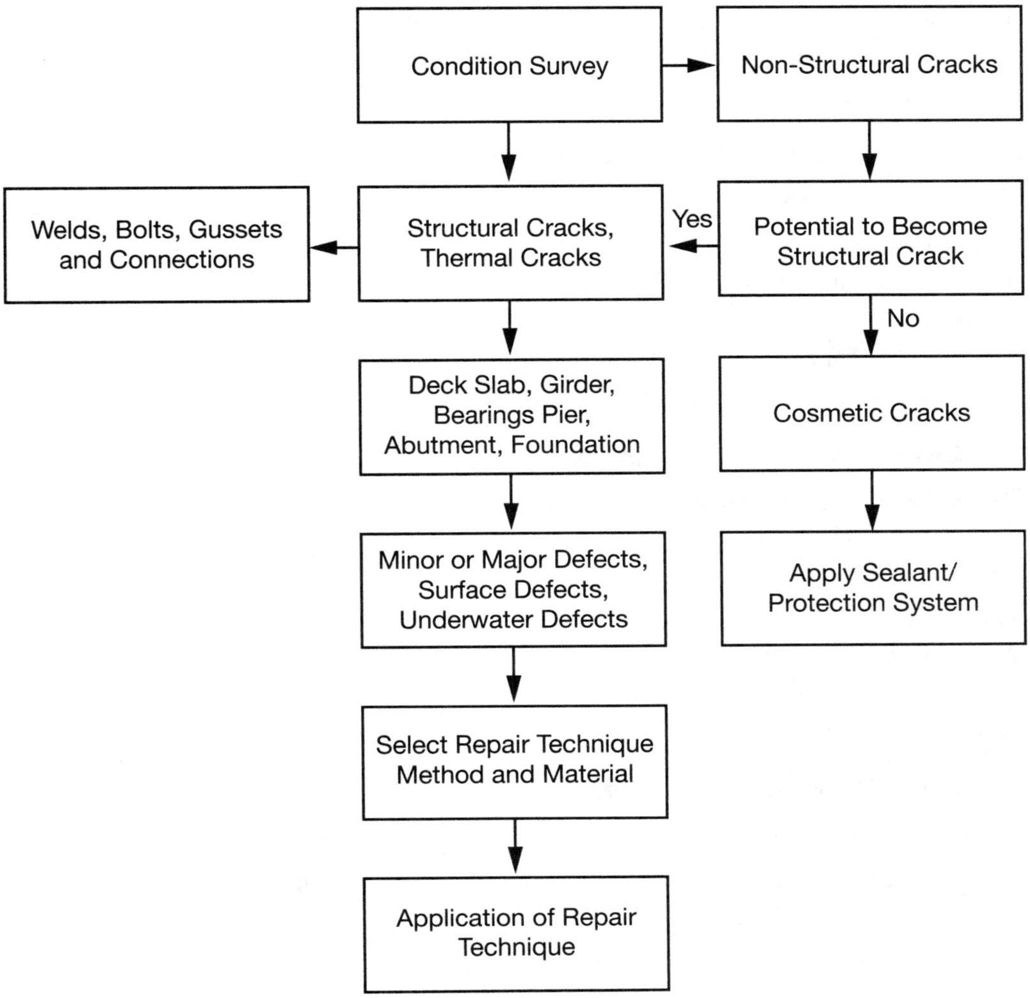

Figure 10.1 Flow diagram for minor and major repairs.

2. Structural repairs: They are required for safety and to prevent costly repairs or replacements.
3. Standard procedures are available for all construction materials such as concrete, steel, and masonry. In selecting a repair technique, feasibility of restoration to original strength and long-term performance of the repaired defect needs to be evaluated. ASTM, ACI, AASHTO, and FHWA repair procedures need to be followed.

10.2.2 Repairs Based on Locations and Access:

1. Visible and invisible defects.
2. Surface defects.
3. Connection defects.
4. Underwater defects.

10.2.3 Repairs Based on Potential Damage

The extent of potential damage needs to be assessed and the consequences and risk of material failures if repairs are unattended. Some of the applications are:

1. Evaluate compressive strength from NDT before repairs.
2. Structural repair and strengthening of concrete elements.
3. Post-tensioning repair.

10.2.4 Factors Important in Reducing Early Cracking

They may be listed as:

1. Low shrinkage.
2. Low modulus of elasticity.
3. High creep.
4. Low heat of hydration.
5. Use of shrinkage compensating cement.

10.2.5 Advanced Concrete Repair Methods

The following methods are a continuation of those discussed in Chapter 9.
Use of a concrete overlay system:

1. LMC.
2. RS LMC.
3. Microsilica fume concrete.
4. Polymer surface treatment—thin epoxy overlay.
5. Corrosion inhibiting admixture.

10.2.6 Crack Width versus Type of Repair

Common repair methods are:

1. Removing damaged concrete by saw cut or jack hammer the damaged areas.
2. Create square vertical sides.
3. Preparing existing concrete for repair.
4. Apply tar and chips.
5. Patching and overlays (partial or full depth patching).

10.2.7 Recommended Repair Techniques Based on Type of Deterioration

Table 10.1 General repair methods and materials.

Type of Deterioration	Repair Methods	Materials	Remarks
Scaling	1. Overlaying	Asphalt cement, epoxy or polymer concrete	Cosmetic reasons/improving appearance.
	2. Grinding	Bituminous coat	Minimize wear and tear of tires.
	3. Shotcrete	Quick-setting mortar	Increasing life of structure.
	4. Coating	Linseed oil coat	
	5. Strengthening	Concrete or steel	Cost effective
	6. Replacement	Concrete or steel	Expensive

(continued on next page)

Type of Deterioration	Repair Methods	Materials	Remarks
Spalling	1. Overlaying	Asphalt cement, epoxy or polymer concrete	Increasing rider comfort
	2. Patching	Concrete, LMC, epoxy or polymer concrete	Increasing life of structure
	3. Shotcrete	Cement mortar	Increasing life of structure
	4. Coating	Bituminous	Increasing life of structure
	5. Strengthening/replacement		Cost effective
Active cracking	Caulking	Elastomeric sealer	Increasing life of structure
	Pressure injection	Flexible epoxy mortar	Increasing life of structure
	Jacketing: strapping overlaying	Steel wire or rod membrane/special mortar	Cost effective
	Strengthening	Steel plate, post tensioning, stitching	Cost effective
Dormant cracking	Caulking	Cement grout or mortar, fast setting mortar	Increasing life of structure
	Pressure injection	Epoxy grouts or mortar	Increasing life of structure
	Coating	Bituminous coating, tar	
	Overlaying	Asphalt overlay with membrane	Increasing rider comfort
	Grinding and overlay	LMC, silica fume concrete	Increasing rider comfort
	Dry pack	Dry pack	Cost effective
	Shotcrete	Cement mortar	Increasing life of structure
	Patching	Cement mortar Epoxy or polymer concrete	Increasing life of structure
	Jacketing	Steel or fiber glass or FRP wrapping	Cost effective
	Strengthening	Post tensioning	Cost effective
	Reconstruction	New concrete construction	Cost effective
Voids	Dry pack	Dry pack	Cost effective
	Patching	Portland cement grout, mortar, cement	Increasing rider comfort
	Resurfacing	Epoxy or polymer concrete	Increasing rider comfort
	Shotcrete	Fast setting mortar	Increasing life of structure
	Pre-placed aggregate	Coarse aggregate, grout	Cost effective
	Strengthening/replacement	Diagnostic design is required	Cost effective

10.2.8 Methods for Reducing Cracks

1. Methods commonly used are as follows:
- Remove unsound concrete
- Full depth patching.
- Partial depth patching.

2. The techniques to reduce cracks are:
- Curing—A process in which concrete is improved through the retention of water sufficient for hydration. Concrete has the ability to redistribute and adjust to shrinkage related

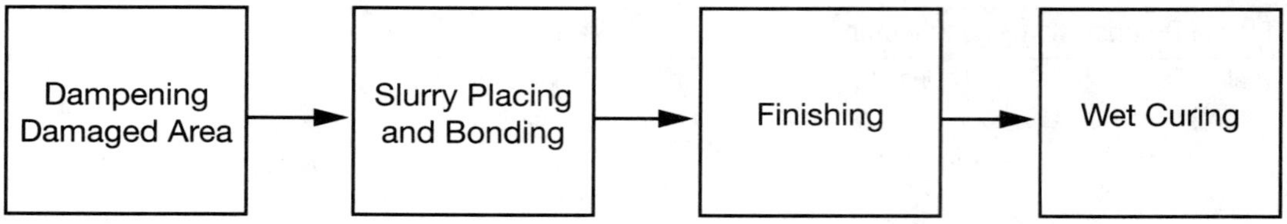

Figure 10.2 Patching procedure for damaged areas.

stresses if properly moistened. Use is made of sprinklers, wet mats, curing compounds (acting as membrane to reduce water evaporation from surface).

- Use of shrinkage compensating concrete.
- Use of stress risers.
- Post-tensioning concrete to introduce compression.
- Use of steel fibers to distribute cracks.
- Existing metal decking: Metal deck is made composite with concrete.

3. Patching procedures (Figure 10.2)
Defective concrete is patched by:

- Portland cement concrete.
- Mortar.
- Epoxy type mortar.
- Dry packing.
- Matching of concrete color with existing concrete is determined by trial use of white cement.
- Cement to washed sand mix proportion is 1:2 to 1:3.
- Repairs apply to areas such as grout holes, bolt holes, and narrow slots. Holes are cleaned and dried before filling.
- Packing is done by wooden tools.
- Use of cathodic protection.
- Use of shotcrete or gunite.

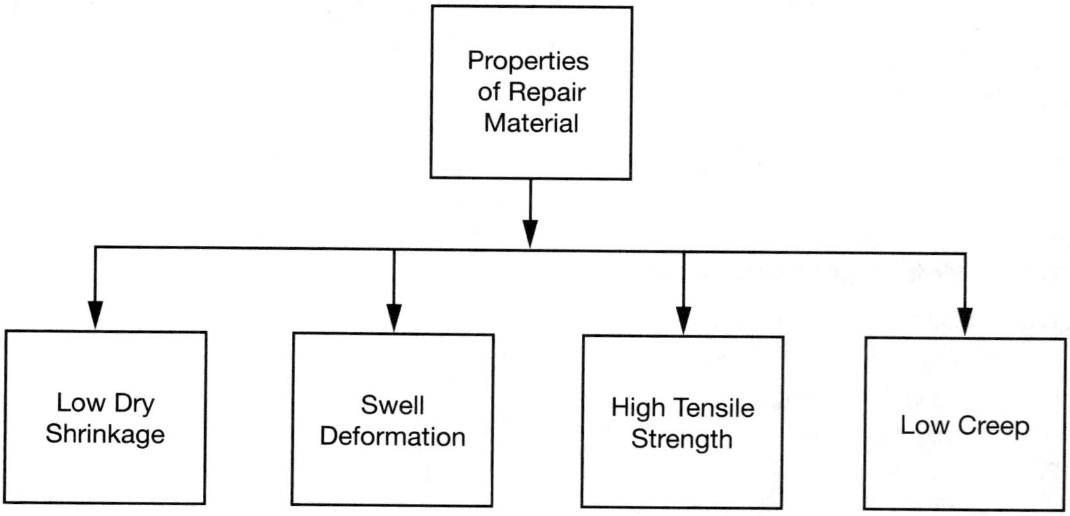

Figure 10.3 Desirable basic properties of concrete repair material.

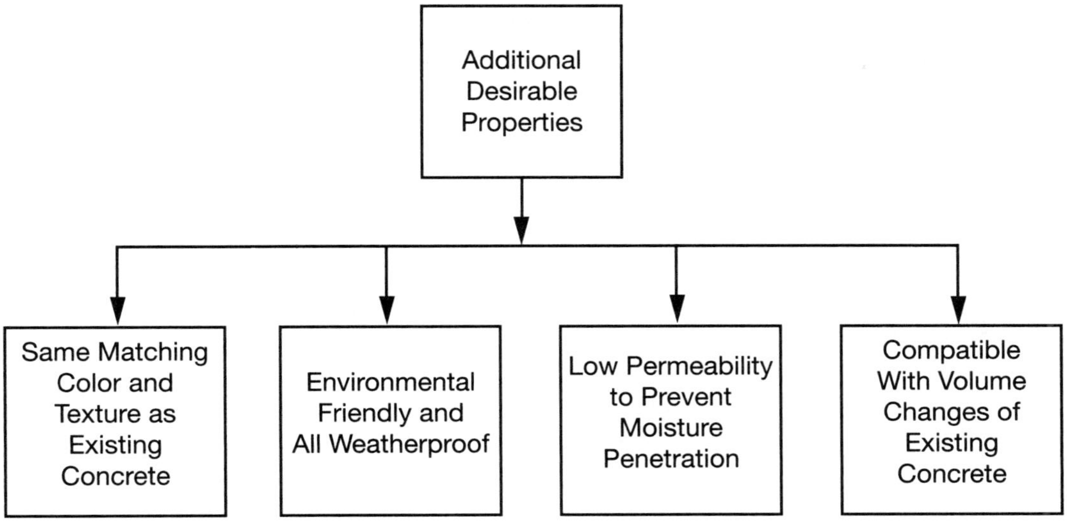

Figure 10.4 Additional properties required for selection of repair materials.

10.2.9 Properties of Repair Materials

Figure 10.3 shows a flow diagram for required properties of the repair material.

Figure 10.4 shows a flow diagram for desirable properties of the repair material in addition to basic requirements.

10.2.10 Repair Methods for Reinforced Concrete

In bridge engineering we are mainly dealing with reinforced or prestressed concrete structures. Figure 10.5 shows a flow diagram for general reinforced concrete repairs including adding extra reinforcing if required.

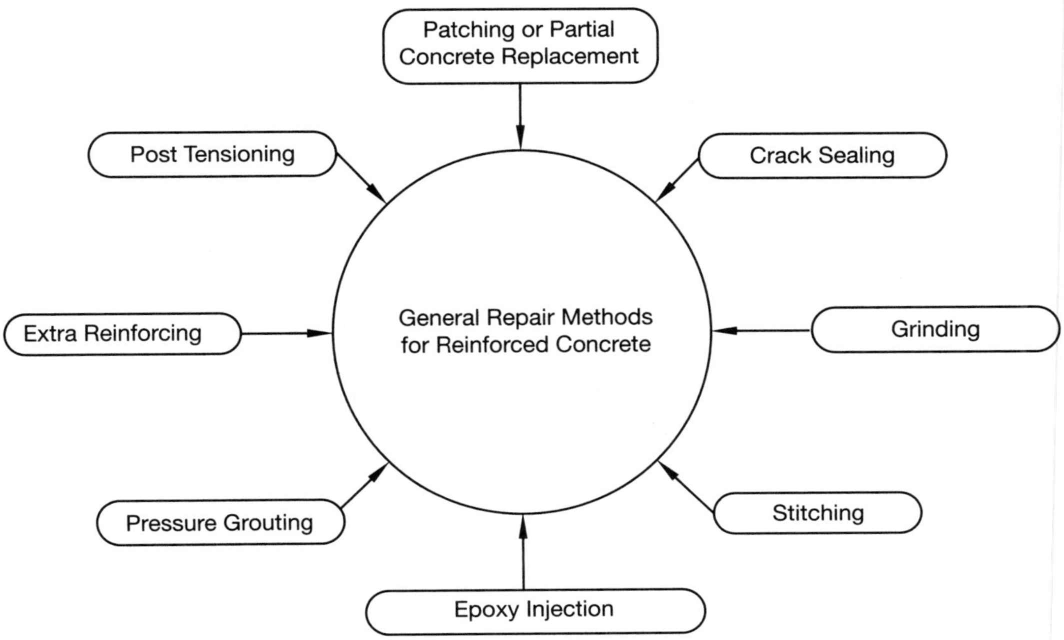

Figure 10.5 Reinforced concrete material repair methods.

10.2.11 Details of Reinforced Concrete Repair Methods

Reinforced concrete repair methods may differ from that of un-reinforced concrete. Common methods are:

- Sealing
- Grinding
- Flexible sealing
- Epoxy injection
- Grouting
- Reinforcing
- Stitching
- External FRP reinforcing and steel plate
- Drilling and plugging
- Jacketing.

10.2.12 Crack Repair Methods (Table 10.1)

Steps to repairs of cracks are:

1. Concrete repair design
2. Patching and packing
3. Shotcrete, carbon fiber reinforcing, chloride extraction, gunite, cathodic protection
4. Improved repair methods using non-shrink hybrid polyurethane mixed with dry silica sand:
   - Thermal resistant
   - Effective bonding
   - Low surface tension.

10.2.13 Suggestions for Designing Durable Concrete for Rehabilitation

1. Experience has shown that in addition to the above specific repair materials, the following guidelines would lead to durable concrete for rehabilitation and replacement. Adverse parameters that affect deterioration such as sharp skew, deck joints, deicing chemicals, etc. need to be addressed in detail:
   - Use slag and pozzolans to reduce concrete permeability and control heat of hydration.
   - Use non-corrosive reinforcement in areas exposed to harsh environments.
   - Use higher percentages of aggregate so that cement pastes are not widespread and may initiate cracks.
   - Use air entrainment to protect concrete from sub-freezing temperatures.
   - Reduce entrapped air voids by compaction and consolidation.
   - Use adequate curing methods to control both temperature and moisture.
   - Use post-tensioning to prevent crack generation.
2. Concrete sealants/sealers: Following are some facts about sealants and sealers:
   - Deck sealants: The major controlling parameter is the depth of penetration. Penetrating sealants prevent concrete deterioration from deicing salts and provide durability and permeability.
   - Silane and siloxane products are used as concrete surface sealers. Moisture on the surface prevents effective application.

Table 10.2 Types of defects in concrete and preventive action.

Type of Deterioration	Description	Preventive Action	Primary Causes
Scaling	1. Surface pitting caused by freeze and thaw cycles. 2. Repeated wetting and drying cycles at splash zone.	Air entrainment in concrete	Salt and deicing chemical agents
Spalling	1. Breaking away of concrete surface from expansive forces caused by corrosion of rebars. 2. Structure movement. 3. High edge pressure.	Adequate concrete cover to rebars	Collision damage Fire
Active cracking/plastic shrinkage (Shallow or full depth)	Evaporation of moisture due to hot, dry or windy weather	1. Use fog nozzles to saturate air 2. Apply curing compound 3. Place plastic cover after finishing	Collision damage Fire
Drying shrinkage cracks	1. Loss of moisture from cement paste constituent and restraints such as subgrade and mass concrete matrix. 2. Surface cracks due to high curing temperature difference in a cross section.	1. Increase amount of aggregate 2. Reduce water content of mix 3. Use high strength concrete	Improper mix design and water: cement ratio
Alkali silica reactive cracks	Carbonate rocks such as dolomite limestone react with alkalis and cause detrimental reactions	1. Proper selection of aggregate 2. Use of pozzolan 3. Use of low alkali cement	Material defects
Dormant Cracking/ settlement cracks	Rebars during curing may restrain concrete from consolidating	1. Use small size bars 2. Concrete with lower slump 3. Increase cover to rebars	Settlement of supports
Delamination	Concrete layer separates at top surface due to expansion force of corroding steel	Increase cover to rebars	Detailing defects
Efflorescence	1. Crystallization of calcium chloride in the form of white deposit on concrete surface 2. Contamination of concrete 3. Increased porosity	Improve mix design by use of admixtures	Material defects
Honeycombing	Formation of voids in column or wall concrete	1. Improve method of vibration 2. Avoid congestion of rebars to permit flow of concrete around rebars	Construction defects

- A design procedure can be formulated for sealant application analogous to concrete mix design for bridge decks with specific exposure conditions. Sealants applied by a flooding procedure will result in additional penetration in a moist concrete deck.

3. Sealers can be a cost-effective means of inhibiting corrosion of uncoated reinforcing steel, steel with too little concrete cover, or steel embedded in concrete which exhibits hairline cracks. However, sealers are not considered a cost-effective means of inhibiting corrosion when applied to mature concrete of standard quality that utilizes other means of corrosion protection, such as epoxy coated steel, specialty overlays, etc.

4. Penetrant sealers: Specify penetrant sealers after grooving existing bridge decks with the following conditions:
   - The existing bridge deck does not conform to the current reinforcing steel cover requirement.
   - The superstructure environment is extremely aggressive due to the presence of chlorides.
   - The existing deck is to be grooved.

Also, sealers cannot be used below grade or below the water line because they provide no protection when submerged.

10.2.14 Structural Repairs

Methods of preventing structural repairs are summarized in Table 10.3.

Table 10.3 Structural defects and preventive action.

Type of Deterioration	Description	Suggested Preventive Action	Primary Causes
Corrosion	Cracks developing in concrete due to corroded rebar steel 1. Carbonation (reduction of pH value). 2. Chloride penetration.	1. Cathodic protection and galvanized rebars. 2. Epoxy and polymer sealer coating.	Detailing defects
Design defects	1. Improper location of expansion and contraction joint. 2. Lack of provision for shrinkage and creep. 3. Insufficient shear reinforcement. 4. Insufficient concrete cover.	1. Locate expansion and contraction joint as per technical specifications. 2. Apply AASHTO design method for shrinkage and creep. 3. Increase shear reinforcement near supports. 4. Increase concrete cover.	Lack of familiarity with AASHTO LRFD code and standard specifications
Skew and curvature effects	Creep effects and stress Concentration at slab edge due to sharp skew angle. Highly curved girder is subjected to St. Venant's and warping torsional stress	1. Improve rebar distribution at skew corners 2. Use internal concrete diaphragms 3. Strengthening required	Design defects
Diagonal tension failure	Structural shear cracks due to foundation settlement, scour, sharp skew and curved girders	1. Increase rebars in tension zone for vertical cracks. 2. Increase transverse reinforcement for diagonal cracks.	Design defects
Shear compression Failure	High principal stress due to flexure and shear	Increase shear and flexural reinforcement near supports by strengthening	Design defects

10.2.15 Some Problems Associated with Concrete Deck Slabs

Some problems associated with concrete deck slabs include:

- Discoloration
- Abrasion loss
- Carbonation
- Curling
- Joint failure
- Honeycombing
- Corrosion of reinforcement
- Leaching
- Delaminations
- Efflorescence.

10.2.16 Types of Corrosion Protection for Concrete

Some of the common methods used for corrosion protection are:

- Corrosion inhibiting admixture
- Calcium nitrate
- Epoxy coated reinforcing bars
- Solid stainless steel bars
- Stainless clad reinforcing bars
- Cathodic protection system
- High performance concrete

1. Treatment for concrete cracking.

Cracking usually results from alkali silica reaction, shrinkage, or settlement. The causes of cracking are similar to concrete spalling, with the addition of drying shrinkage and structural distress.

2. Treatment for concrete spalling.

If the concrete is batched with aggregate that is not chemically inert with the cement, a pattern of map cracking and spalling can develop. Treatments for this condition are:

- To place a thicker cover over the reinforcing bars
- Complete replacement of the concrete.

Another cause of concrete spalling is the combination of freezing temperatures and water penetration into cracks, voids, or porous stone aggregates of the concrete. This cycle of freezing and thawing causes spalls as the water freezes and expands below the surface of the concrete.

Treatments for this condition are:

- Provide proper drainage to prevent ponding of water on the surface
- Patch or replace all cracks and spalls
- Provide a bonded concrete encasement or overlay.

3. Treatments for scaling of hardened concrete.

- Deep scaling: The probable causes of deep scaling are lack of entrained air or an improper water to cement ratio. Treatment options are to either place a bonded concrete encasement around the affected area or to replace the concrete entirely.
- Surface scaling: It is generally caused by improper construction techniques, such as watering the concrete during finishing. Regular sealing of the surface may inhibit the scaling. Otherwise, a bonded concrete encasement can be used or the concrete surface can be ground out and a new surface installed. Another possible option is to ignore the problem until the scaling becomes severe enough to warrant replacement.

When concrete is placed against soil with high sulfate content, the chemical attack causes surface scaling that progresses to deep scaling. The treatment is placing a bonded concrete encasement or complete replacement.

4. When reinforcing steel in concrete corrodes, its volume increases. The expansion causes tensile stresses on the concrete surface which leads to a regular pattern of cracks and spalls over the entire surface.

Treatments for this condition are:

- Patch the surface
- Replace the concrete with a thicker cover
- Completely replace the concrete.

Under drying shrinkage, the volume of the concrete decreases as the concrete cures and the water evaporates from the surface. Tension develops on the surface of the concrete. These cracks can range from singular cracks in thin narrow members, to craze or map cracking for deeper members.

Singular cracks can be treated by:

- Epoxy injection,
- Flexible sealant
- Encasement with reflective crack control
- Complete replacement.

Craze or map cracking can be treated by:

- Surface replacement
- Placement of bonded concrete.

Under structural distress, the concrete produces singular cracks when subjected to:

- Excessive loads
- Unanticipated settlements
- Insufficient reinforcement.

Treatments for this failure are to:

- Reduce the loads
- Correct the settlement
- Add pressure relief joints
- Replace the concrete with proper reinforcement
- Epoxy inject to bond fresh cracks, or stitching.

The two types of cracks may be classified as: working cracks and non-working cracks. The width of a working crack, such as a transverse deck crack, changes due to applied loads or temperature effects. The width of a non-working crack, such as shrinkage cracks in an abutment stem, does not change. The treatment of concrete cracks depends on the type and size of the crack. Silane sealer should be applied to both working and non-working cracks up to 0.30 mm. Cracks greater than 0.30 mm require removal and replacement with a thin bonded concrete overlay.

10.2.17 Optimum Concrete Mix Design

The use of trial mixes needs to be considered for the following additional problems:

- Admixture compatibility
- Plastic shrinkage cracking
- Rapid slump loss
- Variation in air content.

10.2.18 Protecting Reinforcement from Corrosion

Corrosion of reinforcing steel is a major concern for an aging infrastructure. Repairing and replacing damaged concrete caused by rusting reinforcing steel requires time, money, and an imposition on the traveling public.

1. Epoxy coated reinforcement:
   - Epoxy coated reinforcement is the most frequently used type of corrosion protected reinforcement.

- Extra care is required during placement of epoxy-coated reinforcement.
- Repair is required of epoxy coating that is damaged before or during placement.

2. Galvanized reinforcement:

- Galvanized reinforcement can be used in most applications where epoxy is suitable.
- An advantage of galvanized reinforcement over epoxy is shorter development length.
- Galvanized bars shall not be used in prestressed beams.
- The current standard is to use a calcium nitrite corrosion inhibitor in prestressed elements, which negates the need for other corrosion protection measures.
- Uncoated bars and galvanized bars shall not be mixed in the same structural element.

3. Uncoated (plain) reinforcement: In general, uncoated (plain) steel is the most economical choice when the concrete members provide adequate cover, and the reinforcement is not exposed to chlorides or other severe environments.

4. Stainless steel clad reinforcement:

- Bends, development length, and lap splice requirements are similar to plain bars.
- The primary difference between stainless steel clad and solid stainless steel is that stainless steel clad has a plain core that must be protected after cutting, leading to increased time and effort in the field.

5. Solid stainless steel and stainless steel clad reinforcement are appropriate for the following:

- When added durability reduces cost, either long-term or during construction. This can occur:
  - When environmental conditions are particularly severe
  - When the cost of repairs is unusually high due to heavy traffic or construction conditions
  - When design of concrete sections as uncracked under service load is not feasible
  - When cover is less than standard
  - In extreme environments such as in a cap beam beneath an expansion joint
  - A substructure located in or near a body of salt water.

10.2.19 Approximate Reinforcement Cost Comparison

The following table shows comparative cost ratios:

Bar Protection Type	In-Place Cost Ratio	Minimum Expected Service Life
Solid stainless steel	2.0	100
Stainless steel clad	1.6	80
Galvanized	1.1	60
Epoxy coated	1.1	50
Plain	1.0	40

1. A review of the average bid prices (in place costs) indicates that the cost to fabricate, ship, and place plain reinforcing bars is $1.5/kg over the material cost.

2. The cost to fabricate, ship, and place epoxy coated bars is an additional $0.50/kg ($2/kg over the material cost) due to the extra care required during placement and repair to the epoxy coating after placement.

3. Cost to fabricate, ship, and place all types of bars is similar.

10.2.20 Expected Service Life

For the different types of reinforcing bars in conventional concrete (with standard cover) exposed to a corrosive environment, the expected service life is reduced by nearly half. These values are approximate and are based on information obtained from industry sources, university research studies, and professional journals.

The standards for reinforcing bars are given in ASTM A615 and A996. These documents include the minimum dimensions for bending the various diameters and grades of bars. The standard bends for galvanized reinforcing bars are given in ASTM A767.

All substructure components immersed in seawater are considered to be exposed to chlorides on all faces.

10.2.21 Estimated Density of Steel per Unit Volume of Concrete

In lieu of performing an exact calculation from bar bending schedules, approximate values for steel content are given below:

- Abutments, retaining walls and solid piers 45 to 50 kg/m^{3}of concrete volume
- Piers (except solid piers) 55 to 60 kg/m^{3}of concrete volume.

10.2.22 Construction Inspection Requirements for Reinforcement

1. Check minimum cover, lap length, and location and embedment requirements as shown on the contract plans.
2. Verify bar types: Plain, epoxy-coated, galvanized, solid stainless, or stainless clad.
3. Bent bars shall conform to the details shown on the contract plans.

10.3 SUPERSTRUCTURE REPAIRS

10.3.1 Introduction

Repairs are an important aspect of bridge maintenance and constitute a more specialized approach than rehabilitation and restoration. They happen to be the most common maintenance strategy, mainly due to minimum expenditure, duration of implementing repair, and least disruption to traffic flow. The owner will be reluctant to accept major repairs or replacement due to budgetary reasons.

Detailed instructions for each type of steel and concrete repair are spelled out in the standard technical specifications of the highway agency. If, in addition to standard specifications, detailed "special provisions" or "supplemental specifications" are required, a reference note must be added in the contract drawing containing "general notes" to refer to special provisions.

10.3.2 Overview of Bridge Deck Types

Existing decks have survived for decades without replacement. In the U.S., decks are made of concrete, metal, and timber. The life expectancy for composite decks is 25 to 30 years with deck repairs.

Concrete decks:

1. Without overlays
2. With epoxy or galvanized reinforcement bars
3. With bitumenous overlay and without membrane waterproofing
4. With bitumenous overlay and membrane waterproofing—Least expensive of all overlays and significantly extends life of deck.
5. With LMC or concrete overlay—Used extensively and is sensitive to quality of workmanship. They may not have longer service life than waterproofing and bitumenous paving systems. Other types include corrosion inhibitor aggregates.

6. HPC—Ideal for marine environment and for areas where chloride use is prevalent. HPC results in enhanced durability of deck due to decreased permeability. The need for overlay system or even epoxy coated bars is reduced.

   Low viscosity sealants reduce permeability of Portland cement mixes to chloride penetration. They are not required for use with HPC.
7. Fiber reinforced concrete—The amount of reinforcing steel and corrosion can be reduced. Reinforcing steel that is exposed to chlorides from water penetration causes deck cracking and can be further reduced by taking advantage of arching action, HPC, and prestressing.

   Cathodic protection—Effective in controlling corrosion in existing bridge decks but is an expensive method.
8. Prestressed/precast deck has improved quality control.

10.3.3 Open Grid Steel Decks

Open grid steel decks have a poor performance record. Fatigue cracking of welded grid members has been a continual maintenance problem, even on routes with low average daily truck traffic (ADTT). Mechanical connections between the deck and the supporting beams are preferred over welded connections for ease of construction and minimization of weld cracking.

For existing bridges where continuous maintenance welding is experienced, replace deck with concrete filled grid steel deck or fill the existing deck with concrete. Install anti-skid studs when needed.

10.3.4 Concrete Filled Grid Steel Decks

Generally, these decks last for a long time (over 30 years). However, in older designs the filled concrete may crack after 10 years. As reported in manufacturer's literature, consider removing deteriorated concrete and fill the cups with concrete if the full-depth concrete is deteriorating. Use of calcium nitrate may be considered as a rust inhibitor.

10.3.5 Timber Decks

Timber decks are used for pedestrian, equestrian and small span bridges.

1. Replace the deteriorated members with treated lumber. If deck deterioration is over 25 percent, replace the entire deck with treated timber or other material.
2. Bituminous overlay may be provided to improve riding quality. If overlaid, add a leveling course prior to providing the surface course.

10.4 DECK SLAB REPAIRS

10.4.1 Introduction to Deck Concrete Repair

Structural adequacy: When the structural adequacy of a bridge deck to carry current traffic loads is questioned, an in-depth field survey and analysis must be performed. This review should determine the extent of deficiencies as well as the feasibility of rehabilitation.

1. The deck slab is the most vulnerable bridge component due to full exposure to environments and friction from vehicle tires. A deck slab is directly exposed to the elements and to impact from fast moving vehicles. The service life of bridge deck is shorter than other components.

   Use of deicing salts, deposits of oil from car engines, debris deposit from tires, and acids in rain result in deterioration of concrete. Some of the common problems are spall, surface cracks, structural cracks, corrosion of reinforcing steel, and delamination. Deck cracks if not attended to immediately may result in increased impact load and deck replacement.

In addition, cracks may wear and tear vehicle tires. Cracks may also contribute to accidents. For deck repairs, traffic lanes need to be shut down, which causes traffic jams and inconvenience for drivers.

2. Economics and traffic maintenance need to be evaluated when balancing the feasibility of structural restoration against complete replacement.

   Experience, judgment, and research have shown that deterioration often continues in partially rehabilitated decks when only the obviously deteriorated portion of the deck is removed and replaced. To minimize this effect, procedures are required that will determine the extent and type of rehabilitation or reconstruction that should be provided.

3. There has been an increase in the need for concrete rehabilitation on bridge decks. On certain projects, the extent of concrete deterioration actually encountered in the field has far exceeded the dollar amount anticipated in the design stage. This trend needs to be stopped by the selection of adequate materials at the design stage and by performing correct repairs during the structure's life cycle.

   Defects and cracks are most likely to contribute to failure. They need to be analyzed for strength, risk factor, and reduced safety factor. Unlike steel, concrete is a brittle material. The suitability of applying fracture mechanics methods to arrest crack generation and propagation needs to be investigated and included in the design codes.

   As pointed out in Chapter 3, the causes of cracking are usually design defects, overload conditions, and construction defects. A concrete deck is the weakest link in the durability chain. Due to constant friction from braking or abrasion from heavy vehicle tires and due to deicing salts on concrete surface, decks may crack. They need to be repaired every few years or even replaced after 10 to 15 years.

4. The purpose of bridge deck repair projects is to provide a smooth riding surface and improve the bridge deck surface and bridge deck joints so as to extend the useful life of the structure in a cost efficient manner. The procedure is usually to construct interim repairs, which can provide a smooth riding surface and extend the useful life of the structure.

The designer shall evaluate the underside of the deck slab for potential full depth repairs or to check the condition of any stay-in-place forms.

10.4.2 Bridge Deck Repair Work

Bridge deck repair includes the following work items:

1. Full-depth concrete deck replacement.
2. Extensive permanent spall repairs.
3. Parapet, safety walk, and median spall repairs and upgrades.
4. Joint rebuilding.
5. Header reconstruction.
6. Membrane installation.
7. Milling and resurfacing work.

For correct estimation of quantities, the sounding of the concrete deck to detect areas of delamination or cleavage planes should occur close to the time of the actual construction.

Completion of design and contract documents within one construction season of less than one year is preferred.

The following guidelines shall be common to most deck repair projects. Full and partial deck repairs are shown in Figures 10.8.

Type of deck repairs are broadly classified in Figure 10.6. Although there are a host of methods available for repairs, based on practical experience and availability of vendor products and information, the following methods are considered:

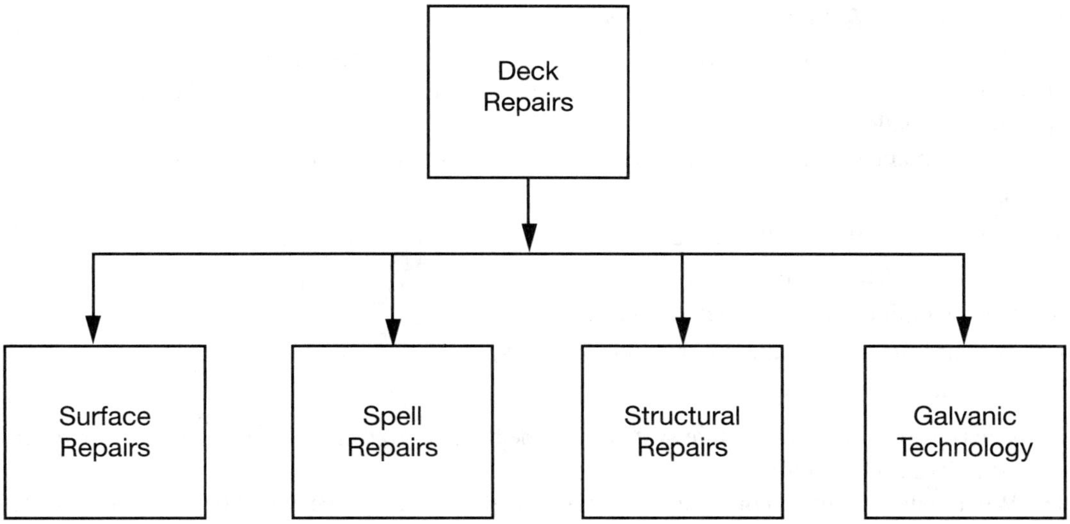

Figure 10.6 Type of deck repairs.

- Preparing surface
- Deck patching
- Remove unsound concrete
- Full depth patching
- Partial depth patching.

1. Class A concrete: For all concrete deck repairs, the coarse aggregate shall be size No. 8. It shall be a maximum of ½ inch in size and be broken stone or crushed gravel.
2. Other materials shall conform to the standard construction specifications:
   - Coarse aggregate
   - Epoxy bonding coat
   - Reinforcement steel, deformed bars
   - Latex emulsion admixture
   - Silica fume admixture
   - Quick-setting patch materials:
   - Membrane waterproofing.
3. Percentage of deterioration: Based on field and laboratory information a decision may be made for the type of rehabilitation. The maximum percentage of deck patching areas that should be included in these projects and remain cost effective is as follows:
4. Treatment for concrete decks with types of overlay:
   - Concrete decks with epoxy or galvanized reinforcement bars, with spalls: Surface deterioration due to loading, abrasion, and other similar activities can be corrected by concrete repairs and latex modified concrete overlays.
   - Concrete decks with bituminous overlay and without membrane waterproofing: Replace such decks when they have a condition rating of three or less.
   - Concrete decks with bituminous overlay and membrane waterproofing: Patch deteriorated concrete and replace affected membrane and overlay until the repaired area is anticipated to be 50 percent or more of the deck area.
   - Concrete decks with latex modified mortar or concrete overlay: Consider latex modified mortar or concrete overlaid decks the same as regular concrete decks when using this table.

Overlay Protective System Commonly Used

If a bridge deck is likely to be exposed to potentially damaging applications of deicing chemicals, salt water, or other hostile environments, a cost effective overlay protective system should be considered.

Design manuals and standard specifications of most highway agencies have listed the following:

1. Latex modified concrete (LMC).
2. Microsilica fume concrete.
3. Corrosion inhibitor aggregate concrete.
4. Polymer surface treatment—thin epoxy overlay.
5. Bituminous overlay with water proofing membrane.

The type of overlay protective system shall be based on the criteria established in approved state standard specifications.

Where latex modified concrete overlay is to be provided, the deck shall be scarified. Scarify the deck in ¼ inch deep passes to avoid structure damage (cracking and spalling) of the remaining deck and to eliminate pulverization of the concrete around the reinforcement due to the high pressures needed for more than a ¼ inch pass.

Repairs of concrete decks shall consist the following:

1. Deck preparation.
2. Membrane waterproofing: Requirements for cleaning and surface preparation of concrete on the existing bridge deck slabs, construction equipment, temperature and weather conditions, application of primer, and other operations pertaining to placing the membrane waterproofing may vary with the proprietary product. Two copies of the manufacturer's technical data sheets shall be submitted at the preconstruction meeting. Construction shall be done in strict conformance to the manufacturer's recommendations.
3. Silica fume concrete: The mix design shall include the sources of fine and coarse aggregates and the composition of silica fume admixture such as fineness, silica content, total chloride ion content, solids content for slurries and moisture content for powders.
4. Saw cut grooving: After completion of the minimum total curing time of 14 calendar days, the overlay shall be grooved according to specifications, provided that the concrete has attained a strength of at least 4000 pounds per square inch as determined from cylinders cast during the placement. Construction equipment needed for saw cutting the overlay is permitted to operate on the overlay. Saw cutting equipment that is to be used shall not overstress the concrete deck or the overlay.

10.4.3 Guidelines for Deck Survey and Deck Condition Evaluation

The following guidelines present procedures that should be considered in determining existing bridge deck conditions and the extent of work required for adequate rehabilitations.

1. A full deck condition survey should only be performed by trained construction personnel.
2. A deck evaluation survey report will be prepared when an existing bridge or structure is to be widened, altered, reconstructed, or rehabilitated.
3. Staging of concrete slab bridges: For single span slab bridges where stage construction is provided, all bridges should be screened to determine whether the main reinforcement is parallel to the centerline of the roadway or is perpendicular to the abutment.

Most state DOTs follow similar procedures. Observe the underside of the deck and record the approximate size and location of all areas exhibiting cracks with or without efflorescence.

Also, record all areas having concrete spalled from the bottom reinforcing for evidence of unsound concrete in the bottom exposed surface of the deck slab (which may indicate structural failure) for:

1. Potential full depth repairs.
2. To check the condition of stay-in-place forms.
3. Where there are indications of delamination, water intrusion, discoloration, spalls, efflorescence or other signs of distress, the underside shall be sounded.
4. The decks shall then be cored in suspicious areas to verify and further define areas of unsoundness. If it is suspected that full depth repair may be required, cores shall be taken full depth or at least to the bottom mat of reinforcing steel in those areas.
5. A description of the core results shall accompany the deck condition survey report.

10.4.4 Deck Testing Steps Commonly Practiced in the U.S.

A limited field condition survey should be made to identify bridge decks that may be structurally inadequate or possibly contaminated with deicing chemicals such that normal maintenance is not expected to provide reasonable service.

Detailed field appraisal: As reported in deatil in listed Bibliography, a detailed evaluation survey should be performed to further define the inadequacies of the existing deck. This appraisal should consider the following:

1. Visual observations: The first step for deck evaluation is to determine the extent of spalling, cracking, and scaling. The top surface of bare concrete decks shall be both visually inspected and sounded for obvious signs of deterioration.
   - Record the deficiencies of either asphalt overlay or the concrete deck wearing surface (e.g., spalling, cracking, scaling, warping, asphalt creep, alligator cracks, etc.).
   - Determine the percentage of spalls and/or patches in the exposed concrete deck wearing surface.
   - Document by photographs visible concrete spalls which have occurred in the deck riding surface.
   - Visual observation does not reveal hidden structural deterioration such as delaminations or corrosion of rebars. Visual surveys are generally expressed in terms of the amount of spalling and patching as a percent of the total deck area.
2. Delaminations: For evidence of delaminations (horizontal fracture planes) or debonding in the concrete deck (to determine extent of internal fractures of the concrete) use infrared thermography, chain dragging, or other approved methods.

   A chain drag is generally used: The chain drag consists of four or five segments of 1 inch link chain about 18 inches long, attached to a 2 foot piece of aluminum or copper tube, to which a 2 to 3 foot piece of tubing is attached at the midpoint, forming a "T."

   Drag the chain in a swinging motion, resulting in a ringing sound while walking along the concrete surface of the deck.

   Outline, with crayon or paint, the areas of the deck over which the chain produces a distinctive "dull" sound. These areas indicate delamination of concrete.
3. Half-cell test to determine the extent of reinforcing steel corrosion: The purpose of half-cell testing is to determine the areas in the deck in which active corrosion is present. Corrosion of the reinforcing bars in concrete decks is detected by electric current flowing from the rebar at one point (the anode) to another point (the cathode). During active corrosion, a potential electrical difference exists between the anode and cathode, which can be measured by copper/copper sulfate half-cells (CSE).

4. Pachometer survey: A pachometer survey is performed to determine the depth of the concrete cover over the reinforcement steel. The equipment shall be calibrated according to the equipment manufacturer's specifications. Locate and expose a reinforcing bar in the deck using a jackhammer. Connect the negative lead of the voltmeter to the reinforcing steel. In order to properly establish the deck condition, establishing the depth of cover over the top reinforcement is necessary. This will provide the evaluator with needed information to properly judge the existing condition versus what is the required minimum depth of cover.
5. Estimate of corrosion level:

   Category 1—Extensive active corrosion: 5 percent or more of the deck area is spalled or 40 percent or more of the deck area is deteriorated or contaminated.

   Category 2—Moderate active corrosion: 0 to 5 percent of the deck area is spalled.

   Category 3—Light to no active corrosion.
6. Chloride content analysis: This provides a quantitative measure of the chloride ion contamination of concrete at selected levels in the deck. Chloride concentrations are significantly greater near the surface of a concrete bridge deck. When rebars have less than specified concrete cover they become appreciably more susceptible to damaging rebar corrosion. Test results have generally established that the corrosion threshold is approximately 2 pounds of chloride per cubic yard of concrete at the level of the rebars for typical bridge deck concrete.

10.4.5 Laboratory Tests

The following practice is commonly used:

1. A sketched plan of the deck area, both top surface and underside, shall be included with the bridge deck condition survey (Figure 10.7). The unsound areas should be plotted on the

Figure 10.7 (A) Bridge closed down for repairs, (B) emergency repairs required on pot holes.

sketch indicating the approximate dimensions that were used to estimate the percentage of total unsound deck area.

2. Core locations shall be determined from conditions detected primarily from the bottom side of the deck; however, the top surface may also indicate areas to be cored. The minimum number of two to four cores to be taken for a bare concrete deck shall be determined by the deck area. For bridge decks with an asphalt overlay additional cores should be taken due to the variability of unknowns hidden under the overlay.
3. At least one core shall be taken from an apparently sound area and the others from questionable areas for comparison. Cores shall be inspected for:
    - Obvious crumbling
    - Stratification or delamination zones
    - Soundness of aggregate
    - Depth and condition of reinforcing steel
    - An estimate of the unsound deck area as a percentage of total deck area shall be made from all of the information gathered from the survey and testing.
    - Deck cores shall be analyzed for chloride content.

10.4.6 Preparing the Deck Condition Report

An important aspect is to prepare a deck condition report. A description of the core results shall accompany the deck condition survey report. The following summary of information is required prior to determining the type of rehabilitation:

1. Concrete strength—Use concrete coring and testing, and Windsor probe results, if permitted.
2. Extent of concrete spalls, in percentage of deck area.
3. Type of deck steel (black versus coated reinforcement bar) concrete cover over the top rebar mat.
4. Salt content in top 1 inch and in the last 1 inch of concrete immediately above the top mat of reinforcing bars.
5. Age of the deck.
6. Air content—The lower the air content, the more important other deck quality factors become.
7. Extent of delamination in percentage of deck area and/or traveled lanes.
8. Half-cell electric potential or other methods of determining active corrosion.
9. Type and location of major cracking indicating superstructure flexibility problems.

10.4.7 Deck Concrete Failure Mechanisms

1. Study the failure mechanism such as:
    - Corrosion due to frequent acid rain and chemical attacks
    - Daily temperature change cycles
    - Fatigue
    - Insufficient cover to rebars

Table 10.4 Patching versus replacement.

1% to 30%	deck deterioration	Deck patching
20% to 60%	deck deterioration	Deck patching with overlay
50% to 100%	deck deterioration	Deck replacement

- Drying shrinkage
- Honeycombing, crack generation, spalls and potholes (bump causes dynamic damage).

2. Study causes of failures:

- Failure to provide timely repairs causes further degradation and increases collateral structural damage
- Concrete wears and tears with age
- Insufficient drainage
- Vehicle impacts
- Wear and tear due to tire friction.

3. Traditional repair techniques may cause more problems (saw cuts create stress risers, jack hammers degrade edges). New repair techniques need to be investigated.

Preparing deck repair details: The defects should be recorded by sketching the defects. Areas of previous repair as well as existing spalls should be sketched.

1. These sheets shall show typical repair details. The method of repairs should be as per state standard details.

2. Repair procedures that the contractor will follow should be outlined in the notes.

3. Deck repair details shall reflect the rebar locations shown in as-built plans. When any portion of the bottom layer of reinforcing bars is exposed, full depth repair is required.

10.4.8 Creating a Schedule

A suggested probable time line for contract from July to the following May is as follows:

1. July to November—Select bridge decks

2. September to February—Perform deck condition surveys.

3. November to March—Complete all construction plan sheets.

4. March to April—Advertising.

5. April to May—Selection of contractor and award.

10.4.9 Case Study of Partial and Full Depth Deck Repairs

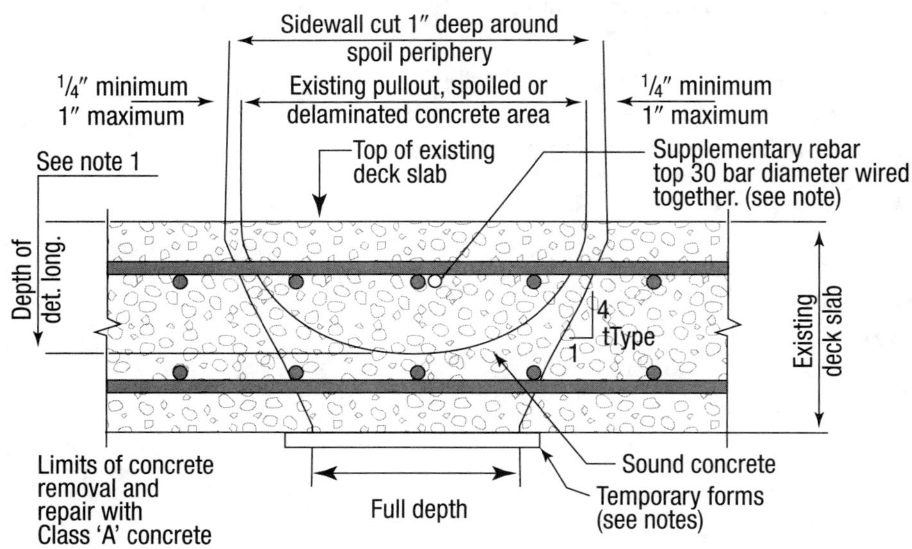

Partial Depth Repair

Deteriorated concrete shall be removed to a minimum depth of 1″ below the bottom of the top layer of reinforcement steel to a minimum depth of the sound concrete. For addition of reinforcement steel due to section losses. See technical specifications.

Full Depth Repair

Full depth repair is required if the bottom mat of reinforcement steel is exposed.

Figure 10.8 Full depth deck repair

10.4.10 Effects of Deck Vibrations

Nowak and Grouni in 1988 showed that deck vibrations, in addition to causing cracking, can affect human perception by reducing comfort level.

1. Factors causing discomfort include acceleration, deflection, and frequency of response.
2. Human reactions to undesirable vibrations are either physiological (due to low frequency and high amplitude vibration such as sea sickness) or psychological (discomfort resulting from unexpected motion).
3. Oehler in a 1957 concluded that cantilever spans were prone to longer periods of vibrations and longer amplitudes than simple or continuous spans.
4. Increasing bridge stiffness does not decrease vibration amplitude sufficiently to remove it from perception range.

   Field studies by Biggs et al have shown that factors influencing dynamic behavior include:

   - Vehicle weight.
   - Bridge geometry.
   - Bridge material.
   - Vehicle/structure interaction.

   Vibrations increase when the natural period of vibration of the span coincides with the time interval between axles passing a reference point on the span.
5. A finite element study by Roeder, Karl, Barth, and Bergman showed that maximum accelerations increased five times for a rough surface compared to a smooth surface. Aramraks observed that vehicle speed greatly influences peak acceleration.

10.4.11 Effects of Deck Joints

Strip seal expansion dams located at abutment backwalls are subjected to impact from vehicles thereby causing cracks in concrete. Approximately 3 feet of deck slab width needs to be replaced with new deck. The top of the back wall is also replaced to accommodate new expansion dams. One of the most frequent replacements is the deck joints including angles, anchor bolts, etc.

To avoid future replacement of deck joints, use of integral abutments needs to be considered.

10.4.12 Need for Effective Deck Drainage

1. With no shoulders and clogged downspouts, the spread of water collects on traffic lanes creating a potentially hazardous condition. Drainage improvements will consist of:
   - Constructing new larger scuppers to take water off the roadway more quickly
   - Replacing existing downspouts with larger diameter downspouts
   - Installing shoulder areas for the water to collect
   - Increasing the roadway cross slope.
2. Cleaning clogged drains and replacing damaged spouts and scuppers would prevent ineffective storm water disposal. This would otherwise slow down traffic. Pools of water on deck are likely to cause accidents.
3. FLOWMASTER or approved software may be used. Based on drainage calculations, drainage inlets will be provided outside according to calculated spacing not exceeding 300 feet (or if possible outside the deck area). Any drainage facilities required for the new decks will be tied into the existing system that carries runoff to the existing low point.

10.5 PAY ITEMS AND QUANTITIES FOR DECK REPLACEMENT PROJECTS

Some of the major items related to deck replacement may be summarized as:

Table 10.5(a) Structural repair contracts (small projects).

Pay Items	Units
Repair Spalled Concrete	Sq Ft
Repair Spalled Concrete—Underwater	Sq Ft
Repair Spalled Concrete—Bearing Pads	Each
Sidewalk, Parapet and Curb Surface Repairs	Sq Ft
Epoxy Resin Injection	Linear Ft
Furnish Epoxy Resin	Gal.
Replace Structural Steel Diaphragm	Each

Table 10.5(b) Bridge deck rehabilitation (intermediate projects).

Pay Items	Units
Concrete Overlays	Sq Yds
Scarification	Sq Yds
Joint Sealers	Linear Ft
Joint Seal Replacement	Each
Joint Preparation and Sealer Installation	Lump Sum
Emergency Concrete Deck replacement	Sq Ft
Emergency Pavement Replacement	Sq Ft
Removal of Existing Surfacing	Sq Ft
Concrete Deck Replacement	Sq Ft

Table 10.5(c) Bridge deck reconstruction (major projects).

Pay Items	Units
Membrane Waterproofing	Sq Yd.
Reinforcement Steel, Epoxy Coated	Lbs.
Drill and Grout Reinforcement Bars	Each
Concrete Deck Replacement	Sq Yd
Permanent Metal Form Removal	Sq Yd
Diaphragms	Each
Joint Reconstruction Types	Linear Ft
Pavement Riser Repair	Linear Ft
Removal of Asphalt Surfacing and Scarify Concrete	Sq Yd

Removal and Replacement of Existing Surfacing	Sq Yd
Spall Repairs	Sq Yd
Deck Haunch Repairs	Sq Yd
Head Block Repairs	Linear Ft
Emergency Joint Reconstruction	Linear Ft
Removal of SIP Metal Forms	Sq Yd
Dowel Bar Removal	Each

Table 10.6 MPT related pay items.

Pay Items	Units
Placing and Removing Concrete Barrier	Each
Install, Maintain and Remove Lane Closings	Each
Temporary Pavement Striping	Linear Ft
Uniformed Flagmen	Man Hours
Variable Message Sign	Each
Truck and Attenuator	Each
Resetting Concrete Barrier	Linear Ft
Furnishing Traffic Control Devices	Lump Sum
Temporary Crash Cushion	Unit
Pavement Striping, White	Linear Ft
Pavement Striping, Yellow	Linear Ft

10.6 ACI RECOMMENDED METHODS OF REPAIR

The following methods developed by the American Concrete Institute (references are in the Bibliography) have emerged for detecting common flaws in concrete decks and are being successfully used:

1. Infrared thermography
2. Impact echo.
3. Ground penetrating radar.

10.6.1 Galvanic Technology

See Chapter 9 for details.

10.6.2 Structural Crack Repairs Using

1. Epoxy injection (ACI RAP-1).
2. Gravity feed with resin (ACI RAP-2).

10.6.3 Spall Repairs Using

1. Low-pressure spraying (ACI RAP-3).
2. Vertical and overhead spall repair by hand application (ACI RAP-6).
3. Horizontal concrete surfaces (ACI RAP-7).
4. Preplaced aggregate method (ACI RAP-9).

10.6.4 Surface Repair Using

1. Form-and-pour techniques (ACI RAP-4).

2. Form-and-pump techniques (ACI RAP-5).

10.6.5 How to Prepare the Surface

Concrete surface preparation: See Figure 10.9. Follow vendor's instruction.

Regardless of the repair method, surface preparation is essentially the same. Concrete is removed until good quality concrete is located. Exposed bars are undercut, and surfaces are cleaned with high-pressure water or are abrasively blasted.

10.6.6 How to Select the Proper Repair Material

Mixtures with high flowability (high slump) will make the placement easier; however, do not sacrifice a low water-cement ratio (< 0.40) for high slump. Use high-range water-reducing admixtures as necessary.

10.6.7 Preconstruction Meeting and Trial Repair

Prior to proceeding with the repair, a preconstruction meeting should include representatives from all participating parties (owner, engineer, contractor, materials manufacturer, etc.). Specifically address the parameters, means, methods, and materials necessary to achieve the repair objectives.

The preconstruction meeting agenda may include the following:

- Site accessibility
- Debris removal and disposal
- Dust, odor, and emissions control
- On-site availability of power, mixing water, control of water runoff
- Confirmation that all materials and equipment are available, especially required quantities of grout and graded, washed aggregate
- Confirmation that all material documentation is on-site
- Noise control
- Methods of curing and time required for curing

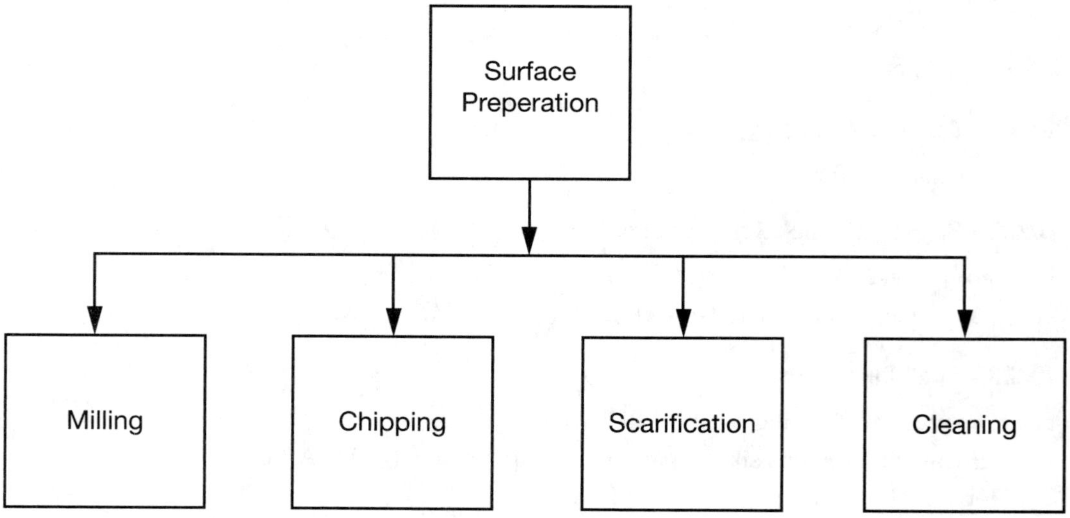

Figure 10.9 Concrete surface preparation prior to repairs.

- Responsibility for quality control and final acceptance
- Possible emergencies and breakdowns
- Safety and all other concerns that could affect the progress of the repair.

Trial repairs using the proposed procedure and materials should follow standard practice for cast-in-place concrete construction except for the calculation of form pressure. Form pressure should be designed for a minimum of 15 psi.

10.6.8 Safety Considerations

Epoxies and HMWMs are hazardous materials. It is the responsibility of the user of the repair methods to establish health and safety practices appropriate to the specific circumstances involved with its use. The user must comply with all applicable laws and regulations, including United States Occupational Safety and Health Administration (OSHA) health and safety standards.

Job site safety practices include, but are not limited to, the following where applicable:

- Available material safety data sheets
- Wearing protective clothing and protective eyewear where required
- Wearing rubber gloves or barrier creams for hand protection
- Availability of eye wash facilities
- Wearing respirators where needed
- Secured storage of hazardous materials
- Availability of necessary cleaning materials
- Notifying occupants of pending repair procedures.
- Providing correct safety guards, maintenance, and warnings for all machinery and equipment to be used
- Providing dust masks for workers operating material mixers
- Confirming that adequate ventilation is available in closed spaces before operating equipment or using products that emit dangerous exhaust, fumes, or dust
- Providing secured storage for all hazardous or flammable materials.

10.6.9 Summary of ACI Repair Methods

Table 10.7 ACI repair methods for deteriorated concrete.

Method	Purpose	Selection of Material
Galvanic technology- Embedded galvanic anodes	Combats the underlying corrosion. Anodes create their protective current internally through natural reaction wherein the anode corrodes to galvanically protect the reinforcing steel.	For chloride-contaminated or carbonated concrete. Galvanic anode is used with cementitious or cementitious-polymer repair materials, which have a low resistivity.
Structural crack repair by epoxy injection (ACI RAP-1)	Epoxy injection can restore structural integrity and reduce moisture penetration through concrete cracks 0.002 in. in width and greater.	ASTM C 881, "Standard Specification for Epoxy-Resin-Base Bonding Systems for Concrete," identifies the basic criteria for selecting the grade and class of epoxies.

(continued on next page)

Table 10.7 ACI repair methods for deteriorated concrete *(continued)*.

Method	Purpose	Selection of Material
Structural crack repairs Gravity feed with resin (ACI RAP-2)	It uses a thin polymer resin to fill the crack. Used for shrinkage cracks and settlement cracks that have stabilized.	Polymer materials used for gravity feed crack repairs are epoxies and high molecular weight methacrylates (HMWM).
Spall repair by low-pressure spraying (ACI RAP-3)	Used for surface/ structural repairs or cosmetic renovation. Spray can be formulated for freeze-thaw durability, sulfate resistance, low permeability	Low-pressure spray-applied repair materials are proprietary, prepackaged cementitious products.
Surface repairs using form-and-pour techniques (ACI RAP-4)	Used on vertical surfaces such as walls, columns, beam sides and bottoms.	All mixture proportions should be optimized to minimize drying shrinkage. Shrinkage testing in accordance with ASTM C 157 measured over a 120-day period.
Surface repair using form-and-pump techniques (ACI RAP-5)	-do-	Separate bonding agents such as grouts or epoxy are not commonly used with this technique.
Vertical/overhead spall repair by hand application (ACI RAP-6)	Ideal for shallow surface repairs. Thin overlays of mortar are cosmetic in nature.	Both Portland cement-based and resin-based repair mortars have been used.
Spall Repair Of Horizontal Concrete Surfaces (ACI RAP-7)	Deterioration caused by corrosion of embedded reinforcement resulting in delamination and spalling.	Water cement ratio (w/c) of not more than 0.40 should be used. Compressive strength should not be less than 4000 psi. Other properties such as low shrinkage
Spall repairs pre-placed aggregate (PPA) method (ACI RAP-9)	1. Uniform aggregate distribution and density are achieved; 2. Ratio of aggregate-to-cement paste is higher than in placeable concrete; 3. Aggregate can be placed in hard-to-reach locations; 4. Shrinkage is reduced by 50 to 100%.	Use sound, tested, properly graded aggregate. Test aggregate for reactivity in accordance with ASTM C 1260. Grade the aggregate in accordance with the recommendations of Table 1 of ACI 304.1R.

10.7 PRESTRESSED CONCRETE BRIDGES

10.7.1 Repair of Prestressed Concrete Bridges

Repair and rehabilitation of damaged or deteriorated prestressed concrete beams, especially the repair of beams that were damaged by oversize vehicles, is required on all rehabilitation projects. The affected structures should be monitored regularly as part of the inspection process, and joints, particularly on prestressed concrete box beam structures, should be kept watertight.

Reference NCHRP Synthesis of Highway Practice 140, "Durability of Prestressed Concrete Highway Structures," when addressing items such as:

1. NCHRP Report 226, "Damage Evaluation and Repair Methods for Prestressed Concrete Bridge Members."
2. NCHRP Synthesis Report 280, "Guidelines for Evaluation and Repair of Prestressed Concrete Bridge Members."

The above reports contain a considerable amount of data on how to repair spalls, splice severed strands, and strengthen beams through the addition of strands.

10.7.2 Rehabilitation of Prestressed Concrete

For defects in prestressed concrete beams post-tensioning is used to eliminate:

- Corrosion of prestressing steel
- Prestress loss
- End block cracks.

Debonding of strands: Spalling of the bearing areas is primarily caused by the infiltration of salt-laden runoff through leaky joints, with subsequent chloride saturation of the beam ends and resulting rusting of mild steel in the beam ends, and worse, rusting and debonding of prestressing strands.

10.8 BRIDGES LOCATED ON RIVERS

10.8.1 Substructure Rehabilitation Issues

Following are examples of substructure rehabilitation issues:

1. Geotechnical issues and foundation design.
2. Foundation erosion.
3. Scour countermeasures design: These have additional issues from flood erosion. In addition to areas for repairs identified in the underwater inspection and evaluation report performed by the diving team, a field inspection and an underwater inspection will be carried out for field verification of the latest conditions.

Methods of repair for deteriorated concrete, repointing mortar joints, applicable design details from AASHTO or a state bridge design manual will be used.

1. If current NBIS rating given in inspection report is say 6 for abutments and 5 for piers, the load rating will be upgraded by performing the recommended repairs.
2. Performing substructure repairs both above and below the water line.
3. Deteriorated expansion joint and back wall elements.
4. Abutment shows deterioration.
5. Abutment back wall has wide cracks at the north and south ends.
6. Concrete aprons at the piers exhibit wide cracks.
7. Tooth dam at abutment is not functioning and needs to be replaced.
8. Remove build up of sand debris at piers.
9. Remove any tree trunks or tree roots between piers.
10. An estimate of cost and repair quantities will be prepared.

10.8.2 Underwater Repairs

Identify defects and perform repair based on underwater inspection reports. Table 10.8 shows the types of defects that may exist for bridges:

Table 10.8 Underwater repairs and debris removal.

No.	Item	Remedy	Alternate Method
1	Spalls in abutment concrete	Surface repairs with approved patch material	
2	Delaminations and cracks in substructure concrete above water	Pressure grouting	
3	Voids in breast wall	Epoxy grouting	Use nonshrink grout
4	Debris accumulation	Cleaning debris	
5	Mortar loss in masonry joints	Repointing mortar	
6	Undermining	Plug with concrete	Use grout bags
7	Broken stone masonry	Place stone and fill with mortar	
8	Cracks in tremie concrete below water	Pressure inject masonry cracks	
9	Spalling in foundation	Pressure grouting	
10	Cracks in apron around piers	Repair concrete apron	Place riprap around piers
11	Corrosion of rebars	Clean, paint, and provide adequate cover	Dowel anchor bars in concrete holes
12	Advance section loss of structural members	Strengthening member	Underpinning

10.9 CONCRETE REPAIR/RESTORATION OTHER THAN FOR DECKS

10.9.1 Field Guide to Concrete Repair Applications

The most common procedure for repairing deteriorated concrete involves the removal of the damaged material and replacement with new concrete or mortar.

Differences in pH, porosity, and chloride content are a few of the factors that may result in corrosion activity. For long-term durability needs, "chip and patch-style" repairs may fail prematurely in chloride exposed structures.

Field practice procedures: All bridge repairs shall be done under the direction of the engineer. The user must be familiar with the applicability of all regulatory limitations and must comply with all applicable laws and regulations. Familiarity with United States Occupational Safety and Health Administration (OSHA) health and safety standards is required. Appropriate health and safety practices must be established.

10.9.2 Patching

To serve as a guide to the designer, the following criteria may be used in the patching selection and evaluation:

1. Patching should be used where the repair depth is 3 inches or greater and the surface can be readily formed and concrete placed.
2. Pneumatically placed mortar should generally be used where the repair surface cannot be readily formed and concrete placed, where the depth of repair is between 1 and 6 inches, and where a repair area greater than 150 square feet is involved.
3. The detail plans shall show and detail the locations of the areas that require patching repairs.

Some modern patching materials include:

- Normal concrete
- Epoxy resin mortar

- Latex modified concrete (LMC)
- Magnesium phosphate concrete
- Other quick setting material.

10.9.3 Crack Sealing and Repairs

1. Critical reinforcement for the pier caps may be vulnerable to corrosion if the concrete is crushed and exposed to contaminants from leaking joints. All such cracks shall be sealed with appropriate epoxy compounds. If the cracking is caused by differential settlement, the situation shall be evaluated and corrected.

Cracks can be repaired by epoxy injection. The location of the cracks shall be shown in the plans and marked in the field.

2. Crack repairs: The bridge inspection report may address needed spall repairs.

Medium to wide cracks in masonry must be sealed by using standard procedures for repairing concrete cracks and spalls such as patching or sealing. Any missing mortar joints will be filled.

3. Spall Repairs: Pier bent or column bent situations can be repaired by jacketing. Repairs will be categorized on the contract documents.

Surface spalls of the concrete elements shall be cleaned to sound concrete and repaired with epoxy mortar. If deteriorated concrete extends beyond the primary reinforcement, the concrete shall be removed to at least 1 inch below the reinforcement and repaired with either concrete (if space permits) or lifts of epoxy mortar. An epoxy bonding compound shall be specified between the old and the new concrete and concrete lifts if needed.

4. Reinforced and prestressed concrete girders: Compared to steel girders, concrete girders are more difficult to strengthen. If reinforcing bars and shear reinforcement is exposed, formwork may be used to recast concrete.

5. A full matrix comparison will be developed to compare the alternatives for costs, optimum sizes of members (weights, depths, etc.), ease for future inspections, constructability, fracture critical requirements, and structural details that will facilitate future widening.

Local noise ordinances will be checked.

6. Nighttime construction will be guided by NCHRP Report # 475 "A Procedure for Assessing and Planning Nighttime Highway Construction and Maintenance" 2002, and NCHRP Report # 476 "Guidelines for Design and Operation of Nighttime Traffic Control for Highway Maintenance and Construction 2002."

10.9.4 Abutment Repairs

Record the abutment type most representative of the taller abutment.
Possible abutment types are as follows:

- Stub
- Semi-stub
- Full height
- Counterfort
- Timber
- Sheet pile
- Masonry
- Open
- Gravity
- Combination.

The following procedures are developed through experience and are reported in literature:

1. Spalling of the abutment stem shall be repaired. Other alternatives, such as casting wall or other support beneath the cantilever, may be feasible and may offer a long-term solution.
2. Back wall deterioration: If the back wall is severely damaged beyond repair, it shall be replaced and a pavement relief joint installed.
3. Structural stability: Abutments on steep slopes are of concern. Such situations can be corrected by slope stabilization and the use of tie backs, replacement of the substructure unit, or by adding a span.
4. Deterioration of concrete at bridge seats requires attention and repairs. Stem and back wall repairs may be required.
5. Reusing existing abutments for new loads should be investigated by using the tieback system.

Existing bridge seats for a two through Girder Bridge may need to be lowered for a slab-beam alternate when using shallower beams. Recasting of abutment cap to provide adequate seat width may be investigated.

When raising of the alignment is required, the total retaining wall lengths at approaches need to be increased in order to make the proposed approach gradient same as the existing gradient.

Use of quick setting concrete and other quick repair techniques at abutment seat, back wall and breast wall may be considered in the interest of saving time and costs.

10.9.5 Pier Repairs

Possible pier repairs are as follows:

1. Hammerhead pier caps: External post-tensioning may be specified for hammerhead pier caps to restore structural integrity. The system shall be designed to carry the entire load assuming the existing cap steel is ineffective.
2. Column types, pile or frame bents.
3. Wall types.

10.9.6 Unknown Foundations

1. Unknown foundations and hidden damage need to be investigated using FHWA procedures.

   If settlement is a problem, the root causes shall be determined. Corrective measures shall be used such as underpinning, replacing superstructure bearings if settlement has ceased, or replacement. If settlement of a substructure unit founded on steel piles is evident, the integrity of the piles could be suspect.
2. Ground penetrating radar (GPR) for concrete evaluation studies: The objectives are to accurately locate rebar, tension cables, grade beams, conduits, voids, and footing slab thickness. Concrete evaluation studies utilizing GPR include the inspection of various foundation systems.

GPR provides an ideal technique for concrete evaluation since it has the highest resolution of any subsurface imaging and is far safer than other methods such as x-ray technology. Recent improvements in hardware and in particular software processing have contributed to the rapidly expanding popularity and usability of this technique.

10.9.7 Foundation Repairs

Table 10.9 shows standard maintenance requirements for foundation repairs.

Table 10.9 Maintenance requirements for foundations.

Substructure Type	Foundation Type	Check Capacity for New Live Loads	Check Capacity for Seismic Loads	Check Capacity for Scour Conditions
Full height abutment Stub abutment Spill through abutment	Unknown	Drill bore holes/perform NDT to determine type and size. Posting for lower load if required	Follow AASHTO/state code requirements	Follow AASHTO/state code requirements
Column bent pier Pile bent pier Wall pier Hammerhead pier	Spread footings	Posting for lower load if required	Retrofit footing with mini piles	Provide countermeasures and retrofit with mini piles if required
	Short piles < 15 Ft.	-do- Geotechnical investigation is required	Geotechnical investigation is required	Geotechnical investigation is required
	Long piles > 15 Ft.	-do-	Geotechnical investigation is required	Geotechnical investigation is required
	Sheet pile	-do-	Geotechnical investigation is required	Geotechnical investigation is required
	Drilled shaft	Check design	Generally okay.	Generally okay.
	Large diameter caisson	-do-	Generally okay.	Generally okay.
Integral abutments/piers	Single row of piles	Check performance. posting for lower load if required	Geotechnical investigation is required	Geotechnical investigation is required
Semi-integral abutments/piers	Two row of piles	-do-	Generally okay.	Generally okay.

BIBLIOGRAPHY

AASHTO/NSBA Steel Bridge Collaboration

Aramraks, T., "Highway Bridge Vibration Studies," Joint Highway Research Project No. JHRP-75-2, Purdue University, 1975.

Aspire, Summer 2007.

Austin, Deron and Ariel Soriano, "Geotextiles for Transportation Projects," *CE News*, April 2005.

Biggs, J. M., H. S. Suer and J. M. Louw, "The Vibration of Simple Span Highway Bridges," *Journal of Bridge Engineering,* American Society of Civil Engineers, Vol. 96, No. EM5, October, pp. 593–620.

Concrete Repair Guide, American Concrete Institute, Committee 546, 2001, Farmington Hills, MI.

Field Guide to Concrete Repair, Repair Application Procedures Bulletins 1–9, American Concrete Institute, Farmington Hills, MI.

Gaidis, J. M. and A. M. Rosenberg, "Estimating Life Cycle Cost of Concrete Using Life-365," *Structure*, March 2004.

Guide to Formwork for Concrete, American Concrete Institute, Committee 347, Farmington Hills, MI, 2001.

Guide for Selecting and Specifying Concrete Repair Materials, No. 03733, International Concrete Repair Institute, 1996.

Guide for Surface Preparation for the Repair of Deteriorated Concrete Resulting from Reinforcing Steel Corrosion, No. 03730, International Concrete Repair Institute, 1995.

Issa, M. A., A. A. Yousif and M. A. Issa, "Investigation of Cracking in Concrete Bridge Decks at Early Ages," *Journal of Bridge Engineering*, American Society of Civil Engineers, May 1999.

Krauss, P. D. and E. A. Rogalla, "Transverse Cracking in Newly Constructed Bridge Decks," National Cooperative Highway Research Program, Report No. 380, TRB, 1996, Washington, DC.

Nowak et al, "Development of a Guide for Evaluation of Existing Bridges," University of Michigan, Ann Arbor, 2000.

Oehler, L. T., "Vibration Susceptibilities of Various Highway Bridge Types," Michigan State Highway Department, 1957.

Park, Sung H., "Bridge Rehabilitation and Replacement (Bridge Repair Practice)," S. H. Park, P.O. Box 7474, Trenton, New Jersey, 08628-0474, 1984.

Park, Sung H., "Bridge Topics," A Power Point Presentation, Hardesty & Hanover, New Jersey.

Roeder, C. W., K. Barth and A. Bergman, "Improved Live Load Deflection Criteria for Steel Bridges," Report 20-7, November 2002.

Seismic Prioritization Program—Seismic Coding Guide, New Jersey Turnpike Association, Garden State Parkway.

Use of Epoxy Compounds with Concrete, American Concrete Institute, Committee 503, 2001, Farmington Hills, Michigan.

Corrosion Control and Galvanic Technology

Bertolini et al., "Cathodic Protection and Cathodic Prevention in Concrete: Principles and Applications," *Journal of Applied Electrochemistry*, 1998, V. 28.

Concrete Repair Guide, American Concrete Institute, Committee 546, 2004, Farmington Hills, MI.

Crack Repair by Gravity Feed with Resin, Mendis and C. McClaskey, eds., American Concrete Institute, Farmington Hills, MI, pp. 113–128.

Glass, G., J. Taylor, A. Roberts and N. Davison, "The Protective Effects of Electrochemical Treatment in Reinforced Concrete," NACE Paper No. 03291, *Corrosion*, 2003.

Glass, G. K., A. M. Hassenein and N. R. Buenfeld, *Corrosion Science*, V. 43, No. 6, 2001, pp. 1111–1131.

Guide for Surface Preparation for the Repair of Deteriorated Concrete Resulting from Reinforcing Steel Corrosion, International Concrete Repair Institute, Technical Guideline No. 03730, 1995.

Hausman, D. A., "Steel Corrosion in Concrete," *Material Protection*, 1967, pp. 19–23.

"Healer for Sika Corporation," Report No. 831298, American Concrete Institute, Committee E 706.

Jerzak, H. P., "High Molecular Weight Methacrylate Resins in California," State of California Department of Transportation, January 1991.

Krauss, P. D., "Bridge Deck Repair Using Polymers," paper presented at International Congress on Polymers in Concrete, San Francisco, CA, Sept. 24–27, 1991.

Monomers, P., "Polymers in Concrete: Advances and Applications," SP-116.

Murray, M. A., and W. O. Eisenhut, "A Concrete Healer and Sealer," *Concrete Repair Digest*, V. 6, No. 3, June/July 1995.

Pfeifer, D. W. and W. F. Perenchio, "Coating, Penetrants, and Specialty Concrete Overlays for Concrete Surfaces," paper presented at NACE seminar, Chicago, IL, Sept. 27–29, 1982.

"Protection of Metals in Concrete Against Corrosion," American Concrete Institute, Committee 222, 2001, Farmington Hills, MI.

Rodler, D. J., D. P. Whitney, D. W. Fowler and D. L. Wheat, *Repair of Cracked Concrete with High Molecular Weight Methacrylate*, American Concrete Institute, Committee E 706, 1989.

Specifications and Special Provisions for Use of High Molecular Weight Methacrylate Monomers to Seal Bridge Decks, California Department of Transportation, Contract No. 04-001754.

Sprinkel, M. M., *Use of High Molecular Weight Methacrylate Monomers to Seal Cracks in Bridge Decks, Retard Alkali-Silica Aggregate Reactions, and Prime Bridge Surfaces for Overlays*, Virginia Department of Transportation Research Council.

Structural Crack Repairs Using Gravity Feed with Resin, American Concrete Institute, RAP-2.

Wiss Janney Elstner Associates, "Corrosion Protection Tests on Reinforced Concrete Treated with Sikaguard 70 and SikaPronto 19 Crack," 1986.

Epoxy Injection (ACI RAP-1)

Causes, Evaluation, and Repairs of Cracks in Concrete Structures, American Concrete Institute, Committee 224, 1993, Farmington Hills, MI.

Concrete Repair Guide, American Concrete Institute, Committee 546, 1996, Farmington Hills, MI.

Emmons, P. H., 1994, *Concrete Repair and Maintenance Illustrated*, R. S. Means Co., Inc., Kingston, MA.

Guide for Evaluation of Concrete Structures Prior to Rehabilitation, American Concrete Institute, Committee 364, 1994, Farmington Hills, MI.

Guide for Repair of Concrete Bridge Structures, American Concrete Institute, Committee 546, 1988, Farmington Hills, MI.

Guide for Verifying Performance of Epoxy Injection of Concrete Cracks, 1998, International Concrete Repair Institute, Technical Guideline No. 03734.

Murray, M. A., 1987, "Epoxy Injection Welds Cracks Back Together," *Concrete Repair*, V. 3.

Promboon, Y., L. D. Olsen and J. Lund, "Nondestructive Evaluation (NDE) Methods for Quality Assurance," *ICRI Bulletin*, V. 15, No. 1, January/February 2002, pp. 12–16.

Standard Specification for Epoxy-Resin Based Bonding Systems for Concrete, American Society for Testing and Materials, C 881-90, 1990,West Conshohocken, PA.

"State-of-the-Art Adhesives for Concrete Construction," *Construction Canada*, May/June 1998.

Trout, J. F., "Epoxy Injection in Construction," the Aberdeen Group, 1998.

Use of Epoxy Compounds with Concrete, American Concrete Institute, Committee 503, 1998, Farmington Hills, MI.

Spall Repairs Using
A. Preplaced Aggregate Method (ACI RAP-9)

Guide for the Use of Preplaced Aggregate Concrete for Structural and Mass Concrete Applications, American Concrete Institute, Committee 304, 1992, Farmington Hills, MI.

Guide to Curing Concrete, American Concrete Institute, Committee 308, 2001, Farmington Hills, MI.

Guide to Surface Preparation for the Repair of Deteriorated Concrete Resulting from Reinforcing Steel Corrosion, International Concrete Repair Institute, Guideline No. 03730.

Guide to Using In-Situ Tensile Pull-Off Test to Evaluate Bond of Concrete Surface Materials, International Concrete Repair Institute, Guideline No. 03739.

King, J. and A. Wilson, "If It's Still Standing, It Can Be Repaired," *Concrete Construction*, July 1988.

King, J. and E. Bush, "Grouting of Granular Materials," *Proceedings, American Society of Civil Engineers Convention*, 1960.

Selecting and Specifying Concrete Surface Preparation for Sealers, Coatings, and Membranes, International Concrete Repair Institute, Guideline No. 03732.

Standard Practice for Making Test Cylinders and Prisms for Determining Strength and Density of Preplaced-Aggregate Concrete in the Laboratory, American Society for Testing and Materials, C 943, 2002, West Conshohocken, PA.

Standard Test Method for Flow of Grout for Preplaced Aggregate Concrete, American Society for Testing and Materials, C 939, 2002, West Conshohocken, PA.

Standard Test Method for Potential Alkali Reactivity of Aggregates (Mortar-Bar Method), American Society for Testing and Materials, C 1260, 2001, West Conshohocken, PA.

Standard Test Method for the Compressive Strength of Grouts for Preplaced-Aggregate Concrete in the Laboratory, American Society for Testing and Materials, C 942, 1999, West Conshohocken, PA.

B. Vertical and Overhead Spall Repair by Hand Application (ACI RAP-6)

Dorsch, D. F., "A Guide to Placement Methods for Repair Materials," *Concrete Repair Digest*, December 1995, pp. 320–325.

Emmons, P., *Concrete Repair and Maintenance Illustrated*, R.S. Means, 1993.

Guide for Evaluation of Concrete Structures Prior to Evaluation, American Concrete Institute, Committee

364, 1994, Farmington Hills, MI.

Guide for Making a Condition Survey of Concrete in Service, (ACI 201.R-92), American Concrete Institute, Committee 201, 1992, Farmington Hills, MI.

Guide for Selecting Concrete Repair Materials, International Concrete Repair Institute, Guideline No. 03733, 1997.

Guide for Surface Preparation for the Repair of Deteriorated Concrete Resulting from Reinforcing Steel Corrosion, International Concrete Repair Institute, Guideline No. 03730, 1995.

Guide for Using In-Situ Tensile Pull-Off Tests to Evaluate Bond of Concrete Surface Materials, International Concrete Repair Institute, Guideline No. 03739, 1997.

Guide to Curing Concrete, (ACI 308R-01), American Concrete Institute, Committee 308, 2001, Farmington Hills, MI.

Guide to the Use of Waterproofing, Damproofing Protective, and Decorative Barriers for Concrete, American Concrete Institute, Committee 515, 1985, Farmington Hills, MI.

Selecting and Specifying Concrete Surface Preparation for Sealers, Coatings, and Polymer Overlays, International Concrete Repair Institute, Guideline No. 03732, 1997.

Smoak, W. G., *Guide to Concrete Repair*, U.S. Department of the Interior, Bureau of Reclamation Technical Service Center, April 1997.

Standard Specification for Liquid Membrane-Forming Compounds for Curing Concrete, American Society for Testing and Materials, C 309, 1998, West Conshohocken, PA.

C. Low-Pressure Spraying (ACI RAP-3)

Standard Specification for Curing Concrete, American Concrete Institute, Committee 308, 1998, Farmington Hills, MI.

Guide to Curing Concrete, American Concrete Institute, Committee 308, 2001, Farmington Hills, MI.

Guide to Shotcrete, American Concrete Institute, Committee 506, 1995, Farmington Hills, MI.

Guide for Selecting Application Methods for the Repair of Concrete Surfaces, International Concrete Repair Institute, Guideline No. 03731, 1996.

Selecting and Specifying Concrete Surface Preparation for Sealers, Coatings, and Membranes, International Concrete Repair Institute, Guideline No. 03732, 1997.

Surface Preparation for Repair of Deteriorated Concrete Resulting from Reinforcing Steel Corrosion, International Concrete Repair Institute, Guideline No. 03730, 1995.

D. Horizontal Concrete Surfaces (ACI RAP-7)

Guide for Evaluation of Concrete Structures Prior to Evaluation, American Concrete Institute, Committee 364, 1994, Farmington Hills, MI.

Guide for Making a Condition Survey of Concrete in Service, American Concrete Institute, Committee 201, 1992, Farmington Hills, MI.

Guide to the Use of Waterproofing, Damproofing, Protective, and Decorative Barriers for Concrete, American Concrete Institute, Committee 515, 1985, Farmington Hills, MI.

Guide to Using In-Situ Tensile Pull-Off Tests to Evaluate Bond of Concrete Surface Materials, International Concrete Repair Institute, Guideline No. 03739, 1997.

Selecting and Specifying Concrete Surface Preparation for Coatings, Sealers, and Polymer Overlays, International Concrete Repair Institute, Guideline No. 03732, 1997.

Standard Specification for Curing Concrete, American Concrete Institute, Committee 308, 1992, Farmington Hills, MI.

Surface Repair Using
A. Form-and-Pour Techniques (ACI RAP-4)

Concrete Repair Guide, American Concrete Institute, Committee 546, 2001, Farmington Hills, MI.

Guide for Selecting and Specifying Concrete Repair Materials, International Concrete Repair Institute, No. 03733, 1996.

Guide for Surface Preparation for the Repair of Deteriorated Concrete Resulting from Reinforcing Steel Corrosion, International Concrete Repair Institute, No. 03730, 1995.

Guide to Formwork for Concrete, American Concrete Institute, Committee 347, 2001, Farmington Hills, MI.

Use of Epoxy Compounds with Concrete, American Concrete Institute, Committee 503, 1998, Farmington Hills, MI.

American Association of State Highway and Transportation Officials (AASHTO)

ASHTO M 160, Standard Specification for General Requirements for Rolled Steel Plates, Shapes, Steel Pilings, and Bars for Structural Use.

AASHTO M 300, Standard Specification for Inorganic Zinc-Rich Primer.

American Society for Testing and Materials (ASTM)

ASTM A 6, Specification for General Requirements for Rolled Structural Steel Bars, Plates, Shapes, and Sheet Piling.

ASTM D 3359, Standard Test Methods for Measuring Adhesion by Tape Test.

ASTM D 4138, Standard Test Method for Measurement of Dry Paint Thickness of Protective Coating Systems by Destructive Means.

ASTM D 4285, Standard Test Method for Indicating Oil or Water in Compressed Air.

ASTM D 4414, Standard Practice for Measurement of Wet Film Thickness by Notch Gages.

ASTM D 4417, Standard Test Methods for Field Measurement of Surface Profile of Blast Cleaned Steel.

ASTM D 4541, Standard Test Method for Pull-Off Strength of Coatings Using Portable Adhesion Testers.

ASTM D 4752, Standard Test Method for Measuring MEK Resistance of Ethyl Silicate Zinc-Rich Primers by Solvent Rub.

Research Council on Structural Connections (RCSC)

Specification for Structural Joints Using ASTM A325 or A490 Bolts, Section 5(b), endorsed by the Research Council on Structural Connections.

American Institute for Steel Construction (AISC)

Sophisticated Paint Endorsement (SPE).

Manual of Steel Construction.

Related Reference Documents
Applicable Ordinances and Regulations
Equipment and Coating Manufacturer's Published Instructions and Product Data Sheets

Best Practices in Context Sensitive Solutions, American Association of State Highway and Transportation Officials, 2006 Competition, Item Code 10-BPCSS-2-OL.

Bridges on Secondary Highways and Local Roads—Rehabilitation and Replacement, National Cooperative Highway Research Program, Report 222, 1980.

Construction Handbook for Bridge Temporary Works, 2008 Interim, Item Code 10-CHBTW-1-11-UL, American Association of State Highway and Transportation Officials.

Detection and Repair of Fatigue Damage in Welded Highway Bridges, National Cooperative Highway Research Program, Report 206, 1978.

Evaluation of Repair Techniques for Damaged Steel Bridge Members: Phase l, National Cooperative Highway Research Program, Project 12-17 Final Report, 1981.

Extending the Service Life of Existing Bridges by Increasing Their Load Carrying Capacity, Federal Highway Administration, FHWA-RD-78-133, 1978.

Guide Design Specifications for Temporary Works with 2008 Interim Revisions, Item Code 10-GSTBW-1-M, American Association of State Highway and Transportation Officials.

Guide Specifications for Applications of Coating Systems with Zinc Rich Primers to Steel Bridges, American Association of State Highway and Transportation Officials, S 8.1-2006, Item Code 10–NSBASBCS-2-OL.

Guide Specifications for Highway Bridge Fabrication with HPS70W Steel, 2003, American Association of State Highway and Transportation Officials.

Guidelines for Evaluation and Repair of Damaged Steel Bridge Members, National Cooperative Highway Research Program, Report 271, 1984.

Guidelines for Evaluating Corrosion Effects in Existing Steel Bridges, National Cooperative Highway Research Program, Report 333, 1990.

Inspector's Guide for Shotcrete Repair of Bridges, American Association of State Highway and Transportation Officials.

LRFD Bridge Construction Specifications, 2nd Edition, with 2008 Interims, Item Code 10-LRFDCONS-2-13, American Association of State Highway and Transportation Officials.

Methods of Strengthening Existing Highway Bridges, National Cooperative Highway Research Program, Report 293, 1987.

Rehabilitation and Replacement of Bridges on Secondary Highways and Local Roads, National Cooperative Highway Research Program, Report 243, 1981.

Shop Detail Drawing Review/Approval Guidelines, American Association of State Highway and Transportation Officials, G 1.1-2000.

Standard Specifications for Transportation Materials and Methods of Sampling and Testing, 28th Edition, American Association of State Highway and Transportation Officials.

Steel Bridge Erection Guide Specification, American Association of State Highway and Transportation Officials, S 10.1-2007.

Steel Bridge Fabrication QC/QA Guide Specification, American Association of State Highway and Transportation Officials, S 4.1-2002.

11

Advanced Repair Methods

Section 3
Repair and Retrofit Methods

11.1 INNOVATIONS MADE IN RECENT YEARS

11.1.1 An Overview

The bridge engineering industry is an evolving world. In the light of recent progress made in repair methods for steel and concrete materials in the U.S. and abroad, new repair techniques have been on the rise. Construction technology has revolutionized modern day repairs significantly. The rapid growth in technology needs to be addressed in detail. It needs to be directed toward redesign and reconstruction. New design standards, new rating standards, and the increasingly competitive marketplace may be challenging, but with the right tools these changes are easier to navigate. They need to be examined and considered for potential use.

In this chapter, market ready technologies, vendor's products, innovative techniques for use of new construction materials, remote health monitoring, and recent developments in repairs and rehabilitation methods are addressed. Use of such efficient methods will cut down the life cycle costs and the duration of maintenance. Some of the newer methods have not been fully tested and precautions may be required in their applications. The latest ideas, ingenuity, and contributions from individual researchers and publications are listed in the Bibliography at the end of this chapter. Also in this chapter, recent developments in new materials and prefabricated concrete elementsfor rapid construction will be discussed.

The motto recommended by FHWA's "Highways for LIFE" initiative is "Get In, Get Out, and Stay Out." In the context of rehabilitation, LIFE initials stand for the following:

L—Long lasting

I—Innovative

F—Fast construction

E—Effective and safe.

Recent innovations may be summarized as:

1. Preventing failures by minimizing the identified deficiencies.
2. Use of advanced methods of analysis and design.
3. Asset management and remote health monitoring methods.
4. Use of modern construction equipment and techniques.
5. New construction materials and systems.
6. Closer interaction between design documents and construction.
7. Continued research efforts in resolving technical issues.

11.1.2 Development of Diverse Repair Technologies

The author has successfully carried out many rehabilitation projects for concrete steel highway and railway bridges. General procedures for repair of old bridges are site specific and include:

1. Field verify applicable items from inspection reports.
2. Prepare a standard checklist of deficiencies.
3. Investigate any mild defects for possible changes in physical conditions.
4. Conduct an in-depth evaluation inspection.
5. Perform emergency repairs.
6. Prepare cost benefit analysis for rehabilitation versus replacement.
7. Perform alternate analysis for selecting the most appropriate method of solution.

For each variety or repair technique, the following approach is needed:

- The basic concept behind the technique.
- Its successful application to specific engineering problems.
- Procedure for field data acquisition.
- Processing.

Other recent innovative developments are:

- Methods to accelerate project completion
- Making accurate information available
- Increase productivity
- Team organization and collaboration with other diciplines
- Automated generation of reports in multiple formats.

The following detailed recommendations are based on current practices and the literature review of a large number of publications, including FHWA, AASHTO, ASTM, and NCHRP. The latest ideas, ingenuity, and contributions from individual researchers, from independent publications and those listed at the end of the chapters are acknowledged.

Rapid developments in transportation technology seen in the past few decades are addressed in the other chapters of this book

11.1.3 Comparative Study of Old and New Methods

The following physical causes are omnipresent in bridge components:

- Deterioration
- Applied direct stress
- Thermal action
- Creep and shrinkage due to changes in atomic bond between constituents.

Table 11.1a shows alternate uses of new methods and technology for rehabilitation. Table 11.1b shows a comparison of steel girder weights for medium and long spans.

Table 11.1a Some examples of old and new methods for rehabilitation.

Type	Old Methods	New Methods
Deck Slab	1. Open steel grid 2. Steel grid filled with concrete 3. Reinforced concrete deck slab	1. Exodermic 2. Fiber reinforced polymer (FRP) concrete 3. Precast and prestressed deck panels
Steel Beams	Cast iron Wrought iron Grade A36 steel	Grade 50 steel Grade 50W steel Grade 70W steel Hybrid grades 50 and 70W steel Grade 100W steel
Concrete Beams	Reinforced concrete beams	Post-tensioned prestressed concrete
Bearings	Rocker & roller	Elastomeric pads Multi-rotational Seismic isolation
Piers	Wall type	Frame bent Pile bent
Abutments	Gravity type Cantilever and full height	Spill through Integral Semi-integral
Wing Walls and Retaining Walls	Reinforced concrete	Mechanically stabilized earth (MSE) walls
Construction Methods	Conventional with formwork	Accelerated bridge construction
Production System	Conventional consultant-contractor team	Design-build-operate
Sign Structures	Bridge mounted and overhead	Variable message sign structures

Table 11.1b Selection of type of Grade 70 W steel for composite steel girders (by unit weight of deck area).

S. No.	Type of Girder System	Superstructure Steel Weight (lbs/ft^2 of deck)	Remarks
1	I-Girders for Single-Span	32	Long spans. Lower weight for HPS 70W. Based on AASHTO LRFD design.
2	I-Girders for Two-Span	19	Medium spans. Lower weight for HPS 70W. Based on AASHTO LRFD design
3	I-Girders for Three-Span	28	Long spans. Lower weight for HPS 70W. Based on AASHTO LRFD design.
4	Box Girders for Three-Span	39	Long spans. Lower weight for HPS 70W. Based on AASHTO LRFD design

11.1.4 Rehabilitation Design Procedures

As discussed in Chapter 9, the following factors affect the selection and design of rehabilitation methods and influence the need for innovations:

- Site layout
- Bridge usage
- Deterrence and access control
- Specific bridge features (moving, suspension)
- Resiliency of bridge components
- Redundancy of the bridge system
- Other bridge specific items

1. Criticality factor for assets: The vulnerability factor for assets can be expressed as a characterization of assets into regions of different levels of criticality and vulnerability:
    - Pre-assessment: Resource identification
    - Assessment: Vulnerability identification
    - Post-assessment: Decision making
2. Plan preparation and presentation: In plan preparation, actual field measurements should be considered more reliable than the drawings.
3. The shop drawings should be considered more reliable than bridge plans.
4. The contract plans shall be sufficiently detailed to provide an overall view of the bridge indicating:
    - The existing and proposed geometric dimensions
    - Limitations and restrictions
    - Extent and type of work to be performed
    - Construction stages
    - Material information
    - All related information needed to rehabilitate the bridge
5. All work shall be accounted for by specific pay items. Pay limits, quantities, and pay items should be adequately defined to eliminate ambiguity or confusion. Where applicable, reasons for critical limitations and restrictions should be explained to assist the contractor and the field inspector in adjusting to the field conditions.
6. For rehabilitation projects, the type, size, and location (TS&L) plans shall have the normal TS&L plan details, plus a complete scope and extent of work.
7. The anticipated bridge rating after the proposed work is incorporated shall be shown.

11.1.5 Common Examples of Innovative Concepts

1. Advanced infrastructure design will utilize ground penetrating radar (GPR) technology for evaluation of the bridge deck under the overlay. This is a high-speed non-destructive evaluation that doesn't require maintenance and protection of traffic. Cost savings per bridge using this method is approximately $0.1 million; time saved is two months.
2. Use of shoulders during the stage construction is inevitable. Therefore, the effective use of the existing shoulders during the stage construction will prove to be a critical factor in the success of the project as measured by the savings in time (schedule) and in cost (design and construction).

    In order to address this critical factor, an innovative approach consists of evaluating the integrity of the existing shoulders at the start of the design phase of this project. Rely on measuring the remaining service life (RSL) of shoulders based on state-of-the-art techniques that are supported by FHWA. This non-judgmental approach has historically resulted in better, more efficient design, since it relies only on visual inspection and coring of the shoulders. Utilizing this method of evaluating the shoulders would allow identifying the exact locations where rehabilitation of the shoulders would be necessary. Substantial savings would be realized in terms of schedule and cost. Cost savings is approximately $0.25 million; time saved is two months.

3. Advanced relocation of O/H utility lines to their permanent position without the use of temporary pole lines. Cost savings is approximately $0.3 million; time saved is three months.
4. Early coordination with regulatory agencies such as EPA and the United States Army Corps of Engineers should be performed to identify potential project permitting requirements; design schedule time saving is three months.
5. Early identification of issues that may be seasonally sensitive, such as presence of endangered and threatened species, will help to reduce project delays; design schedule time saving is one month.
6. To save cost, investigate the possibility of providing an at-grade temporary pedestrian/bicycle access during construction using the existing service road adjacent to the bridge versus a temporary sidewalk carried on structure; time savings is one month; estimated cost savings: $100,000.
7. Road closure versus staging—As an alternate to staging, investigate keeping the bridge open weekdays to accommodate the heavy weekday traffic and partially close the bridge over multiple weekends. Utilizing precast deck/prefabricated superstructure construction will also accelerate schedule; time savings: two months.
8. Reduction in construction staging—Investigate the possibility of eliminating the sub stages required for constructing temporary sidewalks carried by the structure. This would be possible if there is access from below the superstructure to construct the structure-carried temporary sidewalks. As an alternative, consider using a temporary bridge such as ACROW or Mabey type; time savings: one month; cost savings: $50,000.
9. Use of precast concrete modular deck sections in lieu of poured-in-place deck construction, and Inverset type prefabricated superstructure sections for bridge structures may provide considerable cost savings and a couple of month's reduction in construction time.
10. Investigate increasing the strength and service life of the bridge by increasing the beam section properties by making the new deck composite, and making the existing simply supported beams continuous. Potential increase in service life is 25 years.

11.1.5 Introduction of New Topics in Design

In the light of developments in numerical methods and computer techniques a more accurate analysis and design approach can be used:

- AASHTO LRFR rating procedures in place of LFD method
- State codes of practice using ultimate loads and probability approach
- Advanced methods of analysis and non-linear FEM
- New software applications
- Design methods for accelerated bridge construction
- Plan review check list and QC/QA document, e.g., used by NJDOT and Wisconsin DOT
- Use of context sensitive design (CSD).

11.1.6 Application of Innovative Repair and Retrofit Methods

Commercial products and services focus on mitigation technology for retrofit, restoration, and rehabilitation. The innovative techniques listed below are in early stages of development. They cover the use of:

1. Sample project specific guidelines for final plans.
2. Protective coating systems
   - Smart Japanese paint: This paint sends out electrical signals which are picked up by electrodes placed on either side of the paint's resin layer if the structure begins to vibrate. Electrical signal is greater as vibration increases. This paint can evaluate fatigue in old bridges more accurately than strain gauges.
   - The NSF's ATLSS Engineering Research Center has developed "smart paint" with a special dye that outlines a fatigue crack in a bridge as it propagates.
   - Magnified remote cameras: The computer based imaging system can provide spatial measurements and surface analysis. It can detect a surface flaw and determine its size, shape location, and defect details.
   - New vendor products: Seismic isolation bearings, Hilti chemical bolts, etc
   - Improved drainage systems: Examples are Ohio Drainage Guidelines and TN Drainage Manual
   - Innovative techniques include availability of new repair materials, FRP composites to repair overhead sign structures, SIKA CarboDur for general repairs, and SIKAWrap for shear strengthening. Sika Corporation concrete admixture called Sikament 686 can be used in cast-in-place or precast applications as a normal water reducer (ASTM C494 Type A) or as a high-range water reducer (ASTM C494 Type F). Increased workability is provided with no delay in normal set time.
   - Certain AL water-based concrete admixtures, based on aminocarboxylate technology, tolerate extreme cold and hot temperatures. Admixture forms a protective layer on embedded reinforcement that prevents corrosion caused by carbonation, chlorides, and atmospheric attacks. It provides corrosion protection for steel reinforcement, carbon steel, galvanized steel, and other metals embedded in concrete structures.
   - Corrosion prevention and maintenance painting of steel using protective coating systems
   - Corrosion protection for concrete: Bayer Material Science, one of the manufacturers for corrosion protection, has designed thick-film coatings for industrial waterproofing applications, such as concrete and metal bridge decks. Its physical and chemical properties include resistance to wear and abrasion, excellent tensile strength, a homogeneous seal on cracked, porous, corrosion resistance for metal substrates, good weather ability, resistance to microbes, and good chemical resistance. Baytec SPR spray systems can be pigmented to any color.
   - Fluoroethylene vinyl ether (FEVE) resins were developed in the early 1980s in Japan. Fluoropolymers offer a number of desirable properties, among them excellent stability against UV light and the elements, corrosion resistance, and weather, and chemical resistance.
   - Fluoropolymer topcoats now are required for use on all bridges in Japan, both for new construction and for repainting. On one of the longest Japanese bridges (single span of the bridge is 6527 feet) the coating system was a four-coat system consisting of a number of coatings both field- and shop-applied.
   - Deicing overlays: SafeLane surface overlay acts like a rigid sponge, storing the chemicals and automatically releasing them as conditions develop for the formation of ice or snow. This results in safer roads with better mobility and less maintenance because the overlay helps prevent frost or ice from forming on road or bridge surfaces. The final profile is

about ⅜-inch-thick. The recommended method is outlined in AASHTO Task force 34. The SafeLane surface overlay is expected to provide a robust surface for more than 15 years of service, plus the much-needed pavement seal to limit chemical and moisture penetration into the concrete bridge deck.

- Developments in data recording: Examples are data loggers, data acquisition systems, measurement and control products: Data recording computer hardware devices include pen based tablets, notebook computers, and PCs. For field use they are made more rugged than office computers. Techniques to extract data from laser-scan point clouds into 3-D MicroStation drawings have been quite successful.
- Use of laser technology to compute axial forces in cable stayed bridges.
- Asset management using robotic devices: Every transportation agency is faced with management of its assets, which include hundreds of bridges. No two bridges are alike and the vast varieties cover historic masonry arch bridges to the most modern segmental and cable stayed suspension bridges.
- Routine inspection and NDT of bridges can be performed by potential robotic devices using inertial navigation, odometer and laser techniques. A "manipulator" device will fix the sensors on critical bridge locations.
- Isolation bearings and retrofit of bearings: Seismic performance of a movable bridge requires some unique considerations, including the behavior of machinery elements and their tolerance for misalignment during earthquakes, and the issue of the requirement for seismic design when the bridge is in the open position. Surface conditions at the site are highly variable, making site-specific response analysis important.

Seismic retrofit design is based on the seismic deficiencies established through the performance evaluation such as:

- The isolation system consists of lead-core rubber seismic isolation bearings combined with semi-active viscous fluid dampers positioned between the bridge deck and end abutments. Semi-active dampers can offer an effective approach to response modification of seismically isolated highway bridge structures.
- Fiberglass drains for bridges: New fiberglass drains for bridges and elevated highways are light, strong, and easy to install. Fiberglass requires fewer supports than other non-metallic pipe, it resists corrosion, and it does not require painting. This system allows for a high degree of design flexibility.

11.2 DEVELOPMENT IN DESIGN CODES AND MARKET READY TECHNOLOGY

11.2.1 Design Code and Specifications Development

Codes are being revised every few years. Future codes should incorporate principles of "sustainability and context sensitive design." Practice needs to be backed with sound theory. From personal experience and observations the author finds:

1. It may not be accurate enough to use Rankine or Coulomb empirical formulae in seismic zones. Also, most pier caps act as deep beams. Their shear reinforcing details should use the strut and tie model as given in LRFD code.
2. Fabricator and erector shall follow AASHTO construction specifications and "fit-up shall be assumed to be performed under the no-load condition" for which evaluation of erection stress is required.
3. Construction load combinations not covered by AASHTO LRFD bridge design code are summarized:
   - For strength I condition, use all dead load of bridge components (DC), utilities (DW), construction equipment load (CEL) such as screed, and construction live load (CLL).

$\gamma_p = 0.9$ to 1.25. Using AASHTO notations, construction live load combination is

$$\gamma_p\,(DC + DW) + 1.5\,(CEL + CLL).$$

- For strength III condition, use wind on superstructure including forming. Construction wind load combination is

$$\gamma_p\,(DC + DW) + 1.5\,(CEL) + 1.4\,(WS).$$

- For Strength V condition, use construction wind load on equipment (WCEL), load combination is

$$\gamma_p\,(DC + DW) + 1.5\,(CEL) + 1.35\,(CLL) + 0.4\,(WS) + 1.0\,(WCEL);$$

$$\gamma_p = 0.9 \text{ to } 1.25.$$

- Specification developments will ensure quality assurance with ABC.

4. Training of engineers in new technology: It is necessary that construction personnel be given on the job training in the use of new technology:
   - Training aspects should cover tasks such as understanding erection procedures, erection drawings, modern concrete technology and use of new steel, processed timber, etc. in bridge construction.
   - Continuing education in design and construction and rigorous certification requirements needs to be introduced for certain levels of construction personnel.

5. Greater scope of post design activities: On each project a pre-construction workshop shall be conducted to evaluate possible hiccups and constructability issues. Topics would include equipment locations, construction duration, access, right-of-way, and material availability. To resolve constructability issues and any misinterpretation of contract drawings, designers shall be consulted before project construction.

6. Improved traffic control and maintenance and protection of traffic: The objective of using modern rehabilitation technology and innovations is to minimize life cycle maintenance and recurring costs.

11.2.2 Use of Integral Abutment Bridges with Flexible Steel Piles

1. Deck joints: In jointless bridges, by eliminating the deck joints and row of bearings at abutments, long term maintenance costs can be further reduced. Group behavior of a single row of piles at each abutment is of fundamental importance to generate frame actions in both transverse and longitudinal directions. Applied forces result from thermal effects, braking, and earthquakes.

2. Connection detail at the top of piles can be designed for full fixity, partial fixity, and pinned.

3. Orientation: Flexible piles, with the weak axis placed normal to traffic flow, will permit longitudinal movement and rotation, thereby releasing beam stresses.

4. Corrosion: Corrosion of the ends of steel girders is also minimal when girder ends are encased in concrete pile cap. Comparative studies of commonly used piles such as steel H piles, pipe piles, prestressed concrete pile, timber pile, and sheet pile have shown that steel H pile (with weak axis placed normal to traffic flow) is the most flexible type. Use of spread footing, caisson, drilled shaft, or two rows of piles will offer minimum flexibility and will lead to semi-integral abutment conditions.

11.2.3 Long Span Plastic Bridge

Composite plastic bridges are noted for their lightness, allowing the span to carry more weight; less decomposition and rust to damage the environment; and speedy installation that can take hours or days compared with weeks or months for traditional methods.

The world's longest fiberglass-reinforced plastic bridge spans about 175 feet and is at a golf course in Aberfeldy, Scotland. The second-longest, about 140 feet, is in Barcelona, Spain, over a series of railroad tracks.

The nation's longest and world's third-longest fiberglass-reinforced plastic bridge at a northside park in Indiana, built in 2008. The 100-foot-long bridge replaces a concrete one that floodwaters washed out in 2001 at Juan Solomon Park. The bridge provides access to 22 acres of parkland. The Efroymson Fund and the Crooked Creek Community Council bought the bridge, which cost $50,000 for manufacture. Local businesses donated $120,000 in services to help plan and construct the bridge.

11.3 USE OF MODERN CONSTRUCTION MATERIALS

11.3.1 Environmental Benefits of Concrete

1. There is an old saying that *"God made the country and man (the engineer) made the town."* Much of town and bridge making is now possible by the use of longer lasting concrete in the foundations and in the substructure and superstructure.
2. In many structures, exposure to deicing chemicals and marine-sourced chloride is a significant cause of corrosion. As discussed in Chapter 10, the most common procedure for repairing deteriorated concrete involves the removal of the damaged material and replacement with new concrete or mortar. Differences in pH, porosity, and chloride content are a few of the factors that may result in corrosion activity. As a result, "chip and patch-style" repairs may fail prematurely in chloride exposed structures.
3. The concrete industry is continually improving its material quality. Use of concrete in the harshest of environment is an achievement. Limestone, the primary raw material in concrete manufacture is abundantly available in nature. Besides, silica is available in the form of aggregates. Alumina and iron oxide are the other basic ingredients in the manufacture of cement. The concrete industry has reduced its CO_2 emissions in the recent past.

11.3.2 Use of New Concrete Materials

Available concrete materials include:

- Smart concrete
- Accelerators
- Air entraining admixtures
- Water reducers
- Super plasticizers
- Pozzolans
- Emulsions
- Anti-washout admixture for underwater concrete
- Use of HPC.

11.3.3 Use of Concrete Repair Materials

Available concrete repair materials include:

- Composites
- Polymers
- Polymer modified cement
- Epoxy resins
- Polyurethane injection resins
- Shotcrete
- Carbon and glass fiber reinforcement.

11.3.4 General Repair Procedures

The following procedures may be applicable and need to be adopted:

- Assessing damage and deterioration
- Load testing
- Identifying the causes
- Developing reports
- NDT techniques
- Cementitious materials selection process
- Surface preparation
- Placement methods
- Crack and joint repairs
- Protective systems for concrete (membranes and waterproofing, sealers, and coatings)
- Repair of corrosion-related deterioration in concrete structures, which offers unique challenges. In particular, the "ring-anode" effect is a common cause of premature patch failure. It increases corrosion activity adjacent to a repair area. The ring-anode effect is caused by the electrochemical incompatibility between reinforcing steel within a patch and the steel embedded within the surrounding concrete.

11.4 USE OF RECYCLABLE MATERIALS

11.4.1 Recyclable Steel and Waste Products

1. Reinforcing steel and prestressing strands have always been recyclable. Steel used in old ships can be salvaged and is sold as scrap. The steel is removed and recycled.
2. Many different materials are being considered for aggregate substitutes, including:
   - Granulated coal ash
   - Blast furnace slag
   - Various solid wastes including fiberglass waste materials, granulated plastics, paper and wood products wastes, sintered sludge pellets and others

The only two that have been significantly applied for recycling are glass cullet and crushed recycled concrete itself.

11.4.2 Use of Recycled Concrete Aggregate (RCA)

RCA is not a cement substitute. Conventional concrete aggregate consists of sand (fine aggregate) and various sizes and shapes of gravel or stones. The coarse aggregate portion of RCA has no significant adverse effects on desirable mixture proportions or workability.

Recycled fines, when used, are generally limited to about 30 percent of the fine-aggregate portion of the mixture. It is important to note the difference between aggregate and cement, because some materials are used as both a cementitious material and as aggregate (such as certain blast furnace slags).

1. Participation by states: In the U.S. at least 140 million tons of concrete is recycled every year. Several states have high tipping fees for disposal of RCA; this is being done to control landfill usage, thus increasing the reuse of RCA. The Oregon DOT has already taken a lead in this respect on a large scale.

   Another example is the reconstruction of Arthur Ravenet Jr. Bridge in Charleston, South Carolina. Over 248,000 tons of concrete was salvaged from demolition of old structures and reused to create 82 acres of reef habitat.

California, Texas, Virginia, Michigan, Minnesota, and Utah were identified as being among the highest consumers as well as large suppliers of RCA.

The author was associated with a highway embankment project in North Jersey utilizing RCA for structural fill, which is made abundantly available from demolition of debris. This resulted in cost saving from the transportation of tons of wasted but good quality aggregates.

2. Procedure for RCA reuse:

RCA is generally obtained as debris from defunct bridge structures/decks, retaining walls, sidewalks, and curbs that are being removed from service. Concrete can be crushed to a desired gradation. The material is cleaned of unwanted material like bricks, wood, steel, ceramics, and glass. The aggregate must be "clean," without absorbed chemicals, clay coatings, and other fine materials in concentrations that could alter the hydration and bond of the cement paste.

3. Resource conservation: Several factors can be considered in this category.

- Reduced land disposal and dumping: The use of recycled concrete pavement eliminates the development of waste stockpiles of concrete.
- Recycled material can be used within the same metropolitan area, which can lead to a decrease in energy consumption from hauling and producing aggregate.
- It can help improve air quality through reduced transportation source emissions.
- Conservation of virgin aggregate: Many European countries have placed a tax on the use of virgin aggregates. This process is being used as an incentive to recycle aggregates.

4. New concrete made with RCA typically has good workability, durability, and resistance to saturated freeze-thaw action. Aggregate composed of recycled concrete generally has a lower specific gravity and a higher absorption than conventional gravel aggregate. Lack of widespread reliable data on aggregate substitutes can hinder its use.

To design consistent, durable recycled aggregate concrete, more testing is required to account for variations in the aggregate properties. Also, recycled aggregate generally has a higher absorption and a lower specific gravity than conventional aggregate.

5. Aggregate typically accounts for 70 to 80 percent of concrete volume. Aggregates can have a significant effect on the cost of the concrete mixture. As the cost of quarrying for aggregates continues to rise, it makes engineering sense to preserve natural aggregate for future use. Aggregate plays a substantial role in determining concrete's:

- Workability.
- Strength.
- Dimensional stability.
- Durability.

6. Research has revealed that the 7-day and 28-day compressive strengths of RCA are generally lower than values for conventional concrete. Recycled aggregates may be contaminated with residual quantities of sulfate from contact with sulfate rich soil and chloride ions from marine exposure.

Concrete made with RCA has at least two-thirds the compressive strength and modulus of elasticity of natural aggregate concrete. The compressive strength varies with the compressive strength of the original concrete and the water-cement ratio of the new concrete.

7. FHWA conducted a review on the uses of (RCA) and its advantages. The angularity of RCA:

- Helps to increase structural strength in the base, resulting in improved load carrying capacity.
- Residual cementation provides a strong, durable platform from which to build upon.

- Offers better control over gradation. RCA is able to meet gradation and angularity requirements.
- Potential to minimize D-cracking (which is caused by the freeze-thaw expansive pressures of certain types of aggregate) and alkali silica reaction which is caused by the detrimental reaction between silica found in certain aggregate and the alkali (cement) paste. These forms of distress are material related and studies show that the inclusion of RCA in the concrete mix and a suitable fly ash has the potential to reduce these distresses.

8. Recommendations for RCA use:
   - It will be useful to include special provisions in construction contracts to track and implement waste management activities. At the time of bid construction companies can be asked to forecast the percentage of materials they anticipate recycling on their bridge projects. They would be required to include such activities in their monthly progress reports.
   - There is a need to develop mix design and procedures for the reuse of chunks of aggregates with cement mortar coating salvaged from the debris. A sieve analysis and laboratory tests may be carried out to ascertain fc′ value of old mortar coated aggregate and new cement mortar. Such an approach for mix design will result in significant cost savings.

11.4.3 Use of Recycled Glass

1. Some of the specific glass waste materials that have found use as fine aggregate are "non-recyclable" clear window glass and fluorescent bulbs with very small amounts of contaminants. Possible applications for such waste-glass concrete are bike paths, footpaths, gutters, and similar non-structural work.
2. Field testing has shown that crushed and screened waste glass may be used as a sand substitute in concrete. Nearly all waste glass can be used in concrete applications, including glass that is unsuitable for uses such as glass bottle recycling.
3. Glass aggregate in concrete can be problematic due to the alkali silica reaction between the cement paste and the glass aggregate, which over time can lead to weakened concrete and decreased long-term durability. Research has been done on types of glass and other additives to stop or decrease the alkali silica reaction and thereby maintain finished concrete strength. However, further research is still needed before glass cullet can be used in structural concrete applications.

11.4.4 Use of Recycled Plastics

1. Recycled polymers have a range of uses from bridges, footpaths and fences and even flood prevention. It will not chip or splinter and is even vandal proof. The environmental benefits are huge. Despite the alleged versatility the manufacturer cannot use PVC or thermo-set plastics such as polyurethane in its production process.
2. Plastics which are often not usable by most plastic recyclers consequently end up in the waste stream. A vast amount of mixed plastic ends up in landfill. The innovative recycling technology allows the use of material that may otherwise be destined for landfill, or plastic waste which outperforms the traditional alternatives of wood, steel and concrete products.

11.5 USE OF FIBER REINFORCED POLYMER (FRP)

11.5.1 Introduction

1. The use of FRP bars with HPC is another development in the concrete industry. The first FRP bridge was built in Kansas in 1996. Construction followed in New York, West Virginia, Delaware, and Ohio.
2. Fiber reinforced polymer (FRP) bridge decks are used in bridge rehabilitation projects, often because of their relatively low weight and high durability. Related benefits of FRP include

rapid construction and advantages in terms of life cycle costs (e.g., corrosion resistance). FRP is a strong material, lightweight, durable, impervious to deicing salts, weather resistant, and has a low life cycle cost. It is a composite material consisting of high strength thread like fibers embedded in epoxy or polyester resin. There are a variety of fiber types and resins which can be used as composites. The fibers are glass, carbon, or aramid, which are brittle in solid form but have high strength in fiber form. Some of the earliest composite materials were introduced in Egypt for mud bricks with added straw to make the material stronger.

3. FRP is available in a variety of forms such as FRP bars, FRP grids, and tendons for prestressing. The combination of plastic and fibers produces great mechanical properties resulting in a lighter weight material that is durable. It is increasingly being used in strengthening applications and for deck replacements. Longer spans are now possible with lightweight aggregates and FRP, which replaces more expensive steel reinforcing bars.

4. Corrosion is the most common type of distress or deterioration in concrete bridges. Replacing a traditional steel bar with a FRP bar eliminates the expansion that comes with corrosion associated with steel embedded in concrete.

11.5.2 Modern Rehabilitation Methods with FRP

1. The applications are:
- Column wraps to enhance seismic performance.
- Beam wrapping to increase shear capacity.
- Bonded flange plates to increase bending capacity.
- Epoxying FRP rods in grooves cut into the substrate to increase member strength.
- Truss strengthening.
- Wrapping aluminum columns of sign structures.

2. Advantages of FRP:
- Non-corroding, unlike steel rebars.
- Low maintenance.
- Low fatigue.
- Less dead load, high strength to weight ratio.
- Flexural strengthening.
- The benefits of FRP composites have only recently been expanded to include the bridge construction industry. Fiber reinforced polymer (FRP) composites have been providing practical solutions to structural problems in the aerospace, automotive, and manufacturing industries for many years. Research into FRP composites for bridge construction is still in the evolutionary stage. More information is required to provide confidence for the design of FRP bridges and for the development of design standards.

3. Disadvantages of FRP:
- The cost of an FRP bridge is higher than a conventional bridge. The frequency of inspection is greater also. Recommendations for inspection and maintenance are given in ACI 440.2R-02 "The Design and Construction of Externally Bonded FRP Systems for Strengthening Concrete Structures."
- Resistance to fire is much lower than for steel rebars.

4. Visual inspection should record observations for:
- Changes in color
- Debonding
- Cracking
- Deflections

- Peeling
- Blistering
- Crazing.

5. For sign structure repairs, it:
- Costs less than full structural support replacement
- Allows repairs to be done quickly
- Causes less traffic disruption because only lanes beneath repair need to be blocked off.

11.5.3 FHWA Recommendations for the Use of FRP

Appropriate projects to consider include:

- A posted bridge that could benefit from a reduction in dead load and subsequent increase in live load rating.
- A bridge that needs to be widened without imposing additional loads on the substructure.
- Superstructures under 12 m span (and longer spans as technology evolves).
- A historic structure that must be saved (i.e., rehabilitated instead of replaced) due to its cultural value.
- Moveable spans where the light weight can save operating expenses.
- A bridge that needs an improvement in load rating sooner than can be addressed through a capital improvement program. It is often unacceptable to program work for five years in the future when postings present an immediate economic hardship.
- An accelerated schedule to installing decks or superstructures to reduce the cost of maintenance, reduce congeston, and protect traffic.

11.5.4 FRP Deck Selection Criteria

1. Span: A particular deck system is acceptable only up to the support span length at which it has been tested. There are several deck systems that can be used with steel supports spaced at 2.5 m. Superstructures built thus far have been relatively short spans (<14 m). This is due to the cost of controlling deflection. Longer spans will most likely require the use of high performance carbon fiber or a hybrid system which utilizes concrete or another material with FRP.

2. Traffic volume: Though fatigue test data suggests that there is not a concern about using FRP decks on high volume roads, the designer should be cautious about the application in an area that would be difficult to monitor or repair. Low volume roads provide an opportunity for easy access so that more can be learned with little risk.

3. Cost considerations: Since an FRP deck or superstructure typically sells for a premium compared to steel and concrete, part of the scoping process must involve an evaluation of all costs associated with a project. A cost comparison should be made considering all available options. Past price ranges are:

Decks: \$65–\$80 /SF of deck area

Superstructure: \$140–\$150 /SF of deck area

4. Skew and grade: Deck manufacturers have been successful in the fabrication and installation of skewed superstructures and the use of FRP decks on skewed steel bridges. At the present time, there are no general restrictions on such use, but caution is advised in these circumstances. The same is true for percentage grade.

5. Depth: FRP decks are commonly available as a 200 mm (8 in) deep section, although custom depths are available. During preliminary design, a rule of thumb can be used for the depth of an FRP superstructure: i.e., 1 inch for each foot of span. If desired, a slimmer section can be obtained by asking the manufacturer to engineer the fabric architecture accordingly.

6. Specific design criteria are:
- To avoid long term creep, predicted strains under design load shall be less than 20 percent of the FRP composite's minimum guaranteed ultimate strength. The ultimate strength is based on coupon testing and is noted in the approved plans.
- An environmental durability factor (knock down factor) of 0.65 may be applied to the material properties to account for degradation of properties over time.
- Because of the material's typical low modulus of elasticity, most designs will be driven by deflection limitations and not strength requirements. Although the criterion for deflection is somewhat arbitrary, it is typically kept at 1/800 of the supporting span length.

7. Research is required in the areas of long-term environmental effects on FRP:
- In identifying the various failure modes of FRP composites
- In developing design methodologies
- To develop efficient connections to utilize the advantageous properties of FRP composite sections.

11.5.5 Use of Carbon Fiber (CFRP) Pre-impregnated and Steel Reinforced Polymer (SRP) Sheets

For strengthening of decks, reinforcement of columns, and repairs of retaining walls, the use of carbon fiber sheets is gaining popularity (e.g., supplied by Mitsubishi Chemical Corporation).

1. Recently, a comparison of four techniques for strengthening concrete beams was performed by Harrison and Tarek Alkhrdaji.

In addition to externally bonded FRP as a successful technique for strengthening concrete members, other techniques like near surface mounted (NSM) FRP bars have emerged as viable alternatives. Four composite-based strengthening systems were designed to provide equivalent flexural performance. These systems are:
- Externally bonded carbon fiber reinforced polymer (CFRP) sheets
- NSM prefabricated CFRP strips
- Externally bonded steel reinforced polymer (SRP) sheets known as hardwire
- NSM stainless steel bars.

2. Use of laminated fiber reinforced rubber for reducing collision impact: The energy absorption capacity of the laminated fiber reinforced rubber installed at girder ends was studied by Japanese researchers, and compared well to ordinary rubbers. A collision test was carried out between two steel solid bars in order to assess the energy absorption and the impact force during the collision. Then, the numerical simulations of the test results were made by using the mass-spring systems.

It is concluded that both rubbers are effective in the energy absorption during the collision; however, the laminated fiber reinforced rubber can reduce the impact force more than the ordinary rubber. It is also concluded that the energy absorption can be estimated by using the mass-spring systems for the impact force during the collision.

3. New York State installed its first fiber-reinforced-polymer (FRP) composite bridge in October 1998. Since FRPs are relatively new, a comprehensive test program, comprised of load tests and visual inspections every six months and detailed finite-element analysis, is being carried out to monitor the structure and also to evaluate its long-term durability.

4. Dynamic behavior and damage detection of a fiber reinforced plastic bridge superstructure:

Amjad J. Aref and Sreenivas Alampalli studied the dynamic response of a fiber reinforced plastic bridge recently constructed in New York State. The dynamic behavior of the bridge was studied using detailed finite element analysis. These models were then compared with field tests performed on the bridge to validate finite element models.

In a complex structural system such as a bridge made of laminated fiber reinforced plastic materials, several sources of structural degradation may occur during the life span of a bridge. For example, damage due to delamination, cracks, and loss of bond between components, and change in material properties may often take place in various parts of the structures and can not be detected by visual inspection. Thus, a finite element model validated by field testing can be used to foresee several damage scenarios.

The correlation of dynamic testing on the bridge with dynamic analyses of the bridge using finite elements was used to give an indication of the degree of the damage and possible location within the bridge.

5. Smith and Bright investigated the use of fiber reinforced composites for upgrading orthotropic bridge decks. The competitiveness of orthotropic bridge decks depends on their high strength to weight ratio. However, poor durability of paving materials and fatigue failure of welds has contributed to high costs of repair including road user delay costs due to traffic disruption. It is proposed to use a layered surfacing system, combining lightweight asphalt, conventional asphalt, and a layer of glass fiber mesh embedded just beneath the chip-sealed surface. Fatigue tests indicate that glass fiber reinforcement increases the durability by a factor of at least 10.

6. Prestressing with CFRP tendons: Balázs and Borosnyói carried out a comparative experimental study on prestressed concrete beams pretensioned either with CFRP or steel wires. Load-deflection responses were analyzed and pivot point of loading-unloading moduli under repeated loading is defined.

   Considerable corrosion of reinforced concrete members accelerated the research on nonmetallic materials like fiber reinforced polymers (FRP) for structural applications as noncorrosive reinforcement.

7. A case study of using fiber-reinforced polymer decks for bridge rehabilitation: A bridge in Tippecanoe County is the first in Indiana to be rehabilitated with an FRP deck. Among the bridges evaluated, County Road 900E over Wildcat Creek is a three-span continuous steel stringer bridge with two concrete approach spans. The FRP deck replacement would only take place on the three main spans.

11.5.6 Further Studies of Continuous, Long-Term Monitoring of Two Advanced Polymer Composite Bridges

1. Harry W. Shenton III, Michael J. Chajes, William L. Johnson, Dennis R. Mertz, Jack W. Gillespie and their team designed a continuous, long-term monitoring systems for two polymer composite bridges recently built in Delaware. The Magazine Ditch Bridge has been continuously monitored for more than a year. The monitoring system provides data to investigate the effects of sustained load, environmental factors and live load on the bridge. Early results show that daily and seasonal temperature changes can induce strains in the bridge that are equal in magnitude to the maximum live load strains.
2. A similar system has been designed for the first state-owned composite bridge in Delaware. The monitoring system for bridge 1-351 is expected to be on-line by the summer 2000. Presented in the paper is a brief overview of the systems and sample results from the data collected for the Magazine Ditch Bridge.
3. Strengthening of a concrete T-beam bridge using FRP composite laminates: Osman Hag-Elsafi, Sreenivas Alampalli, and their team studied the application of fiber reinforced polymer

(FRP) composite laminates to strengthen a 70-year-old reinforced concrete T-beam bridge in New York.

Leakage at the joints of this single-span bridge led to substantial moisture and salt infiltration in the superstructure as manifested in efflorescence, freeze-thaw cracking, and delamination at some locations. Integrity of the steel reinforcing and overall safety of the bridge was suspected. This concern was heightened by the absence of any documents pertaining to the bridge design, such as bar size, steel type, concrete strength, and design loads. The bridge was strengthened using bonded FRP laminates. Load tests were conducted before and after the FRP laminates were installed to assess effectiveness of the strengthening system. Test results are also compared with those obtained analytically using classical approaches.

4. Fiber reinforced polymer matrix composites (PMCs) are very effective for concrete rehabilitation and deck slab construction. VDOT used it successfully for superstructure replacement of Tom's Creek Bridge.

For injecting hairline and wide cracks in concrete, epoxy injection resins are being effectively used. For waterproofing joints and cracks, sealing leaks and under water repairs to concrete Polyurethane Injection Resins are being effectively used.

11.5.7 Lightweight Truss Bridge Rehabilitation Using Composites and FRP Deck

According to FHWA, there are over 93,000 weight-restricted bridges in the U.S. There are over 19,000 steel trusses which are well suited to lightweight rehabilitation. Existing truss bridges can be made lighter by replacing a heavy concrete deck and numerous asphalt overlays with a lightweight FRP deck. In addition, painting the steel members would increase the life of bridges.

As reported the 140 ft span Warren truss bridge carrying New York State's Route 367 over Bentley Creek in Chemung County built in 1940 found new life in 1999 after the installation of a 13½ in thick, E-glass reinforced vinyl ester resin deck weighing only 32 psf. This reduced the dead load by 80 percent and almost doubled the live load carrying capacity of the bridge: HS12 to HS23 to a level higher than the original design. A 14-ton weight restriction was able to be removed and the bridge reopened to all legal loads.

The deck itself rates much higher than the bridge. It meets L/800 deflection requirement with an inventory load rating of HS85 (154 tons). Load tests indicate that the actual capacity of the deck is even greater than these analytical ratings.

Many innovations were developed:

1. A haunch for each floor beam was uniquely precast to support the deck. This placed the deck at the proper elevation, provided a new load path directly from the deck to the floor beams, raised it above the obsolete steel stringers, and delivered a 2 percent cross slope.

2. The deck was replaced using six discrete deck panels. The modular construction simplified installation and reduced field time.

3. A field splice system was designed to transfer shear and moment between panels. The joints devised rely on FRP cover plates and structural adhesives. The watertight deck protects the steel floor system from the weather.

11.5.8 Use of FRP to Repair Sign Structures

1. Improperly maintained aluminum overhead sign structures create hazards. Lack of inspection during fabrication can yield poor-quality joint welds. Insufficient construction supervision may result in internal stresses in an overhead sign structure before the sign is attached.

Problems cited most frequently were weld defects in sign and pole construction and general fatigue cracking. Fiber-reinforced polymers repair overhead sign structures quickly and economically. FRPs can provide structural integrity to overhead sign supports and prevent them from failing. FRP composite materials have the potential to revolutionize the repair of sign structures with cracked secondary support members. The most common problem is joint cracking between the internal trussing and the main chords of the sign structure.

2. The greatest contributor may be that fatigue design was not a code requirement when many trusses were designed in the 1960s. FRP is used to repair overhead sign structures, and the repair method is relatively quick and economical. It is accomplished by cleaning the damaged area of the sign support thoroughly and wrapping FRP around it. Repairs can be done in place, with only the lanes below the repair area blocked off.
3. Research shows FRP repairs are as strong as welded joints. The specification covers restoration of the tensile capacity of secondary sign structural members, such as internal truss diagonals, and not main members, such as longitudinal truss chords. Benefits include:
   - It costs less than full structural support replacement.
   - It allows repairs to be done quickly.
   - It causes less traffic disruption because only lanes beneath repair need to be blocked off.

11.5.9 Use of Glass Fiber Reinforced Polymer for Lightweight Emergency Bridge

This method is still in the experimental stage, but experimental investigations performed on lightweight bridges made of high performance materials have shown a potential market. A single-lane bridge with a small span was used to achieve cost effectiveness by using pultruded glass-fibre profiles. This material is commonly available in the market in connection with bonded steel reinforcement for bolted joints. A testing program was executed to provide information on the failure criteria and temperature behavior of these bonded hybrid-connections.

11.6 ADVANCEMENTS IN CONCRETE TECHNOLOGY

11.6.1 Salient Features

Concrete bridges are more commonly used for smaller spans. Small steel spans have higher maintenance costs.

1. Ultra HP FRC: Compressive strength reaching 30 ksi is possible and flexural strength of 7 ksi.
2. Using HP lightweight aggregates: Lightweight HPC reduces deadweight, enables longer span lengths, reduces the number of piers in a river, and allows fish travel with the least obstruction.
3. Spliced girders of varying depth enables lightweight concrete to achieve spans of over 200 ft.
4. Overlays: Use of silica fume and high early strength LMC will open deck to traffic within three hours of curing. Silica fume, pozzolans, fly ash, and slag may be used to reduce concrete permeability and heat of hydration. Fly ash and cenospheres are preferred to HPC in bridge decks, piers, and footings. Byproducts of coal fuel such as fly ash, flue gas de-sulfurization materials, and boiler slag provide extra ordinary technical, commercial, and sustainable advantages.
5. Self-consolidating concrete (SCC): Since vibration time is saved SCC helps ABC; more workable concrete with lower permeability than conventional concretes.
6. Rapid setting concrete: Non-shrink, multi-purpose, high strength repair mortar used for concrete repair, plaster repair, mortar bed, formed work, vertical, and overhead applications
7. Blended cement concrete: A blend of Portland cement and a combination of silica fume or fly ash used for enhanced strength and durability. Used in high performance applications

with materials such as slag cement.

8. Fibermesh concrete: Micro-synthetic fibers prevent all early age cracks during concrete's plastic state.
9. Carbon fiber reinforced polymer (CFRP) concrete: For repair and retrofit of concrete structures with glass or FRP.

11.6.2 Prefabrication of Components and Connection Details

The contractor must be knowledgeable about the latest technology and availability of new bridge components. Develop typical connection details for precast deck panels, piers, and abutments. Improve quality of the superstructure by fabricating in a more controlled factory environment.

To facilitate one time shopping for construction products, large supply stores such as Home Depot and Wal-Mart would be helpful. Greater technical service is provided on bridge products such as precast deck units, girders with welded shear connectors, diaphragms, bearings, parapets, precast pier units, etc. Quick assembly is possible by systems such as Conspan, Inverset, Effideck, drop-in deck panels, and post-tensioning in both directions.

Applying segmental construction for long spans to utilize ABC. Balanced cantilever method eliminates the need of formwork.

11.6.3 Use of High Performance Concrete (HPC)

1. Advantages of HPC are wider beam spacing and fewer beams, longer span lengths and fewer piers, increased vertical clearance, less permeable, stronger and more durable concrete, lower initial and life cycle costs, and fewer maintenance requirements. These properties are achieved by special mix design and improved curing.

   With the high performance concrete composite overlay method, the top of the deck is removed to the underside of the top mat of reinforcement and replaced with high performance concrete. Placing the concrete below the top mat forms a composite action between the two layers of concrete.
2. The benefits of a high performance concrete composite overlay include:
   - Addresses any spalling or cracking in the top of the existing deck
   - Composite action helps to strengthen the deck.

   The disadvantages to the system include:
   - Concrete curing time is longer than asphalt or polymer overlays leading to possible MPT issues
   - Assumes the bottom half of the concrete is in acceptable condition to be salvaged
   - Potential cracking of the high performance concrete.

11.6.4 Self Consolidating/Compacting Concrete (SCC)

1. SCC is an improved version of HPC with compressive and tensile strengths exceeding those of normal concrete. The ease of placement and reduced demand for skilled labor are the main advantages of using SCC. It can be used with high range water reducers to achieve high early strength concrete. Proper curing is critical.
2. SCC has very high slump and flowability properties without segregation. It uses specialty admixtures which radically modify slump and flowability. Examples of such admixtures are viscous modifiers and high performance poly carboxylate polymers. Properties of conventional concrete such as compressive strength, cohesiveness, or durability are not affected.
   - It provide superior appearance and long-term durability.

- It eliminates the need for vibration.
- The resulting concrete is totally homogeneous with a uniform surface finish.
- To further enhance durability, a calcium-nitrite corrosion inhibitor may be added to the mix.
- It attains higher quality control resulting with fc' values in excess of 9 ksi.

Other advantages include:

- Voids and honeycombing in many traditional concrete mixes are avoided.
- Successfully used in new construction projects with difficult placement or finishing.
- Effective for concrete repairs of honeycombing.
- Effective for strengthening projects when material must be placed under pressure into confined forms with highly congested reinforcement.
- Concrete is so flowable that instead of measuring the height of the slump cone, the diameter of the circular blob that pours out of a slump test cone is measured.
- Concrete pumping is easier. Forms do not have to be vibrated to consolidate concrete.
- Labor savings because crews can pour large repair areas using a single pump location.
- Architectural finishes are easier to achieve.

Performance of blended cement in high strength self compacting concrete: Rice husk is an agricultural waste generated in massive quantities from rice processing units worldwide. With no worthwhile use, it is a waste material which creates disposal problems. Its high silica content makes it suitable for use with cement. Tahir Kibriya of National University of Sciences and Technology, Pakistan "Investigations in rice husk ash" have suggested improved strength/durability. This experimental study aimed at evaluating the properties of high strength SCC made from blended cements using rice husk ash, Portland cement, natural aggregates, and sand. Wide ranging investigations covering most aspects of mechanical behavior and permeability were carried out for various mixes for compressive strengths of 60N/mm^2, 80N/mm^2 and 100N/mm^2. Compressive strengths of high strength SCC specimen with blended cements were observed to be higher by about 4 to 9 percent than the control specimen, for concrete with 50 percent Portland cement blended with 50 percent rice husk ash.

Higher elastic moduli and reduced permeabilities were observed along with better sulphate and acid resistance. Better strengths and improved durability of such high strength SCC make it a more acceptable material for major construction projects.

Advantages of using SCC:

11.6.5 Use of Shrinkage Compensating Concrete

Concrete Using Shrinkage Compensating Cement

What influences shrinkage compensating cement is the amount of binder, the percentage of silica fume, the water to binder ratio, and the types of super plasticizer, cement and aggregate respectively on both autogenous and total shrinkage of high performance concrete (HPC). Shrinkage can be measured under isothermal conditions and constant humidity of the environment.

Service life can be extended by using modern construction materials. Life cycle costs are also minimized for thousands of bridges. Shrinkage development in HPC can be reduced by using shrinkage compensating cement in lieu of Type 1 cement, to help eliminate drying shrinkage cracking of bridge decks.

FHWA has not participated in the use of this special concrete. This item shall consist of furnishing and placing Portland cement concrete using shrinkage compensating cement for the bridge deck and all parapets, in accordance with the approved specification.

Materials

The cement shall be expansive hydraulic cement. Admixtures used in the concrete mixture must be compatible. Coarse aggregate stockpiles shall be saturated. Saturation shall be completed a minimum of 24 hours prior to use; however, the application of water by sprinkling shall continue as directed by the engineer.

The maximum water/cement ratio shall be between 0.4 and 0.45.

Slump at the time and point of concrete placement shall be 3 to 6 in [75 to 150 mm], except for concrete used to slipform bridge medians and parapets. The contractor may elect to slipform the bridge medians or parapets.

The use of super plasticizers may be required to achieve a placeable concrete within the specified slump range and to meet the maximum water/cement ratio.

Concrete shall contain 6 plus or minus 2 percent of entrained air at time and place of concrete placement.

Maximum ambient temperature at the time of placement of concrete shall be 80°F. Deck formwork beam flanges and reinforcing shall be thoroughly sprinkled with water prior to placement of the concrete. Sprinkled areas shall remain damp until placement of concrete, however, no excess standing water will be allowed.

Concrete shall be placed when the rate of evaporation is less than 0.2 lb/sq ft/hour. Atmospheric conditions shall be monitored throughout the pour. If the evaporation rate is exceeded the pour shall be stopped.

Curing

The finished surface shall be covered with a single layer of clean wet burlap. The fresh concrete surface shall receive a wet burlap cure for seven days. For the entire curing period, the burlap shall be kept wet by the continuous application of water through soaker hoses. Either a 4 mil white opaque polyethylene film or a wet burlap-white opaque polyethylene sheet shall be used to cover the wet burlap for the entire curing period.

Storage tanks for curing water shall be on-site and filled before a pour will be permitted to start. Storage tanks shall remain on-site throughout the entire cure period. They shall be replenished, as required, with a shuttle tanker truck or a local water source such as a fire hydrant.

11.6.6 Ultra-High Performance Concrete (UHPC)

Ultra-high performance concrete is proposed as an innovative new material for the construction of highway bridge superstructures. The path of innovation can be traced back to the 1980s, when the first big gains were made in increasing the resistance or strength of concrete by a factor of 6 to 10. Developed in France during the 1990s, UHPC has seen limited use in North American bridge projects. Consisting of fine sand, cement, and silica fume in a dense, low water-cement ratio mix, this highly moldable material offers a combination of superior properties including compressive strengths up to 30,000 psi and flexural strengths up to 6000 psi, ductility, durability, and a vast range of aesthetic design possibilities.

Advantages

The high-tech versions have different properties that make them more comparable to materials such as stainless steel or aluminum, which are often more expensive still. New types of concrete offer the following advantages:

1. Intrinsically energy-efficient.
2. Excellent insulation against wind and water.
3. Its high density means it stores heat during the day and releases it at night, preventing bridge decks from freezing in winter.

4. UHPC is denser than conventional concrete, which attributes to its remarkable imperviousness and durability.
5. In addition, UHPC is extremely low in permeability and performs better in terms of abrasion and chemical resistance, freeze-thaw, carbonation, and chloride ion penetration.
6. To improve ductility, steel or polyvinyl alcohol (PVA) fibers are added, replacing the need for passive mild reinforcing steel.
7. It sets much faster.
8. Stronger concrete translates into significant gains for the environment. It can be used more thinly, consuming considerably fewer raw materials than regular concrete. The environmental advantage is clear: zero maintenance, zero painting, and a very long life.

Engineers and builders have far greater flexibility to use the material's long-lasting, thermal, and acoustic properties in pedestrian bridges and at bus stations and, in turn, contributing to big energy and other environmental savings. White concrete contains titanium dioxide, which keeps the concrete clean at the same time as destroying ambient pollutants such as car exhaust.

Sustainability

High-tech concrete is just one of the products that has emerged from the research and development labs of cement, steel, and chemicals firms this decade, and it signals a growing commitment by heavy industry to the notion of "sustainability."

UHPC mix designs typically include no aggregates larger than sand, and include steel fibers 0.2 mm in diameter and 13 mm in length. These steel fibers and the special mix design increase the strength and toughness of the UHPC significantly relative to more traditional concretes.

Disadvantages

Low water to cement ratios typically used result in difficult casting and curing conditions. There is a need to evaluate strength and stiffness parameter of as-cast members. The potential for the development of practical quality control techniques for the future implementation of UHPC needs to be considered.

Ultrasonic velocity measurements are used to estimate the bulk elastic modulus, shear modulus, and Poisson's ratio of UHPC and results are compared with traditional destructive methods. The application of ultrasonic testing for the evaluation of early-age material properties and for nondestructive, in-situ materials characterization is extremely useful.

Case Study of New Construction Materials

The new Jakway Park Bridge in Buchanan County, Iowa is one of the first highway bridges in North America to be built with a new generation of ultra-high performance concrete (UHPC) pi-girders. The bridge is 24 ft 3 in wide by 112 ft 4 in long. The UHPC center span is 51 ft 2 in.

It is one of the first North American highway bridge projects to incorporate batching of UHPC in a ready-mix truck. The Iowa Department of Transportation and the Bridge Engineering Center at Iowa State University designed the bridge; a combination of cast-in-place, simple span slabs with a center span consisting of a series of precast UHPC pi-girders. The Jakway Park Bridge has a clean, balanced and symmetrical appearance. The profile is attractive and somewhat unique due to the nature of the approach slab sections meeting the precast pi-girder sections. It can certainly be considered an important technological advancement in the bridge building industry.

Testing of the section by Turner-Fairbank validated the FEM analysis for flexural and shear capacity in the longitudinal direction. The testing also confirmed that the stress in the transverse direction of the deck was unacceptable for service loading and a low transverse, live load distribution between adjacent pi-girders would require stiffening. Future research will address current design and production concerns and develop more efficient beam designs to maximize UHPC's unique structural properties.

11.6.7 Exodermic Bridge Deck

An Exodermic bridge deck is comprised of a reinforced concrete slab on top of, and composite with an unfilled steel grid. Exodermic decks are made composite with the steel superstructure by welding headed studs to stringers, floor beams, and main girders as applicable, and embedding these headed studs in full depth concrete.

This system maximizes the use of the compressive strength of concrete and the tensile strength of steel. The system has overall thicknesses ranging from 6 to 9 ½ inches while weighing 35 to 50 percent less than a standard reinforced concrete deck of the same span. Reducing the dead load of the deck with this system can increase the live load rating. The concrete component of an exodermic deck can be precast or cast-in-place. When the concrete is cast-in-place, the steel grid component acts as a form, and the strength of the grid permits elimination of the bottom half of a standard reinforced concrete slab.

Precast exodermic decks have the benefit of quick installation; they can be erected during short, nighttime work periods, allowing a bridge to be kept fully open to traffic during higher traffic volume hours. This area is poured at the same time as the reinforced concrete deck when the deck is cast-in-place, or separately when the deck is precast. Exodermic decks require no field welding other than that required for the placement (with an automatic tool) of the headed shear studs. While the design of exodermic decks is patented, the availability of the grid from multiple, independent, licensed suppliers allows it to be considered generic and it does not need to be specified as a proprietary product.

The major advantage to the system, besides reduced dead load, is the installation time. Approximately 750 to 1000 sq ft of the cast-in-place system can be installed in a standard eight-hour period. For example, as reported in literature, using a precast exodermic system, 1900 sq ft was installed on the Gowanus Expressway per overnight closing in 2001 and 2000 sq ft per seven hour closing was installed on the Tappan Zee Bridge in 1997.

The system has been in use since the mid-1980s, and sufficient history and performance does not exist to support this prediction. Cracking is dependent on properly cured concrete and the relative stiffness of the superstructure. The manufacturer suggests using an overlay system and prefers an asphalt overlay over LMC or other concrete overlays to increase the speed of construction. Overlays also allow for accommodating vertical geometric variations as opposed to a bare concrete deck.

The major disadvantage to this system is initial cost. Not accounting for savings due to accelerated construction (reduced maintenance and protection of traffic requirements and shorter construction duration), the cost of an exodermic deck is approximately twice the initial construction cost of a cast-in place deck (materials only).

Bridges that do not have simple horizontal and vertical geometry and involve severe skews, small radii curves, or variable cross slopes should use the cast-in-place decks, negating the accelerated construction advantages of the system.

11.6.8 Accelerated Cure Cast-in-Place Concrete

Accelerated cure cast-in-place concrete uses low-slump accelerated concrete in conjunction with the maturity method to obtain design compressive strengths in 12 to 15 hours. The maturity method uses maturity loggers in the deck to analyze the time and temperature profile of in-place concrete to calculate the in-place strength of concrete in real time. Warmer concrete temperatures result in a higher rate of hydration, or more rapid strength gain, and colder concrete hydrates more slowly.

The advantages to this approach include lower traffic impacts due to reduced cure times as compared to standard HPC mix designs. The use of accelerated cure cast-in-place concrete allows for deck replacements to be performed over a series of weekend work cycles or other three-day cycles. The strengths and design life of this system are similar to conventional cast-in-place concrete.

The disadvantage to this system is that high quality control, particularly in regard to the water content and concrete slump, is essential for concrete performance. Although the use of maturity loggers has been successful in New York State, it is not familiar to most contractors; it has a short history and for cast-in-place concrete it is more susceptible to cracking, especially during staged construction.

11.6.9 Reactive Powder Concrete Bridge Girders

The benefits of using reactive powder concrete (RPC) to carry bursting forces in prestressed bridge girders are significant. Tests are reported in literature on three RPC, 150 MPa, deep beams. They have shown significant potential advantages to using RPC in bridge engineering.

11.6.10 Use of Fly Ash, Blast Furnace Slag, and Silica Fume

Concrete can be made more sustainable by use of cementitious materials such fly ash, which is obtained from coal-fired power plants. Use of fly ash in concrete bridges is on the increase. About 70 million tons of fly ash is produced each year in the U.S., but only about 15 million tons of it ends up in concrete products. It's a way to make roads, sidewalks, and bridges stronger and longer-lasting.

Fly ash from power plants and ground-up slag left over from the steelmaking process can replace some of that cement, and it can make the final product stronger.

When mixed with cement, the kind of fly ash produced by burning anthracite and bituminous coal keeps reacting with water to strengthen the concrete nearly a month after pouring.

Some states like Pennsylvania are considering making their use mandatory. During the past five years, the number of concrete suppliers in PA using fly ash and slag in mixes has increased from two plants to 10. Other states such as Minnesota have switched to using more fly ash and less cement. The typical dosage is about 15 to 25 percent in place of cementing materials, but some places like Minneapolis use 40 percent or more.

Like fly ash, slag reduces concrete's susceptibility to being weakened and damaged by water and salt. Blast furnace slag has similar properties to fly ash, but only if it is cooled with water after being taken from the furnace in a molten-hot state. Air-cooled slag doesn't create the same chemical reactions, but it can be used as "aggregate" filler. Up to 15 percent of Portland cement in a typical mix can be replaced with fly ash, or up to 50 percent can be replaced with slag.

Silica fume is a key ingredient in high performance concrete. It significantly increases the life of structures. It is a byproduct of silicon and Ferro-silicon metal production. It is a highly reactive pozzolan and its use decreases silica fume volume in the national waste stream.

The Silica Fume User's Manual and a standard reference manual are available from National Institute of Science & Technology, Gaithersburg, Md.

11.7 ADVANCEMENTS IN STEEL MATERIALS TECHNOLOGY

11.7.1 Revolution in Steel Strength by Using HPS 70W

Modern bridges carry very high moving loads and span over long distances. ASTM A709 HPS 70W (yield strength of 70 ksi) is being used by many states for flanges in hybrid girders exceeding 100 ft spans. Overall costs of construction and life cycle costs are reduced.

The author has used hybrid HPS 70W steel girders on a number of bridge projects.

Other advantages are enhanced resistance to fracture, elimination of maintenance painting, and improved vertical under clearance due to resulting shallow girders. Hybrid systems are possible by keeping web material as 50W steel. Currently, rolled sections are not available in 70W steel and the availability of a range of plate thicknesses is limited. Greater delivery time needs to be allowed in the construction schedule.

High performance and hybrid steel (grade 70):

1. Steel for girder webs and flanges shall be a combination of ASTM A709-00 grade HPS70W manufactured by the thermo-mechanical controlled processing (TMCP) or quenched and tempered heat treatment processing along with ASTM A588/709 grade 50W. All other steel shall be ASTM A709-00 grade 50W.
2. The following plan note establishes specification requirements for fabrication of hybrid girders using high performance grade 70 steel: This work shall be performed as modified by the "Guide for Highway Bridge Fabrication with HPS70W Steel," a supplement to ANSI/AASHTO AWS D1.5. This work consists of furnishing all necessary labor, materials, and equipment to furnish and erect structural steel members, designed as a hybrid/mix of steel materials consisting of ASTM A709, high performance grade HSP70W in combination with grade 50W steel.
3. If projected ADTT exceeds 500, design live load will be HL-93. Fatigue analysis will be based on projected ADTT.

   Survey of CVN toughness requirements: Fracture toughness for steels in civil structures is typically measured using the charpy V-notch (CVN) test. This test is specified where fracture is a concern. The high strain rate during this test is not representative of strain rates applied to structures in service and test temperatures do not match service temperatures.

   Complications occur when different codes, standards, and specifications require significantly different test temperatures and minimum toughness. Steel designated CVN shall be impact tested to exceed the test values of ASTM A709-00 table S1.2 "Non-Fracture Critical Impact Test Requirements" for zone 2, temperature range.
4. Compactness and bracing requirements for high performance steel girders: Earls, Shah, and Thomas experimentally verified nonlinear finite element modeling techniques for the study of high performance steel (HPS) I-shaped bridge girders. An evaluation as to the appropriateness of using the current American bridge specification provisions for cross-sectional compactness and adequate bracing was carried out within the context of applications involving A709 grade HPS483W high performance steel. Based on research, it appears that the current AASHTO flange compactness and bracing provisions are inadequate when applied to A709 grade HPS 483W bridge girders. Intense interactions between local and global buckling, as observed during the failure of the bridge girders, may be the cause.

New flange and web compactness criteria, as well as new bracing criteria, are proposed by Earls et al for use in the design of A709 grade HPS483W bridge girders. Fabrication restrictions include:

1. Application of heat for curving and straightening applications, camber, and sweep adjustment heating is limited to 1100°F maximum, and must be approved.
2. Based on the experience of one fabricator, extreme caution should be exercised when submerged arc welding electrode recommended in appendix A of the "AASHTO Guide for Highway Bridge Fabrication with HPS70W Steel." It has produced weldment containing unacceptable discontinuities in a substantial number of complete penetration groove welds.

   Consideration will be given to other welding processes if qualified and tested and as required by the referenced specifications.
3. In addition to the requirements of ANSI/AASHTO/AWS D1.5 Section 5.17, all procedure qualification tests must be ultrasonically tested in conformance with the requirements of AWS D1.5-95, Section 6, Part c. Evaluation must be in accordance with AWS D1.5-95, Table 9.1, ultrasonic acceptance—rejection criteria—tensile stress.
4. Galvanizing: The plates, hardware, and accessories shall be galvanized after fabrication in accordance with ASTM A 123 or ASTM A 153. Galvanizing damage during erection shall be repaired in an acceptable manner.

11.7.2 Use of HPS 100W Steel in Bridges

AISI/FHWA/U.S. Navy HPS Steering Committee has developed HPS 100W steel grade for 100 ksi minimum yield strength (ultimate tensile strength of up to 130 ksi) in bridges. Currently, plate thickness is manufactured from 3⁄16 to 2.5 inches.

Highway agencies are watching the progress of HPS 70W steel before going on to HPS 100W for their new bridges. However, Nebraska and West Virginia have started using HPS 100W.

HPS 100W has higher toughness, better weldability with a low carbon content, and improved corrosion behavior. It results in lighter weight bridge, reduces substructure costs, and leads to improved seismic performance. It has great potential for special girders when maximum live load deflection criteria are met. A longer span, fewer girders, and wider spacing may result. Also, weathering steel requires minimal painting or maintenance.

11.8 ADVANCEMENTS IN CONSTRUCTION TECHNOLOGY

11.8.1 Modern Construction Equipment

1. Mega movers
2. Record length casting bed in Oregon for girders.
3. Bridge roll-in construction method to avoid lane closures.
4. Innovative construction equipment.

11.8.2 The Design-Build Contractual Procedure

The full system is comprised of design-build-operate-transfer cycle. For bridges the system is still going through a process of development. It is aimed at:

- Improving team work
- Reducing bureaucratic procedures
- Resolving constructability issues quickly. Some of the teething problems are being ironed out.

The traditional approach of client-engineer-contractor is now being gradually replaced by the client opting for design-build contracts and to a "turn key" completion. The pre-qualified contractor hires the consultant or uses his own designers to prepare the contract documents. This method seems to eliminate any bureaucracy and is tailored to give a quick turn out based on the contractor's resources. As a result we need to improvise construction methods which suit the limited resources of the contractor. Design methods therefore revolve around quick turn out requirements.

11.8.3 Enhancing the Environment

The U.S. Congress passed the National Environmental Policy Act in 1969. Its objectives were:

1. To formulate a national policy which will encourage productive and enjoyable harmony between people and environment.
2. To prevent damage to the environment and thereby maintain health and welfare of people.
3. Enrich the understanding of the ecological systems.
4. Establish a council on environmental quality. This resulted in preserving important historic, cultural, and natural aspects of our national heritage. In addition, the quality of renewable resources was enhanced and recycling of resources was made possible. Some of the measures included:

    - Use of precast concrete elements with fewer environmental constraints

- Limiting construction activities to certain months of the year (May to August for example) for the least environmental impact
- No disturbance of the wetlands by adopting an innovative top-down method of construction
- Preserving the natural habitat around the bridge such as providing deer and small animal crossings
- Minimizing damage to flora and fauna
- Minimizing side slope erosion of stream banks.

11.8.4 Roll In–Roll Out Method of Superstructure Construction

The roll in–roll out system has been used for bridges located on airport routes and other essential facilities. The superstructure can be built adjacent to the existing bridge to be replaced on roller supports. During the nighttime window for lane closure, the existing superstructure can be jacked and unhinged off the bearings, rolled out, and the new superstructure rolled in.

To reduce construction impact on local traffic, as an alternative to constructing in stages, Region 11 NYSDOT (presentation given to ASCE by Project Manager Tariq Bashir, P.E.) employed a roll-out, roll-in concept for replacement of the Hillside Avenue and Jamaica Avenue bridges over the Van Wyck Expressway close to JFK Airport.

Conventional staged construction was going to create an inordinate impact to local traffic, due to the high traffic volumes on the expressway and on the avenues as well as on this highly developed area in Queens. The proposed roll-out, roll-in system consists of construction of the new abutments and center pier in stages (mostly performed at night during non-peak hours) and simultaneous construction of the new superstructure adjacent to the existing bridge. Temporary support bents are placed adjacent to the existing structure on both sides and a new superstructure is built next to the existing supported by the temporary bents on one side. The existing bridge is rolled out on the bents on the other side while the new bridge is rolled in place of the existing bridge. This whole operation of roll-out, roll-in was accomplished over one weekend.

Although this system of construction is common for railroad bridges, where extended closures are not practical, it was not tried before in New York State for a highway bridge. The project team relied on its experience with other projects in the midwest and the example of other projects overseas.

Other advantages of this system include no stage construction deck joints, no vibration of live loads onto adjacent green concrete during stages, and accelerated construction duration since the contractor is able to proceed with substructure and superstructure work simultaneously.

Roll-out, roll-in added about $40/sq ft of the deck area, which is 8 percent of the cost related to bridges. The total project cost, including highway elements, was $32 million.

Advantages

Advantages of the system include:

- Significant reduction in construction time
- The same alignment can be maintained for the new bridge
- Least impact on utilities or right-of-waya
- MPT requirements are better met than during staged construction.

Disadvantages

Disadvantages of the system include:

- It requires special construction planning
- It involves design for a temporary support system for the new superstructure
- Additional cost for rollers and temporary substructure is required.

Figure 11.1 Old bridge rolled out.

Work on approaches and bridge deck will require a detour. Investigate alternatives, either to detour one direction of traffic or both directions and provide for pedestrians. An eight- to ten-hour night window is required. A traffic count needs to be performed to assess impact on traffic flow during construction. A new application of the roll-in, roll-out method was successfully applied by NTS DOT (Figures 11.1 and 11.2 was provided by Tariq Bashir, the project manager) to prevent lane closure or a detour very close to the extremely busy JFK Airport. Transportation of a long span curved bridge is shown in Figure 11.3.

An example of a roll-in method was in Badhoevedorp, Netherlands where a preconstructed 390 ft curved girder with a massive weight of 3300 tons was moved within two hours. The road was required to be closed for a week to carry out supporting work of sign structures. Modern technology of transportation, lifting, and erection has made the roll-in, roll-out method possible.

11.8.5 Erection Methods and Precautions to Prevent Construction Failures

A study on bridge failures carried out by the author (Chapter 3) concluded that most failures occur during construction or erection. An ABC system must be put in place to avoid future failures. The study showed that the failure of connections due to overstress from bolt tightening, failure of formwork, local buckling of scaffolding, crane collapse, and overload were some of the causes. Girder stability during stage construction and deck placement sequence need to be investigated and temporary bracing provided. Expansion bearings need to be temporarily restrained during erection. Some flexibility in selecting bolt splice locations may be permitted with approval of the designer. Curved and skew bridges require special considerations such as uplift at supports, achieving cambers, and reducing differential deflections between girders during erection.

Figure 11.2 New bridge rolled in.

Figure 11.3 Transportation of curved bridge.

11.9 ACCELERATED BRIDGE CONSTRUCTION TECHNOLOGY

11.9.1 Understanding Rapid Construction and Associated Needs

Accelerated bridge construction through precasting of bridge components helps to reduce greenhouse gas emissions caused by traffic delays and long-term construction equipment operation. Many innovative concepts are presented here. Each concept is a subject in itself. Modern prefabricated construction materials and methods are vastly different from traditional methods and require innovative ideas for making the system safe and efficient. ABC can be promoted by understanding and analysis.

The owner's requirements are clear: reduction in schedule, wider decks, reduced seismic effects, increased bridge ratings, longer service life, cost savings, and lower maintenance.

Promoting rapid construction/accelerated planning and design:

1. Alternative design-build construction procedure: This approach still has led to faster turn out. For large projects a design-build-finance-operate-maintain process is most likely to be a complete situation. The simpler design-build method is more common by placing builder and designer on one team.
2. Modern construction equipment: The success of ABC is due to powerful equipment. Different erection equipment is required for girders, box beams, trusses, arches, cable stayed, and

Figure 11.4 New bridge under construction.

suspension cable bridges. Timely availability, a leasing facility, and an experienced erection team will be necessary.

3. The erection contractor shall utilize robotics, cranes such as tower cranes (for maximum lightweight pick of 20 tons and heights greater than 400 feet), lattice boom crawler cranes, mobile lattice boom cranes, mobile hydraulic cranes, and lattice ringer cranes for varying heavy weight pickups and accessories. In addition, specialized technology for a superstructure roll-in roll-out method using self-propelled modular transport (SPMT) is provided in a FHWA publication.
4. An evaluation matrix for construction cost, life cycle cost, environmental and social impact, constructability, future maintenance, inspection, and eesthetics needs to be prepared at the planning stage. Planning must address differences in ABC of small, medium, and long spans.

11.9.2 Factors Influencing ABC

Efficient structural planning is required to minimize the constraints in meeting accelerated construction goals such as erection during extreme weather. The following factors may be considered:

1. Improving aesthetics: To improve appearance, an exterior arched beam can be provided to hide the fascia girder.
2. Installing interactive touch screens featuring bridge information for motorists.
3. Increasing rider comfort by using a durable deck overlay protective system to prevent deck cracking; post-tensioning to prevent cracks in concrete and non-corrosive reinforcement; and improving crashworthiness of barriers and parapets.
4. Improving bearings performance: Use of multi-rotational and isolation bearings in seismic zones.
5. Improving bridge lighting and drainage methods by installing solar roof panels at approaches for bridge lighting and signage.
6. Using grouting methods for improving foundation soil including liquefaction mitigation to protect foundations in saltwater against corrosion; resistance to earthquakes, liquefaction, and erosion of soil under footings; preventing fender damage from vessel collision.
7. Constructing bio-retention ponds that collect and filter runoff from the bridge deck.
8. Introducing ground improvement techniques; modifying soil.

11.9.3 Accelerated Bridge Planning (ABP)—A Requirement for ABC

ABP shall promote durability and compliance with environmental/preservation laws.

Accelerated bridge planning (ABP) goes hand-in-hand with ABC. To implement ABC with confidence, applicable code for ABP and relevant specifications is needed. Recommendations are made for adopting innovative approaches for research in emerging construction technology.

As bridges get older, highway agencies have hundreds of bridges to reconstruct and open to the public in any given year. During lengthy construction periods, traffic problems are compounded at multiple bridge sites, which results in a loss of useful man-hours. Traffic jams have adverse effects on the air quality and health of users caused by idle fuel burning. Hence, ailing bridges need to be fixed on a priority basis using rapid construction and with desirable functional improvements.

Practical issues: The original decisions made by pioneer bridge builders need to be revisited. For continuous bridges for example, the use of double rows of bearings is a sign of added security against total bridge collapse. Fully covered bridges with canopies increase the life of the bridge deck, prevent accidents, and have no drainage problems or deicing salts. A return to forgotten fundamentals in some cases may be desirable.

Scope of the ABC approach: ABC is useful for emergency replacement of bridges damaged from construction accidents such as crane collapse, vehicular accidents, ship collision, ice damage, flood or earthquake, for which accelerated planning and design will also be necessary.

Using ABC, formwork, or much of concrete placement and curing is not required.

For rapid construction, such as for busy highways and over rivers and waterways, there is no substitute for ABC.

Activities such as borehole tests, pile driving, shop drawing review, and closure pour are unchanged. While cast-in-place construction is time tested, ABC applications are of recent origin. ABC has a great potential for wider use but many road blocks need to be removed to pave way for a much wider application.

Some limitations of ABP include:

1. An ABC design code based on ABP for alternatives and connections will be required: ABC is an entirely different method of construction. Fabrication, transportation of components, and erection methods are different.

2. Impact on analysis and design: The following design issues need attention:
   - Distribution coefficients in LRFD method—Longitudinal joints alter the aspect ratio of deck panels. While the span to girder spacing ratio may be unchanged, transverse distribution of live load is diminished. Percentage of longitudinal distribution is increased and needs to be evaluated.
   - Load factors—AASHTO's use of load factors for CIP construction will change for precast construction.
   - Changes in boundary conditions compare to conventional construction.
   - The advantage of Poisson's ratio in transverse bending is no longer available due to discontinuity.
   - The orthotropic behavior of the superstructure needs to be improved by providing diaphragms at closer longitudinal spacing.
   - Advantage of arching action at supports will be lost since the deflection pattern has changed.
   - Due to age differences in cast-in-place concrete for closure pour and precast panels, differential shrinkage is likely to take place. Longitudinal joints may crack or open up, requiring repairs and lane closures.
   - Compared to cast-in-place construction, a precast bridge with numerous deck joints is likely to be weaker during earthquakes.
   - The trend in design is to eliminate transverse and longitudinal joints by adopting integral abutment and integral pier approach for unified behavior. Segmental deck construction requires transverse prestressing.
   - Span lengths in concrete bridges are restricted to about 100 feet due to transport restrictions of heavy components.
   - Deck drainage provisions in transverse and longitudinal directions would still require concrete topping with varying thickness in the field.

3. Approach slab construction requires placing precast panels supported on grade. Separation from grade may happen due to a lack of compaction and a raised water table leading to settlement of the approach slab.
   - Future widening is not easy if transverse prestressing of the deck slab has been used.
   - Full-scale testing is required to develop confidence, especially for curved bridges. MPT, approach slab construction, permits and utility relocation, etc. are unavoidable constraints.

- Contractors, in general, have technicians trained in formwork and cast-in-place construction and new training in ABC is required.
- Overemphasis of incentives/dis-incentives pressurizes the contractor into adopting unrealistic schedules at the expense of quality control.
- Also, the manufacturing nature of precast products creates a proprietary system and monopolistic environment which may lead to unemployment of some number of construction workers.

11.9.4 Objectives of ABC/ABP

The primary objectives of ABC/ABP are constructability, erection, serviceability, durability, maintainability, inspectability, economy, and aesthetics. Detailed objectives are as follows:

1. Develop modern codes and construction specifications: Construction mistakes, ship collision, scour, design deficiency, overload, fatigue, and earthquake are the main causes of failure.
2. Introduce specialized training of designers and field staff. Applied mathematical methods and the use of fracture mechanics in design need to be developed and promoted.
3. Monitor construction activities on the "critical path" for saving time and long-term rehabilitation costs.
4. Ensure safety during construction and a safe bridge for users following its completion.
5. Obtain traffic counts for selecting full or partial detours and apply temporary construction staging.
7. Optimize the size and number of girders and use modern material and equipment.
8. Select pleasant bridge colors and aesthetics to keep "America the Beautiful."

11.9.5 The Accelerated Construction Route by Precasting of the Superstructure

Many states have been in the forefront to promote and implement innovative technologies to achieve improved work zone safety, motorist safety and comfort by using joint-less decks and integral abutments, and with minimal environmental disruption.

Audits are in practice to ensure that designers and project managers study alternatives—new manufacture processes, connection details for prefabricated elements, management programs, and quality assurance.

Crews can cut the old bridge spans into segments and remove them, prepare the gaps for the new composite unit, and then set the new fabricated unit in place in an overnight operation. The quicker installation minimizes huge daily delay-related costs and daily traffic control costs.

Construction is usually scheduled for fall months when the weather is more predictable. A single-course deck will save a minimum of six weeks in construction time compared to a two-course deck.

1. On the Route 46 Bridge spanning the Overpeck Creek in Bergen County, the NJDOT decided to use prestressed, precast beams to prevent painting costs.
2. Utilizing a precast superstrucure (Inverset), the NJDOT replaced a structure in South Jersey, Creek Road over Route I-295 SB.
3. Prefabricated deck panels (Inverset, which is no longer proprietary) for three single-span Route 1 bridges over Olden Avenue and Mulberry Avenue in Trenton, NJ were constructed in 2005, over weekends.
4. Besides exodermic and orthotropic decks, new materials used are HPC and corrosion inhibitor aggregate. Precast or steel diaphragms for prestressed beams have been allowed. Precasting has quality control and avoids reinforcement steel placement, concrete pouring, and weeks of curing period.

Figure 11.5 The author's design of precast multi-column pier bent on Route 322-50 in southern NJ.

11.10 PREFABRICATION TECHNOLOGY

11.10.1 Advantages of Prefabricated Bridge Systems

Total prefabricated bridge systems offer maximum advantages for rapid construction and depend on a range of prefabricated bridge elements that are transported to the work site and assembled in a rapid construction process. Prefabricated designs are becoming familiar to the engineers to the extent that they can specify a pre-manufactured bridge in their drawings. The supplier will provide a "cookie cutter" design to satisfy the technical specifications.

As the U.S. interstate highway system approaches the end of its service life, urban congestion continues to grow. Bridge construction extending over several months has become a primary source of congestion. Bridge construction or rehabilitation can be a significant source of congestion because of its sequential nature. In cast-in-place construction, foundations for piers and abutments must be built first, then pier columns and caps must be built before beams and decks are placed. Offsite prefabrication technologies and processes help solve this problem.

Ongoing research is focused on identifying and developing new bridge elements and systems for all materials that would help accelerate bridge construction.

Bridge construction times can be reduced significantly by using precast columns. Columns can be segmented, post-tensioned, reinforced, hollow, or solid concrete.

11.10.2 Prefabricated Bridge Elements and Systems

Prefabricated bridge elements can be manufactured either on site, under controlled conditions, or in a factory and brought to the construction location ready for installation.

Advantages to prefabricated elements include:

- No temporary formwork or stay in place formwork is required
- Construction duration and construction-related traffic disruptions are minimal
- By reducing the number of workers operating near moving traffic, work zone safety is increased
- It reduces environmental impacts by minimizing the site access footprint
- It improves the constructability of bridge designs by controlling manufacturing environments
- It lowers life-cycle costs by increased quality control through controlled fabrication conditions.

Prefabricated superstructure systems, comprised of steel or concrete, are superstructure sections typically constructed off-site at a prefabricator and shipped to the project site. Construction can be completed during limited duration off-peak lane closings.

The ideal span length for this structure is approximately 80 ft using stainless steel reinforcement. The formerly proprietary system known as Inverset is a precast concrete and steel composite bridge superstructure system that utilizes an "upside-down" casting method which takes advantage of the force of gravity to prestress the steel beams. The inverted casting process precompresses the concrete deck, yielding a crack-resistant deck with high durability.

11.10.3 Precast Concrete Steel Composite Superstructure (PCSCS) Units

The advantages to PCSCS units include:

- Reduced beam depth
- Rapid installation: erection times of one hour per unit (after deck removal is complete) allow overnight or weekend installation
- Year-round installation
- Due to precompression in the concrete deck, deck cracking is minimized
- Improved quality due to controlled environment construction.

The primary disadvantages of this system include:

- The initial construction cost of a prefabricated system is approximately 50 percent more than that of a normal superstructure.
- Precompressed concrete cannot be replaced in the field; any future re-decking would require the removal of the entire unit or a reduction in the capacity of the system since the new deck would not be precompressed.

11.10.4 Effideck Bridge Deck System

The proprietary precast concrete deck replacement Effideck System can act compositely when the shear stud connectors are attached to the stringers through pockets in the deck.

The system consists of modular deck panels using a 5-inch thickness supported by closely spaced hollow structural section steel tubes. The panels can span either in the transverse direction between bridge stringers or in the longitudinal direction between floor beams. The panels are bolted to the existing bridge stringers from above the deck and the pockets are then grouted.

Advantages of the Effideck system include:

1. All installation work is performed form the top of the deck, resulting in faster installation without the cost of scaffolding.
2. A steel-on-steel load path ensures that the deck can support load even prior to the completion of the grouting process.
3. A typical panel weighs approximately 75 psf, which is lighter that a conventional precast concrete deck panel.
4. Efficient connection details facilitate overnight installation.

11.10.5 Polymer Concrete Bridge Deck Overlay Systems

Polymer concrete bridge deck overlay systems repair the underlying deck after removal of an existing overlay and placement of a new overlay to seal out moisture and chloride ions from permeating into the deck. Generally, these overlay systems have a 10-year life span. These systems can be used in conjunction with a primer to fill cracks and patch partial depth bridge deck spalls in concrete, such as:

- Polyester polymers
- Epoxy-urethane polymers
- Other engineered polymers that are used to protect the existing deck and improve the riding surface

- Latex modified concrete (LMC) and corrosion inhibitor aggregate concrete are also used.

Polyester polymer concrete overlay systems can achieve 4000 psi in compressive strength and 1600 psi in flexural strength within 24 hours, and the bridge can accommodate traffic due to fast curing system, which also allows traffic to be resumed within a few hours at temperatures lower than 40°F.

Epoxy-urethane co-polymer systems bridge existing cracks, yield an impervious barrier with chloride ion penetration resistance, and provide a high-skid and wear-resistant surface. The system remains flexible throughout its life cycle and at low temperatures. It should be noted that while epoxy-urethane co-polymer systems are promoted to have an expected life span of over 10 years, there are cases where delamination has begun within five years.

The polymer overlay systems have the following advantages:

- Rapid repair of shallow deck spalls and overlaying for quick return of traffic (overnight work)
- Superior adhesion to concrete surfaces
- Material is easily applied.

The disadvantages to these systems include:

- Lower life span than deck replacement
- Requires a sound deck (all spalls repaired and no full depth spall repairs).

Manufacturers of polyester polymer concrete bridge deck systems include Kwik Bond, and manufacturers of epoxy-urethane co-polymer systems include Poly-Carb, Inc.

11.10.6 Examples of Successful ABC Applications

Case Studies

1. The Washington Department of Transportation (DOT) recently minimized traffic disruption on U.S. Interstate 5/South 38th Street interchange in Tacoma, WA, by using partial-depth precast concrete deck panels.
2. When the Virginia DOT needed to keep I-95 open during the James River Bridge replacement, the state used a prefabricated superstructure system for most of the bridge spans. The composite units consisted of a 222 mm concrete deck over steel girders that were fabricated at a nearby casting yard. Crews were able to cut the old bridge spans into segments and remove them, prepare the gaps for new composite unit, and then set new units in place during overnight operation.
3. The New Jersey DOT has used precast columns and caps on Route 322 and Route 50 intersections. In several successful accelerated bridge construction projects in Florida and Texas, pre-fabricated bridge pier components have been utilized, resulting in significant reductions in the construction schedule.
4. The development of high-speed computers, design software, and construction technology have revolutionized modern day repair significantly. The rapid growth in technology therefore needs to be addressed in detail. It needs to be directed toward redesign and reconstruction.
5. Use of HPS: Ali Khan designed bridges with HPS 70W hybrid girders in New Jersey recently. It allows longer spans and lighter girders. Shallower girders improve vertical under clearance, reduce number of girders to be constructed, eliminate painting, and weathering steel provides enhanced resistance to fracture.
6. Parapets: A variety of parapets besides New Jersey barriers are used. The NJDOT permits its contractors to use slip forms for increasing speed of construction, as done successfully on the Route I-195 & I-295 Interchange.

7. Substructure: Integral abutments with fewer piles have been successfully used in New Jersey. They can be constructed more quickly than conventional bridges.
    - An example of an integral abutment bridge using prestressed concrete box beams on Route 46 over Peckman's River was designed by Khan.
    - Currently, a design is in progress for lesser used semi-integral abutments for such bridges on Nottingham Way over Assunpink Creek in Mercer County and Garden State Parkway Bridges over Jakes Branch.
    - Lighter piers or precast concrete pile bents save costs and time as used on Albany Street Bascule Bridge carrying Route 40 & 322 into Atlantic City. Precast post tensioned pier caps were recently used on the Route 9 over the Raritan River and by Khan on Interchange of U.S. Route 322 and NJ Route 50. Drilled shaft foundations and concrete cylinder piles of 36 to 66 in diameter are in use. Precast sheeting has been used for retaining walls and abutments. MSE abutments have performed extremely well. The NJDOT has used RFP material for fender systems for two bridges along the Jersey coast, Route 9 over Nacote Creek, and Route 9 over Bass River. It is environmentally friendly and eliminates marine borers.
8. Scour countermeasures: Minimal marine life disruption and quick construction are achieved by using gabion baskets, articulated concrete, or cable tied blocks in lieu of traditional sheet piles. Khan prepared a "Handbook for Scour Countermeasures" for the NJDOT jointly with CUNY, which was approved by FHWA (7) and in addition helped develop Sections 45 and 46 of the NJDOT Bridge Design Manual.

11.10.7 Extending the Service Life of Bridges

ABC methods were evolved ahead of the design codes. Research is required in many aspects, including:

1. Developing strengthening methods and corrosion mitigation techniques; fabricating stronger girders by eliminating the need for shear stiffeners with the use of folded web plate in steel girders.
2. New methods to monitor and strengthen foundations against scour, earthquake, and impact.
3. Developing and reviving the concept of full canopy on bridges to facilitate mobility, improve drainage, prevent skidding, and eliminate the use of deicing agents.
4. Optimum use of construction materials: Research in the use of FRP composite materials, geo-materials, geo-synthetic products, lightweight, high- and ultra-high-performance concretes and steels. Develop appropriate limit state criteria for the use of these materials, details, components, and optimizing structures for adoption into the LRFD specifications.
5. Reduced duration of bridge construction: Develop contracting strategies such as realistic incentives/disincentives to encourage speed and quality. Mass production management techniques adopted by automobile/aircraft industry may be considered.
6. Mathematical methods, ABC codes and technical specifications: Learn and disseminate emerging technology knowledge and methods such as,
    - A strong back-up of mathematical technique, closed form solutions, and formulae from mathematicians
    - Identify and calibrate the service limit states for unusual construction conditions
    - Begin transition to a performance-based specification, with an accompanying design manual
    - Develop and incorporate security performance standards, especially for long spans
    - Continued development of LRFR provisions is required. Integrate information from

maintenance and operations into code development and vice versa. Promote automated data collection and reporting damage models using the data collected.

11.10.8 Conclusions for ABC

Initiatives taken by FHWA have led to considerable progress in implementing ABC concepts. ABC-related design needs to be made part of AASHTO and state bridge design codes and specifications. To gain confidence in structural behavior, full-scale testing of joints in precast curved deck is required. Analytical methods applicable to discontinuities of components need to be developed. Application of the latest techniques in concrete manufacture, composites, HPS, and hybrid materials is feasible. An integrated software covering all aspects of ABC design and drawings may be developed. Deterrents and bottlenecks such as MPT, construction easement, right-of-way, permit approvals, and utilities relocation need to be resolved and administrative procedures simplified to facilitate ABC. Certification and training of construction personnel, continuing education of engineers in rapid construction techniques, and construction management ABC courses at universities are recommended.

11.11 REHABILITATION OF STEEL BRIDGES

11.11.1 Existing Steel Girders

For each structural component, the following issues will be addressed:

1. Structural steel replacement and/or strengthening
   - Existing steel beams with deck replacement shall be made composite in positive moment regions.
   - In order to determine the remaining service life, a fatigue analysis of existing steel members to be reused or rehabilitated will be carried out in accordance with: AASHTO Guide Specifications for Fatigue Evaluation of Existing Steel Bridge,""2007 AASHTO LRFD Specifications for Bridge Highways," and the current state LRFD bridge design manual.

2. Investigate use of high performance steel: Use of HP 70W for durability, weight, and cost savings, and strengthening fractured floor beams will be considered.

3. Heat straightening: This old technique is used to restore deformed steel members by gradual heating and cooling. The procedure is more of an art than a science and requires experienced craftsmen. Beams or girders that have been struck by trucks or are bent by other causes can often be repaired by heat straightening only, or in combination with field welding to install new sections for the damaged steel member portions. Steel can be bent from overload, collision, earthquake, or fire.

   A repair procedure to straighten plastically deformed regions of damaged steel by applying repetitive heating and cooling cycles is generally used. Each cycle leads to a gradual straightening trend. Maximum temperature is controlled so that thermal stress from heat shall not increase the yield stress of steel. Steel has the capacity to restore to its original condition through heating. The performance of repaired steel does not change. The alternate method of hot mechanical straightening uses an external force by which properties of steel are affected and early fracture can take place. If heat straightening is deemed to be practical, a detail showing the location of the repair and procedures needs to be prepared in the form of a report.

4. Micro composite reinforcing steel bars, corrosion inhibitors and latex modified concrete: Conventional 60 ksi rebars are known to corrode. Micro-composite steel is non-coated and is highly corrosion resistant. Corrosion inhibitors and latex modified concrete can extend the life of bridge decks by more than 20 years.

   The use of corrosion inhibitors with conventional rebars can mitigate the problem by chloride extraction. Corrosion of reinforcing bars is caused by salt penetrating concrete. The chemical corrosion inhibitor additive is designed to protect structural steel reinforcement by preventing oxidation after full or partial depth repairs of bridges. The corrosion inhibitor additive allows its products to chemically counteract the corrosion process that takes place at the interface between the iron and the concrete. ASTM has developed ASTM G109 corrosion inhibitor specifications.

6. Joints, bearings, and devices (JBD): A DOT study investigating joint failures listed "lack of designer's awareness" as a key concern. Topics of critical importance are theory and design of common types of JBD, reliability-based design, installation and maintenance, fatigue and corrosion, modeling, and finite element analysis. Increased awareness of JBDs can enhance the probability that critical components will perform their functions with intended structural control.

11.11.2 Strengthening of Structural Steel Members

1. Welded stud shear connectors: Shear connectors shall be installed full length on all steel beam or girder bridges in which the deck is being removed and replaced. The stud spacing shall be designed in accordance with AASHTO LRFD specifications.

2. Bolted cover plates in tension zones or field welded cover plates in compression zones can be used to increase strength.

3. Jacking the stringers to relieve stresses prior to installing cover plates is desirable so that cover plates will carry dead load and live load stresses. If the plates are installed without relieving the stresses, they will carry live load only.

4. Angles or structural shapes may be attached to the web or flanges.

5. When retrofitting or repairing truss members, temporary support is required since many truss members are non-redundant, and their removal could result in collapse of the structure.

6. Use of slip-critical connections: Refer to "Specifications for Structural Joints Using ASTM A325 or A490 Bolts" for the following four conditions:

- Joints subjected to fatigue load
- Joints with oversized holes
- Joints with slotted holes with loads not perpendicular to slots
- Joints in which slip will be detrimental to performance of structure.

7. Fatigue performance-based analysis: In order to determine the remaining service life, a fatigue analysis of existing steel members to be reused or rehabilitated will be carried out in accordance with "AASHTO Guide Specifications for Fatigue Evaluation of Existing Steel Bridge," current "2007 AASHTO LRFD Specifications for Bridge Highways," and current state bridge design manual. This will help to determine the logic of reuse or replacement.

8. Quality control inspection of welded joints: Rational quality control approaches to fabrication inspection and weld acceptance are required. It ensures that the structure has sufficient fatigue performance. In order to evaluate the effects of weld defects on fatigue performance, fatigue tests of butt-welded joint specimens of 25, 50, and 75 mm thick with various types of weld defects were performed in Japan by Miki and Nishikawa. Acceptable sizes of weld defects are established from these test results and fracture mechanics analysis. A computerized automatic ultrasonic inspection system has been developed and the applicability of these systems has been examined to be satisfactory.

9. Investigate the use of HP 70W for durability, weight, and cost savings of rehab for strengthening fractured floor beams.

10. Removing floor beams with riveted connections to webs: To avoid instability to structural system or local buckling, consider leaving fractured floor beams in place and strengthening with new channel beams.

11. Coupons: A small sample of existing steel may be taken from a redundant location and tested for strength, depending on the available as-built plan and data.

11.11.3 Long-Term Issues

1. As-built plans and/or shop drawings should be reviewed followed by a thorough site inspection making note of material condition, fatigue prone details, utilities, geometry, girder alignment, and possible paint removal and containment considerations.
2. Non-destructive testing should be performed on butt-welded top flange splices to ensure weld soundness.
3. Details of particular importance to check are butt-welded splices, partial length cover plate ends, welded lateral gusset plate connections, connection plate/stiffener welds, and shear connector welds in tension or reversal zones. For example, brittle fracture was observed in the exterior girder of the Blue River Bridge.
4. The designer should consider removing all poor details, fatigue sensitive details, and stress risers of all types. All fatigue sensitive details must be analyzed.
   - Notch effects, such as rivet holes and non-radius cuts, cause stress increases.
   - Lateral connection plates should not be welded to tension flanges.
   - Rivet holes should be made round by reaming to eliminate crack initiation sites.
   - Upgrading of fatigue-sensitive details using bolted over-splicing of partial length cover plate ends should also be considered to meet the allowable fatigue stresses.
   - Riveted girders should not be retrofitted for continuity due to their uncertain fatigue performance and difficult splice detail requirements.
   - Often new load paths are created when widening. The stiffness of new members and strengthening of old adjacent members to carry the new live loads should be considered.
5. Short-term crack sealing and joint repair in steel: Webs are fracture critical members (FCM). Fracture of thin webs is a dangerous scenario and needs to be fixed. Steels used in main members should be ordered to the correct level of strength and toughness. For main members, the material should specify Charpy V Notch (CVN) requirements for the FCM zone and reference the direction of rolling.
6. Bond characteristics of carbon fiber reinforced plastics to structural steels: Fiber reinforced plastics (FRP) as a structural material can provide an opportunity for structural engineers to propose new structural systems instead of conventional steel or concrete.

Researchers in Japan studied the bond strength of carbon fiber reinforced plastics to structural steels in order to develop the composite structure of steel and fiber reinforced plastic. It was concluded that the strong anisotropy of fiber reinforced plastics may cause local shear stress concentration, resulting in premature debonding.

11.11.4 Providing Continuity at Hinge Assemblies

1. Removing hinges and making the members continuous may be desirable. If the hinge cannot be removed, redundancy must be provided in the event of a hinge failure.
2. If a pin and hanger assembly is to be rehabilitated, lubrication and nondestructive testing requirements are desirable.

11.11.5 Connection Designs Based on LRFD Code

1. Formulae for cross frames and diaphragm connection designs based on the latest AASHTO LRFD code were programmed by the author for application to I-95 State Road Viaduct in PA.

2. Fatigue stress evaluation method due to load reversal: The author developed a computer program in MathCAD to evaluate remaining useful fatigue life in connections and steel members. The method was based on AREMA code (and was used for cooper train loads for SEPTA bridges in Philadelphia). Actual train live loads and impact were for silver liner, bombardier, and diesel locomotives used by SEPTA. The yield stress of existing steel was less than 36 ksi. Allowable stress level was based on 2 million cycles capacity. The method can be applied to AASHTO HL-93 Live Loads for Highway Bridges.
3. Newly designed girders need a factor of safety and therefore greater shear resistance than the cumulative shear force from dead and live load plus impact. Web depth at girder ends may need to be increased by using splayed webs.
4. Removing floor beams with riveted connections to webs: To avoid instability to the structural system or local buckling, consider leaving in place fractured floor beams and strengthening with new channel beams.
5. Retrofit of steel beams: The bridge survey and evaluation report will be consulted if the superstructure is in serious condition due to cracks in webs of numerous floor beams or at the truss connections. Drilling of holes at the tip of cracks may be required. The beams need to be retrofitted by installing splice plates at the bottom flange for easy access. Other applicable retrofit details shall be provided as per "AASHTO Guide Specifications for Fatigue Evaluation of Existing Steel Bridges."
6. Coupons: To evaluate strength of old steel, a small sample of existing steel may be taken from a redundant location and tested for strength, depending on the available as-built plan and data.
7. Replace damaged bearings: For proposed construction, fixed and expansion type multi-rotational pot bearings or seismic isolation bearings are likely to be more suitable than other conventional types, such as disc or elastomeric pads, to meet seismic requirements. Bearings will be designed using the LRFD method. Existing rocker and roller bearings have greater height than proposed pot or isolation bearings. The difference in height is handled by placing concrete pads or steel plates under the shallow bearings.

   The superstructure will be jacked to remove and replace bearings. During stage construction, the existing deck slab may be temporarily supported by steel posts and beams prior to demolition. Any underpinning will be removed after the new deck is in position. For existing multi-span steel beam bridges, the designer shall consider providing continuity using flange and web connection plates at pier locations. In such situations, analysis of the girder as a continuous member shall be performed. Replacement of dual bearing lines with a single bearing should be considered, particularly if replacement of the bearings is required.
8. The skew effect behavior of steel bridges and their relative flexibility when compared with prestressed concrete bridges must be evaluated and considered in the analysis.

11.11.6 Rehabilitation of Through Girder Bridges

Past experience has shown that compared to multi-girder system, through-girder systems with floor beams can save up to 2 feet in girder depth. With well-connected closely spaced floor beams acting as a rigid diaphragm between longitudinal girders, this system:

- Will address concerns over the safety of the structure
- Provide much shallower structural depth that would benefit ROW cost
- Will improve sight distance
- Can be designed to carry distributed load even if the through girder had a crack at mid-span
- Is the least expensive alternative

- Will have a shorter repair schedule, with only two longitudinal girders.

As a refinement of the two through-girder system, partial post-tensioning of the main through girders will increase superstructure internal redundancy and serve as a fail-safe system.

A fully redundant four-through girder system may be used with a sidewalk in each end bay. Build the fascia beam strong enough to provide a fail safe system by assuming the main girder has cracked. This will increase redundancy. However, a multi-girder system will have greater redundancy than two through girders.

A study by the author shows that use of 70W steel in multi-girder system can save 12 more inches in girder depth than with 50W steel. This provides the opportunity to utilize high performance steel, reduce structural depth, and lower costs. The example of designing the Route 1 & 9 Magnolia Avenue Bridge using HPS 70W steel girders shows that reduction in structural depth will improve the profile of a bridge. The initial preferred alternative used a 50W multi-girder composite system.

In designing a multi-girder system, Inverset systems or similar prefabricated bridge concepts will accelerate construction.

11.12 RESEARCH IN NEW TECHNIQUES FOR MONITORING, NEW MATERIALS, AND VIRTUAL DESIGN

11.12.1 The University of Michigan Recommends Use of Nanotechnology to Reveal Cracks and Corrosion

A new "skin" for bridges could be a sixth sense for inspectors looking for cracks and corrosion that could lead to a catastrophic failure like the recent Minneapolis bridge collapse.

Bridges are scrutinized every two years and inspectors rely heavily on their eyes to find weak points. If they see red flags, they do more tests. According to a paper published in the journal *Nanotechnology*, researchers at the University of Michigan have recently developed a coating that could be painted or sprayed on structures.

It would allow inspectors to check for damage without physically examining a structure. When it's time to examine the health of the structure, an inspector could push a button and in minutes the skin would generate an electrical resistance map and wirelessly send it to the inspector.

The sensing skin is an opaque, black material made of layers of polymers. Networks of carbon nanotubes run through the polymers. Carbon nanotubes are a fundamental building block of the nanotechnology revolution.

The microprocessor then creates a two-dimensional visual map of that resistance. The map shows inspectors any corrosion or fracturing too small for human eyes to detect.

Each layer of the sensing skin can measure something different. One tests the pH level of the structure, which changes when corrosion is happening. Another layer registers cracks by actually cracking under the same conditions that the structure would. The sensing skin would revolutionize the way current bridge health assessment is conducted, resulting in safer structures and lower-cost of inspection. The skin could be a permanent veneer over strain- and corrosion-prone hot spots of joints in bridges.

11.12.2 University of Maine Developing Lightweight Carbon Arches

The UM Composites Center researchers have estimated their bridge's carbon footprint to be about one-third less than that of a standard concrete bridge and one-fourth less than a standard steel bridge.

The "bridge-in-a-backpack" innovative technology uses carbon-fiber tubes that are inflated, shaped into arches, and infused with resin before being moved into place. The tubes are then filled with concrete, producing arches that are harder than steel yet resistant to corrosion. Finally, the arches are overlaid with a fiber-reinforced decking. The technology has a potential for future use due to its light weight and the portability of its components.

11.12.3 Drexel University Evaluates Concrete Bridges Lacking Documentation

Thirty percent of U.S. bridges lack critical documentation. The lack of documentation for the aging bridges presents a challenge in collecting adequate information about the bridges' materials and reinforcement properties.

Drexel Professors Aktan and Moon therefore used an off-the-shelf, wireless data-acquisition system and falling-weight deflectometer to determine their effectiveness in producing rapid and cost-effective findings.

According to Aktan and Moon, the wireless system is not reliable to be used during load testing but encourage the development of this technology because it could offer time and cost savings for bridge evaluators. The falling-weight deflectometer, which drops weights onto a grid marked on a bridge, could be transformed into a useful tool to complement visual inspections, the common practice of determining bridge safety.

They recommend that the bridges' foundations be inspected annually and that the development of a prototype falling-weight deflectometer be given top priority to expedite testing the large number of bridges without documentation.

11.12.4 The University of Miami Develops a Self-Powered Monitor System

The University of Miami developed a self-powered monitoring system for bridges using wireless sensors. With a scarcity of inspectors and tens of thousands of bridges, the visual inspection process can be long and laborious. Thousands of bridges erected during the 1960s and '70s, when much of the nation's infrastructure was built, do not have sensors installed.

A team of University of Miami College of Engineering researchers are implementing a self-powered monitoring system for bridges that can continuously check their condition using wireless sensors. Sensors can harvest power from structural vibration and wind energy.

They plan to place newly developed wireless sensors, some as small as a postage stamp, others no longer than a ballpoint pen, along strategic points inside the 27-year-old Long Key Bridge in the Florida Keys and on a Northwest 103rd Street quarter-mile steel overpass that leads into Hialeah, Florida.

The sensors, developed by project collaborators Virginia Tech University and New Jersey-based Physical Acoustics Corporation, record all sorts of data, from vibrations and stretching to acoustic waves and echoes emitted by flaws such as cracks. Even the alkaline levels in the concrete of bridge supports are being measured.

The work is part of the National Institute of Standards and Technology Innovation Program and is aimed at developing a more effective system to monitor the health and predict the longevity of bridges. The joint venture is led by Physical Acoustics Corporation.

11.12.5 Innovative Approach for Bridge Inspections

A new imaging program automatically detects irregularities in bridges. Research scientists at the Fraunhofer Institute for Industrial Mathematics ITWM in Kaiserslautern have developed the specialized software jointly with fellow scientists from the Italian company Infracom. The engineers have been using the new software successfully to inspect bridges in Italy.

The software automatically examines the photos of a bridge for certain characteristics and irregularities, for marked discoloration. Even minor damage is identified and signaled.

No two bridges are alike and they differ in terms of their shape, construction material, and surface structure. The color depends on:

- The material
- The dirt or fouling
- The degree of humidity.

The changing effects of weather and temperature, road salt, and the increasing volume of traffic all quickly cause damage such as hairline cracks, flaking concrete, and rust penetration. The software can handle these discrepancies. The researchers have extracted metrics from photographs that include the characteristically elongated shape of a hairline crack, the typical discoloration in damp places, and the structures of the material, which are different for a concrete bridge than for a steel bridge.

All of this information is stored in a database. When the researchers load a photo into the program, the software compares the features of the new image with those of the saved images. If it detects any irregularities, it marks the respective area on the photo. The bridge inspector can decide how serious the damage is and if something needs to be done. The earlier any damage is identified and clearly categorized, the simpler and less expensive it is to repair.

11.12.6 FRP-Retrofitted Reinforced Concrete Beam-Column Joints

Researchers at the University of Queensland, Brisbane, Australia, along with researchers at Iranian universities, have investigated theoretical and experimental behavior of concrete beam-to-column joint specimens (un-strengthened and FRP-strengthened). The ANSYS finite element software was used for modeling RC exterior joints. The specimens were loaded using a step-by-step load increment procedure to simulate the cyclic loading regime employed in testing. Additionally, an automatically reforming stiffness matrix strategy was used to simulate the actual seismic performance of the RC members after cracking, steel yielding, and concrete crushing during the push and pull loading cycles.

The results show that the hysteretic simulation is satisfactory for both un-strengthened and FRP-strengthened specimens. Strengthening FRP sheets in the connection zone move plastic hinges forward from the end of the beam to a reasonable distance, and decrease the maximum concrete strain in the joint region. Furthermore, they enhance the load carrying capacity, ductility, and energy dissipation of the joint.

11.12.7 Recycled Plastic Lumber Bridges

A New Jersey company, Axion International Holdings, of Basking Ridge, NJ, has developed a system in conjunction with researchers at Rutgers University to make bridges from recycled materials that are strong enough to support a U.S. Army tank. The engineers constructed a pair of bridges made entirely from recycled plastic products at Fort Bragg, N.C., and had M1 Abrams tanks driven across the spans.

The M1 Abrams, manufactured by General Dynamics, weighs nearly 70 tons, making it too heavy for the vehicle to use most standard bridges and roads. The tests indicated the structures held up well under both moving and static weight loads and withstood stresses caused when the M1 operator applied the vehicle brakes while on the bridge.

The U.S. Army Corps of Engineers Website states that its construction engineering research laboratory was involved in the design and building of the test structures.

The new bridges can be made entirely from recycled consumer and industrial plastics. The two test thermoplastic spans were made from more than 170,000 pounds (said to be the equivalent of more than 1.1 million 1-gallon milk jugs of recycled plastic). The structures are less expensive to build than traditional wood timber bridges often used on U.S. military bases.

Advantages of recycled structural plastic lumber bridges are speed of installation, reduced costs for construction and maintenance, and eco-friendliness.

11.12.8 Titanium Pedestrian Bridge for the University of Akron

Titanium is as strong and blast-resistant as steel but weighs 40 percent less, the best strength-to-weight ratio of any metal. It is resistant to saltwater as well. It's in ample supply, mined in the southern and western United States, Canada, Australia, and several other countries.

The problem is production cost. The chemical process developed in the 1930s to extract titanium from mineral-laden rocks is expensive. The proposal, though expensive, might be a shot in the arm for Akron's metals industry, and a boon to bridge builders searching for a rust-resistant alternative to steel.

The feasibility study is for constructing the university's proposed pedestrian bridge across the railroad tracks that bisect the UA campus entirely of titanium—a strong, lightweight, virtually corrosion-proof metal. No one has ever built an all-titanium bridge before, according to several metals experts, mainly because of cost concerns.

The Defense Metals Technology Center in North Canton is coordinating with the military solve metals-related technology problems critical to defense and national security. A high-profile venture demonstrating titanium's feasibility in commercial infrastructure projects, especially at a time when governments are having to replace aging steel structures such as Cleveland's corrosion-damaged Inner Belt span, could spark greater demand and open new markets for titanium More demand should spur competition and help drive down production costs.

11.12.9 The Virtual Design Tool (VD) for Visual Management

The virtual design tool (VD) for visual management of large complex projects is an extraordinary tool that construction professionals can employ to collaboratively manage all stages in the design and construction process according to Dimitri Mitchell of Black and Veatch of Kansas City, Missouri.

A walk-through software can read directly from CAD platforms and provide a compact and less hardware-intensive file size by reducing redundant tasks in the pre-construction phase. The usual project phases on a large project are:

1. Conceptual design
2. Preliminary design
3. Detailed design
4. Final design
5. Construction phase.

The objectives of VD are to offer a compressed schedule, alternative design and delivery methods, and reduced costs.

For the successful planning of a project, bridging the gap between design and construction is necessary. Engineering decisions shall be based on the following:

1. Visualization
2. Communication
3. Collaboration
4. Conflict resolution.

A matrix can be generated to embrace process and instrumentation diagrams for major and completed systems, generate individual specifications reports, QA/QC reports, coordinated asset management, renderings for architecture and site aspects, public outreach programs, 3-D walk-throughs of owner and operators, feedback reviews, constructability, safety and interferences studies, training, document management, developing databases for engineering and analysis tools, completion of bid documents, and final site plans and drawings.

Construction professionals can pinpoint potential problems well in advance of the construction phase, which can eliminate cost overruns. Costly misunderstandings and surprises can be avoided when the exact nature of a project is known to all stakeholders. Specialized software needs to be developed to address issues such as:

- Information on bill of materials
- Assets lists and information

- Design data
- Manufacturing details
- Operation and maintenance schedules.

11.12.10 Bridge Construction Using Waste Products

A new pedestrian suspension bridge built with materials generally considered as waste products has opened near the Rattlesnake National Recreation Area in Missoula, Montana. The structure is considered a showcase for innovation in the use of new construction materials, according to the U.S. Forest Service and the Montana Community Development Corp. (MCDC), which helped to secure funding for the bridge. The 90-foot-long, 8-foot-wide bridge was constructed with small-diameter "waste wood" and "waste plastic," as well as recycled tires.

It spans the Rattlesnake Creek and consists of lodge pole pine trees (killed by beetles in Idaho's Nez Perce National Forest) that were debarked and doweled to 6-inch-diameter trusses. The decking material is a fiber-plastic composite (wood flour and PVC plastic). Additionally, rubber mats constructed from recycled tires were laid atop the deck to protect the surface from horse traffic.

BIBLIOGRAPHY

1. Accelerated Bridge Construction

Accelerated Bridge Pier Construction in the U.S.: Seismic Implications.

Capers, Harry, *Rapid Bridge Construction in New Jersey. What Was In It For Us?*, New Jersey Department of Transportation Seminar, Trenton, NJ.

Capers, Harry, *Innovative Ideas for Designing New Jersey's Bridges*, New Jersey Department of Transportation Seminar, Trenton, NJ.

Capers, Harry, Seminar on Accelerated Bridge Construction, MCEER University of Buffalo, SUNY. "High Performance Steel Bridge Concepts," American Iron & Steel Institute, J. Muller International, November 1996.

Glass, G., J. Taylor, A. Roberts and N. Davison, 2003, "The Protective Effects of Electrochemical Treatment in Reinforced Concrete," NACE Paper No. 03291, *Corrosion*, 2003.

Invisible Bridge Construction, Fort Miller Co., 1999.

Khan, Mohiuddin Ali and Richard W. Dunne, "Recent Developments in ABC Technology," paper presented to FHWA Conference on Accelerated Bridge Construction, Baltimore, MD, 2008.

Lushko, Brian, "Accelerated Bridge Construction," *Structure*, August 2007.

"Precast Slab System Workshop Summary Report," Minnesota Department of Transportation/Federal Highway Administration, September 2005.

Prefabricated Bridges, American Association of State Highway and Transportation Officials, Technology Implementation Group, May 2002.

Prefabricated Bridge Elements and Systems, Federal Highway Administration, 2003.

Prefabricated Bridge Technology: Get in, Get Out and Stay Out, Federal Highway Administration, Focus, April 2003.

Ralls, Mary Lou, Ronald Medlock and Sharon Slagle, "Prefabricated Bridge National Implementation Initiative," 2002 Concrete Bridge Conference.

2. New Materials & Technology

Guide for Design-build Procurement, American Association of State Highway and Transportation Officials Item Code 10-DPB-1, 2008.

Guide Specifications for Polymer Concrete Bridge Deck Overlays, American Association of State Highway and Transportation Officials Task Force 34 Report.

Manual for Corrosion Protection of Concrete Components in Bridges, American Association of State Highway and Transportation Officials Task Force 32 Report.

Manual on Uniform Traffic Control Devices (MUTCD), 3rd Edition, American Association of State Highway and Transportation Officials, 2003.

The Use and State of the Practice of Fiber Reinforced Concrete, American Association of State Highway and Transportation Officials, AGC- ARTBA Joint Committee, Task Force 36.

3. Asset Management

Aktan, A. E., "The Integrated Asset Management Paradigm for Highway Transportation Infrastructure."

Asset Management Primer, Florida Highway Administration, 1999.

"Risk-Based Prioritization of Terrorist Threat Mitigation Measures on Bridges," *Journal of Bridge Engineering*, American Society of Civil Engineers, March/April 2007.

Byung, Hwan Oh, Lew Young and Young Cheol Choi, "Realistic Assessment for Safety and Service Life of Reinforced Concrete Decks in Girder Bridges," *Journal of Bridge Engineering*, American Society of Civil Engineers, July/August 2007.

Dewberry, S., "Easing the Way for E-Permitting," *Civil Engineering*, September 2000.

Selecting the Right Bridge Type, National Steel Bridge Alliance, Chapter 7.

Smith, S. "Data Collection—an Advanced Art," *CE News*, April 2002.

Wells, A. "Asset Management—Listen and Learn," *Bridge Design and Engineering*, Third Quarter 2001.

4. Bridge Security

Agrawal, Anil K., Y. Zhihua and Mohammed Ettouney, "Bridge Security and Blast Resistance Design," *American Society of Civil Engineers Seminar*, Phildelphia, PA.

Hinman, Eve "Protecting Infrastructure Against Explosive Attack," *Journal of Bridge Engineering*, American Society of Civil Engineers, July/August 2006.

Lombardi, Francis, "Methodology for Assessing Risk and Prioritizing Security Improvements for Bridges," *Second New York City Bridge Conference*, 2003.

5. Coating System

Attanayake, U., X. Liang, X., S. Ng and H. Aktan, "Penetrating Sealants for Concrete Bridge Decks—Selection Procedure," *Journal of Bridge Engineering*, American Society of Civil Engineers, September 2006.

Guide Specifications for Coating Systems with Inorganic Zinc-Rich Primer, American Association of State Highway and Transportation Officials and National Steel Bridge Association Steel Bridge Collaboration.

6. Constructability

"Design for Constructability," *Steel Bridge Design Handbook*, National Steel Bridge Association, Chapter 13.

"Erection Procedure Effects on Deformation & Stresses in a large radius horizontal curved I-Girder Bridge," *Journal of Bridge Engineering*, American Society of Civil Engineers, July/August 2007.

Schlafly, T. and J. Ericken, "Saying What You Mean—Weld Symbols," *Modern Steel Construction*, July 2007.

Welding Design Procedures and Inspection, U.S. Army Corps of Engineers, 2000.

7. Context Sensitive Design

Barratt, Stephen D., "New York's First Transparent Noise Barrier," *Journal of Bridge Engineering*, American Society of Civil Engineers, July/August 2006.

D'Agostino, Michael J., "Context Sensitive Design Precast Arch Bridges Boost Auto Research Park's Image," *Journal of Bridge Engineering*, American Society of Civil Engineers, July/August 2006.

8. Analysis and Design

Bridge Design Guidelines, Florida Department of Transportation, 2005.

Bridge Design Manual, New York State Department of Transportation, 2005.

Bridge Design Manual, Ohio Department of Transportation, 2005.

Design Manual Part 4, Volume 1, Structures, Pennsylvania Department of Transportation.

International Handbook of Earthquake Engineering: Codes, Programs and Examples (IHEE), Kluwer Academic Publishers, 1995.

Khan, Mohiuddin Ali and K. O. Kemp, "Elastic Full Composite Action in Slab-Beam Systems," *The Structural Engineer*, London, 1970.

Structural Standards Manual, Maryland State Highway Administration, 2006.

9. Inspection

Underwater Evaluation and Repair of Bridge Components, Federal Highway Administration, Demonstration Project, 1995.

Underwater Inspection of Bridges, Federal Highway Administration, Demonstration Project, 1989.

10. LCCA

Economic Analysis Primer, Federal Highway Administration, 2003.

Fiber-Reinforced Polymer, Federal Highway Administration, 2002.

Frangopol, D. and L. Neves, "Probabilistic LCA of Deteriorating Structures under Multiple Performance Constraints," American Society of Civil Engineers.

Frangopol D. and M. Liu, "LCCA for Highway Bridges: Accomplishments and Challenges," Federal Highway Administration.

Huang, K., Y. C. Shen and L. J. Feeser, "An Information System for Bridge Life-cycle Assessment."

Horvath, A., "Life Cycle Environmental Accounting and Decision Making for Bridge Structures," *Structures 2004*, American Society of Civil Engineers.

Life Cycle Cost Analysis Primer, Federal Highway Administration, 2002.

Life Cycle Cost Analysis, Utah Department Of Transportation, Engineering Services Division.

Second New York City Bridge Conference, 2003.

Long Span Structures

Khan, Mohiuddin Ali, "An Experimental Study of Flat Cable Nets," *Proceedings, First Kerensky International Conference on Tension Structures*, organized by Institute of Structural Engineers, American Society of Civil Engineers, and International Association for Bridge and Structural Engineering, London, 1988.

Khan, Mohiuddin Ali, "Factors Affecting Structural Behavior of Suspension Cables Nets," American Society of Civil Engineers/Structural Engineering Institute, *Proceedings of International Conference*, New York, NY, 2005.

Khan, Mohiuddin Ali, "Structural Applications of Suspension Cables and Cable Nets," American Society of Civil Engineers/Structural Engineering Institute, *Proceedings of International Conference*, Seattle, WA, 2003.

Khan, Mohiuddin Ali and H. Tottenham, "Elastic Non-linear Analysis of Prestressed Cable Networks Using Equivalent Continuum Theory," *Proceedings, International Association for Shell and Spatial Structures*, Pacific Symposium, Tokyo.

Khan, Mohiuddin Ali and H. Tottenham, "Static Behavior of Prestressed Cable Cooling Towers," *Proceedings, Second International Conference on Civil Engineering*, Pahlavi University, Sheraz, Iran, 1976.

LRFD Method

Khan, Mohiuddin Ali, Edwin Rossow and S. P. Shah, "Behaviour of Reinforced and Prestressed High Strength Concrete Beams," NED University, Karachi, *Journal of Engineering Research*, Vol. 1, No.

1, pp. 1–29, January 1994.

Khan Mohiuddin Ali, Edwin Rossow and S. P. Shah, "Shear Design of High Strength Concrete Beams," *Structures Congress*, American Society of Civil Engineers/Structural Engineering Institute, Philadelphia, PA, 2000.

Khan, Mohiuddin Ali and H. Nassef, American Society of Civil Engineers seminar, Philadelphia Section, Trenton, NJ, 2005.

Khan, Mohiuddin Ali and M. ElGaaly, New Jersey Department of Transportation seminar, Trenton, NJ, 2001.

LRFD Specifications for the Design of Bridges, American Association of State Highway and Transportation Officials, 1998, Washington, DC.

Rehm, Kelley C., "The Future of Bridge Design, States Implement Load and Resistance Factor Design Specifications to Meet an October Deadline," American Society of Civil Engineers, *Journal of Bridge Engineering*, January/February 2007.

"Strut & Tie Modeling of Disturbed Regions," *Prestressed Concrete Institute Bridge Design Manual*, Chapter 8.

Nondestructive Testing (NDT)

"ASCE, NDT Techniques Tried on George Washington Bridge," *Civil Engineering*, September 2000.

Cao, Honggang and Lim Malcolm, "Nondestructive Testing Methods Help Quickly Diagnose Hidden Problems and Assess Bridge Condition," American Society of Civil Engineers, *Journal of Bridge Engineering*, March/April 2007.

Itani, A., M. Bruneau, L. Carden and I. Buckle, "Seismic Behavior of Steel Girder Bridge Superstructures," *Journal of Bridge Engineering*, American Society of Civil Engineers, Volume 9, Issue 3, May/June 2004.

Ray, James C., S. Yehia, O. Abudayyeh, S. Nabulsi and I. Abdelqader, "Detection of Common Defects in Concrete Bridge Decks Using ND Evaluation Techniques," *Journal of Bridge Engineering*, American Society of Civil Engineers, March/April 2007.

New Materials

AISI-FHWA-Navy, Guide Specifications for Highway Bridge Fabrication with HPS 70W Steel, 2003.

Berke, N. "What's New in Corrosion Inhibitors," *Structure*, July/August 2003.

"Carbon Fiber Prepreg Sheets, Reinforcement and Repair of Concrete Structures," Mitsubishi Chemical Corporation, Tokyo, Japan.

"Dywidag Threadbar Reinforcing Systems," Dywidag-Systems International, Bolingbrook, Illinois.

Fiber-Reinforced Polymer—Priority, Market-Ready Technologies and Innovations, Federal Highway Administration, 2004.

Harry Capers, "High Performers! New Jersey's First High Performance Steel Bridges," *Second New York City Bridge Conference*, 2003.

Hemphill, S., "FRP Composite Decking For Bridge Rehab," *CE News*, August 2001.

High Performance Concrete, Federal Highway Administration, Resource Center, February 2006.

Karbhari, V. M. and F. Seible, "Design Considerations for Use of FRP Composites in the Rehabilitation of Concrete Structures," Division of Structural Engineering, UC, San Diego.

"NDT Tried on George Washington Bridge," *Journal of Bridge Engineering*, American Society of Civil Engineers, September 2000.

New Techniques

Agrawal, Anil, Mohiuddin Ali Khan and Y. Zhihua, *Handbook of Scour Countermeasures*, New Jersey Department of Transporation, Trenton, NJ, 2007.

Automated Column Wrapping Process Overview Manual, XSYS Technologies.

Coletti, D., M. Krueger and R. Renton, "Design Innovation, a Triple Effect," *Roads and Bridges*, August 2007.

"Evaluation of Reinforcing Bars in Old Reinforced Concrete Structures," Concrete Reinforcing Steel Institute, Engineering Data Report Number 48, 2001.

Fielding, Mathew, "High-Tech Instrumentation Helps Optimize Project Schedules and Manage Risks," *Journal of Bridge Engineering*, American Society of Civil Engineers, March/April 2007.

"FRP Finds Place in Composite Mix," *Bridge Builder*, 2001.

Khan, Mohiuddin Ali, Salim Baig and J. Zarriello, " Factors Affecting Stream Stability at Scour Critical Bridges in New Jersey," *First International Conference on Bridge Scour*, Texas A&M University, College Station, Texas, 2002.

Repair and Rehabilitation

Bane, R., "Strengthening Masonry Arch Bridges," *Scanner*, American Society of Highway Engineers, Summer 2007.

Bowman, M., "Brittle Fracture of the Blue River Bridge," *Structures,* 2004, American Society of Civil Engineers.

Dexter, R., K. Johns and R. DiBartolo, "Smart Sign Support," *Civil Engineering*, September 2000.

Dunker, K. and L. Dajin, "Foundations for Integral Abutments," American Society of Civil Engineers, Practical Periodical on Structural Design and Construction, February 2007.

Park, Sung H., *Superstructure Rehabilitation*, Trenton, NJ, 2004.

Steel Bridge Design Handbook, Chapter 23, National Steel Bridge Allliance.

Structural Design Bureau, Movable Bridges, Operating Complex Systems, New Jersey Department of Transportation, Trenton, NJ.

Umphrey, Josh, Derek Beck, G. Ed Ramey and Mary L. Hughes, "Rapid Replacement of Four GDOT Bridge Decks," Practical Periodical on Structural Design and Construction, February 2007.

Zhou, Y. E., "Fatigue Problems in Steel Bridge Structures," *Structures*, 2004, American Society of Civil Engineers.

Structural Health Monitoring

Agrawal, A., Y. Zhihua and Mohiuddin Ali Khan, *Handbook of Scour Countermeasures*, Trenton, NJ, 2007.

Akhradji, Tarek, "FRP: New Life for Bridges and Budgets," *Structure*, March 2002.

Aref, Amjad J. and Sreenivas Alampalli, "Dynamic Behavior and Damage Detection Of A Fiber Reinforced Plastic Bridge Superstructure," *American Society of Civil Engineers Proceedings*, 2004.

Bashir, Tariq M., NYSDOT and G. Lozano, HNTB, "Roll-In Roll-Out Reduces Construction Impact To Local Traffic," *SPAN.*

Blume, John A., Earthquake Engineering Center, Stanford University.

"Bridge Rehabilitation Laser Scanning Sample Projects," Meridian Associates, 2007.

Construction Handbook for Bridge Temporary Works, American Association of State Highway and Transportation Officials, 1995.

Corrosion Protection of Steel Bridges, Chapter 23, National Steel Bridge Alliance.

Detailed Planning Research on Accelerating the Renewal of America's Highway, National Cooperative Highway Research Program, 2003.

Earls, C. J., B. J. Shah and S. J. Thomas, "Compactness and Bracing Requirements for High Performance Steel Girders," *American Society of Civil Engineers Proceedings*, 2004.

Field Manual for Bridge Painting Inspection, FHWA-RD-98-084, Federal Highway Administration.

Frank, W. J., M. Hariharan and J. Wood, "Precast Concrete—A Unique Application," *Structure*, July 2008.

Gehrig, M. D. and J. T. Bryant, "Ground Penetrating Radar for Concrete Evaluation Studies," Bryant Consultants, Carrolton, Texas.

Gold, W. J. and G. J. Blaszak, "Strengthening Concrete Using FRP," *Structural Engineer*, February 2000.

Gu, Xianglin, Zhongyu Chen and Farhad Ansari, "Fiber Optic Distributed Crack Sensor for Concrete

Structures," *Proceedings of Structures Congress 2000,* American Society of Civil Engineers.

Guide Specifications for Design of Pedestrian Bridges, American Association of State Highway and Transportation Officials.

Guide Specifications for Fatigue Evaluation of Existing Steel Bridges, American Association of State Highway and Transportation Officials, 1990.

György L. Balázs and Adorján Borosnyói, "Prestressing with CFRP Tendons," *Proceedings of the International Conference, High Performance Materials in Bridges*, United Engineering Foundation, Inc., July 29–August 3, 2001, Kona, Hawaii.

Harrison, Richard R. and Tarek Alkhrdaji, "Comparison of Four Techniques for Strengthening Concrete Beams Using Performance and Practicality Criteria," *Proceedings of Structures Congress 2006,* American Society of Civil Engineers.

Heat Straightening Repairs of Damaged Steel Bridges, A Technical Guide and Manual of Practice, Federal Highway Administration.

Huston, D. "Structural Sensing, Health Monitoring and Performance Evaluation," *CRC Press*, 2008.

Jakovich, Gary and Vasant Mistry, "National Perspective on Accelerated Bridge Construction," *24th International Bridge Conference*, Pittsburgh, PA, 2007.

Keating, Scott, Thomas N. Claytor, Anthony Davis and Matthew Sheats, "Technological Advancements for High-Speed Industrial Computed Tomography," HYTEC, Incorporated.

Kibriya, Tahir, "Performance of Blended Cement in High Strength Self Compacting Concrete," *Proceedings of Structures Congress 2006*, American Society of Civil Engineers.

Laman, Jeffrey A., Karl M. Reichard and Andrew J. Doller, "Long Gage-Length Interferometric Optical Sensors for Condition Assessment in Bridge Structures," *Proceedings of Structures Congress 2000,* American Society of Civil Engineers.

Los Alamos National Laboratory, ESA-MT, MS.

LRFD Bridge Construction Specifications, 2nd Edition, American Association of State Highway and Transportation Officials, 2004.

LRFD Bridge Design Specifications, 3rd Edition, American Association of State Highway and Transportation Officials, 2004.

Structural Standards Manual, Maryland Department of Transportation, State Highway Administration, Office of Bridge Development, March 2006.

Manual on Use of Self-Propelled Modular Transporters to Remove and Replace Bridges, Federal Highway Administration.

Myers, John J., "A Worldwide Survey of Environmental Reduction Factors for Fiber Reinforced Polymers (FRP)," Department of Civil Engineering, Penn State University, University Park, Pennsylvania.

National Transportation Safety Board, Washington, DC, 20594 Safety Recommendation, (I-35W) highway bridge over the Mississippi River in north Minneapolis, Minnesota, January 2008.

National University of Sciences and Technology, Rawalpindi, Pakistan, *Proceedings of American Society of Civil Engineers 2006.*

PCI Bridge Design Manual, Chapter 15, "Seismic Design," Prestressed Concrete Institute."

Bridge Design Manual, (D.M. 4), Pennsylvania Department of Transportation, Harrisburg, PA.

Precast Concrete Solutions, Fort Miller Company, Schuylerville, New York.

Rahrig, Philip G., "Galvanizing Plate Girders," *Modern Steel Construction*, August 2002.

Rajabipour, Farshad and Jason Weiss, "Linking Health Monitoring in Concrete Structures with Durability Performance Simulations," *Structures*, 2006, American Society of Civil Engineers.

References on the Use of Corrosion Inhibitors in Concrete, Concrete Corrosion Inhibitors Association, Potomac, Maryland.

Rong, Alex, "Structural Analysis of an Ancient Arch Bridge using Modern Technology," *Structures Congress,* American Society of Civil Engineers/Structural Engineering Institute.

Shenton, III, Harry W., Michael J. Chajes, William L. Johnson, Dennis R. Mertz and Jack W. Gillespie, "Continuous, Long-Term Monitoring of Two Advanced Polymer Composite Bridges," *American Society of Civil Engineers Proceedings*, 2004.

Smith, J. W. and S. Bright, "Upgrading Orthotropic Bridge Decks with Fiber Reinforced Composites," *Proceedings of the International Conference, High Performance Materials in Bridges*, United Engineering Foundation, Inc., July 29–August 3, 2001, Kona, Hawaii.

Smith, Z. I., "FRP Strengthening of Beams and Slabs," *Structure*, March 2004.

Steel Bridge Design Handbook, "Design for Constructability," Chapter 13, National Steel Bridge Alliance, 2007.

Straser, Erik, Hoon Sohn, Anne Kiremidjian, and Kincho H. Law.

Teal, Roy, "Fabrication with High Performance Steel Grade HPS 70W," Metals Consultant, Averill Park, NY.

Voo, Jackie, Stephen J. Foster, R. Ian Gilbert and N. Gowripalan, "Design of Disturbed Regions in Reactive Powder Concrete Bridge Girders," *Proceedings of the International Conference, High Performance Materials in Bridges*, United Engineering Foundation, Inc., July 29–August 3, 2001, Kona, Hawaii.

Painting

Cao, H. and M. Lim, "Bridge Check ups," *Bridges*, March/April 2007.

Hemphill, Scott, "Economical Bridge Rehabilitation Using Composites," Construction Innovation Forum, Bloomfield Hills, MI.

"Improving Asset Management, Remote Health Monitoring of the Hawks Falls Bridge," *Bridges*, November/December 2006.

"Latest in Coatings Technology, Long-life Fluoro-Polymer Resins for Bridge Coatings," *Bridges*, September/October 2006.

Steel Bridge Design Manual, Chapter 23, "Corrosion Protection of Steel Bridges," National Steel Bridge Alliance.

12

Protection of Bridges against Extreme Events

Section 3 Repair and Retrofit Methods

12.1 EXTREME EVENT DAMAGE FROM FLOOD SCOUR

12.1.1 Introduction

In this chapter, state-of-the-art scour countermeasures are introduced. A spreadsheet for scour analysis and riprap countermeasures design is available in the Appendix.

According to AASHTO LRFD Specifications (Section C3.7.5) *scour is the most common reason for the failure of highway bridges in the United States.* Scour is the result of the erosive action of running water, excavating and carrying away material from stream beds and banks. Small brooks, streams, rivers and oceans all possess different degrees of kinetic energy. Scour or soil erosion at a bridge is caused by dynamic effects of water in motion. In the U.S., over 36,000 bridges are either scour critical or scour susceptible.

As described in Chapter 3, some examples of recent bridge failures in teh U.S. include:

- Schoharie Creek bridge located on New York State thruway in 1987
- U.S. 51 bridge over Hatchie River in Tennessee in 1989
- Damages to bridges located on the Mississippi River in 1993
- Interstate 5 NB and SB bridges over Los Gatos Creek in California in 1995
- Route 46 bridge on Peckman's River in Passaic County, New Jersey in 1998
- Ovilla Road Bridge located in Ellis County, Texas in 2004.

Applications for FHWA circulars HEC-18 and HEC-23 are discussed. Since scour is a major problem for bridges located on waterways, familiarity with the methods presented will benefit the engineer in terms of safety, economical design of foundations, and solving constructability issues.

The author carried out research on the subject for a joint publication with Anil Agrawal for developing a "Handbook for Scour Countermeasures." The project was sponsored by the New Jersey DOT Bureau of Research and was carried out under the direction of Project Manager Dr. Nazhat Aboo bakr. The detailed report is now available on their Website for general use. For scour analysis methods refer to the NJDOT handbook:

http://www.state.nj.us/transportation/refdata/research/reports/FHWA-NJ-2005-027.pdf

The latest ideas, ingenuity, and contributions from individual researchers are acknowledged and such publications are listed at the end of chapter.

12.1.2 Codes and Design Guidelines

The following FHWA and AASHTO publications serve as major resources for scour analysis and design:

1. HEC-18 "Evaluating Scour at Bridges"
2. HEC-20 "Stream Stability at Highway Structures"
3. HEC-23 "Countermeasures"
4. HEC-25 "Tidal Scour will be Used"
5. AASHTO LRFD "Specifications for Design of Bridges"
6. AASHTO "Model Drainage Manual"
7. MD, NJ, PA, FL and other state codes
8. NCHRP
9. CIRIA (British code)

12.1.3 Design Floods for Bridge Scour

The aim should be to design bridges for all times and for all occasions. AASHTO (LRFD) load combinations for extreme conditions are applicable. The extreme-event limit states relate to flood events with return periods (usually 100 years) in excess of the design life of the bridge (usually 75 years). Foundations of new bridges, bridges to be widened, or bridges to be replaced shall be designed to resist scour for a 100-year flood criteria or even more, which may create the deepest scour at foundations.

12.1.4 AASHTO (LRFD) Load Combinations for Extreme Conditions

There are generally two limit states that may involve consideration of the effects of scour:

1. A flood event exceeding a 100-year flood (the check flood for scour or superflood is used to evaluate scour for this event, a 500-year flood is recommended for the check flood for scour).
2. A vessel collision with the bridge.
3. In addition to the above, there are other conditions relating to scour that the designer may determine are significant for a specified watershed, such as ice loads or debris logging operations, etc. Use load combination "Extreme Event II" as follows:

$$\text{(Permanent Loads)} + WA + FR + CV$$

With all load factors equal to 1.0, nonlinear structural effects must be included and can be significant. It is anticipated that the entire substructure (including piles) may have to be replaced and the superstructure repaired if a bridge is subjected to this design impact load; however, the superstructure need not collapse.

Scour with vessel collision: Substructures must be designed for an extreme vessel collision load by a ship or barge simultaneous with scour. Design the substructure to withstand the following two load/scour (LS) combinations:

Figure 12.1 Flood at Peckman's River bridge on Route 46 in NJ.

1. Load/scour combination
   LS (1) = Vessel collision @ ½ Long-term scour
   Where vessel collision is assumed to occur at normal operating speed.
2. Load/scour combination
   LS (2) = Minimum impact vessel @ ½ 100-year scour
   Where minimum impact vessel is as defined in LRFD with related collision speed.

Required substructure limit states:

1. Limit state 1 (Always required, scour may be zero)
   Conventional LRFD loadings (using load factor combination groups as specified in LRFD Table 3.4.1-1), but utilizing the most severe case of scour up to and including that from a 100-year flood event.
2. Limit state 2 (Applies only if vessel collision force is specified)
   Extreme event of vessel impact (using load factor combination groups as specified in the LRFD) utilizing scour depths for vessel collision.
3. Limit state 3 (Applies only if scour is predicted)
   Stability check during the superflood (most severe case of scour up to and including that from the 500-year flood event).

$$\gamma_p (DC) + \gamma_p (DW) + \gamma_p (EH) + 0.5(L) + 0.5 (EL) + 1.0 (WA) + 1.0 (FR) \qquad (12.1)$$

Where, L = LL + IM + CE + BR + PL, (All terms as per AASHTO LRFD specifications)

The bridge may be inundated at the stage of the design flood for bridge scour.

Check flood for bridge scour: The foundation design shall be checked for a 500-year flood check, or 1.7 times a 100-year flood, if 500-year flood information is not available.

4. Size of opening or degree of obstruction from abutment, pier, and foundations will influence the velocity of water.
5. Catchment area of river, its source of supply, demagogue, storms and melting of glaciers will influence the volume of discharge and erosion.

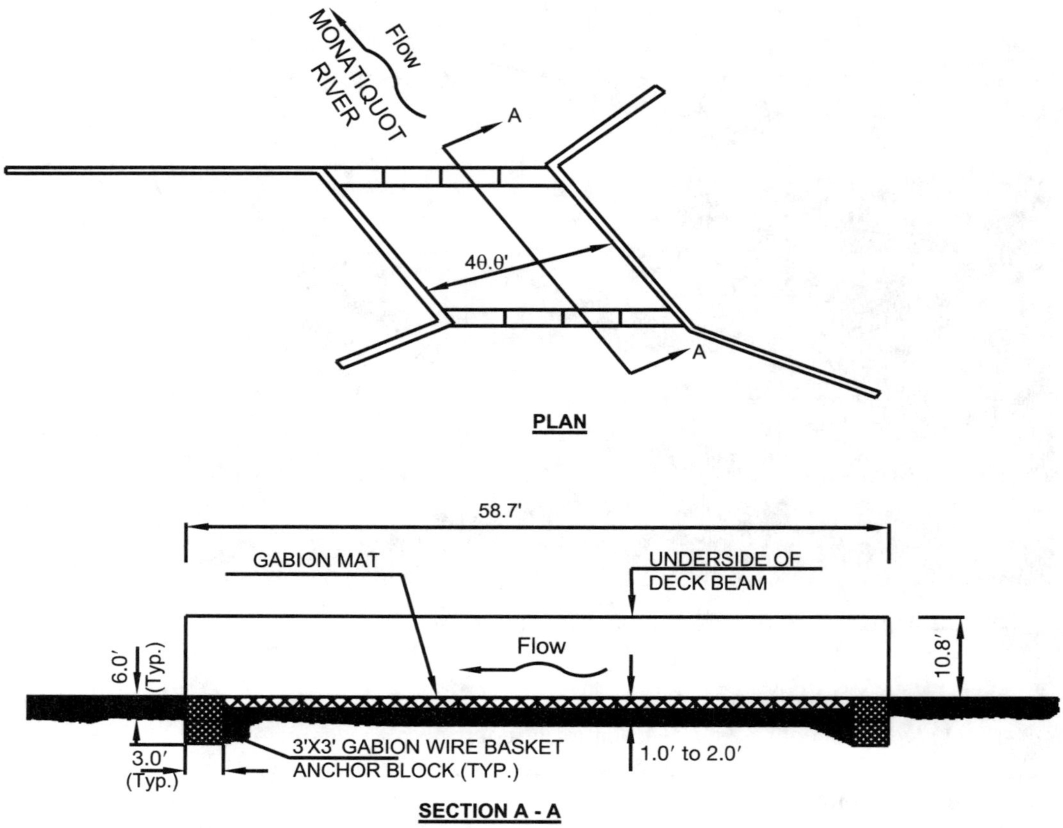

Figure 12.2 Use of Gabion baskets.

12.3 CONDITION OF EXISTING BRIDGES ON RIVERS

12.3.1 Condition Assessment

Refer to the U.S. Department of the Interior, U.S. Geological Survey Guidelines. Requirements for performing scour vulnerability analysis are also given in USGS procedures for scour assessment at bridges in Pennsylvania. Bed material data are required.

Collection and processing of geomorphic, hydrologic, and hydraulic data for assessment of scour at bridges requires borehole information for soil characteristics. Both boring information and grain size analysis are needed for accurate determination of scour.

A scour assessment rating is based on the type of bed material. For example, rock can be classified as erodible, highly erodible, and non-erodible. Observed scour rating can be computed using a computer program.

12.3.2 In-Depth Bridge Inspections of Substructures

NBIS in-depth substructure inspections of the substructure elements need to be peformed. The purpose of the inspections is to identify levels and areas of deterioration of all structural and non-structural substructure elements in order to develop repair recommendations and details.

This effort also includes correlating probing measurements taken near the pier edges with the previous substructure inspections. The FHWA has adopted three diving inspection intensity levels as follows:

Level I: Visual, tactile inspection (100 percent "swim-by" at arm's length).

Level II: Detailed inspection with partial cleaning (100 percent "swim-by" with 10 percent cleaning).

FHWA Level II: Requires that portions of the structure be cleaned of marine growth to identify possible damaged and/or deteriorated areas that may be hidden by surface growth. The cleaning must be performed on at least 10 percent of all underwater elements.

The equipment used to inspect the majority of the bridges consists of a small boat, sounding rod, hand tools, and line tended SCUBA.

Level III diving inspections: Highly detailed inspection with nondestructive testing (highly detailed inspection of a critical structural element where extensive repair is contemplated).

12.3.3 General Procedures for Condition Evaluation

1. The data sought in the literature study includes:
    - Procedures followed in the computation of depth of scour.
    - Models utilized in the computation of the depth of scour.
    - Field observation data on depth of scour of existing structures.
    - Variations used on standard models, such as HEC 18 in computing the depth of scour.
    - Success and failures of models utilized in analysis.
    - Physical parameters used in analysis of the depth of scour.
    - Influence of existing protective measures used to control depth of scour on the computed depth of scour.
2. The first part of the literature search shall discuss experience in bridge scour depths and locations. The second part shall discuss the analytical process to determine various types of scour components.
    - Determination of type of scour
    - Scour depth and scour holes
    - Effects of overtopping
3. Selection and design of scour countermeasures, incuding:
    - Methods for existing bridges
    - Methods for new bridges
    - Methods for unknown foundations.

Alternative evaluation: Based on condition assessment findings, potential repair/remediation recommendations will be developed. Fluctuating river elevations will be taken into account when developing repair recommendations and reviewing the construction feasibility. Evaluation shall include, but is not limited to:

1. Evaluation of constructability.
2. Construction staging.
3. Community impacts during construction: emergency vehicle response, tourist industry, traffic delays, impacts to pedestrians and bicyclists, impacts to local business, and noise.
4. Construction cost estimates shall be developed for each feasible alternative, along with the anticipated construction schedule.
5. Maintenance of traffic schemes for each feasible alternative.
6. Right-of-way requirements.
7. Environmental impact to the waterway and endangered species.

12.3.4 Examples of Findings from Underwater Inspection Reports

Identify defects and perform repair based on the underwater inspection: Based on underwater inspection reports, the following typical defects were found to exist for scour critical bridges:

Table 12.1 Case studies and underwater inspection results.

Item	Remedy	Alternate
Spalls in abutment concrete	Surface repairs with approved patch material	
Delaminations and cracks in substructure concrete above water	Pressure grouting	
Voids in breast wall	Epoxy grouting	Nonshrink grout
Debris accumulation	Cleaning debris	
Mortar loss in masonry joints	Repointing mortar	
Undermining	Plug with concrete	Use grout bags
Broken stone masonry	Place stone and fill with mortar	
Cracks in tremie concrete below water	Pressure inject masonry cracks	
Spalling in foundation	Pressure grouting	
Missing or broken riprap	Replace by R-8 size stone	
Cracks in apron around piers	Repair concrete apron	Place riprap around piers
Silting	Dredging	
Advance section loss of structural members	Strengthening member	Underpinning

12.3.5 Condition Rating

Overall condition of bridge substructures can be assigned the following NBIS condition rating:

- Very good condition—no problems noted.
- Good condition—some minor problems.
- Satisfactory condition—structural elements show some minor deterioration.
- Fair condition—all primary elements are sound, but may have minor section loss, cracking or spalling.
- Poor condition—advance section loss, deterioration or spalling of primary structural elements.

12.3.6 Concept Study Report and Plans

Prepare a draft concept study report, plans, and recommendations to provide a concise aggregation of the important elements of the condition evaluation and an overall synthesis of conclusions and recommendations. The draft concept study report shall include, but is not limited to:

- Condition assessment
- Scour vulnerability assessment
- Discussion of alternatives including a discussion of potential impacts for each alternative
- Recommended repairs/remediation, including supporting justification
- Requirements for design including required survey, geotechnical evaluation, and hydraulic analysis
- A list of applicable permits, including the permit cost and the anticipated duration to obtain the permit in the light of any requirements, including a discussion of required temporary construction easements, if any

- MPT requirements including sketches and/or conceptual plans for construction staging and detours
- Construction costs
- Anticipated construction schedule
- Plans with details to a level sufficient to clearly demonstrate the repair/remediation type and location along with construction staging and construction access.

The next step is value engineering/constructability review. After the draft concept study report is complete, a one- or two-day workshop may be held with team members to review the concept study options. The goal of the workshop is to review comments to advance a preferred design concept.

12.4 PREVENTIVE ACTION AGAINST BANKS AND FOUNDATION SCOUR

12.4.1 Pre-Requisites to Selection of Countermeasures

Selection and design of CMs should be based on hydraulic and scour analyses, geology of the area, and site-specific information, such as the importance and the remaining life of the bridge. A designer must apply engineering judgment in examining the results obtained from scour and hydrologic and hydraulic data. Hydrologic and hydraulic data should include:

1. Performance of the structure during past floods.
2. Effects of regulation and control of flood discharges.
3. Hydrologic characteristics and flood history of the stream and similar streams.
4. Whether the bridge is structurally continuous.

Factors which affect the detailed design of a countermeasure are:

1. Natural issues, such as soil geology.
2. Physical factors, such as width of the bridge opening and traffic volume.
3. Economic considerations such as existing condition of the bridge and the life of the proposed countermeasure compared to the remaining life of the bridge.
4. Available resources for monitoring frequency and underwater inspection.
5. Funding priority for repairs from flood damage and providing adequate countermeasures.

12.4.2 Design of Countermeasures Based on HEC-23 Procedures

Performing scour analysis and use of effective countermeasures such as deep foundations and river training may be required in addition to conventional shielding. Scour countermeasure options include river training measures, dredging, driving sheet piles around the piers and abutments, and filling the gaps with riprap. Alternate measures such as grout bags, cable-tied blocks, and AJAX type blocks will be considered, based on tidal flow conditions.

Wide cracks at piers need to be pressure grouted with epoxy grout, as recommended in the inspection report. Design of required repairs and retrofits to the substructure will consider:

- Modern countermeasures technology requires new construction techniques. The contractor performing such tasks will need properly trained construction crews.
- Providing cofferdams, sheet piling, and bed armoring would require temporary construction works. Alternatives for using quick construction need to be considered.
- The available flow width may be reduced due to construction of cofferdams.
- Velocities through the remaining opening will increase, thereby increasing scour in the channel and around the structure.
- Any pollution of rivers from construction material needs to be monitored. Regular cleaning of the channel may be required.

- Approvals for stream encroachment permits would be necessary.
- The effect of driving sheeting or bed armoring on existing utilities needs evaluation.
- Countermeasures may extend into adjacent property limits. Any construction easement needs to be carefully evaluated and permits obtained.
- Tidal conditions will affect working methods and working hours for construction.
- The health and safety of construction personnel is a concern due to water depth.

12.4.3 Bridge Scour Countermeasures: Categorized by Scour Type

Table 12.2 Type of scour and objectives of each countermeasure.

Scour Type	Countermeasures	Examples	Objectives of Countermeasure
Lateral erosion	Armoring devices (revetment)	Riprap, gabions, cable-tied blocks, tetrapods, precast concrete blocks, used tire, etc. Vegetation planting	Prevention of erosion to the channel bank in the vicinity of the bridge; stabilization of the channel alignment
	-do-	Timber piles, sheet piles jack or tetrahedron fields. Vegetation planting	Reduction of flow velocity near channel bank and inducement of deposition of sediment
	Groynes, Hardpoints	Groynes, spurs, dykes	Reduction of flow velocity near channel bank and inducement of deposition of sediment; stabilization of channel alignment
Degradation	Check dams		Control of channel grade
	Channel lining	Concrete or bituminous concrete pavement	Control of channel degradation
	Bridge modification	Increase of bridge opening width	
Aggradation	Channel improvement	Dredging, clearing of channel Formation of a cut-off wall	Increased sediment transport to reduce sediment deposition at bridge crossing
	Controlled mining		Reducing sediment input at bridge site
	Debris basin		Reducing sediment input at bridge site
Local sour	Armoring devices	Riprap, gabions, cable-tied blocks, etc.	Reduced local scour
	Flow altering devices	Sacrificial piles, deflector vanes, collars	Reduced local scour at piers
	Underpinning of bridge piers		Reduced local scour at piers
	Guide banks		Improved flow alignment at bridge crossing; reduction in local scour at abutments

12.4.4 Description of Substructure Repairs

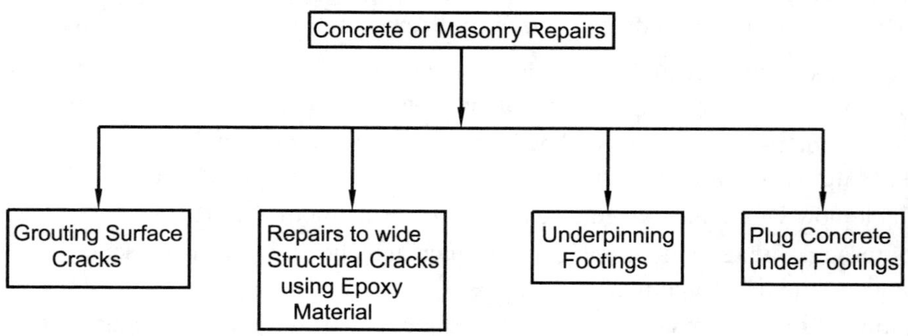

Figure 12.3 Substructure repairs prior to installing structural countermeasures.

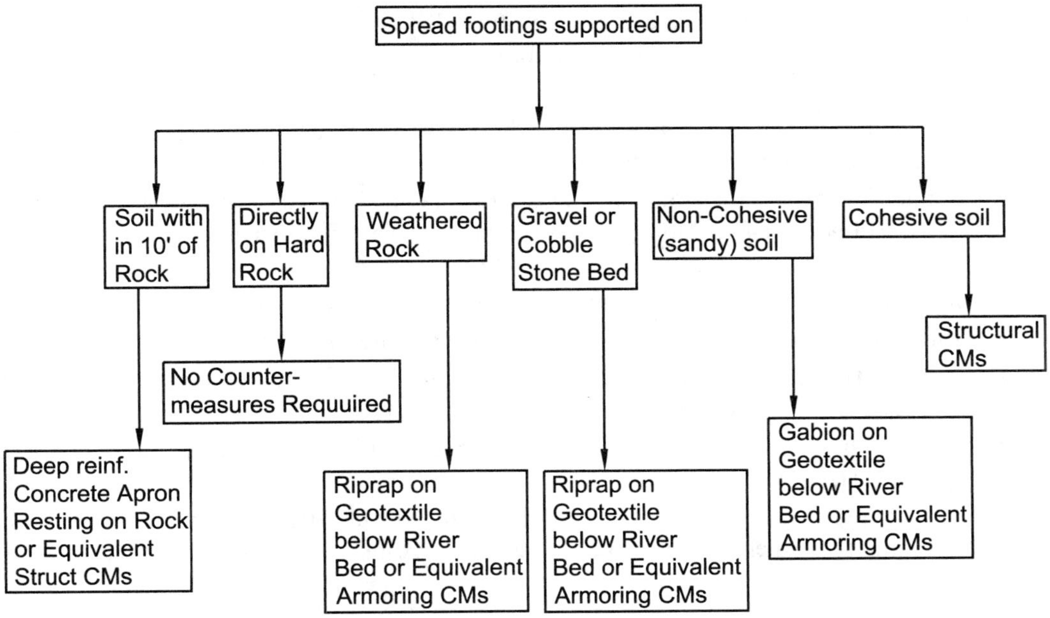

Figure 12.4 Selection of countermeasures (CMs) based on soil types.

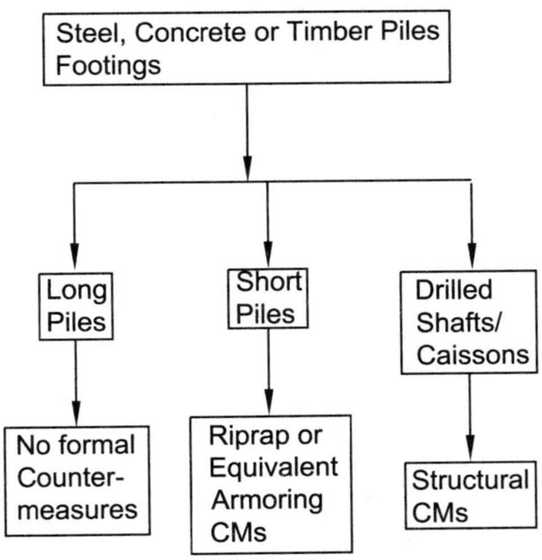

Figure 12.5 Deep foundation types.

12.4.5 Countermeasure Evaluation

1. Bridge footings: Countermeasures are required for scour critical bridges. Ideally, a recommended scour countermeasure will permanently eliminate a bridge's potential vulnerability to scour damage for the peak floods. A state permit is required to install armoring.
   - Installation of riprap, steel sheeting, or a cast-in-place concrete paving to armor the exposed faces of existing footing may be investigated to arrive at the most cost effective method. However, due to confined space under the bridge, driving steel sheeting would not be practical.
   - Constructing a cast-in-place apron wall requires building a temporary cofferdam at times when water elevation is low and when de-watering operations would be least expensive.
   - Countermeasures will be followed by a monitoring program, which includes scour mea-

surements. Scour measurements may be recorded and analyzed following a two-year inspection cycle, with soundings for all bridges.

- Periodic inspections after major floods or coastal storm surge.

2. Channel and banks: Channel improvements by means of channel lining to prevent degradation and armoring of approach banks by revetment materials are recommended.

- Common materials used for shielding of foundations such as standard riprap, stones enclosed in wire nets, gabion mattresses, and artificial riprap such as concrete bags or toskanes are recommended.
- Dredging to increase waterway width and to control aggradation is recommended.
- Filling up of scour holes by dense material, heavy stones or concrete blocks.

3. Detailing of riprap. The author developed the details fiven in Figures 12.6 thru 12.10 for NJDOT Bridge Design Manual.

- Riprap grading—Designating 50 percent of stones in a layer, to be equal or greater than a specified size (D_{50}).
- The specified size can be calculated by hydraulic considerations using FHWA formula. Remaining 50 percent stones can be of smaller size than (D_{50}) to fill the smaller voids between the stones.
- Maximum stone size in a layer $< 1.5\ D_{50}$.
- Minimum thickness of each layer = 300 mm
- Minimum number of layers = 3
- Place riprap around footings at a maximum slope 1 (horizontal): 1 (vertical), with the slope starting at a distance of 450 mm from vertical face of footing
- The top of riprap shall be below the river bed to avoid encroachment of river, or since stones protruding above the bed are likely to be dislodged by floating debris, ice, or currents.

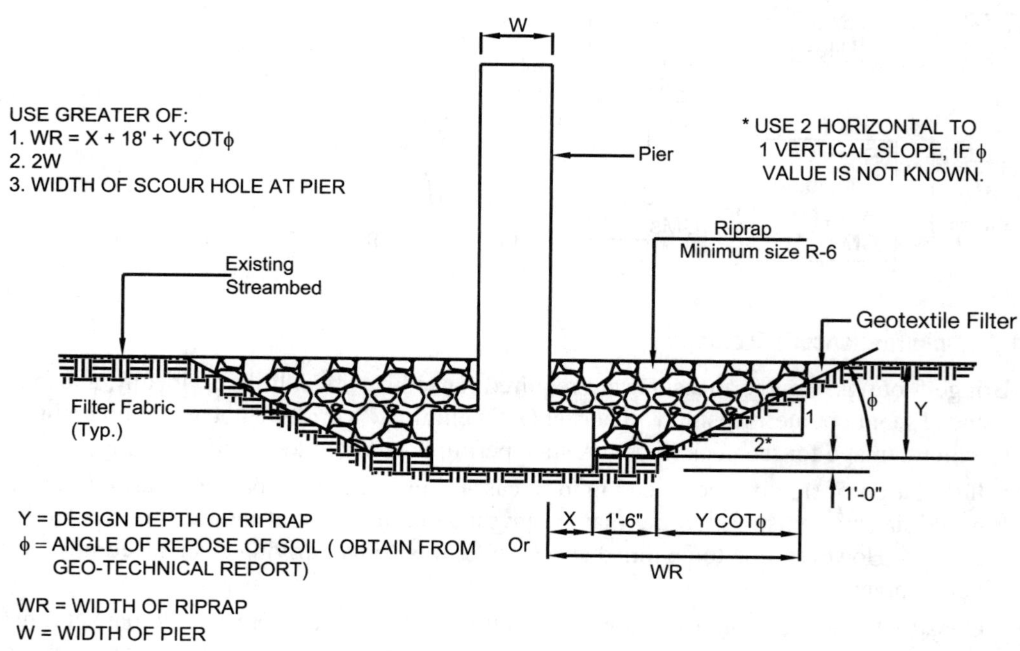

Figure 12.6 Pier cross-section showing riprap details at piers.

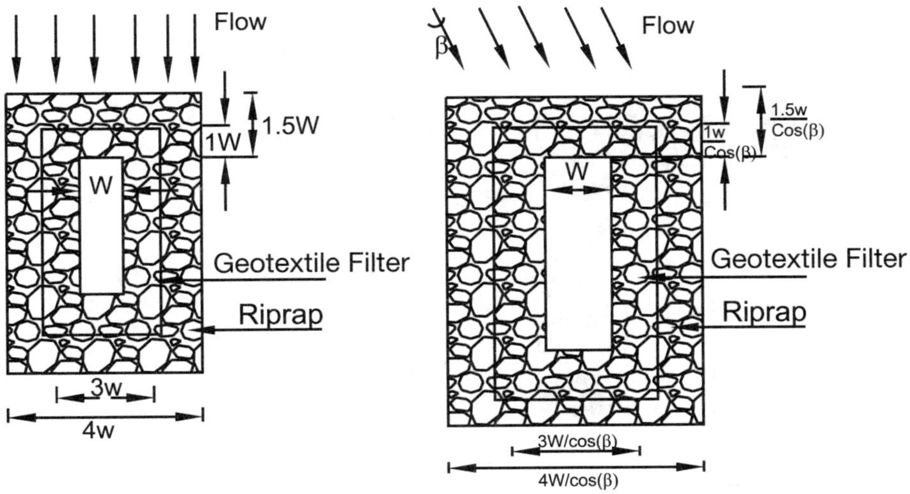

Figure 12.7 Extent of riprap/geotextile filter at pier footings. β = angle of attack.

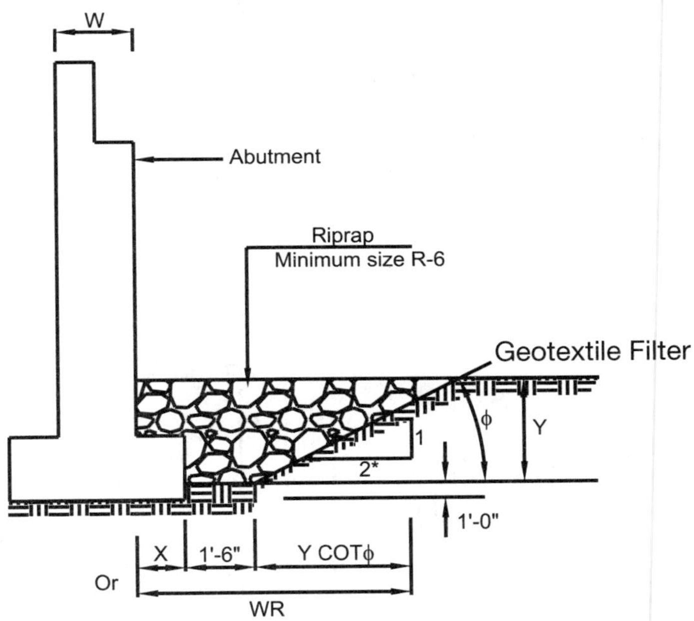

Y = DESIGN DEPTH OF RIPRAP
φ = ANGLE OF REPOSE OF SOIL (OBTAIN FROM GEO-TECHNICAL REPORT)

WR = WIDTH OF RIPRAP
W = WIDTH OF PIER

USE GREATER OF:
1. WR = X + 18' + YCOTφ
2. 2W
3. WIDTH OF SCOUR HOLE AT PIER

* USE 2 HORIZONTAL TO 1 VERTICAL SLOPE, IF φ VALUE IS NOT KNOWN.

Figure 12.8 Riprap details at vertical wall abutments.

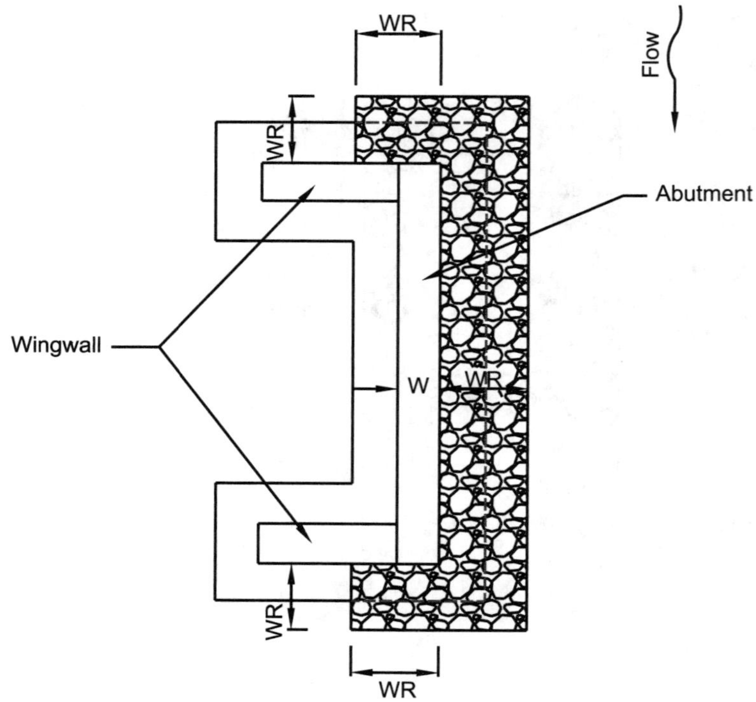

Figure 12.9 Plan of riprap at abutment, wingwalls at 90 degrees.

4. Other agency coordination and permitting

The engineer will be required to contact and meet with representatives of federal, state, local, municipal, and other agencies to review and determine all necessary project requirements and permits.

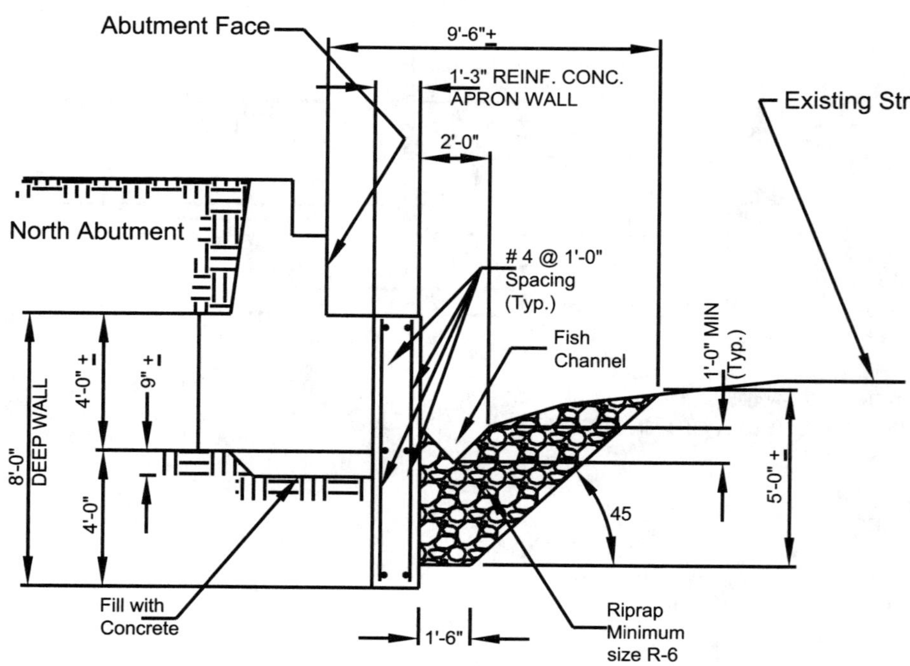

Figure 12.10 Example of concrete apron wall, grouting scour holes with amoring (N.T.S.).

The engineer is required to obtain all applicable permits for the proposed work. Other agencies may include, but are not limited to:

- Army Corp. of Engineers (ACOE)
- State DOT
- State Department of Environmental Protection
- State Department of Conservation and Natural Resources
- Local soil conservation and sediment control
- National Park Service.

Detailed constructability issues: Scour depth should be measured from a reference line 1 ft above the top of footing. If eroded elevation is located at a higher elevation than 1 ft above the top of footing, the higher elevation will be considered. For placing riprap, excavation to the design depth should be carried out. The depth of riprap should be at least below the contraction scour depth. If considerable erosion has already taken place and the riverbed elevation is below the top of the footing, hydraulic analysis should be based on the new channel profile by considering the new opening size.

The same type of armoring countermeasures should be used for abutments and piers for economy and ease of construction. Armoring should not be mixed; i.e., if gabions are selected, then they should be used for the whole bridge site. However, armoring can be combined with a structural CM or river training.

The types of countermeasures to be used will depend upon the bed materials under the bridge.

12.4.6 Structural Countermeasures

For existing bridges that are scour critical, the following types of structural countermeasures will be investigated:

- Foundation strengthening by structural repairs and pumping concrete under the footing
- Driving sheet piling in front of abutment and pier as a shield. Sheet pile can be cut at streambed level
- Constructing a reinforced concrete apron wall attached to exposed face of footing
- Underpinning of foundation
- Providing a curtain wall in front of pile caps
- Providing timber fenders around piers for damage against ice and ship collision
- Sacrificial piles to reduce velocities
- Pier shape modification if required.

12.4.7 Hydraulic Countermeasures

Some of the common types of countermeasures include riprap, gabion baskets, concrete blocks, and sheet piling. Harnessing the river to reduce flood velocities may also be used. Armoring/shield countermeasures such as riprap on geotextile filters, gabion baskets, and articulated or cable-tied blocks on geotextile filters will be considered as alternates.

A filter is required unless the riprap or armor lining has a thickness of at least three times the D_{50} size of the riprap. Detailed design guidelines for geotextiles are available in the following publication: FHWA Publication HI-95-038 by Holtz D.H., Christopher B.R. and Berg R.R., 1995 "Geosynthetic Design and Construction Guidelines", FHWA, Washington D.C.

Flowchart for the design of riprap at bridge piers:

$$D_{50} = \text{Median riprap diameter}$$

12.4.8 Minimum Countermeasure Requirements

If the projected (computed) scour is small or negligible, theoretically a design of a formal countermeasure will not be required. Such cases are:

1. When a spread footing is located or placed on bed rock or when a spread footing is located or placed below the total scour depth.
2. When an additional pile length equal to the projected scour depth is provided.
3. When pile stiffness exceeds the minimum required and the exposed length of pile due to erosion can safely act as a long column.

   Although a minor surface erosion of soil occurrence will not cause a danger to footings, a soil cover or protection to the concrete footing or piles is still required. An adequate soil cover needs to be maintained for:
   - Frost resistance (minimum frost depth requirement)
   - As-built cosmetic appearance
   - Unforeseen error in the scour analysis data or computations.

5. There are several CMs used for the protection of bridge abutments by different state departments of transportation and federal agencies. Two typical approaches for protecting bridge abutments from scour are:
   - Mechanically stabilizing the abutment slopes with riprap, gabions, cable-tied blocks, or grout filled bags, or
   - Aligning the upstream flow by using guide banks, dikes or spurs, or in-channel devices such as vanes and bend way weirs.
   - Underpinning method using mini piles and additional pile cap.

12.4.9 Countermeasures Combined with River Training

Experience has shown that providing armoring CMs alone may not be adequate and a combination of river training measures and armoring is necessary for high velocity rivers. By providing river training measures, less pressure will be put on the armoring mechanism. Accordingly, the effectiveness of the system will be increased. However, since large investments would be involved, economic considerations become important. Hence, cost reductions should be adopted in the design detailing by optimizing the depth and width of armoring mechanisms that are provided as revetment. Using scaling factors discussed later, riprap or gabion blankets may be used.

It is normal practice to protect 100 to 300 feet of riverbanks by revetment at the upstream and downstream of bridges and culverts. They differ from bed armoring in that they have a smaller thickness and are longer. Their sizing takes into account correction factors for stability, gravity, and angle of repose of riprap. In addition to mattresses, continuous framework of articulated concrete blocks and grout bags has been used for revetment. Filters should be used when utilizing concrete blocks and grout bags.

The common types of revetment in use are:

1. Dumped riprap.
2. Wire enclosed riprap mattress.
3. Articulated concrete block system.
4. Grout filled mattresses.
5. Concrete pavement.

It is recommended that some type of river training measure be provided, in addition to armoring countermeasures, when:

1. The flood velocity exceeds 10 ft/sec.
2. The bridge carrying traffic volume exceeds an ADT of 500.

12.4.10 Countermeasures for New Bridges

Considerable research carried out in the U.S. and abroad has resulted in some useful recommendations. The following guidelines are used to minimize different types of scour in design-phase of new bridges.

Contraction scour

1. Use larger bridge openings to allow for debris height accumulation.
2. Use longer span bridges, elevated decks, and crest vertical curves.
3. Use profiles for overtopping during floods and relief bridges.
4. Use reduced superstructure depths. Use open spandrel parapets.
5. Place piers away from the thalweg of a river for minimum height.
6. Excavate waterways to remove debris from smaller floods.
7. Use guide banks on the upstream side to align flow in bridge opening.
8. Use revetments on channel banks to fill slopes at bridge abutments.

Local scour at abutments

1. Place foundations on sound rock.
2. Use deep piling
3. Use stub abutments in lieu of full height abutment.
4. Use sloping walls in place of vertical walls.
5. Use revetments (pervious rock or rigid concrete).
6. Use riprap on spill slopes.
7. Use guide banks at abutments.
8. Monitor and inspect after flood events.

Local scour at piers

1. Place foundations in sound rock or below the total scour line.
2. Use deep piling as foundations.
3. Streamline pier noses (rounded shape).
4. Use pile bents or multiple columns with curtain walls to prevent debris deposit.
5. Use riprap as a temporary measure.
6. Cut cofferdams below contraction scour depths.
7. Monitor and inspect after flood events.

Aggradation

1. Use debris basins.
2. Provide continual maintenance planning.

Degradation

1. Use check dams or drop structures on small to medium streams.
2. Use channel lining.
3. Use deeper foundations.
4. Provide adequate setback of abutments.
5. Use rock and wire mattresses for small channels.

River meander
Locate bridges on straight reaches of streams between bends.

Braided channels

1. Build one long bridge, more than one bridge or a relief bridge.

Requirements of Foundation Design

Spread footings on soil: Place bottom of footings 3 feet below the total scour line. Although minor surface erosion of soil will not cause a danger to the footings, soil cover or protection to the concrete footing or piles should still be provided for the following reasons:

1. Will provide frost resistance (minimum frost depth requirement).
2. Will maintain as-built cosmetic appearance.
3. Will guard against any unforeseen error in the scour analysis data or computations.

A minimum 3 feet depth of riprap or an alternative supplementary countermeasure should be provided adjacent to the footings.

Spread footings on erodible rock: Place the bottom of footing 6 inches below the scour depth. This provision is conservative compared to that for soil conditions in which scour depth may be reduced by 50 percent.

Spread footings on non-erodible rock: This condition is less common in New Jersey since non-erodible rock is not generally found within 10 feet of river beds.

12.4.11 Countermeasures in Ascending Order of Cost

Cost of retrofit is one important consideration in deciding whether to maintain an existing bridge or replace it with a new bridge. The following scale may be used for initial cost consideration when selecting countermeasures:

1. Development of bridge inspection and scour monitoring programs; closing bridges when necessary (least cost).
2. Providing riprap or armoring at piers and monitoring.
3. Providing riprap or armoring at abutments and monitoring.
4. Constructing guide banks (spurs/dikes).
5. Constructing river training countermeasures and channel improvements.
6. Strengthening the bridge foundations, underpinning using minipile.
7. Constructing sills or drop structures (check dams).
8. Constructing relief bridges or lengthening existing bridges (maximum cost).

12.4.12 Selection of Countermeasures as an Alternative to Temporary Riprap

When addressing specific channel conditions, various CMs can be grouped together to provide an overall framework of applications.

The framework would address two nominal categories of channel flow and bed conditions. One category pertains to "moderate flows" when the channel bed is stable and not subject to the passage of large bed forms. The second category pertains to "severe flows" when bed-sediment transport conditions would be disruptive of CMs used in the milder flow and bed conditions.

Selected countermeasure types:

- Scour type to be addressed
  Local scour degradation, lateral erosion
- Description
  Graded broken rock placed below river bed in position by hand or dumped by boats and overlaid with soil

- Advantages
  Familiarity and past experience: Relatively low cost and no maintenance is required. Easy to construct, adjusts to minor scour and is the oldest method in use for shielding footings.

Meets environmental requirements.

- Disadvantages
  Not reliable for stability: Except for large size stone, it can wash out easily in moderate floods; disturbs channel ecosystem until vegetation is reestablished. Maintenance and monitoring required before and after floods.
  Not recommended for general bridge substructure use. Recommended for filling scour holes and bank erosion.

Drawbacks to Riprap

The drawbacks of using riprap are:

- Riprap, though a widely used scour countermeasure, has limited application in conditions marked by very high flow velocities and high intensities of bed-sediment movement.
- In situations where a pier is in close proximity to an abutment, placement of a scour CM may need to take the pier proximity into account as preference.
- In certain situations, required riprap sizes may not be readily available.

Alternatives to Riprap

Hence, the following CMs may be used over the conventional use of riprap.

Armoring countermeasures:

- Gabions and Reno mattresses
- Grout filled bags and mats
- Cable-tied blocks
- Tetrapods, dolos, and related units
- Grade control structures
- Grouted concrete, pavement, and flexible bed armor.

Flow-altering countermeasures (with relatively few field applications)

- Sacrificial piles
- Upstream sheet piles
- Collars and horizontal plates
- Flow-deflecting vanes or plates
- Modified pier shape or texture
- Slots in piers
- Suction applied to bridge pier.

Flow altering CMs should only be used in combination with primary countermeasures to improve their effectiveness.

Selected Countermeasure Type

Sacked concrete/grout filled bags

- Scour type to be addressed:
  Local scour, Degradation, Lateral erosion
- Description
  Fabric bags filled with concrete and stacked to produce a protective layer. Sand filled bags preferred.

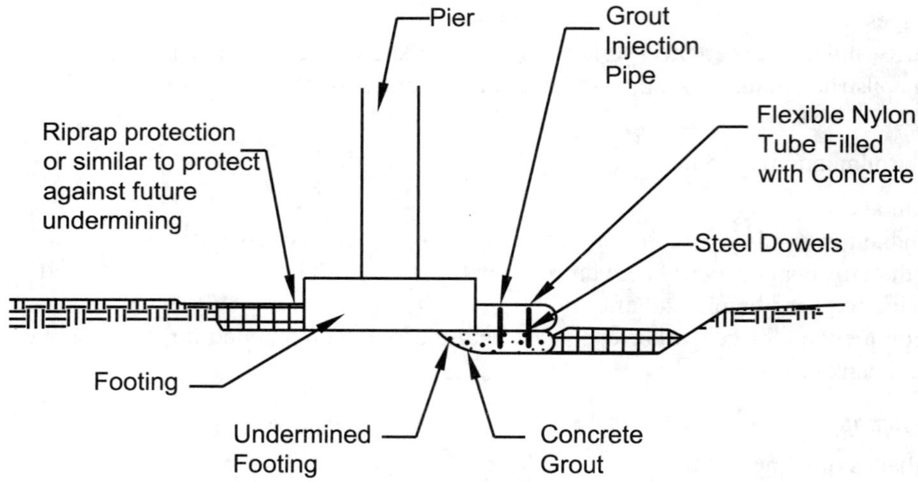

Figure 12.11 Structural repairs, grouting with pipe injection of concrete.

- Advantages
  Suitable for sandy soils only and useful for filling scour holes under footings and elsewhere.
- Disadvantages
  Undermining of toe may result. Likely to prevent vegetation growth or marine life. Risk of pollution from cement washout. Catastrophic failure potential due to lack of interlocking.
- Remarks
  Recommended for filling scour holes under footings subject to meeting environmental requirements.

Interference from Underground Utilities: Installation and maintenance of underground utility pipes close to a bridge foundation cause significant disturbance to the soil and should be checked. If any interference is likely with the proposed CM and utilities need to be relocated or it may result in an environmental issue.

12.4.13 River Training Measures

As discussed in 12.4.9, river training and flow altering CMs can be more important and effective than armoring since they harness the river and control the flood velocities. Due to cost considerations and special applications for higher velocity rivers they will generally take a secondary role when used in combination with primary armoring. They are not attached to the bridge substructure and are used primarily to control floods. River training structures modify a river's flow. They are distinctive in that they alter hydraulics to mitigate undesirable erosion and/or depositional conditions at a particular location or in a river reach. They are more suitable when local scour is a problem.

River training structures can be constructed of various material types. Some of the common types of river training methods are:

- Retard (earth, timber, and steel sheet piles)
- Channel improvements (channelization)
- Guide banks/guide walls
- Groyne (spur/dike/deflector)
- Grade control structure/Check dams

- Collars
- Auxiliary bridge.

River training structures are described as transverse, longitudinal, or areal depending on their orientation to the stream flow.

1. Transverse river training structures are CMs which project into the flow field at an angle or perpendicular to the direction of flow. Groynes are transverse river training structures constructed from stone, earth, sheet piling, or timber cribwork and extend out into the channel from a bank that is at risk of erosion. They are most commonly used on wide braided or meandering channels.
2. Longitudinal river training structures are oriented parallel to the flow field or along a bank line. They use erosion protection systems that include riprap, gabion mattresses, concrete blocks (interlocking or articulated), sheet piling, and bioengineering solutions using soil reinforcement and vegetation cover.
3. Areal river training structures cannot be described as transverse or longitudinal when acting as a system. This group also includes countermeasure "treatments" which have areal characteristics such as channelization, flow relief, and sediment detention.

Examples of areal river training are vertical (bed elevation control) CMs, such as sills or weirs.

12.4.14 Flow Altering Countermeasures

Flow altering countermeasures are recommended for diverting scour away from bridge piers and should be used in combination with riprap, gabions, etc. Sacrificial piles, upstream sheet piles, collars and horizontal plates, flow deflecting vanes or plates, modified pier shape or texture, and slots in piers and pier groups are examples of flow altering CMs.

1. Sacrificial piles: Sacrificial piles are only recommended where the flow is likely to remain aligned with the pile or pier arrangement and for relatively low flow intensities (that is, under clear-water scour conditions).
2. Upstream sheet piles: Upstream sheet piles are placed upstream of bridge piers to arrest scour in the lee of sheet piles. The width of sheet piles should be equal to the width of bridge piers and they should protrude only one third of the depth above the river bed.
3. Selection of river training countermeasures will be based on the following considerations:
   - Flood velocity, medium or high
   - Flow conditions, overtopping or over bank
   - Perennial or seasonal
   - Type of scour, local or contraction, aggradation or degradation
   - Width of waterway, narrow or wide
   - Span length, medium or long
   - Stream alignment, straight, meandering, or braided
   - Environmental requirements
   - Past experience of successful applications

Recommended river training countermeasures: Depending on flood conditions, the following types are recommended:

1. Retard (earth, timber, and steel sheet piles)
2. Channel improvements (channelization)
3. Guide banks/guide walls

The final selection should be made based on project and site specific conditions.

Table 12.3 Comparison of river training measures.

S. No.	Countermeasure Type	Scour Type	Description	Advantages	Disadvantages	Remarks
1	Retard (earth, timber and steel sheet piles)	Local scour, Meandering stream or shifting of thalweg	Permeable or impermeable structure parallel to banks, to reduce flow velocity and induce deposition.	Suitable for maintaining channel alignment. Induce deposition.	Minimum disadvantages since piles are buried below river bed. Expensive	Recommended for high flood velocities.
2	Channel Improvements (channelization)	Contraction and local scour, Aggradation	Channel modifications to increase flow capacity and sediment transport, including dredging, channel clearing.	Suitable for aggradation or if upstream/ down-stream of bridge is clogged.	Minimum disadvantages	Recommended. River encroachment permit requirements apply.
3	Guide banks/guide walls	Local scour at abutments/ channel braiding	Straight or outward curving earth structure/fill to form embankments upstream to align flow through bridge opening and reduce abutment scour.	Improves flow conditions, Moves point of local scour away from abutment, prevent erosion by eddy action.	Minimum disadvantages. Expensive	Recommended for wide rivers with high flood velocity. River encroachment and other permit requirements apply.
4	Groyne (spur/dike/ deflector)	Local scour, upstream lateral erosion and degradation	Impermeable or permeable structure, which projects into flow to alter flow direction, reduces velocity and induces deposition.	Suitable for containment of over bank flow and for braided streams. Proven effective.	Does not prevent downstream lateral erosion of banks or degradation of channel. They project above river bed. Not applicable to streams or narrow channels.	Recommended for wide rivers with high flood velocities. River encroachment permit requirements apply.
5	Grade control structure / check dams	Contraction and local scour, degradation, aggradation, and lateral erosion of banks	Low dam or weir made of concrete, sheet pile, mats, gabions constructed across channel to form debris basin and provide vertical stability of stream bed	Suitable for high flood velocities.	Expensive to install since riprap is required downstream of grade control structure.	Difficult to meet environmental requirements since fish passage is adversely affected.
6	Collars	Local scour	Thin horizontal plate attached to base of pier to deflect flow away from sediment bed.	Suitable for high velocity rivers & for long span bridges. Low cost and maintenance	Debris accumulation for small spans. Does not eliminate scour, not much experience.	Not easy to construct. River encroachment permit requirements apply.

(continued on next page)

Table 12.3 Comparison of river training measures *(continued).*

S. No.	Countermeasure Type	Scour Type	Description	Advantages	Disadvantages	Remarks
7	Relief bridge	Local scour	Constructing an additional or auxiliary bridge adjacent to the scour critical bridge to minimize the discharge and flood velocity.	Suitable for wide rivers and overcomes the problems associated with defects in the original planning and size of opening.	May cost nearly the same as a replacement bridge. Environmentally, it may create additional problems.	It may be difficult to acquire the right-of-way in developed areas.
8	Driving upstream sheeting or using mini piles and additional pile cap.	Scour at pile groups and at bottom of protection.	Driving mini piles and constructing pile cap.	Foundation is stable.	Difficult and expensive.	Bridge needs to be supported during construction.

Size: The median size of riprap D_{50} shall be determined using guidelines in HEC-18. Minimum D_{50} size of riprap shall be R-6 and maximum D_{50} size shall be R-8 as per NCSA rock size and gradation in Table 14.2.

12.5 HEC-18 COUNTERMEASURES MATRIX (CM)

The HEC-18 CM matrix facilitates preliminary selection of feasible alternatives, prior to a more detailed investigation. The matrix lists the CM types in rows, against their characteristics in columns but needs to be modified in the light of specific conditions of state rivers.

The Modified Matrix Table in the Appendix shows a modified CM matrix for simplified conditions. The table is based on engineering factors, environmental factors, and cost. CMs have been organized into groups based on their functionality with respect to scour and stream instability classified into three groups:

Group 1: Hydraulic CMs

Group 1a: River training structures

Transverse structures

Longitudinal structures

Arial structures

Group 1b: Armoring CMs

Revetment and bed armor (rigid, flexible/articulating)

Local armoring

Group 2: Structural CMs

Foundation strengthening

Pier geometry modification

Group 3: Monitoring

Fixed instrumentation

Portable instrumentation

Visual monitoring

Each CM must be selected on the basis of a scour analysis for each specific site. CM characteristics are classified into three groups:

1. Functional applications: Functional applications are the computed or observed scour conditions, such as local, contraction, and stream instability conditions.
2. Suitable river environment: The suitable river environment grouping lists a wide range of physical data for hydraulic and geotechnical conditions related to the river.
3. Maintenance.

12.5.1 Types of Countermeasures in Use

Common types of countermeasures are riprap, gabion baskets, concrete blocks, and sheet piling:

Table 12.4 Distribution of Countermeasures in U.S.

Countermeasure	% Use w/Monitoring	% Use w/o Monitoring
1. Monitoring	76%	—
2. Dumped riprap	16%	67%
3. Extended footings	2%	8%
4. Rock gabions	1.5%	6%
5. Spurs	1.1%	4%
6. Pavements	0.7%	3%
7. Check dams	0.25%	1%
8. Rock bank protection	0.25%	1%
9. Unspecified	2.2%	10%
	Total	100%

Scour depth should be measured from a reference line 1 ft above the top of the footing. If eroded elevation is located at an elevation higher than 1 ft above the top of the footing, the higher elevation is considered. For placing riprap, excavation to the design depth should be carried out. The depth of riprap should be at least below the contraction scour depth.

If considerable erosion has already taken place and the riverbed elevation is below the top of the footing, hydraulic analysis shall be based on the new channel profile by considering the new opening size.

12.5.2 Embedment of Footings for New Bridge Foundations

Footing on non-erodible rock—Minimum 6 inch into bedrock.

Footing on erodible rock—Minimum 3 inch in erodible rock

Footing on soil—Minimum 6 ft in soil (for existing footings minimum 3 ft depth of riprap used).

The same type of armoring CM should be used for abutments and piers for economy and ease of construction.

Armoring should not be mixed, i.e., if gabions are selected, then they should be used for the whole bridge site. However, armoring can be combined with structural CM or river training.

12.5.3 Filter Requirement under Riprap

The use of a filter, or alternatively a geotextile (filter cloth), is of particular importance in ensuring that finer material does not leach through or winnow around the riprap. No filter is required for gravel beds. For sand beds, use a geotextile filter cover equal to the width of a pier (W) from the face of the pier in each direction.

Gradation

The following distribution of riprap sizes should be used.

100 percent finer than 1.5 D_{50}

80 percent finer than 1.25 D_{50}

50 percent finer than 1.0 D_{50}

20 percent finer than 0.6 D_{50}

Effectiveness of Riprap Installations

Effectiveness depends on

- The seal around piers
- Reduced tendency of rock dispersal
- 50 percent less volume of rocks
- Granular filters subject to degradation
- More effective if tied into abutment countermeasure when pier footing is located within three pier diameters of abutment footings.

Constructability of Riprap Installations

- Excavation required
- Sealing geotextile to pile bents is difficult
- Limited ability to pre-excavate due to pier footing and/or pile geometry
- Specialized construction techniques for geotextile placement
- Gravel cushion on geotextile to avoid rupturing
- Performance dependent on construction sequence

Placement of Riprap

Riprap around bridge piers can either be installed from a bridge deck or from the bed/banks of a stream.

- Installation from bridge deck: Riprap is installed from a bridge deck by dumping from trucks and spreading by loader. Since stones are placed irregularly, they are unstable. Also, there are voids between the stones, through which fine particles of soil travel underwater pressure and cause erosion. The success rate with dumped riprap at many existing sites may be misleading since the bridge may not have been subjected to peak design floods of 100 years. For large bridges, installation from a bridge deck can be done by machine placing from a dragline or from buckets.
- Installation from bed/banks of stream: In this method, by hand placing and packing, a compact, mortar less masonry type construction can be achieved. Stones packed into a close interlocking layer will minimize the size of voids. This method of installing riprap results in a stable configuration and in uniformity of stone size distribution. The quality of construction is better than the dumping method. Also, a fish channel made within the top stone layers can be maintained. Although more expensive than other methods, its use is recommended.
- Dumped riprap can be placed by boats when stone sizes are not large.

Durability

The following durability issues should be considered with riprap installations:

- Broad band of failure threshold potential
- Catastrophic failure if riprap is exposed
- Geotextile fails abruptly.

Maintainability

The following maintainability issues should be considered with riprap installations:

- Under-bed installation increases durability
- More maintenance with dumped riprap
- Difficult to repair ripped geotextile or locate damage to riprap
- Clean up difficult after failure
- Gravel filters easier to maintain than geotextile.

Costs

The following cost issues should be considered with riprap installations (may vary according to local conditions):

- Geotextile is more expensive than granular filter
- Pre-excavation costs
- Disposal costs
- Less stone costs
- Traffic disruptions.

12.5.4 Riprap Countermeasure Design for Abutment

Design guidelines for riprap at bridge abutments in this section are based on "Stability of Rock Riprap for Protection at the Toe of Abutments Located at the Floodplain," published in 1991 "Design of Riprap Revetment," Hydraulic Engineering Circular-11, and Design Guidelines 8 and 12 of HEC-23.

1. Side slopes: For preventing slump failure, the side slope is a significant factor in the stability of riprap. It is desirable to decrease the steepness of abutments; thus increasing the stability of the riprap on the slopes.
2. Recommended minimum value for side slopes varies from 1:2 to 1:1.5, (H:V).
3. Extent of riprap protection.
4. Vertical-wall abutments: For vertical wall abutments without wingwalls, wingwalls at 90° or splayed wingwalls, the width of a riprap layer (WR) adjacent to a footing at the river side of an abutment should be the greater of the following:
   - Width of scour hole
   - 2W (W=width of abutment at the base) or

     $2W/\cos(\beta)$ when skew angle of flow $\beta > 15°$

     $X + 18 \text{ in} + y \cot \varphi$

     where X is the width of abutment footing, y is design scour depth at abutment and φ is the angle of natural repose for the soil, as obtained from geotechnical report.

   Place riprap around the footings with the slope starting at a distance of a minimum of 1 ft from the vertical face of the footing.
5. Spill-through abutments: For spill-through abutments, extend the riprap around the abutment and down to the expected scour depth. The launching apron at the toe of the abutment slope should extend along the entire width of the abutment toe and around the sides of the abutment to a point of tangency with plane of embankment slopes.

   The apron should extend from the toe of the abutment into the waterway a distance equal to twice the flow depth in the over bank area near the embankment, not exceeding 25 feet.

   Figures for typical layouts of abutment riprap for abutments near a channel bank, stub abutments near the top of a high channel bank, and abutments near a flood plain are given by the Maryland State Highway Administration.

Minimum riprap blanket thickness, minimum D_{50} size and approximate D_{50} weight for Class 1, 2, and 3 types are shown in Table 12.5.

6. Riprap size distribution for abutments: Riprap size distribution should be the same as that for bridge piers.
7. Flowchart for the design of riprap at bridge abutments: Riprap alone is not recommended as a permanent countermeasure, but as emergency shielding only for a period of five years or longer only after regular evaluation from underwater bridge inspection reports. Riprap can be used as secondary local armoring, in conjunction with primary structural countermeasures or with river training measures.

12.5.5 Traffic and Utility Issues for Riprap Use

1. Site access: Adequate access to the site shall be provided for trucks to deliver riprap.
2. Right-of-way: Construction easements and right-of-way access may be required for the duration of construction.
3. Detours: Detour, lane closure, or nighttime work may be necessary. Coordination with traffic control would be required.
4. Emergency vehicles and school bus services should not be affected by lane closures.
5. Utilities: Relocation of any utilities at the sides of an abutment or a pier may be necessary for the duration of construction. Coordination with utility companies would be required.

12.5.6 Riprap Detailing

1. Construction drawings have to be prepared. Conceptual sketches for the layout of riprap with details for riprap placement at abutments and piers based on this book should be used.
2. In addition to hydraulic data, construction drawings shall show tables summarizing flood elevations, flood velocities, and scour depths.
3. Maximum side slope is 1V:2H although, where excavation is difficult, 1V:1H may be used with fractured rock.
4. Cost: Current estimated cost is $180 to $200 per sq ft. Long distance freight charges for riprap may increase the unit cost by 10 percent.
5. Construction permits: Stream encroachment and other applicable permits must be accounted for. Refer to guidelines that are given for permit applications in the NJDEP Stream Encroachment Technical Manual.

12.5.7 Limitations on the Use of Riprap

1. Monitoring: Riprap shall be used as a countermeasure only if accompanied by field inspection that occurs immediately after floods and by the use of monitoring equipment during floods.
2. Critical velocities: If a 100-year flood velocity exceeds 11 ft/sec., riprap shall not be used.

Table 12.5 Riprap D_{50} size and blanket thickness.

Riprap Class	D_{50} Minimum Size (in)	Approximate D_{50} Weight (Pounds)	Minimum Blanket Thickness (In)
1	9.5	40	19
2	16	200	32
3	23	600	46

Table 12.6 Sizing of gabions based on design velocity.

Gabion Thickness (ft)	Stone Size (inch)	Critical Velocity (ft/sec)	Limiting Velocity (ft/sec)
0.49-0.56	3.3	≤11.5	13.8
	4.3	13.8	14.8
0.75-0.82	3.3	11.8	18.0
	4.7	14.8	20.0
1.0	3.9	13.8	18.0
	4.9	16.4	21.0
1.64	5.9	19.0	24.9
	7.5	21.0	26.2

Based on Agostoni (1988) and CIRIA (2002).

3. Scour depth: If calculated scour depth is high and excavation to place riprap under the riverbed would endanger the stability of soil adjacent to the footing, riprap shall not be used.
4. Economic considerations: If riprap is not available locally or at a reasonable distance it may not be economically feasible. In such situations, other alternates may be considered. Also, if cost of hand placement of riprap is high, other less expensive countermeasures may be considered.
5. Dumped riprap: Truck dumped riprap can easily get dislodged during floods and get washed away due to high velocities. It is less stable compared to hand placed riprap and its use is therefore not recommended.

12.6 DESIGN GUIDELINES

12.6.1 Design of Common Types of Countermeasures

Gabions can also be sized according to NJDOT soil erosion and sediments control standards (using Table 12.7).

1. Articulated concrete blocks: Design guidelines for articulated concrete blocks (ACBs) for abutments are based on Design Guidelines 4 in HEC-23.
2. Design guidelines for articulated concrete blocks (ACBs) for piers are based on NCHRP 24-07. Other recommended sources of information on design of ACBs are HEC-11 and McCorquodale (1993).
3. Concrete armor units/A-Jacks: The basic construction element of A-jacks for pier scour applications is a "module" comprised of 14 individual A-jacks banded together in a densely-interlocked cluster, described as a $5 \times 4 \times 5$ module. The following design procedure for A-Jacks systems for pier-scour protection is based on Design Guidelines 6 of HEC-23.

Table 12.7 NJDOT soil erosion and sediments control standards for sizing gabions.

Gabion Thickness (ft)	Maximum Velocity (ft/sec)
½	6
¾	11
1	14

4. Design procedures for grout bags: The design size of a bag or depth of a layer depends upon the following:
- A design flood velocity of 5 to 10 ft/sec.
- A computed scour depth for contraction and local scour of 3 to 6 ft.
- When hydrostatic pressure builds up, the dead weight of bags should exceed the uplift pressure. The mattresses should be provided with filter drains or drain holes for pressure relief.
- Depending upon the application, bags may vary in capacity, from standard cement bag size, to about 5 ft^3, while mattresses are larger in size up to 15 ft^3 in volume.
- Mats must be bound firmly to the pier itself for a good performance. Mats should be installed with their top surfaces flush to the bed.
- Grout bags should be sized and placed in a manner similar to riprap, and underlain by a geotextile filter with a partial cover or filter layer. Any means to render the surface of bags rough and angular will aid to performance.
- Properly sized bags are more effective when they extend a single layer of protection laterally, rather than if they were stacked. Efforts should be made to avoid stacking of grout bags.
- Flexible bags of sand may be preferable to grout-filled bags.

12.6.2 Case Study of Countermeasures Construction

Where gabions are used as countermeasures, they will be over laid by a soil layer, and vegetation grown for resisting erosion during floods.

Use precautionary measures to keep the river water clean, such as the use of a turbidity dam and silt fence and traffic control during construction. This way DEP underwater construction permit requirements are fully followed with minimum impact on plant and marine life. Table of estimate of quantities shows actual pay and standard items.

If an appropriate method of construction is adopted there will be fewer constructability issues.

Changes in structural details during the long construction process may be necessary due to unforeseen field conditions such as the occurrence of different soils.

Coordination between design and construction teams is essential for answering requests for information (RFIs) and issue agreed design change notices (DCNs) as revisions to contract drawings to finish work within the construction schedule. For delicate repairs a more stringent quality control (QC) procedure may be necessary.

12.6.3 Replacement or New Bridges

1. Use of cofferdams: If the water depth is not high, temporary cofferdams may be required to ensure construction in dry conditions. Without dry conditions the quality of placement of countermeasures will be difficult to monitor or maintain. Figure 12.12 shows an elevation view of a typical cofferdam.
2. Sheet piling left in position: For underwater construction or repairs, temporary sheeting on the stream side is required for installing CMs at the sides of spread footings/pile caps. To prevent long-term scour, temporary sheeting may be left in place after CMs installation is completed.

12.6.4 Design of Timber Fenders to Protect Piers in High Velocity Rivers

Fender systems consist of:
- Transverse pile frames and diagonal bracing

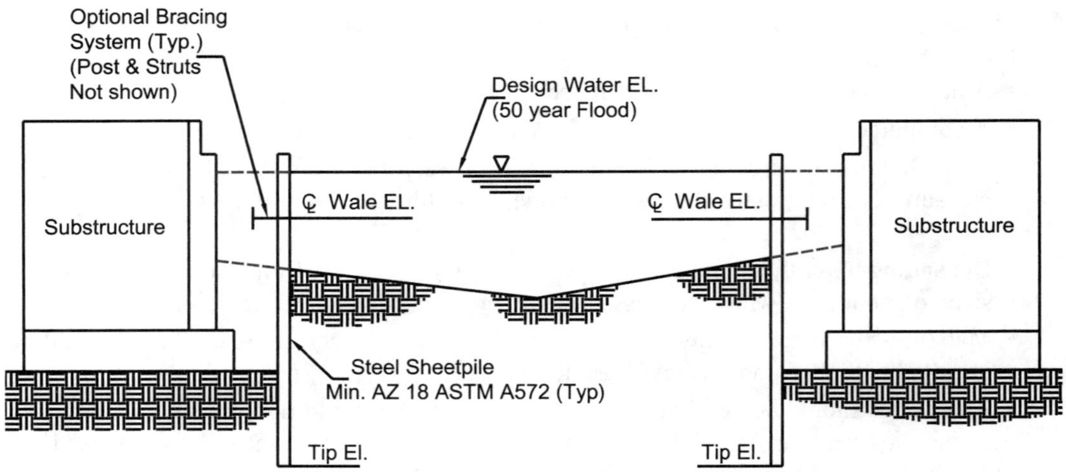

Figure 12.12 Cofferdam location to shield foundation.

- Longitudinal walers supporting vertical sheeting (as T-sections)
- Vertical timber sheeting/planks designed continuous across walers.

Timber space frames with driven piles, longitudinal walers, and vertical planks should withstand:

- Impact forces from floating ice: Ice forces from a 50-year storm will be considered. Data will be provided by coast guards.
- Compression forces from frozen ice.
- Any collision/impact forces from barges.

Timber grades will be assessed from laboratory tests:

Allowable stress in tension ft = Ultimate tensile stress (parallel to grain)/Factor of safety

Allowable bending stress from NDS or AASHTO Bridge Design Specifications may be used.

Example: For southern yellow pine sheeting, typical allowable strengths are: Fb = 1500 psi, Fv = 70 psi

For walers, Fb = 1150 psi, Fv = 65 psi

Members shall be free of any defects or knots.

Space frame/truss analysis: For setting up the model, software such as STAAD-Pro (with capability of designing timber members) can be used. Due to bolted connections between piles and walers, pinned connections may be assumed.

Piles may be assumed fixed at minimum elevation of 5 feet below river bed, if coarse sand layer is present. Calculation of allowable bending and shear stress based on load factors and reduction factors for member shapes and wet conditions.

Computation of bending and shear stress from factored bending moment and shear force diagrams. Vary elevations of ice incidences to produce maximum bending moments. Either LRFD or ASD may be used.

Comparison of computed stresses with the allowable:

- If peak stress is less than allowable, replace timber sheeting with heavier planks or provide additional walers to reduce effective spans.

- If existing piles are weak in bending or shear, place new steel frames (H piles) between existing timber piles and connect to existing walers and planks. H-piles may be spliced with bolted connections. All bolts shall be galvanized.

Dynamic ice forces (refer to AASHTO LRFD specifications):

$$\text{Horizontal force } F = C_n\, p \cdot t \cdot w$$

Example: $C_n = 1.0$ for vertical face

p = Effective ice strength = 200 psi

t = 10 in from 50 year ice storm

w = Width of pier

F = 2.0 kips/inch width

12.6.5 Examples of Existing Underwater Inspected Conditions

As listed in available underwater inspection reports, the two types of remedies are:

1. Repair concrete defects
2. Provide scour/undermining countermeasures for any scour resulting from contraction and local scour.

Debris and silting may reduce river width and increase flood velocity and water pressure on piers. Major design tasks may cover the following tasks associated with the safety of the substructure:

1. Riprap placement/providing effective armoring countermeasures
2. Concrete repairs (patching, spalls, and sealing cracks)
3. Masonry repairs
4. Apron repair around piers
5. Addressing foundation undermining issues
6. Debris removal
7. Strengthening of foundations if required.

12.7 CONSTRUCTABILITY ISSUES

12.7.1 OSHA Recommendations for Slopes of Excavations in Soils

The following maximum values of slopes shall be used for excavation of sloping structures. The angle of repose shall be flattened when an excavation has water conditions.

1. Solid rock—90 degrees
2. Compacted angular gravels—0.5:1 (63 degrees 26 ft)
3. Average soil 1:1— (45 degrees)
4. Compacted sand 1.5:1— (33 degrees 41 ft)
5. Loose sand 2:1— (26 degrees 34 ft).

12.7.2 Computer Software for Hydraulic and Scour Analysis

Construction drawings for scour countermeasures must be based on detailed designs. For detailed design, approved commercial software may be used. A spreadsheet may be developed and used in lieu of software. The following software may be used for detailed analysis and design of scour/scour countermeasures.

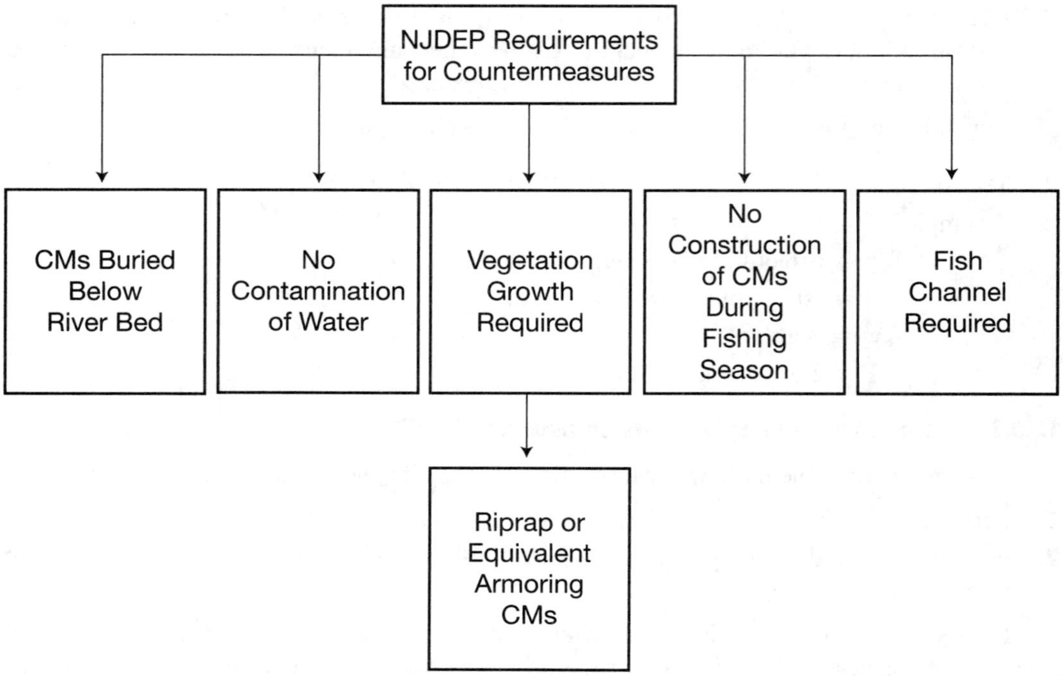

Figure 12.13 DEP general requirements for stream encroachment permit.

Hydrology Programs

- TR55
- PENNSTATE Program
- USGS Methods for different states: Stankowski method is being used in New Jersey.

Hydraulics Programs

The following programs are available to determine the flood intensity, water elevations, scour potential, check FEMA compliance, and help size the proposed structures over waterways.

- HEC-2: This program was developed by the Corps of Engineers and used by FEMA in the past for most of its studies. It has been replaced by HEC-RAS.
- WSPRO is a computer program developed by FHWA which computes water surface profiles and velocities using stream cross sections Manning's variable and Design (Q50) and Basic (Q100) flows.
- HEC-RAS is an Army Corps of Engineers Program. This program can handle variable flows and has a WSPRO subroutine in its water profile routine. In addition, this program computes the possible scour depths at the substructures.
- HYDRAIN is a conglomeration of several programs developed as a pool fund project by several states and FHWA. It includes:
  - HYDRO, a hydrology program;
  - HYDRA, which simulates hydrology and hydraulics on storm drain or sanitary pipes network;
  - HY8, which simulates hydraulic analysis or design for culverts, reservoir routing, and energy dissipators;
  - HYCHL, which analyses and designs channel and rip-rap linings;
  - NFF, which interactively calculates USGS Regression equation flows.

- BRI-STARS is a pseudo two-dimensional hydraulic program that (through the use of stream tubes) provides a time and flow dependent two-dimensional sediment routing (aggradation and degradation) in a bridge cross section.
- UNET: Unlike the riverine flow, this program deals with tidal flow.
- DYNET: Two-dimensional hydraulic program for tidal flow.

Scour Analysis

In-house Excel® spreadsheets based on HEC-18 equations developed by the author (see Appendix for procedure for input data). The sub-routine in the HEC-RAS program does scour analysis. However, in some cases high values of scour depth are obtained for contraction or local scour.

Table 12.8 Comparison of structural countermeasures.

CM	Scour Type	Description	Advantages	Disadvantages	Remarks
Concrete apron/ curtain wall	Contraction and local scour	Concrete walls precast or cast in place against the sides of footing	New wall can rest on hard strata/rock.	Cofferdam is required for construction	Recommended
Local sheet piles	Degradation	Piles driven as shields adjacent to bridge foundations to deflect flow	Suitable for high flood velocities. Stops flow, helpful in dewatering	Scour can occur near sheet piling, construction difficult, rust	Recommended for high scour situations with riprap protection
Extended footing	Local scour	Cast wider concrete slab footing to prevent settlement	Suitable for low scour depths. Acts as curtain wall/apron on side of spread footing	Not suitable for masonry footings. Bridge may be closed to traffic during construction	Recommended for concrete spread footings
Constructing mini piles through spread footings	Degradation	Minipiles driven through footings	Commonly used for footing strengthening	Expensive. Not suitable for old masonry footings	Not recommended for high traffic volume bridges
Under-pinning	Contraction scour Local scour	Lowering the bottom of footing elevation below scour depth	Commonly used for extensive repair or footing strengthening	Expensive. Not suitable for old masonry footings. Bridge needs to be closed to traffic. Disturbance of streambed during construction.	Not recommended for high traffic volume bridges
Use of open parapets or railings	Contraction scour/ pressure flow	Increases flow area and prevents overtopping of flood water	Effective for small openings or where vertical alignment is limited	Additional overflow downstream needs to be checked	Recommended only for overtopping flood situation
Relief bridge	Contraction and local scour	Approach bridge to increase size of waterway opening	Flood water will be discharged rapidly	Expensive. The key scour problem at main bridge may still remain unchanged	Not recommended since utilities need to be relocated

Rating Software

In-house Excel® spreadsheets based on AASHTO Manual for Condition Evaluation of Bridges. A similar approach to SBCI Calculator for Scour Critical Indicator Code as developed by USGS.

Scour Watch Software

The software allows bridge engineers to identify, predict, and record severe floods. Scour watch identifies the occurrence of a flood and processes scour monitoring data from equipment like SHIFLO. The results obtained from the software would expedite maintenance and would facilitate urgent action. Actions required when scour depth is above, within, or below the footing are shown.

12.7.3 Technical Specifications and Special Provisions

Technical specifications deal with description of materials, methods of construction, units of measurement of pay items, and unit costs. For countermeasures that are proprietary in nature they may be obtained directly from the vendors, if not available in state standard specifications. They are required for preparing contract documents and for award of contracts.

The designer should use guidelines provided by the manufacturers only after they have been approved by the engineer-in-charge for the project. Recommended guidelines should be supplemented by design data provided by the manufacturers for designing such countermeasures. Technical specifications for gabions should be used. Gabions are now being frequently used in place of riprap. Any deviations from standard specifications need to be approved as special provisions.

12.8 SCOUR OF PILES, PILE GROUPS, AND CAISSONS

Deep foundations such as piles and caissons are normally required in soft soils, where bearing capacity of soil is not sufficient to support a shallow footing. Such soil has greater scour potential.

Piles are designed as bearing piles or friction piles and in either case must be fully embedded in undisturbed soil at all times. If the surrounding soil is eroded or disturbed the load bearing

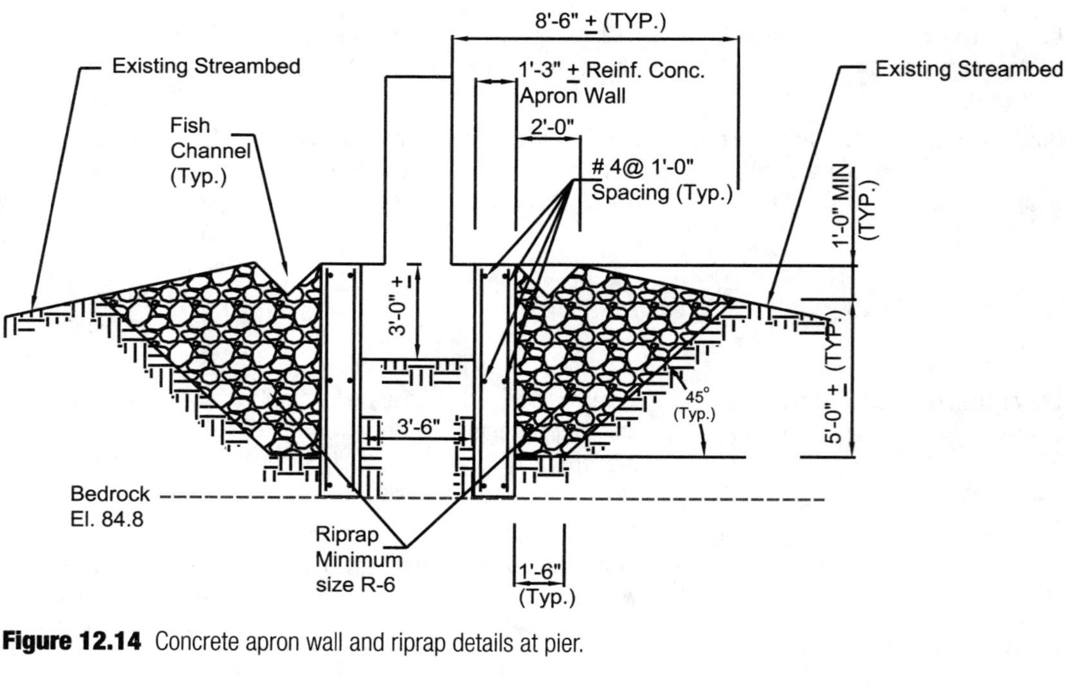

Figure 12.14 Concrete apron wall and riprap details at pier.

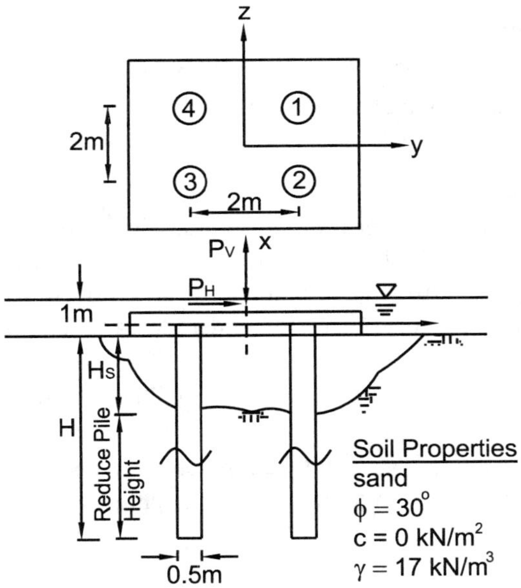

Figure 12.15 Examples sowing scour at piles.

capacity of piles is reduced, leading to settlement and collapse of the pile cap. Figure 12.15 shows failure of piles, resulting from soil erosion and sheet pile protection methods.

Raking piles which are driven closer to the river are more susceptible to erosion than vertical piles. The following steps are required:

1. When using raking piles, they must not be placed at the edge of the river. Instead, the bridge span may be increased.
2. The top 5 to 10 feet of pile, which may be exposed, may be neglected in pile design length. The exposed length of steel or concrete pile should be checked for local buckling.
3. The pile cap should be placed below the calculated scour depth. However, this results in taller and full height abutments, increasing the cost. Shielding of piles by driving sheet piling would prevent erosion.
4. In case of raking piles interfering with sheet piles, sufficient clearance is possible but would reduce the width of waterway, making it difficult to obtain a stream encroachment permit from DEP. Armoring countermeasures or river training measures may be considered.
5. Caissons may be preferred over piles.
6. For long span bridges on harder ground and dense soils, drilled piers will be more resistant to erosion compared to caissons.
7. Integral abutment piles with a single row of piles would require greater protection by providing shielding in the form of sheeting left in place.

12.9 SCOUR AT WINGWALLS

Wingwall normal to abutment:

- No separate scour study needs to be carried out.
- Riprap or armoring needs to be provided for a return length of the maximum of width of scour hole

  2 × width of abutment at base

  (300mm + d cot θ), where d is design scour depth at abutment and θ is the angle of natural repose for the soil, obtained from geotechnical report.

Flared wingwall:

- Scour study will be carried out using computed velocity from HES-RAS analysis.
- Design of riprap will be based on scour depths at wingwall location.

12.10 SCOUR AT CULVERTS

- Hydraulic design: Chapter 9 of AASHTO Model Drainage Manual provides design procedures for the hydraulic design of highway culverts. Included therein are design examples, tables, and charts that provide a basis for determining the selection of a culvert opening. However, no scour analysis needs to be carried out for soil erosion under the culvert floor slab.
- Footings for any flared wingwalls provided at entry and exit of a culvert will be protected by riprap or alternate armoring countermeasures.
- For high velocities exceeding 4 m/sec, riprap at wingwalls will be replaced by a concrete apron, which will extend between opposite wingwalls and the edge of the culvert.

12.11 SCOUR MEASURING EQUIPMENT

Measurement of scour depth using SHIFLO: It consists of a depth finder, which is mounted on a water ski. Depths can be taken at high velocities at locations under the bridge.

12.11.1 Magnetic Sliding Collar

A magnetic sliding collar is a scour-monitoring device that consists of a stainless steel pipe driven into the channel bottom with a sliding collar that drops down the pipe with increase in scour. A magnetic field is produced by magnets located on the collar and the location of the collar can be detected.

12.11.2 Robots for Underwater Inspections

The serpentine robot has multiple joints that provide superior ability to flex, reach, and approach all locations. Robots are still under development by university researchers but have the potential to perform underwater photography and take erosion measurements and perform repairs.

12.12 DEBRIS ACCUMULATION

Debris may accumulate at the upstream or downstream end of a bridge or under a bridge as shown in Figure 12.16.

It changes both the geotechnical and hydraulic characteristics of a bridge. A great majority of bridges are located on narrow stream widths and therefore the effects of debris may be higher.

The work includes classifying the type of debris, studying flow behavior of floating debris, estimating the volume of debris, studying local pier scour associated with debris accumulation, and selecting the type and design of debris sweepers and countermeasures to deflect debris. Recommendations given in FHWA, HEC-9 "Debris Control Structures Evaluation and Countermeasures" will be followed.

Debris consists of indigenous material deposited at the bridge obstruction from:

- Continued floods
- Long-term aggradation—material transported by water flow varies according to demography and the type of terrain
- Sediments, small stones, gravel, and tree leaves and branches
- Broken pieces of timber from furniture, boats, etc. or fragments of metals.

Figure 12.16 Post hurricane Floyd flood scenario. Accumulation of debris at Peckman's River bridge on Route 46 in NJ.

Debris accumulation in bridge openings may result in the following:

- Substantially block a bridge opening, and depending upon the size of debris accumulation, the waterway opening area is reduced
- Hydraulically, reduction in area will increase velocities under the bridge and contraction scour.
- Cause overtopping
- Cause failure of roadway embankments because of increased local scour at piers and/or abutments.

12.12.1 Countermeasures for Debris Accumulation

The most-effective countermeasure against debris accumulation during high-flow flood is monitoring and debris removal. A bridge monitoring crew should closely monitor signs of debris buildup, especially near the lower chord.

If a large difference in the water surface elevation is noticed from the upstream to the downstream side of the bridge, it may be a sign of a severe debris blockage developing beneath the water surface. The structural loading and impact from floating debris may cause cracking of the substructure.

12.13 EXTREME EVENT OF EARTHQUAKE DAMAGE

12.13.1 Recent Developments in Seismic Design

In this chapter, a practical approach to earthquake engineering is presented. An overview of the history and causes of major earthquakes, methods of equivalent static and dynamic analysis, seismic retrofit methods for existing bridges, and available computer software is listed. Latest detailing procedures, bearings retrofit, and examples of disaster management are discussed. Concepts of seismology, mechanisms of earthquake generation and propagation, the difference between various scales used to quantify the size of an earthquake, and its potential to cause damage are reviewed. An attempt is made to examine the design spectrum and conventional methods used to calculate the response of bridges, considering:

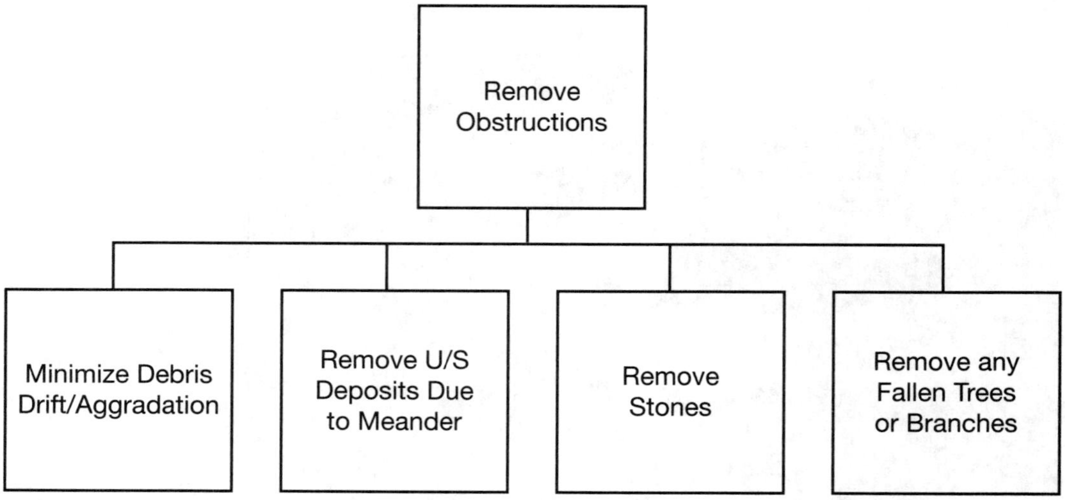

Figure 12.17 Procedures for cleaning up streams for obstructions.

- Base isolation
- Influence of local site conditions
- Seismic protection with energy dissipating devices
- Types of foundations and soil-structure systems to resist earthquake ground motions
- Seismic code provisions.

Bridges are located everywhere, in urban and local areas, on highways, local roads, and on rivers. There is likely to be more than one bridge built within an earthquake zone.

12.13.2 Notable Failures due to Lack of Seismic Design Criteria

There are over 600,000 bridges in the U.S., and of those about 60 percent were constructed before 1970 when AASHTO code did not have a seismic design criteria. There are differences in seismic zoning and bridge architecture between the eastern and western U.S. In the Northeast for example, the majority of older bridges have steel girders, while the majority of California bridges use prestressed concrete superstructures.

Figure 12.18 Collapse of abutment and approach at Balakot bridge on national Route 15 in Hazara, Pakistan, 7.6 magnitude earthquake.

In Chapter 3, notable failure of bridges due to earthquakes were discussed. Seismic events have demonstrated the vulnerability of bridges not designed and detailed to resist seismic loads. After the 1971 San Fernando Earthquake, which displayed the potential for span unseating, CALTRANS instituted a program of seat length retrofitting in California. However, the Whittier Narrows Earthquake (1987), Loma Prieta (1989), and the Northridge Earthquake (1994) have shown that seat length retrofitting is not sufficient to prevent collapse. As a direct result of these events, there is now a standard requirement for retrofit to current standards. Retrofitting methods presented in this chapter are based on research and application in California.

Some historic seismic events in the U.S. include:

- 1906 San Francisco Earthquake: Much of the city was destroyed and was rebuilt.
- 1925 Santa Barbara Earthquake.
- In 1964, all bridges located along Cooper River Highway in Alaska were either destroyed or severely damaged due to an 8.1 magnitude earthquake.
- In 1971, as a result of the San Fernando earthquake (6.4 magnitude), which occurred on the Golden State Freeway in California, more than 60 bridges were severely damaged.
- In 1987, Whittier Narrows had an earthquake event of 5.9 magnitude, which caused extensive damages.
- In 1989, the Loma Prieta earthquake of 7.1 magnitude in California damaged more than 80 bridges, causing scores of deaths and billions of dollars in repair and replacement costs.
- In 1991, a 5.8 magnitude earthquake hit Sierra Madre along the fault zone.
- In 1994, a 6.7 magnitude earthquake struck a 350-square mile area around Northridge.

The above unexpected earthquakes caused severe hardships to the traveling public in terms of loss of time and commerce. In some cases the bridges designed to seismic codes performed well, but secondary causes reduced their carrying capacity, such as washing out of embankments, exposing the concrete piles.

12.13.3 Immediate Effects of Earthquake

While the effects of earthquakes in buildings result from the movement of foundations, the effects on bridges are a combination of both truck moving loads and movement of foundations. In non-seismic design, the load path is applied from the top down (deck slab to vertical members,

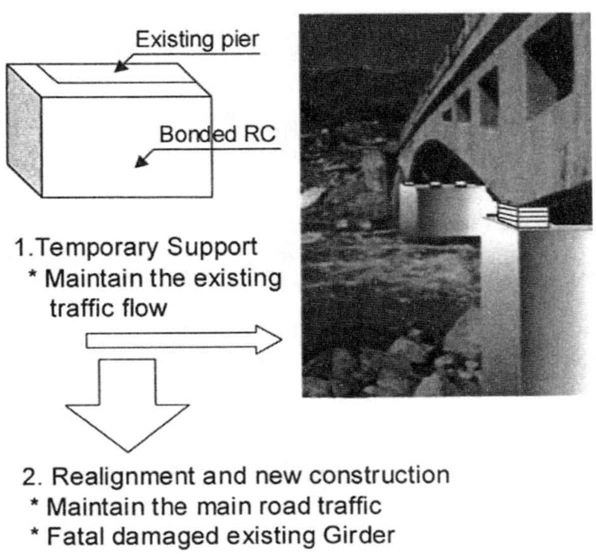

Figure 12.19 Proposed repairs at Balakot bridge in Pakistan.

connections and horizontal footing on grade). In seismic design, the load path is generated from the bottom up (footing displacements first, then transmitted to vertical members, connections, and horizontal members).

Substructure damage during earthquakes generally consists of foundation elements broken in shear, or loss of soil support, or both, caused by liquefaction. Rapid, nondestructive identification of such hidden substructure damage after an earthquake would increase public safety.

Intensive monitoring of earthquakes has revealed basic patterns about size, frequency, and location and scientists have mapped many of the region's faults. But there are a lot of variables when it comes to how earthquakes release energy.

From earthquake damages, the following modes of failure were observed:

- Spalling and cracking of concrete abutments
- Spalling of column-cover concrete
- Shear cracking of column bents
- Settlement of bridge approaches
- Displacements of steel or neoprene bearings
- Large skews and unusual geometries contributed to greater damage
- Soil failure behind abutments
- Foundation failures due to excessive shear or flexural demands
- Ground failure due to liquefaction or excessive soil deformation.

12.13.4 Literature Review on Superstructure Condition Evaluation

Ewins listed an excellent reference on dynamic testing for modal vibration measurement and analysis. For the ultimate load tests, the bridges selected for removal from service were tested to failure. These studies generally provided some insight on the ultimate load capacity and mechanisms of failure that could be used in the future.

Bridge superstructure condition evaluation research programs generally focused on two primary areas: ultimate load tests and dynamic tests.

Laboratory and field studies to evaluate dynamic properties of bridges and relate them to condition assessments have been reported extensively in past years. Some examples include:

1. Baumgartner and Waubke showed frequency measurements in tension hangers under traffic loading can be related to the end fixity of the hangers.
2. Biswas et al. reported a component evaluation technique using a hammer impact using dynamic responses. Results were confirmed with laboratory models, but field verification was limited.
3. Cawley and Adams related changes of successive mode frequencies to the existence and location of structural deterioration in beams.
4. DeWolf et al. and Gregory et al. demonstrated a relationship between dynamic testing results and structural deterioration. Sensitivity of dynamic characteristics to deterioration was shown to depend on the particular modes being observed.
5. Huston et al. reported various full-scale bridge dynamic tests, showing that dynamic characteristics may be revealed using vibratory shakers, impact hammers, and traffic and wind loads.
6. Manning suggested that a localized dynamic analysis might be advantageous because serious loss of strength of a single member may occur before it can be observed on the entire structure.
7. Mazurek and DeWolf showed in field tests that ambient traffic loads could be used as a basis for an automated monitoring scheme based on changes in vibration signatures. Their labora-

tory results encouraged further field investigations. Changes in support condition and crack development affect natural frequencies and modal amplitudes. Changes in modal frequency were up to 30 percent for changes in support condition and up to 10 percent for cracking.

8. The New York State DOT used continuous monitoring of bridge dynamic characteristics. The remote bridge monitoring system is based on measuring dynamic motion (using accelerometers) as well as strain and rotation (using inclinometers). A warning alarm is detected when significant changes in modal frequencies occur. Dynamic response data collected show up to 10 percent scatter in the modal frequency measurement.
9. Salane et al. reported dynamic tests of a bridge for detecting structural deterioration caused by girder fatigue cracks. A concrete deck on steel girders was loaded with an electro-hydraulic actuator system up to 465,000 load cycles. Accelerometers were used to determine damping ratios, frequency contents, and impedance at various stages during the loading. The test results indicated increases in damping ratios with cycles of loading caused by cracking, and a decrease in amplitude at resonant frequencies, as well as a 20 to 40 percent change in computed stiffness coefficients.
10. Salawu and Williams reported a study of the forced vibrations of a bridge before and after repair. The test results demonstrated small changes in natural frequency induced by the repair.
11. Woodward et al. conducted dynamic tests for a full-scale bridge subject to artificially induced fatigue cracking (vertical cuts) in a main girder. Preliminary field test results showed changes in dynamic characteristics due to a maximum amount of damage.

It is possible to conclude from the above studies that simpler interpretations for vibration measurements have not been reported in the literature.

Dynamic measurements can be used to evaluate the distribution of loading in axially loaded members such as cables and truss diagonal braces. The natural frequency of such axially loaded members is highly sensitive to the magnitude of the axial load.

Further work is needed to relate dynamic properties to component deterioration.

12.13.5 Literature Review on Substructure Condition Evaluation

Many state DOTs, such as Connecticut, Florida, and New York, have used instrumentation in their bridge inspection programs. During the North American Workshop on Instrumentation and Vibration Analysis of Highway Bridges in 1995, researchers and practitioners agreed that instrumentation is a viable tool for bridge inspection.

A bridge abutment generally is designed to resist backfill soil pressures; however, for a rigid frame abutment, the thermal deck expansion causes backfill pressures that far exceed the active soil pressures used in design. In addition, bridge skew results in a large horizontal gradient of the backfill pressures, producing local backfill pressures that could exceed the capacity of the abutment walls.

1. Raghavendrachar and Aktan demonstrated that multi-reference impact testing could serve as the main experimental component for comprehensive structural identification of large constructed facilities. Demanding standards are required from modal testing designs for accurate experimental measure of flexibility, which is the inverse of stiffness (also known as displacement divided by force as a function of frequency).
2. Aktan and Helmicki performed a study in instrumented monitoring of a full-scale bridge. For structures subjected to lateral loads impact testing may not be the appropriate method, forced-excitation modal testing using larger vibrators is desirable.
3. California DOT (Caltrans) has an instrumentation program for bridge inspection. It involves monitoring seismic excitations and foundation systems. Practicing bridge engineers recognized the need to evaluate and formalize the use of structural identification and instrumentation.

4. Chen and Kim developed the "bending wave" method for investigating pier conditions and local defects by measuring the velocity dispersion curve of the transverse waves propagated down from the top of a pier. The method proved suitable for short piles in softer soils.
   - The dispersion of the reflected waves from the pier bottom was used to assess overall pier conditions.
   - The dispersion of the directly arrived wave was used to assess local damages.
5. Finno and Prommer studied the impulse response (IR) method for inaccessible drilled shafts under pile caps. Several drilled shafts connected together with concrete grade beams were tested using the nondestructive IR method. Based on the field data, it was found that shaft heads that were more rigid because of larger or several grade-beam connections exhibited greater signal attenuation.
6. Hussein et al. reported the use of compression waves for investigating single pile length and integrity, settlement, and scour.
7. For the Interstate 15 bridges in Salt Lake City, UT, research on dynamic testing for the condition evaluation of bridge bents was performed using vibration tests with horizontal excitation. Modeling and experimental modal vibration test results were compared in terms of mode shapes and frequencies. The estimated location and intensity of the damage or retrofit was identified. Both damaged and repaired substructure states were used to identify the condition of the structure.
8. Pierce and Dowding's method focused on the determination of internal cracking and large local deformations caused by earthquakes for concrete bridge piers using time domain reflectometry (TDR).
9. Warren and Malvar used a falling weight deflectometer to assess structural conditions of reinforced concrete piers. Based on matching dynamic responses, deflected shapes from the FEM and testing results were compared and local stiffness and soft areas of the piers were determined. A software package called Automatic Dynamic Incremental Nonlinear Analysis (ADINA) was used by systematic changing of the stiffness parameters. This method was tested on a real bridge with timber piles in New Jersey. The geometry data were deemed sufficient to identify the effective stiffness of the piers and damaged areas.
10. The Washington State DOT studied lateral load responses of a full-scale reinforced concrete bridge to investigate the seismic vulnerability of bridge piers. This study points out that the seismic design of pier columns in the 1950s and 1960s was not adequate to sustain displacements during earthquakes. This study concluded that the latest pier design was more suitable for earthquakes.
11. The University of California, Berkeley developed a software program in 1996. BASSIN was developed for dynamic analysis of a bridge-abutment-backfill system that is subjected to traveling seismic waves. It can compute 3-D dynamic responses of an arbitrarily configured bridge-abutment-backfill system induced by compression, vertical shear, horizontal shear, and surface waves with arbitrary wavelength, amplitude, and direction of incidence.

12.13.6 The Role of a Structural Engineer in Seismic Vulnerability and Design

It all comes down to handling issues that eventually must fall on the shoulders of a structural engineer. Most of the day-to-day tasks associated with structural engineering include advising the client or highway authority on seismic criteria and costs. The structural engineer is also responsible for structural planning for new bridges, including locations and types of bearings (rotational/translational), as well as ensuring compliance with relevant seismic codes. Selection of computer software for seismic analysis, such as GT-STRUDL, SAP 2000, SEISAB and STAAD-Pro, is another task for the structural engineer.

12.13.7 Seismic/Ductility Detailing

AASHTO Guide Specifications of 1983 introduced ductile detailing practice. For bridges built prior to 1983, the following deficiencies may exist:

- Inadequate anchorage length or embedment
- Inadequate transverse reinforcement in plastic hinge regions
- Inadequate shear reinforcement at joints
- Lap splices located in plastic hinge regions.

The following QA/QC procedures must be followed for the project:

- Keeping seismic design costs to a minimum
- Solving any constructability problems in the field
- Maintaining a professional license and compliance with ethical requirements
- Purchasing liability insurance
- Training design engineers in seismic design procedures.

12.14 SEISMIC ASSESSMENT

12.14.1 Seismic Coding Guide

To estimate seismic vulnerability of the bridge population, it is necessary to collect vulnerability data for each bridge. For 80 percent of the population, vulnerability data is collected as part of the bridge inspection program. The Seismic Coding Guide provides instructions to bridge inspectors for collecting seismic vulnerability data.

Data items related to the superstructure, substructure, bearings, and soils are collected for use in determining bridge vulnerability.

Bridge vulnerability ratings: Component vulnerability ratings are developed to measure a component's ability to resist applied seismic loads. The total bridge vulnerability rating is compiled based upon individual component vulnerability ratings. Bridge vulnerability ratings are based upon the same assessment methods used in a detailed seismic investigation:

- A bridge's dynamic behavior is estimated to determine the magnitude of earthquake induced forces.
- The capacities of critical bridge components (bearing seats, columns, bearing connections, foundations) are evaluated.

Importance ratings: Importance ratings are a quantitative measure of the consequences of bridge damage. The importance rating system developed for the prioritization program considers:

- Financial loss anticipated from bridge damage
- Potential disruption to traffic caused by bridge damage
- Impact to emergency/defense/evacuation routes from bridge damage
- Impact of bridge damage to vehicular traffic on or under the structure
- Potential for utilities service disruption from bridge damage
- Impact of bridge damage to the highway facility considering network redundancy.

The above attributes are combined to determine an importance rating for each bridge, based upon valuations of the highway operations and engineering staff.

12.14.2 Major Bridge Evaluations

As part of bridge evaluations, site-response and liquefaction potential must be assessed using finite element models of each bridge to evaluate seismic response. Critical elements must be identified and conceptual retrofit recommendations are made.

Major bridges are evaluated for two seismic hazard levels:

- A functional evaluation level (450 or 500-year return period)
- A safety evaluation level (2500-year return period).

12.14.3 Selection of Bridges for Seismic Retrofit

To select bridges for seismic retrofit, rehabilitation schedule and cost-effectiveness are used with risk assessment. Key considerations in selecting bridges for seismic retrofit are:

- The schedule for non-seismic rehabilitation
- Cost-effectiveness
- Mobilization and traffic maintenance
- Seismic retrofit should be undertaken at the time of non-seismic rehabilitation since the latter accounts for a major portion of retrofit costs.

Conceptual retrofit measures: Cost estimates on a per square foot basis are developed for each of the retrofit measures. Bridge retrofit costs are estimated for the sample population using these retrofit costs.

Components which require retrofit are identified based upon their vulnerability rating. Conceptual retrofit measures are developed based upon anticipated vulnerabilities of the bridge population. Conceptual retrofit measures include:

- Connection strengthening
- Bearing replacement with isolation bearings
- Column jacketing
- Extending bearing seat lengths
- Foundation strengthening
- Soil improvement.

12.15 THE ROLE OF SEISMIC DESIGN CODES

12.15.1 Need for Comprehensive Specifications

The cost of a bridge increases considerably when seismic effects are included in the design criteria. It makes economic sense to incorporate the latest state-of-the-art in design codes. After the 1971 San Fernando earthquake, a great deal of interest was generated in this discipline and as a result, FHWA published "Seismic Design and Retrofit Manual for Highway Bridges" in 1987. The 1989 Loma Prieta, 1994 Northridge, and the more recent earthquakes in California have revealed safety concerns, even for structures designed in conformity with the latest codes.

Seismic engineering deals with a host of disciplines, such as seismology, earthquake engineering, and foundation design. Case studies highlight the difficulties in applying current procedures. Seismic codes, whether new or old, involve a large number of variables, such as seismic category, dynamic behavior, the geometry of the bridge, and of course, geotechnical considerations.

The topics refer to aspects such as:

- Seismic categories/zoning
- Geotechnical
- Analytical
- Structural detailing issues.

The purpose of a seismic code is:

- To protect life and property

- To develop a quality structure
- To implement uniformity in construction.

As earthquakes continue unabated around the globe, seismic codes have been undergoing a fresh round of development phases every five years or so. Design guidelines are being upgraded for new bridge design and for strengthening or retrofit of existing bridges.

12.15.2 Applicable Structural Codes for Highway and Railway Bridges

The following is a brief timeline of events related to seismic design and retrofit of bridges.

1983: AASHTO Guide Specifications for the Seismic Design of Highway Bridges and Seismic Retrofitting Guidelines for Highway Bridges FHWA RD-83-007

1987: Seismic Design and Retrofit Manual for Highway Bridges FHWA IP-87-6

1991: AASHTO Seismic Design of Highway Bridges is incorporated into AASHTO Standard

Specifications

NEHRP Horizontal Acceleration Maps adopted in AASHTO Seismic Specifications

FHWA Requests State DOT's for retrofit plan of critical and eligible structures

1995: Seismic Retrofitting Manual for Highway Bridges FHWA RD-94-052 issued

The release of FHWA RD-94-052 in 1995 downgraded essential bridges from Seismic Performance Category B to C, necessitating the collection of more detailed bridge seismic vulnerability data.

The progressive development of seismic codes for highway bridges may be summed up as:

1. The "old" code (AASHTO Division 1-A Specifications, 1991).
2. "Current" codes (AASHTO Standard Specification for Highway Bridges, Sixteenth Edition,1996).
3. AASHTO LRFD 2007, Load Resistance Factor Design Specifications for Highway Bridges.
4. Publications ATC 18/23, ATC 32, NCHRP Project 12-49, 2001 and NCHRP Report 472, 2002.

In addition for railway bridges,

5. AREMA 2002 Seismic Design for Railway Structures, Chapter 9 published by American Railway Engineering and Maintenance-of-Way Association.

The AASHTO Division I-A provisions were originally issued as a guide specification in 1983 and were subsequently incorporated with little modification into standard specifications in 1991. The provisions contained in the AASHTO LRFD Bridge Design Specifications are based on provisions and approaches carried over from Division I-A, "Seismic Design," of the AASHTO Standard Specifications for Highway Bridges (AASHTO, 1996). The seismic design LRFD code is similar to LFD, with only minor differences in the following aspects:

- Live loads
- Braking forces
- Return periods of earthquakes
- Load resistance
- Response modification factors.

Table 12.9 Response modification factor.

Response Modification Factor	LRFD	ASD / LFD
Wall type pier	For critical 1.5 essential 2.0 other	2
R.C. pile bents with vertical piles	1.5 for critical 2.0 for essential 3.0 other	3
Single column	1.5 for critical 2.0 for essential 3.0 other	3
Multiple column bent	1.5 for critical 3.5 for essential 5.0 other	5
Foundations	0.5 R but not less than 1.0	0.5 R
Pile bents	1.0 R	1.0 R
Connections-Superstructure to abutment	0.8	0.8
Connections-Columns, piers	1.0	1.0

The current LRFD provisions are based on seismic hazard, design criteria, and detailing provisions that are now considered over 20 years old. In many cases, substructure response modification factors (R) have decreased in the LRFD code, resulting in higher design moments and forces, since

Elastic Seismic Design moment or Force =
Elastic Moment or Force from seismic analysis/R

Substructure Response Modification factors R revised.

12.15.3 Role of Local/State Seismic Codes

AASHTO LRFD seismic specifications are not addressed in sufficient depth or detail to cover every aspect of seismicity, even within the U.S. Current LRFD bias has been primarily for major seismic zones such as 3 and 4, rather than for seismicity of lower seismic zones. Realizing this, independent regional seismic studies were carried out by many states including NJDOT and NYCDOT. The ground motion seismic hazard study for New York City was sponsored by NYSDOT, NYCDOT, and FHWA to develop ground motions in rock, based on a probabilistic seismic hazard study applicable to New York City and the surrounding region. Detailed changes in provisions are also reported in PennDOT Design Manual No. 4.

It is observed that there has been an attempt by state codes to supplement the semi-empirical approach and try to implement only selective provisions. Applications of three state seismic codes from New Jersey (NJ), Pennsylvania (PA), and New York (NY), have shown refinements of AASHTO code with emphasis on different aspects. For example, New Jersey addresses seismic retrofit aspects, Pennsylvania addresses seismic detailing, and New York has multi-modal analysis and soil behavior aspects, in considerable detail.

Modal Superposition: A simplified structural model with lumped mass is used to create the structural model. This method may work for a smaller bridge but does not yield enough information for longer or more complex crossings.

Create a mathematical model to represent the physical structure, which shows the spatial distribution of the mass and stiffness of the structure to the extent necessary to perform adequate calculation. Model common bridges for either a hand single-mode or a computer multi-mode analysis.

For important structures life safety needs to be addressed in the event of a major earthquake. Hence, a two-level earthquake hazard approach needs to be adopted:

Operating design earthquake (ODE) has a return period of 450 years during which there will be no interruption to traffic. When subjected to ODE the structure will be designed to respond essentially in an elastic manner. There shall be no damage to primary members. Only minimal damage to secondary members is permitted.

Maximum design earthquake (MDE) has a return period of 2500 years during which some interruption to traffic is permitted. When subjected to MDE the structure will be designed to respond essentially in an inelastic manner. There shall be no collapse. Any structural damage will be limited to elements that can be easily repaired. The structure will be designed with ductility and strength to survive deformations and loads imposed on the structure during MDE.

Seismic zones versus categories:

LRFD	Code	Seismic	Zones	1 to 4
ASD/LFD	Code	Seismic	Category	A to D

No change in the range or magnitude of acceleration coefficients.

12.15.4 Critical Bridges Importance Category Introduced on Functional Requirement Basis

Table 12.10 Functional classifications of bridges.

Importance Category	E.Q. Return Period	Open To All Traffic	Emergency Vehicles	Security/Defense
Critical	2500 years	Yes	Yes	Yes
Essential	450 years	No	Yes	Yes
Other	<450 years	No	No	No

For vertical non-seismic bridge forces, the primary response is in terms of vertical displacements and to a small extent on the time-dependent varying position of truck live loads. However, for seismic bridge forces, the response is in terms of time-dependent sway and rotations, initiated by soil movements under the footings.

1. In the U.S., seismic design of a bridge is governed by applicable AASHTO LRFD code provisions. AASHTO LRFD seismic zone classifications (zones 1 to 4) have been redefined compared to ASD/LFD code classification (seismic category A to D).
2. Statically determinate (single span bridges) and those located in zone 1 do not require a formal seismic analysis, except for minimum seat width requirement. Also, suspension cable, cable stayed, arch and movable types of bridges are not covered in detail by AASHTO LRFD Bridge Design Specifications.
3. A seismic design is usually not required for buried structures or culvert structures.
4. Critical bridges must remain open to all traffic after the design earthquake of a 2500-year return period event. Essential bridges must be open to emergency vehicles and for security/defense purposes immediately after design earthquake of a 475-year return period event.
5. A percentage of live load may be considered in computing seismic forces, depending on the importance of the bridge.
6. Provide seismic ductility design at the locations where plastic hinges will form, on all new structures.
7. LRFD guidelines recommend use of cracked section properties in considering the rigidity of bridge piers (contrary to the current AASHTO Division 1-A, which used gross properties), and allows for the ability to use soil pressures to resist seismic loads at the abutments.

8. Design issues such as soil-structure interaction, liquefaction, structure displacement verification, P-Delta limitations, and short-period structure response adjustments are also addressed by the LRFD guidelines.
9. Ductility is a very important characteristic of structures because a ductile structure can absorb much more force than a non-ductile structure before it fails. Conversely, non-ductile structures such as un-reinforced masonry or inadequately reinforced concrete are very dangerous because of the possibility of brittle failure.

12.15.5 Load Combinations

For computing elastic design forces, moments and displacements' combined load effects in three directions (transverse, longitudinal, and vertical) will be considered. Inertia effects of design live load without impact shall be included in seismic analysis. The load combinations for extreme events were listed in an earlier chapter in Section 2.

AASHTO—Extreme Event-I load combinations shall be applicable. The guidance provided in Section 3.4 of the AASHTO LRFD Bridge Design Specifications should be followed. Forces from dead load analysis and analysis for other applicable loads for extreme events will be combined with the forces from single or multi-mode analysis as follows:

100 percent of longitudinal seismic forces + 30 percent of transverse seismic forces

100 percent of transverse seismic forces + 30 percent of longitudinal seismic forces

The bridge seat width should be adequate and shall be checked. The guidance provided in Subsection 4.7.4.4 of the AASHTO LRFD Bridge Design Specifications should followed.

Minimum seat width, N: For SPC A & B

$$N = (8 + 0.02L + 0.08H)(1 + 0.000125\,S^2)\text{ inch}$$

For SPC C & D

$$N = (12 + 0.03L + 0.12H)(1 + 0.000125\,S^2)\text{ inch}$$

H = Height of column, pier or abutment in feet

S = Skew angle at support

12.15.6 Computer Models and Implementation of Codes

The common LRFD approach involves an understanding of:

- Basic terminology/glossary
- Definition of seismic hazards
- Deterministic and probabilistic seismic hazard analysis
- Selection of seismic zones.

The theoretical model is based on:

- Multi-degree-of-freedom systems
- Pseudo velocity and acceleration
- Concepts of quasi-static
- Response spectrum
- A time history analysis.

Fundamental applications of D'Alembert's Principle and Newton's second law of motion result in mathematical equations expressed as non-linear matrices, eigenvalues, eigenvectors, and as deformation response spectra.

The dynamic equilibrium equations may be summarized in terms of standard mass and stiffness matrices as $[m]\{\ddot{x}\} + [c]\,\dot{x} + [k]\{x\} = \{P_t\}$

Using the transformation, displacement matrix $\{x\} = \Sigma_{i=(1 \rightarrow p)} \{\varphi\}_i q_i$

When planning a new bridge project, an efficient structural engineer covers the following areas:

- Identifying critical facilities and lifelines
- Determining seismic criteria
- Structural planning
- Selecting the locations and types of bearings (rotational/transnational)
- Selecting computer software for seismic analysis.

Various steps required for the structural analysis are:

- Modeling of the superstructure
- Modeling of the substructure
- Modeling of bearings (guided, unguided, and fixed)
- Idealizing sub-models for connections
- Selecting methods of analysis
- Selecting numerical methods.

Additional software for seismic analysis:

BASSIN—Dynamic analysis of bridges, abutment, backfill systems subjected to traveling seismic waves, Dendrou B. Ababian Associates, El Segundo, CA.

ISADAB—Inelastic static and dynamic analysis of bridges, M. Saidi, University of Nevada, Reno, Nevada.

NEABS—Nonlinear earthquake analysis of bridge systems, University of CA, Berkeley.

MicroSARB—A microcomputer program for seismic analysis of regular highway bridges.

SEISAB—Seismic analysis of bridges. Methods of analysis for zone 2: For computation of seismic forces the magnitude of seismic forces will depend upon:

- Dead weight of the structure
- Ground motion (acceleration coefficient)
- Type of soil
- Fundamental period of vibration
- Importance classification.

Single span bridges: With a single span, the ability to resist an earthquake is increased, due to a higher relative stiffness of abutments compared to piers.

1. No formal analysis is required for seismic forces. The guidance provided in Subsection 4.7.4.2 of the AASHTO LRFD Bridge Design Specifications should be referred to.
2. Width of the bridge seat should be adequate and shall be checked. The guidance provided in Subsection 4.7.4.4 of the AASHTO LRFD Bridge Design Specifications should be referred to.
3. Abutment does not need to be designed for seismic forces from the superstructure except for bridges with integral abutments. For bridges with semi-integral abutments, only the bridge seat width shall be checked.
4. Abutment shall be designed for static earth pressure and additional seismic induced forces earth pressure, using Mononobe-Okabe method, which is an extension of Coulomb's method for soil pressure on retaining walls. Backfill is assumed unsaturated so that liquefaction effects are negligible. The backfill is assumed cohesion-less.

5. Seismically induced active and passive pressures will be considered.

6. Long multi-frame bridges shall be analyzed with multiple elastic models. A single multi-frame model may not be realistic since it cannot account for out-of-phase movement among the frames and may not have enough nodes to capture all of the significant dynamic modes. Each multi-frame model should be limited to five frames plus a boundary frame or abutment on each end of the model. Adjacent models shall overlap each other by at least one useable frame. The boundary frames provide some continuity between adjacent models but are considered redundant and their analytical results are ignored. A mass less spring should be attached to the dead end of the boundary frames to represent the stiffness of the remaining structure.

7. Engineering judgment should be exercised when interpreting the deformation results among various sets of frames since the boundary frame method does not fully account for the continuity of the structure.

8. In the future, the classification of bridge categories will be performance based. Emphasis will be placed on pushover analysis, ductility, use of new soil site factors, liquefaction effects, and loss of bearing capacity during a seismic event.

The structural engineer needs training in structural dynamics, geotechnical engineering, computer software, and seismic detailing. This may require developing new courses at the university level, continuing education courses, and holding suitable examinations for licensure.

12.16 SEISMIC DESIGN

12.16.1 Need for Monitoring Earthquake Effects

An earthquake event is not just localized to a bridge. It is widespread, sometimes over hundreds of miles. Factors affecting bridge foundation behavior are:

- Movements of natural ground in the vicinity of the bridge
- Land slides
- Man-made embankments
- Cut slopes
- Drainage structures
- Utilities
- Bridge approaches.

It is observed that following a major earthquake event, not all the bridges located in the seismic zone behave exactly the same way. At each site the type of soil is different. Due to uncertainty of the location of an epicenter, there are not enough studies available of the unique soil response in each case. It has been difficult, though not impossible, to determine in advance the magnitude of an earthquake and the response of a bridge foundation. Also, bridges need to be monitored for follow-up damage from aftershocks, which may continue for several weeks.

The condition of bridges in post seismic event may be determined as:

- Those subjected to collapse
- Those subjected to severe structural damage
- Those subjected to minor structural damage
- Those subjected to little damage.

Regional geologic and tectonic history needs to be reviewed. Interaction between earth structures and bridges needs to be considered.

Selecting deep versus shallow foundations for seismic resistance: Deep foundations and those well anchored into the soil are more stable against seismic forces than shallow foundations.

Piles and caissons are likely to perform better than spread footings supported on soils that are subjected to differential settlement.

Improving performance of approach slabs: Settlement at approaches may occur depending upon structural backfill material and embankment height. During earthquakes, the dynamic response of each element is different. Longer and stronger approach slabs (10 feet to 25 feet long) are preferred. The following action is needed to improve the combined structural behavior of bridges and approaches:

1. Replacing or rehabilitating a bridge approach slab helps to minimize seismic impact on a bridge.
2. Providing a pavement relief joint for all concrete pavement types, when one has not been provided previously or when the expansion length of the concrete pavement is less than 1000 ft apart.
3. Servicing the existing pavement relief joint as needed to ensure a smooth ride.

12.16.2 Bridges That Are Located on Rivers

Soils around bridges that are located on rivers are likely to have a high water table, requiring an increased drainage provision behind abutments. Submerged water pressure exerted on an abutment is greater than dry soil pressure. During a seismic event, riverbeds and banks, unless on rock, footings are subject to movements resulting in increased erosion and scour of foundations. Use of the Mononobe-Okabe Method, which is an extension of Coulomb's Theory for seismic effects, does not consider subsurface or above riverbed water elevations.

Piers in deep rivers may be subjected to hydro-dynamic forces in addition to the inertia forces as a result of violent shaking of ground. Hydro-dynamic forces need to be considered as well.

12.16.3 Recommendations for New Bridge Design Based on Investigation of Failures

1. Provide adequate seat widths, especially for simply supported spans to prevent girders from becoming unseated (Section 12.15.5).
2. Use continuous spans in place of a series of simply supported spans. This will reduce the need for expansion joints.
3. Use approach slabs with positive ties to abutments such as in integral and semi-integral abutment bridges. This will reduce the effect of soil slumping behind the abutment.
4. Bearings for simply supported spans should have adequate strength in restrained directions, i.e., able to resist lateral seismic loads. Check stability of bearings in unrestrained directions to prevent anticipated displacement.
5. Column reinforcement: Place laps or splices away from locations of plastic hinges (at column to cap or column to footing connections). Provide adequate confinement of transverse ties or spiral reinforcement in columns.
6. Footings: Design footings to resist full shear and moment demands. Use soil improvement techniques to reduce potential for soil failure or liquefaction.

12.16.4 Recommendations for Existing Bridge Design Based on Investigation of Failures

Identify and prioritize those bridges first for repairs or replacement, based on vulnerability.

1. Either extend seat width (Figure 12.23) or provide cable restrainers (Figure 12.20) if seat widths are inadequate.
2. Provide continuity by casting a continuous deck slab and by structural modifications.
3. Eliminate expansion joints to improve seismic performance.

4. Consider bearings retrofit or replacement. Use seismic isolation technology (Section 12.20) to retrofit bridges with short stiff columns.
5. Provide column jacketing if there is inadequate confinement (i.e., insufficient transverse ties or hoops) in the column or if there are splices in or laps in hinge zones.
6. Provide footing extensions or overlays if there is a lack of footing reinforcement, inadequate shear strength, or soil bearing strength.

12.17 COMMON RETROFIT CONCEPTS AND CODE APPLICATIONS

12.17.1 Scope of Work

The retrofitting of bridge structures is primarily to prevent loss of life due to collapse of bridges in large seismic events. It is much less expensive to repair hinges than it is to replace the entire structure. The most destructive earthquake failure mode for bridges is the unseating of in-span hinges.

The second most seismically vulnerable aspect of a bridge is the columns. These members hold the structure up and if they shear or are weak in flexure, the structure may also experience a catastrophic failure.

Other retrofits such as in-fill walls, catcher blocks, foundation work, etc. will be considered.

The use of cable restrainers as a primary means of resisting seismic forces is not permitted.

1. The seismic retrofit design of existing highway structures shall follow the guidelines of the FHWA publication titled "Seismic Retrofitting Manual for Highway Bridges" currently numbered as FHWA-RD-94-052, May, 1995.
2. Highway structures shall be retrofitted to SPC B earthquake.

Evaluating vulnerability ratings: Structural component vulnerability ratings are determined at different seismicity levels, consistent with the range of anticipated seismic activity. The need for retrofit is based on other considerations, including:

- Component seismic vulnerability ratings
- Inspection report showing overall condition of bridge
- Live load rating.

Currently, bridges retrofitted to Caltran's (California Department of Transportation) standards are designed to survive a seismic event of the same magnitude as that used for new design. Emphasis is placed upon ensuring that any sustained damage is repairable and that collapse is prevented.

In general, the Japanese program bears a number of similarities to that adopted by Caltran.

Bridge retrofit in Japan: Japan began retrofitting bridges in the 1970s, at about the same time as Caltran's Phase I retrofit program began. The scale of the Japanese program was a great deal larger than that of Caltran's program, with several billion dollars spent on seismic retrofit during the 1970s and 1980s. The brunt of this effort was directed toward the installation of restrainers across movement joints and increasing seat lengths (similar to Caltran's Phase I) followed by column retrofitting.

Structural components: The scope of work is as follows,

1. Upgrading existing bridges to meet new seismic criteria—Screening of seismically deficient bridges is required. Priority of repairs needs to be established. Only bridges with higher Importance Factor will have top priority.
2. Retrofitting bridges damaged in an earthquake—There are three categories, including:

- Bridges with minor repairs
- Bridges with major repairs
- Bridges which require replacement.

3. Replacing high rocker and roller bearings.

4. Extending bearing seats (Figure 12.23).
Bearing seat lengths must meet the minimum support lengths as per the design specifications. This must be addressed on rehabilitation projects. Seat extensions should be provided.

5. Add shear blocks and/or pedestals: Structures which are deficient in areas such as seat length and bearing instability, or have inadequate superstructure to substructure connections, may be retrofitted by addition of shear blocks and/or dowel bars, or by construction of concrete pedestals which will act as shear blocks and alleviate bearing instability.

12.17.2 Nonstructural Components

Although nonstructural components are lighter weight, they can participate as structural components. Usually 25 perent of combined weight is considered as the required minimum. Such components are likely to include:

- Stay-in-place forms used for deck construction
- Large diameter utility pipes
- Bridge mounted signs
- Light poles
- Railings
- Sidewalks.

12.17.3 Back Walls of Integral and Semi-Integral Abutments

Back walls of integral and semi-integral abutments have pin connected approach slabs and influence the seismic response of such bridges. No remedial action is required for bridge types that are exempt from a ranking process. These include:

- Bridges designed to seismic standards
- Bridges with seismic performance category (SPC) A that is not a critical facility
- For bridges with SPC B, only a vulnerability rating for restrainers, support lengths, and liquefaction vulnerability rating (LVR) for certain sites is required.
- For bridges with SPC C or D, in addition to a vulnerability rating for restrainers, support lengths, and LVR for certain sites, a vulnerability rating for columns, abutments, and foundations is required.

12.17.4 Seismic Hazard Rating

Simply supported spans with high skew and inadequate seat width are most vulnerable to overturning. Continuous superstructures with integral abutments are most stable.

A reconstructed bridge (replacing a damaged or completely-destroyed bridge) should:

- Include seismic features
- Be an assessment of:
  - What types of repairs are necessary, and
  - What type of financial/material assistance will be provided
- Include an assessment of what types of seismic features should be included in the design
- Include site-specific features and modifications

- Be repaired if required in a seismically-improved manner
- Be based on information about how to repair it in a seismically-improved manner.

According to Penn DOT Design Manual, Part 4, all bridges that are scheduled for rehabilitation shall be evaluated with regard to seismic failure vulnerability. The purpose of this evaluation is to assess seismic retrofit measures and to incorporate into the rehabilitation plans those measures deemed warranted to eliminate or mitigate such failure vulnerability. (Bridge Safety Assurance Seismic Vulnerability Manual.)

At the time of original bridge construction, there were no seismic criteria in effect. Hence, it is important to evaluate by analysis the seismic vulnerability of the bridge, using three-dimensional finite element software, such as SAP or ADINA.

The bridge R-factor is used as the principal parameter describing the bridge structure. A method for computing column damage fragilities for the following three damage states, based on the bridge force reduction factors (R-factors), can be derived:

1. Column longitudinal bar buckling
2. Column wrapping
3. Column concrete cover spalling

High quality carbon composite jackets or casings can be wrapped around columns by wrapper machines. Jacket material is made up of carbon fibers pre-impregnated with epoxy resin.

The methods outlined below are recommended procedures and are not intended to restrict the ingenuity and creativity of the design engineer. If it is found through a seismic analysis that the substructure is in need of seismic retrofit, it will probably be economically advantageous to study bearing replacement as part of a retrofit.

Most U.S. states require seismic analysis and complete seismic assessment on all rehabilitation projects. As part of this assessment, the designer must indicate all deficient seismic items and provide preliminary details for any needed seismic retrofits. The structure needs to be completely upgraded to handle seismic loads. Standard retrofit details may be applicable.

FHWA Research Reports FHWA-IP-87-6, FHWA-RD-83-007, and FHWA-RD-94-052 also contain acceptable references for retrofit details.

12.17.5 Seismic Retrofit Report

A seismic retrofit report shall be prepared to provide a determination as to a bridge structure's eligibility for a seismic retrofit.

A flow chart to provide guidance in determining if a bridge structure qualifies as a seismic retrofit candidate can be found in Section 1.45.12. The results of the analysis, performed in accordance with the flow chart, shall be provided in the seismic retrofit report.

In preparing the seismic retrofit report, the following guidance shall be followed. Initially, seismic retrofitting of a bridge structure shall only be considered under one or more of the following conditions:

1. The planned work will involve widening of a deck by more than 30 percent of its area
2. The planned work will involve an entire deck replacement.
3. The planned work will involve superstructure rehabilitation or replacement, major abutment or pier repairs to bearing seat areas, or bearing repairs or replacement.

Several methods of seismic retrofit are outlined for bearings and expansion joints within the FHWA Retrofit Manual that is referenced above. Of these methods the following are recommended for consideration in order of preference.

- Modify existing bearings to resist seismic loads or to prevent toppling of existing bearings by installing longitudinal displacement stoppers

- Longitudinal joint restraints as outlined in Subsection 5.2.1 of FHWA Retrofit Manual
- Modifications to steel bearings, such as:
  - Increase size, number, or embedment of anchor bolts.
  - Increase the outer diameter of the pin head.
  - Increase the width of the expansion rocker.
  - Increase the top and bottom dimension of the pintle detail for increased movement.
- Modifications to elastomeric nearings:
  - Secure bearing against horizontal and vertical movement.
  - Modify the plan area and/or thickness of the elastomeric bearing to reduce seismic forces to the substructure.

12.17.6 Determine Seismic Forces per Latest Criteria for Replacement Bearings

1. The current AASHTO LRFD Specifications for Highway Bridges should also be referred to for guidance in providing designs for pot, disc, and elastomeric type bearings.
2. The AASHTO Guide Specifications for Seismic Isolation Design shall be used for designing isolation bearings when they have been deemed necessary for accommodating seismic loads. These bearings have special performance characteristics, which will alter the dynamic response of a bridge. Superstructure forces can be reduced by factors of 2 to 5 in the lower seismic zones and there are corresponding reductions in the forces transferred to the piers and abutments.
3. Columns: The second most seismically vulnerable aspect of a bridge is the columns. These members hold the structure up and if they shear or are weak in flexure, the structure may also experience a catastrophic failure. Columns are retrofit utilizing welded A36 steel casings ranging from 1/4 to 5/8 inch in thickness. Steel casings are made circular or as close to circular as practical using an elliptical shape for rectangular columns. The shells are pre-manufactured in two or more sections then field welded to encircle the column. For circular casings, a minimum 1 inch gap is left all around and the void is filled with grout. Grout is pumped from the bottom up to guarantee no air gaps will be created. Injection ports for the pump will need to be spaced vertically depending on the capacity of the pump used.
4. Catcher blocks: The simplest and least expensive means of allowing the superstructure to "float" over the substructure in a large earthquake is to place seat extenders (catcher blocks) under the girders. Once the bearings fail or shear, the weight of the girders transfers to the catcher blocks to ride out the earthquake without a drop failure.
5. Seismic hooks: Where possible, the new top and bottom mats should use seismic hooks to tie the two mats together. Seismic hooks are usually #4 bars with hooked ends that grab the top and bottom mats of a footing (Figure 12.28).
6. Anchor slabs: If a situation arises in which work to a vulnerable substructure is difficult and costly (say a deep canyon), an anchor slab can be used to stabilize the structure at road level by driving the loads to the abutments. If the superstructure has a rigid transverse configuration (such as box girder bridges), stability both longitudinally and transversely can be achieved by creating a fixed moment connection at the abutments.

An anchor slab replaces the approach slab in which the new slab is anchored firmly with new piles or is simply a large concrete block. As the superstructure moves away from the abutment or tries to rotate in plane due to the transverse swaying of the middle of the structure, the new anchor slab is engaged and resists this motion. Thus, the stiffness of the superstructure, by being fixed at the ends, will increase the transverse stiffness of the system.

Jacking method for bearing replacement: The bridge deck and supporting girder needs to be lifted. This is done in small increments using a hydraulic jack. New bridges have transverse girders and jacking positions shown for future bearing replacement. However, existing beams may not have a provision for lifting. The uplift force is calculated and an analysis is required for jacking stresses induced in the girder and deck slab. The travel lane needs to be shut down so that live load is zero.

12.17.7 Alternate Retrofit Concepts

Cable restrainers: Restrainers are designed to limit the relative displacement between superstructures at movement joints to prevent span unseating (Figure 12.20). The most common form of restrainer retrofit used in the Caltran's Phase 1 program consists of installing cable restrainers anchored to the webs or diaphragms of concrete bridges or the bottom flanges of steel girders. High strength steel bars may be used alternatively.

Restrainers are typically installed with enough slack to allow normal movements from temperature/creep and shrinkage without engaging. In California they are periodically checked and adjusted when necessary.

Restrainers used in Japan: The Japanese use similar strategies to prevent support length failure, although the restrainer configurations are different. To allow rotation/translation, steel chains and steel plates with pins are utilized in addition to cable strategies.

It is noted that web connections for restrainers are only recommended when the web has sufficient strength to transmit simultaneously seismic and dead load forces in this region. Bottom flanges of girders offer more suitable locations for restrainer attachments.

Footing retrofits: This type of retrofit is expensive and difficult to perform.

Fluid viscous dampers provide complete protection for bridges against column vulnerability due to shear failure.

Column additions or replacements: In extreme situations, column can be added or replaced to gain ductility and rigidity. This work is difficult and costly. Column replacement is an extreme retrofit measure. A bridge is jacked up to relieve the load into an existing column and the column removed. Typically, 10 percent DL lateral capacity is used for the support system in moderate seismic areas.

- Column vulnerability due to foundation deficiencies (for single column bents on piles)
- Pier vulnerability due to flexure failure at column reinforcement splices
- Abutment vulnerability: Abutment failure is linked indirectly with settlement of approach fill, such as for spill through.

12.18 STEEL AND CONCRETE BRIDGE DETAILING REQUIREMENTS

12.18.1 Ductility Requirements

Ductility at plastic hinge locations will be provided so as to dissipate energy during a seismic event. A comprehensive set of special detailing requirements for steel components, which are expected to yield and dissipate energy in a stable and ductile manner during earthquakes, were developed in AASHTO specifications. These include provisions for ductile moment-resisting frame substructures, concentrically braced frame substructures, and end-diaphragms for steel girder and truss superstructures.

For concrete bridge design, the minimum amount of longitudinal steel was reduced to 0.8 percent. This is expected to result in significant material cost savings when used with capacity design procedures.

Splices in reinforcing steel: Splicing of flexural reinforcement is not permitted in critical locations of ductile elements.

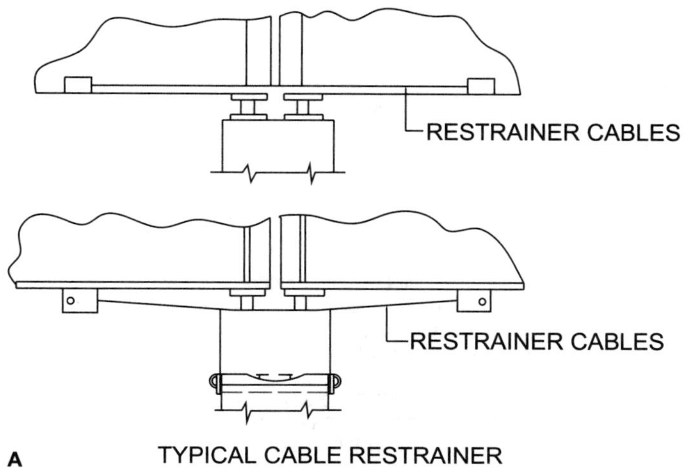

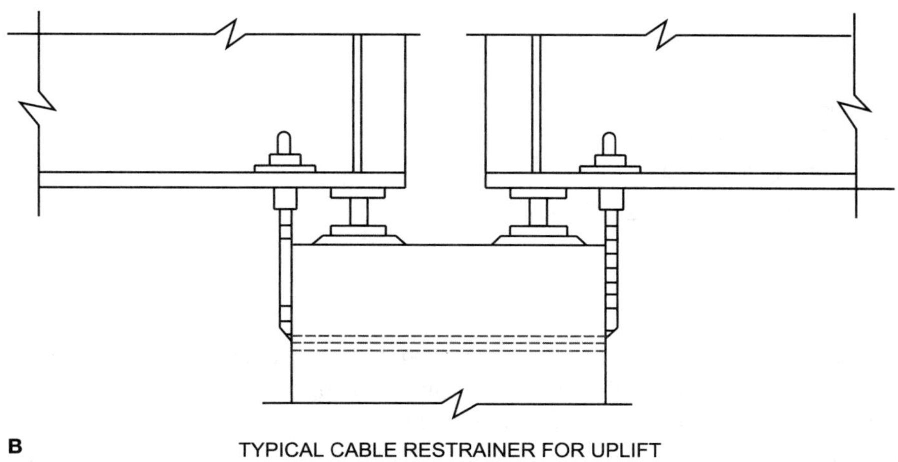

Figure 12.20 (A) Steel girders retrofit. (B) Alternate use of cable restrainers for steel girders.

Hoop and spiral reinforcement splices: Ultimate splices are required for all spiral and hoop reinforcement in ductile components.

Detailing of reinforcing bars:

1. Reinforcing bars provided in foundation and column/wall shall be adequate to resist the design moments and shear forces.
2. Standard details for ductility will be followed. Whenever appropriate, the design detailing requirements recommended by Applied Technology Council, 1996, Publication ATC-32—Improved Seismic Design Criteria for California Bridges should be followed.

12.18.2 Guidelines for Reinforcement

Footing Reinforcement

Footing reinforcement shall be designed for the applied loads, but the following minimum requirements shall be provided to maintain the integrity of the footing in the event of seismic loading.

1. Top and bottom reinforcement in footings in both the transverse and longitudinal directions shall be provided with hooks (180 degrees or 90 degrees) at both ends.

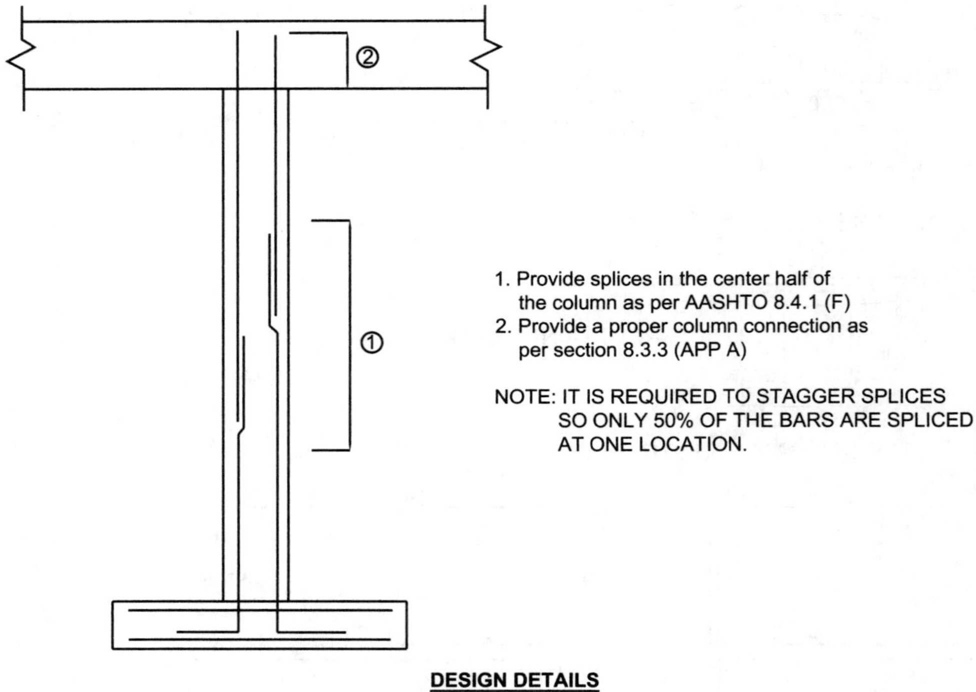

Figure 12.21 Column detailing.

2. Vertical stirrups using #13 bars with alternated 135 degree hooks at one end and 90 degree hooks at the other shall be used in all footings to connect the top and bottom reinforcement mats. Spacing shall be a maximum of 1.2 m in both directions.
3. The bottom reinforcement mat in footings with piles shall be placed 50 mm clear above the tops of the piles. In special cases, where design requirements dictate and the pile pattern permits, the bars may be located between piles. In this case, a minimum clear distance of 75 mm shall be maintained between the reinforcing bars and the piles.
4. The vertical compression reinforcement of all abutment stems and walls shall be doweled into the footing with #16 bars. These dowels should have 180 degree hooks on both ends. Minimum reinforcement shall be #16 bars at 450 mm.
5. The minimum top reinforcement for a continuous pier footing shall be #19 bars at 300 mm in both the transverse and longitudinal directions.
6. The minimum top reinforcement for an individual pier footing shall not be less than 50 percent of the area of the designed bottom reinforcement or #19 bars at 300 mm in both the transverse and longitudinal directions.

Abutment Reinforcement

The top layer of bridge seat reinforcement for steel girder, prestressed concrete I-beams, and spread prestressed concrete box beams shall be #25 bars at 150 mm. For adjacent prestressed concrete box and slab unit structures, the top layer of bridge seat reinforcement shall be #25 bars at 200 mm.

The minimum vertical reinforcement shall be #16 bars at 450 mm. The entire capacity of these bars shall be developed by embedment or lapping the bar.

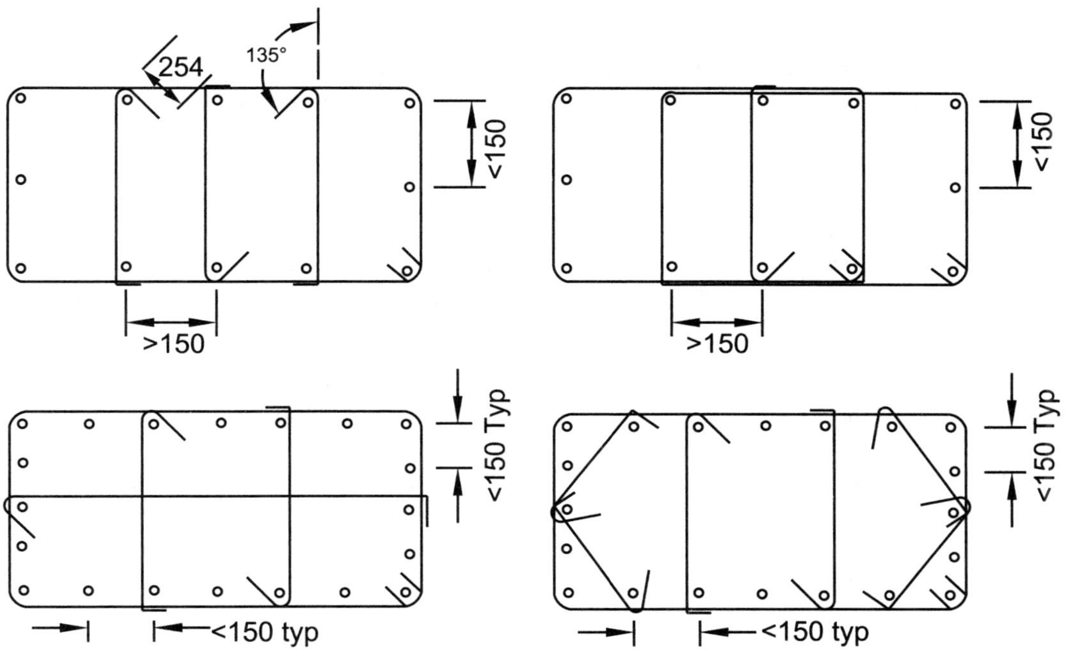

Figure 12.22 Alternate methods for detailing of transverse reinforcements in columns/piers (dimensions in mm).

Column Reinforcement

All lap splices shall be located within the middle half of the column height. Dowels shall extend at least one-quarter of the column height or 3.0 m, whichever is greater. Splices in the vertical design reinforcement shall be staggered. Vertical reinforcement shall be extended into the pier cap for the full embedment length. Continuous ties shall surround the vertical reinforcement. Ties shall be not less than #13 bars. The spacing of lateral ties in the interior length of pier columns shall not exceed the least plan dimension of the compression member or 300 mm, whichever is less. For seismic reasons, when a plinth is provided at the base of a column, the design vertical reinforcement of the columns shall extend into the footing. Additional reinforcement in the plinth may be required due to other design forces.

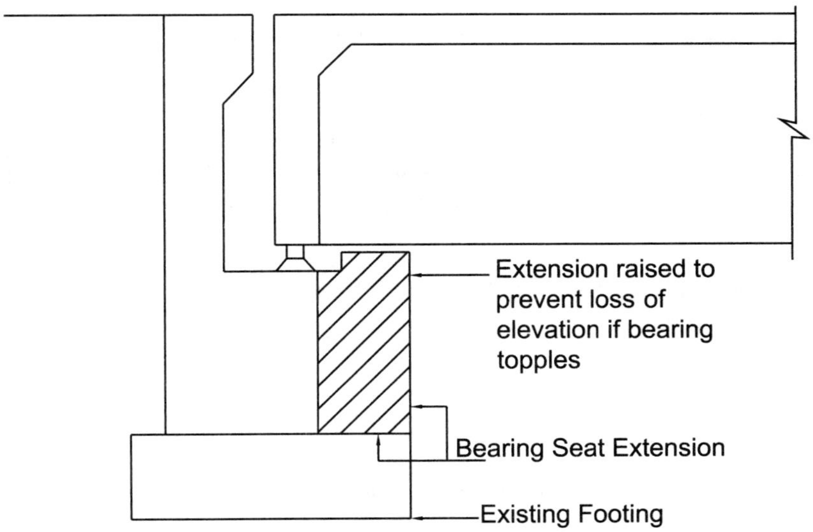

Figure 12.23 Bearing seat extensions.

Pier Cap Reinforcement

The splices of top bars in the cap beam shall be staggered so no more than 50 percent of the bars are spliced at one location. The splices shall be located in areas of low negative moment. The splices of bottom bars in the cap beam shall be staggered so no more than 50 percent of the bars are spliced at any one location. The splices shall be located in areas of low positive moment. Both the bracket and corbel and the strut and tie methods recognize that direct shear is the primary behavioral mode instead of flexure, and is resisted by tension reinforcement across the shear plane. As a result of these methods, more reinforcement may be required in the top of the overhang than would be required if a normal cantilevered beam is assumed.

Temperature and Shrinkage Reinforcement

Exposed faces of abutments, walls, and solid piers shall be provided with a minimum reinforcement of #16 bars at 450 mm placed vertically and #16 bars at 300 mm placed horizontally to resist temperature and shrinkage stresses.

The rear faces of abutments and walls shall be provided with a minimum reinforcement of #16 bars at 450 mm in both directions.

12.19 RETROFIT AND STRENGTHENING

12.19.1 Description

An important aspect of maintenance is retrofit. An overview of retrofit methods is presented here. Retrofit may be defined as *strengthening or upgrading the structural quality of a bridge to enhance its performance.* For example, when there is a need for seismic criteria to be implemented for a bridge that was originally designed for wind forces only, it needs to be retrofitted with multi-rotational or seismic isolation bearings. Similarly, foundations may be retrofitted by underpinning.

Common retrofit concepts apply to seismic retrofit or upgrading underwater countermeasures. Uncommon retrofit concepts are foundation under pinning using mini-piles. Bearing retrofit: Problems with some of the damaged bearings appear to be:

- Locking up or freezing such that no movement is possible. This situation results in overstress at the end of the girder, such that a failure may occur
- Shear cracks may be caused by a frozen bearing
- Similarly, roller action may be blocked due to dirt
- Water and salt penetration gives rise to rust and corrosion. Eventually, the girder may settle.

Any existing damaged or seismically deficient bearing which does not allow rotation and translation during a seismic event will need to be retrofitted or replaced.

Steel retrofits or replacement: It is recommended to replace existing A7 steel with ASTM A709M Grade 250 or Grade 345 whenever possible. FCM zone 2 steel should be used for FCM members.

- Seismic retrofit by replacing rocker and roller bearings or extending bearing seats
- Upgrading scour countermeasures
- Painting structural steel.

Available options:

1. No build option
   - Structure is in advanced stage of deterioration—condemn the bridge and just post the load limits
   - Low traffic volume and/or no money available

- Too much local opposition
- Inadequate safety
- Unacceptable delay and detour.

2. Rehabilitation schemes—major repairs
- Improve geometry
- Correct joints and bearing problems
- Improve deck and drainage
- Increase live load carrying capacity
- Add members.

3. Reconstruction schemes—major improvement
- Post-tensioning.

4. Total replacement schemes
- On-site
- Off-site
- New bridge schemes on new alignment.

5. Safety improvement
- Crash tested parapet and railing.

6. Substructure stabilization/repairs.

12.19.2 Proposed Solutions

Perform a cost/benefit analysis offered by alternate retrofit schemes and recommend the appropriate seismic rehabilitation/retrofit measures. If a cost analysis shows that repair is cost effective, and the repair will restore the joint, every effort should be made to schedule the maintenance.

Substructure retrofit measures:

- Bearing strengthening—use of restrainers
- Foundation improvement
- Bearing replacement
- Elastomeric bearing
- Use of Isolation bearing

Seismic Retrofit

1. Seismic retrofit goals
- Minimize the risk of unacceptable damage
- Unacceptable damages
- Loss of life
- Collapse of all or part of bridge
- Loss of use of vital transportation route (essential route)

2. Seismic retrofit process
- Evaluate and upgrade the seismic resistance of existing bridges
- Preliminary screening—inventory
- Detailed evaluation
  - Vulnerability rating
  - Seismic bridge ranking
- Design retrofit measures

3. Retrofit measures
    - Strengthening members: Bearing strengthening by restrainers
    - Seat width improvement
    - Column strengthening: FRP wrapping or jacketing of column
    - Foundation improvement
    - Bearing replacement
    - Elastomeric bearing
    - Isolation bearing
    - Dampers
    - Retrofitting for continuity.

12.19.3 Retrofit and Strengthening Methods

Retrofit and strengthening methods are distinct from repairs and may include a host of components. The deck slab is most vulnerable and is likely to require repair, strengthening, or retrofit every 15 years for most bridges. Bearings for existing bridges require retrofit for increased seismic resistance and scour countermeasures for footings.

Underpinning: During stage construction, the existing deck slab may be temporarily supported by steel posts and beams prior to demolition. Any underpinning will be removed after the new deck is in position.

Strengthening methods may be summarized as:

1. Concrete deck repairs by patching
2. Epoxy injection
3. Deck protection by HDC

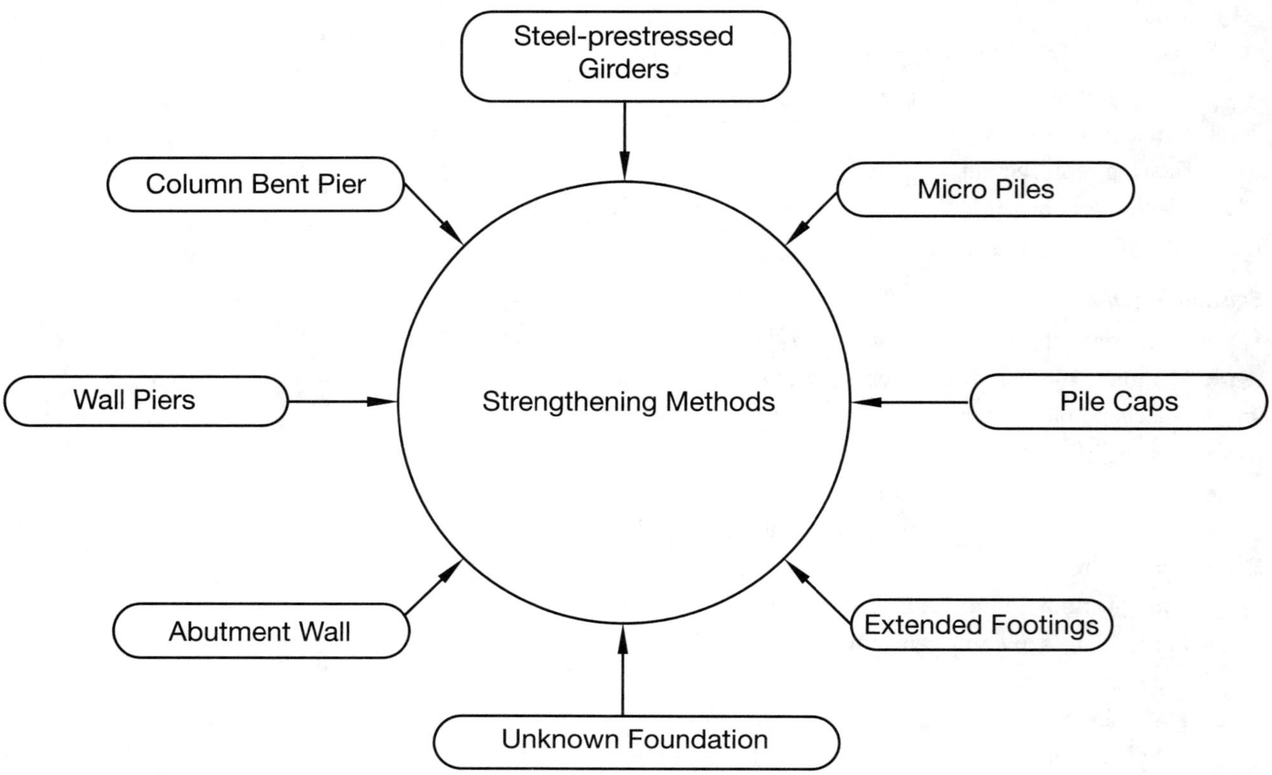

Figure 12.24 Strengthening methods of bridge components.

4. Silica fume
5. Waterproofing membrane, polymer surface treatment, CP system, ECE, protective sealant and coatings, LMC, CIA, improving skid
6. Various DOT memorandums on specialized strengthening topics.

Figure 12.24 shows various strengthening methods for bridge components, which include:

1. Adding members—longitudinal or clutch beams.
2. Deck grooving: For widened superstructures where at least one traffic lane is to be added, specify grooving for the entire deck area.
3. For projects with shoulder widening only, specify that the bridge floor finish matches that of the existing bridge deck surface. New construction utilizing C.I.P bridge decks will not be surfaced with asphalt concrete.
4. Salvaging of existing beams and railings: For superstructure replacement projects, it is customary to retain the existing beams and railings. Demolition in such cases needs to be carried out with extra care so that such features required for continuity are not damaged. FHWA would encourage retaining the existing beams whenever it would serve a useful purpose. Previous procedures of salvage value accreditation have been discontinued.

12.19.4 Continuity Retrofit

During a rehabilitation project, the expansion joint at a pier can be eliminated by splicing the simple spans together to form a continuous girder. This has several benefits, including reducing the possibility of deterioration of the girder and substructure due to a leaky joint, increasing the bridge's resistance to seismic displacements, and slightly improving the load carrying capacity of the superstructure.

However, continuity retrofit can result in undesirable structural performance characteristics that must be addressed in the design. Increased vulnerability to fatigue may result due to areas of the existing beams being subjected to stress reversals and higher stress ranges compared to simple span behavior. The end regions of retrofitted girders originally designed for simple span positive moments of small magnitude are subjected to larger magnitude negative moments.

While the deck joints over the interior supports are eliminated, the deck in this area is subjected to tension under service loads and crack control measures must be the considered. Continuity can also increase seismic loads on individual piers depending on bearing fixity configurations.

The scope of a project may help determine when it is appropriate to retrofit two or more simple spans into one continuous span. For a rehabilitation that involves a deck overlay only, the extra cost of concrete removal required to retrofit the simple span may be beyond the project scope. However, if some deck scarification, deep removal, and joint replacement are also scheduled as part of the rehabilitation, a cost assessment should be done to determine if retrofitting the simple span girders to be continuous is reasonable.

Complete deck replacement projects provide excellent opportunities to include girder retrofit since the girders will be readily accessible and the future costs of maintaining the joints will be eliminated.

The cost of providing continuity retrofits for full deck replacement projects should be compared to the cost of replacing the girders. This is particularly relevant when the cost of cleaning and painting the existing steel is required for the retrofit alternate.

12.19.5 Typical Retrofit Details and Guidelines

Compared to continuous for live load only designs, fully continuous retrofits require more complex splice and retrofit details.

However, a retrofit that provides full continuity for both dead and live loads is advantageous because the combined girder should behave like a conventional continuous girder. Since this retrofit requires so much more of the girder to be exposed in the area of the splice, a fully continuous retrofit should only be done in conjunction with a full deck replacement. Another benefit is that the existing two lines of bearings at the pier can be replaced by a single bearing line.

Retrofit Design Guidelines

There are a number of ways to retrofit a bridge structure to prevent collapse. These physical restoration or strengthening techniques have a unique style of design and construction. Alternative methods of retrofitting should be investigated to comply with the unique situations and compared. The method chosen must be economical and constructible with least impact to the traveling public.

1. The retrofitted span should be analyzed as fully continuous or continuous for live load to determine the new moments and increased shears induced over the interior support. The bolted flange splices shall be designed to carry the new moments, while the web splice shall be designed to carry the increased shear. The existing piers and bearings (if being retained) shall be analyzed for the increased reactions due to continuity.
2. Continuity retrofits should be designed for an MS18 (HS 20) live load. Upgrading the superstructure to an MS23 (HS 25) design is not required.
3. One method of increasing the design moment capacity of the continuous girder is to increase the girder's section properties by adding bolted cover plates to the flanges of the existing girders.
4. The negative moment capacity of the girder may be enhanced by considering the girder over the pier as a composite section. If not damaged, the stud shear connectors for the simple span beams may be left in place during deck removal operations. In most cases, the existing shear connectors are adequate to provide composite action in the negative moment region between the girder and the longitudinal reinforcing steel in the deck.
5. For both fully continuous and continuous for live load retrofits, additional longitudinal reinforcing steel must be installed in tension regions of the continuous deck. For fully continuous retrofits, provisions of Standard Specifications for Highway Bridges should be applied.
6. Filler plates may be used to make up differences in thickness between flanges or between webs to be spliced.
7. Bolts through the bottom flanges must be arranged to avoid interfering with the bearing(s).
8. Installing the splice may require removing the existing end diaphragms and bearing stiffeners. A new line of diaphragms and bearing stiffeners should be placed over the centerline of the new bearings. Rolled beams may not have bearing stiffeners.
9. Expansion joints on the structure should be checked to verify that they can handle the thermal expansion of the continuous girders. If it is determined that new joints are required, they should be designed with the current design procedure for expansion joints.
10. The existing pier shall be analyzed for any increased longitudinal or seismic loading caused by the continuity retrofit. Current seismic retrofit criteria should be reviewed.
11. Field confirmation of dimensions and steel condition is essential.
12. Caution is advised when using continuity retrofits with stage construction. The design must carefully consider construction sequencing. In no case shall two simple spans be attached to a deck continuous over a pier.
13. When replacing the bearings, care must be taken to ensure that the elevation of the superstructure remains the same.

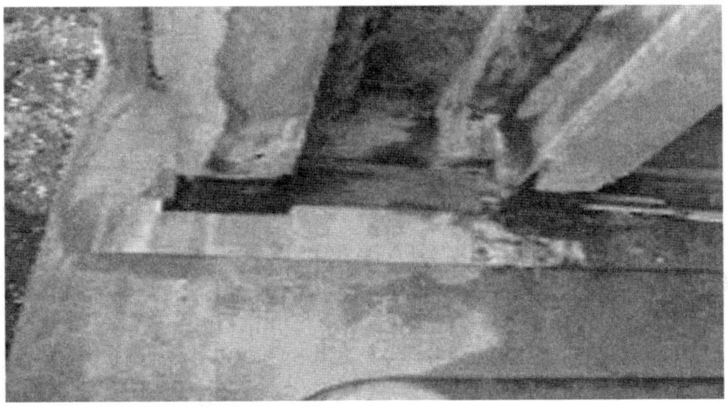

Figure 12.25 Corroded bearings due for replacement.

12.19.6 Substructure Retrofit

From as-built drawings it is observed that none of the existing abutment, pier walls, and footing rebar details seem to conform to the seismic rebar detailing given in AASHTO LRFD Design Manual. Hence, in a seismic event, the substructure performance may not be as safe as required. If the piers are found inadequate to resist seismic forces and moments for zone 2, one or more of the following measures is recommended:

1. Jacketing the pier
2. Flexural steel strengthening
3. Use of seismic isolation bearings to release seismic forces (energy dissipating force).

12.19.7 Bearing Replacement

Bearing replacement will not be restricted to replacing a single damaged bearing, but will be applicable to replacement of all the fixed or expansion bearings located on the same bearing line on the abutments or piers.

1. For bearing replacements, jacking of the bascule span superstructure will be considered for minimum dead load and no live load combination. Care will be taken to prevent instability of the suspended span during the jacking operation.
2. New bearings will be designed for seismic loads using current seismic criteria for non-vulnerable bearings. Proposed bearings will be multi-rotational pot bearings or elastomeric bearings. Elastomeric bearings are relatively shallow and differential height built-up is required.
3. Pot bearings will be guided, non-guided, or fixed types. They offer a better compatibility with the existing bearings in terms of stiffness characteristics than elastomeric bearings.
4. Alternate schemes: Due to the large number of existing rocker type bearings on the abutments and piers, it may be expensive to replace all the bearings. It is proposed to replace only the damaged, frozen, or contracted bearings. Selective replacements will be considered based on seismic evaluation. The following schemes will be considered as alternates:
   - Scheme 1: Replace corroded expansion (high rocker type) bearings and fixed bearings with appropriately guided, non-guided, or fixed pot bearings.
   - Scheme 2: Replace fixed bearings with elastomeric bearings, and replace all expansion bearings with an elastomeric bearing with or without a sliding plate.
   - Scheme 3: Maintain and retrofit bearings as required and provide a keeper and/or catcher system to prevent toppling of rocker bearings and to limit movement of all bearings in the event of a shear failure.

- Scheme 4: Maintain and retrofit bearings as required and provide longitudinal restraints/ties as appropriate to reduce the risk of toppling of the rocker bearings.

At some of the expansion bearings, existing shims under stringers may be replaced. The need for using multi-rotational or isolation bearings will be investigated. Isolation bearings may be considered for higher seismic zones. Multi-rotational pad bearings or simpler elastomeric pads, fixed, and expansion types will be considered. Bearing design will be based on the LRFD method. The bridge may be upgraded for seismic resistance using isolation bearings to replace both fixed and expansion bearings. Expansion bearings can cause a malfunction due to lack of maintenance. Bearings are usually replaced by jacking by hydraulic jacks placed on abutment seats where expansion bearings are located. Jacking beams need to be connected to webs by welding stiffener plates. Bolted connections at the end of jacking beams need to be designed for the dead load. Jacking is usually not done under live load and the traffic lane needs to be closed.

At least one constructible scheme for bearing replacement must be shown in the contract documents. All related analyses, including the effects of the jacking on all connections and elements, must be performed.

Thought shall be given to any required jacking procedure and constraints. Girders shall be raised uniformly in a transverse direction in order to avoid inducing stresses into the superstructure.

The following jacking design guidelines need to be considered:

- Deck should be closed to traffic until jacking is done and bearings are completed
- Do not include impact load to design jacking force requirements
- If shims and blocks are used for temporary support under traffic, design must include impact
- If strengthening during jacking is required, all details must be shown in contract documents.
- A special provision or notes on contract drawings will be added for the contractor to submit alternate schemes
- Seismic retrofit.

Minimum support widths for bridge seat: For minimum width requirements, a spreadsheet will be developed based on AASHTO formula, and available bearing widths at each abutment and pier will be checked.

12.19.8 Procedures for Bearing Retrofit

If any deficiencies exist, corrective measures must be incorporated into the contract plans. Temporary bearing repairs will be good for five to 10 years. Typical modifications to steel bearings to withstand loadings should include:

- Increase size, number, or embedment of anchor bolts
- Increase the outer diameter of the pin head
- Increase the width of the expansion rocker
- Increase the top and bottom dimension of the pintle detail for increased movement.

Options available for upgrading the support system for seismic resistance are bearing retrofit or bearing replacement including:

- Modification of the existing bearings to resist seismic loads
- Prevent toppling of existing bearings by installing longitudinal displacement stoppers
- Installation of longitudinal joint restraints, as outlined in Subsection 5.2.1 of FHWA Retrofit Manual
- Bearing replacement with bearings described in AASHTO LRFD Specifications 2007 for Bridges and Structures.

In retrofitting rocker type bearings, the strength of the assembly would have to be increased. Strengthening could be achieved by adding keeper bars to the lower components and by replacing existing upper bolts with high strength bolts. The weak links in bearing assemblies are the girder to sole plate bolts or welds, pin heads, dowels, and the anchor bolts, all of which are capable of shear failure during a seismic event.

Performance during seismic events: The expansion rocker bearings or pin/hinge type bearings located on bridge abutments and piers have been known to perform poorly due to bearing toppling and out-of-plane rotation. A solution is a catcher system that is designed to prevent loss of elevation based on bearing toppling. A built-up assembly of welded steel plates would be attached to the girder or the bridge seat. The assembly would allow normal movements to occur, but would support the girder in the event of excessive longitudinal movement.

12.19.9 Neoprene Bearings

Problems such as major uneven deformation or walk-out shall be corrected.

1. Bearings for temporary construction condition: Appropriate bearing type and restraining connections shall be designed to endure construction and traffic loads.
2. Rocker and roller bearings: Readjust all rocker and roller bearings to restore their required function. Clean, paint, and lubricate (roller bearings only) as warranted. For deck replacement or other major bridge rehabilitation projects, rocker bearings should be replaced and roller bearings should be considered for replacement unless seismic criteria is met.
3. Other metal bearings: Restore the required function of these bearings, as warranted, by repairing or replacing worn-out parts.
4. Pot bearings: Ensure that the neoprene material is adequately contained in the pot and the gap between the top of the pot and the piston bearing plate is fairly uniform under dead load. Also, sufficient end distance should exist to the stainless steel plate (mirror plate) for expansion and contraction at extreme temperatures.
5. Other multi-rotation bearings: If adverse functional conditions exist, corrective measures must be incorporated into the contract plan.

   The types of bearings include:

   - Disk type
   - Elastomeric
   - Pad
   - Pot
   - Rocker
   - Roller
   - Sliding plate
   - Pin
   - Spherical
   - Rocker plate
   - Isolation
   - No bearing (integral or continuous frame-type bridge)

   Alternatives to be considered are:

1. Replace high rocker and roller bearings.

   The vulnerability of rocker and other bearings has become obvious in past earthquakes. The replacement of rocker bearings in regions of moderate seismicity is recommended in most current retrofitting guidelines and state standards. In addition, the resetting of rocker bearings represents a rather substantial maintenance cost. For both seismic and life cycle cost perspectives, rocker bearing replacement is warranted for the more vulnerable and important bridges.

2. Extend bearing seats. Bearing seat lengths must meet the minimum support lengths as per the design specifications. This must be addressed on rehabilitation projects.
3. Provide cribbing for vulnerable bearings. While it is desirable to eliminate vulnerable bearings (i.e., rocker and roller bearings), this is not always possible or cost effective. Cribbing is used as a temporary measure until an economical bearing replacement can be performed (i.e., during a deck or bridge replacement).
4. Add shear blocks and/or pedestals—Structures which are deficient in areas such as seat length and bearing instability, or have inadequate superstructure to substructure connections, may be retrofitted by addition of shear blocks and/or dowel bars, or by construction of concrete pedestals which will act as shear blocks and alleviate bearing instability.

12.19.10 Rocker Bearings

During an earthquake, bearings are subjected to displacements, rotation, and lateral forces in various directions, resulting in brittle failure of the unidirectional steel high rocker and low sliding bearings. Replacing these bearings with ductile, multi-rotational, and multi-directional bearings provides safety against potential unseating of the superstructure. During rehabilitation work, existing structures are jacked to remove the existing steel rocker or steel sliding bearings.

Due to the height differences between the elastomeric bearings and the existing rocker bearings, existing pedestals are built up to a higher elevation, as recommended in the Seismic Retrofitting Manual for Highway Bridges (FHWA RD-94-052).

Recent earthquakes, including Kobe, indicate that the failure of the steel bearings caused substantial damage to the superstructure, which closed down the highway system.

Depending upon the capacity of the existing substructure, it is prudent to investigate seismic demand due to a design earthquake on the substructure. When warranted, the seismic demand can be reduced by adjusting the bearing configurations in one of the following manners:

1. Providing all expansion bearings with transverse restraints, thus reducing the transverse force demand by distributing it to all the substructures through expansion and fixed bearings. Providing conventional laminated elastomeric bearings at the expansion supports and using a lead core base isolation bearing at the fixed support will reduce the seismic demand.
2. Providing guided pot bearings at the expansion supports and a lead core base-isolation bearing at the fixed support. Since the coefficient of friction for expansion pot bearings is less than for elastomeric bearings, it will further reduce the demand on the substructure.
3. Providing a fixed bearing at one of the abutments and expansion bearings at the pier and the other abutment, which will help to reduce the seismic demand at the pier.

12.19.11 Multiple Simple Span

By providing continuity coupled with replacement of bearings, the lateral resistance of a superstructure will be enhanced and seismic loads will be distributed among all the substructure elements. The length of the superstructure that can be made continuous is a function of thermal movement.

When connecting the unrestrained ends of adjacent girder spans, it is important to provide a complete splice between the flanges and the webs. Similarly, shear blocks should be designed to resist the movement of the superstructure beyond the anticipated thermal movement. A detailed method is provided in the FHWA Retrofitting Manual for designing the restrainers and shear blocks.

Connections: All connections and anchor bolts should be designed for a minimum lateral force of 19 percent of the dead load plus live load reactions at the support. Anchor bolts should be anchored into the bridge seat to resist uplift.

12.20 DEVELOPMENTS IN PASSIVE DAMPING SYSTEMS

12.20.1 Seismic Dampers

Seismic dampers are increasingly being utilized to dissipate energy from an earthquake. The seismic dampers can be positioned within a seismic isolation system to limit the isolation system displacement. The isolation system consists of sliding isolation bearings in combination with a controllable fluid damper and limits the response of the isolation system and the superstructure for earthquake ground motions.

The efficiency of various dissipation mechanisms to protect structures from pulse-type and near-source ground motions needs to be studied. The response of structures with low to moderate isolation periods is substantially affected by the high frequency fluctuations that override the long duration pulse. The concept of seismic isolation is beneficial even for motions that contain a long duration pulse.

A semi-active electromagnetic friction damper for response control of structures: The smart damper used is a magneto-rheological (MR) damper. It is shown that the smart MR dampers can reduce displacements and forces in the piers further than the passive dampers. While these displacement reductions can be achieved by increasing the passive damping further, it can only be done at the expense of greater forces in the piers. One example is the isolation of the south side span and damping of the main cables of the Golden Gate Bridge.

12.20.2 Use of Fluid Viscous Dampers

Fluid viscous dampers can provide complete protection for bridges in seismic events. The systems have been proven in extensive tests at the Multidisciplinary Center for Earthquake Engineering Research (MCEER) at the State University of New York at Buffalo, and the Earthquake Engineering Research Center (EERC) at the University of California at Berkeley.

Isolation Bearings and Dampers

Seismic isolation bearings isolate a structure from the ground motion produced by an earthquake. The energy absorption devices are designed to absorb the energy associated with an earthquake. The seismic devices are designed to have both existing bridge retrofit and new bridge applications. Isolation bearings and dampers (shock absorbers) have limited use in bridge retrofit but are being used more for very vulnerable structures or extremely important structures such as toll bridges.

Due to the cost to jack the bridge up to replace bearings, typically simple seat extenders or catcher blocks are placed to take the load if bearings fail. Once the bearings fail, the bridge is isolated. Dampers are used for important structures in which much energy can be dissipated with large movements, thus limiting the extreme movements and preventing more damage.

To replace an existing bearing with an isolation bearing, adequate headroom or space is required between the bottom of the girder and the seat face. Low profile isolation bearings are used if this space is limited.

There are a number of ways to retrofit a bridge structure to prevent collapse. Various methods of retrofitting a bridge should be compared and the best method chosen based on economics and limited impact to the traveling public during reconstruction.

Bridges that are located in seismic zones 3 and 4 worldwide are using fluid viscous dampers.

12.20.3 Use of Isolation Bearings

Isolation bearings should be considered when column and footing design is substandard to resist seismic forces and moments and when rocker bearings are used in seismic zones. Column and footing retrofit are more expensive than seismic isolation. Isolation bearings have the following two functions:

1. To dissipate seismic energy by increasing the damping of the structural system.
2. To create a more flexible structural system which experiences less acceleration.

Three broad categories of isolation bearings are available in the U.S. market. They are:

- Lead rubber bearings (LRB)
- High damping rubber bearings (HDR)
- Sliding bearings.

The use of isolation bearings offers force reduction and is on the increase: Force reductions of the order of 3 to 6 are achievable using isolation strategies. Even for longer return period seismic events the need for retrofitting substructure elements would be eliminated.

Multi-rotational isolation bearing characteristics allow the distribution of lateral loads in proportion to the lateral strength of the existing substructure units. However, sufficient rigidity should be provided under wind and braking loads to keep displacements to within tolerable limits. Caltran and the Seismic Retrofitting Manual for Highway Bridges (FHWA RD-94-052) discourage using isolation bearings in bridge foundations resting on soft soil or for bridges with tall flexible piers.

The development of appropriate performance specifications is provided by the vendors and major manufacturers of isolation bearings. Prices for 100K to 250K bearings range from $2500 to $5000 each, respectively, without the sole plates and masonry plates, which are not supplied by manufacturers.

Sliding isolation systems can be many times more effective in damping as compared to lead rubber isolators. Due to release of seismic forces, foundation and column design are more economical.

Friction pendulum bearings are seismic isolators that are installed to lengthen the isolated structure and minimize the strongest earthquake forces.

12.21 FOUNDATION RETROFIT

12.21.1 Seismic Performance of Foundations

Footing work tends to be the most difficult and costly of all retrofit measures. It should be avoided at all costs. Usually only single column bent structures need a footing upgrade because multi-columns can allow the base of the column top in thus increasing the structure's natural period and reducing the shear to the column.

A number of potential failures are possible depending upon type of foundation. Only when there is little ductility do capacity failure modes need to be considered. These failures are structural only and do not take into account liquefaction effects. Since footings are hidden below ground, not enough failure data exists, unlike the superstructure.

Spread footings, retrofit will be required for the following conditions:

1. Footing shear or flexure failure.
2. Bearing capacity exceeded.

Pile foundations: Pile work will be the next order of work. The pile type used will reflect the most practical situation for the soil conditions. In many cases driven piles are selected for low overhead situations that may be encountered due to the height of the bridge. Low overhead pile driving is very expensive but can be achieved with clear distances down to 15 feet.

Figure 12.26 Liquefaction after an earthquake. Sand ejected through a crack forming a series of sand boils along the railroad tracks adjacent to Deschutes Parkway in Olympia. (Photo courtesy of Geomatrix).

All piles should be positively anchored to the new footing to create uplift resistance. The bottom mat reinforcement (if widened) can be extended by welding or the use of mechanical couplers.

- Inadequate pile capacity
- Inadequate soil capacity
- Shear failure of pile section
- Pile Pullout failure
- Pile flexure or tension failure

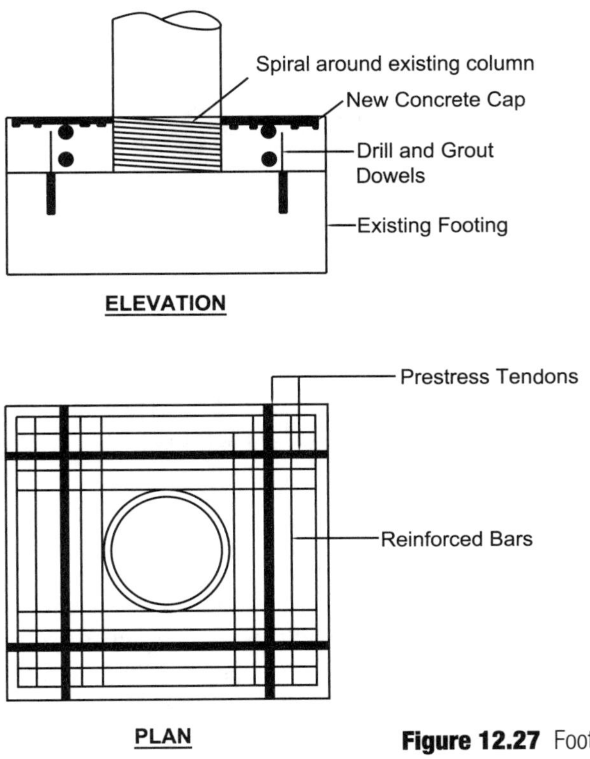

Figure 12.27 Footing retrofit by increasing thickness.

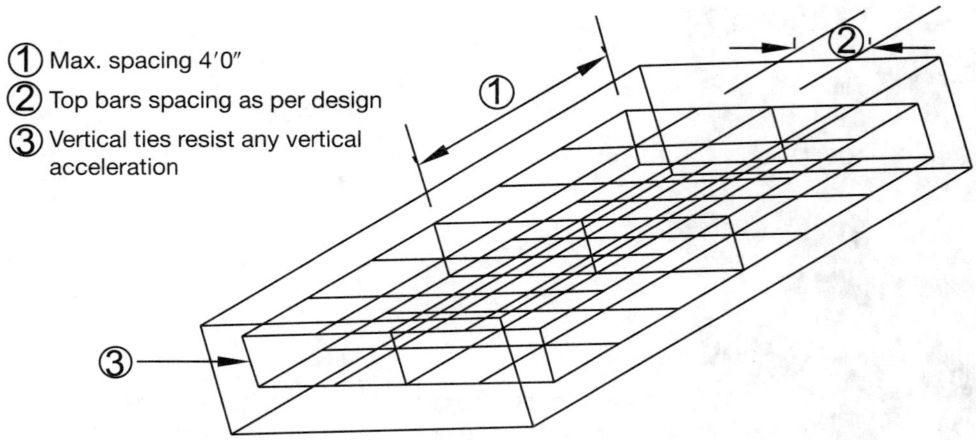

Figure 12.28 Seismic detailing for footings (vertical ties are required) to form a 3-D grillage.

12.21.2 Liquefaction

Liquefaction has been one of the most significant causes of damage to bridge structures during past earthquakes. The damage has been related to lateral movement of soil at the bridge abutments and the loss of lateral and vertical bearing support of foundations of bridge piers. Lateral flow or spreading of approach fills and settlement of liquefied soils would result. For seismic hazard intensity less than 6.0, the effects of liquefaction can be neglected.

Liquefaction vulnerability: Bridges with discontinuous superstructure and non-ductile supporting members have a higher vulnerability than continuous superstructures. It will depend upon severe, moderate, and low liquefaction. Liquefaction may lead to:

- Settlements of foundations
- Bearing capacity failure of shallow foundations
- Loss of support of friction piles
- Slope failures at abutments
- Large lateral ground movements on mildly sloping ground.

The probability of liquefaction is determined based on:

- Earthquake induced bedrock accelerations
- The probability that these accelerations will occur
- Soil profile type
- Properties from soil borings
- The remaining life of the bridge.

Seismic design using liquefaction assessment methodology: In areas of moderate and high seismicity, liquefaction hazard at a bridge site can be can be assessed by field study of:

- Site response spectra
- Seismological parameters
- Soil susceptibility.

Use of geotechnical field-testing methods such as cone penetration test (CPT), standard penetration test (SPT), and seismic crosshole logging and seismic downhole tests can be investigated. Better liquefaction assessment will lead to improved performance during seismic events.

The consequences of liquefaction affect the performance of the bridge. Hence, it is necessary to perform remedial or retrofit measures, which consist of:

- Strengthening the foundations to resist the consequences of liquefaction
- Performing measures to reduce the consequences of liquefaction
- The pile cap lateral capacity to resist displacements can be taken into account. The primary changes in the proposed provisions include using spring constants for driven piles, drilled shafts, and spread footings.

12.22 DISASTER MANAGEMENT

12.22.1 Introduction

While engineers try to grapple with seismic resistant designs, post-earthquake relief and reconstruction needs to be tackled on an emergency basis. The author was associated in disaster management aspects for earthquakes in Erzurum, Turkey (1987) and more recently in the north of Pakistan (2005). The post-disaster process must include:

- Loss estimation: Loss of life, property, and commerce
- Disaster planning and management
- Implementation of disaster recovery procedures
- Short-term and long-term needs for recovery
- Investigation of vulnerable populations
- Seismic risk assessment
- Maintaining hazard insurance

12.22.2 Preparing a Damage Inventory

For bridge inspection after an earthquake, the following field inventory chart may be used:

Table 12.11 Format of the inventory of damage following an earthquake.

Bridge Components and Highway Structures	Mildly Damaged	Moderately Damaged	Severely Damaged	Remarks
Approach Slab				
Embankment				
Utilities/Light Poles				
Deck Slab				
Parapet and Railing				
Inlets and Scuppers				
Sidewalks				
Girders				
Bearings				
Sign Structures				
Retaining Walls				
Abutments				
Piers				
Footings				
Scour Countermeasures				

12.22.3 Recommendations for Reconstruction

1. Use of lightweight aggregate concrete

   Seismic forces are directly proportional to the mass of the superstructure, i.e., reinforced deck slab, parapet, sidewalk, and median barrier. It is recommended to reduce unit weight of concrete by using lightweight aggregate concrete. It is possible to reduce density from 150 pcf to approximately 110 pcf by using commercially produced lightweight aggregates. Seismic forces and moments will be proportionately reduced by 30 percent. Loads on foundations will be reduced and lighter bridges have a better chance of withstanding earthquakes.

2. Use of ultra high performance concrete (UHPC)

   In the U.S., there has been a trend in the last decade to use high strength concrete (HSC) and HPC in bridges. It has resulted in lighter prestressed concrete girders and in some cases in thinner decks reducing concrete mass by over 10 percent. Optimization of member sizes tends to reduce seismic forces and moments proportionately.

   Also, an increase in concrete strength has resulted in reducing the cross sectional area of reinforcement steel. Steel cost has risen considerably recently compared to the cost of concrete. For economic design it is desirable to use even higher strengths of concrete such as ultra high strength concrete for deck slabs. Drawbacks found in normal strength concrete such as micro-cracking can be avoided. Ultra high strength concrete can be manufactured by:

   - Optimization of grain size distribution
   - Use of pozzolans such as slag, silica fumes, and fly ash for a dense cement-sand matrix and partial replacement of cement in mix design. High silica content increases the rate of hydration of cement and high early strength can be achieved at early ages.
   - Use of superplasticizers such as melamine formaldehyde condensates, resins, and third generation superplasticizers such as carbonate chains surrounding cement grains help reduce the water:cement ratio.
   - Use of fiber reinforced polymer (FRP) concrete to reduce micro-cracking in decks, columns, and walls.

3. Seismic design aspects are dependent upon local geology and seismicity of the area. In the site selection for a bridge or highway, care shall be taken to avoid locating bridges in a fault zone.
4. For the structural engineer, familiarity with seismic terminology and conventional dynamic analysis is necessary. For important bridge structures located in zone 2, use of time history or response spectrum analysis shall be considered. In addition, the latest methods of performance-based seismic engineering (PBSE) shall be applied. Pushover analysis and capacity spectrum design (CSD) methods are studied. Emphasis is rightly placed on pushover analysis, ductility, use of new soil site factors, liquefaction effects, and loss of bearing capacity during a seismic event.
5. Awareness of the natural disaster mitigation process and familiarity with shake table tests, to determine structural behavior of a complex bridge, is desirable.
6. AASHTO LRFD code does not cover all aspects of seismic design. A separate seismic code is required as a companion to AASHTO LRFD code. NCHRP 12-49 is an important step.
7. State codes should be updated to reflect the latest developments based on NCHRP 12-49. They should deal with local seismicity and offer a simplified explanation of complex aspects.
8. Use of isolation bearings: Both tall buildings and bridges need to be retrofitted with seismic isolation (pendulum) bearings to increase seismic resistance capacity. Moments on footings are reduced considerably. A non-isolated bridge would sustain structural damage during a safety level design earthquake event.

The author designed similar bearings for Schuylkill River Bridge in Philadelphia. The bearings had an 8 inch displacement capacity and support dead and live load of up to 500 kips. The bearings were installed pre-displaced.

BIBLIOGRAPHY

EXTREME EVENT OF FLOOD SCOUR

Baig, Salim M., Anello F. Monaco and J. C. Patel, "Screening of Bridges in New Jersey for Scour," *Journal of Hydraulic Engineering,* 1993.

Best Management Practices for Erosion and Sediment Control, FHWA- FLP-94-005, Federal Highway Administration, 1995.

Bridge Design Manual, New Jersey Department of Transportation, 2002.

Bridge Inspectors Reference Manual, NHI 03-001, Federal Highway Administration, 2002.

Bridge Scour and Stream Instability Countermeasures, HEC No. 23, Federal Highway Administration, 1997.

Bureau of Structural Evaluation, New Jersey Bridge Scour Evaluation Program Stage II, New Jersey Department of Transportation, 1994.

Chow, V. T., *Open Channel Hydraulics,* McGraw-Hill Book Co, New York, NY, 1959.

Circular 9, 3rd Edition, Report No. FHWA-IF-04-016, Federal Highway Administration, 2005.

Commentary to AASHTO Manual for Condition Evaluation of Bridges.

"Countermeasures," *Journal of Hydraulic Engineering,* Circular, Number 23, Federal Highway Administration, 1998.

Davis, S. R., J. E. Pagan, Ortiz and Linda Kelbaugh, "Streamlining the Permit Approval Process for the Installation of Scour Countermeasures at Bridges," *Journal of Hydraulic Engineering,* 1993.

"Debris Control Structures Evaluation and Countermeasures," *The Journal of Hydraulic Engineering,* Circular No. 9, 2006.

"Design of Riprap Revetment," *Journal of Hydraulic Engineering,* Circular Number 11, Federal Highway Administration.

Determination of Unknown Subsurface Bridge Foundations, NCHRP Report No. 21-5, Geotechnical Guideline No. 16, Federal Highway Administration, 1998.

Elias, Victor, *Strategies for Managing Unknown Bridge Foundations,* FHWA-RD-92-030, Federal Highway Administration, U.S. Department of Transportation, February 1994, Washington, DC.

Evaluating Scour at Bridges, 4th Edition, HEC No. 18, Federal Highway Administration, 2001.

"Evaluating Scour at Bridges," *Journal of Hydraulic Engineering,* Circular Number 18, Federal Highway Administration, 1996.

Federal Coding Guide, Federal Highway Administration, 1994.

Hamill, Les, *Bridge Hydraulics,* E & FN Spon, London & New York, 1999.

History of Bridge Scour Research and Evaluations in the U.S., E. V. Richardson, American Society of Civil Engineers, 1999.

Horne, W. A., "Evaluating Unknown Bridge Foundations, Is It Worth It?" *Proceedings of the International Water Resources Engineering Conference,* Memphis, Tennessee, August 3–7, 1998.

Hunter, D. S., M. A. Hixon and Salim M. Baig "Comparisons of Two Methods of Screening Bridges for Scour," *Journal of Hydraulic Engineering,* 1993.

Lagasse, P. F., L. W. Zevenbergen, J. D. Schall and P. E. Clopper, *Bridge Scour and Stream Instability Countermeasures Experience,* Federal Highway Administration, 2001.

LRFD Design Manual, Part 4, Chapter 7, Pennsylvania Department of Transportation.

Load Resistance Factor Design (LRFD) Bridge Design Specifications, American Association of State Highway and Transportation Officials, 1998.

Manual for Condition Evaluation of Bridges, 1994 including all Interims, American Association of State Highway and Transportation Officials.

Melville, B. W. and S. E. Coleman, *Bridge Scour*, Water Resources Publication, Colorado, 2000.

Model Drainage Manual, American Association of State Highway and Transportation Officials.

Parker, G., C. Toro-Escobar and R. L. Voigt, "Countermeasures to Protect Bridge Pier from Scour," University of Minnesota, National Cooperative Highway Research Program, Project 24-7, 1998. *Recording and Coding Guide for the Structure Inventory and Appraisal of the Nation's Bridges*, Federal Highway Administration Report FHWA-ED-89-044, 1989.

Park, Sung H., *Bridge Inspection and Structural Analysis*, Trenton, NJ, 1999.

Procedures Manual, New Jersey Department of Transportation.

Richardson, E. V. and P. F. Lagasse, *Instrumentation for Measuring Scour at Bridge Piers and Abutments*, Final Report, Phase III, NCHRP Project No. 21-3, Transportation Research Board, Washington, DC, 1994.

River Analysis System, Users Manual, U.S. Army Corps of Engineers, HEC-RAS, 1995.

Ruff, J. F. and L. M. Fotherby, *Bridge Scour Protection Systems Using Toskanes*, Engineering Center, Colorado State University, 1995.

Scour at Bridges, Technical Advisory 5140.20, Federal Highway Administration, 1988.

Sheppard, D.M., *Bridge Scour in Tidal Waterways*, Transportation Research Board, Washington, DC, 1995.

"Stream Stability and Scour at Highway Bridges," NHI Course No. 13046, Publication No. FHWA HI-91-011, United States Department of Transportation, Federal Highway Administration, October 1999.

Stream Stability and Scour at Highway Bridges, Water Resources Eng., American Society of Civil Engineers, 1999.

"Stream Stability at Highway Structures," *Journal of Hydraulic Engineering,* Circular, Number 20, Federal Highway Administration, 1997.

Technical Manual for Land Use Regulation Program, Bureau of Inland and Coastal Regulations Stream Encroachment Permits, New Jersey Department of Environmental Protection.

Tidal Hydraulic Modeling for Bridges Users Manual, Ayres Associates, December 1997.

Underwater Evaluation and Repairs of Bridge Components, FHWA-DP-98-1, Final Report, Federal Highway Administration, November 1995.

Underwater Inspection and Evaluation of New Jersey Bridges, New Jersey Department of Transportation, May 1994.

Underwater Inspection of Bridges, FHWA-DP-80-1 Final Report, Federal Highway Administration, November 1989.

Wojcik, P., "New Jersey Bridge Scour Evaluation Program," *Journal of Hydraulic Engineering,* 1993.

Additional References Cited

Agostini, R., L. Cesario, F. Ferraialo and A. Papetti, *Flexible Gabion and Reno Mattress Structures in River and Stream Training Works*, Officine Maccaferri (Bologna), 1988.

Ashmore, P. E. and G. Parker, "Confluence Scour in Coarse Braided Streams," *Water Resources Research*, 19, (1983).

Briaud, J. L., H. C. Chen, P. Nurtjahyo and J. Wang, "Shear Stress Approach to Pier Scour Predictions," *Proceedings of First International Conference on Scour of Foundations*, Texas A&M University, November 2002.

Briaud, J. L., H. C. Chen, Y. Li, P. Nurtjahyo and J. Wang, *Complex Pier Scour and Contraction Scour in Cohesive Soils*, NCHRP Report 24-15, Transportation Research Board, National Research Council, Washington, DC, 2003.

Bridge Design Manual, No. 4, Chapter 7, "Scour," Pennsylvania Department of Transportation, 2006.

Bridge Scour Evaluation Program, "Stage II—In-depth Scour Evaluation," Phase 3, New Jersey Department of Transportation, June 2000.

Brown, S.A. and E.S. Clyde, "Design of Riprap Revetment," *Journal of Hydraulic Engineering* Circular 11 (HEC-11), Report No. FHWA-IP-89-016, Federal Highway Administration, U.S. Department of Transportation, Washington, DC, 1989.

Brunner, G. W., "Computer and Physical Modeling of Bridge Scour,"ASCE Compendium of Papers on "Stream Stability and Scour at Highway Bridges," 1998.

Dumas, C. and J. Krolack, "A Geotechnical Perspective: Design and Construction of Highway Bridge Foundation for Scour," *Proceedings of First International Conference on Scour of Foundations*, Texas A&M University, November 2002.

Ettema, R., "Flume Experiments on Abutment Scour, Hydroscience and Engineering," University of Iowa, Volume 1, *Proceedings of First International Conference on Scour of Foundations*, Texas A&M University, November 2002.

Fotherby, L. M., "Alternatives to Riprap for Protection Against Local Scour at Bridge Piers," Transportation Research Record, Vol. 1420, TRB, National Research Council, Washington, DC, 1993.

Gao, D., Lilian Posada and Carl Nordin, "Pier Scour Equations Used in China," ASCE Compendium of Papers on "Stream Stability and Scour at Highway Bridges," 1998.

Guidelines for Bridge Maintenance and Rehabilitation, by Secretary, U.S. Department of Interior, Washington, DC.

Khan, Mohiuddin A., J. Zarriello and S. A. Baig, "Stream Instability Study for New Jersey Bridges," *Proceedings of First International Conference on Foundation Engineering*, Texas A&M University.

Klaassen, G. J. and K. Vermeer, "Confluence Scour in Large Braided Rivers with Fine Bed Material," *Proceedings of International Conference on Fluvial Hydraulics*, Budapest, (1988).

Kosicki, A. and H. Johanesson, "The Challenges of Scour Prediction," *Proceedings of First International Conference on Scour of Foundations*, Texas A&M University, November 2002.

Kumar, Virender, K. Raju and N. Vittal, "Reduction of Local Scour Around Bridge Piers Using Slots and Collars," *Journal of Hydraulic Engineering*, Vol. 125, No.12, Federal Highway Administration, Washington, DC, 1999.

Lagasse, P. F., P. L. Thompson and S. A. Sabol, "Guarding Against Scour," *Civil Engineering*, 65(6), 1995, 56–59.

Manual on the Use of Rock in Hydraulic Engineering, CUR/RWS Report 169, Centre for Civil Engineering Research and Codes, Balkema, Rotterdam, 1995.

May, R. W. P., J. C. Ackers and A. M. Kirby, *Manual on Scour at Bridges and Other Hydraulic Structures*, Construction Industry Research and Information Association, London, 2002.

May, R., "Comparison of General Scour Prediction Equations for River Crossings,"' *Proceedings of First International Conference on Scour of Foundations*, Texas A&M University, November 2002.

Maynord, S. T., "Gabion-Mattress Channel-Protection Design," *Journal of Hydraulic Engineering*, Vol. 121 No. 7, Federal Highway Administration, 1995, pp. 519–522.

McCorquodale, J. A. "Cable-tied Concrete Block Erosion Protection," *Proceedings of Hydraulic Engineering Conference*, San Francisco, CA, 1993.

Melville, B. W. and Anna Hadfield, "Use of Sacrificial Piles as Pier Scour Countermeasures," *Journal of Hydraulic Engineering*, Vol. 125, No. 11, Federal Highway Administration, 1999.

Melville, B. W. and R. Ettema, "Bridge Abutment Scour in Compound Channels," *ASCE Compendium of Papers on "Stream Stability and Scour at Highway Bridges,"* 1998.

Melville, Bruce W. and Stephen E. Coleman, *Bridge Scour*, Water Resource Publications, LLC, Highlands Ranch, Colorado, 2000.

Parker, Gary, Carlos Toro-Escobar and Richard L. Voigt, *Countermeasures to Protect Bridge Piers from Scour*, Users Guide and Final Report, NCHRP 24-7, National Cooperative Highway Research Program, Transportation Research Board, National Research Council, 1998.

Pennsylvania Department of Transportation, 1993, Bureau of Design, Bridge Management System (BMS), Coding Manual, office version: Pennsylvania Department of Transportation Publication 100A.

"Procedures for Bridge Scour Assessments at Bridges in Pennsylvania," Open File Report 00-64, by Peter J. Cinotto and Kirk E. White (In cooperation with PENNDOT), U.S. Department of the Interior, U.S. Geological Survey, Lemoyne, PA 2000.

"Procedures for Estimating Scour at Bridge Abutments, Parts 1 and 2—Live Bed Scour and Clear Water Scour," Maryland State Highway Administration, ASCE Compendium of Papers on "Stream Stability and Scour at Highway Bridges," 1998.

Racin, J. A., *Gabion Facilities Along the Pacific Coast Highway*, Report No. FHWA-CA-TL-93-17, California Department of Transportation, Sacramento, California, 1993.

Reihsen, G. and L. J. Harrison, "Debris-Control Structures," *Journal of Hydraulic Engineering,* Circular No. 9, Federal Highway Administration, Washington, DC, 1971.

Richardson, E. V. and P. F. Lagasse, *Instrumentation for Measuring Scour at Bridge Piers and Abutments*, Final Report, Phase III, NCHRP 21-3, National Cooperative Highway Research Program, Transportation Research Board, National Research Council, Washington, DC, 1994.

Richardson, E.V. and S.R. Davis, "Evaluating Scour at Bridges," *Journal of Hydraulic Engineering,* Circular 18, 4th Edition, FHWA NHI 01-001, Federal Highway Administration, Washington, DC, 2001.

Soil Erosion and Sediment Control Standards, New Jersey Department of Transportation, 1989.

"Software User's Guide for Determining the Pennsylvania Scour Critical Indicator Code and Streambed Scour Assessment Rating for Roadway Bridges," Open-File Report 01-446, by Mark F. Henneberg and Jeffrey L. Strause, (In cooperation with PENNDOT), U.S. Department of the Interior, U.S. Geological Survey, New Cumberland, PA, 2002.

Tedrow, J. C. F., *Soils of New Jersey*, Krieger Publishing Company, 1986.

Wire Mesh Gabions, U.S. Army Corps of Engineers, Report CW-02541, 1991.

EXTREME EVENT OF EARTHQUAKE DAMAGE

Codes and Publications Related to Seismic Design

Basoz, N. and A. Kiremidjian, "Prioritization of Bridges for Seismic Retrofit," Technical Report NCEER-95-00, Standford Univeristy, January 1995.

Bridge Seismic Retrofit Manual and Its Applications in Missouri Highway Bridges, Federal Highway Administration.

Bridges and Structures Design Manual, 3rd Edition, New Jersey Department of Transportation, 1998.

Chen W. F. and C. Seawthorn, *Earthquake Engineering Handbook*, CRC Press, 2003.

Chopra, A. K. and R. K. Goel, "Direct Displacement-Based Design: Use of Inelastic vs. Elastic Design Spectra," *Earthquake Spectra*, Volume 17, No. 1, February 2001.

Chopra, A. K., *Dynamics of Structures, Theory and Application to Earthquake Engineering*, Prentice-Hall, 2nd Edition.

Journal of Earthquake Engineering, Vol. 8, USA, 2004.

Khan, M. A., "Seismic Analysis of Bridge Substructures using AASHTO Code Methods," *Structures Congress*, 15th Analysis and Computation Specialty Conference, ASCE, Denver, Colorado, 2002.

Leelataviwat, S. et al, "Toward Performance Based Seismic Design of Structures," *Earthquake Spectra*, Volume 15, No. 3, August 1999.

LRFD Specifications for Highway Bridges, American Association of State Highway Transportation Officials (AASHTO), 1998.

LRFD Specifications for the Design of Bridges, American Association of State Highway and Transportation Officials, Washington, DC, 1998.

Mackie, Kevin R. and Božidar Stojadinovi, "R-Factor Parameterized Bridge Damage Fragility Curves," *Journal of Bridge Engineering*, ASCE, July/August 2007.

"Manual for Railway Engineering," The American Railway Engineering and Maintenance-of-Way Association.

Memari, A., H. Harris, A. Hamid and A. Scanlon, *Seismic Column Reinforcement Study*, Final Report, prepared for the Pennsylvania Department of Transportation, 2001.

Practice of Earthquake Hazard Assessment (PEHA), Robin McGuire, International Association of Seismology, IASPEI, 1993.

Pradeep, Kumar T. V. and D. K. Paul, "Force-Deformation Behavior of Isolation Bearings," *Journal of Bridge Engineering*, ASCE, July/August 2007.

Proceedings of International Conference on Earthquake Engineering, Lahore, 2006.

Regulations for Seismic Design: A World List (RSD), International Association for Earthquake Engineering (IAEE), Tokyo, 1996.

Standard Specifications for Highway Bridges, 16th Edition, American Association of State Highway and Transportation Officials, 1996.

"Seismic Evaluation and Retrofit of Steel Bridges," First U.S. Seminar, Hyatt Regency, San Francisco, CA, October 1983.

Saadeghvaziri, M. Ala, A. R. Yazdani-Motlagh and S. Rashidi, "Seismic Response and Evaluation of MSSS Bridges in New Jersey," Federal Highway Administration.

Supplement 2000 to RSD (SUP), IAEE, Tokyo 2000.

USEFUL REFERENCES

Applied Technology Council, Publication 18/23.

Applied Technology Council, 1996, Publication ATC-32, Improved Seismic Design Criteria for California Bridges: Provisional Recommendations.

Bozorgnia, Y. and V. V. Bertero, *Earthquake Engineering, From Engineering Seismology to Performance-Based Engineering*, CRC Press, 2004.

Bridge Software Development International Ltd, Allentown, PA, 1998.

Garden State Parkway Seismic Prioritization Program, New Jersey, Executive Summary.

Seismic Design and Retrofit Manual for Highway Bridges, Federal Highway Administration, 1987.

PUBLICATIONS RELEVANT TO EARTHQUAKE DESIGN

Seismic Behavior

ATC-32, *Improved Seismic Design Criteria for California Bridges: Provisional Recommendations*, Applied Technology Council (ATC).

Caltrans Seismic Design Criteria, Version 1.2, December 2001.

Greene, M., "Skopje, Yugoslavia: Seismic Concerns and Land Use Issues During the First Twenty Years of Reconstruction Following a Devastating Earthquake," *Earthquake Spectra*, Volume 3, Number 1, February 1987.

"Investing in a Safer Future," U.S. Geological Survey, Earthquake Hazards Program in National Earthquake Hazards Reduction Program, 2007.

Khan, Mohiuddin Ali, "Comparative Study of Seismic Design Codes with the Proposed NCHRP Code 12-49," *Proceedings of Second International Bridge Conference*, New York, NY, 2003.

Khan, Mohiuddin Ali, "Modeling and Seismic Analysis of Integral Abutments," *Proceedings of International Conference*, ASCE-SEI, Nashville, TN, 2004.

Khan, Mohiuddin Ali, "Seismic Analysis of Bridge Substructures Using AASHTO and AREMA Code Methods," *Structures Congress*, Denver CO, 2002.

Khan, Mohiuddin Ali, "Seminar on Earthquake Engineering," New Jersey Institute of Technology, Newark, NJ, October 2006.

Khan, Mohiuddin Ali, "Seminar on Earthquake Engineering," Temple University, Philadelphia, PA, March 2007.

Nikolaou S., P. Deming and P. Edinger, "Liquefaction Assessment Methodologies for Seismic Bridge Design and Their Applications to New York City Projects," *Proceedings of Second International Bridge Conference*, 2003.

Practice of Earthquake Hazard Assessment (PEHA), Robin McGuire, International Association of Seismology, IASPEI, 1993.

Regulations for Seismic Design: A World List (RSD), International Association for Earthquake Engineering (IAEE), Tokyo, 1996.

Applications to Lightweight Beam Design

Khan, Mohiuddin Ali, Edwin Rossow and Surendra Shah, "Behaviour of Reinforced and Prestressed High Strength Concrete Beams," NED University, Karachi, *Journal of Engineering Research*, Vol. 1, No. 1, pp. 1–29, January 1994.

Khan, Mohiuddin Ali, Edwin Rossow and Surendra Shah, "Shear Design of High Strength Concrete Beams," *Proceedings of International Conference*, ASCE-SEI, Philadelphia, PA, 2000.

Khan, Mohiuddin Ali and K. O. Kemp, "Elastic Full Composite Action in Slab-beam Systems," *The Structural Engineer*, London, 1970.

Murata, Masahiko, "The Great Hanshin-Awaji Earthquake, The Experience and Reconstruction," *Proceedings of Gujarat State Disaster Management Authority International Conference on Earthquake Recovery*, September 30, 2003.

Technical Resource Materials

Agrawal A., Y. Zhihua and M. Ettouni, *Bridge Security and Blast Resistance Design*, CUNY, 2004.

Ballard, T., A. Krimotat and R. Mutobe, *Utilizing Pushover Analysis for Seismic Performance of Steel Bridge Structures*, SC Solutions, Inc., Santa Clara, CA.

"Blue Pages," New York State Department of Transportation, March 2000.

Bozorgnia, Yousef and V. V. Bertero, *Earthquake Engineering*, CRC Press, 2004.

Bridge Design Code, New York City Department of Transportation, 1996.

Bridge Design Manual, 2003, New Jersey Department of Transportation.

Bridge Design Manual (DM 4), Pennsylvania Department of Transportation, 1998.

Bridge Seismic Retrofit Manual and Its Applications in Missouri Highway Bridges, Federal Highway Administration.

Buchheit, John A. and Harry Capers, *Constructible and Economical Seismic Retrofits for Existing Bearings*, New Jersey Department of Transportation, Trenton, NJ.

Buckle, I. G. and I. M. Friedland, "Improved Screening Procedure for Seismic Retrofitting of Highway Bridges," *Fourth International Bridge Engineering Conference*, Transportation Research Board, San Francisco, CA, 1995.

Comprehensive Specification for the Seismic Design of Bridges, National Cooperative Highway Research Program, TRB, 2002.

"Computer Software for Earthquake Engineering," National Information Service for Earthquate Engineering, University of California, Berkeley, 1993.

Cooper, et al, *The Northridge Earthquake*, Federal Highway Administration.

Dekker, David R., *The Repair and Strengthening of Reinforced Concrete Bridge Piers*, Department of Civil Engineering, University of Canterbury, Christchurch, New Zealand, March 1992.

Farrar, Charles R. and Thomas A. Duffey, "Bridge Modal Properties Using Simplified Finite Element Analysis," *Journal of Bridge Engineering*, American Society of Civil Engineers, February 1998.

Freeman, S. A., *Performance Based Seismic Engineering: Past, Current and Future*, WJE Associates.

Garden State Parkway Seismic Prioritization Program, Section 3, Bridge Vulnerability Ratings, Appendix A, New Jersey Turnpike Authority.

Hipleyoffice, Patrick, Earthquate Engineering, "Bridge Retrofit Construction Techniques," *Proceedings of the Second National Seismic Conference on Bridges and Highways*, 1997.

Idriss, I. M., *An Update of the Seed-Idriss Simplified Procedure for Evaluating Liquefaction Potential*, Transportation Research Board Workshop, Washington, DC, January 1999.

Kawashima, K., Seismic Strengthening and Repair of Highway Bridges in Japan, *Proceedings of the First U.S.-Japan Workshop on Seismic Retrofit of Bridges*, Public Works Research Institute, Ministry of Construction, Tsukaba Science City, Japan, December 17–18, 1990.

Liao, S. S. C., et al, "Regression Models for Evaluating Liquefaction Probability," *Journal of Geotrechnical Engineering*, Vol. 114, No. 4, pp. 389–411.

Mackie, A.M., Kevin, ASCE and Božidar Stojadinovi, A.M., ASCE.

Maffei, J., "The Seismic Evaluation and Retrofitting of Bridges," Ph.D. Thesis, University of Canterbury, CALTRANS Memos to Designers, Memo 20-4, March 1995.

Maharashtra Emergency Earthquake Rehabilitation Programme, Revenue and Forests Department, Mumbai, India.

Malik, Ayaz H., "Seismic Retrofit Of Steel Bridges," *Modern Steel Construction*, March 1997.

National Cooperative Highway Research Program (NCHRP)/Multidisciplinary Center for Earthquake Engineering Research (MCEER), 2001, Project 12-49.

National Cooperative Highway Research Program Report 12-49, 2002.

PCI Bridge Design Manual, Chapter 15 "Seismic Design."

Priestley, MJN, Seible F. and Y. H. Chai, "Design Guidelines for Assessment, Retrofit, and Repair of Bridges for Seismic Performance," SSRP 92-01, August 1992, p. 33.

Seed, H. B. and I. M. Idriss, "Simplified Procedure for Evaluating Soil Liquefaction Potential," *Journal of Soil Mechanics and Foundations Division*, Vol. 97, SM 9, pp. 1249–1273.

Seed, H. B., et al, "Influence of SPT Procedures in Soil Liquefaction Resistance Evaluations," *Journal of Geotechnical Engineering*, Vol. 111, No. 12, December 1985, pp. 1425–1445.

"Seismic Analysis of Bridges 1998," SEISAB software, Imbsen Associates, CA, 1996.

Seismic Design for Railway Structures, American Railway Engineering and Maintenance-of-Way Association (AREMA, 2002), Chapter 9.

"Seismic Design of Bridges," Prepared by Berger/Abam Engineers Inc., National Cooperative Highway Research Program, NCHRP 12-49, July 2001.

Seismic Energy Products, Earthquake Protection Products, Athens, Texas.

Seismic Isolation of American River Bridge, Earthquake Protection Systems, Inc., Richmond, CA.

Seismic Retrofitting Manual for Highway Bridges, 1994, FHWA Contract DTFH61-92-C-00106, p. 22.

Setting the Standard in Earthquake Protection, DIS, Lafayette, CA.

"Structural Analysis and Design," STAAD-PRO software 2002, Research Engineers Inc., CA.

Structural Engineering Institute, Denver, 2002 and Seattle, 2003 *Proceedings of Structural Congress*.

University of Central Florida and Dept. of Civil and Environmental Engineering, University of California Berkeley.

Villaverde, Roberto, F*undamental Concepts of Earthquake Engineering*, CRC Press, 2008.

Williams, Alan, *Seismic Design of Buildings and Bridges*, Oxford University Press, 2003.

Wilson, C. A., "New Deformation Patterns Improve Bond of Reinforcing Bars," *Civil Engineering*, March 1951, pp. 34–37.

Yen, W. Phillip, John D. O'Fallon, James D. Cooper and Jeffrey J. F. Ger, Missouri Department of Transportation.

Yost, J. R., S. Ratanalert and R. Rasnasinghe, "Seismic Analysis and Design of Continuous Prestressed Concrete Bridges," *ASCE Proceedings*, 2002.

Appendices

Appendix 1 SCOUR COUNTERMEASURES

Appendix 2 ESTIMATE OF UNIT CONCRETE REPAIR COSTS (2009)

Appendix 3 QUICK REFERENCE TO AASHTO LRFD SPECIFICATIONS

Appendix 4 QUICK REFERENCE TO AASHTO LRFR MANUAL

Appendix 1

Scour Countermeasures

Table A1 Modified HEC-18 bridge scour and stream instability countermeasurers matrix.

Countermeasures Group	Countermeasures Characteristics									
	Functional Applications			Suitable River Environment						Maintenance
	Local Scour[1]		Contraction Scour	Stream Instability		River Type	River Size	Velocity	Bed Material	Estimated Allocation of Resources
	Abutment[1]	Pier	Flood Plain and Channel[3]	Vertical/ Aggradation,[4] Degradation[5]	Lateral Erosion/ Meander[6]	Braided = B Meandering = M Straight = S	Wide = W Moderate = M Small = S	Moderate = M Slow = S	Course = C Sand = S Fine = F	High = H Moderate = M Low = L
GROUP 1. HYDRAULIC COUNTERMEASURES										
GROUP 1A. RIVER TRAINING STRUCTURES										
TRANSVERSE STRUCTURES	o = Unsuitable		O = Well Suited	D = Secondary Use						
Impermeable Spurs (Groins, Wing Dams)	D	D	o	o	O	B M	W M	M S	C S F	M L
Drop Structures (Check Dams, Grade Control)	D	D	D	O	o	B M S	W M S	M S	C S F	M
LONGITUDINAL STRUCTURES										
Retards	D	o	o	o	O	B M S	W M S	M S	S F	H M
Plantation	o	o	o	O	O	M S	W M S	M S	S F	L
Bulkheads	O	o	o	o	O	B M S	W M S	M S	C S F	M
Guide Banks	O	D	D	o	D	B M S	W M	M S	C S F	M L
GROUP 1B. ARMORING COUNTERMEASURES										
REVETMENTS and BED ARMOR										
RIGID										
Concrete Pavement	O	D	O	D	O	B M S	W M S	M S	C S F	M
Rigid Grout Filled Mattress/ Concrete Fabric Mat	O	D	D	D	O	B M S	W M S	M S	C S F	M
FLEXIBLE/ARTICULATING										
Riprap on Textile	O	D	D	D	O	B M S	W M S	M S	C S F	M
Riprap Fill Trench	D	o	o	o	O	B M S	W M S	M S	C S F	M
Gabion/Gabion Mattress	O	D	D	D	O	B M S	W M S	M S	S F	M
Articulated Concrete Blocks (Interlocking/Cable Tied)	O	D	D	D	O	B M S	W M S	M S	C S F	M L

	Countermeasures Characteristics									
	Functional Applications			Suitable River Environment						Maintenance
	Local Scour[1]		Contraction Scour	Stream Instability		River Type	River Size	Velocity	Bed Material	Estimated Allocation of Resources
Countermeasures Group	Abutment[1]	Pier	Flood Plain and Channel[3]	Vertical/ Aggradation,[4] Degradation[5]	Lateral Erosion/ Meander[6]	Braided = B Meandering = M Straight = S	Wide = W Moderate = M Small = S	Moderate = M Slow = S	Course = C Sand = S Fine = F	High = H Moderate = M Low = L
LOCAL SCOUR ARMORING										
Riprap (Fill/Apron)	0	D	X	X	X	B M S	W M S	M S	C S F	H M
Concrete Armor Units (Toskanes, Tetrapods)	D	D	X	X	X	B M S	W M S	M S	C S F	M L
Grout Filled Bags/Sand Cement Bags	0	D	X	X	X	B M S	W M S	M S	C S F	H M
Gabions	0	D	X	X	X	B M S	W M S	M S	C S F	M L
Articulated/Hollow Concrete Blocks	0	D	X	X	X	B M S	W M S	M S	S F	M L
Sheet Pile	D	D	X	X	X	B M S	W M S	M S	C S F	M L
GROUP 2. STRUCTURAL COUNTERMEASURES										
Crutch Bents/Underpinning	o	0	0	0	D	B M S	W M S	M S	C S F	L
Pumped Concrete/Grout Under Footing	0	0	D	D	D	B M S	W M S	M S	C S F	M
Lower Foundation/Curtain Wall	0	0	0	0	0	B M S	W M S	M S	C S F	L
PIER GEOMETRY MODIFICATION										
Extended Footings	X	0	X	X	X	B M S	W M S	M S	C S F	M
Sacrificial Piles	X	0	X	X	X	B M S	W M S	M S	C S F	M
GROUP 3. MONITORING										
FIXED INSTRUMENTATION										
Sonar Scour Monitor	D	0	0	0	D	B M S	W M S	M S	C S F	M
Magnetic Sliding Collar	0	0	0	0	D	B M S	W M S	M S	C S F	M
PORTABLE INSTRUMENTATION										
Physical Probes	0	0	0	0	0	B M S	W M S	M S	C S F	L
Sonar Probes	0	0	0	0	0	B M S	W M S	M S	C S F	L
VISUAL MONITORING										
Periodic Inspection	0	0	0	0	0	B M S	W M S	M S	C S F	H
Flood Watch	0	0	0	0	0	B M S	W M S	M S	C S F	H

Appendix 2

Estimate of Unit Concrete Repair Costs (2009)

ESTIMATE OF UNIT CONCRETE REPAIR COSTS (2009)

Item	Cost
1. Minor repairs	
Type 1: Routing and sealing cracks	\$15 per ft.
Type 2: Epoxy Injection	\$70 per ft.
2. Major repairs	
Type 3: Demolition, cleaning rebars, putting repair material area over 50% of beam length.	\$60 per sq. ft.
Type 4: For smaller areas less than 50% of beam length	\$55 per sq. ft.
Costs include: Material Labor Equipment High reach rental	
3. Corrosion Inhibitor for rebars	\$3.5 per sq. ft.
4. Painting and coating of concrete surfaces	\$6.0 per sq. ft.

Appendix 3

Quick Reference to AASHTO LRFD 2008 Specifications

QUICK REFERENCE TO AASHTO LRFD 2008 SPECIFICATIONS

The references made to the section numbers in LRFD Specifications are briefly summarized here for ready reference. There are 14 sections in the voluminous handbook covering a variety of subjects.

Section Number	Description
1	**Introduction:** Service; Fatigue; Strength; and Extreme Limit States; References
2	**General Design and Location Features:** Design Objectives; Hydrology and Hydraulics; Bridge Security
3	**Loads and Load Factors:** Load Factors and Combinations; Permanent Loads; Live Loads; Water Loads; Wind Loads; Ice Loads; Earthquake Effects; Earth Pressure; Force Effects due to Superimposed Deformations; Friction Forces; Vessel Collision; Blast Loading
4	**Structural Analysis and Evaluation:** Acceptable Methods of Structural Analysis; Mathematical Modeling; Static Analysis; Dynamic Analysis; Analysis by Physical Models
5	**Concrete Structures:** Material Properties; Limit States; Compression members; Design Considerations; Design for Flexural and Axial Effects; Shear and Torsion; Prestressing and Partial Prestressing; Details of Reinforcement; Development and Splices; Durability; Specific Members; Provisions for Culvert Types
6	**Steel Structures:** Materials; Limit States; Fatigue and Fracture; General dimensions and Details; Tension Members; Compression Members; I-Section Flexural Members; Box-Section Flexural Members; Miscellaneous Flexural Members; Connections and Splices; Provisions for Structure Types; Piles
7	**Aluminum Structures:** Materials; Limit States; Fatigue and Fracture; Design Considerations; General Dimensions and Details; Tension Members; Compression Members; Flexural Members; Torsion; Combined Force Effects; Connections and Splices; Provisions for Structure Types
8	**Wood Structures:** Materials; Limit States; Components in Flexure; Components Under Shear; Components in Tension Parallel to Grain; Components in Combined Flexure and Axial Loading; Bracing Requirements; Camber Requirements; Connection Design
9	**Decks and Deck Systems:** General Design Requirements; Limit States; Analysis; Concrete Deck Slabs; Metal Decks; Wood Decks and Deck Systems

10	**Foundations:** Soil and Rock Properties; Limit States and Resistance Factors; Spread Footings; Driven Piles; Drilled Shafts; Micro Piles; Seismic Analysis and Design of Foundations
11	**Abutments, Piers and Walls:** Soil Properties and Materials; Abutments and Conventional Retaining Walls; Piers; Nongravity Cantilevered Walls; Anchored Walls; Mechanically Stabilized Earth Walls; Limit States and Resistance Factors; Prefabricated Modular Walls; Seismic Design of Abutments
12	**Buried Structures and Tunnel Liners:** Soil and Material Properties; Limit States and Resistance Factors; General Design Features; Metal Pipe, Pipe Arch and Arch Structures; Long Span Structural Plate Structures; Structural Plate Box Structures; Reinforced Concrete Pipe; Reinforced Concrete Cast-in-place and Precast Box Structures and Reinforced Cast in Place Arches; Thermoplastic Pipes; Steel Tunnel Liner Plates; Precast Reinforced Concrete Three Sided Structures
13	**Railings:** General; Materials; Limit States and Resistance Factors; Traffic Railing; Pedestrian Railing; Bicycle Railings; Combined Railings; Curbs and Sidewalks.
14	**Joints and Bearings:** Movement and Loads; Bridge Joints; Requirements for Bearings; Special Design Provisions for Bearings; Load Plates and Anchorage for Bearings; Corrosion Protection.

Appendix 4

Quick Reference to AASHTO LRFD Manual

QUICK REFERENCE TO AASHTO LRFR MANUAL

For ready reference the Contents of AASHTO Manual for Condition Evaluation and LRFR of Highway Bridges are presented here:

Section 1 Introduction; Definitions and Terminology; Important References

Section 2 Bridge File (Records)

Section 3 Bridge Management Systems

Section 4 Inspection

Section 5 Material Testing

Section 6 Load and Resistance Factor Rating

Section 7 Fatigue Evaluation of Steel Bridges

Section 8 Nondestructive Load Testing

Section 9 Special Topics – Evaluation of Unreinforced Masonry Arches; Direct safety assessment of bridges; Historic bridges

Appendix A Illustrative Examples (A1 to A9)

Members for steel stringers and truss bridges; reinforced concrete slab and beam bridges; prestressed concrete and timber bridges are rated for design; legal and permit live loads.

Limit states for evaluation:
Strength I for design live load for all bridges
Strength I for legal live load for all bridges
Strength II for permit load for all bridges

References are made throughout the Manual to AASHTO LRFD Specifications.

Index

A

AASHTO codes, 191
AASHTO load combinations, 216–217, 226–227
AASHTO procedures, 349
abbreviations, 415
ABC factors, 406
above water inspection, 384
abutment type, 326–327
abutments, 579
accelerated construction, 23–24, 421–422, 505–509
accelerated planning, 506–508
accelerated repair, 405–406
accelerated schedules, 412
acoustic emissions, 379–380
additional inspections, 350
advanced concrete repair methods, 442
advanced repair methods, 477–521
advancements in concrete technology, 494–500
alternate analysis, 419
analysis method, 61
analysis procedures, 161
analytical load rating, 188–189
application modules, 286
applied loads, 136–137
approach slab, 152–159, 328
approach to rehabilitation, 40
asset management, 381–383
avoid failures, 117–124

B

bank scour, 535–579
beam corrosion, 433
beam modeling, 143
bearing replacement, 591–594
bearing seat elevations, 298
bearing stiffeners, 260
bearings isolation, 596
    bid documents, 11
blast load, 103–104
BMRR Practices, 56–57
BMS, 335–344
bolts, 265–266
bomb blast, 104
box beams, 141
bridge
    abutments, 579
    accelerated schedule, 412
    aesthetics, 34–35
    analysis, 61
    applications, 51
    approach slab, 152–159
    beams, 155–158, 426–427
    bearings, 357–358
    behavior, 143–146
    blast load, 103–104
    bolts, 265–267
    box beams, 141
    BRIDGIT, 353–354
    case studies, 87–90

bridge *(continued)*
- clearances, 19–21, 102
- codes, 67, 191
- collision, 99–101
- components, 495
- composite systems, 143–151, 175–180
- concrete T-beam, 274–276
- condition assessment, 532
- connections, 264–266, 495
- construction, 95–99, 505–509
- continuous span, 269–270
- cost, 51, 290–291
- critical members, 341
- debris, 562–563
- deck, 50–51, 225, 266–274, 377–378, 415–417, 440, 452–461, 499
- deflection, 258–260
- design, 49, 52–59, 92–94, 99–101, 115, 180–182, 191–193, 228–232, 240–247, 408–409, 480, 577–578
- detailing requirements, 582–586
- deterioration, 75–76
- disaster management, 390–391
- disasters, 75–78
- during construction, 28–31
- elastic analysis, 137–138
- elasticity, 135
- emergency repairs, 389–390
- engineering inspection, 346
- engineering, 78
- engineers, 47
- evaluations, 569–570
- expected life, 11
- extreme event, 529–601
- extreme events, 368–372, 529–601
- failure, 23, 75–81, 85–92, 94–95, 107–110, 112–114, 116–124, 169–171, 564–565
- bridge fatigue, 94–95, 204–208
- FHWA ratings, 359–363
- fire damage, 104–105
- foundation scour, 85–87
- foundation, 344
- fracture, 60, 83–85, 204–208
- fragility, 107
- girders, 163–169, 225–226, 283–284, 338–339, 500
- hazards, 116, 338–339
- highway, 571
- historic, 26, 66–73
- inspection 47–48, 110, 347–348, 344–357
- inspectors, 364–366, 391
- legal loads, 240–241
- legal requirements, 78
- lightweight emergency, 494
- lightweight truss, 493
- live load analysis, 161–163
- live loads, 136, 161–163, 185–234, 244–254
- load resistance, 261–266
- long spans, 35
- maintenance, 16–22 56–57, 110, 115, 383–387
- management, 25, 335–344, 383–387, 407–413
- material, 50
- members, 135–136
- models, 151–152
- monitoring, 336–337, 380–383
- moveable, 107
- neglect, 110
- new, 543
- not on river, 85
- obsolete, 414
- on river, 368, 383–387, 467–468, 532–535
- overlay, 50
- overload, 366–367
- planning, 26, 47–48, 294, 506–508
- PONTIS system, 353–354
- prestressed concrete girder, 276–279
- prestressed concrete, 466–467
- preventative actions, 92–94
- preventative measures, 124–128
- prioritization, 339
- production errors, 115
- protection, 51
- quality control, 95
- railway, 571
- rating, 191–193, 255–260
- redundancy, 60
- rehabilitation, 41–43, 54–57, 285–294, 389, 426–431, 513–517
- repair, 47, 56–57, 387–394, 397–343, 403–406, 413–415
- replacement, 295–303, 285–303, 317–318, 339
- resistance factors, 239–240
- retrofits, 47

risk assessment, 339–341
risk management, 127–128
rural settings, 35
science, 78
scope of repairs, 417–419
scour countermeasures, 535–579
security, 26
shear design, 180–182
sign structures, 141–142
single span, 269–270
slab beam, 138–139
steel sections, 171–175
steel, 50, 158–160, 255–260
stiffeners, 260–261
strategic topics, 5–6
stress analysis, 135–137
structural behavior, 158
structural components, 185–190
structural evaluation, 344–347
structural health monitoring, 372–380
structural planning, 121–122
structural solutions, 121–122
structure, 48
substructure inspection, 532–535
supervision, 47
surface elements, 18–19
technology, 66–73
temporary design, 223–224
train accidents, 101–102
trusses, 140–141, 426–427
two-girder steel, 261–262
type classification, 16–18
vehicle impact, 102–103
vibrations, 204
voided slab, 283–284
widening, 282–284, 339

BRIDGIT, 353–354

C

carbon fiber, 491–493
case studies, 318–321, 324–325
cast-in-place concrete, 499–500
changing technology, 4–5
closed forms, 153
coating systems, 429–431
code, 28–29, 67
applications, 578–582
implementing, 574–576
communication errors, 115–116
compact sections, 222
compliance, 28–29
composite material, 50
composite systems, 143–151
computational models, 151–152
computer software concepts, 133–135
computing issues, 4
concept study report, 534–535
concrete
abutment repair, 469–470
aggregate, 486–488
alternative material, 500
benefits, 485
bridge girders, 500
cast-in--place, 499–500
corrosion protection, 449–450
crack repair, 469
deck slabs, 449
durable, 446–448
failure, 459–460
filled steel decks, 453
high performance, 495
materials, 485–486
mix design, 450
patching, 468–469
reinforced, 445–446
repair, 49–50, 397–343, 439–471
self consolidating, 495–496
shrinkage, 496–497
T-beam bridge, 274–275
technology, 494–500
ultra-high performance, 497–498
condition assessment, 399, 532
condition evaluation, 533
condition rating, 359–362, 368–372, 534
connections, 264–266, 515–516
constructability, 4, 304–306, 402–403, 557–560
construction
deficiency, 95–99
equipment, 502
failures, 95–99, 504–505
loads, 223–228
materials, 485–486
methods, 503–504
products, 433–434
related activities, 125–126
route, 508

construction *(continued)*
- scheme, 303–304
- stages, 30
- supervision, 95
- systems, 23–24
- technology, 502–505

continuity retrofit, 589
continuous span bridge, 269–270
contract documents, 11
contractual procedure, 502
corrosion prevention, 426–431
corrosion protection, 51, 432–433, 449–450
cost effective strategies, 51
cost-benefit analysis, 290–291
countermeasures, 544–557
covered bridges, 73
crack reduction, 443–445
crack repair, 440–441
crack repair, 446
crack width, 442
critical members, 341
critical problems, 398–406
curved girder, 322–323

D

damage inventory, 599
dampers fluid viscous, 595
dampers seismic, 595
dead loads, 229
debris accumulation, 562–563
deck
- applications, 377–378
- components, 10
- condition, 456–457, 459
- cracking, 51
- drainage, 50, 461
- efflorescence, 51
- flood scour, 529–532
- geometry, 10
- joints, 461
- materials, 10
- pour, 225
- reconstruction summary, 415
- rehabilitation methods, 419–421
- repair work, 454–456
- repairs, 440
- replacement, 271–274, 462–463
- selection, 490–491
- slab design, 266–274
- survey, 456–457
- testing, 457–459
- types, 452–453
- vibrations, 461

deep beams, 155
defect repair, 440–441
deficient structures, 33
deflection, 196–204, 231, 306
design
- checks, 222
- codes, 483–485, 570–576
- countermeasures, 535–536
- deficiency, 92–94
- details, 237–238
- diagnostic, 27, 58–59
- errors, 115
- guidelines, 554–557
- method, 228, 237–238
- related actions, 124–125
- seismic, 576–579
- shear, 260
- topics, 481
- vessel, 99–101

design-build construction (DBC), 23–24
development in design codes, 483–485
diagnostic analysis, 133–134
diagnostic case studies, 87–90
diagnostic design, 27, 52–53
diagnostic methods, 134
disaster history, 75–78
disaster management, 390–391, 599
diverse repair technologies, 478
diving inspections, 385–386
drainage design, 411
drainage, 434
ductility requirements, 582–583
durable concrete, 446–448
dynamic testing results, 381

E

early cracking reduction, 442
earthquake damage, 105–108, 563–569
earthquake failures, 108
earthquake immediate effects, 565–566
education, 26
elastomeric material, 50
emergency repairs, 389–390, 415–416

engineering
- activities, 295
- forensic, 78–81
- hurricane, 109–110
- inspection, 346
- maintenance, 39–44, 40–41, 133
- wind, 109–110

enhanced concrete, 23–24
environment enhancing, 502–503
environmental
- concerns, 36–37
- engineer, 36
- issues, 4
- permits, 36–37, 412–413

equestrian traffic, 35
erection methods, 504–505
erection stresses, 306
erosion science, 23
evaluation of deflection, 258–260
excessive fatigue, 94–95
existing bridge design, 577–578
exodermic bridge deck, 499
extreme conditions, 227–228, 530–532
extreme events, 368–372, 563–569

F

failure
- aspects, 81
- avoidance, 117–124
- causes, 116–117
- diagnostics, 81–85
- history, 85–92
- material, 169–171
- rating, 372
- solutions, 124
- studies, 81–83

fathometer survey, 386–387
fatigue evaluation, 401
fatigue, 204–208
fender system reconstruction, 440
FHWA
- deterministic approach, 291
- policy, 26–27
- procedures, 349
- ratings, 359–363
- recommendations, 790

fiber reinforced polymer, 488–494
final design submission, 408
finite element method, 142
fire damage, 104–105
fire preventative, 105
floating ice, 99–101
flood scour damage, 529–530
floor beams, 426–427
flow altering countermeasures, 547–549
fluid viscous dampers, 595
forensic engineering, 78–81
forms of failure, 116, 123–124
foundation repairs, 470–471
foundation retrofit, 596–599
foundation scour, 90–92, 535–579
foundation, 328, 344
fracture causes, 83–85
fracture critical members, 60–61
fracture, 204–208
functional obsoleteness, 59
fundamental equations, 137–138

G

galvanic technology, 431–433
geometric issues, 59
girder, 159, 255–257
- design data, 228–229
- materials, 8–9
- replacement, 322–323
- selection, 163–169
- shapes, 11
- systems, 516–517
- steel, 513–514

government agencies, 61–65
gusset plates, 261–265

H

hazard control principles, 116
hazard resiliency, 338–339
heat straightening, 427
HEC-18 countermeasures matrix, 549–554
high performance concrete, 495
high priority repairs, 416–417
hinge assemblies, 515
historic bridges, 26, 67–69
historic footings, 69–70
historic masonry arch, 70–71
historical perspective, 31–33
history of bridge inspections, 347–348

history of disasters, 75–78
hurricane engineering, 109–110
hydraulic countermeasures, 541
hydraulic inadequacy, 32

I

infrastructure government agencies, 61–65
innovative concepts, 480–481
inspection, 344–357
  access, 365
  data, 350
  frequency, 355–356
  in-depth, 356–357
  reports, 44
inspectors, 364–366
instrumentation, 375

J

judicial reports, 81

K

key issues, 398–406
knowledge base, 25
knowledge databases, 126–127

L

laboratory tests, 458–459
lack of maintenance, 110–112
legal loads, 240–246
lighting design, 411
lightweight emergency bridge, 494
limit state design equation, 237–240
limit state for substructure, 217–223
live load analysis, 161–163, 213–216
live loads, 189, 209–220, 229, 244–254
load
  behavior, 190–194
  combinations, 229, 223–228, 530–532, 574
  effects, 238–239
  factors, 190–191
  rating, 401
local seismic codes, 572–574
long spans, 35
long-term concrete repairs, 440
low priority repairs, 417
LRFD
  application, 367–368
  load requirements, 194–196
  specifications, 209–210
  T-beam bridge design, 228–232
LRFD/LRFR method, 193–194
LRFR application, 367–368

M

maintaining the environment, 35–36
maintenance, 56–57, 383–387
  activities, 48–49
  protection of traffic (MPT), 30–31
  engineering, 7–8, 39–44
  errors, 115
  knowledge database, 23
  manual, 22–23
  principles, 39–40
  theoretical approach, 23
major repairs, 418–419
malfunction of bearings, 112
management systems, 335–344
management triple constraint, 407–413
management, 383–387
market ready technology, 483–485
material deterioration, 75–76
materials recyclable, 486–488
mathematical issues, 4
mechanics of failure, 116
medium priority repairs, 417
member sizing, 135–136
metal corrosion, 428–429
methods advanced repair, 477–521
minimum countermeasure requirements, 542
mitigation solutions, 43–44
modeling methods, 140
modern construction materials, 485–486
modern rehabilitation, 1
monitoring plan, 373–374
monitoring techniques, 517–521
monitoring, 336–337, 380–383
moveable bridges, 107
moving loads, 203–204
multi-hazard, 338

N

national policy, 26–27
NCHRP studies, 400–401
near-term concrete repairs, 429–440
neglect, 110–112

new applications, 51
new bridge countermeasures, 543–544
new construction products, 433–434
new materials, 23–24, 517–521
new techniques, 517–521
new topics in design, 481
non-compact sections, 223
nondestructive evaluation, 377
nonlinear analysis in concrete, 159–161
nonlinear analysis in steel, 159–161
nonstructural components, 579
numerical models, 151–152

O

obsolete bridges, 414
open grid steel decks, 453
optic sensors, 383
outreach program, 409–410
overlay, 50
overload prevention, 366–367

P

painting inspection, 431
parametric study of deflection, 196–198
parametric study, 320–321
parapet design, 12–13
passive damping systems, 595–596
pedestrian traffic, 35
permit loads, 210–214
pier repair, 470
pier type, 327
piling failures, 123–124
plan preparation, 294
planning issues, 44
planning, 26, 321
plastic behavior of composite, 175–180
plastic behavior of steel, 171–175
plastic behavior, 194
policy, 26–27
PONTIS system, 353–354
post-design tasks, 12
post-design services, 409
potential damage repairs, 441–442
prefabrication technology, 509–513
preliminary design submission, 408
preserving aesthetics, 34–35
prestressed concrete girder, 276–279
prestressed girders, 158–159
preventative
    actions, 92–95
    design, 52–53
    fire damage, 105
    measures, 92–94
    wind failure, 109–110
preventing construction failures, 504–505
production errors, 115
progress maintenance, 23–28
progress reports, 413
progressive design phases, 57
project management, 286–287
public involvement, 409–410
public outreach, 30–31
published references, 11

Q

quality control, 95, 367, 413
quality planning, 44–45

R

railing design, 12–13
rapid construction, 505–506
rating factor, 186–190
rating procedures, 255–260
rating structural members, 260
reconstruction duties, 294
reconstruction parameters, 21
reconstruction recommendations, 600–601
recyclable materials, 486–488
recycled concrete aggregate, 486–488
recycled glass, 488
recycled plastics, 488
redundancy, 60–61
rehabilitation, 56–57
    approach, 40
    benefits, 1–4
    decisions, 285–289
    deficiencies, 54–56
    design, 480
    funding, 21
    methods, 489–490
    planning, 44–46
    process, 6–7, 44–51
    report, 44, 389
    structural components, 185–190
    tasks, 11
    technologies, 66–67

reinforced concrete protection, 450–452
reinforced concrete repair, 445–446
reinforcement guidelines, 583–585
remote health monitoring, 374–375
remote sensors, 376–376
REP/REH ratios, 288–289
repair, 56–57, 397–343
    alternative, 419–426
    comparative study, 478–480
    costs, 422–426
    critical problems, 398–406
    deck, 453–461
    items, 422–426
    key issues, 398–406
    materials, 419, 445, 485–486
    methods, 419, 445, 463–466, 482–483
    plans, 403–404
    procedures, 486
    report, 387–394
    sign structures, 493–494
    techniques, 442–443
    technologies, 478
replacement, 339
    alternatives, 317–318
    case studies, 324–325, 328–331
    constructability issues, 303–308
    decisions, 285–289
    deflection, 306
    erection stresses, 306
    girder, 322–323
    guidelines, 301
    issues, 295
    methodology, 296–297
    over rivers, 306–308
    planning, 297–298, 321
    practical considerations, 301–303
    projects, 296
    span, 298–301
    strategy, 295–296
    structural components, 185–190
    superstructure, 308–317
    type, 297
research in new techniques, 517–521
resistance factors, 190–191, 229, 239–240
resource applications, 11
retaining walls, 327–329
retrofit, 586–595
    concepts, 578–582
    continuity, 589
    details, 589–591
    foundation, 596–599
    guidelines, 589–591
    methods, 482–483, 588–589
    substructure, 591
risk assessment, 339–341
risk management, 127–128
river training, 542, 546–547
rocker bearings, 594–595
routine design, 52–53
routine inspections, 349–350
rural settings, 35

S

safety management, 75–78
salient features, 494–495
scheduling, 460
scope of repairs, 417–419
scope of work, 578–579
scour
    analysis software, 557–560
    caissons, 560–561
    culverts, 562
    design, 530, 535
    guidelines, 530
    measuring equipment, 562
    piles, 560–561
    preventative action, 535–579
    safety evaluation, 387
    type, 536
    vulnerability, 369
    wingwalls, 561–562
security, 26
seismic
    assessment, 569–570
    coding guide, 569
    dampers, 595
    design, 563–564, 568–569, 576–579
    failures, 107–108
    hazard rating, 579–580
    rating, 369–372
    replacement bearings, 581–582
    resistance, 23, 33, 107
    retrofit report, 580–581
    retrofit, 106, 570, 587–588
self compacting concrete, 495–496
service life, 24–25
service, 228
shallow beams, 155

shear connectors, 260
shear design, 180–182, 260
shear stress, 155–158
short-term repairs, 413–415
shrinkage compensating concrete, 496–497
sign structures, 141–142, 154, 493–494
signs of failure, 117–119
simplified structural options, 48
single span bridge, 268–269
slab beam bridges, 138–139
slopes ISHA recommendations, 557
software substructure, 233–234
software superstructure, 232–233
software, 283, 557–560
soil conditions, 32
span classification, 9–10
span selection, 321–322
special access equipment, 384
special provisions, 415–416
stakeholders, 30–31
standards, 28–29
state seismic codes, 572–574
steel
    bridge rehabilitation, 513–517
    fatigue, 232–234
    girders, 225–226, 513–514
    material technology, 500–502
    members, 514–515
    rehabilitation, 426–431
    repair, 397–343
    selection, 50
    trusses, 426–427
strategic plan, 21–23
strength limit, 230–230
strength, 227
strengthening methods, 588–589
strengthening steel members, 514–515
strengthening, 586–595
structural
    behavior, 158
    countermeasures, 541
    deficiencies, 54–56, 58–59
    engineer, 568
    evaluation, 344–347, 399–400
    health monitoring, 372–380
    items, 401–402
    members, 260
    planning and solutions, 121–122, 228–229
    repairs, 448
    solutions, 44
    steel painting, 427–428
    systems, 137
structure deficient, 33
structure success, 31
study of failures, 81–83
substructure, 217–223
    alternatives, 326
    components, 357–358
    condition evaluation, 342–344, 267–568
    damage, 105–108
    failure, 123
    inspection, 532–535
    items, 358
    repair, 404–405, 536–537
    retrofit, 591
    software, 233–234
    stiffness methods, 142–143
    type, 137
    underwater, 384–385
sufficiency ratings, 363
superstructure
    alternatives, 325–326
    analysis, 140–143
    components, 220–221, 351
    condition evaluation, 341–342, 566–567
    construction, 503–504
    failure avoidance, 122–123
    fatigue, 222
    inspection, 351–352
    load capacity, 367–368
    load, 222
    maintenance, 12, 13–15
    repairs, 452–453
    replacement, 12, 13–15, 308–317, 404–405
    service loads, 221
    software, 232–233
    widening, 284–285

T

T-beam bridge, 274–275
technical specifications, 560
technology, 483–485
temporary riprap, 544–546
theoretical concepts, 135
theory of elasticity, 135
thermal stresses, 112

through girder bridges, 516–517
timber decks, 453
traffic
  count, 28–29
  equestrian, 35
  loads, 6
  maintenance, 28–31, 403
  pedestrian, 35
  protection, 28–31
train accidents, 101–102
train live load, 153–154, 211–213
training, 26
transportation inspection facilities, 119–121
traverse stiffeners, 260
triple constraint management, 407–413
trusses, 140–141
two girder steel bridge

U

ultra-high performance concrete, 497–498
unanticipated studies, 413
underwater inspection, 384–385, 533–534
unexpected failures, 112–115
unforeseen failure causes, 112–114
unknown foundations, 470
utility relocation, 410–411

V

vehicle accidents, 103
vehicle impact, 102–103
vertical clearance, 102
vessel collision, 99–101
vibration parameters, 204
virtual design, 517–521
voided slab bridge, 283–284
vulnerability evaluation, 363

W

weathering steel, 427
weight posting, 367
well-known historic bridges, 71–72
widening, 282–285, 339
  activities, 284–285
  design methods, 283
  planning, 285–289
  practice, 282–283
  project management, 286–287
  requirements, 284
  structural components, 185–190
wind engineering, 109–110
wingwalls, 327–329